Eco-efficient concrete

Related titles:

Non-destructive evaluation of reinforced concrete structures: Deterioration processes and standard test methods (Volume I) (ISBN 978-1-84569-560-6)

Failure, distress and repair of concrete structures (ISBN 978-1-84569-408-1)

Understanding the rheology of concrete (ISBN 978-0-85709-028-7)

Details of these books and a complete list of titles from Woodhead Publishing can be obtained by:

- visiting our web site at www.woodheadpublishing.com
- contacting Customer Services (e-mail: sales@woodheadpublishing.com; fax: +44 (0) 1223 832819; tel.: +44 (0) 1223 499140 ext. 130; address: Woodhead Publishing Limited, 80, High Street, Sawston, Cambridge CB22 3HJ, UK)
- in North America, contacting our US office (e-mail: usmarketing@woodheadpublishing.com; tel.: (215) 928 9112; address: Woodhead Publishing, 1518 Walnut Street, Suite 1100, Philadelphia, PA 19102-3406, USA

If you would like e-versions of our content, please visit our online platform: www.woodheadpublishingonline.com. Please recommend it to your librarian so that everyone in your institution can benefit from the wealth of content on the site.

We are always happy to receive suggestions for new books from potential editors. To enquire about contributing to our materials series, please send your name, contact address and details of the topic/s you are interested in to francis.dodds@woodheadpublishing.com. We look forward to hearing from you.

The Woodhead team responsible for publishing this book:
Commissioning Editor: Jessica Rowley
Publications Co-ordinator: Lucy Beg
Project Editor: Sarah Lynch
Editorial and Production Manager: Mary Campbell
Production Editor: Richard Fairclough
Cover Designer: Terry Callanan

Eco-efficient concrete

Edited by

F. Pacheco-Torgal, S. Jalali, J. Labrincha and V. M. John

WOODHEAD PUBLISHING

Oxford Cambridge Philadelphia New Delhi

Published by Woodhead Publishing Limited,
80 High Street, Sawston, Cambridge CB22 3HJ, UK
www.woodheadpublishing.com
www.woodheadpublishingonline.com

Woodhead Publishing, 1518 Walnut Street, Suite 1100, Philadelphia,
PA 19102-3406, USA

Woodhead Publishing India Private Limited, G-2, Vardaan House, 7/28 Ansari Road,
Daryaganj, New Delhi – 110002, India
www.woodheadpublishingindia.com

First published 2013, Woodhead Publishing Limited

British Library Cataloguing in Publication Data
A catalogue record for this book is available from the British Library.

Library of Congress Control Number: 2012954993

ISBN 978-0-85709-424-7 (print)
ISBN 978-0-85709-899-3 (online)

The publisher's policy is to use permanent paper from mills that operate a sustainable forestry policy, and which has been manufactured from pulp which is processed using acid-free and elemental chlorine-free practices. Furthermore, the publisher ensures that the text paper and cover board used have met acceptable environmental accreditation standards.

Typeset by Replika Press Pvt Ltd, India
Printed and bound in the UK by MPG Books Group

Contents

Contributor contact details

(* = main contact)

Editors

Dr F. Pacheco-Torgal* and Professor S. Jalali
C-TAC Research Centre
Sustainable Construction Group
University of Minho
Campus de Azurem
4800-058 Guimarães
Portugal

E-mail: torgal@civil.uminho.pt

Dr J. A. Labrincha
Ceramics and Glass Engineering Department (CICECO)
University of Aveiro
Campus Universitário de Santiago
3810-193 Aveiro
Portugal

E-mail: jal@ua.pt

Dr V. M. John
University of Sao Pãolo
Brazil

E-mail: vmjohn@lme.pcc.usp.br

Chapter 1

Professor Guillaume Habert
Swiss Federal Institute of Technology (ETH Zürich)
Wolfgang Pauli Strasse, 15
CH-8092, Zürich
Switzerland

E-mail: habert@ibi.baug.ethz.ch

Chapter 2

Mr Bruno Luís Damineli, Professor Rafael Giuliano Pileggi and Professor Vanderley M. John*
Department of Construction Engineering
Escola Politécnica
University of São Paulo
05508-900 São Paulo - SP
Brazil

E-mail: bruno.damineli@lme.pcc.usp.br; rafael.pileggi@lme.pcc.usp.br; vmjohn@lme.pcc.usp.br

Chapter 3

Professor Snežana B. Marinković
Faculty of Civil Engineering
University of Belgrade
Bulevar kralja Aleksandra 73
11000 Belgrade
Serbia

E-mail: sneska@imk.grf.bg.ac.rs

Chapter 4

M. I. Sánchez de Rojas Gómez* and M. Frías Rojas
Eduardo Torroja Institute for Construction Science (IETcc-CSIC)
Department of Cement and Materials Recycling
c/ Serrano Galvache, 4
28033 Madrid
Spain

E-mail: srojas@ietcc.csic.es

Chapter 5

M. Frías Rojas* and M. I. Sánchez de Rojas Gómez
Eduardo Torroja Institute for Construction Science (IETcc-CSIC)
Department of Cement and Materials Recycling
c/ Serrano Galvache, 4
28033 Madrid
Spain

E-mail: mfrias@ietcc.csic.es

Chapter 6

A. R. Pourkhorshidi
Department of Concrete Technology
Road, Housing and Urban Development Research Center
Tehran
Iran

E-mail: alip_208@yahoo.com

Chapter 7

Assistant Professor Dr Burak Uzal
Department of Civil Engineering
Abdullah Gul University
38039 Kayseri
Turkey

E-mail: burakuzal@gmail.com; burak.uzal@agu.edu.tr

Chapter 8

Professor Martin Cyr
Laboratoire Matériaux et Durabilité des Constructions (LMDC)
INSA/UPS Civil Engineering
Université de Toulouse
135 Avenue de Rangueil
31077 Toulouse Cedex
France

E-mail: Martin.Cyr@insa-toulouse.fr

Chapter 9

Associate Professor Dr E. Güneyisi,* Associate Professor Dr M. Gesoğlu and Z. Algin
Civil Engineering Department
Gaziantep University
Gaziantep 27310
Turkey

E-mail: guneyisi@gantep.edu.tr

Chapter 10

Professor Dr İlker Bekir Topçu
Department of Civil Engineering
Engineering-Architecture Faculty
Eskişehir Osmangazi University
26480, Eskişehir
Turkey

E-mail: ilkerbt@ogu.edu.tr

Chapter 11

Associate Professor Keren Zheng
School of Civil Engineering
Central South University
Jian'gong Building
22# Shaoshan Nan Road
Changsha 410075
China

E-mail: krzheng_1@yahoo.com.cn

Chapter 12

Dr Mark Tyrer
University of Coventry
Coventry, UK
and
Mineral Industry Research Organisation
Concorde House
Trinity Park
Solihull
Birmingham B37 7UQ
UK

E-mail: mark.tyrer@miro.co.uk

Chapter 13

Dr F. Pacheco Torgal*
C- TAC Research Centre
Sustainable Construction Group
University of Minho
Campus de Azurem
4800-058 Guimarães
Portugal

E-mail: torgal@civil.uminho.pt

Professor Yining Ding
State Key Laboratory of Coastal and Offshore Engineering
Dalian University of Technology
116024 Dalian
China

E-mail: ynding@hotmail.com

Chapter 14

Dr Antonio Eduardo Bezerra Cabral
Department of Structural Engineering and Civil Construction
Federal University of Ceará
Campus do Pici
Bloco 710
60.455-760 Fortaleza
Ceará
Brazil

E-mail: eduardo.cabral@ufc.br

Chapter 15

Dr F. Pacheco Torgal*
C- TAC Research Centre
Sustainable Construction Group
University of Minho
Campus de Azurem
4800-058 Guimarães
Portugal

E-mail: torgal@civil.uminho.pt

S. Miraldo
University of Aveiro
Portugal

E-mail: ssmiraldo@gmail.com

Dr J. A. Labrincha
Ceramics and Glass Engineering Department (CICECO)
University of Aveiro
Campus Universitário de Santiago
3810-193 Aveiro
Portugal

E-mail: jal@ua.pt

J. de Brito
ICIST
DECivil-IST
Technical University of Lisbon
Av. Rovisco Pais
1049-001 Lisbon
Portugal

E-mail: jb@civil.ist.utl.pt

Chapter 16

Professor Dr Mariaenrica Frigione
Department of Engineering for Innovation
University of Salento
73100 Lecce
Italy

E-mail: mariaenrica.frigione@unisalento.it

Chapter 17

Dr I. García-Lodeiro,* Dr A. Fernández-Jiménez, and Professor A. Palomo
Instituto Eduardo Torroja, CSIC
Serrano Galvache 4
28033-Madrid
Spain

E-mail: iglodeiro@ietcc.csic.es

Chapter 18

Professor M. A. G. Aranda and Dr A. G. De la Torre
Departamento de Química Inorgánica
Universidad de Málaga
Campus de Teatinos, s/n
29071-Málaga
Spain

E-mail: g_aranda@uma.es

Chapter 19

Dr Abir Al-Tabbaa
Department of Engineering
University of Cambridge
Cambridge
UK

E-mail: aa22@cam.ac.uk

Chapter 20

Professor Maria S. Konsta-Gdoutos
Department of Civil Engineering
Democritus University of Thrace
12 Vas. Sofias
GR-67100 Xanthi
Greece

E-mail: mkonsta@civil.duth.gr

Chapter 21

Dr F. Pacheco-Torgal*
C-TAC Research Centre
Sustainable Development Group
University of Minho
Campus de Azurem
4800-058 Guimarães
Portugal

E-mail: torgal@civil.uminho.pt

Dr J. A. Labrincha
Ceramics and Glass Engineering Department (CICECO)
University of Aveiro
Campus Universitário de Santiago
3810-193 Aveiro
Portugal

E-mail: jal@ua.pt

Foreword

The sustainability of the use and construction with concrete has been in the limelight for some years and the issues involved with it will be further intensified in the 21st century. Concrete is the most important and highly used construction material for setting the infrastructure and housing of our society, and the demand for it is growing rapidly in view of the increase in world population and the impressive advancement in the standard of living in emerging countries. Thus, it is a formidable challenge to explore and apply new ways for reducing its ecological and environmental impact.

The scientific and technological challenges are particularly high if one takes into account that concrete is a relatively 'green' material, since its embodied energy and carbon per unit volume are much smaller than any other construction material. Because of the already low environmental imprint, it is hardly feasible to achieve a breakthrough by a single step or technology. Considering the fact that concrete is made from locally available raw materials, it would hardly be possible to come up with a single solution which could be applicable globally to all parts of the world. Therefore, the more practical way to go, at least for the near future, is to develop and advance a range of technologies which could be implemented selectively, depending on local conditions and availability of raw materials, and could be combined to provide additive influences that, when summed will result in significant improvement in terms of ecological impact.

Over the years numerous scientific approaches and technologies have been developed and applied to ease the ecological impact of concrete by providing means to produce cements and concretes with reduced energy and carbon imprints and at the same time maintaining and even improving long-term durability performance, to provide additional ecological and environmental benefits. This know-how and the state of art of advancement over a wide front of technologies need to be available to any researcher and technologist dealing with sustainability of construction with concrete. The current book, which compiles reviews of a wide range of technologies and strategies, will enable the concrete technologist to be in a better position for producing and developing ecologically friendly concrete, based on bringing together and synergizing several strategies, which could be applicable to local conditions and availability of materials.

The book presents a very useful input for producing more ecological friendly concrete. It achieves this goal by dealing with the impact of the concrete from a variety of viewpoints: the concrete constituents themselves, cement and concrete, the concrete as a whole, the application of wastes and by-products, the use of natural materials of low environmental impact as well as exploring new types of binders. It thus provides a sound basis for developing and application of ecologically friendly concretes.

Professor Arnon Bentur
Faculty of Civil and Environmental Engineering
National Building Research Institute
Technion, Israel Institute of Technology

Introduction

F. PACHECO TORGAL, University of Minho, Portugal

During the last century, global materials use increased 8-fold and as a result Humanity currently uses almost 60 billion tons (Gt) of materials per year (Krausmann *et al.*, 2009). The more important environmental threat associated to its production is not so much the depletion of non-renewable raw materials (Allwood *et al.*, 2011), but instead the environmental impacts caused by its extraction, namely extensive deforestation and top-soil loss. In 2000 mining activity worldwide generated 6000 million tons of mine wastes to produce just 900 million of raw materials (Whitmore, 2006). This means an average use of only 0.15%, resulting in vast quantities of mineral waste, the disposal of which represents an environmental risk in terms of biodiversity conservation, air pollution and pollution of water reserves. For instance, in 2003 alone US mining activities were responsible for releasing 3 billion pounds of toxic substances. As a result, since the 1970s there have been 30 serious environmental accidents in mines, 5 of them occurring in Europe (Pacheco-Torgal and Jalali, 2011).

Concrete is the most used construction material on Earth, currently about 10 km^3/year (Gartner and Macphee, 2011). For comparison, the amount of fired clay, timber, and steel used in construction represent, respectively about 2, 1.3 and 0.1 km^3 (Flatt et al., 2012). The main binder of concrete, Portland cement, represents almost 80% of the total CO_2 emissions of concrete, which in their turn are about 6–7% of the planet's total CO_2 emissions (Shi et al., 2011). This is particularly serious in the current context of climate change and gets even worse because Portland cement demand is expected to increase almost 200% by 2050 from 2010 levels reaching 6000 million tons/year (Pacheco-Torgal and Jalali, 2011).

The concept of eco-efficiency was first introduced in 1991 by the World Business Council for Sustainable Development (WBCSD) and includes

> the development of products and services at competitive prices that meet the needs of humankind with quality of life, while progressively reducing their environmental impact and consumption of raw materials throughout their life cycle, to a level compatible with the capacity of the planet.

For concrete this concept means producing more with fewer resources and less waste and emissions. For instance replacing Portland cement by supplementary cementitious materials (SCMs) (Lothenbach et al., 2011) and virgin aggregates by industrial wastes (Meyer, 2009).

This book covers several aspects related to the eco-efficiency of concrete. The first chapter analyses the environmental impact of Portland cement production. A detailed description of the cement production process is made and an inventory data on energy use, CO_2, PM10, SOx and NOx emissions is presented.

The second chapter covers the efficiency of binder use, suggesting two indicators for it (binder intensity índex and the CO_2 intensity index). The life cycle assessment (LCA) of concrete will be the subject of Chapter 3. The two different types of impact assessment methods are described (damage-oriented approach like Eco-Indicator 99 and the 'midpoints' approach like the Centrum voor Milieukunde Leiden (CML) methodology). This chapter also includes a case study related to ready-mixed concrete production. Since the use of SCMs leads to a significant reduction in CO_2 emissions per ton of cementitious materials it is then no surprise that this issue is the theme of the following nine chapters.

Chapter 4 is concerned with the major natural pozzolans, along with the technical, economic and environmental advantages of using such materials in cement manufacture. The characterization, pozzolanic properties, reaction kinetics and mechanical strength of pozzolans of different origins are described. The use of new pozzolans obtained from fired clay waste, which can be classified as a natural calcined pozzolan, is also addressed.

The availability of artificial pozzolans generated in industrial processes and from agro-industrial waste is the subject of Chapter 5. This chapter addresses the scientific (characterization, pozzolanic properties, reaction kinetics) and technical (physical and mechanical) aspects of new blended cement matrices covering the cases of silicon–manganese slag, copper slag, pulverized coal combustion bottom ash, fluid catalytic cracking catalyst, sewage sludge ash, electric arc furnace steel slag, rice husk ash, sugar cane ash, activated paper sludge and palm oil fuel ash.

The sixth chapter discusses various methods available for evaluating pozzolans. It includes the direct methods (thermo-gravimetric analysis, determination of insoluble residue, mineralogical analysis, saturated lime test, EN 196-5 and Frattini test) and indirect methods (ASTM C618, ASTM C1240 and electrical conductivity). The merits and disadvantages of the methods are elaborated, and some existing anomalies are highlighted.

Chapter 7 focuses on the properties and durability of high volume natural pozzolan (HVNP) concrete. It includes workability, air content, setting time, mechanical strength and modulus of elasticity, resistance to chloride-ion penetration, resistance to sulfate attack and alkali–silica reaction.

The influence of SCMs on concrete durability is analysed in Chapter 8. The cases of the alkali–silica reaction, delayed ettringite formation, sulfate attacks, acid attacks, frost resistance, abrasion, carbonation and chloride ingress are dully assessed.

Chapter 9 addresses the theme of self-compacting concrete (SCC) with high volume SCMs. The properties of fresh and hardened concrete as well as their durability are analyzed. Chapter 10 deals with concrete containing ground granulated blast furnace slag (GGBFS) as partial replacement for Portland cement. Properties and durability of this material are reviewed.

One of the most important environmental features of concrete is the ability to recycle wastes from other industries. This avoids the use of large areas for landfill disposal which is a major threat for biodiversity. That is why Chapters 11 to 14 are related to the reuse of wastes in concrete.

The practice of recycling glass containers is growing in many countries; however, a significant portion is not recycled partially because the bulk of waste glass can be collected in mixed colors. The European Container Glass Federation statistics refer that the recycling rate of glass in EU27 is just 68%. Chapter 11 summarizes properties of fresh and hardened recycled glass concrete. The influences of recycled glass on the properties and durability of concrete are discussed. Special attention is given to alkali–silica reaction.

The subject of Chapter 12 relates to the reuse of municipal solid waste incineration (MSWI) ashes in concrete. Because of their content in heavy metals, high levels of chlorides and dioxins, this kind of waste is considered hazardous by the European Waste Catalogue. This chapter covers the composition of MSWI ashes, their hydration in concrete, its performance in service and the environmental implications of incorporating such residues.

Chapter 13 covers the reuse of polymeric wastes (tyre rubber and polyethylene terephthalate (PET) bottles). An estimated 1000 million tyres reach the end of their useful life every year and it is expected that by the year 2030 the number of tyres from motor vehicles is expect to reach 1200 million representing almost 5000 millions tyres to be discarded on a regular basis. Tyre landfilling is responsible for a serious ecological threat. Mainly waste tyre disposal areas contribute to the reduction of biodiversity, also the tyres hold toxic and soluble components. The implementation of the Landfill Directive 1999/31/EC and the End of Life Vehicle Directive 2000/53/EC banned the landfill disposal of waste tyres, creating the driving force behind the recycling of these wastes. Annual consumption of PET bottles represents more than 300,000 million units and since this waste is not biodegradable it can remain in nature for hundreds of years. Therefore, recycling PET waste in concrete is an option with undoubted environmental benefits.

Chapter 14 concerns the performance of concrete with construction and demolition wastes (CDW) – another important group of wastes that is suitable to be used in concrete. These wastes are often used in roadfill which constitutes a down-cycling option. Although the use of CDW has been studied for almost 50 years, today we still see too many structures that are made with virgin aggregates. The reasons lie in its low cost, the lack of incentives, low landfill costs and even sometimes the lack of technical regulations. On 19 November 2008 the EU approved the Revised Waste framework Directive No. 2008/98/EC. According to this Directive the minimum recycling percentage for CDW by the year 2020 should be at least 70% by weight. This Directive will therefore increase the reuse of this particular waste.

Chapter 15 reviews current knowledge on concrete carbonation when SCMs and/or recycled aggregates concrete (RAC) are used. Although a wide range of literature has been published on this field, the different conditions used by different authors limit comparison and in some cases contradictory findings are noticed. Besides, since most investigations are based on the use of the phenolphthalein indicator, which provides a poor estimate of the real concrete carbonation depth, there is a high probability that past research could have underestimated the corrosion potential associated with concrete carbonation. Some remedial actions to minimize concrete carbonation when recycled aggregates are used are described.

The subject of Chapter 16 is concrete with polymers. The use of polymers in concrete goes back as far as 1923 when for the first time a patent was issued for a concrete floor with natural latex, so that Portland cement was used only as filler. This material has superior durability over ordinary Portland cement concrete, assessed by resistance to acid attack, resistance to action of ice-melting, resistance to diffusion of chlorides. Three different cases are assessed (polymer modified concrete, polymer impregnated concrete and polymer concrete). The importance of water reducing admixtures, polymeric coatings and polymeric adhesives to concrete durability is also reviewed.

Chapters 17, 18 and 19 are related to the development of 'greener' binders respectively comprising alkali-activated binders, calcium sulfoaluminate cement and magnesia cement. Alkali-activated cements (which include the geopolymers group) are synthesized from aluminosilicate materials after being mixed with alkaline solutions. This binder is associated with lower carbon dioxide emissions than Portland cement. It shows high mechanical performance and also superior resistance to acid attack and abrasion than current Portland cement concretes. This new binder can use fly or bottom ashes from power stations or mining and quarrying wastes as raw materials and can be used for the immobilization of radioactive and toxic wastes which gives it an undeniable environmental value.

Sulfoaluminate cements use lower calcination temperatures than Portland cement, and also use limestone with a lower carbonate content, thus representing a reduction in CO_2 emissions. Reactive magnesia cement is also manufactured at much lower temperatures than Portland cement and has the additional advantage of being able to uptake significant quantities of CO_2.

Chapter 20 covers nanotechnological developments of concrete, including nano-modification with the addition of carbon nanotubes (CNTs), single-walled nanotubes (SWCNTs), multi-walled CNTs (MWCNTs) and nanofibers (CNFs). The use of ultrasonication energy in order to obtain proper dispersion of CNTs and CNFs within the cementitious matrix is analyzed. The effect of the addition of silica nanoparticles to minimize the degradation by calcium leaching, thus increasing concrete durability, is also discussed. This feature will allow structures with a longer service life thus needing less maintenance and conservation operations thus having a lower environmental impact.

Finally Chapter 21 shows how biotechnology is able to enhance concrete durability by means of biomineralization, a phenomenon by which organisms form minerals first used for crack repair by Gollapuddi et al. (1995). This chapter covers the use of bacteria as admixture in concrete and also biomineralization as a concrete surface treatment. This technique seems to be an environmental friendly alternative to current concrete surface treatments which are based on organic polymers (epoxy, siloxane, acrylics and polyurethanes), all of which have some degree of toxicity.

This book has no intent to be even close to a treatise on the subject of concrete eco-efficiency. Treatises on concrete are only within the reach of grand engineers like Adam Neville, Pierre-Claude Aitcin and Arnon Bentur. Be that as it may, they lived in a time when a single engineer could gather the enormous amount of knowledge required for a treatise. No longer. This is the time for the democratization of the investigation process. Around the world thousands of individual contributors produce each year several hundred important papers on concrete science and technology. In 2011 around 1500 papers on Portland cement concrete were published in international journals. Unfortunately, although much knowledge has been and still is generated, only a little is used by the concrete industry. An excellent proof of that can be found in the fact that 31 years after Professor Roger Lacroix coined the expression 'high performance concrete', still only 11% of the concrete ready-mixed production corresponds to the HPC strength class target (ERMCO, 2011). The sole purpose of this book is therefore to display recent investigations on eco-efficient concrete in order to highlight the importance of this subject. It is to be hoped that 31 years from now ready-mixed concrete production has turned itself entirely eco-efficient and not just 11%.

References

Allwood, J.; Ashby, M.; Gutowski, T.; Worrell, E. (2011) Material efficiency: A white paper. *Resources, Conservation and Recycling* **55**, 362–381.

ERMCO (2011) *Statistics of the year 2010*. Boulevard du Souverain 68, B-1170 Brussels, Belgium.

Flatt, R.; Roussel, R.; Cheeseman, C.R. (2012) Concrete: An eco material that needs to be improved. *Journal of the European Ceramic Society* (in press).

Gartner, E.; Macphee, D. (2011) A physico-chemical basis for novel cementitious binders. *Cement and Concrete Research* **41**, 736–749.

Gollapudi, U. K.; Knutson, C. L.; Bang, S. S.; Islam, M. R. (1995) A new method for controlling leaching through permeable channels. *Chemosphere* **30**, 695–705.

Krausmann, F.; Gingrich, S.; Eisenmenger, N.; Erb, K.-H.; Haberl, H.; Fischer-Kowalski, M. (2009) Growth in global materials use, GDP and population during the 20th century. *Ecological Economics* **68**, 2696–2705.

Lothenbach, B.; Scrivener, K.; Hooton, R. (2011) Supplementary cementitious materials. *Cement and Concrete Research* **41**, 1244–1256.

Meyer, C. (2009) The greening of the concrete industry. *Cement and Concrete Composites* **31**, 601–605.

Pacheco-Torgal, F.; Jalali, S. (2011) *Eco-efficient construction and building materials.* Springer Verlag, London, UK.

Shi, C.; Fernández Jiménez, A.; Palomo, A. (2011) New cements for the 21st century: The pursuit of an alternative to Portland cement. *Cement and Concrete Research* **41**, 750–763.

Whitmore A (2006) The emperor's new clothes: Sustainable mining. *Journal of Cleaner Production* **14**, 309–314.

Part I

Eco-efficiency of Portland cement concrete

1
Environmental impact of Portland cement production

G. HABERT, ETH Zürich, Switzerland

DOI: 10.1533/9780857098993.1.3

Abstract: With the current focus on sustainability, it is necessary to evaluate cement's environmental impact properly, especially when developing new 'green' concrete types. Therefore, this chapter investigates the available literature on every process concerned during the production of cement. Inventory data on energy use, CO_2, PM10, SOx and NOx emissions are collected. Alternatives and improvement are briefly described regarding energy performance and mineralogy changes that it induces for clinker.

Key words: life cycle assessment (LCA), CO_2, cement kiln dust (CKD), energy efficiency.

1.1 Introduction

Cement production has undergone a tremendous development from its beginnings some 2000 years ago. While the use of cement in concrete has a very long history, the industrial production of cements started in the middle of the 19th century, first with shaft kilns, which were later on replaced by rotary kilns as standard equipment worldwide. Today's annual global cement production has reached 2.8 billion tonnes, and is expected to increase to some 4 billion tonnes per year (Schneider et al., 2011). Major growth is foreseen in countries such as China and India as well as in regions like the Middle East and Northern Africa. At the same time, the cement industry is facing challenges such as cost increases in energy supply, requirements to reduce CO_2 emissions, and the supply of raw materials in sufficient qualities and amounts. The World Business Council for Sustainable Development (WBCSD) and its Cement Sustainability Initiative, comprising cement producers worldwide, has initiated the project 'Getting the Numbers Right' which for the first time provides a good database for most of the global cement industry with respect to CO_2 and energy performance (WBCSD, 2008). Cement production is responsible for 5% of global anthropogenic CO_2 emissions and 7% of industrial fuels use (IEA, 2007).

In this chapter the process of cement production will be described regarding its contribution in terms of energy consumption and associated environmental impacts. The chapter will then focus on the differences that exist between the different cement production plants. It will be highlighted that these differences are due to technology differences but also to internal variability in the supply chain. To conclude, the main perspective in term of reduction of cement production's environmental impacts will be discussed.

1.2 Description of the cement production process

As a first approximation, Portland cement clinker can be described as a four-component system consisting of the four major oxides: CaO, SiO_2, Al_2O_3 and Fe_2O_3 (Fe^{2+} being neglected). Raw materials should be mixed precisely to provide appropriate amount of elements. All these raw materials together with the fuel ash are then combined to form the typical clinker composition. In the best available cement manufacturing technology, limestone and silicon, aluminum and iron oxides are crushed and then milled into a raw meal. This raw meal is blended in blending silos and is then heated in the pre-heating system which dissociates carbonate to calcium oxide and carbon dioxide. The meal is then passed through the kiln for heating. A reaction takes place between calcium oxide and other elements to produce calcium silicates and aluminates at about 1500 °C. Primary fuel is used to keep the temperature high enough in the burning zone for the chemical reactions to take place. When the clinker cools to a subsolidus temperature, an assemblage is formed consisting of C_3S, C_2S, C_3A and C_4AF. The clinker will then be inter-ground with gypsum, limestone and/or ashes to a finer product called cement. A comprehensive cement manufacturing process can be found in the European Cement Association (CEMBUREAU, 2010). Details of cement manufacturing process for few selected countries around the world can be found in Bastier et al. (2000), JCR (2000), Kaantee (2004) and Sogut et al. (2009). Figure 1.1 shows a comprehensive brief description of cement manufacturing process, the different steps of which are described very accurately, for instance by Madlool et al. (2011) are briefly explained as follows.

1.2.1 Limestone quarrying and crushing

Naturally occurring calcareous deposits such as limestone, marl or chalk provide calcium carbonate ($CaCO_3$) and are extracted from quarries, often located close to the cement plant. Very small amounts of 'corrective' materials such as iron ore, bauxite, shale, clay or sand may be needed to provide extra iron oxide (Fe_2O_3), alumina (Al_2O_3) and silica (SiO_2) to adapt the chemical composition of the raw mix to the process and product requirements.

An open mining process inducing the use of drilling, blasting and heavy earth moving equipment is common for quarrying operations.

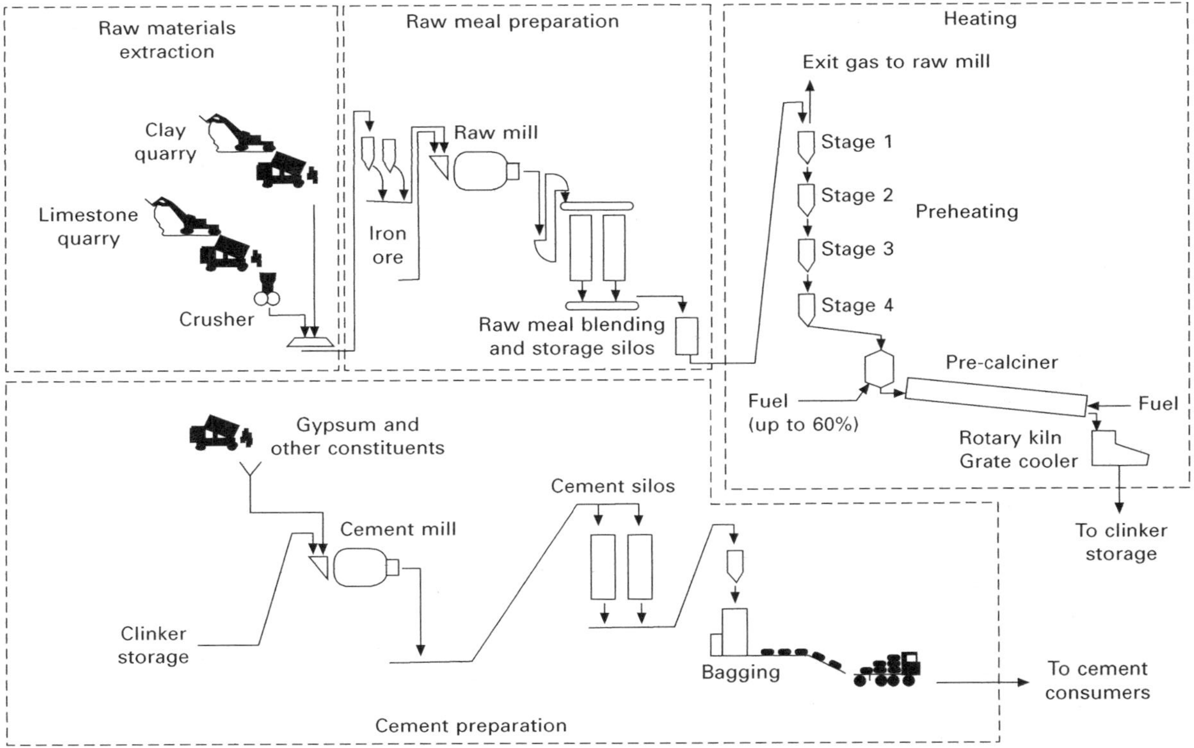

1.1 Overview of cement manufacturing process.

1.2.2 Raw mill

The grinding process impacts the cement manufacture in at least two ways. First, an increase of the fineness and homogeneity of the raw mix accelerates the clinkering reactions, leading to less variability of the clinker due to the enhanced stability of the kiln and better energy efficiency in producing a clinker of better consistency. The second effect concerns the fineness of the cement that enhances mechanical properties of the final product.

The cement industry utilizes four types of mills: ball mill, vertical roller mill, roller press (high-pressure grinding roll, HPGR) and horizontal roller press mill. Different mill feeds, such as raw material, coal, clinker and slag, have different grindabilities, feed particle size ranges and moisture content as well as different throughput rates, fineness data and other quality parameters. Each mill type associated with the grinding process is more suitable for some applications and requirements over others. Raw material grinding has the objective of producing a homogeneous raw meal with a fineness requirement (less than 10–15% residue on the 90 μm sieve) from a raw material containing moisture (3–8%) with a feed particle size of 100–200 mm. A vertical mill is installed in 80% of all new grinding plants, which has lower power consumption and allows simultaneous grinding, drying and separation. A ball mill is still used for 12% of all raw material applications.

1.2.3 Blending and storing silo

To reduce the natural chemical variation in the various raw materials, it is necessary to blend and homogenize the raw material efficiently. This is done through continuous blending silos.

1.2.4 Pre-heater and kiln

A preheater is a series of vertical cyclones through which the raw meal is passed, coming into contact with swirling hot kiln exhaust gases moving in the opposite direction. In these cyclones, thermal energy is recovered from the hot flue gases, and the raw meal is preheated before it enters the kiln. Depending on the raw material moisture content, a kiln may have up to six stages of cyclones with increasing heat recovery with each extra stage.

After the preheater, modern plants have a precalciner, where limestone is decomposed to lime and carbon dioxide. Here, the chemical decomposition of limestone typically emits 60–65% of total emissions. Fuel combustion generates the rest, 65% of which occur in the precalciner. Gartner (2004) has highlighted the fact that decarbonation process is the most energy-consuming process during the chemical reaction.

The precalcined meal then enters the kiln. Fuel is fired directly into the kiln to reach temperatures of up to 1450 °C. As the kiln rotates, about 3–5 times per minute, the material slides and tumbles down through progressively hotter zones towards the flame. The intense heat causes chemical and physical reactions that partially melt the meal into clinker. More details can be found in Sorrentino (2011). The clinker is then discharged as red-hot from the end of the kiln and passed through different types of coolers to partially recover the thermal energy and lower the clinker handling temperature.

Different types of kilns are briefly explained below but more details can be found for instance in Szabó et al. (2003). Wet rotary kilns are used when water content of the raw material is within 15–25%. Usually wet slurry is produced to feed into the kiln. The wet kiln feed contains about 38% water. This makes the meal more homogeneous for the kiln, leading to lower electrical energy use for the grinding. However, overall energy consumption will be high in order to evaporate water in the slurry. This process is still in use in some countries. However, many countries are shifting from wet kilns to dry kilns to reduce the overall energy consumption.

Semi-wet rotary kilns are used when the wet raw material is processed in a filter after homogenizing to reduce moisture content. It is an improved version of the wet process. This is mainly used for retrofitting the existing wet kilns. This process can reduce energy consumption by 0.3 GJ/tonne of clinker.

In semi-dry rotary kiln, waste heat recovered from the kiln is used to remove moisture content. Then the dried meal is fed into the kiln. As the existing long dry kilns without preheaters still consume more energy than new technologies, they are often improved by the addition of a preheater. Dry kilns with preheaters include kilns with 4–6 multistage cyclone preheaters. The raw materials are passed through the cyclones. As one part of the calcinations already takes place in the preheater, it is possible to reduce the energy consumption due to reduction in the length of the kiln.

With dry kilns with preheater and pre-calciner, an additional combustion chamber is installed between the preheater and the kiln. This pre-calciner chamber consumes about 60% of the fuel used in the kiln, and 80–90% of the calcinations take place here. This reduces energy consumption by 8–11%. Low-temperature waste heat from the combustion chamber can also be recovered for other purposes.

Finally, a number of shaft kilns can be found in China and India. In India their share is 10%, while in China it is over 80% of the capacities. Their usual size is between 20 and 200 tonnes/day, and many of them are operated manually. Clinker quality is highly dependent on the homogenization of pellets and fuel, and on the air supply. Inadequate air supply or uneven air distribution makes combustion incomplete, resulting in low-quality clinker and high CO and volatile organic compound (VOC) emissions. We show the kiln's energy consumption from Madlool et al. (2011) in Table 1.1.

1.2.5 Cooler

From the kiln, the hot clinker (1500 °C) falls onto a grate cooler where it is cooled down to 170 °C by incoming combustion air, thereby minimising energy loss from the system (Zeman, 2009).

1.2.6 Cement mill

This is the final step in a cement manufacturing process. In this step, the clinker is ground together with additives in a cement mill. All cement types contain around 4–5% gypsum to control the setting time of the product.

Traditionally, ball mills have been used for grinding, although more efficient technologies of roller presses and vertical mills are used in many modern plants today. As for raw material, four types of grinding systems are used: ball mill, vertical roller mill, high-pressure roller press and horizontal roller press mill. The ball mills are still used in clinker grinding even though its application has decreased in new plant projects. A vertical roller mill uses approximately 40% less power than a traditional ball mill. Vertical mills with an integral separator are used for finish grinding with a mill capacity of 350 tonne per hour. An high-pressure roller press (HPRP) is used in various systems with ball mills for primary grinding and semi-finish grinding or as a single-stage grinding process for finish grinding. Horizontal roller press mills are suitable for use as a single-stage grinding process. Energy savings are 50% compared to ball mills and 20% compared to vertical mills, but the throughput rates are limited at 120 tonne per hour (Madlool et al., 2011). By comparing ball mill and HPRP, it has to be noted that even if the particle size distribution is similar (between 90 μm and 20 μm), higher Brunauer–

Table 1.1 Specific thermal energy consumption in a clinker manufacturing process

Kiln process	Thermal energy consumption ($GJ/t_{clinker}$)
Wet process	5.85–6.28
Long dry process	4.60
Shaft kiln	3.70–6.60
1-stage cyclone pre-heater	4.18
2-stage cyclone pre-heater	3.77
4-stage cyclone pre-heater	3.55
4-stage cyclone pre-heater plus calciner	3.14
6-stage cyclone pre-heater plue calciner	< 2.93

Emmett–Teller (BET) surface area analysis is obtained with HPRP leading to higher water demand than the ball meal products (Celik et al., 2007). It may be presumed that a single-stage grinding process with a vertical roller mill or roller presses will continue to expand at the cost of the ball mill. Chemical compounds are used to improve the particle comminution during the grinding of materials (grinding aids). The most commonly used grinding aids include propylene glycol, triethanolamine, triethanolamine acetate and tri-isopropylamine. The mechanism of action of grinding aids is not known precisely and their efficiency possibly varies with the type of grinder. The toxicity associated with the use of such chemicals must be taken into account (Bensted and Smith, 2009).

1.2.7 Storing in the cement silo

The final product is homogenized and stored in cement silos and dispatched from there to either a packing station (for bagged cement) or to a silo truck.

1.3 Main impacts

Many environmental studies have detailed the impacts of the different processes involved in the cement manufacturing (Josa et al., 2004, 2007; Valderrama et al., 2012; von Bahr et al., 2003). Recent studies show that calcination has a predominant role in the environmental impacts of cement production (Cagiao et al., 2011; Chen et al., 2010a). Figure 1.2 shows that processes involved for the preparation of raw materials and those involved after the calcination (milling) represent not more than 20% of the impact for all impact categories. Furthermore, Fig. 1.2 highlights the fact that environmental impacts are due either to the direct emissions on the cement kiln or to indirect impact associated to the preparation of the fuels.

1.3.1 Global scale

Over the last decades, the emphasis has clearly shifted towards a global focus on climate change. A recent study has gathered the global impacts of cement production. Table 1.2 presents a summary of values found in literature (Febelcem v.z.w. 2006; Humphreys and Mahasenan 2002; Hendriks et al., 2011; Josa et al., 2004; Van den Heede and De Belie, 2012) for cement-related CO_2 emissions. They are usually the sum of the CO_2 emitted during the calcination process (raw material CO_2: RM-CO_2) and the CO_2 associated with energy use (energy-bound CO_2: EB-CO_2) (Gartner, 2004). With respect to the latter, some authors have made a distinction between indirect and direct energy bound CO_2 (IEB- and DEB-CO_2). IEB emissions comprise the CO_2 emissions associated with the generation of electrical power to operate the

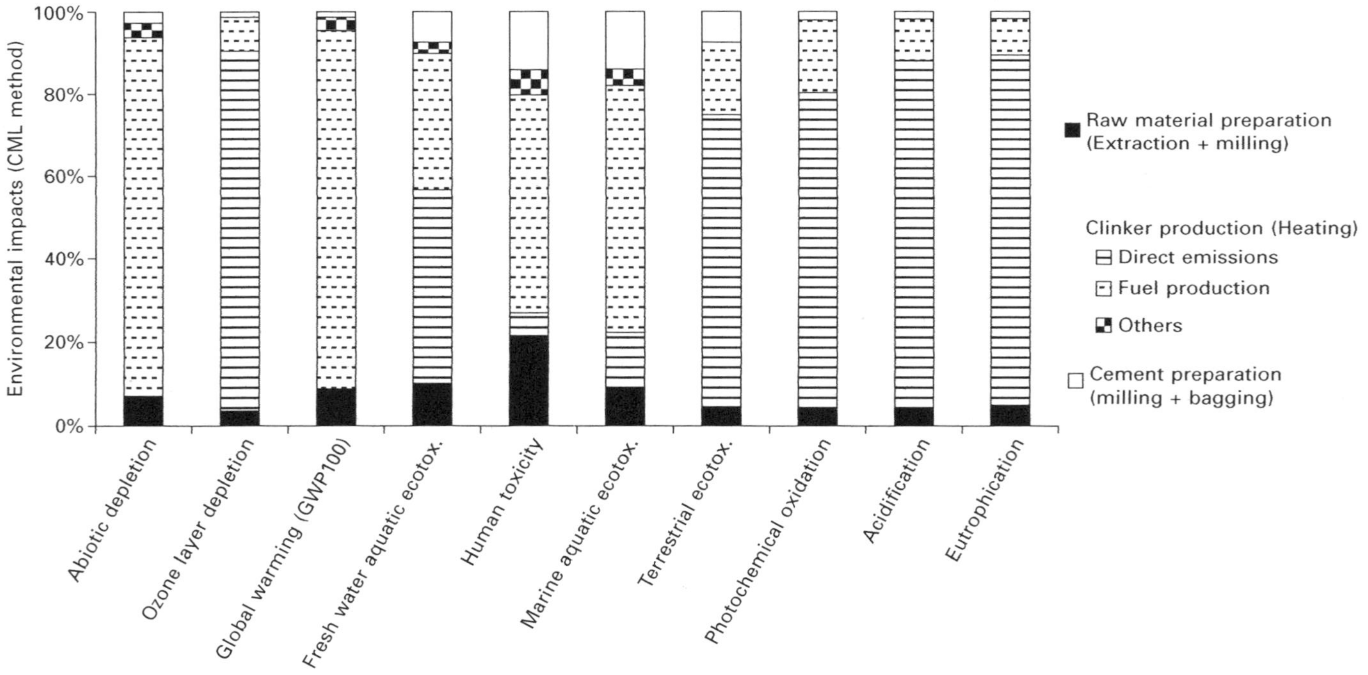

1.2 Relative environmental impact assessment of the different processes involved in cement production. CML = Centrum voor Milieu kunde Leiden. (Adapted from Chen et al., 2010a.)

Table 1.2 Summary of CO_2, SO_2, NOx and cement kiln dust (CKD) emissions for Portland cement production expressed in g/kg cement* or g/kg clinker**

CO_2			SO_2			NOx			CKD/PM-10		
Ref	Values		Ref	Values		Ref	Values		Ref	Values	
[1]	870	g/kg*	[3]	0.40–0.60	g/kg*	[3]	2.4	g/kg*	[3]	0.1–10	g/kg*
[2]	810	g/kg*	[5]	0.82	g/kg*	[5]	1.2	g/kg*	[5]	0.49	g/kg*
[3]	800	g/kg*	[6]	0.58	g/kg*	[6]	1.5	g/kg*	[6]	0.04	g/kg*
[4]	820	g/kg*	[7]	0.27	g/kg**	[7]	1–4	g/kg*	[7]	200	g/kg**
[5]	690	g/kg*	[8]	0.54	g/kg**				[9]	0.1–0.3	g/kg*
[6]	810	g/kg*	[9]	2.5	g/kg*				[10]	150–200	g/kg*
[7]	900	g/kg*									
[10]	895	g/kg*									
Mean value	814	g/kg*	Mean value	0.53	g/kg*	Mean value	3.65	g/kg*	Mean value	25	g/kg*

References [1]: Humphreys and Mahasenan, 2002; [2]: Hendriks et al., 2011; [3]: Josa et al., 2004; [4]: Flower and Sanjayan, 2007; [5] Chen et al., 2010a; [6] ATILH, 2002; [7]: Van Oss and Padovani, 2003; [8]: Huntzinger and Eatmon, 2009; [9]: Febelcem v.z.w., 2006; [10]: Gartner, 2004 (adapted from Van den Heede and De Belie, 2012).

cement plant, while the direct energy bound emissions are associated with the fuel combustion in the cement kiln. Regarding DEB-CO_2, the efficiency of the cement kiln plays an important role. Under optimum conditions heat consumption can be reduced to less than 2.9 GJ/ton clinker. A typical modern rotary cement kiln with a specific heat consumption of 3.1 GJ/ton clinker emits approximately 0.31 kg DEB-CO_2, while this amount equals about 0.60 kg/kg clinker for an inefficient long rotary kiln burning wet raw materials with an extra heat consumption of around 0.6 GJ/ton clinker (Damtoft et al., 2008).

Possibilities to reduce RM-CO_2 emissions are rather limited. Partially replacing the traditional raw materials by blast-furnace slag (BFS) or class C fly ash (FA) with a higher calcium content is one option. In practice, replacement levels of about 10% are commonly reported (Habert et al., 2010). For a limestone replacement of 10%, the total CO_2 reductions can in theory be as high as 25% (Damtoft et al., 2008). To reduce the emissions any further, alternative clinker chemistries need to be considered. This could be the development of belite type cement or sulfo-aluminate clinker for instance. A brief summary of these alternative chemistry can be found in Flatt et al. (2012).

1.3.2 Regional scale

Regional environmental impacts include SO_2 and NOx emissions which contribute to acid rain. Table 1.2 includes an overview of the estimated SO_2 and NOx emisions for Portland cement according to literature (Josa et al., 2004; Van Oss and Padovani, 2003).

The majority of SO_2 emitted is derived from fuel combustion and processing of raw materials in the kilns. However, the majority of the SO_2 leaves the kiln with the clinker (EPA, 1994) as it is absorbed due to the high alkalinity of clinker (Houghton et al., 1996).

The generation of NOx and CO is mainly as outputs from fuel usage during clinker production. Their emissions are highly dependent on temperature and oxygen availability and rotary kilns produce much more NOx and less CO because of their higher operation temperature and stable ventilation compared to shaft kilns (Lei et al., 2011). This technology difference can explain the rapid increase in rotary kiln installation in China.

1.3.3 Local scale

Cement kiln dust (CKD) emissions are the main contributors to the local impact. The size of CKD (0.05 to 5 μm) is within the size range of respirable particles (EPA, 1994). Since the diameter is smaller than 10 μm, CKD is classified as PM10. According to the EPA (1999), these fine particulates of unburned and partially burned raw materials present in the combustion gases of the cement kiln, are considered as a potential hazardous waste due to their caustic and irritative nature. As mentioned in Table 1.2, the amount of CKD generated per kg of clinker produced equals about 15–20% (by mass) (Van Oss and Padovani, 2003). Nowadays, both the environmental and health risks associated with CKD can be reduced significantly by means of mineral carbonation. Sequestering carbon in CKD stabilizes the waste. The reduction in pH reduces health risks and the generation of harmful leachate (Huntzinger and Eatmon, 2009). In addition, the utilization of CKD for sequestration appears to have an advantage on the global scale, since about 7% of the carbon emissions can be captured this way. On the local scale, attention should be paid to the chromium content of cement. For instance, the sale of cement containing more than 2 ppm of soluble Cr(VI) when hydrated is prohibited by European Directive 2003/53/EC (European Union, 2003). Hexavalent chromium or Cr(VI) is not stable. When dissolved, Cr(VI) can penetrate the unprotected skin and be transformed into to Cr(III) which combines with epidermal proteins to form the allergen to which some people are sensitive. The Cr(VI) content can originate from (i) raw materials and fuel entering the system, (ii) magnesia–chrome refractory blocks, (iii) wear metal from crushers containing chromium alloys and (iv) additions of gypsum, pozzolans, ground granulated BFS, mineral components, CKD and set regulators (Hills and Johansen, 2007).

Finally a number of additional pollutants such as polychlorinated dibenzo-*p*-dioxins, dibenzofurans and heavy metals can be potentially released (Abad et al., 2004; Schumacher et al., 2003). Recent studies that evaluate the potential health risk for populations living in the neighbourhood of a cement plant show that a seasonal pattern were observed with higher values recorded during the colder periods. However, the carcinogenic and non-carcinogenic risks derived from human exposure to metals and polychlorinated dibenzodioxins/Furans (PCDD/Fs) were within the ranges considered acceptable by international regulatory organisms (Rovira et al., 2011). Furthermore, the intensive use of alternative fuels such as sewage sludge or municipal solid waste which would otherwise be disposed somehow/somewhere or refuse-derived fuels allow a significant decrease in PCDD/Fs levels as well as in some metal concentrations (Rovira et al., 2010).

1.4 Future trends

1.4.1 Alternative materials

Alternative fuels

The use of alternative fuels and raw materials (AFR) for cement and clinker production is certainly of high importance for the cement manufacturer but also for society as a whole. Alternative fuel utilization began in the mid-1980s. Starting in calciner lines, up to almost 100% alternative fuel firing at the pre-calciner stage was very quickly achieved. Alternative fuels are mainly used tires, animal residues, sewage sludges, waste oil and lumpy materials. The last are solid recovered fuels retrieved from industry waste streams, and to a growing extent also from municipal sources. These refuse-derived fuels are pre-treated light fractions processed by mechanical or air separation. Waste-derived fuels consist of shredded paper, plastics, foils, textiles and rubber and also contain metal or mineral impurities.

While in some kilns up to 100% substitution rates have been achieved, in others, local waste markets and permitting conditions do not allow for higher rates of AFR. In any case, AFR utilization requires the adaptation of the combustion process. Modern multi-channel burners designed for the use of alternative fuels allow control of the flame shape to optimize the burning behavior of the fuels and the burning conditions for the clinker (Wirthwein and Emberger, 2010). In a conventional preheater kiln (without pre-calciner), it is only possible to burn fuels in the kiln inlet with substitution rates of up to 25–30%. This is different in pre-calciner kilns, as usually up to 65% of the total fuel energy input is fired into the calciner and a minimum of 35% through the main kiln burner. As a consequence, when alternative fuels are used in the pre-calciner, it does not change the nature of the fuels introduced in the kiln and does not change therefore the kiln performance. Most operators then first increase the alternative fuel substitution in the pre-calciner. After this, they start to increase the proportion of alternative fuels in the sintering zone firing.

When alternative fuels are used in the cement kiln, these alternative fuels are mixed with the raw meal and can have an influence on the clinker properties. A recent paper made a review of the chemical consequences of the minor elements added to the clinker by the use of alternative fuel (Sorrentino, 2011). For example, phosphorus is present in meat and bone meal, and chlorides are present in refuse-derived fuels. These minor elements affect the burning process (clinkering, cooling, and emission) both thermodynamically by modifying the phase assemblages in the system $CaO–SiO_2–Al_2O_3–Fe_2O_3$ and kinetically by modifying the chemical and physical properties of the interstitial melt. Studies of industrial clinker show that phosphorus (up to 0.7%) is distributed into belite grains without any structural modification (Moudilou et al., 2007). In laboratory experiments, it is known that a small amount of P_2O_5 is added to suppress the 'dusting effect' due to the transformation of β-C_2S to γ-C_2S (Fukuda et al., 2008). Phosphorus replaces silicon in both alite and belite and, therefore, increases the content of belite with phosphorus contents in the clinker of up to 1%. The distribution of P^{5+} in the silicate phases is closer to 1 (Herfort et al., 2010). The increased belite contents that are often reported then have to be compensated by increasing the calcium content of the clinker. Burnability is improved when HPO_4^{3-}, HPO_4^{3-} with F^-, or with F^- and SO_3^{2-} are added into the raw meal and the strength is increased with the addition of HPO_4^{3-} and F^- (Guan et al., 2007).

The addition of chloride-containing components increases the burnability of the raw meal and allows higher alite contents at the same clinkering temperature. Halogens have a great capacity for reducing the viscosity of the liquid phase and CaO is highly soluble in liquid phases rich in halogen (Maki, 2006).

Sulfur (S^{6+}) has a strong preference for belite. By increasing SO_3 in the clinker, the alite becomes richer in Al and Fe, particularly in Al, with an increase in the Al/Fe ratio. The C_3A content decreases and C_4AF increases with increasing SO_3 content. An increase in the SO_3 content causes a reduction of the primary phase volume of C_3S (decrease of C_3S/C_2S ratio); 2.5% of SO_3 seems to be the threshold of formation of C_3S. SO_3 and P_2O_5 decrease both the viscosity and surface tension of the liquid and the polymorphic form of C_3S, M1, is predominant as the constituent phase. MgO and SO have opposite effects on the size and the phase constitution of alite. MgO favors the occurrence of the M3 polymorph, whereas SO_3 favors M1 (Maki, 2006). The addition of SO_3 or SO_3 + HPO_4^{3-} simultaneously reduces the burnability, whereas it is improved with the addition of SO_3 + HPO_4^{3-} and F^-.

The increase in Na_2O content modifies the polymorphism of C_3A (orthorhombic) with a limit of 4%. Na is partially incorporated in C_2S, stabilizing the form α and α' at room temperature. When Na_2O is added, C_2S decreases and C_3S increases with a possible occurrence of free lime (Gotti et al., 2007). Alkali metal oxides (Na_2O and K_2O) increase the viscosity and decrease the surface tension of the liquid phase. The nucleation and growth decrease in rate and large alite crystals are grown with distinct facets. A small content of chromium is found in cement. It arises primarily from the raw material and the wear of steel devices. When it is present as chromate $CrO_4{}^{2-}$, it provokes skin allergies. Reducing agents such as ferrous iron sulfate are added to prevent this shortcoming (Bensted, 2006).

The maximum amount of Cu, Ni, Sn, and Zn that could be incorporated in a laboratory clinker has been determined experimentally. This threshold limit is found to be 0.35% of Cu, 0.5% of Ni, 1% of Sn and 0.7% for Zn (Gineys et al., 2011).

A recent study has evaluated the perspective of alternative fuel used and has shown that in developed countries a ratio of 40–60% of alternative fuel in 2050 can be achieved while in the developing countries this ratio will be around 25–35% (WBCSD, 2009). Technically, much higher substitution rates are possible. In some European countries, the average substitution rate is over 50% for the cement industry and up to 98% as yearly average for single cement plants. As fuel-related CO_2 emissions are about 40% of total emissions from cement manufacture, the CO_2 reduction potential from alternative fuel use can be significant.

Although, technically, cement kilns could use up to 100% of alternative fuels, there are some practical limitations. The physical and chemical properties of most alternative fuels differ significantly from those of conventional fuels. While some (such as meat-and-bone meal) can be easily used by the cement industry, many others can cause technical challenges. These are related to, for example, low calorific value, high moisture content, or high concentration of chlorine or other trace substances. For example, volatile metals (e.g., mercury, cadmium, thallium) must be managed carefully, and proper removal of CKD from the system is necessary. This means pre-treatment is often needed to ensure a more uniform composition and optimum combustion. However, the achievement of higher substitution rates has stronger political and legal barriers than technical ones:

- Waste management legislation significantly impacts availability: higher fuel substitution only takes place if local or regional waste legislation restricts land-filling or dedicated incineration, and allows controlled waste collection and treatment of alternative fuels.
- Local waste collection networks must be adequate.
- Alternative fuel costs are likely to increase with high CO_2 costs. It may then become increasingly difficult for the cement industry to source significant quantities of biomass at acceptable prices.
- The level of social acceptance of co-processing waste fuels in cement plants can strongly affect local uptake. People are often concerned about harmful emissions from co-processing, even though emissions levels from well-managed cement plants are lower with alternative fuel use (Rovira et al., 2010). In addition, alternative fuel use has the potential to increase thermal energy consumption, for example when pre-treatment is required as outlined above.

Alternative cementitious materials

Another option to reduce the environmental impact of cement production is to reduce the amount of clinker in the cement. This option allows a cost reduction in the cement manufacturing process by saving burning costs and is a highly efficient CO_2 reduction way. The next chapters will describe the different materials that can be used to develop these low clinker cements. In this section we will discuss the question associated with the environmental evaluation of these alternative cementitious materials. According to ISO standards, when a production system produces more than one product, it is necessary to attribute an environmental burden to each product. This is the case for most of the supplementary cementitious materials (SCM) and specifically to fly ash and blast furnace slags that are by-products from other industries. The question of the value of the environmental load for SCM has been emphasized by many authors and has not yet been solved. The current practice is to consider them as waste and with that assumption to assign a null environmental impact to them, but this situation will evolve toward a practice where the different industrial sectors will have to share these environmental loads. Particularly in Europe where, a recent European Union Directive (EU, 2008) notes that a waste may be regarded as by-product if the following conditions are met:

- Condition (a) further use of the substance or object is certain;
- Condition (b) the substance or object is produced as an integral part of a production process;
- Condition (c) the substance or object can be used directly without any further processing other than normal industrial practice;
- Condition (d) further use is lawful, i.e. the substance or object fulfils all relevant product, environmental and health protection requirements for the specific use and will not lead to overall adverse environmental or human health impacts.

This Directive is very relevant to the use of SCMs such as blast furnace slags and fly ash. Therefore in Europe these two materials can no longer be considered as waste but instead as by-products. Hence, the question is what is the environmental cost of these by-products?

Different scenarios can be promoted. Dividing the impact by the relative mass value of the different products (steel/slags) induces too high an environmental load for by-products and a division in relation to the relative economic value of the products seems more in accordance with the perception of what could be the environmental load of these SCMs. The main problem with the economic allocation is the question of price variability as illustrated in Fig. 1.3 where blast furnace slag is varying between 40 and 90 €/t and steel price is varying between 150 and 1500 €/t. For further reading on this subject see Van den Heede and De Belie (2012).

Among alternative cementitious materials, it should also be pointed out that rather than partly substituting the clinker, it is also possible to develop new binders. Chapters 16 to 18 focus on these new chemistries. The environmental concerns associated with these binders are similar to the SCMs as they are often based on the use of waste/by-product which induces allocation problems. If raw materials are used, the energy for heating is most of the time lower than for clinker except for the production of alkaline solution which can change the environmental balance depending on the amount that is needed (Habert et al., 2011; McLellan et al., 2011).

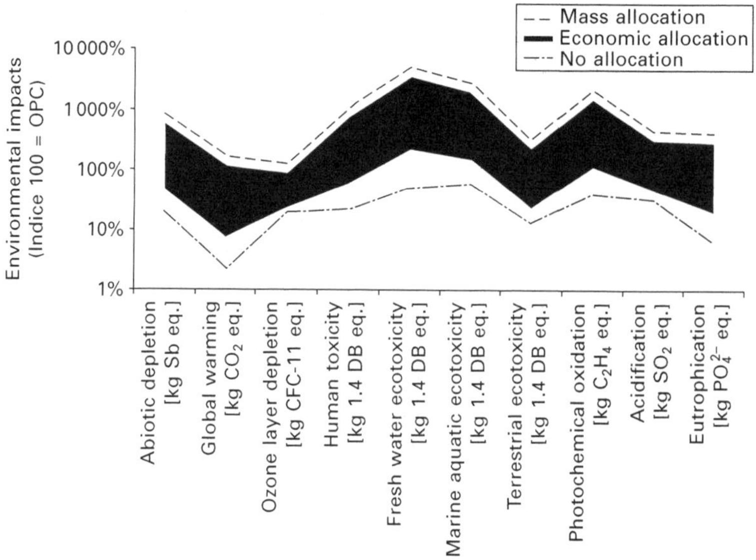

1.3 Environmental impact of a mass of GBFS (1.11 kg) equivalent to the replacement of 1 kg of cement CEM I, for the different CML indicators with different allocation procedures. (Adapted from Chen et al., 2010b.)

1.4.2 Energy efficiency

Energy demand in clinker production has been significantly reduced over the last few decades. The theoretical minimum primary energy consumption (heat) for the chemical and mineralogical reactions is approximately 1.6–1.85 GJ/t (Klein and Hoening, 2006). However, there are technical reasons why this will not be reached, for example unavoidable conductive heat loss through kiln/calciner surfaces. For reduction of the specific power consumption (electricity), other barriers are also preventing the industry from reaching this minimum, for example:

- A significant decrease in specific power consumption will only be achieved through major retrofits. These have high investment costs and so retrofits are currently limited.
- Strengthened environmental requirements can increase power consumption (e.g., dust emissions limits require more power for dust separation regardless of the technology applied).
- The demand for high cement performance, which requires very fine grinding and uses significantly more power than low-performing cement.

As a consequence the best available techniques (BAT) levels for new plants and major upgrades are 2900 to 3300 MJ/t clinker, based on dry process kilns with multistage-preheaters and precalciners (European Commission,

2010). A critical review on energy use and savings in the cement industries can be found in Madlool et al. (2011), but the factors involved to further reduce this demand are plant-specific, which will involve large investment. As moisture content of the raw materials determines the heat consumption (Bauer and Hoenig, 2010; Klein and Hoenig, 2006), the main driver to reduce energy consumption on a global average is the kiln size, which is, in most cases, not applicable for existing installations. An increase in the cement plant capacity could still enhance the productivity and therefore the efficiency of cement plants. However, cement plant capacities will remain in the typical range of between 1.5 and 2.5 million t/yr, resulting in typical single clinker production lines between 4000 and 7000 t/day and very large cement and clinker lines of 10,000 or even 12,000 t/d will generally be the exception (Nobis, 2009).

As the change in cement kiln size or capacity is difficult, waste heat recovery may play an important role. Actually 30–45 kW h/t of clinker is becoming feasible for recovered energy and the waste heat utilization industry itself is developing technologies to widen the potential for energy recovery. Depending on the volume and temperature level of waste heat, a range of specific technologies can be applied. A large quantity of low temperature waste heat (below 350 °C), approximately 30% of the total heat consumption of the system, is still not recovered. Several different low temperature waste heat power generation technologies for cement production have been developed including the steam Rankin cycle with three main patterns: single and dual pressure or flash steam generation system (Jintao et al., 2009).

1.4.3 Carbon capture and storage

The use of pure oxygen instead of air can in theory result in a very significant improvement in thermal efficiency, because it reduces the volume of the exhaust gases (and their associated heat losses) by a factor of about 3. It also leads to exhaust gases that are essentially a simple mixture of CO_2 and water vapor, which could then easily be separated by condensation, the resulting pure CO_2 then being readily transportable or directly injectable into underground aquifers or other such potential disposal sinks. This type of approach is currently under consideration by the electric power generating industries for a new generation of coal burning power plants, and the cement industry could in theory try to apply the same approach. However, the electrical energy required to produce pure oxygen from air with current technology is about 420 kW h/t-O_2 (ECRA, 2009). If we consider that the minimum O_2 requirement is 1 mol per mol of exhaust CO_2, this energy already represents 10–15% of the energy needed to produce clinker. Based on this, oxygen enrichment would not actually save a lot of energy or CO_2 generation in cement manufacture. This situation will evidently improve as the primary energy efficiency of electric power generation plant and air separation plants improves, but this is likely to be a slow process.

1.4.4 Cement production

Finally, it should not be forgotten that the main impact of cement is due to its tremendous demand. Actually, 1 kg of cement emits only 0.6–0.8 kg of CO_2 which is negligible compared to other material production emissions such as aluminum or insulation materials. However, the impact of cement industries is more important that other energy intensive industries because of the volume of cement production. A recent study showed that the main driver for CO_2 emission reduction in the French context from 1990 to 2005 has been the reduction in cement production rather than the improvement in technology (Habert et al., 2010). This is due on one hand to the fact that investment costs are important and on the other hand that technology improvements on the clinker production technology are limited compared to the mitigation objective.

Cement production evolution is linked with economic activity and the levels of industrialization and infrastructure development of the country. These parameters can be expressed as an intensity of cement use that refers to the amount of cement used per unit of gross domestic product (GDP) (kg/unit of GDP). Note that a unit of GDP is here adjusted to 1000 constant dollars (base year: 2000) and expressed in term of purchasing power parities (PPP) which are the rates of currency conversion that eliminate the differences in price levels between countries. Cement intensities differ between countries according to economic growth (GDP) and economic structure. Different studies attended to demonstrate that this intensity follows an inverted U-shape curve (Lafarge, 2006; Scheubel and Nachtwey, 1997; Szabó et al., 2003; Vuuren, et al., 1999). Intensity of cement demand will then decline in developed countries and increase in many developing countries (Taylor et al., 2006).

In Western Europe Organization for Economic Co-operation and Development (OECD) countries, the intensity of cement use is currently estimated at 21 kg of cement per unit of GDP (1000 USD, PPP 2000) and will be around 17 kg of cement per unit of GDP at the 2050 horizon (Taylor et al., 2006). To give a comparison, the intensity of cement demand in China is today around 131 kg of cement by unit of GDP (1000 USD, PPP 2000) and is expected to be reduced to the usage of Western European countries by the 2050 horizon (Taylor et al., 2006).

With assumptions on GDP evolution and on cement demand intensity, it is then possible to evaluate the cement production evolution. A model described as VLEEM 2 (Very Long Term Energy Model) has been used to make assumption on future cement production (Chateau, 2005). It expects a strong increase of cement production in developing countries (Szabó et al., 2003) and a limited increase in developed OECD countries in 2050 (Chateau, 2005). Other evaluations, following a business as usual (BAU), scenario expect an increase in cement production in Western European countries until 2020 and stagnation afterward (Szabó et al., 2003). In a recent study of the International Energy Agency (IEA 2007), global cement production is expected to rise from the current situation of 3 to 4 Gt (between 3.86 and 4.38) in 2050 and then remain stable. These evolutions seem irrespective of fuel and CO_2 emission prices (Pardo et al., 2011).

1.5 Conclusion

In this chapter, it has been shown that an improvement in cement plant efficiency is possible, particularly in developing countries where the majority of the cement production is located and where there are still cement plants with wet rotary kilns. However, an increase in energy efficiency incurs a large investment cost, which will then probably not be achievable, except for waste recovery systems. Furthermore, an increase in alternative fuel use has great potential as soon as cement plants have preheaters and that the waste supply chain is organised. Another improvement of the environmental impact of cement production can be achieved with clinker substitution. This solution is very efficient from technological, economic and environmental points of view as no drastic change in the production process is needed. Finally, the development of new low-carbon binders is needed to reach carbon reduction objective. Many possibilities are currently under development: magnesia and sulfo-aluminate cements or alkali activated binders.

1.6 References

Abad E., Caixach J., Rivera J., Gustems L., Massague G., Puig O. 2004. Temporal trends of PCDDs/PCDFs in ambiant air in Catalonia (Spain). *Science of the Total Environment*, **334–335**, 279–285.

ATILH. 2002. *Environmental inventory of French cement production*. Association Technique des Liants Hydrauliques (Hydraulic Binder Industries Union), Paris.

Bastier R. J., Bocan A., Gilbert B., Regnault A. 2000. Cement kiln: clinker workshops (in French), Techniques de l'ingénieur, BE 8844.

Bauer K., Hoenig V. 2010. Energy efficiency of cement plants. *Cement International*, **8**, 148–152.

Bensted J. 2006. Significance of chromate VI reducing agent in cements. *Cement Wapno Beton*, **1**, 29–35.

Bensted J., Smith J.R. 2009. Grinding aids during cement manufacture. *Cement Wapno Beton*, **4**, 179–188.

Cagiao J., Gomez B., Domenech J.L., Gutierrez Mainar S., Gutierrez Lanza H. 2011. Calculation of the corporate carbon footprint of the cement industry by the application of MC3 methodology. *Ecological Indicators*, **11**, 1526-1540.

Celik I.B., Oner M., Can N.M. 2007. The influence of grinding technique on the liberation of clinker minerals and cement properties. *Cement and Concrete Research*, **32**, 1334–1340.

CEMBUREAU 2010. www.cembureau.be/about-cement/cementmanufacturing-process. Accessed 24/09/2011.

Chateau B. 2005. VLEEM 2 (Very Long Term Energy Model), European Commission.

Chen C., Habert G., Bouzidi Y., Jullien A. 2010a. Environmental impact of cement production: detail of the different processes and cement plant variability evaluation. *Journal of Cleaner Production*, **18**, 478–485.

Chen C., Habert G., Bouzidi Y., Jullien A., Ventura A. 2010b. LCA allocation procedure used as an incitative method for waste recycling: an application to mineral additions in concrete. *Resources, Conservation and Recycling*, **54**, 1231–1240.

Damtoft J.S., Lukasik J., Herfort D., Sorrentino D., Gartner E.M. 2008. Sustainable development and climate change initiatives. *Cement and Concrete Research*, **38**, 115–127.

ECRA (European Cement Research Academy) 2009. Carbon Capture Technology: ECRA's approach towards CCS; Communication Bulletin. www.ecra-online.org/fileadmin/redaktion/files/pdf/ECRA_CCS_Communication_Bulletin.pdf. Accessed 18/03/2011.

EPA 1994. Emission factor documentation for AP-42, section 11.6: Portland Cement Manufacturing, Final report, EPA Contract 68-D2-0159, MRI Project No. 4601-01.

EPA 1999. Management standards proposed for cement kiln dust waste. Environmental Fact Sheet, EPA 530-F- 99-023.

European Commission 2010. Reference Document on Best Available Techniques in the Cement, Lime and Magnesium Oxide Manufacturing Industries.

European Union (EU) 2003. Directive 2003/53/EC of the European Parliament and the Council of 18 June 2003. *Official Journal of European Union*, **L178**, 24–27.

European Union (EU) 2008. Directive 2008/98/EC of the European Parliament and of the Council on waste and repealing certain directives. *Official Journal of European Union*, **L312**, 3–30.

Febelcem v.z.w. 2006. *Environmental report of the Belgian cement industry 2006* (in French). Brussels: Febelcem v.z.w.

Flatt R.J., Roussel N., Cheeseman C.R. 2012. Concrete: An eco material that needs to be improved. *Journal of the European Ceramic Society*, in press.

Flower D.J.M., Sanjayan J.G. 2007. Greenhouse gas emissions due to concrete manufacture. *International Journal of Life Cycle Assessment*, **12**, 282–288.

Fukuda K., Iwata T. Yoshida H. 2008. Melt differentiation induced by crystallisation of clinker minerals in a $CaO–SiO_2–Al_2O_3–Fe_2O_3$ pseudo quaternary system. *Journal of American Ceramic Society*, **91**, 4093–4100.

Gartner E. 2004. Industrially interesting approaches to low-CO_2 cements. *Cement and Concrete Research*, **9**, 1489–1498.

Gineys N., Aouad G., Sorrentino F., Damidot D. 2011. Threshold limits for trace elements in Portland cement clinker. *Cement and Concrete Research*, **41**, 1177–1184.

Gotti E., Marchi M., Costa U. 2007 Influence of alkalis and sulphates on the mineralogical composition of clinker, 12th International Congress on the Chemistry of Cement, Montreal, 20078 M3-01-1.

Guan Z., Chen Y., Qin S., Guo S. 2007. Effect of phosphorus on the formation of alite rich Portland clinker. 12th International Congress on the Chemistry of Cement, Montreal, 20078 M3-01-3.

Habert G., Billard C., Rossi P., Chen C., Roussel N. 2010. Cement production technology improvement compared to factor 4 objectives. *Cement and Concrete Research*, **40**, 820–826.

Habert G., d'Espinose de Lacaillerie J.B., Roussel N. 2011. An environmental evaluation of geopolymer based concrete production: reviewing current research trends. *Journal of Cleaner Production*, **19**, 1229–1238.

Hendriks C.A., Worrell E., deJager D., Block K., Riemer P. 2011. Emission reduction of greenhouse gases from the cement industry. www.wbcsdcement.org/pdf/tf1/prghgt42. Accessed 08/11/11.

Herfort D., Moir H.K., Johansen V., Sorrentino F., Bolio Arceo H. 2010. The chemistry of Portland cement clinker. *Advances in Cement Research*, **22**, 187–194.

Hills L., Johansen V.C. 2007. Hexavalent chromium in cement manufacturing: literature review. *PCA R&D Serial No. 2983*. Skokie: PCA.

Houghton J.T., Meira Filho L.G., Lim B., Tréanton K. (Eds.) 1996. *Revised 1996 IPCC Guidelines for National Greenhouse Gas Inventories, Greenhouse Gas Inventory Reference Manual* (Volume 3). Bracknell: Meteorological Office.

Humphreys K., Mahasenan M. 2002. *Toward a sustainable cement industry. Substudy 8: Climate change*. Geneva: WBCSD.

Huntzinger D.N., Eatmon T.D. 2009. A life-cycle assessment of Portland cement manufacturing: comparing the traditional process with alternative technologies. *Journal of Cleaner Production*, **17**, 668–675.

IEA 2007. *Tracking industrial energy efficiency and CO_2 emissions*. international Energy Agency: Paris.

JCR 2000. *Integrated Pollution Prevention and Control (IPPC): Reference Document on Best Available Techniques in the Cement and Lime Manufacturing Industries*. European Commission.

Jintao H., Qishou L., Jiaotong X., Yang P., Jinzhou H. 2009. Waste heat recovery. *World Cement*, **40**, 62–66.

Josa A., Aguado A., Heino A., Byars E., Cardim A. 2004. Comparative analysis of available life cycle inventories of cement in the EU. *Cement and Concrete Research*, **34**, 1313–1320.

Josa A., Aguado A., Cardim A., Byars E. 2007. Comparative analysis of the life cycle impact assessment of available cement inventories in the EU. *Cement and Concrete Research*, **37**, 781–788.

Kaantee U. 2004. Cement manufacturing using alternative fuels and the advantages of process modeling. *Fuel Processing Technology*, **85**, 293–301.

Klein H., Hoening V. 2006. Model calculations of the fuel energy requirement for the clinker burning process. *Cement International*, **3**, 45–63.

Lafarge 2006. *Sustainable report*. Technical report. Lafarge.

Lei Y., Zhang Q., Nielsen C., He K. 2011. An inventory of primary air poluutants and CO_2 emissions from cement production in China, 1990–2020. *Atmospheric Environment*, **45**, 147–154.

Madlool N.A., Saidur R., Hossain M.S., Rahim N.A. 2011. A critical review on energy use and savings in the cement industries. *Renewable and Sustainable Energy Reviews*, **15**, 2042–2060.

Maki I. 2006. Formation and microscopic textures of Portland cement clinker minerals, part 1. *Cement Wapno Beton*, **2**, 65–85.

McLellan B.C., Williams R.P., Lay J., van Riessen A., Corder G.D. 2011. Costs and carbon emissions for geopolymer pastes in comparison to ordinary Portland cement. *Journal of Cleaner Production*, **19**, 1080–1090.

Moudilou E., Amin F., Boullotte B., Thomassin J.H., LeCoustumer P., Lefrais Y. 2007. Phosphorus effect on physical and mechanical properties: Relation between clinker micro structure and hydration, *12th International Congress on the Chemistry of Cement*, Montreal, 2007 8 T1-04-1.

Nobis R. 2009. Burning technology. *Cement International*, **7**, 52–71.

Pardo N., Moya J.A., Mercier A. 2011. Prospective on the energy efficiency and CO_2 emissions in the EU cement industry. *Energy*, **36**, 3244-3254.

Rovira J., Mari M., Nadal M., Schumacher M., Domingo J.L. 2010. Partial replacement of fossil fuel in cement plant: risk assessement for the population living in the neighborhood. *Science of the Total Environment*, **408**, 5372–5380.

Rovira J., Mari M., Schumacher M., Nadal M., Domingo J.L. 2011. Monitoring environmental pollutants in the vicinity of a cement plant: a temporal study. *Archives of Environmental Contamination and Toxicology*, **60**, 372–384.

Scheubel B., Nachtwey W. 1997. Development of cement technology and its influence on the refractory kiln lining, in: *Refra Kolloqium*, Berlin, Germany, pp. 25–43.

Schneider M., Romer M., Tschudin M., Bolio H. 2011. Sustainable cement production – present and future. *Cement and Concrete Research*, **41**, 642–750.

Schumacher M., Agramunt M.C., Bocio A., Domingo J.L., de Kok H.A. 2003. Annual variation in the levels of metals and PCDD/PCDFs in soil and herbage samples collected near a cement plant. *Environment International*, **29**, 415–421.

Sogut M.Z., Oktay Z., Hepbasli A. 2009. Energetic and exergetic assessment of a trass mill process in a cement plant. *Energy Conversion and Management*, **50**, 2316–2323.

Sorrentino F. 2011. Chemistry and engineering of the production process: State of the art. *Cement and Concrete Research*, **41**, 616–623.

Szabó L., Hidalgo I., Ciscar J. C., Soria A., Russ P. 2003. *Energy consumption and CO_2 emissions from the world cement industry*. European Commission.

Taylor M., Tam C., Gielen D. 2006. Energy efficiency and CO_2 emissions from the global cement industry, in: International Energy Agency (Ed.), *Energy Efficiency and CO_2 Emission Reduction Potentials and Policies in the Cement Industry*. Paris: IEA.

Valderrama C., Granados R., Cortina J.L., Gasol C.M., Guillem M., Josa A. 2012 Implementation of best available techniques in cement manufacturing: a life-cycle assessment study. *Journal of Cleaner Production*, **25**, 60–67.

Van den Heede P., De Belie, N. 2012. Environmental impact and life cycle assessment (LCA) of traditional and 'green' concretes: literature review and theoretical calculations. *Cement and Concrete Composites*, **34**, 431–442.

Van Oss H.G., Padovani A.C. 2003. Cement manufacture and the environment, part II: environmental challenges and opportunities. *Journal of Industrial Ecology*, **7**, 93–127.

von Bahr B., Hanssen O.J., Vold M., Pott G., Stoltenberg-Hansson E., Steen B. 2003. Experiences of environmental performance evaluation in the cement industry. Data quality of environmental performance indicators as a limiting factor for benchmarking and rating. *Journal of Cleaner Production*, **11**, 713–725.

Vuuren D.P., Strengers B.J., Vries H.J.M.D. 1999. Long-term perspectives on world metal use – a system dynamics model. *Resources Policy*, **25**, 239–255.

Wirthwein R., Emberger B. 2010. Burners for alternative fuels utilisation: optimisation of kiln firing systems for advanced alternative fuel co-firing, *Cement International*, **8**, 42–46.

WBCSD 2008. The cement sustainability initiative: getting the numbers right, [online], Cement Industry Energy and CO_2 Performance. World Business Council for Sustainable Development, www.wbcsdcement.org/pdf/csi-gnr-report-with%20label.pdf Accessed 18/03/2011.

WBCSD 2009. *Cement technology road map 2009. Carbon emissions reductions up to 2050.* International energy agency, 36pp.

Zeman F. 2009. Oxygen combustion in cement production. *Energy Procedia*, **1**, 187–194.

2
Lower binder intensity eco-efficient concretes

B. L. DAMINELI, R. G. PILEGGI and V. M. JOHN,
University of São Paulo, Brazil

DOI: 10.1533/9780857098993.1.26

Abstract: This chapter discusses strategies for mitigating CO_2 emissions from the concrete chain. Current strategies, which focus on cement production processes such as clinker replacement, kiln efficiency, fuel substitution and CO_2 capture, among others, are not effective enough to compensate for the increase in production that is expected to occur in the future. Binder use efficiency is therefore a crucial strategy for decreasing environmental impacts. Exploratory data from literature, real market and lab research indicates that the potential for decreasing emissions from the concrete chain is very high, and will require changes to the formulation and processing of concrete, as well as the production of aggregates, cement and fillers.

Key words: binder efficiency, clinker replacement, benchmark, packing, dispersion.

2.1 Introduction

Anthropogenic CO_2 emissions from cement production have been rising steadily in both absolute and relative terms since the 1940s. The actual numbers are contested, though, varying from 5% of the total manmade CO_2 emissions in 2003 (Bernstein et al. 2007) to 8% in 2006 (Müller and Harnisch 2008).

This growth has occurred despite some success in reducing specific (t/t) CO_2 emissions (or CO_2 intensity), mostly due to an increase in energy efficiency in the kilns and the replacing of clinker with other materials[1]. As an example, between 2000 and 2006 total cement production increased 54% whereas absolute emissions increased only 42% (IEA and WBCSD 2009). The fact remains that current mitigation strategies have not been able to

[1]A significant share of the actual benefits of replacing clinker by materials such as blast furnace slag and fly ash is because these materials were considered to be CO_2 free (no allocation) since they were considered to be waste. This allocation criterion is under discussion (Chen et al. 2010; Birat 2011) and will probably be changed, causing a further increase in cement industry CO_2 emissions.

compensate for the increasing global cement demand, and it is clear that this trend cannot continue indefinitely (IEA and WBCSD 2009).

The demand for cement-based materials is expected to continue growing, and under current technology will require a continuous increase in cement production. The 2050 forecast for global cement production varies between 3.69×10^9 t/year in a low-growth scenario (IEA and WBCSD 2009) and 5×10^9 t/year in a high-growth scenario (Müller and Harnisch 2008), representing a 2.5 increase factor in relation to the 2010 production. This growth is expected to happen mostly in Kyoto's so-called 'non Annex I' developing countries that are under no obligation to reduce total CO_2 emissions.

If conventional mitigation strategies remain the same, a significant increase in total CO_2 emissions from cement production (IEA and WBCSD 2009; Müller and Harnisch 2008) can be expected. Simultaneously, due to the Kyoto protocol, global CO_2emissions are expected to be reduced. The report by Müller and Harnisch presents a scenario that combines the 450 ppm mitigation path, which implies a 50% reduction in global CO_2 emissions through 1990 with business as usual production of 5 Mt of cement with an increase of 260% cement-related CO_2 emissions in the same period of time (Müller and Harnisch 2008). In consequence, by the year 2050 cement production will be responsible for between 20% (IEA and WBCSD 2009) and 30% (Müller and Harnisch 2008) of the anthropogenic CO_2 emissions, a situation that will have enormous costs, both in political terms as well as in carbon taxes. This can be further aggravated if allocation of CO_2 for blast furnace slag and fly ash becomes a rule, or due to a future shortage of these residual materials.

The Blue scenario of the IEA and WDBSC (2009) roadmap proposes a total emission reduction of 0.79 Gt of CO_2 from a baseline of 2.34Gt – a 33% cut. Clinker substitution and energy efficiency potential are estimated to account for 10% each of the total reduction. The use of alternative fuels to CO_2 intensive pet-coke, especially renewable biomass, has the potential to reduce emissions even further, to 24%. But 56% of the reduction will come from carbon caption and storage (CCS), requiring from US$474 billion to US$593 billion in investments (the high-demand scenario) and, therefore, significantly increasing the cost of cement. Cement prices are due to increase, and will produce a high social impact on developing countries that need to improve and expand their built environment, including housing stock, at low costs.

Therefore, the cement industry needs to find new mitigation strategies that are environmentally and economically viable. So far, none of the new binders have shown the potential to scale up and take a significant share of the market (IEA and WBCSD 2009; Müller and Harnisch 2008; Habert et al. 2010). A little-explored alternative is to focus on the efficiency of cement applications, as suggested by John (2003) and Müller and Harnisch

(2008). This opens an entirely new field to investigation, such as strategies to promote the use of industrialized cement solutions, increase concrete strength, and reduce concrete production variability. Among all options, the reduction of the binder content in cement products, especially concrete, has progressively attracted researchers' attention (Damineli et al. 2010; Habert and Roussel 2009; Wassermann et al. 2009). This is especially attractive because aggregates and inert fines have much lower environmental impacts and costs than binders. These options depend on the capacity of the concrete research community to develop new innovative solutions related to the characteristics of aggregates, the formulation of methods, cementitious materials processing technologies as well as more reactive clinker and new binders (Damtoft et al. 2008; Popovics 1990).

This chapter will discuss the potential and limits of clinker replacement in terms of the availability of supplementary cementitious materials, present the concept, and further explore the potential of reducing binder intensity in concrete as a tool to mitigate CO_2 and other environmental impacts resulting from cement production.

2.2 Supplementary cementitious materials: limits and opportunities

The increase in the use of so-called supplementary cementitious materials, especially wastes, like blast furnace slag and fly ash pozzolans, has received much attention from the research community (Papadakis and Tsimas 2002) as a tool to mitigate environmental impacts (Damtoft et al. 2008), improve durability, and even to reduce costs. A substantial effort has been made to increase the replacement rates (Bilodeau and Malhotra 2000; Fu et al. 2000), a strategy that has become part of many green building certifications methods. But cement production is increasing, and a simple but crucial question needs to be answered: do we have mineral admixtures enough to make all the cement (and concrete) in the world more sustainable?

2.2.1 Blast-furnace slag (BFS)

Blast-furnace slag (BFS) use in cement dates from more than 100 years back, being part of the cement industry wherever it is available. Smithers Apex (2009) mapped demands and trends of the ferrous slag market. It was concluded that owing to low-carbon and other environmental policies, the demand for ferrous slag in cement industries worldwide stands to outstrip supply. The total supply of BFS is equal to about 13% of world cement production, which is a shortfall in comparison with demand. These by-products will become increasingly marketable, with the following price trends (Smithers Apex, 2009; USGS, 2010b).

Figure 2.1 shows that the total slag available to the cement market is expected to increase, from around 180×10^6 t/year (2008) to 290×10^6 t/year (2020). Humphreys and Mahasenan (2002) estimate that BFS intended for cement production in 2020 will be 123×10^6 t/year. A third source (WBCSD, 2009) provides a total BFS of 200×10^6 t/year in 2006.

Considering all sources, the availability of blast furnace slag will not surpass 300×10^6 t/year. A probable scenario for 2020 is around 4×10^9 t/year of cement, resulting in a maximum average clinker replacement with BFS lower than 10%. Humphreys and Mahasenan (2002) estimate that BFS will replace a maximum of 7% clinker by 2020. In some places, BFS will even decrease: according USGS (2010b), annual BFS production, in the USA, reached a peak of 21.6×10^6 t in 2005, but decreased to 12.5×10^6 t in 2009. Additionally, the potential use of slag in clinker-free alkali activated cements was explored (Pacheco-Torgal et al. 2008).

A new dilemma into this equation is the allocation of CO_2 to BFS. This will be a complex issue with several possible options, all giving very different CO_2 intensity values for the slag (Birat 2011; Chen et al. 2010). The solution will probably result from negotiation between the two industrial sectors. There is no obvious interest to develop cements with higher amounts of slag than the current standardized limits, except where a technical reason might exist (e.g. heat of hydration or durability) or in places that have localized surpluses due to logistic restrictions. In general, BFS should be considered a scarce material that must be explored in a very efficient way.

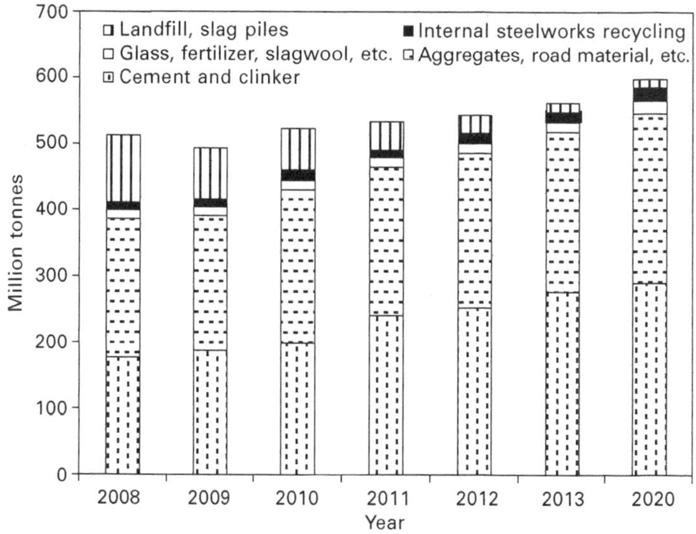

2.1 Forecast for the global use of ferrous slag, 2008–2020 (Smithers Apex, 2009).

2.2.2 Silica fume (SF)

Total world silica fume (SF) production was estimated to be 900 × 10^3 t/year in 2006 (ACI, 2006). In the USA the estimated SF production in 2004 was between 100 and 120 × 10^3 t (EPA, 2008). This accounts for less than 0.01% of total cement production making the analysis of silica availability unnecessary. Due to this reason, and the fact that it is a very expensive material, approximately US$800/t (Malhotra, 2005), it is only used in some special concretes.

2.2.3 Fly ash (FA)

Coal plants are still the world's most important energy source at present. Humphreys and Mahasenan (2002) evaluated total world fly ash (FA) availability and concluded that, by 2020, the average annual production will be 205 × 10^6 t. WBCSD (2009) estimates a total of 500 × 10^6 t/year of FA in 2006. USGS (2010a) shows that the FA production in the USA was near 56.9 × 10^6 t in 1999, increasing to a peak of 65.7 × 10^6 t in 2006 and, decreasing to 57.2 × 10^6 t in 2009. India produced about 100 × 10^6 t in 2005 (Malhotra, 2005). Considering that both countries are among the six largest coal producers in the world and that not all FA produced winds up in cement production, the rough estimate of Humphreys and Mahasenan (2002) seems to provide a good insight as to the magnitude of the amount of FA worldwide available for concrete.

Considering that the total annual world cement production expected for the same year (2020) is more than 3 × 10^9 t, the maximum average clinker replacement content by FA will be 7% or so. As for making this expectation worse, if it is true that coal production is increasing and will increase in the medium term, it is also true that a significant and quick decrease in its production from 2020 onward should be expected (Höök et al. 2008), as is shown in Fig. 2.2. So, a world FA production decrease is only a matter of time.

Allocation of CO_2 to fly ash is also under discussion (Chen et al. 2010) since for every ton of fly ash produced, from 2.9 to 5.2 t of CO_2 are released into the atmosphere (Yamamoto et al. 1997). Again, except for technical or localized reasons, there seems to be no need to promote the use of higher amounts of pozzolans in cement mixtures. Fly ash must also be considered a scarce material.

2.2.4 The need for additional supplementary cementitious materials

In considering the relative scarcity of traditional supplementary cementitious materials, the development of new supplementary materials becomes a

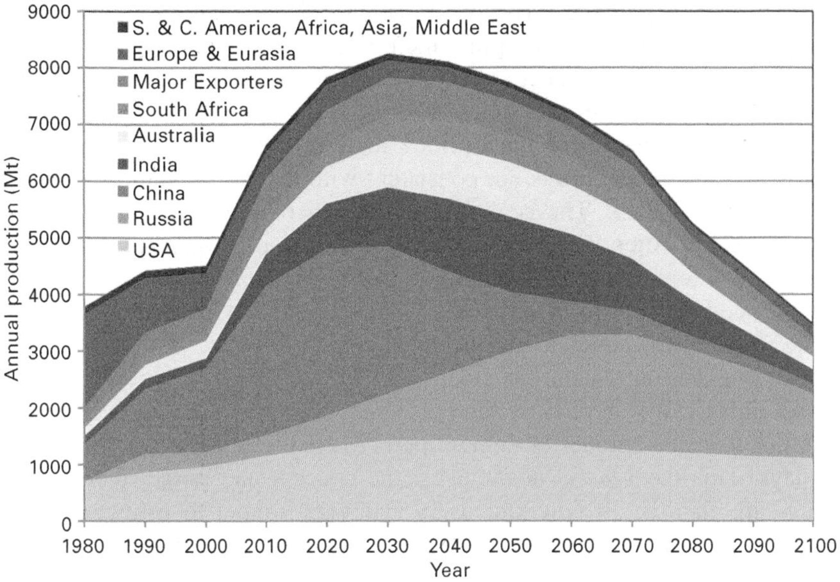

2.2 The forecast for world coal production divided into regions, from 1980 up to 2100 (Höök et al., 2008).

priority, as does as a better understanding of their chemical and physical effects (Damtoft et al. 2008; Gartner 2004; Lothenbach, et al. 2011).

Considering the new trend in allocating environmental impacts for waste that has reached a consistent market, the production of artificial pozzolans, such as calcined clays (Fernandez et al. 2011; He et al. 1995) and other new binders has become even more important (Damtoft et al. 2008; Gartner, 2004; Scrivener and Kirkpatrick 2008; WBCSD 2007). Probably, though, it will be difficult to provide new supplementary cementitious materials from waste flows in large quantities.

2.3 Binder efficiency in concrete production

Another strategy for reducing CO_2 emissions is to improve the efficiency of binder use (Damtoft et al. 2008; John 2003; Müller and Harnisch 2008). This is already being done by the cement industry to some extent, by replacing binders with almost inert limestone filler (Lothenbach et al. 2008), with current European standards allowing above 30%. But it must also be done during the actual formulation and production of cement products.

2.3.1 Binder use efficiency indicator

Simple and robust performance indicators are needed to allow the evaluation of use efficiency. Popovics (1990) defined efficiency as the strength developed by one unit of cement mass. The author developed a procedure for estimating the f_{eco}, the compressive strength which gives the highest efficiency for a given set of materials. Isaia et al. (2003) also adopted the strength developed by 1 kg of Portland cement as an indicator. However, these authors did not count supplementary cementitious materials (pozzolans) as cement. Nowadays, this procedure is not consistent with the global scarcity of such high-energy products. The economic efficiency of a concrete mix design was defined by Aïtcin (2000) as the cost of 1 MPa, or 1 year of service life of the concrete structure. This is a very advanced concept and service life is rather complicated to forecast and might depend on factors that are not in the scope of the concrete producer.

Our proposition (Damineli et al. 2010) is to adapt a concept that is almost universally adopted by the ready-mix concrete industry to measure its efficiency: the amount of cement per 1 MPa of compressive strength at 28 days. This has several advantages: the concept and range of resulting values are familiar to potential users and very simple to estimate, so it is easy to develop a benchmark of current performance. However, since the aim was to access the efficiency of use of scarce energy-intensive material, the amount of 'cement' was replaced by an equal amount of binder, by removing the amount of limestone filler from the calculation. Therefore, we define binder intensity (bi_{cs}) as the amount of binder (B, in kg m^{-3}) needed to provide 1 MPa of compressive strength (CS) at a given age:

$$bi_{cs} = \frac{B}{CS}$$

Since the CO_2 footprint of cement and even aggregate are available in many countries, we also proposed a CO_2 intensity of concrete (ci_{cs}), defined as the amount of CO_2 (C, in kg.m^{-3}) released to provide 1MPa of compressive strength.

$$bi_{cs} = \frac{C}{CS}$$

Using the same concept, all series of indicators can be easily defined, even the expected service life, were possible and relevant. Actually, the concept is to change the functional unit of concrete from one cubic meter to a relevant performance indicator, such as compressive strength or bending strength.

A benchmark on both *bi* and *ci* is presented in Fig. 2.3. This includes laboratory data from Brazil, plus another 28 countries, as well as data from two ready-mix concrete companies in São Paulo. It is worth mentioning that all the datasets overlap. It is also visible that there are significant differences in the binder efficiency for any given strength, evidencing the potential to increase eco-efficiency. From the available data it seems that a minimum *bi* with current technology is around 5 kg m^{-3} MPa^{-1} for concretes above 50 MPa. Below that threshold, minimum *bi* seems to follow the 250 kg m^{-3} binder content line, which corresponds roughly to the minimum cement content in many national standards (ABNT, 2006; Grube and Kerkhoff, 2004). Since most of the concrete is below 50 MPa class, the actual binder intensity is much higher than the possible minimum.

As an example, we estimated a *ci* based on very simple information. BFS, FA and SF were considered to be CO_2 free. The clinker was considered an average value of 1 kg CO_2 by 1 kg of clinker. Aggregate emissions, mostly related to transportation distances, were also disregarded. Minimum *ci* values are around 2 kg m^{-3} MPa^{-1}, and correspond to high fractions of typical CO_2 intensity which is between 3 and 9 kg m^{-3} MPa^{-1}. And, unlike *bi*, CO_2 intensity seems not to be as sensitive to compressive strength.

The combination of *bi* and *ci* allows demonstrating the risks of judging environmental impact based only on one indicator. The data shows that it is possible to produce concrete with very low *ci* values – which could imply an efficient concrete in response to global warming concerns – both using very little pure Portland cement (ci = 4 kg m^{-3} MPa^{-1}) or with a very high *bi* due to the heavy use of 'zero CO_2 emissions' supplementary materials. This strategy is even cited in the LEED certification as a way for 'green' concretes (US Green Building Council, 2009). However, the foreseen change in CO_2 allocation procedures for the supplementary cementitious materials will obviously change this equation and certainly will raise the CO_2 intensity level. It will make CO_2 intensity more dependent on binder intensity. For more detailed discussions about *bi* and *ci*, see Damineli et al. (2010).

As can be seen in Fig. 2.3, average data from two ready-mix concrete (RMC) producers were obtained. Data from the first one (triangles) are representative of a 3 month production interval, and from the second one (squares) represents 6 months. Since these ready-mix concrete producers are among the best ones in the Brazilian market, it is assured that the analysis can be considered as a good parameter for assessing the current state of technology in the market. RMC1 works with concretes in an interval of 20–40 MPa, with a *bi* varying from 8 to 12 kg m^{-3} MPa^{-1} and average near 10 kg m^{-3} MPa^{-1} – almost double the best *bi* found in literature. On the other hand, RMC2 reached *bi* near 6.5 to 7 kg m^{-3} MPa^{-1}. But maximum values are also higher than RMC1. Market averages seems to be something near 10 kg m^{-3} MPa^{-1}.

2.3.2 Strategies to increase binder use efficiency

Binder use efficiency is strictly dependent on two concepts: (1) maximization of particle mobility with lowest water content possible; and (2) an increase of

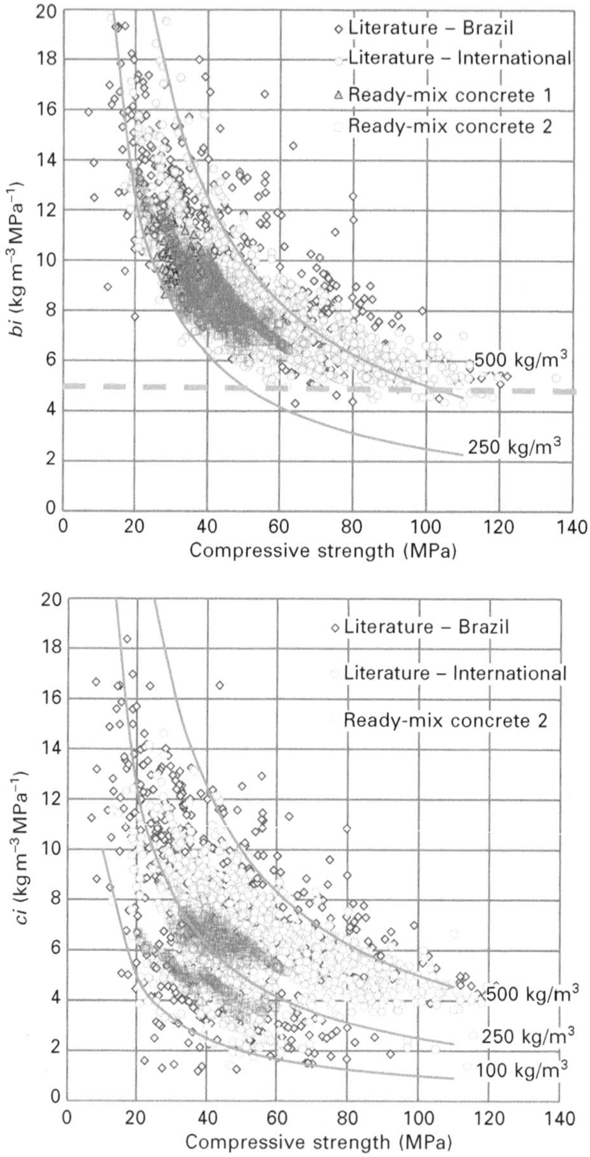

2.3 Benchmark of *bi* (left) and *ci* (right). This includes laboratory data from Brazil, plus 28 other countries (Damineli et al. 2010) as well as comprehensive unpublished dataset from two ready-mix companies in São Paulo. CO_2 intensity is very dependent on cement composition.

binder particle reactivity by exposing the highest surface area to hydration. To achieve this, two different strategies are important: dispersion of particles and packing of particles.

Dispersion of particles

Fine particles have a tendency to agglomerate since the smaller the particle, the lower the effect of the gravitational force and the higher the effect of the surface force that act on them (physical-chemical attraction and repulsion). These interactions are determined by van de Waals forces and by different chemical charges for different phases, making them attract. So, in normal conditions, cementitious suspensions are agglomerated. For the suspension these agglomerates may result in large particles, which: (1) modify particle size distribution; and (2) mobility of flow lines makes increasing the viscosity of the system difficult (Pandolfelli et al. 2000); (3) increase water consumption since there are voids between the particles where water is located; and (4) decrease binder reactivity since the surface area available to chemical hydration reactions is decreased by the contact between particles. So, the dispersion is crucial for increasing binder efficiency. A concrete made with high binder efficiency needs to be completely dispersed.

Packing with flowability

Packing of particles is another crucial concept that needs to be controlled to increase binder use efficiency. Concrete is a material that should have compressive strength and adequate rheology that is conducive to that required by the project. The problem occurs when the efficiency of one is increased leading to a decrease in efficiency of the other. So, in a fixed concrete mix an increase in strength is responsible for a decrease in rheological parameters such as flowability and/or viscosity, and vice versa.

In a system of granular particles, flowability is produced by the mobility proportioned by the fluid inserted between the grains. First, the fluid fills the voids between particles, then it covers the surface area of the particles. After these two steps, the fluid starts to detach the particles (Vogt, 2010), creating an increasing distance of separation that allows them to move independently. The greater the distance, the easier the movement.

When the voids between particles are high, the fluid content required for allowing the flow is higher. So, the particle packing design needs to decrease the voids between particles as much as possible, because with lower fluid content in the voids, the fluid content needed to make the system flow is lower (Pileggi et al. 2003). So, a good packing design for a granular system is one that that allows the greatest flowability with the lowest content of fluid since the efficiency of binder depends strictly on lower fluid contents.

From the rheological point of view, concretes can be described, in the fresh state, as heterogeneous suspensions composed of: (1) aggregates with diameters above 125 μm (predominance of mass forces); (2) fines, reactive or inert, with diameters under 125 μm (predominance of surface area forces)

(Hunger and Brouers 2009); and (3) water. Due to a problem of component mass and density, water is the fluid that can move the fines, and the union of water plus fines creates a paste; and the paste becomes the fluid for the aggregates. It can be inferred that two different pack models are necessary for controlling concrete rheological behavior, one for the paste and another for the aggregates. Because of this, the distance between particles for concrete suspensions is usually calculated by two indexes: the MPT (maximum paste thickness, which measures the distance between aggregates and, consequently, the mobility of them and is determined by the paste); and the IPS (inter-particle spacing, measuring distance between fines, is determined by the water) (Pandolfelli et al. 2000):

$$\mathrm{IPS} = \frac{2}{\mathrm{VSA}} \times \left[\frac{1}{V_s} - \left(\frac{1}{1 - P_{of}}\right)\right]$$

where VSA = volumetric surface area, calculated by the product of specific surface area and solid density; V_s is the volumetric solid fraction on suspension; and P_{of} is the pore fraction on system when all particles are in touch in the maximum packing condition:

$$\mathrm{MPT} = \frac{2}{\mathrm{VSA_c}} \times \left[\frac{1}{V_{sc}} - \left(\frac{1}{1 - P_{ofc}}\right)\right]$$

where $\mathrm{VSA_c}$ = VSA of coarse fraction; V_{sc} is the volumetric coarse solid fraction on suspension; and P_{ofc} is the porosity of coarse fraction on system when all particles are in touch in the maximum packing condition.

A concrete design with high binder efficiency is one that uses the lowest water content while allowing the highest IPS possible, since it is the condition when the flowability of the paste is increased at the same time as the compressive strength is the highest possible due to the lowest water content. Conversely, a high-binder efficient concrete is also one that uses the lowest paste content while allowing the highest possible MPT, because in this condition the rheology requirements are the same, even though there is a low paste content – which implies the lowest binder content for the same compressive strength.

Adequate concrete rheology is currently reached by increasing paste content to increase the MPT, where, for current concretes, pastes are composed of water and cement. This implies the use of a binder for guaranteeing rheological parameters which, consequently, decreases binder use efficiency since the w/c ratio is normally kept constant, which, however, does not change compressive strength, but the increase in paste content increases binder content.

The use of inert fines for providing rheology for the concrete is an effective way of decreasing binder content by replacing it with a very low impact product – and also, one that is largely available since inert fillers are made from materials easily found throughout the world (Craeye et al. 2010; Esping, 2008; Hunger and Brouers, 2009; Moosberg-Bustnes et al. 2004; Petit and Wirquin, 2010). Inert fines increase the packing of the paste by filling the voids between cementitious particles; therefore, the water content required to provide the flowability is decreased since less water is needed to fill these voids, thus, more water becomes available for detaching particles. But the fines also have high surface area, which can increase water demand and subsequently, decrease strength. Due to this, together with the packing project, the characteristics of the fines need to be controlled carefully to allow for good rheology without increasing water demand too much. The advance in dispersant technologies also allows increasing use of these fines in the paste.

Exploratory results

By adapting concepts for increasing packing and complete dispersion used in refractory concretes (Bonadia et al. 1999; Ortega et al. 1999; Pileggi and Pandolfelli 2001; Pileggi et al. 2003) to make Portland cement concretes – which use much different and cheaper materials – and, by using fillers for the design of high performance pastes for concrete (Lagerblad and Vogt, 2004; Moosberg-Bustnes et al. 2004; Vogt, 2010), important binder intensity reductions were achieved as presented by the large dark circular dots on Fig. 2.4.

Our preliminary results show it is possible to mix concretes with compressive strengths between 20 and 40 MPa (same as market requirements) and with *bi* between 5 and 7 $kg\,m^{-3}\,MPa^{-1}$, filling an empty area of the graphic – the area with binder consumption under 250 $kg\,m^{-3}$. This applies in mixtures with binder contents lower than that required by national standards and still having good strength, which proves that national standards are actually a big hindrance as regards the increase of binder efficiency. It is also fundamental to know that most of these concretes are self-compacting, which means, in practice, that lower *bi* results could be achieved if the rheological requirements were decreased to produce concretes with higher yield stress.

Just as an example, Fig. 2.5 presents the full particle size distribution of one of the concretes presented in Fig. 2.4. The concrete shown in Fig. 2.5 used 194.4 $kg\,m^{-3}$ of total binder content to reach a 28-day compressive strength of 29.5 MPa ($bi = 6.6\,kg\,m^{-3}\,MPa^{-1}$), slump 280 mm (self-compacting concrete), IPS of 0.23 μm and MPT of 5.6 μm. Rheological tests also indicated that it would be appropriate for pumping. The very low viscosity and yield stress of this mix is achieved with a *bi* much lower than market standards for its strength. For applications that do not need to have such a fluid rheology, this *bi* could be easily decreased.

A serious limitation to implement packing strategy is the existence of

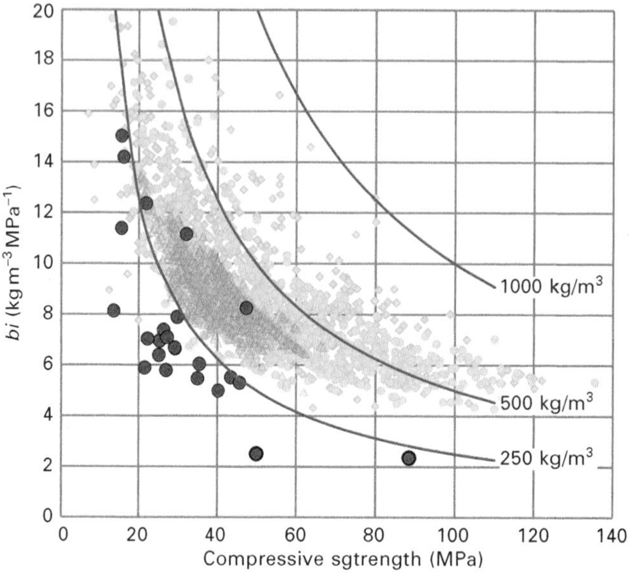

2.4 Examples demonstrating the potential to reduce *bi* of concretes. Concretes formulated and produced in our lab at USP as well as in collaboration with CBI/Sweden. Most concretes are self compacting. The two with *bi* below 4 have slump above 180 mm and were produced at CBI.

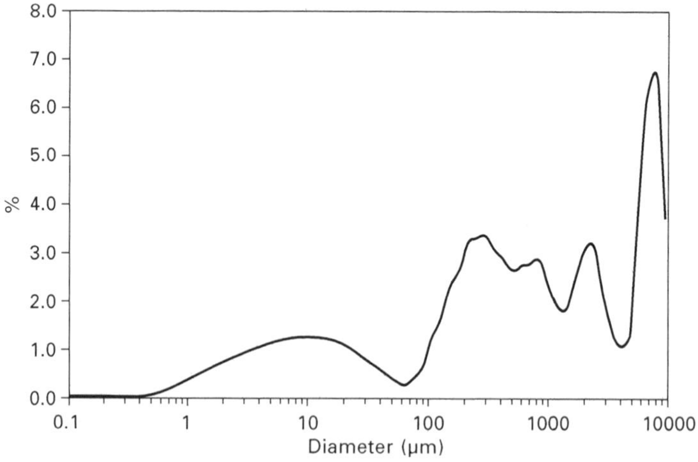

2.5 Particle size distribution of a concrete mixed in the laboratory. The distribution is a combination of eight different materials, six of them below 2 mm.

variety of controlled fillers. Exploring these raw materials in Sweden[2], it was possible to produce concretes even more efficiently. A formulation using only 210 kg m^{-3} of total binder (including 10 kg of SF), slump of 180 mm reached 88.4 MPa (100 × 200 mm cylinder) at 28 days. This corresponds to a *bi* of 2.37 kg m^{-3} MPa^{-1} – less than one-half of the best practice found in the literature benchmark. Comparing with marketable concretes, this means cutting total binder consumption by a factor of 4^2. In the same project, another concrete having only 126.3 kg m^{-3} of total binder content was also designed and achieved 190 mm of slump and 28 day-compressive strength of 50 MPa, which delivered a *bi* of 2.53 kg m^{-3} MPa^{-1}.

This shows that the difficulty of designing concretes with very low cement content can be overcome by scientific methods of packing and dispersion of particles, requiring lower binder content for achieving rheological parameters.

2.3.3 In-use performance and robustness of low binder concretes

The need for reductions in binder intensity questions the need to minimize cement content. Wassermann et al. (2009) show that concrete durability performance indicators are not affected (carbonation depth, shrinkage) or even improved (chloride penetration and capillary absorption coefficient) when binder content is reduced within the limits investigated. Dhir et al. (2004) conducted a comprehensive investigation and concluded that specifying minimum cement content for concrete durability was not necessary. Popovics (1990) shows that for the same water/cement (W/C) ratio an increase in binder content results in a decrease in compressive strength. So the authors believe that standards concerning minimum cement content in concrete should be revised. This is a fundamental strategy for allowing the decrease of *bi* for concretes with less than 50 MPa – the most widely used at present. A *bi* of 5 or something near could be feasible even for low strength concretes.

However, the published results deal with binder content above 221 kg m^{-3} (Wassermann et al. 2009) and 215 kg m^{-3} (Dhir et al. 2004). So there is a need to investigate the effect of further reducing the binder content to values as low as 120 kg m^{-3} in the most important degradation mechanisms and, eventually, develop new protection strategies.

[2]Research in collaboration with Prof. Björn Lagerblad (CBI-KTH). One of the concretes formulated in this research won the 'StarkastBetong' (the strongest concrete) international competition, which was held in CBI Betonginstituten, 2012. This competition established a maximum cement content of 200 kg m^{-3} and did not limit the use of supplementary materials. However, the winner used just 200 kg m^{-3} of cement and 10 kg m^{-3} of silica fume as total binders.

Another important issue that needs to be better comprehended for the improvement of binder use efficiency is the robustness of the concrete mix. Low binder content implies that small variation on water content will have larger impact on water/cement ratio affecting the porosity and mechanical strength. Also, this concretes demands efficient dispersants and the robustness of the system cement-dispersants must be improved (Nkinamubanzi and Aïtcin 2004).

The effect of binder content reduction on mechanical behavior, including fracture, dimensional stability and creep behavior also need to be investigated.

2.4 Conclusion and future trends

Current strategies for decreasing CO_2 emissions in cement production are not enough to mitigate environmental impacts, as expected by society. Carbon capture and storage, as under discussion by the cement industry, is not a desirable option due to cost impact that leads to social and economic implications in developing countries. On the otherhand, binder use optimization has enormous potential to supply the cement-based materials that society needs,with much lower environmental impacts.

Our results show that it is possible to produce in laboratory conditions concretes using less than $3\,kg\,m^{-3}\,MPa^{-1}$ of binder. This binder intensity is almost a third of the current market and lab benchmarks. This can be done using well-established packing engineering, selected fillers and commercial Portland cement and dispersants. Scaling up this solution will require substantial R&D effort, facing problems related to robustness of this systems and its long-term performance. It also might demand new technologies for production of better controlled aggregates, especially ultra-fine particles, more reactive clinkers, new admixtures, at compatible cost.

The authors believe that a *bi* between 5 and $6\,kg\,m^{-3}\,MPa^{-1}$ can be a feasible target for the market. This will allow a reduction of 30–50% of total binder intensity, which make possible doubling the cement-based material production without investing in new kilns and using much additional fuel. This strategy seems to be economically attractive to the industry since it means replacing expensive investment and the operational cost of kilns by new and more sophisticated mills. It will also avoid the need for huge investment and operational costs of carbon caption and storage.

2.5 Acknowledgements

Bruno Damineli's research is supported by CNPq and CAPES. The authors wish to acknowledge the collaboration of CBI Betonginstituten team and Prof. Björn Lagerblad (KTH-CBI).

2.6 References and further reading

Aitcin, P. C. 2000. Cements of yesterday and today: *Concrete of tomorrow. Cement and Concrete Research, New York*, n. 30, p. 1349–1359.

ACI 2006. 234R-06: *Guide for the Use of Silica Fume in Concrete*, 2006, 63p. American Concrete Institute. Available: <http://www.concrete.org/bookstorenet/ProductDetail.aspx?ItemID=23406>.

ABNT 2006. NBR 12655: *Concreto de cimento Portland – Preparo, controle e recebimento – Procedimento*. Associação Brasileira de Normas Técnicas Rio de Janeiro, 18p (in Portuguese).

Bernstein, L. et al. 2007. Chapter 7: Industry. In: Metz et al. (ed). *Climate Change 2007: Mitigation – Contribution of Working Group III to the Fourth Assessment Report of the Intergovernmental Panel on Climate Change*, Cambridge University Press, Cambridge, United Kingdom and New York, USA, pp. 447–496 (IPCC Report).

Bilodeau, A., and Malhotra, V. M. 2000. High-volume fly ash system: Concrete solution for sustainable development. *ACI Structural Journal* **97** (1): 41–48.

Birat, J.P. 2011. The sustainability footprint of steelmaking by-products. *Steel Times International*, September. <http://www.steeltimesint.com/contentimages/features/Web_Birat.pdf>.

Bonadia, P. et al. 1999. Applying MPT principle to high-alumina castables. *American Ceramic Society Bulletin, United States*, 78, (3): p. 57–60.

Cembureau 1999. Environmental benefits of using alternative fuels in cement production – a life-cycle approach, 25p. Available :<http://www.cembureau.be>.

Chen, C., Habert, G., Bouzidi, Y., Jullien, A., Ventura, A. 2010. LCA allocation procedure used as an incitative method for waste recycling: an application to mineral additions in concrete. *Resources, Conservation and Recycling* **54** (12) (October): 1231–1240.

Craeye, B., et al. 2010. Effect of mineral filler type on autogenous shrinkage of self-compacting concrete. *Cement and Concrete Research, New York*, **40**: 908–913.

Damineli, B. L. et al. 2010. Measuring the eco-efficiency of cement use. *Cement and Concrete Composites*, **32**: 555–562.

Damtoft, J. S. et al. 2008. Sustainable development and climate change initiatives. *Cement and Concrete Research, New York*, **38**: 115–127.

Dhir, R.K., McCarthy, M.J., Zhou, S.; Tittle, P.A.J. 2004. Role of cement content in specifications for concrete durability: cement type influences. *Proceedings of the ICE – Structures and Buildings*, **157**(2): 113–127.

EPA (Environmental Protection Agency) 2008. Study on Increasing the Usage of Recovered Mineral Components in Federally Funded Projects Involving Procurement of Cement or Concrete to Address the Safe, Accountable, Flexible, Efficient Transportation Equity Act: A Legacy for Users, 225p. Available: <http://www.epa.gov/osw/conserve/tools/cpg/pdf/rtc/report4–08.pdf>.

Esping, O. 2008. Effect of limestone filler BET(H_2O)-area on the fresh and hardened properties of self-compacting concrete. *Cement and Concrete Research*, **38**: 938–944.

Fernandez, R., Martirena, F., Scrivener, K. L. 2011. The origin of the pozzolanic activity of calcined clay minerals: A comparison between kaolinite, illite and montmorillonite. *Cement and Concrete Research* **41**(1): 113–122.

Fu, X., Hou, W., Yang, C., Li, D., Wu, X. 2000. Studies on Portland cement with large amount of slag. *Cement and Concrete Research* **30**(4): 645–649.

Gartner, E. 2004. Industrially interesting approaches to 'low CO_2' cements. *Cement and Concrete Research*, **34**: 1489–1498.

Grube, H., Kerkhoff, B. 2004. The new German concrete standerds DIN EN 206-1 and DIN EN 1045-2 as basis for the design of durable constructions. *Concrete Technology Reports*, p. 19–27.

Habert, G., Roussel, N. 2009. Study of two concrete mix-design strategies to reach carbon mitigation objectives. *Cement and Concrete Composites*, **31**: 397–402.

Habert, G., Billard. C., Rossi, P., Chen, C., Roussel, N. 2010. Cement production technology improvement compared to factor 4 objectives.*Cement and Concrete Research* **40**(5): 820–826. doi:10.1016/j.cemconres.2009.09.031.

He, C., Osbaeck, B., Makovicky, E. 1995. Pozzolanic reactions of six principal clay minerals: activation, reactivity assessments and technological effects. *Cement and Concrete Research* **25**(8): 1691–1702.

Höök, M. et al. 2008. A supply-driven forecast for the future global coal production. Contribution to ASPO, 2008, 48p. Available: <http://www.tsl.uu.se/uhdsg/Publications/Coalarticle.pdf>.

Humphreys, K., Mahasenan, M. 2002. Substudy 8: Climate Change. In: *Battelle. Toward a Sustainable Cement Industry*, 34p (WBCSD Report). Available: <http://wbcsd.org>.

Hunger, M., Brouers, H. J. H. 2009. Flow analysis of water–powder mixtures: Application to specific surface area and shape factor. *Cement and Concrete Composites*, **31**: 39–59.

IEA (International Energy Agency) 2006. *Energy Technology Perspectives – Scenarios & Strategies to 2050*, 486p.Available :<http://www.iea.org/Textbase/techno/etp/index.asp>.

IEA and WBCSD (International Energy Agency) and (World Business Council For Sustainable Development) 2009. *Cement Technology Roadmap 2009 – Carbon Emissions Reductions up to 2050*, 36p (CSI Report). Available: <http://wbcsd.org>.

Isaia, G. C., Gastaldini, A. L. G., Moraes, R. 2003. Physical and pozzolanic action of mineral additions on the mechanical strength of high-performance concrete. *Cement and Concrete Composites* **25**(1) 69–76. doi:10.1016/S0958-9465(01)00057-9.

John, V. M. 2003. On the sustainability of concrete. *UNEP Industry and Environment, Paris*, **26**, (2–3): 62–63. Available: <http://www.uneptie.org/media/review/archives.htm>.

Lagerblad, B., Vogt, C. 2004. *Ultrafine Particles to Save Cement And Improve Concrete Properties*. Swedish Cement and Concrete Research Institute.

Lothenbach, B., Saout, G. L., Gallucci, E., Scrivener, K. 2008. Influence of limestone on the hydration of portland cements. *Cement and Concrete Research* **38**(6): 848–860. Available doi:10.1016/j.cemconres.2008.01.002.

Lothenbach, B., Scrivener, K., Hooton, R.D. 2011. Supplementary cementitious materials. *Cement and Concrete Research* **41** (3): 217–229. Available G:\wwf_Cimento. G:\wwf_Cimento.

Malhotra, V. M. 2005. Availability and management of fly ash in India. *The Indian Concrete Journal*, 5p. Available: <http://www.icjonline.com/forum/Point_of_view.pdf>.

Moosberg-Bustnes, H., Lagerblad, B., Forssberg, E. 2004. The function of fillers in concrete. *Materials and Structures*, **37**: 74–81.

Müller, N., Harnisch, J. 2008. *A Blueprint for a Climate Friendly Cement Industry*. Gland: WWF Lafarge Conservation Partnership, 2008, 94p. (WWF-Lafarge Conservation Partnership Report). Available <http://assets.panda.org/downloads/english_report_lr_pdf.pdf>.

Nkinamubanzi, P. C., Aïtcin, P. C. 2004. Cement and superplasticizer Combinations:

Compatibility and robustness. *Cement, Concrete and Aggregates*, **26**(2): p. 102–109

OCC (Office for Climate Change) 2005. *Stern Review: the Economics of Climate Change*, 27p (Executive Summary). Available: <http://www.occ.gov.uk/activities/stern.htm>.

Ortega, F. et al. 1999. Optimizing particle packing in powder consolidation. *American Ceramic Society Bulletin, United States*, **79**(2): 155–159.

Pacheco-Torgal, F., Castro-Gomes, J., Jalali, S. 2008. Alkali-activated binders: A review. Part 1. Historical background, terminology, reaction mechanisms and hydration products. *Construction and Building Materials* **22**(7): 1305–1314.

Pandolfelli, V. C. et al. 2000. Dispersão e empacotamento de partículas – princípios e aplicações em processamento cerâmico.São Paulo: Fazendo Arte Editorial, 224 p (in Portuguese).

Papadakis, V.G., Tsimas, S. 2002. Supplementary cementing materials in concrete: Part I. Efficiency and design. *Cement and Concrete Research* **32**(10): 1525–1532.

Petit, J-Y., Wirquin, E. 2010. Effect of limestone filler content and superplasticizer dosage on rheological parameters of highly flowable mortar under light pressure conditions. *Cement and Concrete Research*, **40**: 235–241.

Pileggi, R. G. et al. 2003. High-performance refractory castables: Particle size design. *Refractories Applications and News, United States*, **8**(5): 17–21.

Pileggi, R. G., Pandolfelli, V. C. 2001. Rheology and particle size distribution of pumpable refractory castables. *American Ceramic Society Bulletin, United States*, **80**(10): 52–57.

Popovics, S. 1990. Analysis of the concrete strength versus water-cement ratio relationship. *ACI Materials Journal*, **87**(5): 517–529.

Scrivener, K. L., Kirkpatrick, R. J. 2008. Innovation in use and research on cementitious material. *Cement and Concrete Research*, **38**: 128–136.

Smithers Apex 2009. *Future of Ferrous Slag: Market forecasts to 2020*, 144p (Market Report). Available: <http://www.smithersapex.com/Future-of-Ferrous-Slag-Market-Forecasts-to-2020.aspx>.

Stewart,B. R., Kalyoncu, R. S. 1999. Materials flow in the production and use of coal combustion products. In: 1999 International Ash Utilization Symposium, Center For Applied Energy Research University of Kentucky, paper 46, 1999, Proceedings. Kentucky, 1999, 9p. Available: <http://www.flyash.info/1999/econom/kalyon2.pdf>.

USGS (United States Geological Survey) 2007. *USGS Minerals Yearbook: Cement Annual Report 2005*. Available: <http://minerals.usgs.gov/minerals/pubs/commodity/cement/cemenmyb05.pdf>.

USGS (United States Geological Survey) 2010a. *Coal and Combustion Production Statistics*. 3p. Available: <http://minerals.usgs.gov/ds/2005/140/>.

USGS (United States Geological Survey) 2010b. *Iron and Steel Slag Statistics*, 3p. Available: <http://minerals.usgs.gov/ds/2005/140/>.

US Green Building Council 2009. LEED 2009 for New Construction Version 2.2 Reference Guide.

Vogt, C. 2010. Ultrafine particles in concrete: influence of ultrafine particles on concrete properties and application to concrete mix design, 155 p. PhD Thesis, Royal Institute of Technology, Stockholm, Sweden.

Wassermann, R., Katz, A., Bentur, A. 2009. Minimum cement content requirements: A must or a myth? *Materials and Structures*, **42**(7): 973–982.

WBCSD (World Business Council for Sustainable Development) 2007. *The Cement Sustainability Initiative (CSI)*, 8p (CSI Report). Available: <http://wbcsd.org>.

WBCSD (World Business Council For Sustainable Development) 2009. *Cement Industry Energy and CO_2 Performance – Getting the Numbers Right*, 44p (CSI Report). Available: <http://wbcsd.org>.

Yamamoto, J. K. et al. 1997. Environmental impact reduction on the production of blended portland cement in Brazil. *Environmental Geosciences*, **4**(4): 192–206.

3
Life cycle assessment (LCA) aspects of concrete

S. B. MARINKOVIĆ, University of Belgrade, Serbia

DOI: 10.1533/9780857098993.1.45

Abstract: The concrete industry is considered to be a large consumer of energy and natural resources, and is one of the main sources of greenhouse emissions and waste generation. The production and utilization of concrete and concrete structures have a large impact on the environment, and so the environmental assessment of concrete is of great importance in terms of achieving a sustainable society. This chapter includes instructions on how to apply the life cycle assessment (LCA) methodology to concrete, including a general description, life cycle inventory and life cycle impact assessment of concrete, future trends and sources of further information.

Key words: concrete, life cycle assessment, (LCA) environmental product declaration, tools, rating systems, sustainability assessment.

3.1 Introduction

Concrete is the most widely used building material in the construction industry. It is estimated that today's world concrete production is about 6 billion tonnes per year, i.e. one tonne per person per year (ISO/TC 71, 2005). The concrete industry is regarded as a large consumer of natural resources such as natural aggregates and non-renewable fossil fuels. Moreover, consumption is rising constantly and rapidly as the production and utilization of concrete increases. For example, three billion tonnes of aggregate are produced each year in the countries of the European Union (European Environment Agency, 2008). Waste arising from the construction sector, so-called construction and demolition waste (C&D waste), is also a relevant concern in the protection of the environment. For example, about 850 millions tonnes of C&D waste are generated in EU per year, which represent 31% of the total waste generation (Fisher and Werge, 2009). After production, concrete 'lives' through various types of buildings, bridges, roads, dams, etc., which have their own impact on the environment. Among them, buildings are considered as large energy consumers in the course of their service life. According to the European Commission, they are responsible for 40% of energy consumption and 36% of CO_2 emissions in the European Union (European Commission, 2011).

The most important environmental impacts of the production and utilization of concrete and concrete structures are:

- large consumption of natural resources;
- large consumption of energy (mostly for cement production and reinforcement steel production; in addition, for operation and maintenance of buildings and other structures; finally for transportation, construction, demolition and recycling to a lesser extent);
- large emissions of greenhouse gasses, primarily CO_2 which are responsible for climate change and originate mostly from cement production and energy consumption; to a lesser extent, emissions of SO_2 which is responsible for acidification and mostly originates from the transportation phase;
- large amount of construction and demolition waste produced.

Therefore, concrete has a large impact on the environment because of its enormous production and utilization. That is why the environmental assessment of concrete is of great importance with regard to the efforts towards creating a sustainable society. There are many methodologies for evaluating the environmental loads of processes and products during their life cycle, but the most widely acknowledged (and standardized) is life cycle assessment (LCA).

3.2 General description of life cycle assessment (LCA) methodology

LCA is a methodology for evaluating the environmental loads of processes and products during their life cycle. According to ISO standards 14040–14043 (ISO, 2006a), LCA consists of four steps: (1) goal and scope definition, (2) creating the life cycle inventory (LCI), (3) assessing the environmental impact (LCIA) and (4) interpreting the results (Fig. 3.1).

The first step of an LCA involves the goal and scope definition. The goal of an LCA study must be clearly defined including the intended application, the intended use of the results and users of the results (intended audience). The goals of an LCA study are, for example, to compare two or more different products fulfilling the same function, to identify improvement possibilities in further development of existing products or in innovation and design of new products, etc. The definition of the scope of an LCA study sets the borders of the assessment: what is included in the modeled system and what assessment methods are to be used. The following items must be clearly described within the scope definition: the system to be studied and its function, the functional unit, the system boundaries, the types of impact and the methodology of impact assessment, data quality requirements, assumptions and limitations, the type of critical review, if any (Jensen et al., 1997). Detailed information on each of these items is given in Section 3.3.

Inventory analysis is the second step in an LCA. It involves data collection

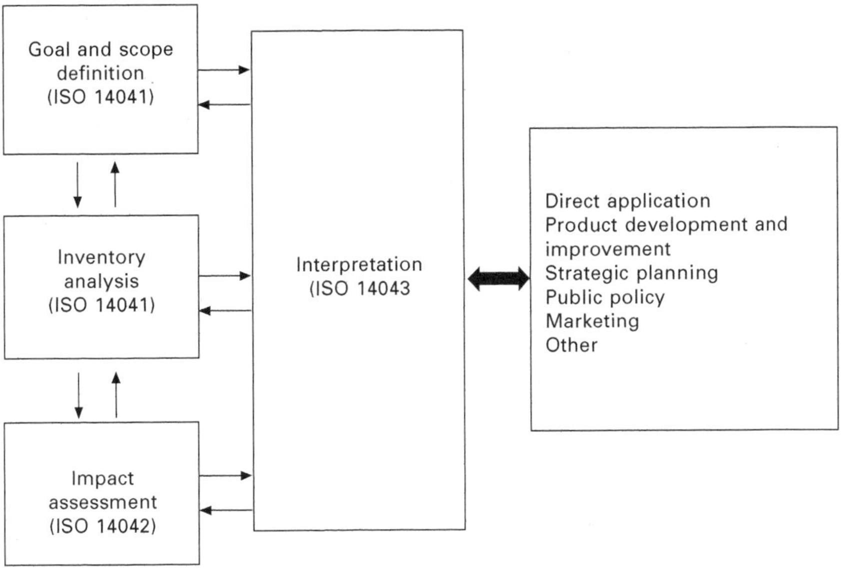

3.1 LCA phases.

and calculation procedures to quantify relevant inputs and outputs of a product system. These inputs and outputs include the use of resources, emissions to air, water and soil, and waste generation associated with the system. The inventory analysis is supported by a process tree (process diagram, flow tree) which defines the phases in the life cycle of a product, (Fig. 3.2). Each of the different phases can be made up from different unit processes, i.e. production of different kinds of raw materials to be combined in the material production phase. The different phases are often connected by transport processes. The data regarding material and energy consumption, waste and emissions have to be collected from all unit processes in a product life cycle. The data can be site specific (from specific companies, specific areas or specific countries) or more general, from more general sources like trade organizations or public surveys. For detailed LCA studies, site-specific data should be used. The average data (from trade organizations, from previous investigations of the same or similar product, etc.) may be used for conceptual or simplified LCA (Jensen et al., 1997). However, this step is often the most work-intensive part of an LCA, especially if the site-specific data are needed for all the unit processes in the life cycle.

Impact assessment is the third phase in an LCA, in which the potential environmental impacts of the modeled system are evaluated. This step consists of three mandatory elements: (1) selection of impact categories, category indicators and characterization models, (2) classification and (3) characterization.

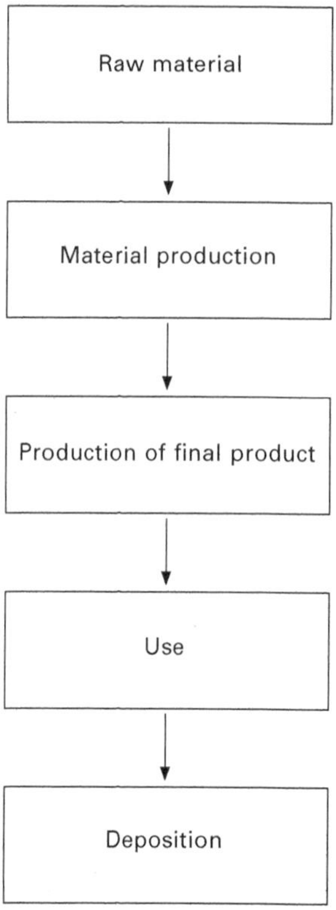

3.2 Example of a simple flow diagram to be used as a support in the LCI.

The impact categories are selected in order to describe the impacts caused by the analyzed product or product system. This is a follow-up of the decisions made in the goal and scoping phase. Examples of the impact categories usually considered are: depletion of abiotic and biotic resources, land use, climate change, stratospheric ozone depletion, ecotoxicological impacts, human toxicological impacts, photo-oxidant creation, acidification, eutrophication, etc. The impact categories relevant for concrete production and utilization are described in detail in Section 3.5.

Classification of the inventory input and output data is a qualitative step based on scientific analysis of relevant environmental processes (Jensen et al., 1997). The classification has to assign the inventory input and output data (emissions, resources and waste) to chosen environmental impacts, i.e. impact categories.

The aim of the characterization step is to model each impact category based on scientific knowledge, wherever possible (Jensen et al., 1997). This means that the category indicator and the relationship between the inventory input and output data and the indicator must be defined. In fact, the converted LCI results are aggregated into a category indicator result, which is the numerical value of the indicator in appropriate units. The indicator result is the final result of the mandatory part of a LCIA. The examples of the characterization models are shown in Section 3.5. To date, consensus has not been reached either for one single default list of impact categories or for one single characterization model.

Normalization, grouping, weighting and additional LCIA data quality analysis are the optional steps within the LCIA phase (ISO, 2006a). Normalization is calculating the magnitude of the category indicator results relative to reference information. Grouping is assigning impact categories into one or more sets as predefined in the goal and scope definition, and it may involve sorting and/or ranking. Weighting is the process of converting and possibly aggregating indicator results across impact categories using numerical factors based on value choices. Since weighting may include aggregation of the weighted indicator results, the outcome of this step can be one number. This score, or index, represents the environmental performance of the product system under study. It should be noted that according to ISO 14040 there is no scientific way to reduce LCA results to a single overall score or number, hence it cannot be used for comparative assertions. Data quality analysis is an additional technique aiming at better understanding the reliability of the collection of indicator results, i.e. the LCIA profile.

Interpretation is the fourth phase in LCA containing the following main issues: identification of significant environmental issues, evaluation of the results with the aim of establishing their reliability, and conclusions and recommendations.

The application of this methodology in the case of concrete production and utilization is discussed in the following sections.

3.3 Life cycle assessment (LCA) of concrete: goal and scope definition

Within this step, the following items must be defined:

- The goal of the study, for example: comparison of two or more different types of concrete or concrete structural elements, assessment of the whole building or bridge, determination of the sources of the relevant impacts and possible optimization potentials, etc.
- The function of the product or product system, i.e. functional unit.
- The system boundaries.
- The relevant impacts.
- The necessary data to reach the goal of the study.

The functional unit is defined by the function of the product or product system (by services it provides with certain quality and for a certain period). For a material, the functional unit can include the quantity (m^3), the mechanical properties (in appropriate units) and durability related properties (in years). For a building, the functional unit can include the size (m^2), the lifetime (years), the need for maintenance or the materials used. The results of the study will directly relate to the chosen functional unit. For example, if the functional unit is one cubic meter of concrete of certain quality, the result will be the amount of resources, waste and emitted pollutants per each cubic meter of such a concrete.

The system boundaries determine what part of the life cycle will be considered. The complete life cycle of the concrete structure is shown in Fig. 3.3. It includes the production of raw materials, the production of concrete, construction and service phase, demolition or dismantling, and disposal or recycling of waste materials.

Generally, there are two possible different situations when assessing the environmental burdens of the concrete. The first is the assessment of the environmental impact of the concrete as a building material or a specific structural element (for example, a precast bridge girder), i.e. the assessment of the production phase only. This is a cradle-to-gate type of analysis (Fig. 3.3) and the result of the study is called environmental product declaration (EPD). The second is the assessment of the environmental impact of the concrete structure or structural element as a part of a building, bridge, road or some other kind of the construction work. This is a cradle-to-grave type of analysis (Fig. 3.3). The type of assessment and system boundaries are chosen depending on the goal of an LCA study: whether the goal of the study is to evaluate the environmental impact of the concrete as a building material (cradle-to-gate) or the goal of the study is to evaluate the environmental impact of the building, bridge, road, etc., as a whole (cradle-to-grave).

Among various environmental impacts, for a specific product or process, some impacts are more and some impacts are less important. At this point, the relevant impact for concrete should be determined, which is discussed in detail in Section 3.5. Then, the necessary data for calculation of the chosen environmental impacts must be collected. This is done within the next step of LCA – life cycle inventory phase.

3.4 Life cycle assessment (LCA) of concrete – life cycle inventory (LCI)

Based on the decisions made in the goal and scope phase, the process tree is drawn, as shown in Fig. 3.3. Each of the different phases in the life cycle of concrete is made up from different unit processes. For example, the cement production within the raw or constituent material production phase consists of many unit processes (Fig. 3.4).

To calculate the environmental impacts, energy and materials flows as

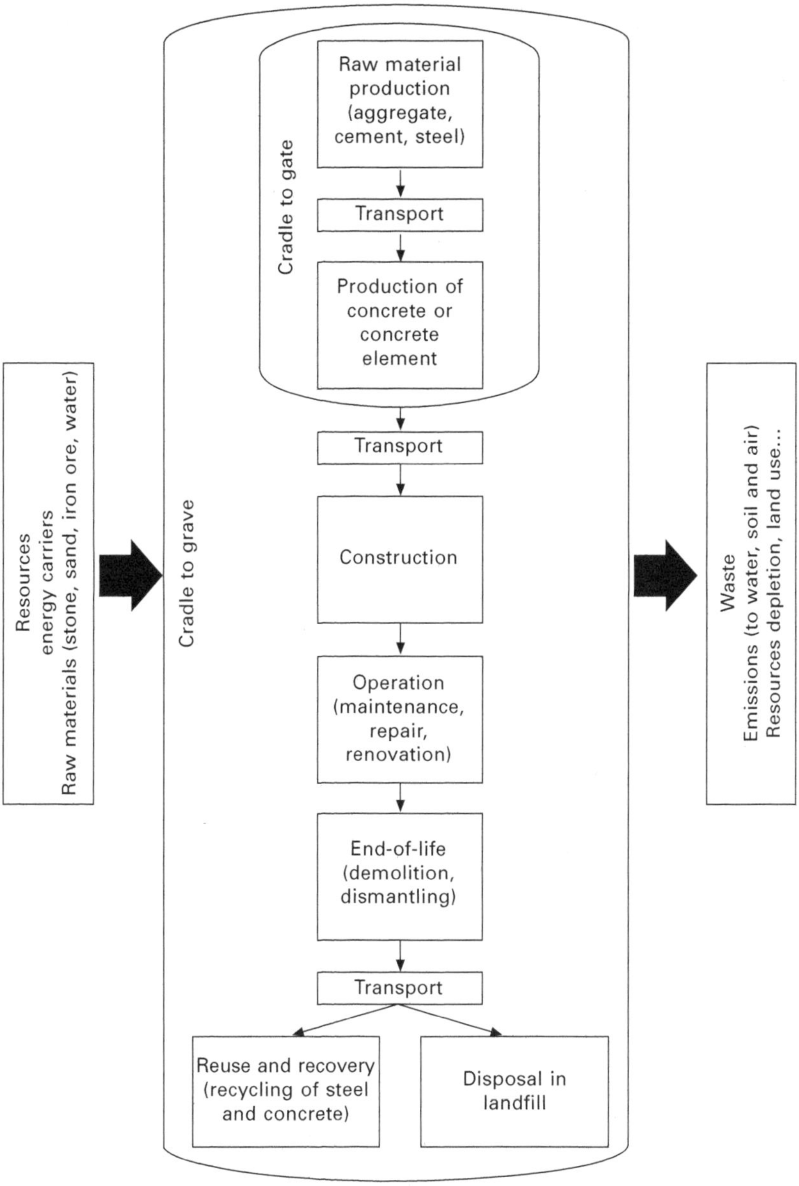

3.3 Life cycle of a concrete structure.

well as emissions must be estimated for each unit process (Fig. 3.5). Now the question is how to get the necessary data. Energy and material flows data can be obtained from bills of quantities and usually it is not the problem. However, the information about the type and amount of pollutants emitted into the air, water and soil must be measured or calculated. Looking at the process tree in Fig. 3.3, it is easy to conclude that in the case of a building, for

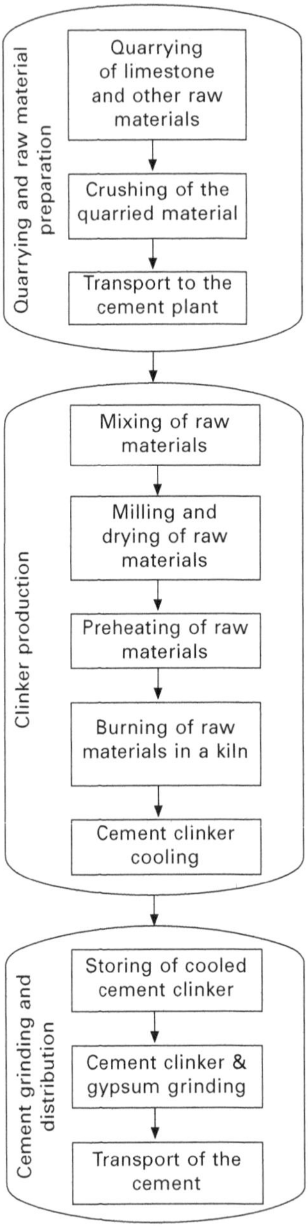

3.4 Phases of cement production.

example, there will be a hundreds of unit processes for which this operation has to be performed. This is obviously very time-consuming and expensive work. That is why, especially in the case of cradle-to-grave assessment, the necessary emission data are taken from existing databases.

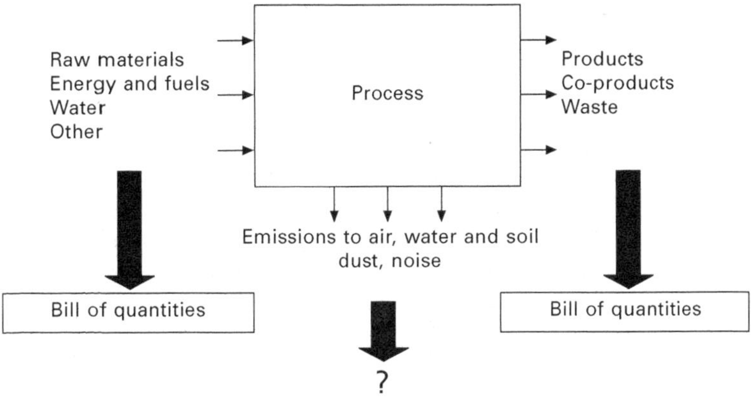

3.5 Energy and material flows and emissions of a unit process.

3.4.1 Databases

In order to facilitate the LCI phase, many databases have been developed. These include public national or regional databases, academic databases, industrial databases and commercial databases (Khasreen et al., 2009). Some of them are offered in combination with LCA software tools. Databases provide inventory data on a variety of products and basic services that are needed in every LCA, such as raw materials, electricity generation, transport processes and waste services, as well as complex products sometimes (Finnveden et al., 2009).

Some of the most famous databases and LCA software tools are listed in Table 3.1 (Bribián et al., 2009, Khasreen et al., 2009). The majority of databases are based on average data representing average production and supply conditions for products and services. It should be pointed out again that average data from databases can be used for simplified LCA studies; for detailed LCA studies site-specific data (measured or calculated for each unit process) should be used.

The most important problem with databases is that the data for the same product or process can differ from one database to another. The reasons for these variations are different boundaries assumed, different energy source and supply assumptions, different transportations types and distances assumptions, production differences, etc. (Khasreen et al., 2009). All these issues are dependent on geography, i.e. they differ from country to country.

The availability and geographical representativeness of data on concrete in some databases is shown in Table 3.2. The ELCD database is the European Life Cycle Database that is developing under the European Platform on LCA, which is a project of the European Commission (European Commission – JRC, 2011). Ecoinvent is a database developed by the Ecoinvent Centre, Switzerland, originally called the Swiss Centre for Life Cycle Inventories

Table 3.1 List of some LCA databases and tools

Database	Country	Function	Software	Level	Web site
Athena	Canada	Database + Tool	Impact Estimator Eco Calculator	Specific building	www.athenaSMI.ca
BEES	USA	Tool	BEES	Specific building	www.nist.gov/el/economics/BEESSoftware.cfm
Boustead	UK	Database + Tool	Boustead	General	www.boustead-consulting.co.uk
Ecoinvent	Switzerland	Database	No	General	www.pre.nl/ecoinvent
Eco-Quantum	Netherlands	Tool	Eco-Quantum	Specific building	www.ivam.uva.nl/?id=2&L=1
Envest 2.0	UK	Tool	Envest	Specific building	http://envest2.bre.co.uk/
ELCD	EU	Database	No	General	http://lca.jrc.ec.europa.eu
EQUER	France	Database + Tool	EQUER	Specific building	www.izuba.fr/logiciel/equer
Gabi	Germany	Database + Tool	Gabi	General	www.gabi-software.com
GEMIS	Germany	Database + Tool	GEMIS	General	http://www.oeko.de/service/gemis/en/
JEMAI	Japan	Database + Tool	JEMAI	General	www.jemai.or.jp/english/index.cfm
SimaPro	Netherlands	Database + Tool	SimaPro 7	General	www.pre.nl
Spin	Sweden	Database	No	General	http://195.215.251.229/Dotnetnuke/
TEAM	France	Database + Tool	TEAM 4.0	General	www.ecobilan.com
Umberto	Germany	Database + Tool	Umberto	General	www.umberto.de
US NREL	USA	Database	No	General	www.nrel.gov/lci

Table 3.2 LCI data on concrete in some databases

Data on:			ELCD	Ecoinvent	USNREL
Aggregate production	Crushed	Location	European average	Switzerland	NA[1]
	River	Location	NA	Switzerland	NA
Cement production		Location	European average	Switzerland	USA
Concrete production		Location	NA	Switzerland	NA
Transport		Location	European average	Switzerland, European average	USA

[1]Not available.

(Ecoinvent Centre, 2011). NREL is the US Life-Cycle Inventory Database developed mainly by National Renewable Energy Laboratory, USA (NREL, 2011). The representative Japanese database JEMAI is not listed in Table 3.2 because it is available only in Japanese.

As can be seen from Table 3.2, only Ecoinvent offers all the necessary LCI data for concrete production. However, the data is geographically dependable. Ecoinvent is an internationally accepted database, although it is mostly based on Swiss data regarding concrete. This is a potentially misleading situation because the types of raw materials, energy and transport data and even technologies may be different in different countries. That is why each country should have its own database according to its construction industry resources and traditions.

Unlike general databases, LCA tools for building assessment usually contain databases with 'final' LCI data on specific building materials and products. For example, ATHENA database contains LCI data on several concrete structural products in various mixes, size and strength designations: 20 and 30 MPa ready-mixed concrete, with 25% and 35% fly ash content, 60 MPa ready-mixed concrete, precast double T beams, precast hollow deck, concrete blocks (Athena Sustainable Materials Institute, 2011). Data are representative for North America.

Table 3.3 provides, as an indication of the general range of values, CO_2 emissions and energy consumption for concrete raw materials and reinforcement steel production (FIB TG 3.8, 2011). The data should be considered only as rough estimates and should not be used in detailed LCA studies.

3.4.2 LCI data on cement

The research performed so far in the area of LCA of concrete (Flower and Sanjayan, 2007; Marinkovic et al., 2008; Oliver-Solà et al., 2009; Šelih

Table 3.3 Value ranges of CO_2 emissions and energy consumption for concrete raw materials and steel reinforcement production (FIB TG 3.8, 2011)

	Inventory data, cradle-to-gate			Remarks
	CO_2 emissions ($kgCO_2/kg$)		Energy use (MJ/kg)	
	Lower	Upper		
Portland cement	0.80	1.0	4–7	
Natural aggregate	0.004	0.008	0.05–0.07	Lower: river aggregate Upper: crushed stone aggregate
Reinforcement steel	0.70	2.90	10–40	Lower: electric arc furnace; only scrap Upper: blast furnace; no recycled material
Prestressing steel	0.80	3.00	10–40	Lower: electric arc furnace; only scrap Upper: blast furnace; no recycled material

and Sousa, 2007) showed that because of the large CO_2 emissions, cement production is the biggest contributor to all concrete environmental impacts. Moreover, it is estimated that cement production is responsible for 5–7% of total world CO_2 emissions. These emissions are due to CO_2 release from the calcination process of limestone (to produce cement clinker), combustion of fossil fuels used for achieving the necessary temperature in the kiln (1400–1500 °C), as well as from power generation. According to Josa et al. (2004), who conducted the comparative analysis of several different types of cement produced in Europe, the average emissions per 1 kg of Portland cement type I (95% of clinker and 5% of gypsum) are: approximately 800 g of CO_2, 2.4 g of NO_x, 0.5 g of SO_2 and 0.2–0.3 g of dust. The average total energy consumption is about 4.3 MJ. Similar figures are obtained for cement production in Serbia (Marinkovic et al., 2008) except for SO_2 emissions that are much bigger. This is a consequence of large SO_2 emissions from electricity production (electricity is produced mostly in thermo-power plants in Serbia).

An obvious solution to this problem is to replace a part of the cement clinker with industrial waste materials that have pozzolanic activity, such as blast furnace slag (BFS) and fly ash (FA). Two benefits are gained in this way: application of waste material in the new product, so decreasing the amount of waste to be disposed of in landfills, and lowering the CO_2 emissions from clinker production. This procedure is already included in the cement production technology in many European countries (in Serbia, the replacement of clinker with BFS and/or FA up to 30% by mass has been done for last 40 years). However, the replacement ratio should be limited to a level that does not significantly affect the basic properties of cement and, consequently, the concrete.

LCA of cement with supplementary cementitious materials that are by-products or waste from another industrial process (such as BFS or FA) is always connected to allocation procedures. To be used as a substitution for cement clinker, these by-products must be additionally processed (so-called secondary process) and this has an impact on the environment, too. The following question is then raised: should the environmental impacts of the primary production process (production of the main product) be associated with the production of the by-product, and how? If BFS and FA are considered as waste, no environmental impacts from the primary production process are allocated to the secondary process. If they are considered as by-products, then part of the environmental impacts from primary production process are allocated to the secondary process and it can be done using two different allocation procedures. According to the mass allocation procedure, the part of the primary production impacts allocated to by-product production is determined from the ratio of the mass of the by-product and the total mass of the main and by-product.

According to the economic allocation procedure, the part of the primary production impacts allocated to by-product production is determined from the ratio of the price of the by-product and the sum of the prices of the main product and by-product. The choice of the allocation procedure is very important because it significantly affects the environmental impacts of the by-product. Chen et al. (2010) have shown that a mass allocation procedure imposes large environmental impacts on BFS and FA that can jeopardize the decision to apply these by-products as substitutes for cement clinker. However, the economic allocation procedure, which imposes smaller environmental impacts on BFS and FA, has the disadvantage of being unstable because of the potential price fluctuations on the market. They have concluded that no allocation procedure appears to be unquestionable.

3.5 Life cycle assessment (LCA) of concrete: life cycle impact assessment (LCIA)

3.5.1 Methodology

Choice of impact category and category indicator

Generally, there are two different types of impact assessment methods. The first one is called the damage-oriented approach (also the top-down approach or 'endpoints') and the second one is called the problem-oriented approach (also the bottom-up approach or 'midpoints').

When defining the impact category, the category indicator must be chosen somewhere in the environmental mechanism. The task of LCIA is to establish a relation between the inputs, e.g. fossil fuels or minerals, and outputs, e.g. emissions, of the LCI phase with the impacts on the environment. For this reason, for each impact category an indicator should be chosen in the environmental mechanism, which as far as possible represents the totality of all impacts in the impact category. This indicator can in principle be located at any position in the mechanism, from the LCI results down to the category indicators. However, the environmental relevance is typically higher for indicators chosen later in the environmental mechanism (ISO, 2003). For example, the environmental mechanism of acidification is shown in Fig. 3.6 (ISO, 2006a). In the damage-oriented approach, the category indicator is chosen at the endpoint of the environmental mechanism. In the problem-oriented approach, the category indicator is chosen at an intermediate level somewhere along the mechanism – at midpoint. Examples of category midpoints and endpoints for some impact categories are shown in Table 3.4 (ISO, 2003).

The most famous example of damage-oriented approaches is Eco-indicator 99 methodology, developed at Pré Consultants B.V., The Netherlands (Goedkoop and Spriensma, 2001). The impact categories are defined at the endpoint of the environmental mechanism and express the damage to Human Health, Ecosystem Quality and Resources (Table 3.5). Definitions at this level are much easier to comprehend than the rather abstract definitions of midpoints such as infrared radiation, UV-B radiation or proton release. However, the problem with this approach is in fact that it is not easy to establish a clear relationship between the LCI results and damage categories. In the top-down approach, it is not possible to avoid normalization, grouping and weighting.

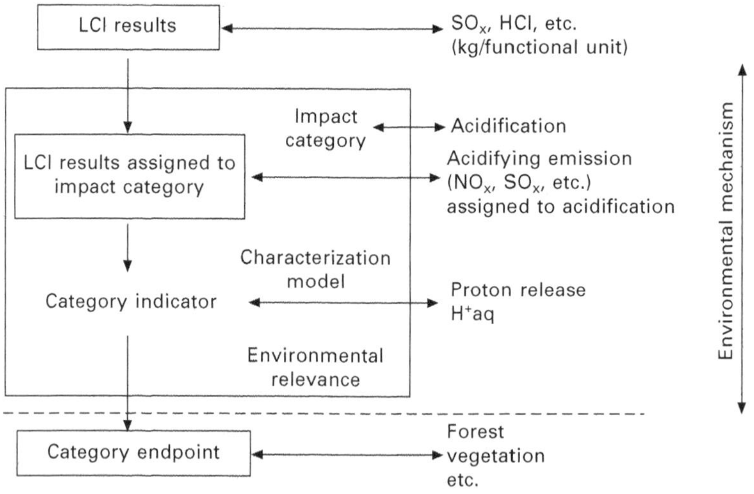

3.6 Concept of category indicators – acidification example (ISO, 2006a).

Table 3.4 Examples of intermediate variables (midpoints) and endpoints for some impact categories (ISO, 2003)

Impact category	Choice of indicator level	
	Examples of intermediate variables	Examples of category endpoints
Climate change	Infrared radiation, temperature, sea-level	Human life expectancy, coral reefs, natural vegetation, forests, crops, buildings
Stratospheric ozone depletion	UV-B radiation	Human skin, ocean biodiversity, crops
Acidification	Proton release, pH, base-cation level, Al/Ca ratio	Biodiversity of forests, wood production, fish populations, materials
Eutrophication	Concentration of macronutrients (nitrogen, phosphorus)	Biodiversity of terrestrial and aquatic ecosystems
Human toxicity	Concentration of toxic substances in environment, human exposure	Aspects of human health (organ functioning, human life expectancy, number of illness days)
Ecotoxicity	Concentration or bio-availability of toxic substances in environment	Plant and animal species populations

Table 3.5 Impact categories according to Eco-indicator 99 and CML methodologies

Eco-indicator 99		CML
Impact category	Sub-categories	Impact category
Damage to human health	Caused by carcinogenic substances	Depletion of abiotic resources Impacts of land use
	Caused by respiratory effects	Climate change
	Caused by climate change	Stratospheric ozone depletion
	Caused by ionizing radiation	Human toxicity
	Caused by ozone layer depletion	Ecotoxicity
Damage to ecosystem quality	Caused by ecotoxic substances	Photo-oxidant formation
	Caused by acidification and eutrophication by airborne emissions	Acidification Eutrophication
	Caused by land use	
Damage to resources	Caused by depletion of minerals and fossil fuels	

Weighting is not a scientifically based operation, but it relies upon the opinion and attitude of experts towards different environmental effects. On the other hand, the relationship between midpoint category indicators and LCI results is easily established through appropriate, scientifically based, characterization models. That is why the 'midpoints' approach is often used to quantify the results in the early stage in the cause and effect chain to limit the uncertainties (Mateus and Bragança, 2011). The most representative example of this type of approach is CML methodology, developed at the Institute of Environmental Sciences (CML) of the Faculty of Science, Leiden University in The Netherlands. Their choice of impact categories (only baseline) is shown in Table 3.5 (Guinée et al., 2002).

Characterization model

The link between the LCI results (extraction of resources, emissions and waste), and the category indicator is normally given by clear modeling algorithms (ISO, 2003). These modeling algorithms are called characterization models. Many characterization models for different impact categories have been developed. In the following, characterization models proposed by CML methodology (Guinée et al., 2002) are briefly presented.

- Climate change: the characterization model of the IPCC (Intergovernmental Panel on Climate Change). The IPCC provides characterization factors, global warming potentials (GWPs), for three different time horizons: 20, 100 and 500 years in terms of CO_2 equivalents. The indicator result is estimated by calculating the product of the amount of emitted greenhouse gas per functional unit of produced material (m_i) and the GWP given in CO_2-equivalents for each gas (GWP_i) . Finally, the contribution to the indicator result from each gas is summarized:

$$\text{Climate change} = \sum_i GWP_i \times m_i \tag{3.1}$$

- Stratospheric ozone depletion: the characterization model of the WMO (World Meteorological Organization). This model provides characterization factors, stratospheric ozone depletion potentials (ODPs) for a steady state in terms of CFC-11 equivalents:

$$\text{Stratospheric ozone depletion} = \sum_i ODP_i \times m_i \tag{3.2}$$

  The indicator result is expressed in kg of the reference substance, CFC-11 equivalent. ODP_i is the steady state ozone depletion potential for substance i, while m_i (kg) is the quantity of substance i emitted.

Similarly, the category indicators (midpoints) are proposed for other impact categories, and together with other relevant data, shown in Table 3.6. The

Table 3.6 Summary of impact categories data according to CML methodology

Impact category	Description	LCI results	Category indicator	Characterization model	Characterization factor	Indicator result	Category endpoint
Climate change	Temperature increase in the lower atmosphere	Emissions of greenhouse gases	Increase of infrared radiative forcing (W/m^2)	IPCC	Global warming potential (GWP) for each emission (kg CO_2-eq./kg emission)	Kilograms of CO_2-equivalents	Years of life lost (YLL), coral reefs, crops, buildings
Stratospheric ozone depletion	Decomposition of the stratospheric ozone layer which causes an increased UV radiation	Emissions of ozone-depleting gases	Increase of stratospheric ozone breakdown	WMO	Ozone depletion potential (ODP) for each emission (kg CFC-11-eq./kg emission)	Kilograms of CFC-11-equivalents	Illness days, marine productivity, crops
Photo-oxidant formation	Creation of 'smog' as a local impact and 'tropospheric' ozone as a regional impact	Emissions of substances (VOC, CO) to air	Quantity of tropospheric ozone formed	CML	Photochemical ozone creation potential (POCP) for each emission (kg ethylene-eq./kg emission)	Kilograms of ethylene equivalents	Illness days, crops
Acidification	Acidification of aquatic and terrestrial ecosystems (acid rains, acid lakes…)	Emissions of acidifying substances to air	Maximum release of protons (H^+)	CML	Acidification Potential (AP) for each emission (kg SO_2-eq./kg emission)	Kilograms of SO_2 equivalents	Biodiversity, natural vegetation, wood, fish, monuments
Eutrophication	Enrichment of aquatic and terrestrial ecosystems with nutrients	Emissions of nutrients to air, water and soil	Deposition increase divided by N/P equivalents in biomass	CML	Eutrophication potential (EP) for each emission (kg PO_4^{3-}- eq./kg emission)	Kilograms of PO_4^{3-} equivalents	Biodiversity, natural vegetation, algal bloom

Table 3.6 Continued

Impact category	Description	LCI results	Category indicator	Characterization model	Characterization factor	Indicator result	Category endpoint
Human toxicity	The toxic impacts of chemical and biological substances on humans	Emissions of toxic organic substances to air, water and soil	Environmental toxic substances concentration increase	CML	Human toxicity potential (HTP) (kg 1,4-dichlorobenzene-eq./kg emission)	Kilograms of 1,4-dichlorobenzene equivalents	Aspects of human health (organ functioning, human life expectancy, number of illness days)
Ecotoxicity	The toxic impacts of chemical and biological substances on aquatic and terrestrial ecosystems	Emissions of toxic organic substances to air, water and soil	Environmental toxic substances concentration increase	CML	Ecotoxicity potentials (ETPs) for each emission (kg 1,4-dichlorobenzene-eq./kg emission)	Kilograms of 1,4-dichlorobenzene equivalents	Biodiversity
Depletion of abiotic resources	Depletion of mineral resources which are considered as non-renewable	Extraction of resources, including fossil fuels	Extraction of material in the ore as a function of estimated supply horizon of the reserve base	CML	Abiotic depletion potential (ADP) for each resource (kg antimony-eq./kg of resource)	Kilograms of antimony equivalents	Availability of resources

characterization factors (potentials) for relevant substances and each impact category, according to the characterization models previously explained, are given in Table 3.7.

The list given in Table 3.6 cannot be regarded as complete. Other categories may, for instance, focus on radiation, noise and odor, working environmental impacts or land use, but for these categories as yet no widely accepted characterization models are available (ISO, 2003). However, characterization models for impacts of land use (competition), ionizing radiation and odor exist in CML methodology, see Guinée et al. (2002).

Case study: ready-mixed concrete production in Belgrade, Serbia

The case study presented here is the example of the LCIA of ready-mixed concrete production in the capital of Serbia – Belgrade. The impact assessment follows the standard protocol of LCA (ISO, 2006a). Therefore, the goal of the study is the LCIA of ready-mixed concrete production in Belgrade. This goal determines the system boundaries: the analyzed part of the life cycle includes production and transport of aggregate and cement, and production of concrete (the production of admixtures and water is not included as their contribution is very small) – cradle-to-gate type of assessment (Fig. 3.7). The construction, service and demolition phases are excluded because the impacts for these phases significantly depend on the type of the concrete structure to be made. The functional unit of 1 m^3 of ready-mixed concrete is used in this work.

The production of ready-mixed concrete is located in Serbia, so all the LCI data for aggregate, cement and concrete production were collected from local suppliers and manufacturers (Marinkovic et al., 2008). Emission data for diesel production and transportation, natural gas distribution and transport that could not be collected for local conditions were taken from GEMIS database (Öko-Institut, 2008). Data were taken from no earlier than 2000, so the processes analyzed in this work are based on recent technologies and normal production conditions.

The problem-oriented (midpoints) methodology is chosen for the impact assessment. The impacts are evaluated using the CML method (Guinée et al., 2002), only the total energy consumption expressed in MJ (energy use) is calculated instead of the depletion of abiotic resources. The impact categories included in this work are: climate change, eutrophication, acidification and photo-oxidant creation (POC).

Transport types and distances are estimated as typical for the construction site located in Belgrade. Cement is transported by heavy trucks from cement factory to concrete plant and assumed transport distance is equal to 150 km. River aggregate is most often used in Serbia for concrete production and is therefore transported by medium-sized ships; assumed transport distance is equal to 100 km.

Table 3.7 Characterization factors according to CML methodology

Impact categories	Substance	Characterization factors									
		Climate change (kg CO_2– eq./kg)	Stratosph. ozone depletion (kg CFC-11-eq./kg)	Photo-oxidant formation (kg ethylene-eq./kg)	Acidification (kg SO_2– eq./kg)	Eutrophication (kg PO_4^{3-}- eq./kg)		Human toxicity[1] (kg 1,4-DCB- eq./kg)	Ecoxicity, FAETP[2] (kg 1,4-DCB-eq./kg)		Depletion of abiotic resources (kg antimony-eq./kg)
		Air emissions	Air emissions	Air emissions	Air emissions	Air emissions	Water emissions	Air emissions	Air emissions	Water emissions	Extraction of resource
Climate change	carbon dioxide	1									
	dinitrogen oxide	310									
	methane	21									
Stratospheric ozone depletion	CFC-11		1								
	HALON 1301		12								
	tetrachlorometh.		1.2								
Photo-oxidant formation	ethylene			1							
	methane			0.006							
	ethane			0.123							
	propane			0.176							
Acidification	sulfur dioxide				1						
	ammonia				1.88						
	nitrogen dioxide				0.70						
Eutrophication	ammonia					0.35					
	nitrogen dioxide					0.13					
	phosphate						1				
	phosphorus						3.06				
	nitrogen						0.42				
Human toxicity	1,4 dichloroben.							1			
	sulfur dioxide							0.096			
	nitrogen dioxide							1.2			
	arsenic							3.5E+5			

	lead	4.7E+2			
	nickel	3.5E+4			
Ecoxicity	1,4 dichloroben.		2.4E-3	1	
	phenol		1.50	240	
	cadmium		290	1500	
	lead		2.40	9.60	
	chromium		1.90	6.90	
	copper		220	1200	
Depletion of abiotic resources	antimony				1
	iron				8.43E-8
	crude oil				0.0201
	natural gas				0.0187
	hard coal				0.0134
	fossil energy				4.81E-4

[1]Characterization factors given as an example only for air emissions.
[2]Characterization factors given as an example only for FAETP and air and water emissions.

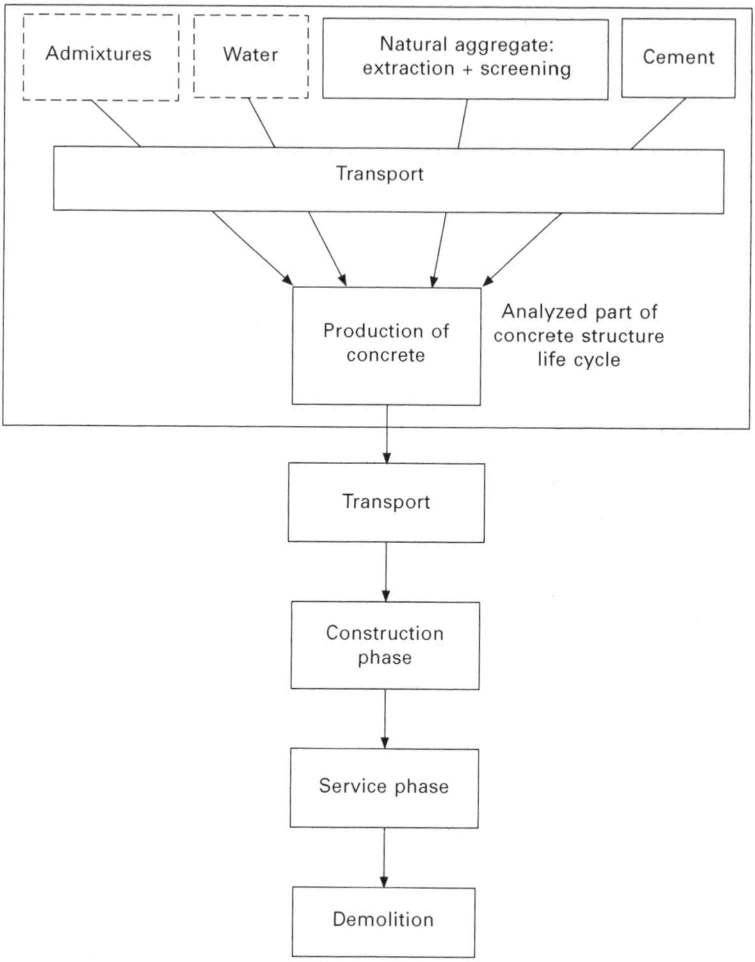

3.7 Analyzed part of the concrete structure life cycle.

The type and amount of component materials used for concrete production are shown in Table 3.8. The mix proportion of concrete was determined from two conditions: the target concrete strength class was C25/30 (characteristic compressive cylinder/cube strength equal to 25/30 MPa), nomenclature according to Eurocode 2 (CEN, 2004) and the target slump 20 minutes after mixing was 6±2 cm.

Table 3.9 shows the collected LCI data for aggregate, cement, concrete production and for transport. Calculated cumulative energy requirement and emissions to air for the production of 1 m^3 of ready-mixed concrete (so-called inventory table) are presented in Table 3.10. For chosen impact categories, category indicator results (per functional unit) are calculated according to CML methodology and are shown in Table 3.11. Figure 3.8 shows the

Table 3.8 Concrete mix proportions and tested properties for concrete strength class C25/30

Components		Unit	Amount
Cement CEM I 42.5 R		(kg/m^3)	315
Aggregate	0/4 mm	(kg/m^3)	658
(river aggregate,	4/8 mm	(kg/m^3)	338
Morava river)	8/16 mm	(kg/m^3)	282
	16/31.5 mm	(kg/m^3)	601
Water		(kg/m^3)	180
w/c[1]		/	0.571
a/c[2]		/	5.965
Properties			
Density		(kg/m^3)	2396
Slump after 20 minutes		(cm)	5.5
Compressive strength at 28 days		(MPa)	39.2

[1]Water-to-cement ratio.
[2]Aggregate-to-cement ratio.

Table 3.9 LCI data for various phases of the concrete life cycle

	Production of cement (kg)	Production of aggregate (kg)	Production of concrete (1 m^3)	Transport (tkm)	
				Heavy truck	Medium-sized ship
Energy (MJ)					
Coal	3.370140				
Natural gas	0.083178				
Diesel	0.024369	0.014780		1.540900	0.599850
Electricity	0.507672		20.06894		
Emissions to air (g)					
CO	4.203224	0.003475	0.722680	0.318850	0.155420
NO_x	2.279068	0.015579	13.22440	0.984380	0.426770
SO_x	3.646948	0.005447	98.75360	0.430940	0.171510
CH_4	1.002748	0.001296	0.433290	0.123860	0.046390
CO_2	861.2028	1.377926	5698.210	110.7700	43.38800
N_2O	0.000756	0.000055	0.029100	0.002950	0.001290
HCl	0.067800		2.680210		
HC	0.000580		0.023080		
NMVOC	0.034732	0.000392	0.071040	0.124710	0.075790
particles	0.711981	0.001455	11.99120	0.193270	0.152050

contribution to the total indicator results of various phases in raw material extraction and material production part of the concrete life cycle.

The results show that the cement production is by far the largest contributor to all impact categories (Fig. 3.8). It causes approximately 84% of the total energy use, 93% of the total climate change, 81% of the total eutrophication, 86% of the total acidification and 83% of the total photo-oxidant creation.

Table 3.10 Inventory table per 1 m^3 of concrete

	Cement (kg) 315.00	Aggregate (kg) 1879.00	Concrete 1 m^3	Transport 1 m^3	Total
Energy (MJ)					
Coal	1061.594				1061.594
Diesel	7.676	27.772		185.519	220.967
Natural gas	26.201				26.201
Electricity	159.917		20.069		179.986
Emission to air (g)					
CO	1324.016	6.530	0.723	44.269	1375.537
NO_x	717.906	29.273	13.224	126.702	887.106
SO_x	1148.789	10.235	98.754	52.589	1310.366
CH_4	315.866	2.435	0.433	14.569	333.303
CO_2	271278.882	2589.123	5698.210	13386.488	292952.703
N_2O	0.238	0.103	0.029	0.382	0.752
HCl	21.357		2.680		24.037
HC	0.183		0.023		0.206
NMVOC	10.941	0.737	0.071	20.133	31.882
particles	224.274	2.734	11.991	37.702	276.701

Table 3.11 Category indicator results per 1 m^3 of concrete

Category Indicator result	Energy use MJ	Climate change g CO_2-eq.	Eutrophication g PO_4^{3-}-eq.	Acidification g SO_2-eq.	POC g C_2H_4-eq.
Cement	1255.388	279251.727	93.328	1670.117	49.054
Aggregate	27.772	2683.073	3.805	30.726	0.527
Concrete	20.069	5718.354	1.719	110.369	0.055
Transport	185.519	13872.883	16.471	141.280	9.753
Total	1488.748	301526.038	115.324	1952.493	59.390

The main reason for such a situation is a large CO_2 emission during the calcination process in clinker production and fossil fuel usage, as explained. The contributions of the aggregate and concrete production phases are very small, while the contribution of transport phase lies somewhere in between. The contribution of transport depends on the transport scenario: assumed transport distance and type. It can be significant, especially to POC and to a minor extent to eutrophication, for larger distances or other transport vehicles than assumed in this study.

It can be seen from the inventory table of concrete (Table 3.10) that the relevant impact categories are: energy consumption, fossil fuel depletion, and categories related to emissions released mostly from the cement production and transport such as climate change, acidification, eutrophication and photo-oxidant creation (smog). Human toxicity can also be included because of the sulfur dioxide and nitrogen dioxide emissions. In addition, solid waste

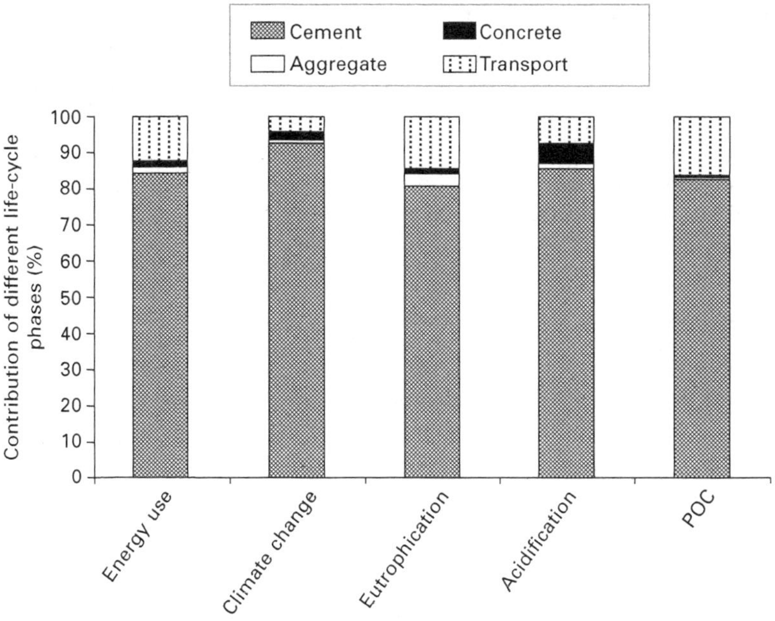

3.8 Contribution of different life-cycle phases to category indicator results.

production and mineral resources (sand and stone) depletion are one of the major impacts of concrete, although they were not treated in this case study.

Unfortunately, most of the proposed methodologies do not include solid waste production as an impact category, or consider sand and stone as abiotic resources that can be depleted. Besides the fact that sources of quality sand and stone for aggregate production are not endless, especially at regional level, their unlimited extraction also has strong impact on the environment and led to direct local devastation of the natural environment, whether crushed stone or river aggregate. The possible way to account for these issues is to develop a special indicator for natural bulk resources (that are used in concrete production, i.e. sand and gravel) depletion. Habert et al. (2010) proposed a methodology for development of such an indicator.

3.5.2 EPD

EPD is defined by international standard ISO 14025 (ISO, 2006b) as ISO Type III label. It contains specific results from an LCA of a product according to ISO 14040 (ISO 2006a), presented in a formalized and comparable way. Besides LCA results, EPDs declare other environmentally relevant issues, which are not covered by LCA such as technical data or information on special substances.

According to ISO 14025 (ISO 2006b), three different steps are needed for the generation of EPD. Firstly, the standard sets the framework for EPD programs and the validation of EPDs. Secondly, the program holder defines detailed rules for LCA of defined product categories, so-called product category rules (PCR). These rules determine the format and content of EPD for specific product categories. And finally, according to these rules, the data for the EPD for the respective product category is gathered and documented. All data within the EPD must be verified by an independent third party.

EPD as a Type III declaration differs from other environmental labels and declarations that can be found in the market today in the following ways:

- EPD contains quantitative information on a number of standardized environmental effects which is obtained by application of generally accepted and standardized LCA methodology;
- within the specific product category, the same format and content of EPD enables easy comparison between products;
- all data in EPD must be verified by an independent third party.

In that way, EPDs solve the problems associated with manufacturers of products publishing unverified, misleading, un-comparable and often incomprehensible environmental data derived through non-standardized methodologies.

One of the most important applications of EPDs in the construction sector is the use for environmental assessment of construction works. Looking at the process tree in Fig. 3.3, it is easy to conclude that in the case of a building, for example, there will be a hundreds of unit processes for which LCI data should be collected. This is obviously very time-consuming and expensive work even for the simplest construction work. The EPDs for building materials, precast structural elements, various building products (for example, facade elements) significantly facilitate this work because they contain all the necessary environmental data collected and processed in a formalized and comparable way. EPD programs can be found in several European countries, many of them specialized for the building sector. However, EPDs are still a very young communication tool. Even in the countries where EPD programs are under developement, the number of existing EPDs is fairly small.

For example, in Germany the holder of the EPD program for building products is AUB- Association of Building Product Producers and Distributors (Braune et al., 2007). Up to now, 14 PCRs for different construction products have been developed (IBU, 2011). It is interesting to note that there are EPDs for structural steel, for some timber and masonry products, but an EPD for structural concrete is still missing.

According to the AUB program, EPD contains a summary (usually of one page) and the following chapters: product definition, raw materials, manufacturing of the building product, working with the building product, building product in use, singular effects, end-of-life phase, LCA, evidence and verification, and references. The summary of EPD contains the results of the LCA. The example of tabular overview of category indicators in summary is given from the EPD for structural steel, manufacturer CELSA Barcelona, Table 3.12 (IBU, 2011). The values of indicators in Table 3.12 are calculated per 1 kg of structural steel. In the end-of-life phase, 88% recycling, 11% reuse and 1% loss are assumed.

As another example, the part of the EPD for Portland cement (CEM I) issued by the European Cement Association in 2008 is given in Table 3.13 (CEMBUREAU, 2008). It was developed as business-to-business communication tool with the prime intention to provide measurable and verifiable input for the environmental assessment of construction works. The functional unit is 1000 kg of cement.

3.5.3 LCA tools and rating systems

LCA tools are software designed for calculation of the environmental impacts of products and services based on LCA methodology. The most recognized and sophisticated LCA tools are listed in Table 3.1. Some of these tools are general and therefore can be used for the environmental assessment of various types of construction works (SimaPro, Gabi, TEAM, Umberto). Besides them, there are tools developed specifically for buildings and building products, such as Athena, BEES, Eco-Quantum, Envest, EQUER.

The application of LCA in the environmental assessment of buildings and other construction works is a very complex task as a construction incorporates

Table 3.12 Category indicators in an AUB EPD (CELSA Barcelona) for structural steel (IBU, 2011)

Structural steel: sections, rebars and marchant bars				
Indicator	Unit per kg	Production	End-of-life	Total
Primary energy, non-renewable	(MJ)	11.26	0.26	11.52
Primary energy, renewable	(MJ)	1.09	–0.15	0.94
Global warming potential (GWP)	(kg CO_2-eq.)	0.68	0.10	0.76
Ozone depletion potential (ODP)	(kg CFC-11-eq.)	7.68E-08	–1.26E-08	6.60E-08
Acidification potential (AP)	(kg SO_2-eq.)	4.22E-03	4.97E-05	4.27E-03
Eutrophication potential (EP)	(kg PO_4^{3-}–eq.)	2.56E-04	–6.82E-06	2.49E-04
Photochemical oxidant creation potential (POCP)	(kg C_2H_4-eq.)	2.95E-04	5.73E-05	3.52E-04

Table 3.13 EPD for Portland cement CEM I (CEMBUREAU, 2008)

Impact category	Unit per 1000 kg	Pre-factory	Cement factory	Total
Raw materials				
Natural resources				
Renewable	(kg)	0	0	0
Non-renewable	(kg)	18	1447	1465
Secondary resources				
Renewable	(kg)	0	0	0
Non-renewable	(kg)	0	44	44
Energy resources				
Natural resources				
Renewable	(MJ)	132	0	132
Non-renewable	(MJ)	1370	2501	3871
Secondary resources				
Renewable	(MJ)	0	157	157
Non-renewable	(MJ)	0	638	638
Use of water	(kg)	1467	226	1693
Global warming	(kg CO_2-eq.)	118	781	899
Acidification	(kg SO_2-eq.)	1.10	1.30	2.40
Ozone depletion	(kg CFC-11-eq.)	0.000043	0	0.000043
Photochemical oxidant formation	(kg C_2H_4-eq.)	0.13	0.12	0.25
Eutrophication	(kg PO_4-eq.)	0.05	0.20	0.25
Waste for disposal				
Non-hazardous	(kg)	665	not relevant	665
Hazardous	(kg)	1.2	not relevant	1.2

hundreds and thousands of different products and furthermore, the expected service life of construction works compared to the service life of other products is very long. For that reason, LCA tools mentioned in Table 3.1 are not widely used by most stakeholders (designers, contractors, occupants) but much more by experts and very often at academic level (Mateus and Bragança, 2011).

During the last two decades various assessment and rating systems (mostly for buildings) which simplify LCA for practical use, have been developed. The first commercially available environmental assessment framework and rating system for buildings was BREEAM (Building Research Establishment Environmental Assessment Method), developed in UK in 1990 by Building Research Establishment (BREEAM, 2011). Since then many different tools of this type have been launched around the world. The most widely used, besides BREEAM, are: LEED developed by US Green Building Council (LEED, 2011), SBTool – previously known as GBTool – developed by International Initiative for a Sustainable Built Environment (iiSBE, 2011), DGNB developed by German Sustainable Building Council (DGNB, 2011).

Although the sets of indicators and consequently the assessment methods integrated in these rating systems are not based on LCA methodology comprehensively and consistently, they play an important role in improving the sustainable performance of buildings (Mateus and Bragança, 2011). Besides, unlike LCA tools, these systems contain some kind of a rating system that ends in awarding a certain certification (credit point, grade) to assessed building (Ding, 2008).

There is also one LCA tool developed specifically for reinforced concrete structures. It is called EcoConcrete (EcoConcrete, 2011). EcoConcrete is developed and promoted by the 'Joint Project Group on the LCA of concrete' (JPG), set up by and composed of the European associations of the different components of concrete: BIBM (International Bureau for Precast Concrete), CEMBUREAU (European Cement Association), EFCA (European Federation of Concrete Admixtures Associations), ERMCO (European Ready Mixed Concrete Organisation) and EUROFER (European Confederation of Iron and Steel Industries). It is a tailor made and peer reviewed MSExcel-based software for environmental assessment of European ready-mixed and precast concrete products. EcoConcrete provides results of LCA of ten selected concrete applications, according to three different methodologies (CML, EDIP and Eco-Indicator) and ISO standards.

The ten functional units correspond to the following concrete applications: a flat slab, a continuous beam, a foundation pile, a motorway pavement, a bridge pylon, a separation floor, a load-bearing wall, the elements of a solid wall, a column and pavement blocks. The first five correspond to ready-mixed concrete applications and the last five correspond to precast concrete applications. These ten functional units are analyzed from cradle to grave, using inventory data provided by co-owners of software and taking into account the use and maintenance phase and the end-of-life scenario of different applications considered (Josa et al., 2005). EcoConcrete is available under license including a training program. The limitations of EcoConcrete tool are in its database: data is European average and limited to ten specific products, so it cannot be applied in detailed LCA or for the assessment of elements that involve different components (for example, recycled aggregate instead of natural aggregate) or different execution procedures or plant processes. Some of the applications of this tool have been published. For example, the assessment of concrete with fine recycled concrete aggregate (Evangelista and de Brito, 2007), concrete sidewalks (Oliver-Solà et al., 2009) and concrete walls (Šelih and Sousa, 2007).

3.6 Conclusion and future trends

As is well known, three pillars of sustainability are environmental, social and economic aspects. Therefore, the integrated approach to design and evaluation of the overall performance of sustainable construction works should include the assessment of all three sustainability aspects in addition to technical and functional requirements. The inclusion of social and economic aspects is considered as a step from 'green' to 'sustainable'.

The social aspects are related to wellbeing and comfort. These are the most subjective aspects, but for certain types of construction work such as buildings, they are fundamental to the perception of the user. The social aspects are already integrated in some rating systems. For example, the following categories and indicators are integrated in the Portuguese version of SB Tool – SB Tool^{PT} (Bragança and Mateus, 2011):

- occupant's health and comfort (indicators: natural ventilation efficiency, toxicity of finishing, thermal comfort, lighting comfort, acoustic comfort);
- accessibility (indicators: accessibility to public transportations and accessibility to urban amenities);
- awareness and education for sustainability (indicator: education of occupants).

The economic aspects are related to the cost, but the cost analysis should comprise all phases of the life cycle. As with social aspects, the cost analysis is already included in some rating systems. In SB Tool^{PT} (Bragança and Mateus, 2011) for example, this category is called life-cycle costs, and the indicators are capital cost and operation cost. On the other hand, the diversity of so far proposed environmental and sustainability assessment methods makes the comparison between results obtained by different tools and especially, rating systems not comprehensively based in LCA methodology, practically impossible. To overcome this problem at the European level, the European Commission mandated the CEN (European Centre of Normalization) to develop a standard method for the assessment of construction works. For this purpose, the Technical Committee CEN/TC 350, 'Sustainability of Construction Works', was created in 2005. The standards being developed by this committee are to provide a voluntary, harmonized and horizontal (applicable to all products/building types) method for the environmental assessment of new and existing construction works across their entire life cycle. In addition, CEN/TC 350 aims at developing the standards for the environmental product declarations of construction products. In the meantime, the scope of these standards has been extended to include all sustainability aspects as social and economic performance of buildings. To date, following CEN/TC 350 standards and prestandards have been produced (Dias and Ilömaki, 2011):

- EN 15643-1:2010, Sustainability of construction works – Integrated assessment of building performance. Part 1: General Framework;

- prEN 15643-2:2009, Sustainability of construction works – Integrated assessment of building performance. Part 2: Framework for the assessment of environmental performance;
- prEN 15643-3:2008, Sustainability of construction works – Integrated assessment of building performance. Part 3: Framework for the assessment of social performance;
- prEN 15643-4:2008, Sustainability of construction works – Integrated assessment of building performance. Part 4: Framework for the assessment of economic performance;
- prEN 15978:2010, Sustainability of construction works – Assessment of environmental performance of buildings – Calculation method;
- CEN/TR 15941:2010, Sustainability of construction works – Environmental product declaration – Methodology for selection and use of generic data.

Regarding the environmental aspect, prEN 15643-2:2009 sets the environmental indicators that should be used in the European building sustainability assessment methods, Table 3.14 (Mateus and Bragança, 2011). In future, all standardized European sustainability assessment should consider the same list of indicators, the new sustainability rating systems should be consistent with it and it is expected that the existing ones will be adapted to this new approach (Mateus and Bragança, 2011).

Therefore, future work should be focused on developing environmental product declarations for various materials and construction products and technologies, which is important especially for concrete and concrete products as they are the most widely used. Besides, a lot of work remains

Table 3.14 Impact categories for environmental impacts/aspects assessment according to prEN 15643-2:2009

Environmental impacts expressed with the impact categories of LCA	Environmental aspects expressed with data derived from LCI and not assigned to the impact categories of LCA
Climate change expressed as global warming potential	Use of non-renewable resources other than primary energy
Destruction of the stratospheric ozone layer	Use of recycled/reused resources other than primary energy
Acidification of land and water resources	Use of non-renewable primary energy
Eutrophication	Use of renewable primary energy
Formation of ground level ozone expressed as photochemical oxidants	Use of fresh water resources
	Non-hazardous waste to disposal
	Hazardous waste to disposal
	Nuclear waste (separated from hazardous waste)

to be done in the area of social and economic aspects, in their inclusion in the assessment of construction works.

Here it should also be mentioned that there is an ISO standard regarding environmental management for concrete and concrete structures that is currently under development. This is ISO/FDIS 13315: Environmental management for concrete and concrete structures, Part 1: General principles and Part 2: System boundary and inventory data. It is currently prepared by ISO TC71/SC8 (ISO TC71/SC8, 2011).

3.7 Sources of further information and advice

The websites of the most frequently used LCA tools and databases are given in Table 3.1. Detailed research on LCI data of Portland cement concrete produced in United States, which include ready-mixed concrete, concrete masonry and precast concrete, is performed by Portland Cement Association (Marceau et al., 2007). In addition, for slag cement concrete and the same geographical area, LCI data are available in Prusinski et al. (2004). Comparative analysis of LCA tools and rating systems can be found in Ding (2008), Haapio and Viitaniemi (2008), Bribián et al. (2009) and Erlandsson and Borg (2003). A very thorough and detailed information about most frequently used rating systems is available in Mateus et al. (2011) and at SB Alliance web site (SB Alliance, 2011).

Some examples of the application of LCA in concrete and concrete structures are published: FIB TG 3.8 (2011), Hájek et al. (2011), Kawai et al. (2005), Marinković et al. (2010), Oliver-Solà et al. (2009), Sakai (2005), Šelih and Sousa. (2007), although some of the studies regarding concrete are focused only on energy consumption and CO_2 emissions. Further information on environmental design of concrete structures can be found in FIB TG3.3 (2004), FIB TG3.6 (2008) and FIB TG 3.8 (2011).

3.8 Acknowledgement

The work reported here is a part of the investigation within the research project TR36017 'Utilization of by-products and recycled waste materials in concrete composites in the scope of sustainable construction development in Serbia: investigation and environmental assessment of possible applications', supported by the Ministry for Science and Technology, Republic of Serbia. This support is gratefully acknowledged.

3.9 References

Athena Sustainable Materials Institute (2011), LCI Databases-Database details. Available from: http://www.athenasmi.org/our-software-data/lca-databases/products/ [Accessed 30 November 2011].

Bragança, L., Mateus, R. (2011), 'Improving the design of a residential building using the Portuguese rating system SBTool^{PT}', in Bragança, L., Koukkari, H., Blok, R., Gervásio, H., Veljkovic, M., Borg, R.P., Ungureanu, V., Schaur, C. (eds.), *Sustainability of Constructions – Towards a better built environment.* Proceedings of the International Conference: Innsbruck, 3–5 February 2011. Malta, Gutenberg Press Ltd, 197–204.

Braune, A., Kreiβig, J., Sedlbauer, K. (2007), 'The use of EPDs in building assessment – Towards the complete picture', in Bragança, L., Pinheiro, M., Jalali, S., Mateus, R., Amoêda, R., Guedes, M.C. (eds.), *Portugal SB07 Sustainable construction, Materials and Practices – Challenge of the Industry for the New Millenium*, Amsterdam, IOS Press, 299-304.

BREEAM (2011), Homepage of BREEAM. Available from: http://www.breeam.org/ [Accessed 30 November 2011].

Bribián, I.Z., Usón, A.A., Scarpellini, S. (2009), 'Life cycle assessment in buildings: State-of-art and simplified LCA methodology as a complement for building certification', *Building and Environment* **44**, 2510–2520.

CEMBUREAU (2008), Environmental Product Declaration – CEMBUREAU EPD CEM I. Available from: http://www.cembureau.be/topics/sustainable-construction/environmental-product-declaration [Accessed 12 July 2011].

CEN (2004), Eurocode 2: Design of Concrete Structures – Part 1-1: General Rules and Rules for Buildings, EN 1992-1. Brussels, CEN.

Chen, C., Habert, G., Bouzidi, Y., Jullien, A., Ventura, A. (2010), 'LCA allocation procedure used as an incitative method for waste recycling: An application to mineral additions in concrete', *Resources, Conservation and Recycling* **54**, 1231–1240.

DGNB (2011), Certification system. Available from: http://www.dgnb.de/_en/certification-system/index.php [Accessed 12 November 2011].

Dias, A.B., Ilömaki, A. (2011), 'Standards for Sustainability Assessment of Construction Works', in Bragança, L., Koukkari, H., Blok, R., Gervásio, H., Veljkovic, M., Borg, R.P., Ungureanu, V., Schaur, C. (eds.), *Sustainability of Constructions – Towards a better built environment.* Proceedings of the International Conference: Innsbruck, 3–5 February 2011. Malta, Gutenberg Press Ltd, 189–196.

Ding, G.K.C. (2008), 'Sustainable construction – The role of environmental assessment tools', *Journal of Environmental Management* **86**, 451–464.

EcoConcrete (2011), Homepage of EcoConcrete. Available from: http://www.therightenvironment.net/EcoConcrete.htm [Accessed 08 December 2011].

Ecoinvent Centre (2011), Database. Available from: http://www.ecoinvent.org/database [Accessed 10 August 2011].

Erlandsson, M., Borg, M. (2003), 'Generic LCA-methodology applicable for buildings, constructions and operation services-today practice and development needs', *Building and Environment* **38**, 919–938.

European Commission (2011), Energy efficiency in buildings. Available from: http://ec.europa.eu/energy/efficiency/buildings/buildings_en.htm [Accessed 15 July 2011].

European Commission - JRC (2011), Life cycle thinking and assessment. Available from: http://lct.jrc.ec.europa.eu/assessment [Accessed 02 December 2011].

European Environment Agency (2008), *Effectiveness of environmental taxes and charges for managing sand, gravel and rock extraction in selected EU countries*. EEA Report. No 2/2008, Copenhagen, Schultz Grafisk. Also available from: http://www.eea.europa.eu/publications/eea_report_2008_2 [Accessed 17 August 2009].

Evangelista, L., de Brito, J. (2007), 'Environmental life cycle assessment of concrete made with fine recycled concrete aggregates', in Bragança, L., Pinheiro, M., Jalali, S.,

Mateus, R., Amoêda, R., Guedes, M.C. (eds.), *Portugal SB07 Sustainable construction, Materials and Practices – Challenge of the Industry for the New Millenium*, Amsterdam, IOS Press, 789–794.

FIB TG3.3 (2004), 'Environmental design', *fib (International Federation for Structural Concrete) bulletin 28*, Stuttgart, Sprint-Digital-Druck.

FIB TG3.6 (2008), 'Environmental design of concrete structures-general principles', *fib bulletin 47*, Germany, DCC Siegmar Kästl e.K.

FIB TG 3.8 (2011), 'Guidelines for Green Concrete Structures', *fib bulletin* 67, Lausanne, International Federation for Structural Concrete (fib).

Finnveden, G., Hauschild, M.Z., Ekvall, T., Guinée, J., Heijungs, R., Hellweg, S., Koehler, A., Pennigton, D., Suh, S. (2009), 'Recent developments in life cycle assessment', *Journal of Environmental Management* **91**, 1–21.

Fisher, C., Werge, M. (2009), EU as a Recycling Society. ETC/SCP working paper 2/2009. Available from: http://scp.eionet.europa.eu.int. [Accessed 14 August 2009].

Flower, D.J.M., Sanjayan, J.G. (2007), 'Green house emissions due to concrete manufacture', *International Journal of Life Cycle Assessment* **12**, 282–288.

Goedkoop, M., Spriensma, R. (2001), The Eco-indicator 99 A damage oriented method for Life Cycle Impact Assessment, Methodology Report. Available from: http://www.pre.nl [Accessed 16 May 2009].

Guinée, J.B., Gorrée, M., Heijungs, R., Huppes, G., Kleijn, R., Koning, A. de, Oers, L. van, Wegener Sleeswijk, A., Suh, S., Udo de Haes, H.A., Bruijn, H. de, Duin, R. van, Huijbregts, M.A.J. (2002), *Handbook on life cycle assessment. Operational guide to the ISO standards. I: LCA in perspective. IIa: Guide. IIb: Operational annex. III: Scientific background*, Dordrecht, Kluwer Academic Publishers.

Haapio, A., Viitaniemi, P. (2008), 'A critical review of building environmental assessment tools', *Environmental Impact Assessment Review* **28**, 469–482.

Habert, G., Bouzidi, Y., Chen, C., Jullien, A. (2010), 'Development of a depletion indicator for natural resources used in concrete', *Resources, Conservation and Recycling* **54**, 364–376.

Hájek, P., Fiala, C., Kynčlová, M. (2011), 'Life cycle assessments of concrete structures – a step towards environmental savings', *Structural Concrete Journal of the fib* **1**, 13–22.

IBU (2011), Homepage of the Institute Construction and Environment IBU. Available from: www.bau-umwelt.de [Accessed 22 September 2011].

iiSBE (2011), SB Method and SB Tool. Available from: http://www.iisbe.org/sbmethod [Accessed 24 October 2011].

ISO (2003), 'Environmental management – Life cycle impact assessment – Examples of application of ISO 14042', ISO/TR14047, Geneva, International Organization for Standardization.

ISO/TC 71 (2005), Concrete, reinforced concrete and prestressed concrete. Business plan. Available from: http://isotc.iso.org/livelink/livelink/fetch/2000/2122/687806/ISO_TC_071__Concrete__reinforced_concrete_and_pre-stressed_concrete_.pdf?nodeid=1162199&vernum=0 [Accessed 21 August 2009].

ISO (2006a), 'Environmental Management – Life Cycle Assessment', Set of International standards: ISO 14040-14043, Geneva, International Organization for Standardization.

ISO (2006b), 'Environmental labels and declarations, Type III environmental declarations', ISO 14025, Geneva, International Organization for Standardization.

ISO TC71/SC8 (2011), Environmental management for concrete and concrete structures.

Available from: http://www.iso.org/iso/iso_catalogue/catalogue_tc/catalogue_tc_browse.htm?commid=548367 [Accessed 11 December 2011].

Jensen, A. A., Hoffman, L., Møller, B.T., Schmidt, A. (1997), Life Cycle Assessment. A guide to approaches, experiences and information sources. Environmental Issues Series no.6. European Environment Agency. Available from: http://www.eea.europa.eu/publications/GH-07-97-595-EN-C [Accessed 14 September 2011].

Josa, A., Aguado, A., Heino, A., Byars, E., Cardim, A. (2004), 'Comparative analysis of available life cycle inventories of cement in EU', *Cement and Concrete Research* **34**, 1313–1320.

Josa, A., Aguado, A., Gettu, R. (2005), 'Environmental assessment of cement based products: Life cycle assessment and the EcoConcrete software tool', in Dhir, R.K., Dyer, D.T., Newlands, M.D. (eds.), *Achieving Sustainability in Construction*, Proceedings of International Conference: Dundee, 5–6 July 2005. Thomas Telford, 281–290.

Kawai, K., Sugiyama, T., Kobayashi, K., Sano, S. (2005), 'Inventory data and case studies for environmental performance evaluation of concrete structure construction', *Journal of Advanced Concrete Technology* **3**, 435–456.

Khasreen, M.M., Banfill, P.F.G., Menzies, G.F. (2009), 'Life-cycle assessment and the environmental impact of buildings: A review', *Sustainability* **1**, 674–701. doi:10.3390/su1030674

LEED (2011), Homepage of LEED. Available from: http://www.usgbc.org/ [Accessed 16 October 2011].

Marceau, M.L., Nisbet, M.A., VanGeem, M.G. (2007), Life Cycle Inventory of Portland Cement Concrete, PCA R&D Serial No. 3011, Skokie, PCA. Available from: http://assets.ctlgroup.com/aea962c9-279b-4cf2-9dac-9706094e408e.PDF [Accessed 14 June 2011].

Marinkovic, S., Radonjanin, V., Malesev, M., Lukic, I. (2008), 'Life cycle environmental impact assessment of concrete', in Bragança L., Koukkari, H., Blok, R., Gervasio, H., Veljkovic, M., Plewako, Z., Landolfo, R., Ungureanu, V., Silva, L.S., Haller, P. (eds.), *Sustainability of Constructions – Integrated Approach to Life-time Structural Engineering*. COST action C25. Proceedings of Seminar: Dresden, 6–7 October 2008. Possendorf, Herstellung, Addprint AG, 3.5–3.16.

Marinković, S., Radonjanin, V., Malešev, M., Ignjatović, I. (2010), 'Comparative environmental assessment of natural and recycled aggregate concrete', *Waste Management* **30**, 2255–2264.

Mateus, R., Bragança, L. (2011), 'Life-cycle assessment of residential buildings', in Bragança, L., Koukkari, H., Blok, R., Gervásio, H., Veljkovic, M., Borg, R.P., Ungureanu, V., Schaur, C. (eds.), *Sustainability of Constructions – Towards a better built environment*. Proceedings of the International Conference: Innsbruck, 3–5 February 2011. Malta, Gutenberg Press Ltd, 255–262.

Mateus, R., Bragança, L., Blok, R., Glaumann, M., Wetzel, C., Bikas, D., Giarma, C., Kahraman, I., Aktuglu, Y. (2011), 'Use of rating systems in the process towards sustainable construction', in Bragança, L., Koukkari, H., Blok, R., Gervásio, H., Veljkovic, Plewako, Z., M., Borg, R.P. (eds.), *Sustainability of Constructions – Integrated Approach to Life-time Structural Engineering. COST action C25.* Volume 1 – *Summary report of the Cooperative Activities.* Malta, Gutenberg Press Ltd, 51–97.

NREL (2011), U.S. Life-Cycle Inventory Database. Available from: http://www.nrel.gov/lci/database/ [Accessed 04 July 2011].

Öko-Institut (2008), Global Emission Model for Integrated Systems GEMIS. Available on line at http://www.oeko.de/service/gemis/en/index.htm. [Accessed 20 January 2008].

Oliver-Solà, J., Josa, A., Rieradevall, J., Gabarerell, X. (2009), 'Environmental optimization of concrete sidewalks in urban areas', *International Journal of Life Cycle Assessment* **14**, 302–312.

Prusinski, J.R., Marceau M.L., VanGeem, M.G. (2004), Life cycle inventory of slag cement concrete. Available from: http://www.slagcement.org/Sustainability/pdf/Life%20Cycle%20Inventory%20of%20Slag%20Cement%20Concrete.pdf [Accessed 10 December 2011].

Sakai, K. (2005), 'Environmental design for concrete structures', *Journal of Advanced Concrete Technology* **3**, 17–28.

SB Alliance (2011), Tools & research. Available from: http://www.sballiance.org/ [Accessed 18 November 2011].

Šelih, J., Sousa, A.C.M. (2007), 'Life cycle assessment of construction processes', in Bragança, L., Pinheiro, M., Jalali, S., Mateus, R., Amoêda, R., Guedes, M.C. (eds.), Portugal SB07 *Sustainable construction, Materials and Practices – Challenge of the Industry for the New Millennium*, Amsterdam, IOS Press, 366-372.

Part II
Concrete with supplementary cementitious materials (SCMs)

4
Natural pozzolans in eco-efficient concrete

M.I. SÁNCHEZ DE ROJAS GÓMEZ and M. FRÍAS ROJAS, Eduardo Torraja Institute for Construction Science (IETcc-CSIC), Spain

DOI: 10.1533/9780857098993.2.83

Abstract: This chapter discusses the major natural pozzolans, along with the technical, economic and environmental advantages of using such materials in cement manufacture. The characterisation, pozzolanic properties, reaction kinetics and mechanical strength of pozzolans of different origins are described. The use of new pozzolans obtained from fired clay waste, which can be classified as a natural calcined pozzolan, is also addressed.

Key words: natural pozzolans, natural calcined pozzolans, characterisation, pozzolanic activity, blended cement matrices, properties.

4.1 Introduction

The term pozzolan, originally applied to a kind of volcanic tuff found at Pozzuoli, at the foot of Mount Vesuvius, is now used generically to define materials which, while not cementitious per se, have constituents that at ambient temperature combine with lime in the presence of water to form permanently insoluble and stable compounds that behave like hydraulic binders. The $Ca(OH)_2$ needed for the pozzolanic reaction may come directly from hydrated lime or a hydrating Portland cement.

Pozzolans have been in use since antiquity. The works erected with pozzolan- and lime-based 'Roman cement and mortars', pantheons, coliseums, stadiums, basilicas, aqueducts, bridges and a wide variety of other structures (cited by Vitruvius and Pliny in their writings), have endured to our times as invaluable relics of Roman civilisation.

Pozzolans continue to be used today, as attested to by international standards that cite various types of pozzolanic materials among the constituents of cement. Roy and Langton (1989) suggest that calcined clays mix with slaked lime (calcium hydroxide) were the first hydraulic binder to be made. Malinowsky (1991) reports ancient constructions from 7000BC in the Galilei area (Israel) using this type of binder. The eruption of Thera in 1500BC, which destroyed part of Santorin island was responsible for the appearance of large amounts of ashes used by the Greeks to make mortars that reveal having hydraulic properties. However, the Romans already knew that calcined clay was needed to produce mortars with a high performance, so their use was not conditioned by the availability of natural pozzolans (Hazra and Krishnaswamy, 1987).

The use of pozzolans is justified from a number of perspectives. Technically, depending on their nature, pozzolans afford cement special characteristics, in particular by enhancing its durability, understood here to mean mortar and concrete resistance to chemical attack by aggressive external agents. This is because pozzolanic additions set off a chemical reaction in cement paste in which the lime hydrolysed from portlandite, a calcium silicate present in clinker, combines with active acid components in the pozzolan, silica and alumina to form silicates and aluminates. This raises the tobermorite content in the hydrated cement paste, with beneficial results. The various forms and structures of tobermorite (calcium silicate hydrate – CSH) are primarily responsible for cement paste (and consequently mortar and concrete) bonding and strength. Therefore, any increase in the (secondary) tobermorite formed by pozzolanic action over the (primary) amount formed directly during hydration of the silicates present in clinker enhances the mechanical strength of the binder. This effect is more significant in the medium and long term, since pozzolanic reactions tend to be slow, although that depends on the reactivity of the pozzolan in question. Given the same fineness, then, and the same clinker, pozzolan-free cements generally exhibit higher short-term mechanical strength than cements with pozzolan additions, but in the medium and long term, the latter have substantially higher strength than the former.

Moreover, portlandite fixation in the form of secondary tobermorite affords cement paste greater chemical resistance or durability, primarily for two reasons. On the one hand, portlandite is considerably more water soluble and vulnerable to acid media than tobermorite, which is less liable to be leached or carried away. This conversion from portlandite to tobermorite makes cement pastes more resistant to attack by pure mountain and granitic soil water, which is a powerful solvent, and to aggressive acid or carbonic solutions. Pure water dissolves portlandite physically; acid solutions convert it chemically into soluble calcium salts, while carbonic solutions turn it into soluble and leachable calcium bicarbonates. The concomitant porosity weakens cement paste, mortars and concretes. Pozzolanic action prevents or mitigates that effect.

Secondary tobermorite gel formation densifies cement paste, generally lowering total porosity, while increasing the proportion of micropores at the expense of mesopores and especially macropores, even where total porosity rises slightly. This has very beneficial effects, since micropores are less interconnected and more difficult to penetrate, making cement paste less permeable to water and aggressive ionic solutions, primarily sulphates and chlorides, and less permeable also to ion penetration by diffusion. Pozzolanic cement paste is ten times less chloride ion-penetrable than Portland cement without such additions (Calleja, 1992).

The presence of pozzolans in cements helps dilute clinker components, including tricalcium aluminate, C_3A, and the aluminates, ferrite-aluminates and calcium ferrites in general that are most sensitive to aggressive external agents such as sulphates in the soil or sea water or mist.

Taylor (1997) noted that the inclusion of active additions in Portland cement improves the performance of the resulting matrices. This improvement is due to both the so-called filler effect in which filler particles act as nuclei for the formation of hydrated Portland cement phases, and the pozzolanic effect, whereby pozzolanic reaction products fill the voids left by the excess water. These developments change the pore structure and pore size distribution substantially.

Economically, pozzolans are added at the end of the process, i.e. in the cement mill with no need for prior drying, dehydration or calcination, with the concomitant savings in energy. As a rule, the cost of purchasing, quarrying and transporting pozzolans is amply offset by the aforementioned energy savings.

Environmentally, the reduction in the fuel consumed in the furnace entails smaller volumes of combustion gases and hence lower carbon dioxide, sulphur, nitrogen oxide and carbon monoxide (as appropriate) emissions. This not only favours compliance with the Kyoto Treaty, but is also in line with worldwide environmental policy in the struggle against climate change and the effort to reach the EU's 80–90% greenhouse gas reduction target by 2050 (Infocemento, 2011; Isrcer, 2011; Oficement, 2011).

Nonetheless, the availability of these traditional pozzolans is declining due to less intensive and less extensive quarrying, as well as a downward trend in the establishment of new sites, primarily to minimise the impact on the landscape. Environmental policies also seek to eliminate or reduce spoil banks, prioritising industrial waste and by-product recycling for use as prime materials in cement manufacture (Chapter 5). For all these reasons, waste or by-products of natural origin that after activation can be classified as natural or natural calcined pozzolans are being actively sought. This is the case of the waste generated by the fired clay industry.

This chapter discusses the most common traditional pozzolans and the studies conducted on natural calcined pozzolans. The research conducted by the Eduardo Torroja Institute's Material Recycling Team on the viability of using brick and roof tile waste as pozzolanic additions is also described. Such fired materials, given their origin and processing, can be classified as natural calcined pozzolans.

4.2 Sources and availability

The term pozzolan is used for a number of different materials, all of which exhibit a high reactivity with lime. Establishing a single and precise classification of pozzolans is no easy task, however, for the word covers a wide variety of materials with different origins and different chemical and mineralogical compositions. Unsurprisingly, several classifications can be found in the literature, proposed by Calleja (1969), Massazza (1974, 1993), Sersale (1980), Soria (1983) and Malhotra (1987), to name a few. All, however, concur in grouping pozzolans as either natural or artificial. The former require no processing that would prompt major chemical or mineralogical changes, for they are inherently active. The latter are the outcome of the chemical or structural transformation of materials initially devoid of pozzolanicity. The typically pozzolanic components in this group need heat to be activated. Natural calcined pozzolans form part of this group.

The origin of natural pozzolans may be purely mineral or purely organic, although some mixed materials can also be found. When of a mineral origin, pozzolans are usually materials deposited in the proximity of the volcano after an explosive volcanic eruption. Depending on its viscosity, cooling rate and gas content, the original magma generates (powdery) ash, pumice (fragments with small, regular alveoli separated by a thin film of lava), scoria (regular pores and higher density) and bombs (dense material).

When the incoherent fragments of these original pyroclasts are exposed to diagenetic cementing processes, they convert into compact rocks known as tuffs. Attendant upon such cementation are the chemical-mineralogical transformations that have a direct effect on the pozzolanicity of these materials.The incoherent materials include the traditional Italian pozzolans from Campania and Latium (respectively the regions of Naples and Rome); Spanish rock such as the pumice found at Campo de Calatrava in the province of Ciudad Real, Olot in the province of Gerona and Almería, in the province of the same name; Greek materials such as Santorin earth, pozzolans from the French Massif Central and vitreous rhyolites found in the United States and India. Tuffs include German trass, used in Roman times, and materials from the Canary Islands, Naples, Romania and Crimea.

Non-volcanic materials come from rocks that have been simply deposited, such as clay, or have an organic origin such as diatomaceous earth, which contains the siliceous skeletons of microorganisms that were deposited in seawater. They may also contain clay or sand, such as in the Danish möler, which is of mixed origin. In Spain the largest beds are located in the provinces of Albacete and Jaén. Materials of this type found in Central Asia are known as Gliezh, originally schists that were calcined during underground combustion. In Spain, sedimentary materials known as opaline rock, whose main component is non-crystalline silica, are found in the provinces of Salamanca and Zamora.

Another material, amorphous silica is basically a sedimentary rock formed in surface. Actually, it forms in volcanic hydrothermal systems in which fluids, containing colloidal silica particulates having a definite temperature, cool by reaching to the Earth's surface and they also reach to excessive saturation. Amorphous silica is also called silica sinter or geo-silica (Davraz and Gündüz, 2005, 2008). Other materials have a diverse origin (volcanic, sedimentary and organic), such as Sacrofano soil (north of Rome) and gaize (Ardennes, France), a rock rich in siliceous organic remains and clay.

Materials with a high clay content, which are initially inert, exhibit significant pozzolanicity when calcined at temperatures of 600 to 900 °C and ground to the same fineness as cement. Due to their origin, these materials consist mainly of silica and alumina. The loss of chemically combined water during calcination destroys the crystalline network of the clay constituents, rendering their components amorphous, poorly defined and unstable. One example of such materials is srkhi, found in India. Another material is the bentonite clay that is found in many different areas of Khyber Pakhtunkhwa province of Pakistan.

Calcined clay products include fired clay manufactures, such as roof tile and brick, whose processing entails the use of temperatures able to activate the clay used as a prime material.

4.2.1 Natural pozzolans

According to the chemical analyses found in the literature (Schwiete et al., 1968; Calleja, 1969; Mehta, 1981; Soria, 1983; Sánchez de Rojas, 1986; Malhotra, 1987; Sánchez de Rojas et al., 1993; Uzal and Turanli, 2003, 2012; Turanli et al., 2004; Moropoulou et al., 2004) the various types of natural pozzolans do not differ substantially in this respect. They are all highly acidic, with a predominance of silica and alumina as well as iron oxide, which together generally amount to over 70% of the total, although in some cases the silica may account for 90%. The difference between mineral and organic pozzolans is fairly straightforward, inasmuch as the former usually contain less silica and more alumina and alkalis. The loss on ignition (LOI) values tend to vary widely among pozzolans regardless of their origin.

Pumice stone (natural volcanic pozzolan) from Gerona, for instance, exhibits 0.9% LOI, whereas in the pumice from Ciudad Real the value is 4.9%, even though the sums of their silica, alumina and iron oxide contents come to 69 and 69.5%, respectively.The most significant differences among tuffs lie in the percentage of alkalis (Na_2O and K_2O): Canary tuff contains 11.5%, while the material from Crimea has only 3.7%; and the LOI is 6% in Canary and 11.7% in Crimea tuff, even though their acid contents are very similar (80 and 81.8%, respectively) (Calleja, 1969; Soria 1983).

Two volcanic tuffs from different regions in Turkey (Uzal and Turanli, 2003) have very similar chemical compositions and LOI values of 6.3 and 5.9%; their silica, alumina and iron oxide contents are 82.4 and 81.4% and their alkali contents, 5.7 and 4.2%. Other volcanic materials such as volcanic ash formed during volcanic eruptions (Siddique, 2011) exhibit a low LOI (1%), along with an 80% acid, and around a 5% alkali content.

Silica, alumina and iron oxide account for around 80% of natural zeolitic tuffs (Ahmadi and Shekarchi, 2010; Uzal and Turanli, 2012), which also have a low alkali content (3–5%) and medium LOI values (6–9%).

Chemical analyses of trasses reveal that the Rhine variety has LOI values of 4.6–11.8%, the Bavarian variety 6.1–8.8% and the Austrian variety 15%. The sum of their acid components comes to around 75%, although differences exist in the alkali content: 9% in the first, 4% in the second and 3% in the third (Schwiete et al., 1968). Spanish opaline has high silica contents, from 87 to 90%, barely more than traces of alkalis (0.5%) and 3% LOI (Sánchez de Rojas et al., 1993).

The Greek pozzolans, such as Santorin earth, have silica, alumina and iron oxide contents of 65.1, 14.5 and 5.5%, respectively, and 6.5% alkalis with a LOI of 3.5% (Mehta, 1981). These values do not differ substantially from the pozzolanic material studied by Moropoulou et al. (2004), who reported silica, alumina and iron oxide values of 69.7, 12.3 and 2.3%, along with 4% alkalis and 7.4% LOI, although the literature also contains data on other Greek materials with higher LOI (Turanli et al., 2004).

Of the organic pozzolans, möler has a LOI of 5.6% and a combined silica, alumina and iron oxide content of 86%, whereas the LOI values for diatomaceous earth range from 7 to 15%, even though both materials have 80–90% acid components and an alkalic content of under 1% (Calleja, 1969; Soria, 1983; Sánchez de Rojas, 1986; Sánchez de Rojas et al., 1993).

The amorphous silica rock in the Isparta Keciborlu region have silica and alumina contents of 92.48% and 2.60%, respectively, and 1.12% alkalis with a LOI of 1.85% and sulphur (SO_3) contents are low (Davraz and Gündüz, 2005, 2008). Also, the bentonite clay of Pakistan meets the requirements of chemical composition of natural pozzolans (silica and alumina contents of 54% and 20% and iron oxide content of 8.6%), with a LOI of 5.4% and an alkalic content of under 5% (Memon et al., 2012).

Mineralogical composition, which is a determinant for material activity, also varies widely from one pozzolan to another. The main crystalline component of pumice is augite, a pyroxene and a major constituent of eruptive rocks. Magnetite and hematite may also be present in this stone (Sánchez de Rojas, 1986; Sánchez de Rojas et al., 1993).

The chief constituents of volcanic tuff are feldspars and zeolites. Ordinary trachytic, pyroxenic or augitic tuff consists of alkaline feldspars and may therefore contain high percentages of alkalis and iron- and magnesium-bearing minerals (Calleja, 1969).

Diatomaceous earth comprises calcite, quartz and traces of opal-CT (cristobalite-tridymite) and usually has high loss on ignition values (up to 15%) (Sánchez de Rojas, 1986; Sánchez de Rojas et al., 1993).

Opaline rocks consist primarily of silica, found in a number of metastable forms: opal A, which is practically amorphous, and opal-CT. Quartz, kaolinite and smaller amounts of micas are also present (Sánchez de Rojas, 1986; Sánchez de Rojas et al. 1993).

The Keciborlu amorphous silica rocks generally contain more than 90% silicon dioxide, 68.90% of which are amorphous (Davraz and Gündüz, 2005, 2008). Bentonite at a temperature of 200 °C contains anorthite, bentonite and sanidine among other crystalline compounds (Memon et al., 2012).

4.2.2 Natural calcined pozzolans

In its discussion of common cement manufactured with natural calcined pozzolan (CEM II/A&B-Q), Spanish and European Standard EN197-1 (2011) defines natural calcined pozzolans as thermally activated volcanic, clay, schist or sedimentary materials. Pozzolans must consist essentially of reactive silicon dioxide (SiO_2) and aluminium oxide (Al_2O_3), and secondarily of iron (Fe_2O_3) and other oxides. Reactive silicon dioxide must account for at least 25 wt% of the total.

As a rule, clay minerals, which are initially inert, exhibit significant pozzolanicity when calcined at temperatures of 500–900 °C and ground to the same fineness as cement. The loss of chemically combined water during calcination destroys the crystalline network of the clay constituents, rendering their components amorphous or vitreous. This thermodynamic instability is largely responsible for the pozzolanicity of these calcined materials (Hea et al., 1995).

In the phyllosilicate family, research has focused mainly on kaolinite, one of whose two structural sheets consists of tetrahedral silica and the other of octahedral alumina. Its tetrahedral ('t') and octahedral ('O') sheets together form 't-O' layers that are electrically neutral and inter-connected by weak van der Waals bonds. When kaolinite is heated to 450–600 °C the resulting dehydroxylation gives rise to a disorderly phase (13.76% mass loss) known as metakaolin (MK).

4.3 Pozzolanic activity

'Pozzolanic activity' or 'pozzolanicity' is the capacity of a material to react with lime. The many methods in place to assess this property from the chemical or mechanical perspective (Donatello et al., 2010) are discussed in depth in Chapter 6 – Tests to assess pozzolanic activity. The results described below were obtained with an accelerated chemical method which, applied to pozzolan/calcium hydroxide systems, establishes the pozzolanicity of the test material (Sánchez de Rojas et al., 1993).

Lime fixation rates in natural pozzolans vary depending on their origin. The graphs in Fig. 4.1 show the 7- and 28-day lime fixation values for Spanish volcanic materials: pumice from Ciudad Real (Pumice-CR), pumice from Gerona (Pumice-Ge). Tuff from the Canary Islands (Tuff-Ca), tuff from Murcia (Tuff-Mu), tuff from Huelva (Tuff-Hu) and tuff from Almería (Tuff-Al). The findings for non-volcanic materials are shown in Fig. 4.2: opaline

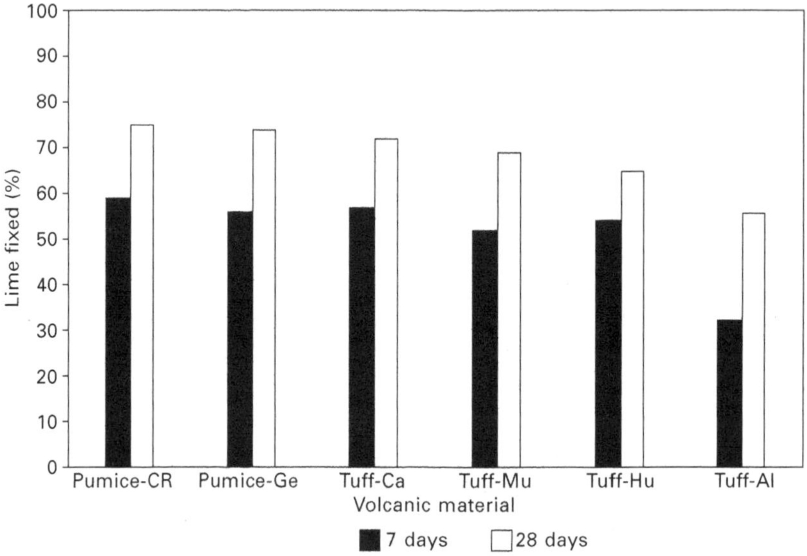

4.1 Lime fixed over time: volcanic materials.

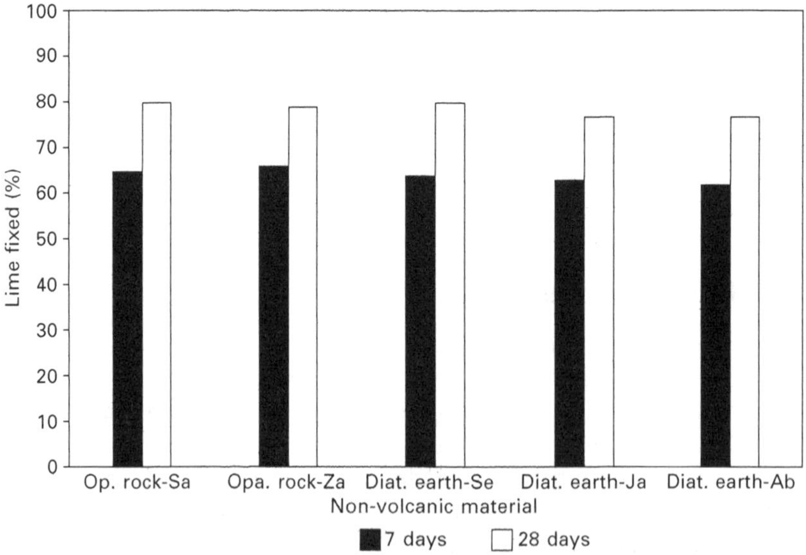

4.2 Lime fixed over time: non-volcanic materials.

from Salamanca (Op. Rock-Sa), opaline from Zamora (Op. Rock-Za) and diatomaceous earth from Seville, Jaen and Albacete (Diat. Earth-Se, Diat. Earth-Ja, and Diat. Earth-Ab). According to these figures, opaline is the most active pozzolan (fixation of the greatest amount of lime over time), followed by diatomaceous earth, pumice and tuff, in that order. Of the tuffs, the best performers were the materials from the Canary Islands (Sánchez de Rojas, (1986). Similar test findings show that certain natural zeolites can fix 94% of the lime in 14 days (Ahmadi and Shekarchi, 2010), while others (Uzal and Turanli, 2012) are able to fix only 75% in 28 days.

MK has long been known to exhibit such high pozzolanicity (Murat, 1983; de Silva and Glasser, 1990; Kostuch et al., 1993; Coleman and Page, 1997; Badogiannis et al., 2005). MK fixes 70% of the available lime in the first 7 days and 81% in the first 28 (Frías et al., 2000).

Doubts nonetheless existed about the stability of the hydrated phases formed during the pozzolanic reaction, for these phases were associated with volume changes at high curing temperatures, in turn believed to favour the conversion of metastable hexagonal hydrated phases into stable cubic phases (de Silva and Glasser, 1993).This transformation would entail a loss of strength and durability due essentially to the release of large amounts of water and the presence of a smaller volume of stable phase. Certain studies (Cabrera and Frías, 2001; Frías and Cabrera, 2001, 2002; Frías and Sánchez de Rojas, 2003; Frías, 2006), however, have shown that at a curing temperature of 60 °C, the pozzolanic reaction between MK and calcium hydroxide in an MK/$Ca(OH)_2$ system simultaneously generates metastable phases ($C_4AH_{10}$and C_2ASH_8) as well as a cubic phase (C_3ASH_6) of the hydrogarnet family, with no sign of the feared conversion reaction. Moreover, the pore structure is refined in this process, improving cement performance when blended with MK (Pera et al. 1998; Cabrera and Nwaubani, 1993; Chabannet et al. 2000; Ramlochan et al. 2000; Sabir et al., 2001). Consequently, MK, as a highly active natural calcined pozzolan, exhibits ideal properties for use in cement manufacture.

4.4 Properties of pozzolan-blended cement

4.4.1 Heat of hydration

Portland cement hydration reactions are so highly exothermal that they heat the cement paste. Heat develops rapidly during setting and initial hardening and gradually declines and finally stabilises as hydration slows. Hence, 50% of the heat is generated in the first 3 days and 80% in the first 7 (Soria, 1980). Moreover, the substantial temperature variations recorded in the first few hours may cause shrinkage, generating the cracks observed in some construction works involving large masses of concrete or structures with cement-rich mortar or concrete (Springenschmid, 1991). All the cement constituents participate in the generation of this heat: most prominently tricalcium aluminate with 207 cal/g and free lime with 279 cal/g, while dicalcium silicate contributes the least, with 62 cal/g. Since cement heat of hydration depends on the proportion of its constituents, it is particularly important to quantify its composition, as well as the proportion of any additions.

The use of pozzolans in cement reduces the heat released during hydration, although the pozzolanic material–lime reaction also generates heat, which would explain why the decline is not proportional to the amount of clinker replaced in the blended cement (Sánchez de Rojas et al., 1993, 2000; Sánchez de Rojas and Frías, 1996; Frías et al., 2000).Consequently, early age pozzolanic behaviour can be estimated by measuring heat of hydration, which can also be used to establish its impact in the development of low or very low heat of hydration cements (European Standards EN 197-1, 2011, and EN 14216, 2005).

The Langavant calorimeter is the procedure recommended in Spanish legislation (European standard EN 196-9, 2011).This semi-adiabatic method consists of quantifying the heat generated during cement hydration using a Dewar flask, i.e. a thermally insulated vessel, as a calorimeter (Alegre, 1961). According to the aforementioned standard, measurements are to be taken only up to the age of 41 hours, for heat rises so slowly at later ages variations are on the order of measurement error.

Figure 4.3 shows heat of hydration over time up to 14 hours for cements prepared with different materials. In the studies discussed, since the batching was the same in all cases (cement : addition, 70 : 30), the effect of each material on the heat of hydration generated by the base Portland cement, which was the same in all the mixes.

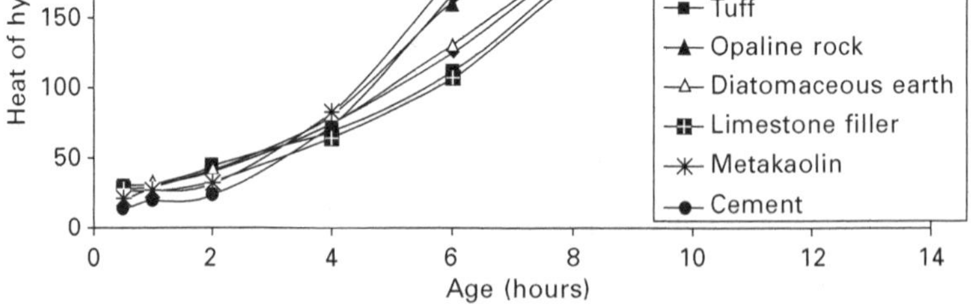

4.3 Heat of hydration versus time: mortar prepared with different additions (cement : addition ratio, 70 : 30).

In the first few hours of the test, the pozzolanic materials raised the heat of hydration as a result of the filler effect (Taylor, 1997), which favours hydration and pozzolanicity. Between 5 and 12 hours, however, the reactions were highly exothermal and the upward slope on the curves was very steep, with the base cement exhibiting greater heat of hydration than the blended materials. Over time, the pozzolanic activity of the material was instrumental in the development of heat of hydration.

This effect is shown more clearly in Fig. 4.4 which depicts the variation in heat of hydration in mixed mortars normalised to the base mortar (assigned a value of 0) in the first 14 hours of the trial. Note that the heat of hydration declined with rising addition pozzolanicity. Hence, opaline lowered the heat of hydration less than the other additions. Given its high pozzolanicity, the MK blend even exhibited greater heat of hydration than the control due to its specific reactions with calcium hydroxide, for the formation of aluminium hydrate compounds plays an important role in the development of the heat released (Frías et al., 2000).

4.4.2 Mechanical properties

The existing European Standard (EN 197-1, 2011) lists the mechanical specifications (compressive strength) to be met by commercial Portland

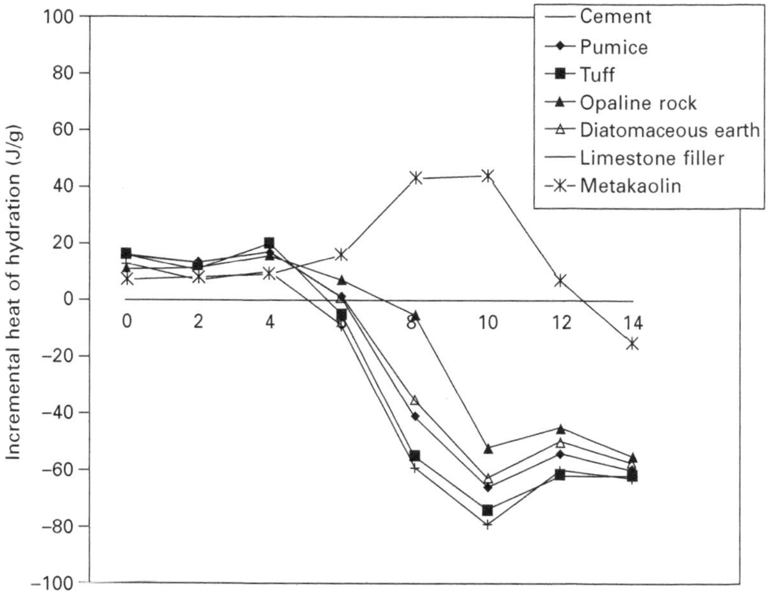

4.4 Incremental heat of hydration in mortars with different additions (cement : addition ratio, 70 : 30) referred to base cement (100 : 0).

cement, by strength category. In the research discussed here, mechanical trials were conducted with mortars prepared with a CEM I 42.5N Portland cement (defined in standard EN 196-1, 2005), blended with a series of natural pozzolans: pumice from Ciudad Real, volcanic tuff from the Canary Islands and Almería and opaline from Salamanca. Two replacement ratios were used: 20 and 35%.

The mechanical findings (Fig. 4.5) showed that the mortars containing 20% pumice from Ciudad Real had a 28-day mean strength of 45 MPa, or 90% of the strength found for Portland cement alone (100/0). When 35% of the cement was replaced, however, compressive strength was 19% lower than in the control. Similarly, Hossain (2003, 2005) and Siddique (2011), studying volcanic ash and pumice powder, reported that the compressive strength of mortar containing blended cement declined with rising percentages of pozzolan. Mehta (1981), in a study with blended Portland cements containing 10, 20 and 30% Santorin earth, found that the 7-day strength was proportional to the cement content in the blend. In the 28-day specimens, however, the cement containing 10% pozzolan was 6% stronger than the control, while strength in the 20 and 30% pozzolan blends was only 7 and 18% lower than in pure Portland cement.

Canary Island tuff raised mortar mechanical strength. When the ratio was 80/20, the 28-day values were similar to the findings for cement (98.8%). When the ratio was 65/35, strength in the samples was 90% of the value for the unblended cement. By contrast, the Almería tuff exhibited deficient behaviour, with 28-day values for the 65/35 blend of 24 MPa or 50% of Portland cement strength. By contrast, other authors reported good results for cements with a high with tuff content. Uzal and Tutanli (2003, 2012)

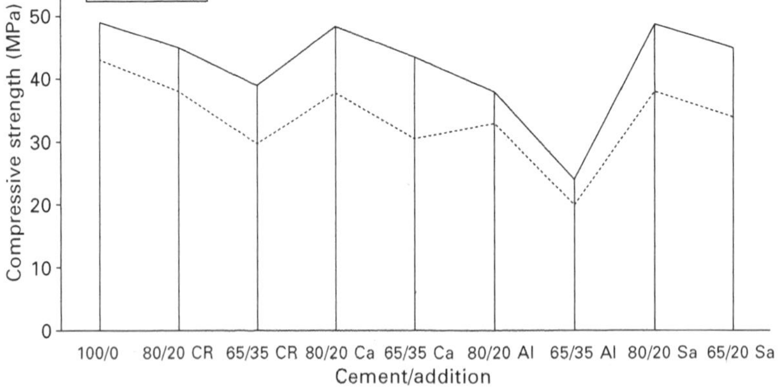

4.5 Compressive strength in mortars with different cement : addition ratios.

observed that cement containing 55% zeolitic tuff and 45% Portland exhibited high strength, with 28-day values similar to the strength found for 100% Portland cement.

Compressive strength has also been reported to rise at all test ages with respect to the control for concrete made with cement containing natural zeolite (Ahmadi and Shekarchi, 2010). The best results were obtained with a replacement ratio of 15%. Opaline also yielded good results, even better than those obtained with volcanic tuff. When the ratio was 80/20, the 28-day values were similar to the findings for cement (99.6%). When the ratio was 65/35, strength in the samples was 92% of the value for the unblended cement. The excessive amount of mixing water required for diatomaceous earth constitutes a substantial drawback to its use as a pozzolan, which adversely affects its use as a pozzolanic material.

The Keciborlu amorphous silica rocks generally show strength enhancement with the addition of AS at all curing ages and higher strengths are obtained in comparison with that of control mix (Davraz and Gündüz, 2005) and the comparative compressive strength analysis of bentonite mixes showed higher strength than the reference cement (Memon et al. 2012).

The effect of MK has been studied in pastes, mortars and concretes. Poon et al. (2001) and Badogiannis et al. (2005) explored the effect of the presence of MK as a partial cement replacement (5–20 wt%) on cement paste compressive strength. Mechanical behaviour was observed to improve in both these studies, according to which the optimal replacement ratio is 10%. Other authors have focused on the effect of using metakaolin as an addition (10–30 wt%) on the mechanical properties of mortars (Curcio et al., 1998; Li and Ding, 2003; Potgieter-Vermaak and Potgieter, 2006), likewise observing a rise in compressive strength.Wild et al. (1996) and Brooks and Johari (2001) reported that the greatest rise in compressive strength was obtained with replacement ratios of 20 and 15%, respectively.

The effect of adding varying percentages (5–30 wt%) of MK to cement on the properties of new concretes has also been researched (Wild et al., 1996; Brooks and Johari, 2001; Qian and Li, 2001; Roy et al., 2001; Sabir et al., 2001; Badogiannis et al., 2004; Güneyisi et al., 2008; Abbas et al., 2010). All these studies revealed the beneficial effect of this material on concrete compressive and tensile strength (Qian and Li, 2001).

4.5 Conclusion and future trends

As noted earlier, the future of natural pozzolans will depend on the availability of quarries, i.e. the maintenance and enlargementof existing sites or the discovery of new ones. In this regard, a review of the literature shows that papers have been published recently on volcanic tuffs from Turkey (Uzal and Turanli, 2003; Turanli et al., 2004), natural zeolites (Uzal and Turanli,

2012), natural pozzolans from the island of Milos in Greece (Velosa and Cachim, 2009), and volcanic ash from Rabaul in the Papua New Guinean province of East New Britain, where the source was a volcano known as Mount Tavurvur (Hossain, 2005). Such papers attest to the topicality of research on natural pozzolans.

Nonetheless, the difficulties in exploiting natural pozzolan quarries continue to grow, primarily due to concerns around their impact on the landscape. Alternative materials must therefore be sought. Furthermore, the Kyoto protocol has established demanding reduction targets for greenhouse gas emissions, while the use of eco-friendly materials has been driven by the growing awareness of the contribution of energy consumption to global warming and climate change. These materials may be industrial by-products, as explained in Chapter 5, or materials which, while not active in their original natural state, have undergone industrial processing that affords them pozzolanic characteristics. This is the case of the waste generated by the fired clay industry.

The inclusion of ceramic waste in Chapter 4 on natural pozzolans rather than in Chapter 5 on artificial pozzolans is justified by the origin of the material. Ceramic waste comes from calcined natural clay. While this thermal treatment is the result of a manufacturing process, it nonetheless constitutes thermal activation. As a result, if these materials were included in a possible future standard, they would be regarded as natural calcined pozzolans (Q), further to the definition of that material in European Standard (EN 197-1, 2011), natural calcined pozzolans are materials of volcanic origin, clay, shales or sedimentary rocks active by thermal treatment.

Thirty million tonnes of fired clay products such as bricks, roof tiles and block were manufactured in Spain in 2006, although with the economic crisis, this figure has since tumbled to ten million tonnes (Hispalyt, 2011). Nonetheless, the percentage of discards continues to make this material attractive for use in construction, as a source of pozzolan for cement production. In addition, according to the National Construction and Demolition Waste Plan (Official State Journal, 2009), 54% of the 40 million tonnes of construction and demolition waste produced annually in Spain comprise fired clay materials, which are presently being studied for their aptitude for the above purpose[1].

While the polluting power of this inert waste is low, it poses a severe environmental problem because it must be stockpiled, a practice that mars the landscape. During the manufacture of brick and similar products, which involves dehydration followed by firing at controlled temperatures ranging

---

[1]Sánchez de Rojas M.I., Head researcher for research project BIA 2010-21194-C03-01, funded by the National Plan for Scientific Research, Development and Technological Innovation. Ministry of Science and Innovation.

from 700 to 1000 °C, the high proportions of clay minerals present in the natural materials used acquire properties characteristic of 'fired clay'. As a result, these minerals may become activated, i.e., pozzolanic.

The pozzolanic properties of waste clay discards have been emphasised by the authors, who noted that hydrated products similar to those obtained with other pozzolanic materials are formed in the respective pozzolanic reactions. The impact of the temperature at which the waste is obtained has likewise been studied. When the firing temperature is inappropriate (under- or overfired material), the chemical and mineralogical composition of the waste varies significantly with respect to the product obtained under optimal firing conditions. But in any event the temperature used (around 900 °C) is sufficient to activate the clay and confer pozzolanic properties on the discarded material. Moreover, studies have been conducted on the feasibility of using waste clay brick as a raw material in the manufacture of concrete roofing tiles, either as a replacement for cement, capitalising on its pozzolanic properties, or as part of the aggregate (Rivera et al., 2001; Sánchez de Rojas et al., 2001a, 2001b, 2003, 2006, 2007a, 2007b; Senthamarai and Devadas, 2005; Lavat et al., 2009; Pacheco-Torgal and Jalali, 2010, 2011; Medina et al., 2012a, 2012b).

This type of waste, like other pozzolanic materials, is very acidic, with silica (53.88%) prevailing in its composition, followed by alumina (16.80%) and iron oxide (5.29%). It also contains CaO (12.41%) and the alkalis Na_2O and K_2O (0.58% and 3.10%), respectively. The main crystalline compounds in its mineralogical composition are quartz, muscovite, calcite, microcline and anorthite.

Pozzolanic activity has been studied in brick and roof tile, ground to different Blaine fineness values.The results are shown in Fig. 4.6. The lime fixation values for the brick and tile tested are shown in Fig. 4.7. The findings reveal that fired clay waste exhibits acceptable pozzolanicity, for the percentage of

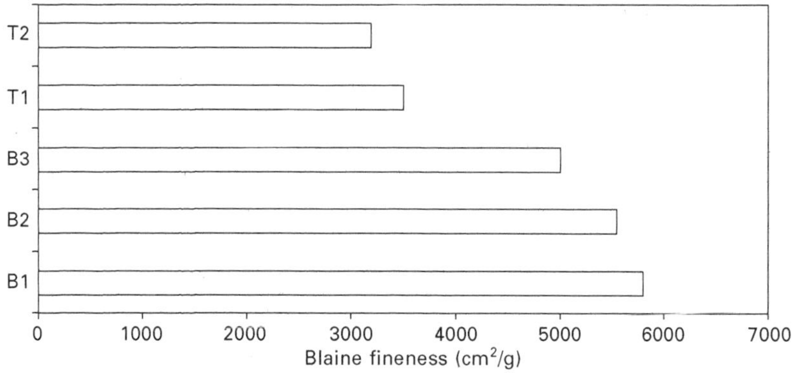

4.6 Blaine fineness for different types of fired clay waste.

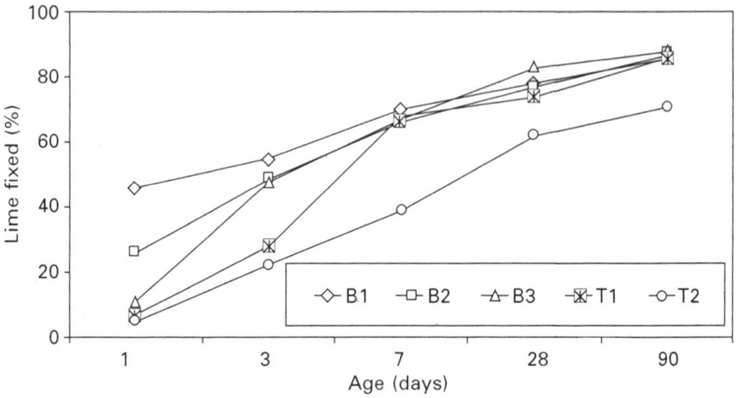

4.7 Lime fixed over time by fired clay material.

lime fixed in the 1-day specimens may be as high as 40% of the total lime available under the most favourable (finest material) conditions and 5% in the least (coarsest material). These results highlight the importance of fineness in early age specimens. Over time, lime fixation tends to even out, with similar 90-day results for all the samples. These findings indicate that the firing temperature used to manufacture these materials (around 900 °C) suffices to activate the pozzolanic properties of clay. Note that these fired clay materials behave much like the most active natural pozzolans and as MK.

The effect of including fired clay waste in mortars was studied for this review by running mechanical strength trials and comparing the results to the findings for standard mortar. The waste was used as a pozzolan at a cement replacement ratio of 15%. The fired clay waste used for the trial was obtained by mixing ground brick and roof tile ground to a Blaine fineness of 3500 cm^2/g.

The findings for bending and compressive strength in percentage of the values for the control mortar (with no fired clay additions) in 24-hour and 28-day specimens are shown in Fig. 4.8. The 24-hour bending and compressive strength values were similar in the test and control specimens. The 28-day values were slightly lower than in the control, however, but in all cases, the percentage decline was smaller than the cement replacement ratio. This is an indication that these waste materials act as pozzolans and contribute to mechanical strength. The above studies confirm that additions have a beneficial effect on cement characteristics, since they contribute to enhancing mechanical strength in the medium to long term.

In this same vein, i.e. the reuse of fired clay industry waste, Toledo et al. (2007) studied material from plants in Brazil and Velosa and Cachim (2009) a residue from expanded clay production. Both teams confirmed that this waste is apt for use as a pozzolanic addition in concrete with a hydraulic

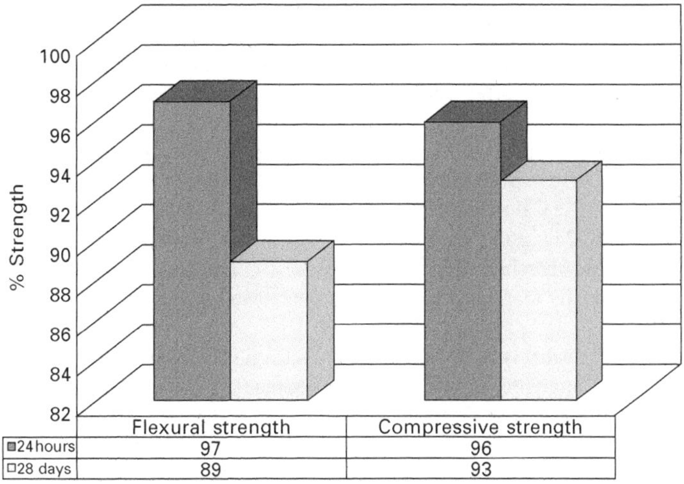

4.8 Flexural and compressive strength in mortars prepared with 15% fired clay waste (cement:waste ratio, 85 : 15) referred to unblended cement (100 : 0) mortars (=100%).

lime binder. Other authors (Moropoulou et al., 2004) have reported that waste from clay fired at low temperatures (<900 °C) has lower pozzolanicity than MK.

4.6 Sources of further information and advice

Nano-science and nanotechnology have recently drawn considerable attention from the scientific community in light of the potential applications of nano-scale particles, whose high surface area-to-volume ratio translates into exceptional chemical reactivity as well as unique physical properties.

Improvements in concrete performance have also been sought by resorting to these new disciplines (Korpa and Trettin, 2008; Sanchez and Sobolew, 2010; Cardenas et al., 2011; Nina et al., 2012). One possible application of nanotechnology is the manufacture of nano-pozzolans to raise the reactivity of the original materials. Authors such Askarinejad et al. (2012) are studying the manufacture of pozzolannano-structures from bulk natural pozzolans, opening a new line of research based on natural pozzolans.

4.7 Acknowledgements

The research conducted at the Eduardo Torroja Institute for Construction Science and cited in this chapter was funded by several Spanish Ministries (Project refs:AMB96-1095 and BIA2010- 21194-C03-01).

4.8 References

Abbas R, Abo-El-Enein SA, Ezzat ES (2010), 'Properties and durability of metakaolin blended cements: Mortar and concrete', *Materiales de Construcción*, **60**, 300, 33–49.

Ahmadi B and Shekarchi M (2010), 'Use of natural zeolite as a supplementary cementitious material', *Cement and Concrete Composites*, **32**, 134–141

Alegre R (1961), 'La Calorimétrie des Ciments au CERILH', *Revue des Matériaux*, **547**, 218–229; **548**, 247–262.

Askarinejad A, Pourkhorshidi AR, Parhizkar, T (2012), 'Evaluation the pozzolanic reactivity of sonochemically fabricated nano natural pozzolan', *Ultrasonics Sonochemistry*, **19**, 119–124.

Badogiannis E, Papadakis VG, Chaniotakis E, Tsivilis S (2004), 'Exploitation of poor Greek kaolins: strengths' development of metakaolin concrete and evaluation by means of k-value', *Cement and Concrete Research*, **34**, 6, 1035–1041.

Badogiannis E, Kakali G, Dimopoulou G, Chaniotakis E, Tsivilis S (2005), 'Metakaolin as a main cement constituent: Exploitation of poor Greek kaolins', *Cement and Concrete Composites*, **27**, 2, 197–203.

Brooks JJ and Johari MMA (2001), 'Effect of metakaolin on creep and shrinkage of concrete', *Cement and Concrete Composites*, **23**, 6, 495–502.

Cabrera J and Frías M (2001), 'Mechanism of hydration of the MK-lime-water system', *Cement and Concrete Research*, **31**, 2, 177–182.

Cabrera J and Nwaubani O (1993), 'Strength and chloride permeability of concrete containing red tropical soils', *Magazine of Concrete Research*, **45**, 164, 169–178.

Calleja J (1969), 'Las Puzolanas'. Monografía no. 281. Instituto Eduardo Torroja de Construcción y Cemento (in Spanish), (*Pozzolans*. Monograph No. 281. Eduardo Torroja Institute for Construction Science).

Calleja J (1992), 'Los nuevos cementos europeos para hormigones'. *Cemento y Hormigón*, **709**, 1157–1181 (in Spanish) (New European cements for concrete).

Cardenas H, Kupwade-Patil K, Eklund S (2011), 'Corrosion mitigation in mature reinforced concrete using nanoscale pozzolan deposition', *Journal of Materials in Civil Engineering*, **23**, 6, 752–760.

Chabannet M, Girodet C, Bosc JL, Pera J (2000), 'Effectivenees of MK on the freezing resistance of mortars', *CANMET ACI*, Barcelona, **I**, 173–186.

Coleman NJ and Page CL (1997), 'Aspects of the pore solution chemistry of hydrated cement pastes containing MK', *Cement and Concrete Research*, **27**, 1, 147–154.

Curcio F, Deangelis BA, Pagliolico S (1998), 'Metakaolin as pozzolanic micro filler for high-performance mortars', *Cement and Concrete Research*, **28**, 6, 803–809.

Davraz M and Gündüz L (2005), 'Engineering properties of amorphous silica as a new natural pozzolan for use in concrete', *Cement and Concrete Research*, **35**, 1251–1261.

Davraz M and Gündüz L (2008), 'Reduction of alkali silica reaction risk in concrete by natural (micronised) amorphous silica', *Construction and Building Materials*, **22**, 1093–1099.

De Silva PS and Glasser FP (1990), 'Hydration of cements based on MK thermochemistry', *Advances in Cement Research*, **3**, 12, 167–177.

De Silva PS and Glasser FP (1993), 'Phase relations in the system CaO-Al_2O_3-SiO_2-H_2O relevant to MK-calcium hydroxide hydration', *Cement and Concrete Research*, **23**, 3, 627–639.

Donatello S, Tyrer M, Cheeseman C (2010), 'Comparison of test methods to assess pozzolanic activity', *Cement and Concrete Composites*, **32**, 121–127.

European Standard EN 196-1 (2005), 'Method of testing cement. Part 1: Determination of strength'.

European Standard EN 196-9 (2011), 'Method of testing cement. Part 9: Heat of hydration semi-adiabatic method'.

European Standard EN 197-1 (2011), 'Cement – Part-1: Composition, specifications and conformity criteria for common cements'.

European Standard EN 14216 (2005), 'Cement – Composition, specifications and conformity criteria for very low heat special cement'.

Frías M (2006), 'The effect of metakaolin on the reaction products and microporosity in blended cement pastes submitted to long hydration time and high curing temperature', *Advances in Cement Research*, **18**, 1, 1–6.

Frías M and Cabrera J (2001), 'Influence of MK on the reaction kinetics in MK/lime and MK-blended cement systems at 20 °C', *Cement and Concrete Research*, **31**, 4, 519–527.

Frías M and Cabrera J (2002), 'The effect of temperature on the hydration rate and stability of the hydration phases of MK–lime-water systems', *Cement and Concrete Research*, **32**, 133–138.

Frías M and Sánchez de Rojas MI (2003), 'The effect of high temperature on the reaction kinetics in MK/lime and MK-blended cement matrices at 60 °C', *Cement and Concrete Research*, **33**, 643–649.

Frías M, Sánchez de Rojas MI, Cabrera J (2000), 'The effect that the pozzolanic reaction of MK has on the heat evolution in MK-cement mortars', *Cement and Concrete Research*, **30**, 2, 209–216.

Güneyisi E, Gesoğlu M, Mermerdaş K (2008), 'Improving strength, drying shrinkage, and pore structure of concrete using metakaolin', *Materials and Structures*, **41**, 5, 937–949.

Hazra, PC and Krishnaswamy, VS (1987), 'Natural pozzolans in India, their utility, distribution and petrography', *Records of the Geological Survey of India*, **87**, 675–706.

Hea Ch, Osbzckb B, Makovicky E (1995), 'Pozzolanic reactions of six principal clay minerals: activation, reactivity assessments and technological effects', *Cement and Concrete Research*, **25**, 8, 1691–1702.

Hispalyt (2011), http://www.hispalyt.es, Asociación Española de Fabricantes de Ladrillos y Tejas de Arcilla Cocida, (in Spanish).

Hossain KMA (2003), 'Blended cement using volcanic ash and pumice', *Cement and Concrete Research*, **33**, 1601–1605.

Hossain KMA (2005), 'Volcanic ash and pumice as cement additives: pozzolanic, alkali-silica reaction and autoclave expansion characteristics', *Cement and Concrete Research*, **35**, 1141–1144.

Infocemento (2011), http://www.infocemento.com (in Spanish).

Isrcer (2011), http://www.isrcer.org, Instituto para la Sostenibilidad de los Recursos) (in Spanish).

Korpa A and Trettin R (2008), 'Very high early strength of ultra-high performance concrete containing nanoscale pozzolans using the microwave heat curing method', *Advances in Cement Research*, **20**, 4, 175–184.

Kostuch JA, Walters GV, Jones TR (1993), 'High performance concretes incorporating MK: A review', *Concrete 2000*, Ravindra K. Dhir and M.R. Jones (Edts), University of Dundee, 2, 1799–1810.

Lavat A E, Trezza MA, Poggi M (2009), 'Characterization of ceramic roof tile wastes as pozzolanic admixture', *Waste Management*, **29**, 1666–1674.

Li Z and Ding Z (2003), 'Property improvement of Portland cement by incorporating with metakaolin and slag', *Cement and Concrete Research*, **33**, 4, 579–584.

Malhotra VM (1987), *Supplementary cementing materials for concrete*, Ottawa, CANMET.

Malinowsky R (1991), 'Prehistory of concrete', *Concrete International*, **13**, 62–68.

Massazza F (1974), 'Chemistry of pozzolanic addition and mixed cements', *6th International Symposium on Chemistry of Cement*, Moscow.

Massazza F (1993), 'Pozzolanic cements', *Cement and Concrete Composites*, **15**, 185–214.

Medina C, Sánchez de Rojas MI, Frías M (2012a), 'Reuse of sanitary ceramic wastes as coarse aggregate in eco-efficient concretes', *Cement and Concrete Composites*, **34**, 48–54.

Medina C, Frías M, Sánchez de Rojas MI (2012b), 'Microstructure and properties of recycled concretes using ceramic sanitaryware industry waste as coarse aggregate', *Construction and Building Materials*, **31**, 112–118.

Mehta PK (1981), 'Studies on blended Portland cements containing Santorin earth', *Cement and Concrete Research*, **11**, 507–518.

Memon SA, Arsalan R, Khan S, Yiu Lo T (2012), 'Utilization of Pakistani bentonite as partial replacement of cement in concrete', *Construction and Building Materials*, **30**, 237–242.

Moropoulou A, Bakolas A, Aggelakopoulou E (2004), 'Evaluation of pozzolanic activity of natural and artificial pozzolans by thermal analysis', *Thermochimica Acta*, **420**, 135–140.

Murat M (1983), 'Hydration reaction and hardening of calcined clays and related minerals. I. Preliminary investigation on metakaolinite', *Cement and Concrete Research*, **13**, 259–266.

Nina F, Abang AAA, Ramazan D (2012), 'Development of nanotechnology in high performance concrete', *Advanced Materials Research*, **364**, 115–118.

Official State Journal (2009), 'National construction and demolition waste plan for 2008–2015', *BOE no. 49*, 26/02/09.

Oficemen (2011), http://www.oficemen.com, Agrupación de Fabricantes de Cemento de España (in Spanish).

Pacheco-Torgal F and Jalali S (2010), 'Reusing ceramic wastes in concrete', *Construction and Building Materials*, **24**, 5, 832–838.

Pacheco-Torgal F and Jalali S (2011), ' Compressive strength and durability properties of ceramic wastes based concrete', *Materials and Structures*, **44**, 1, 155–167.

Pera J, Bonnin E, Chanannet M (1998), 'Immobilization of wastes by MK-blended cements', *6th CANMET ACI, Inter. Conf. on the fly ashes, S.F., slag and Natural pozzolanas in concr*ete, II, Bangkok, Thailand, 997–1005.

Poon CS, Lam L, Kou SC, Wong R (2001), 'Rate of pozzolanic reaction of metakaolin in high-performance cement pastes', *Cement and Concrete Research*, **31**, 9, 1301–1306.

Potgieter-Vermaak SS and Potgieter JH (2006), 'Metakaolin as an extender in South African cement', *Journal of Materials in Civil Engineering*, **18**, 4, 619–623.

Qian X and Li Z (2001), 'The relationship between stress and strain for high-performance concrete with metakaolin', *Cement and Concrete Research*, **31**, 11, 1607–1611.

Ramlochan T, Thomas M, Gruber A (2000), 'The effect of MK on alkali-silica reaction in concrete', *Cement and Concrete Research*, **30**, 339–344.

Rivera J, Sánchez de Rojas MI, Frías M (2001), 'Properties of cement pastes containing calcined clay from waste ceramic tiles as pozzolana', *7th CANMET/ACI Int. Conference on Fly ash, silica fume, slag and natural pozzolans in concrete, Madras, India*, Supplementary paper, 357–370.

Roy DM and Langton C (1989), 'Studies of ancient concretes as analogs of cementitious sealing materials for repository in Tuff', *LA-11527-MS*, Los Alamos Nacional Laboratory.

Roy DM, Arjunan P, Silsbee MR (2001), 'Effect of silica fume, metakaolin, and low-calcium fly ash on chemical resistance of concrete', *Cement and Concrete Research*, **31**, 12, 1809–1813.

Sabir BB, Wild S, Bai J (2001), 'Metakaolin and calcined clays as pozzolans for concrete: a review', *Cement and Concrete Composites*, **23**, 6, 441–520.

Sanchez F and Sobolew K (2010), 'Nanotechnology in concrete – A review', *Construction and Building Materials*, **24**, 11, 2060–2071.

Sánchez de Rojas MI (1986), 'Estudio de la relación estructura-actividad puzolánica de materiales silicios españoles (origen: natural y artificial) y su utilización en los conglomerantes hidráulicos', PhD. Thesis. Autonomous University of Madrid, Spain (in Spanish) (Study of the relationship between structure and pozzolanic activity of (natural and artificial) Spanish siliceous materials and their use in hydraulic binders).

Sánchez de Rojas MI and Frías M (1996), 'The pozzolanic activity of different materials, its influence on the hydration heat in mortars', *Cement and Concrete Research*, **26**, 2, 203–213.

Sánchez de Rojas MI, Frías M, García N (1993), 'The influence of different additions on portland cement hydration heat', *Cement and Concrete Research*, **23**, 1, 46–54.

Sánchez de Rojas MI, Frías M, Rivera J (2000), 'Studies about the heat of hydration developed in mortars with natural and by-product materals', *Materiales de Construcción*, **50**, 260, 39–48.

Sánchez de Rojas MI, Frías M, Rivera J, Escorihuela MJ, Marín FP (2001a), 'Research about the pozzolanic activity of waste materials from calcined clay', *Materiales de Construcción*, **51**, 261, 45–52.

Sánchez de Rojas MI, Marín FP, Frías M, Rivera J (2001b), 'Viability of utilization of waste materials from ceramic products in precast concretes', *Materiales de Construcción*, **51**, 263, 149–161.

Sánchez de Rojas MI, Frías M, Rivera J, Marín FP (2003), 'Waste products from prefabricated ceramic materials as pozzolanic addition', *11th International Congress on the Chemistry of Cement Durban, South Africa*, II, 935–944.

Sánchez de Rojas MI, Marín FP, Rivera J, Frías M (2006), 'Morphology and properties in blended cements with ceramic waste materials recycled as pozzolanic addition', *Journal of the American Ceramic Society*, **89**, 12, 3701–3705.

Sánchez de Rojas MI, Frías M, Marín FP, Rivera J (2007a), 'Microstructure of concrete made with blends of fly ash cement and waste clay by backscattered electrón', *9th CANMET/ACI Inter. Conf. Fly Ash, Silica Fume, Slag and Natural Pozzolans in Concrete, Warsaw, Poland*, Supplementary paper, 79–90.

Sánchez de Rojas MI, Marín FP, Frías M, Rivera J (2007b), 'Properties and performances of concrete tiles containing waste fired clay materials', *Journal of the American Ceramic Society*, **90**, 11, 3559–3565.

Schwiete HE, Kastanja P, Ludwing U, Otto PA (1968), 'Investigation on the behaviour of natural and artificial puzzolans', *The Fifth Inter Congress on the Chemistry of Cement, Tokyo*, IV, Supplementary paper IV-63, 135–139.

Senthamarai RM and Devadas MP (2005), 'Concrete with ceramic waste aggregate', *Cement and Concrete Composites*, **27**, 910–913.

Sersale R (1980), 'Structure et caracterérisation des pozzolanes et des cendre volantes', *7th Inter Congress on the Chemistry of Cement, Paris*, 1, IV, IV-1/3-21.

Siddique R (2011), 'Effect of volcanic ash on the properties of cement paste and mortar', *Resources, Conservation and Recycling*, **56**, 66–70.

Soria F (1980), 'Estudio de Materiales: IV – Conglomerantes hidráulicos', *Instituto Eduardo Torroja de la Construcción y del Cemento. CSIC*, Madrid, 186pp (in Spanish), (Materials studies: IV. Hydraulic binders).

Soria F (1983), 'Las puzolanas y el ahorroenergético en los materiales de construcción', *Materiales de Construcción*, **190–191**, 69–84, (in Spanish),' (Pozzolans and energy savings in construction materials).

Springenschmid R (1991), 'Cracks in concrete caused by the heat of hydration', *Zement Kalk Gips*, **3**, 132–138.

Taylor HFW (1997), '*Cement Chemistry*', 2nd edition. Thomas Telford Publishing, Thomas Telford, London.

Toledo Filho, RD, Gonçalves JP, Americano BB, Fairbairn EMR (2007), 'Potential for use of crushed waste calcined-clay brick as a supplementary cementitious material in Brazil', *Cement and Concrete Research* **37**, 1357–1365.

Turanli L, Uzal B, Bektas F (2004), 'Effect of material characteristics on the properties of blended cements containing high volumes of natural pozzolans', *Cement and Concrete Research*, **34**, 2277–2282.

Uzal B and Turanli L (2003), 'Studies on blended cements containing a high volume of natural pozzolans', *Cement and Concrete Research*, **33**, 1777–1781.

Uzal B and Turanli L (2012), 'Blended cements containing high volume of natural zeolites: Properties, hydration and paste microstructure', *Cement and Concrete Composites*, **34**, 101–109.

Velosa AL and Cachim PB (2009), 'Hydraulic-lime based concrete: Strength development using a pozzolanic addition and different curing conditions', *Construction and Building Materials*, **23**, 2107–2111.

Wild S, Khatib JM, Jones A (1996), 'Relative strength, pozzolanic activity and cement hydration in superplasticised metakaolin concrete', *Cement and Concrete Research*, **26**, 10, 1537–1544.

5
Artificial pozzolans in eco-efficient concrete

M. FRÍAS ROJAS and M. I. SÁNCHEZ DE ROJAS GÓMEZ, Eduardo Torroja Institute for Construction Science (IETcc-CSIC), Spain

DOI: 10.1533/9780857098993.2.105

Abstract: This chapter aims to stress the importance of finding new artificial pozzolans for the manufacture of future, more eco-efficient cements and concretes, given the social and environmental benefits they deliver. It also addresses the scientific (characterisation, pozzolanic properties, reaction kinetics) and technical (physical and mechanical) aspects of new blended cement matrices.

Key words: new artificial pozzolans, characterisation, pozzolanic activity, blended cement matrices, properties.

5.1 Introduction

The industrial development attendant upon human social evolution has greatly enhanced social welfare. Nonetheless, that development has also led to the generation of huge volumes of industrial by-products and waste, which in most cases are stockpiled in landfills and, depending on their nature, incinerated under uncontrolled conditions; which poses severe environmental, economic and technical problems.

For a number of decades, some of these industrial by-products have been successfully recycled in the manufacture of cements as artificial pozzolans, as evinced by the existing European legislation (EN 197-1, 2011). The scientific, technical and environmental advantages of adding silica fume (SF), fly ash (FA) or granulated blast furnace slag (GBS) to cement matrices (pastes, mortars and concretes) are well known, while interest in the subject has focused in particular on improving durability (Malhotra, 1987; Sánchez de Rojas et al., 1989, 1999; Frías and Sánchez de Rojas, 1997; Taylor, 1997; Siddique, 2008; Lothenbach et al., 2011). The availability of these traditional pozzolans has been observed to decline in recent years, however, primarily as a result of technological changes in industrial processes and the use of alternative fuels. This has raised certain doubts or uncertainties about the future of commercial cements made with these materials. Consequently, the forthcoming revision of European Standard EN 197-1 is expected to address the need for Member States to accept new pozzolans, affording the industries concerned greater flexibility as they rise to the challenges facing 21st-century society.

For this reason, recent research has focused on the identification of new wastes whose characteristics make it apt for use as a supplementary cementitious material in future blended cements. This line of endeavour is one of the cement industry's research priorities and worldwide environmental policy guidelines, which stress the need to recycle rather than stockpile industrial waste. It also forms part of the European Commission premises on waste co-processing in the cement industry as a key element in combating climate change and reducing greenhouse gas emissions between 80 and 90% by 2050, in keeping with Community targets (Infocemento, 2011 Isrcer, 2011; Oficemen, 2011).

As a result of the foregoing and in light of the present global economic crisis, the urgent need to find alternative materials is a worldwide priority, given the substantial losses incurred by the construction industry, mainly in Spain. That, however, calls for a firm commitment by all concerned: waste-generating companies, research institutions, the cement and concrete industry and government at all levels.

This chapter discusses the main scientific-technical features of the alternative industrial wastes and by-products studied by the 'Materials Recycling' research team at the Eduardo Torroja Institute (Spanish National Research Council), which may serve as alternative artificial pozzolans in the future. For a clearer understanding of the findings, industial wastes have been divided into two main groups: artificial pozzolans from industrial processes (SiMn slag, Cu slag, coal combustion bottom ash, fluid catalytic cracking) and artificial pozzolans from agro-industrial waste (sugar cane, paper sludge, rice husks, bamboo leaves).

5.2 Sources and availability

5.2.1 Artificial pozzolans generated in industrial processes

Silicon–manganese slag (SiMn)

This kind of pozzolan is obtained during SiMn production as a combination of the non-profitable part of the raw materials and the fluxes such as quartz or lime that are added to the ore in ferroalloy production furnaces to endow the cast-iron products with certain physical properties. Due to the difference in density, this slag can be separated from the ferroalloy (SiMn) during casting and poured onto a slag bed, from where it is transported, classified and stored, if necessary.

Slag generation the world over comes to around 10 Mt per year, 2.2 Mt of which are produced in Europe. In Spain, 150 000 t are generated by the Ferroalloys División of Ferroatlántica at its Boo de Guarnizo Plant, located in the province of Cantabria. Table 5.1 shows the X-ray fluorescence (XRF) results on the chemical composition of the Spanish SiMn slag. The main oxides are SiO_2 and CaO (which summed over 67%), followed by Al_2O_3 (2.2%) and MnO (9.9%). The remaining compounds account for less than 5% each. The sulphide content comes to 0.42% and loss on ignition (LOI) is – 0.91% (weight gain). According to these data, the SiMn slag would be classified as acidic (Ca/Si = 0.59), because its Ca/Si chemical modulus is less than 1 (Pera et al., 1999). The X-ray diffraction (XRD) mineralogical composition of the slag obtained shows akermanite ($Ca_2MgSi_2O_7$) to be the main crystalline phase. Other compounds identified are $CaAl_2Si_2$, $K_2MgSi_5O_{12}$, Ca_2SiO_4, manganese oxides, and sulphides (mainly manganese). Figure 5.1 depicts a fragment of the SiMn slag analysed. This slag exhibits five clearly differentiated layers, in which the greenish coloration and the textural morphology varied substantially with depth (1-5).

Copper slag (Cu)

For decades, the metallurgical industry has used non-ferrous slag (Pb, Zn, Cu, Ni) as an alternative source for obtaining raw materials for construction

Table 5.1 SiMn slag chemical composition

Oxides	SiO_2	Al_2O_3	Fe_2O_3	CaO	MgO	Na_2O	K_2O	MnO	TiO_2	SO_3
(%)	42.6	12.2	1.0	25.2	4.2	0.36	2.2	9.9	0.36	0.12

5.1 Morphological aspect of the SiMn slag.

and, more specifically, as artificial additions in cement manufacture (Douglas and Malhotra, 1987). Nonetheless, research has focused primarily on slag generated in copper metallurgy via fusion of primary minerals or copper-based industrial waste. The end product is a shiny black, stable, vitreous, compact, abrasive slag with variable particle size distribution (primarily under 10 mm) (Fig. 5.2). Further to the data available worldwide, 27 Mt of Cu slag are generated yearly, 700 000 t of which by Spanish industry.

The chemical composition of copper slag varies within a normal range, depending primarily on the process and raw materials used. Cu slag consists of Fe_2O_3 (45–60%) and SiO_2 (20–36%), followed by Al_2O_3 (2–7%) and ZnO (1–8%). None of the remaining oxides present (CuO, CaO, MgO) comes to 2% of the total weight of the sample. The sulphide content is lower than 0.5%, with and LOI of 4 to 7% (weight gain), due to the presence of oxidisable compounds. The main crystalline component, fayalite (SiO_4Fe_2), is a member of the olivine family. XRD analysis has also identified traces of iron oxides (FeO, Fe_2O_3 and F_3O_4).

Pulverised coal combustion bottom ash (BA)

In addition to FA, well known for its pozzolanic properties (Sánchez de Rojas and Frías, 1996) and included the world over in legislation on commercial cement and concrete manufacture, other types of combustion wastes (bottom ash, BA) are generated at coal-fired steam power plants and stockpiled in open spaces, with the resulting adverse impact on the environment. Given its characteristics, BA (Fig. 5.3) is presently being studied from the scientific, technical and environmental standpoints for use alone or in combination

5.2 Morphological aspect of the Cu slag.

5.3 Morphological aspect of the bottom ash.

with FA (Cheriaf et al., 1999; Bai et al., 2003; Sanjuan and Menéndez, 2011). Generation of this type of waste has declined drastically worldwide, particularly in Europe, in the wake of restructuring in the coal industry, the growing use of gas as an alternative to coal and new less polluting electric power generation approaches such as combined cycle plants. The USA is estimated to have generated around 18 Mt of bottom ash in 2007, 10% less than in 2002. Spanish production in 2009 was estimated to come to 150 000 t, a drastic decline from the 1 Mt recorded in the late 1990s.

Substantial discrepancies can be found in the scant literature on the subject with respect to the chemical composition of bottom ash, which can be attributed prmarily to differences in the type of coal used (Cheriaf et al., 1999; Sanjuan and Menéndez, 2011). As a rule, the sum of SiO_2+Al_2O_3+Fe_2O_3 accounts for 85%, a value comparable to the findings for fly ash, which has been used in cement manufacture for many years. The CaO content in bottom ash is under 6%, i.e., this siliceous ash can be classified as EN type V and ASTM type F. The XRD pattern for this material reveals a sizeable vitreous phase, which is responsible for its pozzolanic activity. While quartz and mullite are the main crystalline compounds in BA, smaller proportions of calcite and haematite have also been detected.

Fluid catalytic cracking catalyst (FCC)

FCC is a substance that increases the rate of a chemical reaction by reducing the activation energy and is often used in oil refinery process for the rupture of high molecular weight hydrocarbon chains, a process needed to optimise the proportion of gasoline produced. FCC has two particle sizes (Fig. 5.4): one, at 69 μm, is drawn from the spent catalyst, and the other, at around 22 μm,

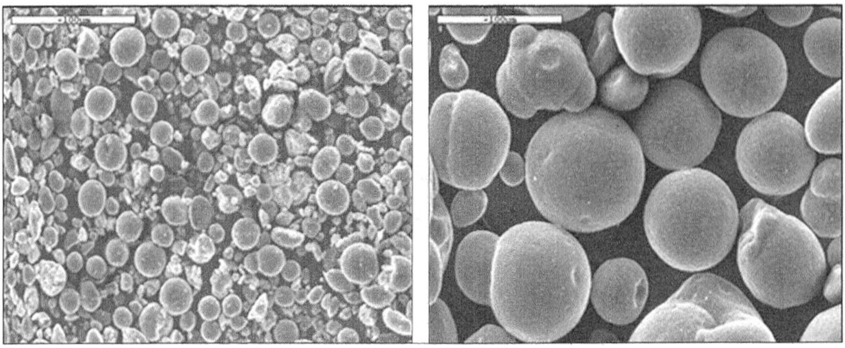

5.4 Morphological aspect of two FCC.

present in smaller proportions is collected in electrostatic precipitators in FCC units. The world production of this waste, which is estimated to amount to around 1100 t/day, is highly concentrated in specific areas. XRF chemical analysis of FCC shows that it is a silico-aluminium, consisting primarily of SiO_2 (43.62%) and Al_2O_3 (51.57%) (García de Lomas et al., 2006, 2007). Other standardised and highly reactive pozzolans such as metakaolin (Frías, 2006; Rojas, 2006; Banfill and Frías, 2007; Siddique, 2008) are similar in nature. XRD mineralogical studies reveal a high amorphous matter content (around 80%) and the presence of a single crystalline phase, a hydrogen aluminium silicate very similar to faujasite ($Na_2O.Al_2O_3.4.7SiO_2.xH_2O$).

5.2.2 Artificial pozzolans from agro-industrial wastes

As a rule, agro-industrial wastes are associated with technical, economic and especially environmental problems, due to the large quantities of wastes generated and the concomitant handling and shipping complexities involved. Consequently, the final solution is often uncontrolled combustion. These problems will intensify in the near future, according to the United Nations Food and Agriculture Organization (FAO). The planet's projected nine billion inhabitants in 2050 will raise the demand for food by 70% over today's figures. Suitable measures should therefore be taken to process and recycle the resulting waste as raw materials for other industries. One such alternative, in light of the rising price of oil, is to use agro-industrial wastes as an alternative fuel to generate electric power (rice husk and bagasse have heating capacity values of 16.3 and 19.3 MJ/kg, respectively) (Armesto et al., 2002). At the same time, the characteristics of these wastes when fired under controlled temperature and retention time conditions would appear to be an inexhaustible source of new pozzolans that could be used in the manufacture of future commercial cements.

Rice husk ash (RHA)

Rice husk, one the most abundant types of agricultural waste, has been used for decades in construction. Worldwide production is estimated to be 140 Mt, 700 000 t of which are generated in Spain. Incinerating this husk under controlled conditions (500–900 °C)(Fig. 5.5), which vary depending on the source of the ash (laboratory or cogeneration), produces a practically amorphous material (Copra, et al., 1981; Armesto et al., 2002), with different morphologies (Borrachero et al., 2007).

RHA comprises mainly SiO_2 (85–90%), with much smaller K_2O (2.5–5.5%), Na_2O (0.1–1.2%) and P_2O_5 (0.8–3.7%) contents. Its LOI ranges from 5 to 9%. Its reactive silica content is upward of 80% of the total silica in the ash. RHA contains some crystalline compounds, which may differ with the calcining temperature. At temperatures of over 800 °C (cogeneration), reflection peaks typical of cristobalite, at 21.95°, 28.45° and 36.15° (2θ), are detected and the material is found to be practically amorphous. Quartz and non-burnt coal may also be present.

Sugar cane ash (CSA)

The sugar cane industry generates two types of waste. The first, sugar cane straw (SCS), is usually burnt in open landfills (or even on the plantation itself) and practically unknown to the scientific community. Only Frías et al. (2005, 2007) and Villar et al. (2008) have reported the potential of this type of ash (SCSA) as an active addition. The second is sugar cane bagasse (SCB), which is what remains after the juice has been extracted from the cane. Research has focused on recycling this latter type of waste as an artificial pozzolan. According to the data available, world cane production is in the order of 1.5 billion tonnes, which translates into 375 Mt of bagasse and, once

5.5 Morphological aspect of the rice husk ash.

incinerated, 15 Mt of ash (SCBA) (Fig. 5.6). Since this is a recent line of research, very few papers on the subject have been published (Payá et al., 2002; Rainho et al., 2008; Chusilp et al., 2009). Ash particle morphology, texture and composition are affected by calcining temperature, which ranges widely, from 600 to 1 000 °C (Morales et al., 2009).

The chemical compositions of the two types of sugar cane wastes (SCS and SCB) when calcining at 800 °C are given in Table 5.2. Both contain SiO_2, Al_2O_3, Fe_2O_3 and CaO, which together account for over 85% of the total. The rest of the oxides are present in concentrations of under 3.5%. Bagasse ash has a smaller silica and higher alumina and iron oxide content than straw ash. Depending on the origin of the bagasse waste and the calcining process, however, ash chemical composition may vary. The table shows the differences in the chemical compositions of two types of bagasse ash obtained with different processes (laboratory and cogeneration). Although these differences are not conclusive because the bagasse waste used also differed, the BA was highly contaminated by quartz particles (Fig. 5.7), illustrating the importance of optimising agro-industrial waste-fired cogeneration and the re-use of the ash as a pozzolanic material.

Both types of sugar cane ashes exhibit very low crystallinity after calcining at temperatures of ≤ 800 °C. Calcite, quartz, mullite, iron oxides and carbon

5.6 Morphological aspect of the sugar cane bagasse ash.

Table 5.2 Chemical compositions of sugar cane ashes (%)

Oxides	SiO_2	Al_2O_3	Fe_2O_3	CaO	MgO	Na_2O	K_2O	P_2O_5	TiO_2	SO_3	LOI
SCSA	70.20	1.93	2.09	12.20	1.95	0.50	3.05	1.40	0.02	4.10	1.81
SCBA	58.61	7.32	9.45	12.56	2.04	0.92	3.22	2.09	0.34	0.53	2.73
SCBA*	66.61	9.46	10.08	1.43	0.92	0.22	3.19	1.04	2.44	0.10	4.27

*Bottom ash from a cogeneration process.

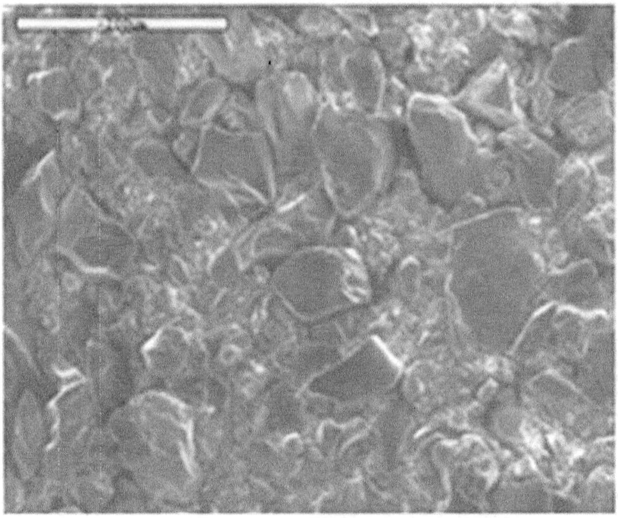

5.7 Particles of quartz in cogenerated sugar cane bagasse ash.

may be present in the ashes as crystalline compounds. When the calcining temperature was raised to 1000 °C, however, cristobalite was detected as a result of the recrystallisation of amorphous to crystalline silica.

Activated paper sludge (APS).

The paper industry, which uses recycled paper as a raw material, generates great volume of paper sludge wastes, which presently constitutes an alternative source of metakaolinite, a highly pozzolanic product classified in class Q in European legislation (Bai et al. 2003). Europe generates on the order of 2.5 Mt of such sludge yearly, and Spain alone 800 000 t. The dry sludge (35–40% moisture) consists primarily of organic matter (30%), calcite (35%) and kaolinite (20%) (Fig. 5.8).

Recent studies conducted by Frías et al. (2008a, 2008b, 2010a), García et al. (2007) and Vigil et al. (2007) established the scientific, technical and environmental basis for its reuse as a new artificial pozzolan. These authors showed that the optimum activation conditions to eliminate all the organic matter and completely convert the kaolinite into reactive metakaolinite consisted of calcination at 650–700 °C for two hours. The product obtained is depicted in Fig. 5.9: its particles are normally smaller than 90 μm with luminosity (whiteness index) of over 90%, an important property for the manufacture of white blended cements. Chloride ions (<0.02%) may or may not be present in paper sludge, depending on the industrial bleaching process used by each individual paper manufacturer (Frías et al., 2011a). The chemical composition of this activated sludge may vary substantially with

5.8 Morphological aspect of the paper sludge waste.

5.9 Aspect of the calcined paper sludge.

the activating conditions, nature of the recycled paper, industrial process and chemical composition. Studies conducted on this type of activated waste reveal that it consists primarily of silica (30%), lime (30%) alumina (18%) and magnesium <5%), with calcite, talc, chlorite and phyllosilicates (illite) as crystalline phases.

5.3 Pozzolanic activity in waste

For a waste to be regarded as an active addition to cement it must exhibit pozzolanic properties. This is defined to be the reactivity observed in siliceous or silicoaluminous substances or a combination of the two with calcium hydroxide to form hydrated pastes similar to the pastes obtained during Portland cement hydration. This property can be determined by a number of methods, which adopt different chemical or mechanical approaches

(Donatello, et al. 2010) (also see Chapter 6 – Tests to evaluate pozzolanic activity). An accelerated chemical method is used in order to evaluate this property in waste/calcium hydroxide systems, which predicts subsequent behaviour in cement matrices (Frías et al., 2008c).

The findings on the amount of fixed lime with reaction time show that all the types of waste considered here react with the portlandite in the medium (Fig. 5.10). Nonetheless, they differ in terms of reaction rate, which has a direct effect on the final application of the blended cement (mass or high performance concretes). Copper slag pozzolanic behaviour is similar to the reactivity observed in FA, i.e. with nil early age activity, which rises sharply in 28- to 90-day specimens. FCC exhibits higher 24-h activity than SF. SiMn slag fixes lime at an intermediate rate between SF and FA in the first 28 days of the reaction. Thereafter its activity is practically nil. The differences in reactivity between the acid components of the waste and the lime in the medium are closely related to factors such as fineness, industrial process, cooling conditions, degree of crystallinity and reactive silica content.

Analyses of the ash obtained by calcining the agro-industrial wastes show that lime consumption is very high at first hours of the reaction, different behaviour to that observed for the waste from industrial process mentioned above (Fig. 5.11). The percentages of lime fixed exceed the values obtained with pure metakaolin, defined as a calcined pozzolan (Q) in the existing legislation. From 25 to 40% of the total lime is fixed in the first 24 hours; the subsequent reaction rate is so high that it outpaces the findings for SF. After 28 days, the pozzolanic activity in RHA, SCBA and APS is negligible. The good performance observed in agro-industrial wastes is due primarily

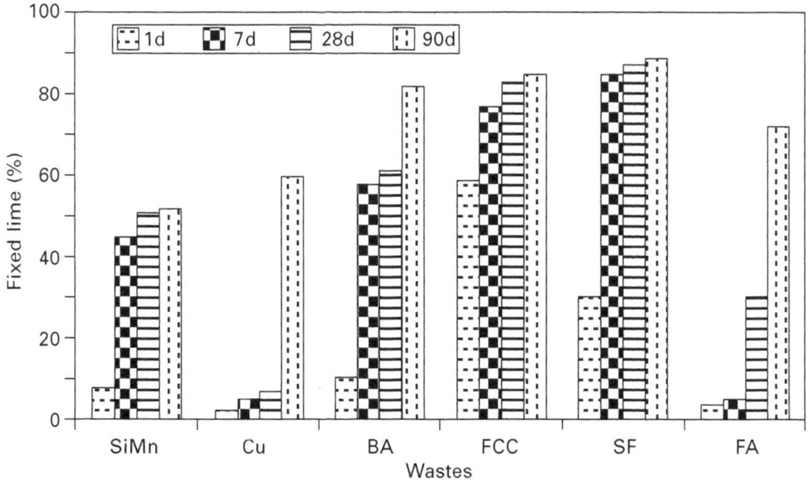

5.10 Fixed lime values of industrial wastes versus reaction time.

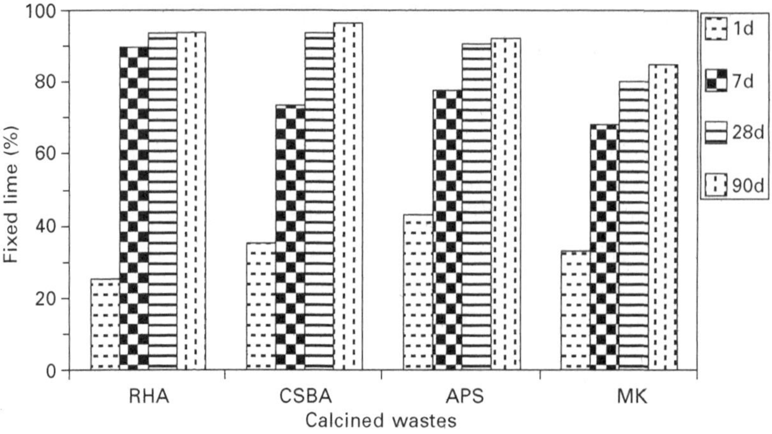

5.11 Fixed lime values of calcined agroindustrial wastes versus reaction time.

to its amorphous nature and ash particle fineness, with low density and high surface area.

5.4 Physical and mechanical properties

The existing European legislation (EN 197-1) specifies physical (initial setting time and soundness) and mechanical (compressive strength) requirements for commercialising Portland cements by strength category. These requirements are discussed below.

5.4.1 Physical properties

Table 5.3 shows the relative variation in initial setting (IS) time and soundness (expansion) values for the artificial pozzolans studied here for several replacement percentages. As a rule, the inclusion of non-ferrous slag (SiMn and Cu) and BA retard the initial setting time (positive values) with respect to the reference cement paste, more effectively at higher addition contents. This development is the result of the presence of certain minority elements known to retard anhydrous cement particle hydration (Pb, Zn, Mn). However, the addition of FCC accelerates cement hydration by 12 to 19% as a result of its high fineness.

Agro-industrial pozzolans such as RHA and SCBA can retard initial setting time by 25–50% with respect to the control paste, while APS accelerates initial setting by a substantial 11–23%, primarily as a result of the dual (physical and chemical) effect of calcite in the starting pozzolan (Banfill et al., 2009).

Table 5.3 Physical and mechanical properties of blended cements

Pozzolans		IS (%)	Expansion (mm)	Relative compressive strength (%)		
				7 days	28 days	90 days
SiMN	5%	+6.5	≤ 0.5	−7.8	−3.1	0.0
	15%	+6.5		−13.4	−11.4	-3.2
Cu	30%	+35.0	≤ 0.5	-37.0	-30.0	-20.0
BA	6%	0	≤ 0.5	−5.0	−4.0	-3.5
	20%	+14.5		−18.5	−7.5	-7.2
FCC	10%	−12.0	≤ 0.5	+1.0	+8.2	+7.3
	20%	−19.3		−12.6	+0.8	+1.6
RHA	10%	+25	≤ 0.5	−10.0	-6.0	+10.0
CSBA	10%	+25	≤ 0.5	+23.0	+25.0	+27.0
	20%	+50		+9.5	+12.6	+13.0
APS	10%	−11.5	≤ 1.0	+5.0	+2.5	0.0
	20%	−23.0		3.1	−1.5	+2.0

Expansion tests conducted on blended cement pastes reveal that these pozzolans do not expand, for all the values obtained are under 1 mm, a value comfortably below the 10-mm ceiling specified in the standard.

5.4.2 Mechanical properties

Table 5.3 also gives the 7, 28 and 90-day relative compressive strength values for the blended cement mortars. It shows that each pozzolan affects compressive strength differently, as would be expected, depending on pozzolanic activity, reaction rate and replacement percentage. The SiMn, Cu and BA pozzolans show lower strength values than the control (negative values), although the difference narrows with reaction time (Frías et al., 2006; Frías and Rodríguez, 2008; Sánchez de Rojas et al., 2008; Najimi et al., 2011). By contrast, adding FCC to the cement matrix improves compressive strength, with values higher than in the control after 7 days of hydration (positive values) (Payá et al., 2003; García de Lomas et al., 2007). This same pattern was observed for the agro-industrial pozzolans RHA and APS, whose 28- to 90-day strength values are similar to or greater than the findings for the control mortar (Vegas et al., 2006; Siddique, 2008; Ferreiro et al., 2009). The presence of limestone in APS leads to positive 7-day strength, which accelerates hydration in the anhydrous C_3A particles (Taylor, 1997). Finally, thc SCBA blended cement mortars exhibit compressive strength values up to 27% higher than the reference mortar (Singh et al., 2000; Ganesen et al., 2007; Chusilp et al., 2009; Frías et al., 2011b).

Despite the strength loss, all the new artificial pozzolans discussed in the present chapter are compliant with the present European legislation on physical and mechanical requirements (standard 28-day strength), an indication of their scientific and technical viability as pozzolanic additions.

5.5 Conclusion and future trends

Artificial additions, used as supplementary cementitious materials in commercial cement and concrete manufacture, play an important role in social, economic and environmental development. For different reasons, traditionally used pozzolans (silica fume, granulated blast furnace slag and fly ash) and even natural pozzolans pose serious short-term availability problems that are obliging cement and concrete producers to seek alternatives to enhance industry sustainability.

Artificial pozzolans such as mentioned above are being explored today for their potential to improve blended cement matrix performance. For most, however, the results reported in the literature are contradictory due to the novelty of this line of research, the variability of the parameters analysed (fineness, nature of the waste, calcining temperature, industrial process, cooling, replacement grade, type of cement), which may distort the scientific and technical advantages. Moreover, the effect of these new additions on durability, contraction-expansion, and fresh and hardened concrete properties is nil or virtually unknown.

For that reason and in light of the present worldwide economic crisis, research on possible artificial pozzolans is likely to be considerably more intense in the future to establish the scientific, technical and normalisation bases for future blended cements and concretes. The research conducted on agro-industrial wastes merits special mention, for its use as a co-generation fuel will lead to huge volumes of a type of ash that is an ideal active addition (due to its high reactive silica content) for cement used in construction.

5.6 Sources of further information and advice

In addition to the artificial pozzolans set out above, the literature addresses research related to other industrial wastes with pozzolanic properties, such as discussed below.

- *Sewage sludge ash (SSA)* is a waste obtained from the incineration of waste water sludge. According to available statistics, the USA and EU alone generate around 1.2 Mt of SSA yearly. Its immediate effect when used as an addition is to lower blended cement mortar mechanical strength, more intensely as the replacement ratio rises. Certain doubts remain around the possible use of this kind of ash as an active addition in view of its high P_2O_5 (0.3–26.7%), SO_3 (0.1–12.4%), Na_2O (0.01–6.8%)

and MgO (0.02–23.4%) content, as well as the presence of heavy metals such as chromium, copper and zinc at concentrations of under 2700, 5500 and 10 000 mg/kg, respectively. These elements have an adverse effect on reaction kinetics, microporosity, fresh and hardened concrete properties and durability (Monzó et al., 1999; Bai et al., 2003; Cyr et al., 2007; Jamshidi et al., 2012).

- *Electric arc furnace steel slag (EAFS)* generated during steel manufacture-related mining and refining. In the late 1990s, Europe generated 12 Mt/year of steel slag, 35% of which was stockpiled. That slag, primarily black slag, is known to have no pozzolanic properties and hence is primarily recycled as an aggregate for concrete and road building (Frías and Sánchez de Rojas, 2004; Frías et al., 2010b). Nonetheless, a recent study (Muhmood et al., 2009) showed that it may be recycled as an active addition to cement if subjected to prior remelting and water quenching, after which its mechanical behaviour is similar to performance in the control.
- *Palm oil fuel ash (POFA)*, obtained when palm shell oil, is used as a fuel to generate electric power (Borrachero et al., 2007). Production is estimated at 1.1 t per crop hectare which, after incineration, generates ash (5% of the waste) rich in silica (58%) and potassium oxide (8.3%) and with excellent pozzolanic properties (Tangchirapat et al., 2007). Research has shown that replacing ordinary Portland cement (OPC) with up to 20% of (finely ground) POFA raises concrete compressive strength, although at higher percentages its inclusion in cement lowers strength values compared to the reference concrete.

5.7 Acknowledgements

The research discussed in this chapter was funded by Spanish ministries (Project refs: MAT2003-06479-CO3; CTM2006-12551-CO3; MAT2009-10874-CO3), the Spanish National Research Council (CSIC) (Project refs: 200460E617 and 2003CU0009) and the following companies: Elmet, Ferroatlántica (Ferroalloys Division), Repsol, Altuna and Uria, Holmen Paper Madrid and Maicerías Española.

5.8 References

Armesto L., Bahillo A., Veijonen K., Cabanillas A., Otero J., (2002), 'Combustion behaviour of rice husk in a bubbling fluidised bed'. *Biom Bioe*, **23**, 171–179.

Bai, Y., Bashaer, P.A., (2003), 'Influence of furnace bottom ash on properties of concrete'. *Proceedings of the Institution of Civil Engineers. Struc Build*, **156**, 85–92.

Bai J., Chaipanich A., Kinuthia JM., O'Farrell M., Sabir BB., Wild S., Lewis MH., (2003), 'Compressive strength and hydration of waste paper sludge ash-ground granulated blastfurnace slag blended pastes'. *Cem Concr Res*, **33**, 1189–1202.

Banfill P., Frías M., (2007), 'Rheology and conduction calorimetry of cement modified with calcined paper sludge'. *Cem Concr Res*, **37**, 184–190.

Banfill P., Rodríguez O., Sánchez de Rojas MI., Frías M., (2009), 'Effect of activation conditions of a kaolinite based waste on rheology of blended cement pastes'. *Cem Concr Res*, **39**, 843–848.

Borrachero MV., Payá J., Monzó J., (2007), 'Reutilización de Residuos agrícolas en mezclas de cementantes', in Frías M., Sánchez de Rojas MI, Azorín V, *Seminario 12 Reciclado de Materiales en el Sector de la Construcción (CEMCO),* Madrid, 137–152.

Cheriaf, M., Rocha, JC., Pera, J., (1999), 'Pozzolanic properties of pulverized coal combustion bottom ash'. *Cem Concr Res*, **29**, 1387–1391.

Chusilp N., Jaturapitakkul C., Kiattikomol K., (2009), 'Effect of ground bagasse ash on the compressive strength and sulfate resistance of mortars'. *Construct Build Mater*, **23**, 3523–3531.

Copra, SK., Ahluwalia SC., Laxmi S., (1981), 'Technology and manufacture of rice husk ash masonry cement', Proceedings of ESCAP/RCTT Workshop on Rice-Husk Ash Cement, New Delhi.

Cyr M., Coutand M., Clastres P., (2007), 'Technological and environmental behaviour of sewage sludge ash in cement based materials'. *Cem Concr Res*, **37**, 1278–1289.

Donatello, S., Tyrer, M., Cheeseman, C., (2010), 'Comparison of test methods to assess pozzolanic activity. *Cem Concr Comp* **32**, 121–127.

Douglas E., Malhotra VM., (1987), 'A review of the properties and strength development of non-ferrous slags-portland cement binders', in Malhotra VM, *Supplementary cementing materials for concrete*, Canada, CANMET, 373–428.

European Standard EN 197-1 (2011), 'Cement – Part-1: Composition, specifications and conformity criteria for common cements'.

Ferreiro S., Blasco T., Sánchez de Rojas MI., Frías M., (2009), 'Influence of activated art paper sludge-lime ratio on hydration kinetics and mechanical behavior in mixtures cured at 20 °C'. *J Am Cerm Soc*, **92**, 3014–3021.

Frías M., (2006), 'The effect of MK on the reaction products and microporosity in blended cement pastes submitted to long hydration time and high curing temperature'. *Adv Cem Res*, **18**, 1–6.

Frías M., Sanchez de Rojas, MI., (1997), 'Microstructural alterations in fly ash mortars: Study on phenomena affecting particle and pore size'. *Cem Concr Res*, **27**, 619–628.

Frías M., Sánchez de Rojas, MI., (2004), 'Chemical assessment on the electric arc furnace slag as construction material: Expansive compounds'. *Cem Concr Res*, **34**, 1881–1888.

Frías M., Rodríguez C., (2008), 'Effect of incorporation ferroalloy industry wastes as complementary cementing materials on the properties of blended cement matrices'. *Cem Concr Comp*, **30**, 212–219.

Frías M., Villar E., Sánchez de Rojas MI., Morales E., (2005), ' The effect that different pozzolanic activity methods has on the kinetic constants of the pozzolanic reaction in sugar cane straw-clay ash/lime'. *Cem Concr Res*, **35**, 2137–2142.

Frías M., Sánchez de Rojas MI., Santamaría J., Rodríguez C., (2006), 'Recycling of silicomanganese slag as pozzolanic material in Portland cements: Basic and engineering properties'. *Cen Concr Res*, **36**, 487–491.

Frías M., Villar E., Morales E., (2007), 'Characterization of sugar cane straw waste as pozzolanic material for construction: Calcining temperature and kinetic parameters'. *Waste Manag*, **27**, 533–538.

Frías M., Sánchez de Rojas MI., Rodríguez O., García R., Vigil R., (2008a), 'Characterization of calcined paper sludge as an environmentally friendly source of MK for manufacture of cementitious materials'. *Adv Cem Res*, **20**, 23–30.

Frías M., Rodríguez O., Garcia R., Vigil R., (2008b), 'Influence of activation temperature on reaction kinetics in recycled clay waste-calcium hydroxide systems'. *J Am Ceram Soc*, **91**, 4044–4051.

Frías M., Rodríguez O., Vegas I., Vigil R., (2008c), 'Properties of calcined clay waste and its influence on blended cement behaviour'. *J Am Ceram Soc*, **91**, 1226–1230.

Frías M., Rodríguez O., Nebreda B., Villar E., (2010a), 'Influence of activation temperature of kaolinite based clay wastes on pozzolanic activity and kinetic parameters'. *Adv Cem Res*, **22**, 135–142.

Frías M., San José JT., Vegas I., (2010b), 'Steel slag aggregate in concrete: the effect of ageing on potentially expansive compounds'. *Mater Construc*, **60**, 33–46.

Frías M., Vegas I., García R., Vigil R., (2011a), 'Recycling of waste paper sludge in cements: Characterization and behaviour of new eco-efficient matrices', in Kumar S., *Integrated Waste Management*, vol. II, Intech, Croatia, 301–318.

Frías M., Villar E., Savastano H., (2011b), 'Brasilian sugar cane bagasse ashes from the cogeneration industry as active pozzolans for cement manufacture'. *Cem Concr Comp*, **33**, 490–496.

Ganesan K., Rajagopal K., Thangavel K., (2007), 'Evaluation of bagasse as supplementary cementitious material'. *Cem Concr Comp*, **29**, 515–524.

García R., Vigil R., Vegas I., Frías M., (2007), 'The pozzolanic properties of paper sludge waste'. *Constr Build Mater*, **22**, 1484–1490.

García de Lomas, M., Sánchez de Rojas, MI., Frías, M., Mújika, R., (2006), 'Comportamiento científico-técnico de los cementos portland elaborados con catalizadores FCC'. *Monografía* del IETcc (CSIC), no 412, Madrid.

García de Lomas, M., Sánchez de Rojas, MI., Frías, M., (2007), 'Pozzolanic reaction of a spent fluid catalytic cracking catalyst in FCC-cement mortars'. *J Therm Anal Cal*, **90**, 443–447.

Infocemento (2011), http://www.infocemento.com

Isrcer (2011), http://www.isrcer.org

Jamshidi, A., Jamshidi, M., Mehrdadi, N., Pacheco Torgal, F., (2012), Mechanical performance and water absorption of sewage sludge ash concrete (SSAC). *International Journal of Sustainable Engineering*, **5**, 228–234.

Lothenbach, B., Scrivener, K., Hooton, R., (2011), Supplementary cementitious materials. Cem Concr Res, **41**, 1244–1256.

Malhotra VM., (1987), *Supplementary cementing materials for concrete*, Ottawa, CANMET.

Monzó J., Payá J., Borrachero MV., Peris-Mora E., (1999), 'Mechanical behaviour of mortars containing sewage sludge ash and Portland cements with different tricalcium aluminate content'. *Cem Concr Res*, **29**, 87–94.

Morales E., Villar E., Frías M., Santos SF., Savastano H., (2009), 'Effect of calcining conditions on the microstructure of sugar cane waste ashes: Influence in the pozzolanic activation'. *Cem Concr Comp*, **31**, 22–28.

Muhmood L., Vitta S., Venkateswaram D., (2009), 'Cementitious and pozzolanic behaviour of electric arc furnace steel slags'. *Cem Concr Res*, **39**, 102–109.

Najimi M., Sobhani J., Pourkhorshidi AR., (2011), 'Durability of copper slag contained concrete exposed to sulfate attack'. *Constr Build Mater*, **25**, 1895–1905.

Oficemen, (2011). http://www.oficemen. com

Payá J., Borrachero MV., Díaz-Pinzón L., Ordoñez LM., (2002), 'Sugar cane bagasse ash

(SCBA): Studies on its properties for using in concrete production'. *J Chem Technol Biotechnol*, **77**, 321–325.

Payá J., Monzó J., Borrachero MV., Velázquez S., Bonilla M., (2003), 'Determination of pozzolanic activity of fluid catalytic cracking residue', *Cem Concr Res*, **33**, 1085–1091.

Pera, J., Ambroise, J., Chabannet, M., (1999), 'Properties of blast furnace slags containing high amounts of manganese'. *Cem Concr Res*, **29**, 171–177.

Rainho S., Eunice A., Tadeu G., Fidel A., (2008), 'Sugarcane bagasse ash as a potential quartz replacement in red ceramic'. *J Am Ceram Soc*, **91**, 1883–1887.

Rojas MF., (2006), 'Study of hydrated phases present in a MK-lime system cured at 60 °C and 60 months of reaction'. *Cem Concr Res*, **36**, 827–831.

Sánchez de Rojas, MI., Frías, M., (1996), 'The pozzolanic activity of different materials, its influence on the hydration heat in mortars'. *Cem Concr Res*, **26**, 203–213.

Sánchez de Rojas, MI., Luxán, MP., Frías, M., García, N., (1989), 'The influence of different additions on portland cement hydration heat'. *Cem Concr Res*, **23**, 46–54.

Sánchez de Rojas, MI., Rivera, J., Frías, M., (1999), 'Influence of the microsilica state on pozzolanic reaction rate'. *Cem Concr Res*, **29**, 945–949.

Sánchez de Rojas, MI., Rivera J., Frías, M., (2008), 'Use of recycled copper slag for blended cements'. *J Chem Technol Biotechnol*, **83**, 209–217.

Sanjuan, MA., Menéndez, E., (2011), 'Experimental analysis of pozzolanic properties of pulverized coal combustion bottom ash compared to fly ash in Portland cements with additions', in Palomo A, Zaragoza A, López Agüí JC. *13th Inter Congress on the Chemistry of Cement*, Madrid, 58.

Siddique R., (2008), *Waste materials and by-products in concrete*, Berlin, Springer.

Singh NB., Singh VD., Rai S., (2000), 'Hydration of bagasse blended Portland cement', *Cem Concr Res*, **30**, 1485–1488.

Taylor HFW., (1997), *Cement chemistry*, 2nd ed., London, Thomas Telford Publishing.

Tangchirapat W., Saeting T., Jaturapitakkul C., Kiattikomol K., Siripanichgorn A., (2007), 'Use of waste ash from palm oil industry in concrete'. *Waste Manag*, **27**, 81–88.

Vegas I., Frías M., Urreta J., San José JT., (2006), 'Obtaining a pozzolanic addition from the controlled calcination of paper mill sludge. Performance in cement matrices'. *Mater Construc*, **56**, 49–60.

Vigil R., Frías M., Sánchez de Rojas MI., Vegas I., García R., (2007), 'Mineralogical and morphological changes of calcined paper sludge at different temperatures and retention in furnace'. *App Clay Sci*, **36**, 279–286.

Villar E., Frías M., Valencia E., (2008), 'Sugar cane wastes as pozzolanic materials: application of mathematical model'. *ACI Mater*, 105, 258–264.

6
Tests to evaluate pozzolanic activity in eco-efficient concrete

A. R. POURKHORSHIDI, Road, Housing and Urban Development Research Center, Iran

DOI: 10.1533/9780857098993.2.123

Abstract: This chapter discusses various methods available for evaluating pozzolans as cement replacement materials. The merits and disadvantages of the methods are elaborated, and some existing anomalies are highlighted. Guides on applying a combination of these methods are presented in order to evaluate pozzolans in a way that complies with pozzolan performance in concrete.

Key words: natural pozzolan, fly ash, silica fume, pozzolanic activity, strength activity index.

6.1 Introduction

Human beings have used natural pozzolans for 6000–7000 years. Malinowski and Frifelt (1993) reported that the oldest example of hydraulic binder, dating from 5000-4000 BC, was a mixture of lime and natural pozzolan, a diatomaceous earth from the Persian Gulf. The next oldest report of using pozzolans was in the Mediterranean region. The pozzolan was volcanic ash produced from two volcanic eruptions: one sometime between 1600 and 1500 BC.

In recent decades environmental considerations and energy efficiency requirements have ushered researchers to focus on sustainable development. One of the most pollutant industries is cement production; inasmuch as producing 1 tonne of cement results in the emission of about 1 tonne of CO_2 (Mehta, 1999; Uzal et al., 2007). Therefore, the application of blended cements, in lieu of ordinary Portland cement, is rapidly increasing, which reaps great environmental and economical benefits while improving concrete properties (Meyer, 2009).

Contemporary surveys reveal that the majority of European cement production is allocated to blended cements (CEMBUREAU, 2001). EN 197-1 (2000) designates 27 different cement types, from which 26 are categorized as blended cements. Blended cements consist of different supplementary cementitious materials (SCM), such as fly ash, silica fume, blast furnace slag, limestone and natural pozzolans. Natural pozzolans, owing to their abundance and relatively low costs, present considerable potential for employment in the cement and concrete industries. Additionally, their application generally results in decreases in pollutant emissions and increases in concrete durability properties (Shannag and Xeginobali, 1995; Colak, 2002, 2003; Targan et al., 2003; Ezziane et al., 2007; Najimi et al., 2008). For a long time the use of natural pozzolans has been mostly restricted to Italy, where considerable reserves of natural pozzolans are found (Kogel et al., 2006). Nowadays, ample resources of natural pozzolans capable of being used in binary cements exist in Italy, China, the USA, Chile, Greece, Cameroon, Algeria, France, Turkey, Iran, Saudi Arabia, Honduras, etc. (Massazza, 1993; Poon et al., 1999; Türkmenoğlu and Tankut, 2002; Feng and Peng, 2005; Kogel et al., 2006; Çavdar and Yetgin, 2007, 2009; Jana, 2007; Kaid et al., 2009).

Heretofore, numerous investigations on cement replacement materials, such as natural pozzolans, silica fume and fly ash, have resulted in various methods for the assessment and classification of pozzolans (Diamond, 1982; Mehta, 1986; Luxan et al., 1989; Sybertz, 1989; Wesche et al., 1989; Tashiro et al., 1994; Shi and Day, 2001; Paya et al., 2001; Gava and Prudencio, 2007a,b). These methods are principally based on testing the chemical characteristics of pozzolans, and physical properties of cement–pozzolan mixtures. The complicated and multifaceted interaction of pozzolans in concrete mixtures depends on a battery of parameters. Hence, testing methods normally interpreting from few properties and observations have limited ability in foreseeing real performances of pozzolans in concretes. Recent researches assert that pozzolanic activity should be assessed in conjunction with real concrete/pozzolan mixture performance; consequently, a simple and versatile procedure, reflecting real pozzolan performance in concrete mixtures would be invaluable (Luxan et al., 1989; Sybertz, 1989; Tashiro et al., 1994; Paya et al., 2001; Gava and Prudencio, 2007a,b).

6.2 Methods for evaluating pozzolanic activity

The combination of pozzolanic materials with hydraulic cements has long been utilized in construction practices. Research on specifications and testing methods of fly ash started in the 1930s when fly ash, from coal-burning electric power plants, was used in cement concretes (ACI 232.2R, 2003). The first classification of natural pozzolans, based on pozzolanic constituents, was introduced in the 1950s (ACI 232.1R, 2000). Also, investigations of silica fume performance in concrete began in the Scandinavian countries around 1950 (ACI 234R, 1996).

Hitherto, extensive research on pozzolanic materials has resulted in various standards and testing methods. Nonetheless, it is apparent that these methods are unable to project all the aspects of pozzolan performance in concrete. Peer reviews of the literature suggest that some standard procedures totally neglect the importance of pozzolan performance in the concrete mixture, inasmuch as some methods have ambiguities.

The methods and standard procedures for the evaluation of specifications and requirements for pozzolans are designated in two main categories: *direct methods* and *indirect methods*. Direct methods, 'monitor the presence of $Ca(OH)_2$ and its subsequent reduction in abundance with time as the pozzolanic reaction proceeds' (Donatello et al., 2010). That is to say, some chemical (e.g., insoluble residue) and mineralogical (e.g., amourphous or crystalline minerals) properties of the pozzolan are traced (e.g., via X-ray diffraction, XRD) and measured (e.g., thermo-gravimetry) which provide a basis for evaluating the pozzolan quality.

Indirect methods 'measure a physical property of a test sample that indicates the extent of pozzolanic activity' (Donatello et al., 2010). Hence, some characters of the pozzolan in combination with cement (e.g., compressive strength of mortar containing pozzolan as cement replacement) are measured, and pozzolanic quality is assessed. In the following an overview of these methods is presented.

6.3 Direct methods

Direct methods stem from techniques usually applied in powder technology applications.

6.3.1 Thermo-gravimetric analysis

The pozzolanic activity, lime binding capacity, of pozzolans can be measured by thermo-gravimetric analysis. This method is based on the thermal decomposition of crystalline calcium hydroxide in a temperature range of 400–500 °C to calcium oxide and water. By applying the above-mentioned temperatures, the calcium hydroxide is decomposed and the resulting water is evaporated. The weight reduction resulting from water evaporation in the test is very low for suitable pozzolans and high in weak pozzolans (Moropoulou et al., 2004; Pourkhorshidi et al., 2010a, b).

6.3.2 Determination of insoluble residue

The insoluble residue (IR) test, according to EN 196-2 (2005), determines the amount of insoluble residue ingredients. This test can be carried out by two methods or better described in two conditions. In the first (here named method I), hydrochloric acid and sodium carbonate are used, while the second (here termed method II) utilizes hydrochloric acid and potassium hydroxide as a stronger condition for solution (Pourkhorshidi et al., 2010a, b, 2011).

Although these methods are not directly applied for recognition of amorphic minerals, results have significant compatibility in case of amorphic minerals in natural pozzolans. Therefore, measurement of insoluble residue according to the EN196-2 procedure is recommended for evaluation of amorphic and non-amorphic phases that are at play in pozzolanic activity (Pourkhorshidi et al., 2010a, b).

6.3.3 Mineralogical analysis

Mineral components of pozzolans are identified semi-quantitatively by the XRD method. The amount of non-crystalline (amorphous) silica having a major role in pozzolanic activities is determined through identification of the amount of crystalline minerals (Pourkhorshidi et al., 2010a, b). It should be noted that there is a remarkable relation between IR and XRD; as the insoluble residue content indicates the amount of soluble and insoluble minerals detected in XRD.

Regarding the evaluation of major oxides, assessment of amorphic and crystaline minerals for natural pozzolans is of prime importance. Results of numerous antecedent studies have revealed that amorphic constituents (such as cristopolite, zeolite and calcite) are the most significant active alkali phases. There is no universal deterministic limit to the amount of amorphic minerals in natural pozzolans considered for cement replacement. However, investigations show that natural pozzolans, with less than 15% amorphic minerals, would not have good performance as a cement replacement. Another method being used for the identification of crystalline and non-crystalline minerals in pozzolans is petrography which can be performed using a petrography-microscope.

6.3.4 EN 196-5 and Frattini test

EN 196-5 (2005) evaluates pozzolanic activity of pozzolan cements designated in EN 197-1 (2000). According to EN 197-1 (first published in 1995) pozzolanic cements are classified in two groups, namely; CEM IV/A, and CEM IV/B. The former type comprises 65–89% clinker plus a mixture of silica fume, natural pozzolan and fly ash. The latter type has 45 to 64% clinker.

In EN 196-5, pozzolanicity is assessed based on the concentration of calcium ion Ca^{2+} (expressed as calcium oxide or CaO), present in the aqueous solution in contact with hydrated blended cement after a fixed period. As seen in Fig. 6.1, a unique predefined curve demarcates pozzolanic and non-pozzolanic areas. The blended cement is satisfactory, if the concentration of CaO in the solution falls below the curve.

In this method, test samples are prepared consisting of cement and natural pozzolan and mixed with distilled water. After preparation, samples are

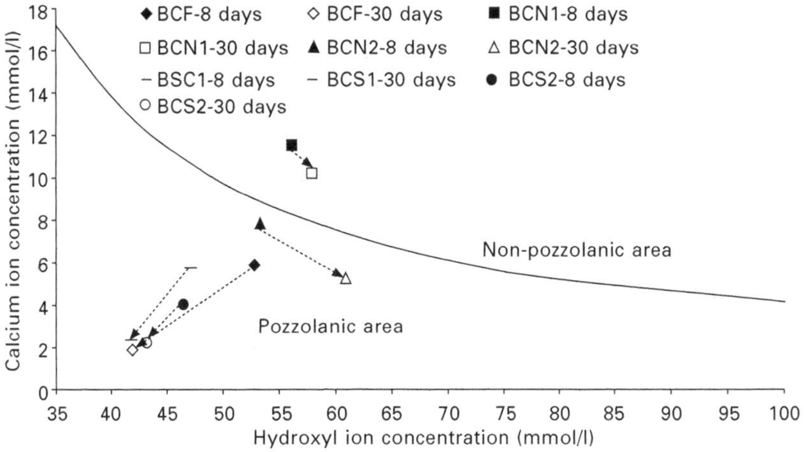

6.1 Treatment results for pozzolan cement pastes at 8 and 30 days in 40 °C (acc. EN 196-5).

left for 8 days in a sealed plastic bottle in an oven at 40 °C. It should be noted that the number of days spent in the oven can be increased depending on the pozzolans. After the above-mentioned times, samples are vacuum filtered through a 2.7 μm nominal pore size filter paper and allowed to cool to ambient temperature in sealed Buchner funnels. The filtrate is analysed for $[OH^-]$ by titration against dilute HCl with methyl orange indicator and for $[Ca^{2+}]$ by pH adjustment to 12.5, followed by titration with 0.03 mol/l EDTA solution using Patton and Reeders indicator (Donatello et al., 2010; Pourkhorshidi et al., 2010a, b).

Results are presented as a graph of $[Ca^{2+}]$, expressed as equivalent CaO, in mmol/l versus $[OH^-]$ in mmol/l. Test results lying below the curve (see Fig. 6.1), indicate removal of Ca^{2+} from solution which is attributed to pozzolanic activity. Results lying on the line indicate zero pozzolanic activity and results above the line correspond to no pozzolanic activity. It should be noted that this procedure assumes no other source of soluble calcium is present in the system, as leaching of calcium would invalidate this approach (Donatello et al., 2010; Pourkhorshidi et al., 2010a, b).

6.3.5 Saturated lime test

The saturated lime method is similar to the Frattini test, wherein the pozzolan is mixed with saturated lime (slaked lime; $Ca(OH)_2$) solution instead of cement and water (Donatello et al., 2010). Measurement of the residual dissolved calcium shows the amount of lime fixed by the pozzolan. During cement hydration, $Ca(OH)_2$ is precipitated; however, in the presence of pozzolan $Ca(OH)_2$ reacts with the pozzolan, and this allows more solid $Ca(OH)_2$ to dissolve. In the saturated lime test a fixed quantity of $Ca(OH)_2$ is available in solution, thus the quantity of Ca^{2+} ions is accurately known at the start of the test, and since Ca^{2+} ions only interact with pozzolan or water, the amount of lime fixed by pozzolan is quantifiable. Test results indicate mmol CaO fixed or % total CaO fixed per gram of test pozzolan (Donatello et al., 2010). The saturated lime test method has been used for paper sludge waste (Garcia et al., 2008), sugar cane straw waste (Frías et al., 2005, 2007) and ferroalloy industry wastes (Frías and Rodriguez, 2007).

6.4 Indirect methods

Amongst the indirect methods ASTM C618 (2003) and ASTM C1240 (2003) are more reliable and usually implemented in current design practice.

6.4.1 ASTM C618

The most common and simplest method for assessing pozzolanic activity is the ASTM C618 standard practice. ASTM C618 presents chemical and physical requirements and specifications of fly ash and natural pozzolans for cement replacement (see Table 6.1), where the standard tests procedure of ASTM C311 (2002) is incorporated. ASTM C618 was published in 1968 to combine and replace ASTM C350 on fly ash and ASTM C402 on other pozzolans (ACI 232.1R, 2000).

Briefly, in this method, the foremost important criteria for pozzolanic activity are: (1) the sum of chemical components, that is, SiO_2+ Fe_2O_3+Al_2O_3 and (2) the strength index defined by ASTM C311. In ASTM C311 the strength index is determined as the ratio of compressive strength for mortar with 20% pozzolan replacement to compressive strength of control mortar. For the control mortar, w/b = 0.485, and water in the test mortar is regulated to give same consistency as the control mortar. In fact in this method the role of natural pozzolan in mobilization of the concrete strength is determined. Note that w/b stands for water/binder, where binder consists of cement plus cement replacement material (if any).

ASTM C618 has disadvantageous and vague points. For example the blain of some natural pozzolans affect test results. As the pozzolan's blain increases, the amount of water required to ensure the target slump (i.e., slump identical to control mortar) increases. This in turn increases w/b and leads to a smaller strength index (Pourkhorshidi et al., 2010a, b). Many investigators have suggested that both the trial and control mortars should possess the same w/b ratio, in order to produce authentic results. It should be noted, however, that the effects of blain on pozzolanic activity cannot be generalized for all pozzolan types, that is to say in some pozzolans there is a limit, beyond which blain does not affect pozzolanic activity. Notwithstanding

Table 6.1 Chemical/physical properties and standard specification of pozzolans

Requirements		Standard specification		
		Natural pozzolan, Class N, ASTM C618	Fly ash, Class F, ASTM C618	Silica fume, ASTM C1240
Chemical requirements	SiO_2+Al_2O_3+Fe_2O_3, %	min, 70	min, 70	–
	SiO_2	–	–	min, 85
	Sulphur trioxide (SO_3), %	max, 4	max, 4	–
	Moisture content, %	max, 3	max, 3	max, 3
	Loss on ignition, %	max, 10	max, 10	max, 6
Physical requirements	Amount retained when wet-sieved on 45 μm sieve, %	max, 34	max, 34	max, 10
	Strength activity index, at 7 days, percent of control	min, 75	min, 75	–
	Strength activity index, at 28 days, percent of control	min, 75	min, 75	–
	Accelerated strength activity index, at 7 days, percent of control	–	–	min, 105
	Water requirement, percent of control	max, 115	max, 105	–
	Autoclave expansion or contraction, %	max, 0.8	max, 0.8	–
	Specific surface, m^2/g	–	–	min, 15

this, the ASTM C311 abandons discussion of the pozzolan PSD (particle size distribution) and only demarcates the percentage passing 45 μm.

6.4.2 ASTM C1240

Originally it was intended to consider silica fume in ASTM C618; however, this suggestion was rejected and silica fume is covered in ASTM C1240 – first introduced in 1993 (ACI 234R, 1996). ASTM C1240 addresses physical and chemical requirements of silica fume for cement replacement (see Table 6.1). The prominent requirements are: (1) the amount of SiO_2, and (2) the accelerated strength activity index, defined in ASTM C311, considering that; the amount of silica fume replacement in binary cement is 10%, the amount of water in tested mortar is regulated with super-plasticizers to be identical with control mortar, and specimens are stored at 65 °C for 6 days.

6.4.3 Electrical conductivity

The electrical conductivity method was first introduced by Rassk and Bhaskar (1975) who measured the electrical conductivity of amount of silica dissolved in a solution of hydrofluoric acid (HF) with dispersed pozzolan. Luxan et al. (1989) suggested a pozzolanic index given by the variation between initial and final electrical conductivity of a calcium hydroxide–pozzolan suspension for 120 seconds. Feng et al. (2004) used Luxan's methods to assess pozzolanic properties of rice husk ash. Tashiro et al. (1994) suggest that only 72 h testing is required to assess pozzolanic behavior with electrical conductivity for lime-pastes cured under steam at 70 °C. McCarter and Tran (1996) alluded that calcium hydroxide consumption after 72 h correlates with electrical resistance of several artificial pozzolans. Paya et al. (2001) recommend a technique predicated on electrical conductivity which assesses pozzolanic activity in less than one hour. Sinthaworn and Nimityongskul (2009) studied pozzolans dispersed in an ordinary Portland cement using a temperature of 80 °C and suggested that is possible to assess pozzolanic activity by measuring changes in electrical conductivity in a 28 h test.

6.5 Comparison and guidelines

Attempts have been made to compare pozzolanicity and pozzolanic activity indices measured by various methods. The results of ASTM C618 show the least conformity with concrete mixture results, whereas the Frattini test, insoluble residue and recognizing crystalline and amorphous minerals presented suitable agreement with concrete mixture results (Pourkhorshidi et al., 2010a, b).

Moropoulou et al. (2004) compared pozzolanic activity by differential thermal analysis/thermo-gravimetric (DTA/TG) analysis for metakaolin and ceramic powder (artificial pozzolans) and earth of Milos (natural pozzolan). The metakaolin has a higher $Ca(OH)_2$ consumption than the earth of Milos or the ceramic powder (Fig. 6.2). The metakaolin has the lowest overall $Ca(OH)_2$ consumption although at 3 days curing its consumption is higher than earth of Milos.

Das (2006) suggests predicting pozzolanic activity of fly ash by strength index using a model based on fly ash chemical composition and fineness. Franke and Sisomphon (2004) and Sisomphon and Franke (2011) state that DTA is not accurate for low calcium hydroxide content and suggest a chemical method to determine the calcium hydroxide content. Uzal et al. (2010) suggested that compressive strength of lime–pozzolan paste specimens (2.5 cm^3) cured at 50 °C is a more reliable indicator than conventional pozzolanic tests. Gava and Prudencio (2007a, 2007b) present a comparison of pozzolanic activity index results obtained from several test procedures

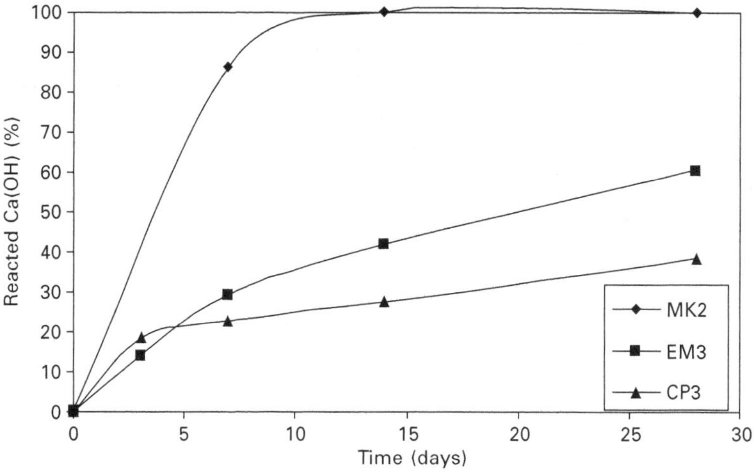

6.2 Calcium hydroxide reaction vs curing time for metakaolin (MK2), earth of Milos (EM3) and ceramic powder (CP3) (Moropoulou et al., 2004).

recommended by Brazilian, American and British Standards to evaluate three types of pozzolan: fly ash, silica fume and rice-husk ash. They assert that different standards may sometimes generate conflicting results. Pourkhorshidi et al. (2010a, b) mention that strength activity index in ASTM C618 sometimes classifies as high reactive pozzolans, some supplementary material that led to concrete with unsatisfactory performance, and that EN 196-5 had superior compatibility with real concrete performance. Donatello et al. (2010) stipulate that strength activity index of ASTM C618 and of EN 196-5 are not suitable for evaluating pozzolanic activity of incinerator sewage sludge ash. Accordingly sludge ash is a low reactive pozzolan, hence the 28-day curing period is not sufficient, thus suggesting the Frattini test to evaluate its pozzolanic activity. Donatello et al. (2010) compared pozzolanic activity of metakaolin, silica fume, coal fly ash, incinerated sewage sludge ash using the Frattini test, the saturated lime test and the strength activity index test. It was observed that saturated lime test is not a reliable test because it does not correlate with other tests that could be explained by the different activator to pozzolan ratio. They recommend application of the Frattini and SAI methods in combination with an independent determination of $Ca(OH)_2$ content (thermal or diffraction methods) for the assessment of the pozzolanic activity.

Although the electrical conductivity test correlates with other standard methods it is not valid for high calcium fly ashes. Villar-Cocina et al. (2003), Frías et al. (2005), Rosell-Lam et al. (2011) and Dalinaidu et al. (2007) have suggested a mathematical model predicated on electrical conductivity measurements to describe the process in a kinetic–diffusive or kinetic regimes.

Dalinaidu et al. (2007) also describe a procedure to assess the pozzolanic activity by measuring electrical conductivity of a pozzolan–lime solution.

After the initial evaluation for selecting the appropriate natural pozzolans, its performance in concrete should be evaluated. The performance of pozzolan in concrete is complicated and dependent on a host of parameters and conditions, thus evaluating pozzolan performance in concrete is sensitive and requires proficient insight. A battery of different parameters such as pozzolan PSD and blain, cement type, amount of replacement, w/b and so on, have prominent influence on the pozzolan behaviour in concrete. Hence, available standard procedures cannot always accurately define the amount of replacement and projected concrete performance.

One can think of high-tech and expensive methologies for the assessment of pozzolanic activity; however, simple and versatile methods that are applicable in modest laboratories and agree well with the pozzolan behavior in concrete are beneficiary and popular. A reliable assessment of pozzolans should adress two main phenomena; (a) recognition of pozzolanicity in contrast to non-active fillers, and (b) evaluating pozzolan performance as cement replacement in concrete. As a guide to reliable evaluation of pozzolans the procedure illustrated in Fig. 6.3 is recomended (Parhizkar et al., 2010). Therefore, this guideline consists of steps which start from simple and prelimenary methods, towards the accurate and more expensive methods. Considering Fig. 6.3 the steps for evaluating pozzolans are as follows:

- Step 1, *Comparing with ASTM C618.* Since all pozzolans and some non-active fillers satisfy ASTM C618 requirements, materials that do not satisfy the requirements are not a pozzolan.
- Step 2, *Determining insoluble residue and minerals.* In this step, firstly IR of natural pozzolans is measured according to method I of EN 196-2. If a pozzolan cannot satisfy the limitations, IR is evaluated again by method II of EN 196-2.
- *Step 3, Frattini test.* At this step, a Frattini test is carried out according to EN 196-5, except that if a pozzolan does not show pozzolanic activity, the time of the test should be continued until 30 days.
- *Step 4, Evaluation of natural pozzolans in concrete.* The optimum percentage of replacement, depending on concrete condition, can be selected in this step. Moreover, if some pozzolans show little differences in former steps, the quality of them can be judged in this step.

6.6 Conclusion and future trends

The proposed and approved test methods are mainly based on testing the chemical characteristics of pozzolans, and physical properties of cement–pozzolan mixtures. These test methods measure only one or few properties

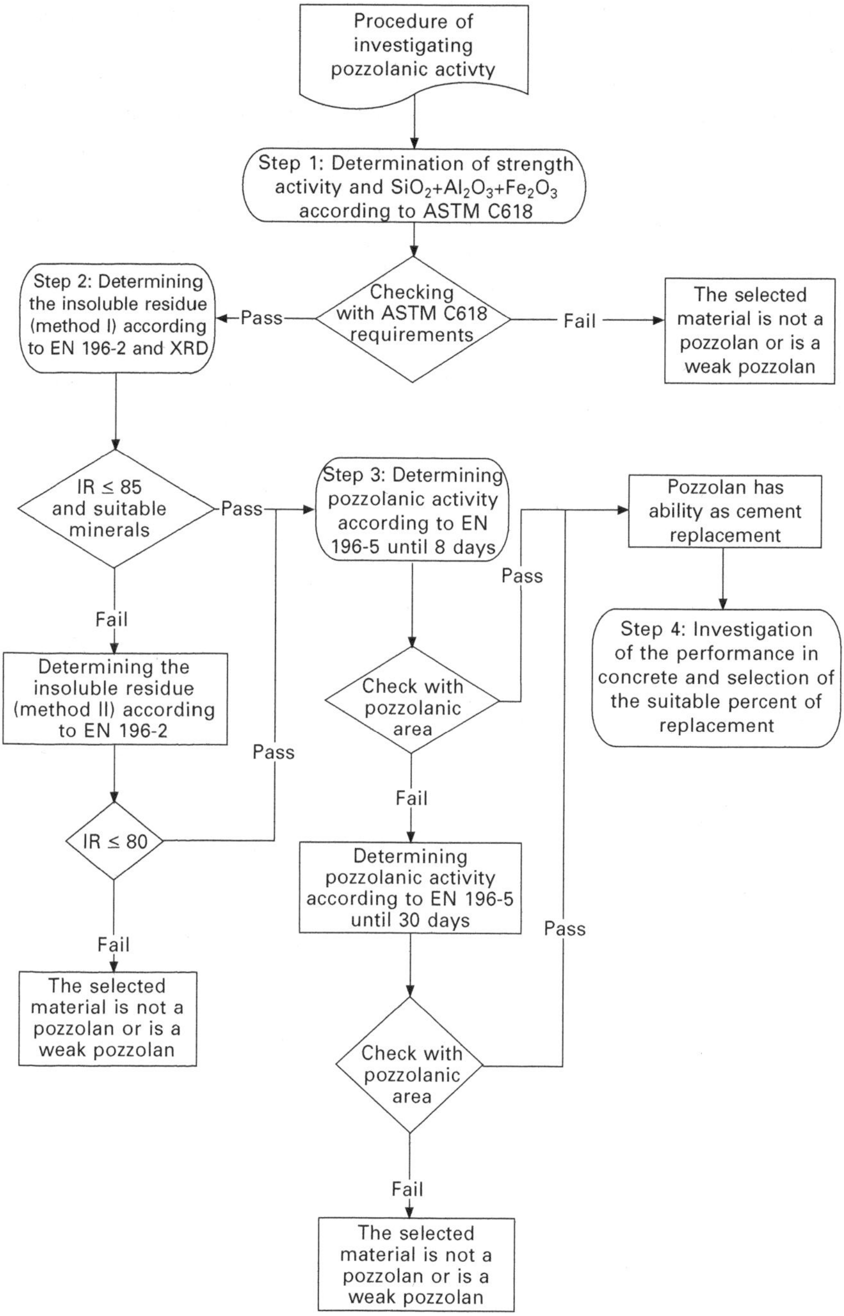

6.3 Flowchart of guideline (procedure of investigating pozzolans).

and have limited ability in foreseeing real performances of pozzolans in concretes.

These days most of the designs are based on the performance of materials and structures at older ages. The extensive use of pozzolanic materials to reduce permeability and enhance the durability of concretes reveals the fact that performance tests are necessary for evaluation of concrete mixtures containing pozzolanic materials. Simple test methods for the permeability of mortars and concretes to assess the durability of pozzolanic materials are among the performance tests which it is essential are considered in the future.

Heretofore, a combination of different methods can determine/assess the suitable pozzolan material for use as cement replacement in blended cements. Yet in the near future improvements in techniques and methods for evaluating the properties of powders and different materials may enhance the ability to precisely evaluate poozzolans, before combining them with cement.

In order to evaluate pozzolans from the prospect of chemical composition and crystalline matter the high-tech techniques of XRF (X-ray fluorescence), XRD (X-ray diffraction), mass spectrometry, ICP (inductively coupled plasma) and EDAX (energy despersive analysis X-ray) are applicable. Concerning the morphology and particle sizes of pozzolans SEM (scanning electron microscopy), TEM (transmission electron microscopy) and BET (Brunauer Emmet Teller) can be deployed.

Besides improvements in physio-chemical analysis methods, such as tracing and determination of chemical reaction rates by HPLC (high-performance liquid chromatography), GC (gas chromatography) and NMR (nuclear magnetic resonance spectroscopy) provide a more accurate and faster assessment of pozzolans.

6.7 References

ACI 232.1R. (2000), Use of Raw or Processed Natural Pozzolans in Concrete, American Concrete Institute, Farmington Hills, MI.

ACI 232.2R. (2003), Use of Fly Ash in Concrete, American Concrete Institute, Farmington Hills, MI.

ACI 234R. (1996), Guide for the Use of Silica Fume in Concrete American Concrete Institute, Farmington Hills, MI.

ASTM C618 (2003), Standard Specification for Coal Fly Ash and Raw or Calcined Natural Pozzolan for Use in Concrete, American Society for Testing and Materials, ASTM International, West Conshohocken, PA, USA.

ASTM C311 (2002), Standard Test Methods for Sampling and Testing Fly Ash or Natural Pozzolans for Use in Portland-Cement Concrete, American Society for Testing and Materials, ASTM International, West Conshohocken, PA, USA.

ASTM C1240 (2003), Standard Specification for Silica Fume Used in Cementitious Mixtures, American Society for Testing and Materials. ASTM International, West Conshohocken, PA, USA.

Çavdar, A., Yetgin, Ş. (2007), Availability of tuffs from northeast of Turkey as natural pozzolan on cement, some chemical and mechanical relationships. *Construction and Building Materials* **21**, 2066–2071.

Çavdar, A., Yetgin, Ş. (2009), The effect of particle fineness on properties of Portland pozzolan cement mortars, *Turkish Journal of Science & Technology* **4**, 17–23.

CEMBUREAU (2001), *The European Cement Association (CEMBUREAU), annual report, 2001.*

Colak, A. (2002), The long-term durability performance of gypsum–Portland-cement–natural pozzolan blends. *Cement and Concrete Research* **32**, 109–115.

Colak, A. (2003), Characteristics of pastes from a Portland cement containing different amounts of natural pozzolan. *Cement and Concrete Research* **33**, 585–593.

Das, S. (2006), Yudhbir – A simplified model for prediction of pozzolanic characteristics of fly ash, based chemical composition. *Cement and Concrete Research* **36**, 1827–1832.

Dalinaidu, A., Das, B., Singh, D. (2007), Methodology for rapid determination of pozzolanic activity of materials. *Journal of ASTM International* **4**. doi: 10.1520/JAI100343

Diamond, S. (1982), Characterization and classification of fly ashes in terms of certain specific chemical and physical parameters. *International Symposium on the Use of PFA in Concrete* (Cabrera J. G. and Cuseus A. R. (eds)). Leeds, 9–22.

Donatello, S., Tyrer, M., Cheeseman, C.R. (2010), Comparison of test methods to assess pozzolanic activity. *Cement and Concrete Composites* **32**, 121–127.

EN 197-1 (2000), Cement; Part 1: Composition, Specifications and Conformity Criteria for Common Cements. European Committee for Standardization.

EN 196-5 (2005), Methods of Testing Cement; Part 5: Pozzolanicity Test for Pozzolan Cements, European Committee for Standardization.

EN 196-2 (2005), Methods of Testing Cement; Part 2: Chemical Analysis of Cement. European Committee for Standardization.

Ezziane, K., Bougara, A., Kadri, A., Khelafi, H., Kadri, E. (2007), Compressive strength of mortar containing natural pozzolan under various curing temperature. *Cement and Concrete Composites* **29**, 587–593.

Feng, N., Peng, G. (2005), Applications of natural zeolite to construction and building materials in China. *Construction and Building Materials* **19**, 579–584.

Feng, Q., Yamamichi, H., Shova, M., Sugita, S. (2004), Study on the pozzolanic properties of rice husk ash by hydrochloric acid pretreatment. *Cement and Concrete Research* **34**, 521–526.

Franke, L., Sisomphon, K. (2004), A new chemical method for analyzing free calcium hydroxide content in cementing material. *Cement and Concrete Research* **34**, 1161–1165.

Frías, M., Rodríguez, C. (2007), Effect of incorporating ferroalloy industry wastes as complementary cementing materials on the properties of blended cement matrices. *Cement and Concrete Research*, **30**, 212–219.

Frías, M., Villar-Cocina, E., Sánchez de Rojas, M.I., Valencia-Morales, E. (2005), The effect that different pozzolanic activty methods has on the kinetic constants of the pozzolanic reaction in sugar cane straw-ash/lime systems: application of a kinetic-diffusive model. *Cement and Concrete Research* **35**, 2137–2142.

Frías, M., Villar-Cocina, E., Valencia-Morales, E. (2007), Characterisation of sugar cane straw waste as pozzolanic material for construction: calcining temprature and kinetic parameters. *Waste Management*, **27**, 533–538.

Garcia, R., Vigil de la Villa, R., Vegas, I., Frias, M., Sanchez de Rojas, MI. (2008), The pozzolanic properties of paper sludge waste. *Construction and Biulding Materials* **22**, 7, 1484–1490.

Gava, G.P., Prudencio Jr, L.R. (2007a), Pozzolanic activity tests as a measure of pozzolans performance: Part 1. *Magazine of Concrete Research* **59**, 729–734.

Gava, G.P., Prudencio, Jr, L.R. (2007b), Pozzolanic activity tests as a measure of pozzolans performance: Part 2. *Magazine of Concrete Research* **59**, 735–741.

Jana, D. (2007), A new look to an old pozzolan, clinoptilolite – A promising pozzolan in concrete, *Proceedings of the 29th ICMA conference on cement microscopy*, Quebec City, West Chester: Curran Associates Inc., 168–206.

Kaid, N., Cyr, M., Julien, S., Khelafi, H. (2009), Durability of concrete containing a natural pozzolan as defined by a performance-based approach. *Construction and Building Materials* **23**, 3457–3467.

Kogel, J.E., Trivedi, N.C., Barker, J.M., Krukowski. S.T. (2006), *Industrial minerals & rocks: commodities, markets, and uses*. Society for Mining, Metallurgy, and Exploration Inc, Littleton, Colorado, USA, 7th edition.

Luxan, M.P., Madruga, F., Saaverda, J. (1989), Rapid evaluation of pozzolanic activity of natural products by conductivity measurement. *Cement and Concrete Research* **19**, 63–68.

Malinowski, R., Frifelt, K. (1993), *Prehistoric Hydraulic Mortar: The Ubaid Period 5–4000 years BC: Technical Properties*, Document D12, Swedish Council for Building Research, Stockholm, Sweden, 16 pp.

Massazza, F. (1993), Pozzolanic cements. *Cement and Concrete Composites* **15**, 185–214.

McCarter, W., Tran, D. (1996), Monitoring pozzolanic activity by direct activation with calcium hydroxide. *Construction and Building Materials* **10**, 179–184.

Mehta, P.K. (1986), Standard specifications for mineral admixtures: an overview. *Proceedings of First International Conference on the Use of Fly Ash, Silica Fume, Slag and Natural Pozzolans in Concrete* (Malhotra V. M. (eds)). American Concrete Institute, Detroit, 637–658.

Mehta, P.K. (1999), Concrete technology for sustainable development. *Concrete International* **21**, 47–52.

Meyer, C. (2009), The greening of the concrete industry. *Cement and Concrete Composites* **31**, 601–605.

Moropoulou, A., Bakolas, A., Aggelakopoulou, E. (2004), Evaluation of pozzolanic activity of natural and artificial pozzolans by thermal analysis. *Thermochimica Acta*, **420**, 135–140.

Najimi, M., Jamshidi, M., Pourkhorshidi, A. R. (2008), Durability of concretes containing natural pozzolan. *Proceedings of the Institution of Civil Engineers, Construction Materials* **161**, 113–118.

Parhizkar, T., Najimi, M., Pourkhorshidi, A.R., Jafarpour, F., Hillemeier, B., Herr, R. (2010), Proposing a new approach for qualification of natural pozzolans. *International Journal of Science and Technology; Scientia Iranica* **17**, 450–456.

Paya, J., Borrachero, M. V., Monzo, J., Peris-Mora, E., Amahjour, F. (2001), Enhanced conductivity measurement techniques for evaluation of fly ash pozzolanic activity. *Cement and Concrete Research* **31**, 41–49.

Poon, C.S., Lam, L., Kou, S.C., Lin, Z.S. (1999), A study on the hydration rate of natural zeolite blended cement pastes. *Construction and Building Materials* **13**, 427–432.

Pourkhorshidi, A.R., Najimi, M., Parhizkar, T., Hillemeier, B., Herr, R. (2010a), A comparative study of the evaluation methods for pozzolans. *Advances in Cement Research* **22**, 157–164.

Pourkhorshidi, A.R., Najimi, M., Parhizkar, T., Jafarpour, F., Hillemeier, B. (2010b), Applicability of the ASTM C618 standard for evaluation of natural pozzolans. *Cement and Concrete Composites* **32**, 794–800.

Rassk, E., Bhaskar, M. (1975), Pozzolanic activity of pulverized fuel ash. *Cement and Concrete Research* **5**, 363–376.

Rosell-Lam, M., Villar-Cocina, E., Frias, M. (2011), Study on the pozzolanic properties of a natural Cuban zeolitic rock by conductometric method: Kinetic parameters. *Construction and Building Materials* **25**, 644–650.

Shannag, M.J., Yeginobali, A. (1995), Properties of pastes, mortars and concretes containing natural pozzolan. *Cement and Concrete Research* **25**, 647–657.

Shi, C., Day, R.L. (2001), Comparison of different methods for enhancing reactivity of pozzolans. *Cement and Concrete Research* **31**, 813–818.

Sinthaworn, S., Nimityongskul, P. (2009), Quick monitoring of pozzolanic reactivity of waste ashes. *Waste Management* **29**, 1526–1531.

Sisomphon, K., Franke, L. (2011), Evaluation of calcium hydroxide contents in pozzolanic cement pastes by a chemical extraction method. *Construction and Building Materials* **25**, 190–194.

Sybertz, F. (1989), Comparison of Different Methods for Testing the Pozzolanic Activity of Fly Ashes. *Proceedings of the 3rd International Conference on Fly Ash, Silica Fume, and Natural Pozzolans in Concrete* (Malhotra V. M. (eds)) (ACI SP 114-22). Trondheim, American Concrete Institute, Detroit, 477–497.

Targan, S., Olgun, A., Erdogan, Y., Sevinv, V. (2003), Influence of natural pozzolan, colemanite ore waste, bottom ash, and fly ash on the properties of Portland cement. *Cement and Concrete Research* **33**, 1175–1182.

Tashiro, C., Ikeda, K., Inome, Y. (1994), Evaluation of pozzolanic activity by the electric resistance measurement method. *Cement and Concrete Research* **24**, 1133–1139.

Türkmenoğlu, A.G., Tankut, A. (2002), Use of tuffs from central turkey as admixture in pozzolanic cements assessment of their petrographical properties. *Cement and Concrete Research* **32**, 629–637.

Uzal, B., Turanli, L., Mehta, P.K. (2007), High-volume natural pozzolan concrete for structural applications. *ACI Materials Journal* **104**, 535–538.

Uzal, B., Turanli, L., Yucel, H., Goncuoglu, M., Culfaz, A. (2010), Pozzolanic activity of clinoptilotite: A comparative study with silica fume, fly ash and a non-zeolitic natural pozzolan. *Cement and Concrete Research* **40**, 398–404.

Villar-Cocina, E., Valencia-Morales, E., Gonzalez-Rodriguez, R., Hernandez-Ruiz, J. (2003), Kinetics of the pozzolanic reaction between lime and sugar cane straw ash by electrical measurement: A kinetic-diffusive model. *Cement and Concrete Research* **33**, 517–524.

Wesche, K., Alonso, I.L., Bijen, I., Schubert, P., Von Berg, W. Rankers, R. (1989), Test methods for determining the properties of fly ash for use in building materials. *Materials and Structures* **22**, 299–308.

7 Properties of concrete with high-volume pozzolans

B. UZAL, Abdullah Gul University, Turkey

DOI: 10.1533/9780857098993.2.138

Abstract: This chapter focuses on the materials and properties of high-volume natural pozzolan (HVNP) concrete. The characteristics of natural pozzolans used in high-volume pozzolan mixtures are discussed, together with the fresh and hardened properties of HVNP cementitious systems, their hydration characteristics and their microstructures.

Key words: concrete, durability, high volume, natural pozzolan.

7.1 Introduction

The use of pozzolanic mineral admixtures in concrete mixtures improves the properties of fresh and hardened concrete, providing better workability, low heat of hydration, increased ultimate strength, and improved durability performance. The same advantages can be applied to concrete mixtures made with blended Portland cement by including pozzolan as a clinker-replacement material and, as an additional benefit, CO_2 emissions resulting from Portland cement production will be reduced.

In order to get the most benefit from incorporating pozzolanic materials into concrete systems, it is reasonable to use as high a proportion as possible. High-volume pozzolan cementitious systems, containing 50% or more pozzolanic material by weight of total binder content, efficiently reduce Portland cement consumption and improve the durability of structural concrete mixtures. Greenhouse gas emissions are also reduced as a consequence of lower cement consumption. Ground granulated blast furnace slags, fly ashes and natural pozzolans are among the most suitable for high-volume incorporation into concrete mixtures, due to their physical properties and comparatively low cost. Silica fume and metakaolinite are not appropriate because of their relatively high cost and water demand.

Many reports have already been written on the use and properties of high-volume fly ash concrete. The main focus of this chapter is on the composition and properties of high-volume natural pozzolan (HVNP) concrete mixtures, which are a new type of high-volume pozzolan concrete. Future trends and sources for further information about high-volume pozzolan concrete are also presented in individual sections. Section 7.3 covers the characteristics of natural pozzolans used in HVNP concrete mixtures and typical mixture proportions of HVNP concrete. In Sections 7.4–7.6, properties of fresh and hardened HVNP concrete are discussed, such as workability, superplasticizer requirement, air content, setting time, strength development, modulus of elasticity, and resistance to chloride-ion penetration. In Section 7.7, future trends and sources of further information on high-volume pozzolan concrete (HVPC) are presented.

7.1.1 Definition and history of development of HVPC systems

The term high-volume pozzolan concrete dates back to the late 1980s, when Malhotra first used high volumes of fly ash (50% by mass of total binder) in Canada CANMET (Malhotra and Ramezanianpour, 1994; Malhotra and Mehta, 2002). This kind of concrete was called high-volume fly ash (HVFA) concrete, and was characterized by very low water content, and having 50% or more of the Portland cement replaced by ASTM Class F fly ash. In order to provide excellent workability with low water contents, superplasticizers must be used for HVFA concrete mixtures. Comprehensive studies on HVFA concrete have shown that it is a high-performing material, exhibiting sufficient early-age strength, very high ultimate strength, low shrinkage and superior durability (Malhotra and Mehta, 2002).

The use of natural pozzolans instead of fly ashes in high-volume pozzolan cementitious systems was first proposed in a preliminary study conducted by Uzal and Turanli (2001). HVNP systems were defined similarly to HVFA, with 50% or more natural pozzolan by mass, and reduced water content aided by superplasticizers. Although HVNP systems initially showed relatively poor performance in terms of strength development, further studies have indicated that HVNP systems could perform comparably to HVFA mixtures, if the appropriate natural pozzolanic material and superplasticizer were selected (Uzal and Turanli, 2003, 2012; Turanli et al., 2004, 2005; Uzal et al., 2007).

7.1.2 Significance of HVPC in the sustainable development of concrete technology

The concrete industry is not sustainable due to its consumption of natural resources, the huge amounts of greenhouse gases generated by Portland cement production, and lack of durability of concrete structures. HVPC systems provide reasonable solutions to these sustainability deficiencies.

The HVFA concrete system, which uses solid waste material from thermal power plants, affords solutions to the all the aforementioned sustainability issues. Fly ash is a by-product of energy production and so can be obtained with no additional resource consumption. The HVNP concrete system, by contrast, consumes natural pozzolanic materials. However, HVNP concrete systems improve sustainability in other ways, by reducing greenhouse gas emissions and increasing the durability of concrete structures.

The climatic changes that have occured in recent years are believed to be the result of the increased atmospheric content of greenhouse gases, mainly CO_2. The Portland cement industry is responsible for approximately 7% of total global CO_2 emissions (Mehta, 1999). Reducing Portland cement production using pozzolanic materials could therefore make the concrete industry more sustainable. In addition, the improved durability conferred by pozzolanic minerals results in longer service life of concrete structures and reduced repair and renewal costs. Clearly, incorporating high quantities of pozzolanic materials into concrete mixtures results in more sustainable concrete systems and structures. This is the main rationale driving efforts to develop HVPC systems.

7.2 Composition of natural pozzolans for high-volume natural pozzolan (HVNP) systems

Natural pozzolans are inorganic materials with no inherent binding abilities. However, they harden when mixed with water and calcium hydroxide, or with materials that release calcium hydroxide, such as Portland cement clinker (Massazza, 2001). This hydraulic reactivity with slaked lime is called pozzolanic activity.

Most natural pozzolans are volcanic in origin, while some others originate from sedimentary deposits such as diatomaceous earth. The chemical composition of natural pozzolans depends on the nature and the origin of the deposits. SiO_2, Al_2O_3, Fe_2O_3 are the major oxides present in natural pozzolans. Other oxides such as CaO, Na_2O, K_2O, and SO3 are also part of their chemical composition. Oxide compositions of natural pozzolans, which were formerly used in studies on HVNP cementitious systems, are shown in Table 7.1. As shown in Table 7.1, the oxide composition of natural pozzolans may vary considerably. In fact, the effect of chemical composition on the performance of cementitious systems is not considerable. The mineralogical composition and physical characteristics, including particle size distribution, specific surface area, etc., of pozzolans are more influential.

It is commonly believed that the pozzolanic activity of natural pozzolans is mainly governed by the quantity and composition of glassy phases present. However, some mainly crystalline natural minerals with very little glassy phase, for instance zeolitic minerals such as clinoptilolite, can exhibit high pozzolanic activity and performance in cementitious systems (Uzal et al., 2010).

Table 7.1 Chemical composition of natural pozzolans formerly tested in HVNP systems (% by weight)

Pozzolan no.	SiO_2	Al_2O_3	Fe_2O_3	CaO	MgO	Na_2O	K_2O	SO_3	LOI*	Ref.
1	62.3	15.0	5.1	4.9	1.2	2.2	1.8	0.2	6.3	Uzal and Turanli (2003)
2	61.7	16.2	3.5	5.0	1.5	3.5	2.4	0.1	5.9	
3	61.5	18.2	3.8	4.6	0.9	3.9	4.1	0.1	2.7	Turanli et al. (2004)
4	54.2	15.2	6.4	8.8	2.4	1.5	2.1	0.5	11.6	
5	62.2	15.2	3.3	4.9	1.5	3.5	2.4	0.5	6.1	
6	69.3	11.4	1.3	1.7	0.8	1.0	4.0	n.a.**	6.2	Uzal and Turanli (2012)
7	66.1	11.0	1.1	3.7	1.0	0.2	1.8	n.a.**	9.4	
8	71.0	13.4	1.2	1.7	0.3	3.2	4.7	–	3.3	Uzal et al. (2007)
9	54.2	14.8	5.6	7.3	2.7	1.0	1.2	0.1	12.7	

*LOI: Loss on ignition.
**n.a.: Not available.

Table 7.2 Qualitative mineral composition of the natural pozzolans formerly tested in HVNP systems

Pozzolan no.	Glass	Quartz	Cristobalite	Albite	Calcite	Clinoptilolite	Ref.
1	+	–	–	+	+	+	Uzal and Turanli (2003)
2	+	–	–	+	–	–	
3	+	+	–	+	–	–	Turanli et al. (2004)
4	+	+	–	+	–	+	
5	+	–	–	+	–	–	
6	–	–	–	–	–	+	Uzal and Turanli (2012)
7	–	–	–	–	–	+	
8	+	+	+	+	–	–	Uzal et al. (2007)
9	+	+	–	+	+	+	

The common minerals present in natural pozzolans are glass, quartz, feldspar, mica and analcime. In some deposits, glassy phases in volcanic tuffs transform into crystalline zeolitic minerals. The nature of zeolitization and the type of the resulting zeolitic mineral depend on hydrothermal conditions in the geology of the deposit. On the other hand, geological transformation of glassy phases in natural materials of volcanic origin may result in formation of clay minerals (argillation process). Zeolitization improves the pozzolanic activity of volcanic materials whereas argillation decreases it (Massazza, 2001). The mineral composition of some of the natural pozzolans formerly used in studies on HVNP cementitious systems as determined by X-ray diffraction (XRD) analysis is shown in Table 7.2.

7.3 Physical characteristics of finely ground natural pozzolans

The physical properties of finely ground natural pozzolans, such as particle size distribution and specific surface, have considerable effects on fresh and hardened concrete properties, especially in the case of high-volume incorporation. The water demand and workability of pozzolan-containing concrete mixtures are dependent upon particle size distribution and smoothness of surface texture (Malhotra and Mehta, 2002). Particle size distribution is also a significant parameter for the extent of pozzolanic activity. It is believed that pozzolan particles larger than 45 µm exhibit little or no pozzolanic action, and the particles smaller than 10 µm are most effective in pozzolanic reactions.

In contrast to fly ashes obtained in particulate form, natural pozzolans are received in bulk and, then processed in order to give finely divided pozzolanic material. Their fineness and particle size distribution therefore depend on the types and parameters of processing equipment used. Figure 7.1 shows the particle size distribution curves for some finely ground natural pozzolans processed with laboratory equipment (jaw crusher and ball mill) (Uzal et al., 2010). Particle size distribution of a Turkish fly ash is also shown in Fig. 7.1 for the purpose of comparison.

Fly ash particles are spherical while finely ground natural pozzolans are composed of angular particles. Spherical particles of fly ash result in lower water requirement and improved workability when compared with natural pozzolan incorporated concrete mixtures. The difference between fly ash and

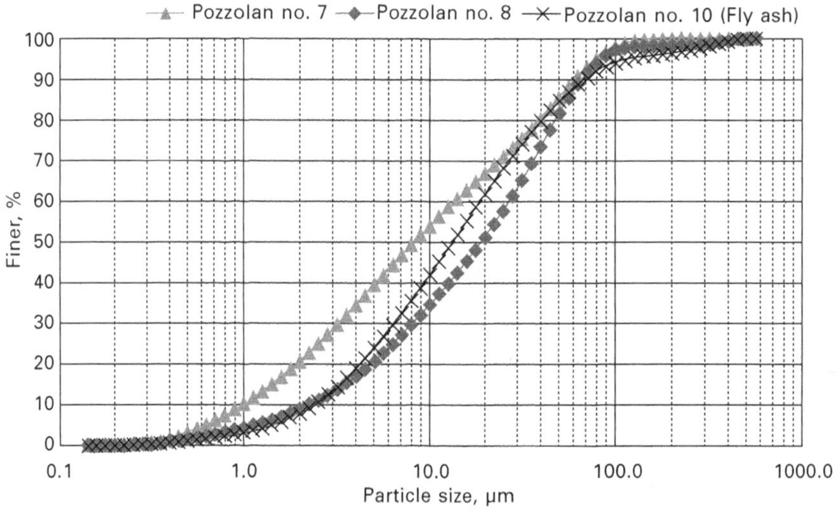

7.1 Particle size distribution curves of finely ground natural pozzolans and a Turkish fly ash (adapted from Uzal et al., 2010).

natural pozzolans is more significant than conventional concrete mixtures containing mineral admixtures, especially in HVPC applications. Another particle characteristic that distinguishes the fly ash and natural pozzolans is surface texture. Fly ash particles have smooth surfaces whereas the surface texture of natural pozzolans is generally porous, resulting in relatively high specific surface area. Table 7.3 shows some physical properties including particle characteristics such as specific surface area of a fly ash as well as a zeolitic and a non-zeolitic natural pozzolan as determined by the Brunauer–Emmett–Teller (BET) method (Uzal et al., 2007, 2010). As seen from Table 7.3, specific surface area of the fly ash is considerably lower than natural pozzolans, although their median particle sizes are similar. This fact is one of the reasons for the difference in water requirement of fly ashes and natural pozzolans.

7.4 Hydration characteristics and microstructure of high-volume natural pozzolan (HVNP) cementitious systems

HVNP cementitious systems display specific hydration processes and microstructures as a result of low Portland cement content and high volume of natural pozzolan. Uzal and Turanli (2012) investigated the hydration and paste microstructure of HVNP cementitious systems containing 55% natural zeolitic pozzolan and 45% Portland cement by mass. They reported that the $Ca(OH)_2$ content of hardened HVNP cement pastes is significantly lower than that of control Portland cement paste, which are below the detectable limits of thermal analysis by 91 days (Fig. 7.2). They also determined that HVNP cement pastes contain fewer pores larger than 50 nm when compared with the reference Portland cement paste at the end of 91 days, which results

Table 7.3 Some physical properties of finely ground natural pozzolans and a Turkish fly ash

Pozzolan no.	Specific gravity	Median size of particles (μm)	Blaine fineness (m^2/kg)	BET surface area (m^2/kg)	SAI*, %		Ref.
					7 days	28 days	
6	2.16	13.9	995	35500	71	84	Uzal and Turanli (2012)
7	2.19	8.2	1287	26870	79	98	
8	2.38	19.1	413	n.a.	78	80	Uzal et al. (2007)
9	2.26	15.4	670	24650	79	94	
10 (Fly ash)	2.37	13.2	388	3380	86	93	Uzal et al. (2010)

*SAI: Strength activity index as determined in accordance with ASTM C 311.

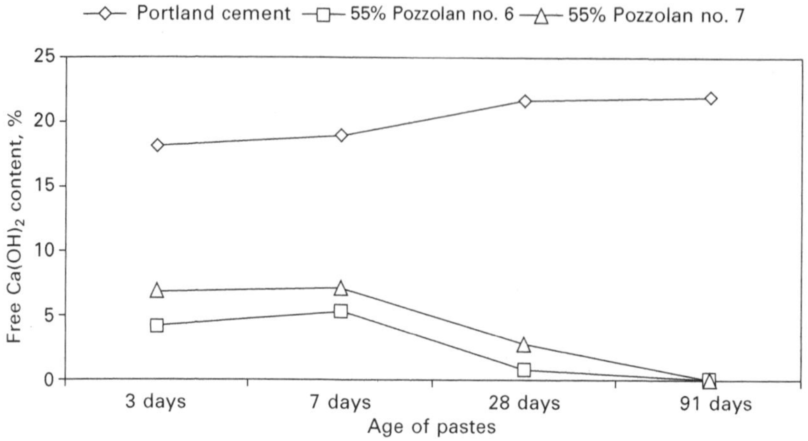

7.2 Free $Ca(OH)_2$ content of HVNP cement pastes containing 55% by weight of zeolitic pozzolan (adapted from Uzal and Turanli, 2012).

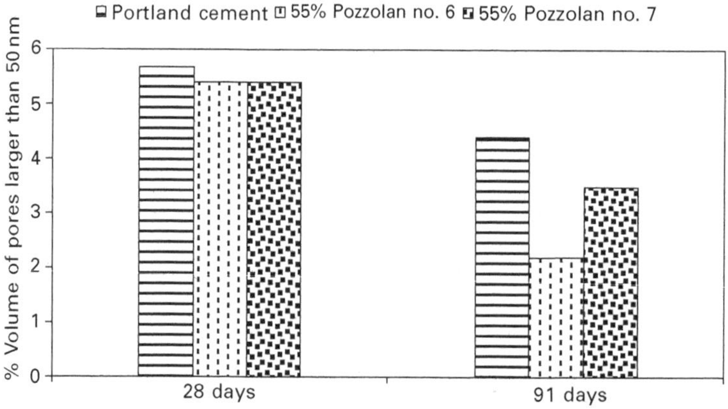

7.3 Percent volume of pores larger than 50 nm in hardened cement pastes (adapted from Uzal and Turanli, 2012).

in higher mechanical strength and lower permeability in HVNP cement pastes (Fig. 7.3). As shown in Fig. 7.3, a pozzolanic reaction progressing beyond 28 days refines larger pores and reduces their percent amount more drastically when compared to the ordinary Portland cement paste. Uzal and Turanli (2012) also reported that XRD analysis and scanning electron microscopy (SEM) observations on hardened cementitious pastes indicated the dissolution of the crystalline zeolite phase in natural pozzolan to exhibit a pozzolanic reaction.

Sato and Trischuk (2008) investigated the hydration process of cementitious pastes containing 50% natural pozzolans (chabazite and clinoptilolite types of natural zeolite) using a conduction calorimetry technique. They reported that the rate of heat development accelerates during the first hours of hydration of cement pastes containing 50% chabazite when compared with the reference Portland cement paste. They attributed the acceleration of hydration to the high BET surface area of the natural zeolite, rather than its fine particle sizes.

7.5 Mixture proportions for high-volume natural pozzolan (HVNP) concrete

HVNP concrete should be proportioned in accordance with the required levels of workability, strength and durability, as with conventional concrete mixtures. Due to some differences in the nature and particle characteristics of fly ashes and natural pozzolans, typical mixture proportions show some dissimilarity between HVFA and HVNP concrete mixtures. HVFA concrete, in general, has 100–180 kg/m^3 Portland cement content and 150–270 kg/m^3 fly ash content, with approximately 0.30–0.40 water to cementitious materials (cement + fly ash) ratio (Malhotra and Ramezanianpour, 1994; Malhotra and Mehta, 2002). The slump values of >150 mm are provided in approximately 115 kg/m^3 water content with the help of superplasticizers in HVFA concrete. On the other hand, Uzal et al. (2007) showed that HVNP concrete for structural applications could be proportioned with 200 kg/m^3 Portland cement and 200 kg/m^3 natural pozzolan, with 0.45 ratio of water to cement + pozzolan by weight w/(c + p) and slump values of 100–150 mm (Uzal et al., 2007). Superplasticizer is a mandatory component of HVNP concrete, and its dosage is approximately 1.5–2% by weight of total cementitious materials content. The maximum coarse aggregate size is 19 mm in typical examples of HVNP concrete mixtures. Typical mixture proportions for HVNP concrete are shown in Table 7.4 (Uzal et al., 2007).

7.6 Properties of fresh and hardened high-volume natural pozzolan (HVNP) concrete

7.6.1 Slump and workability

The incorporation of mineral admixtures generally improves the workability of fresh concrete, which can be defined as ease of mixing, transporting,

Table 7.4 Typical mixture proportions for HVNP concrete for a slump of 100–150 mm (Uzal et al. 2007)

Portland cement (kg/m^3)	Natural pozzolan (kg/m^3)	Water (kg/m^3)	Aggregates		Superplasticizer (kg/m^3)
			Fine (kg/m^3)	Coarse (kg/m^3)	
200	200	180	820	850	2.6–6

placing and compaction without segregation. In the case of high-volume replacement of Portland cement by finely ground natural pozzolans, the concrete mixtures exhibit better cohesiveness and workability compared to ordinary concrete. Other advantages include improved pumpability, compactibility and finishability. On the other hand, high-volume incorporation of pozzolanic materials into concrete mixtures makes use of superplasticizers, which are necessary in obtaining high slumps due to the low water content of HVPC mixtures. The type and dosage of superplasticizer vary depending on the nature and particulate characteristics of the pozzolan, as well as the compatibility between the superplasticizer and cementitious materials. Selecting a superplasticizer for use in HVPC mixtures is critical to obtaining a fresh concrete with satisfactory slump and workability.

7.6.2 Air content

In published literature, it has been reported that HVFA concrete requires a somewhat higher dosage of air-entraining agent compared to the control concrete mixture without fly ash. For high-volume fly ash concrete mixtures, relatively high carbon content in fly ash (>6%) may cause difficulties in air entrainment (Malhotra and Ramezanianpour, 1994).

The air content of non-air-entrained HVNP concrete is somewhat lower than the control concrete mixture for the same water-to-cementitious materials content (w/cm) (Uzal et al., 2007). This is actually true of all types pozzolan incorporated into concrete mixtures. Similarly to HVFA concrete, HVNP concrete might require a higher dosage of air-entraining agent for a certain level of air content compared to a plain concrete mixture.

7.6.3 Setting time

HVNP concrete mixtures exhibit quite different setting times compared with HVFA concrete. HVFA concrete mixtures typically show comparable initial setting time to the control concrete made with the same water content and the same ratio of water to cement + fly ash by weight w/(c + f), and a setting time approximately 3 hours longer that the control (Malhotra and Ramezanianpour, 1994). However, published data on HVNP systems has demonstrated that HVNP replacement may cause accelerated initial and final set in paste and concrete specimens when compared to the control specimens without replacement (Uzal and Turanli, 2003, 2012; Turanli et al., 2004, 2005; Uzal et al., 2007). Table 7.5 shows the setting times of HVNP cement pastes and concrete mixtures in comparison to reference mixtures without natural pozzolan. It is believed that the accelerated setting times of HVNP cementitious systems are related to self-desiccation of the hydrating system

Table 7.5 Setting time of HVNP pastes and concretes

Cementitious system	Pozzolan	Setting times (minute)		Reference
		Initial	Final	
Reference PC paste	–	155	180	Uzal and Turanli (2003)
HVNP paste	Pozzolan no. 1	157	190	
HVNP paste	Pozzolan no. 2	98	140	
Reference PC paste	–	150	188	Turanli et al. (2004)
HVNP paste	Pozzolan no. 3	197	295	
Reference PC paste	–	175	215	
HVNP paste	Pozzolan no. 4	103	157	
Reference PC paste	–	160	189	
HVNP paste	Pozzolan no. 5	115	144	
Reference PC paste	–	218	245	Uzal and Turanli (2012)
HVNP paste	Pozzolan no. 6	112	185	
HVNP paste	Pozzolan no. 7	54	164	
Reference PC concrete	–	265	375	Uzal et al. (2007)
HVNP concrete	Pozzolan no. 8	280	425	
HVNP concrete	Pozzolan no. 9	185	395	

Note: PC = Portland cement

due to incessant absorption of free water by the natural pozzolan, due to their high specific surface area.

7.6.4 Strength

Strength development in HVPC mixtures varies significantly depending on the type (fly ash or natural pozzolan) and source of the pozzolan used. HVFA and HVNP concrete mixtures exhibit sufficient strength values for structural applications both at early and later ages (Malhotra and Ramezanianpour, 1994; Uzal et al., 2007). Since HVFA concrete mixtures are typically made with lower w/cm values compared to HVNP mixtures, HVFA concrete generally shows a higher rate and extent of strength development, especially in the early stages. However, for the same w/cm, some HVNP concrete mixtures exhibit comparable or higher compressive strength development for up to 28 days when compared with the identical HVFA concrete mixture. Beyond 28 days, HVFA concrete is again stronger than the HVNP concrete (Uzal et al., 2007). Compressive strength development of the HVFA and HVNP concrete mixtures made with the same w/cm is illustrated in Fig. 7.4 (Uzal et al., 2007). Splitting-tensile strengths of the HVFA and HVNP concrete at 28 days are approximately 10–15% and 8–10% of the corresponding 28-day compressive strength, respectively (Malhotra and Ramezanianpour, 1994; Uzal et al., 2007).

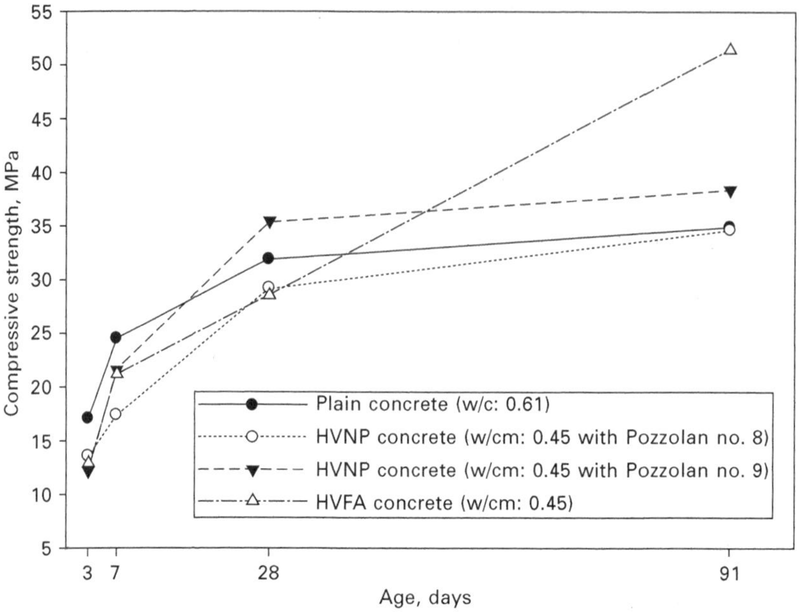

7.4 Compressive strength development of HVNP and HVFA concrete (adapted from Uzal et al., 2007).

7.6.5 Modulus of elasticity

The modulus of elasticity of HVNP concrete made with a zeolitic natural pozzolan is somewhat lower than that of the control concrete made with the same w/cm ratio (Uzal, 2007). However, it is known that the modulus of elasticity of HVFA concrete is higher than that of Portland cement concrete with comparable strengths (Malhotra and Ramezanianpour, 1994). This difference between HVFA and HVNP concrete mixtures could be related to differences in microstructure of the interfacial transition zone, but it requires further study. In both types of high-volume pozzolan concretes, the modulus of elasticity is close to or higher than 30 GPa at 28 days, and thus they are suitable for structural applications.

7.6.6 Resistance to chloride-ion penetration

Investigations into the resistance of HVFA and HVNP concretes to penetration by chloride ions, as measured by ASTM C 1202 test method, indicated that their chloride ion permeability is very low (Malhotra and Ramezanianpour, 1994; Bouzoubaa et al., 2000; Uzal et al., 2007; Uzal, 2007). ASTM C 1202 (2005) test method is based on the measurement of the total electrical charge, in coulombs, passed through a concrete disc subjected to a potential difference. This test is the most commonly accepted method for evaluating the resistance of concrete specimens against chloride-ion penetration. The chloride-ion penetrability is considered to be low and very low for the values of total charge passed of 1000–2000 coulombs and 100–1000 coulombs, respectively, in accordance with ASTM C 1202 (2005).

In HVFA concrete, the charge generally varies from 500 to 2000 coulombs at 28 days, and lower than 1000 coulombs at 3 months depending on mixture proportions and curing conditions (Bouzoubaa et al., 2000, Malhotra and Mehta, 2002). Reports on HVNP concrete demonstrated that charge values are comparable to the values for HVFA concrete, for the same w/cm (Fig. 7.5) (Uzal et al., 2007). It is possible to obtain charge values lower than 500 coulombs at 28 days in HVNP concrete made with a highly reactive zeolitic natural pozzolan at 0.45 w/cm, whereas the corresponding value is approximately 4500 coulombs for the control concrete mixture (Uzal, 2007).

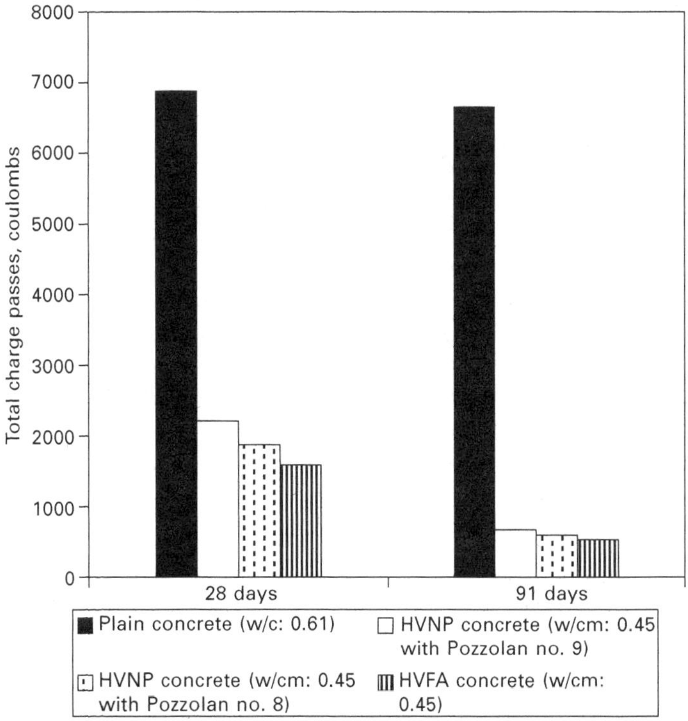

7.5 Chloride-ion penetration of HVNP and HVFA concrete (adapted from Uzal et al., 2007).

7.6.7 Resistance to sulfate attack and alkali–silica reaction

Turanli et al. (2005) investigated the sulfate resistance performance of HVNP blended cement mortars in accordance with ASTM C 1012 (1994) sulfate immersion test, by measuring the expansion of mortar bars. HVNP blended cement mortars showed considerably lower expansion (about 0.04%) than the control mortar (about 0.13%) at the end of 36 weeks of sulfate immersion. This excellent durability against sulfate attack is attributable to low Portland cement content and thus low content of $Ca(OH)_2$, which is a key factor in deterioration caused by sulfate attack.

Reported investigations have demonstrated that HVFA and HVNP cementitious systems are quite successful in controlling the expansions caused by alkali–silica reactions. Ramachandran et al. (1992) demonstrated that high volume fly ash replacement efficiently reduces alkali–silica reaction expansions, so the mortar bars of HVFA system exhibited negligible expansion when subjected to the accelerated mortar bar test.

Uzal and Turanli (2003) and Turanli et al. (2005) used the accelerated test method ASTM C 1260 (1994) to evaluate the alkali–silica reaction expansions of HVNP cementitious systems. They showed that the expansion of HVNP mortars remained at an insignificant level whereas the control mortar had expanded by more than 0.35% at the end of 30 days of testing.

7.7 Future trends

The importance of high-volume pozzolan cementitious systems in the sustainable development of cement and concrete technology is clearly accepted, not only by researchers but also by members of the concrete construction industry. The cement and concrete industry undeniably increases global warming and climatic changes due to the huge amounts of greenhouse gases, mainly CO_2, released into the atmosphere during production of Portland cement. High volume pozzolan cementitious systems with low Portland cement content provide an effective solution to the problem of meeting increasing construction needs in a sustainable manner.

The published literature contains many reports on the properties, durability characteristics and field performance of high-volume fly ash concrete. However knowledge about high-volume natural pozzolan cementitious systems is so limited and the literature is lacking in data concerning their field performance. HVNP cementitious systems should be investigated in a more detailed manner, especially in countries that do not have access to quality fly ashes. Where good quality fly ash that conforms to the standard requirements is unavailable, finely ground natural pozzolans could be evaluated as alternative pozzolanic materials for making high volume pozzolan concrete.

Recent trends have shown the future of high-volume fly ash cementitious systems. Bentz et al. (2011) have recently presented a new approach to optimizing fly ash blended cements by adjusting the particle sizes of cement and fly ash, in order to solve problems related to the rate of strength development at early ages. They showed that equivalent 1-day and 28-day compressive strength may be achieved with 35% fly ash replacement when the particle size distributions of cement and fly ash were selected appropriately. Therefore it could be anticipated that research efforts will focus on the property improvement of high-volume pozzolan systems by optimizing the system components. The type and dosage of superplasticizers as a key component of HVPC mixtures will probably be one of the variables of the optimization approaches. Furthermore, research should be focused on the development of new kinds of superplasticizers, which are more compatible with HVPC systems.

Ternary HVPC systems could be a future research topic. The use of two different pozzolanic materials together with Portland cement in a concrete mixture may generate more improvement than if only one types of pozzolanic material is incorporated (Bentz, 2010; De Weerdt et al., 2011; Bentz et al., 2012). The use of fly ash and natural pozzolans together in HVPC mixtures could also provide a synergistic effect on concrete properties and durability characteristics, especially on water requirement and workability properties, and should therefore be investigated.

7.8 References

ASTM C 1012 (1994), 'Standard test method for length change of hydraulic-cement mortars exposed to a sulfate solution', *Annual Book of ASTM Standards.*

ASTM C 1202 (2005), 'Standard test method for electrical indication of concrete's ability to resist chloride ion penetration', *Annual Book of ASTM Standards.*

ASTM C 1260 (1994), 'Standard test method for potential alkali reactivity of aggregates (mortar-bar method)'. *Annual Book of ASTM Standards.*

Bentz, D.P. (2010), 'Powder additions to mitigate retardation in high-volume fly ash mixtures'. *ACI Materials Journal* **107**, 508–514.

Bentz, D.P.; Hansen A.S.; Guyn, J.M. (2011), 'Optimization of cement and fly ash particle sizes to produce sustainable concretes'. *Cement and Concrete Composites* **33**, 824–831.

Bentz, D.P.; Sato, T; Varga, I.; Weiss, W.J. (2012), 'Fine limestone additions to regulate setting in high volume fly ash mixtures'. *Cement and Concrete Composites* **34**, 11–17.

Bouzoubaa, N.; Zhang, M.H.; Malhotra, V.M. (2000), 'Laboratory-produced high-volume fly ash blended cements: Compressive strength and resistance to the chloride-ion penetration of concrete', *Cement and Concrete Research* **30**, 1037–1046.

De Weerdt, K.; Kjellsen, K.O; Sellevold, E.; Justnes, H. (2011), 'Synergy between fly ash and limestone powder in ternary cements'. *Cement and Concrete Composites* **33**, 30–38.

Malhotra, V.M.; Mehta, P.K. (2002), *High-performance high-volume fly ash concrete: materials, mixture proportioning, properties, construction practice, and case histories.* Supplementary Cementing Materials for Sustainable Development Onc., Ottawa, Canada.

Malhotra, V.M.; Ramezanianpour, A.A. (1994), *Fly ash in concrete.* CANMET Natural Resources Canada.

Massazza, F. (2001), 'Pozzolona and pozzolanic cements', *Lea's Chemistry of Cement and Concrete*, fourth edition, edited by P.C. Hewlett. Elsevier Butterworth Heinemann.

Mehta, P.K. (1999), 'Concrete technology for sustainable development'. *Concrete International* **21**, 47–52.

Ramachandran, S.; Ramakrishnan, V.; Johnston, D. (1992), 'The role of high volume fly ash in controlling alkali-aggregate reactivity'. *Proceedings of 4th CANMET/ACI International Conference on the Use of Fly Ash, Silica Fume, Slag, and Natural Pozzolans in Concrete, Edited by V.M. Malhotra.* Turkey.

Sato, T; Trischuk, K. (2008), *Use of high volumes of zeolite in cement and mortar systems.* Institute for Research in Construction National Research Council, Report No:B1371.

Turanli, L.; Uzal, B.; Bektas, F. (2004), 'Effects of material characteristics on properties of blended cements containing high volumes of natural pozzolans', *Cement and Concrete Research* **34**, 2277–2282.

Turanli, L.; Uzal, B.; Bektas, F. (2005), 'Effect of large amounts of natural pozzolan addition on properties of blended cements'. *Cement and Concrete Research* **35**, 1106–1111.

Uzal, B. (2007), *Properties and hydration of cementitious systems containing low, moderate and high amounts of natural zeolites.* PhD Thesis, Middle East Technical University. Turkey.

Uzal, B.; Turanli, L. (2001), 'High volume natural pozzolan blended cements: Physical properties and compressive strength of mortars'. Supplementary Proceedings of Three-Day CANMET/ACI International Symposium on Sustainable Development and Concrete Technology, San Francisco, USA.

Uzal, B.; Turanli, L. (2003), 'Studies on blended cements containing a high volume of natural pozzolans'. *Cement and Concrete Research* **33**, 1777–1781.

Uzal, B.; Turanli, L. (2012), 'Blended cements containing high volume of natural zeolites: Properties, hydration and paste microstructure'. *Cement and Concrete Composites* **34**, 101–109.

Uzal, B.; Turanli, L.; Mehta, P.K. (2007), 'High-volume natural pozzolan concrete for structural applications'. *ACI Materials Journal* **104**, 535–538.

Uzal, B.; Turanlı, L.; Yücel, H.; Göncüoğlu, M.C.; Çulfaz, A. (2010), 'Pozzolanic activity of clinoptilolite: A comparative study with silica fume, fly ash and a non-zeolitic natural pozzolan'. *Cement and Concrete Research* **40**, 398–404.

8
Influence of supplementary cementitious materials (SCMs) on concrete durability

M. CYR, University of Toulouse, France

DOI: 10.1533/9780857098993.2.153

Abstract: This chapter presents the effects of supplementary cementitious materials (SCMs) on properties affecting the durability of concrete (porosity, permeability), and on the major factors affecting concrete durability (alkali–silica reaction, delayed ettringite formation, sulphate attacks, acid attacks, frost resistance, abrasion, carbonation and chloride ingress). General trends and specific effects of ground-granulated blast-furnace slag, fly ash, silica fume and metakaolin, when used as cement replacement, are reported and discussed.

Key words: pozzolans, mineral admixtures, supplementary cementitious materials, ground-granulated blast-furnace slag, fly ash, silica fume, metakaolin, porosity, permeability, alkali–silica reaction, delayed ettringite formation, sulphate attacks, acid attacks, frost resistance, abrasion, carbonation, chloride ingress, corrosion.

8.1 Introduction

For many engineers, concrete materials are specified principally on the basis of their compressive strengths at 28 days. Current standards, such as EN 206-1, correct this point of view by taking the durability of concretes into account and classifying them according to exposure classes related to environmental actions (corrosion induced by carbonation or by chlorides, freeze/thaw attacks and chemical attacks).

According to ACI Committee 201's Guide to Durable Concrete (ACI, 2008), the 'durability of hydraulic-cement concrete is defined as its ability to resist weathering action, chemical attack, abrasion, or any other process of deterioration. Durable concrete will retain its original form, quality, and serviceability when exposed to its environment.' Supplementary cementitious materials (SCMs) greatly affect the durability of concrete and an extremely large number of research papers have been published, increasing the difficulty of generalizing the effect of a given mineral admixture on a durability property. Nevertheless, this chapter, which is in three main sections, tries to present general trends concerning the effects of SCMs on the principal topics regarding the durability of concrete (Fig. 8.1):

- the major properties governing durability, i.e. porosity and permeability;

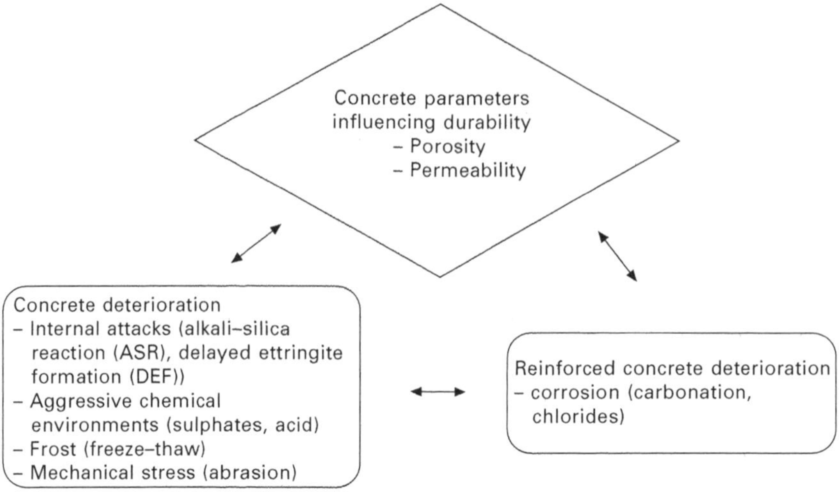

8.1 Principal topics treated in this chapter regarding the durability of concrete.

- the principal mechanisms of concrete deterioration, i.e. internal, chemical, physical and mechanical attacks;
- the two durability aspects related to reinforced concrete corrosion, i.e. carbonation and chloride ingress.

It should be noted that this chapter is intentionally limited to the principal artificial supplementary cementitious materials (SCMs) most commonly used, i.e. fly ash (FA), silica fume (SF), metakaolin (MK), and ground-granulated blast-furnace slag (GGBS). Ternary and quaternary binders are omitted. Moreover, only the results and effects concerning the replacement of Portland cement are treated.

8.2 Influence of supplementary cementitious materials (SCMs) on moisture transfer properties of concrete

The durability of cement-based materials is greatly influenced by the transport properties of the concrete, mainly by the difficulty that aggressive agents (chlorides, sulphates, etc.) have in penetrating the porous network of the concrete. The capacity of the concrete to exchange with its external environment depends strongly on the porous structure of the material. The formulation of a durable concrete especially requires a reduction of the open porosity (Section 8.2.1) and permeability (Section 8.2.2). SCMs are known to play a significant role in the modification of these two properties.

8.2.1 Porosity

The porous structure of a cement paste is made up of capillary pores (~200–700 nm), which are residues of the water-filled spaces in the fresh paste, and gel pores (~3–30 nm), which are the interstitial voids in the hydrates (e.g. C-S-H) (Verbeck and Helmuth, 1968). The presence of aggregates in mortars and concretes creates an interfacial transition zone (ITZ), which is a porosity having a size comparable to that of capillary pores (Maso, 1980; Ollivier et al., 1995). The orders of magnitude of the total porosity at 28 days are around 15% for an ordinary concrete, 10–12% for a high-strength concrete, and 7–9% for a very high-strength concrete.

The capillary pores, which are the largest pores in size, play a preponderant role in the transfer properties of the concrete, and thus on the durability performances, especially when they are interconnected (open pore structure). The capillary porosity is strongly reduced when:

- The water–cement ratio decreases. According to the results of Mehta and Manmohan (1980), no interconnected pores of 100 nm or more remain in cement paste after 28 days of curing when the water–cement ratio is smaller than 0.50. Interconnection between ITZs can, however, still exist if the thickness and number of ITZs are too great (Ollivier and Torrenti, 2008).
- The degree of hydration increases. Mehta and Manmohan (1980) showed a reduction of the size and volume of capillary pores for prolonged curing times, leading to a significant decrease in the open porosity. The minimum relative humidity in the concrete must be at least 75% to allow hydration to continue (Ollivier and Torrenti, 2008).

The presence of SCMs affects the porosity of cement-based materials: refinement of the pore size distribution compared to control mixtures, decrease of the thickness and porosity of the ITZ (Carles-Gibergues, 1981; Scrivener et al., 1988; Poon et al., 1999), etc. The total porosity of cement-based materials with SCMs can be affected by chemical (pozzolanic reaction) and physical (compactness) influences. In the latter case the porosity is decreased by using very small particles such as silica fume (if correctly dispersed with a superplasticizer), since the minimal porosity of a solid particle arrangement is proportional to $\sqrt[5]{\frac{d}{D}}$ (Caquot, 1937), where d and D are the minimal and maximal sizes of grains, respectively. An optimization of the grading curve can also help decrease the porosity, for instance by use of the Fuller and Thompson (1907) approach, which permits a maximum density gradation of the solid particles to be attained.

A schematic representation of pozzolan effects on the porosity of hardened cement pastes is presented in Fig. 8.2. In this figure, the pore size distribution is represented as a cumulated curve giving, for a particular equivalent diameter

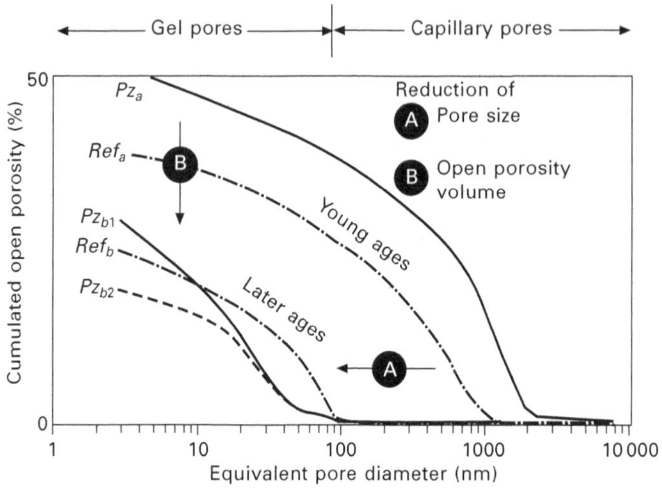

8.2 Schematic representation of the pore size distribution of cement pastes with (Pz_a, Pz_{b1} and Pz_{b2}) and without (Ref_a and Ref_b) mineral admixture.

of pore, the total volume of pores having a dimension greater than or equal to this diameter. These cumulated curves highlight possible reductions of the maximum pore size and of the open porosity, as illustrated by arrows A and B respectively (Fig. 8.2). These reductions are achieved for any paste with (Pz_a to Pz_b) or without (Ref_a to Ref_b) pozzolan by a decrease in the water–binder (w/b) ratio or an increase in the paste curing time or degree of hydration.

Specific effects of SCMs on the pore structure of cement pastes, when compared to reference pastes with Portland cement only, are as follows:

- At young ages, SCMs lead to an increase in paste porosity and the size of pores (Pz_a vs. Ref_a) because a smaller quantity of hydrates is formed in the paste. At that time, most pozzolans have not yet reacted to produce pozzolanic C-S-H and the replacement of a fraction of cement by a mineral admixture has merely the effect of diluting the cement. With fly ash and GGBS, this situation can last for several days (more than 7 days for some fly ashes (Feldman, 1983; Poon et al., 2001)) and can even be prolonged if the SCMs are used at high doses or cured for particularly short periods of time. For very active pozzolans such as silica fume and some metakaolins, this state can be short, e.g. less than 3 days (Khatib and Wild, 1996; Poon et al., 2001, 2006).
- At later ages, the pozzolanic reaction leads to the production of hydrates in addition to those of Portland cement. The main consequences are as follows:
  - A refinement of the open porosity (arrow A on Fig. 8.2), leading to

a reduction in the size of the larger pores and in the average pore diameter. Most works found in the literature confirm this characteristic for all usual SCMs: Feldman (1983) and Uchikawa (1986) for GGBS, Feldman (1983), Nagataki and Ujike (1986) and Poon et al. (1999, 2001) for fly ash, Nagataki and Ujike (1986), Batrakov et al. (1992), Hooton (1993) and Poon et al. (1999, 2001, 2006) for silica fume, and Khatib and Wild (1996), Frias and Cabrera (2000), Poon et al. (2001, 2006), Güneyisi et al. (2008) and Badogiannis and Tsivilis (2009) for metakaolin.

- A decrease of the volume of the open porosity with age, which sometimes allows the paste/concrete with pozzolan to reach values lower (Pz_{b2} vs. Ref_b on Fig. 8.2) than that of the reference paste/concrete. This situation has been seen, especially for low w/b (e.g. 0.30), for very active pozzolans such as silica fume (5, 10, 20%) and metakaolin (10, 20%) (Yogendran et al., 1987, Poon et al., 2001, 2006; Güneyisi et al., 2008), but also for fly ash (20, 40, 50%) and GGBS (60%) (Al-Amoudi et al., 1993; Shafiq and Cabrera, 2004). Higher contents of less reactive SCMs are necessary to reach the same efficiency.
- Otherwise the open porosity can be equal to or higher than that of comparable plain Portland cement pastes (Pz_{b1} vs. Ref_b on Fig. 8.2) for high w/b or mineral admixture contents: Feldman (1983) (fly ash and GGBS), Gonen and Yazicioglu (2007) (fly ash), Cwirzen and Penttala (2005) (silica fume), Khan (2003) (silica fume), Khatib and Wild (1996) and Frias and Cabrera (2000) (metakaolin). It should be noted that a few authors (Hooton, 1993, for silica fume, Poon et al., 1999, for silica fume and fly ash) obtained higher porosity in the pozzolan pastes compared to reference pastes, but the contrary in concrete. This situation shows the efficiency of SCMs in improving the ITZ.

8.2.2 Permeability

The permeability refers to the ease with which liquids or gases can enter and move through the concrete due to a pressure head. The literature reports measurements obtained by various methods based on the use of water, air or other gases (oxygen, nitrogen…).

The usual range of permeability for a typical concrete is of the order of $10^{-18} - 10^{-20}$ m^2 (water permeability), and $10^{-16} - 10^{-18}$ m^2 (gas permeability) (Baroghel-Bouny, 2004) but lower values can be obtained for high-strength concretes or for concretes with SCMs. The dispersion of values found in replicate measurements may be significant, and the general scatter of the results can be very high. This is the reason why the results are often presented on a logarithmic scale and why the difference between the permeability of two distinct concretes must be substantial (a factor of about 10) to be statistically significant.

General trends

The water content and the curing time greatly influence the permeability of cement-based materials. Generally speaking, these two parameters are more important regarding permeability variations than the effect of cement replacement by a mineral admixture. In the latter case, the measured permeability usually varies by a factor of 1 to 10 (see Fig. 8.3), while it has been shown that the variations of w/b (e.g. 0.30 vs. 0.70) combined with different curing times (e.g. > 7 days vs. < 1 day) can lead to permeability variations of several orders of magnitude (Dhir et al., 1989).

The ITZ is sometimes regarded as a key feature governing the permeability of mortars and concretes (e.g. (Winslow et al., 1994; Garboczi and Bentz, 1996). The beneficial effect of pozzolans on the ITZ (Bentur and Cohen, 1987; Xu et al., 1993; Gao et al., 2005), i.e. the densification of the microstructure and the improvement in the mechanical properties of the bond by a decrease in the quantity of portlandite crystals and a modification of their orientation arrangement at the ITZ, could partly explain the reduction of permeability when pozzolans are used.

Although results in the literature about the effect of SCMs on permeability are quite dispersed, some general trends, schematized on Figs 8.3 and 8.4, can be highlighted. On the one hand, some rare results, involving mainly fly ash or GGBS, imply an increase in the permeability with the use of mineral admixture (1 on Fig. 8.3) (Kasai et al., 1983; Shi et al., 2009; Elahi et al.,

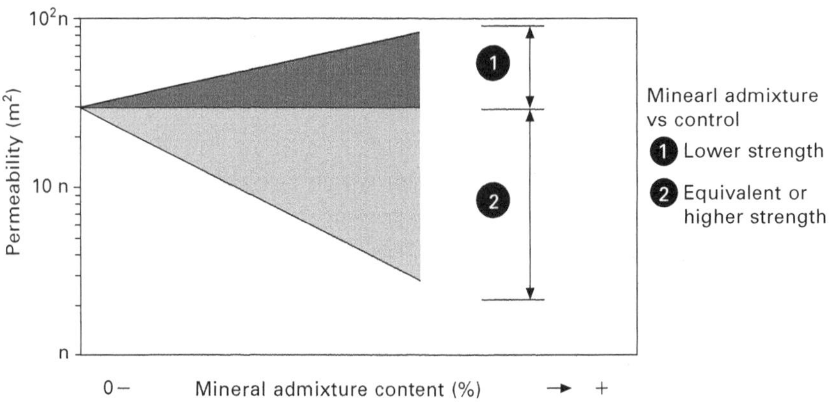

8.3 Schematic representation of the mineral admixture effect on concrete permeability ($n = -18$ (gas permeability) or -20 (water permeability).

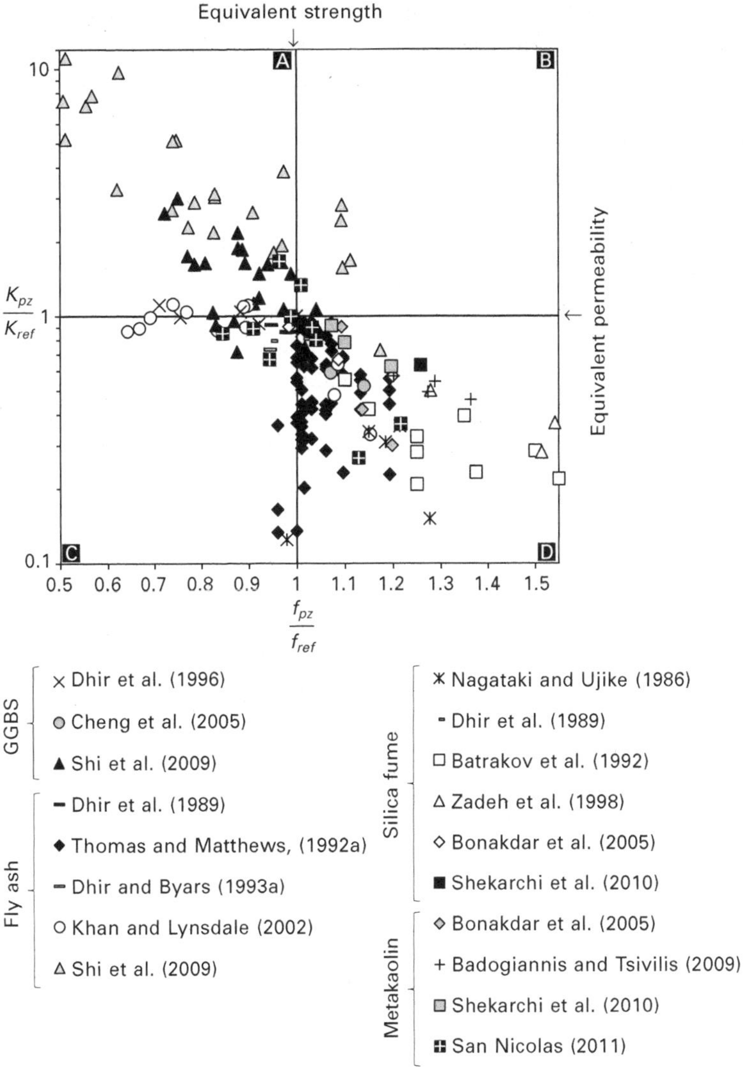

8.4 Relation between relative permeability (K_{pz}/K_{ref}) and relative compressive strength (f_{pz}/f_{ref}) for concretes containing GGBS, fly ash, silica fume and metakaolin. A, B, C and D delimit the effects of SCMs on permeability (K) and compressive strength (*f*).

2010). This situation, often associated with pozzolan concretes having much lower strength than the control concrete (zone A on Fig. 8.4), could be due to the use of weakly reactive mineral admixture, to high replacement rates, or to an insufficient curing time (as stated earlier, SCMs are very sensitive to curing time).

On the other hand, several studies report a beneficial effect of GGBS, fly ash, silica fume and metakaolin on the permeability of concrete compared to a control mixture without mineral admixture (2 on Fig. 8.3), with different efficiencies depending on many parameters (quality and quantity of pozzolan, time of curing, concrete properties...). Most of the studies concern mixtures having slightly or much higher strength than control concretes (zone D on Fig. 8.4). However, the increase in compressive strength alone does not explain the reduction of permeability of pozzolan concrete, since significant decreases can be obtained even for equivalent concrete strengths (region near the interface between zones C and D on Fig. 8.4) (Thomas and Matthews, 1992a). These reductions are often attributed to the formation of smaller and less permeable capillaries (Manmohan and Mehta, 1981), i.e. a pore structure refinement due to the pozzolanic reaction, and to a reduction of the pore continuity (Feldman, 1984; Hooton, 1986). It can be noted that very few increases in permeability are found when the compressive strengths of pozzolan concrete are higher than those of control concretes (zone B on Fig. 8.4).

Particular effects of SCMs

GGBS contents up to 67% are able to maintain permeabilities (±10%) similar to that of control concrete, even for decreases in strength of up to 30% (zone C on Fig. 8.4) (Dhir et al., 1996). Significant reductions in permeability can also be found. For instance, Cheng et al. (2005) obtained 41 and 48% decreases for GGBS contents of 40 and 60% respectively.

Low-lime and high-lime *fly ashes* used up to the range of 30–50% can lead to reductions in permeability, especially when the concretes are correctly cured (Nagataki and Ujike, 1986; Dhir and Byars, 1993a; Naik et al., 1994; Khan and Lynsdale, 2002).

For *metakaolin* (MK), Bonakdar et al. (2005) showed a continuous decrease of permeability (–70%) with up to 15% MK. This trend was confirmed by Shekarchi et al. (2010), who found similar results. Badogiannis and Tsivilis (2009) and San Nicolas (2011) found an optimum at 10 and 20% respectively. These differences in the quantity of metakaolin to be used may be related to the characteristics of the product (e.g. purity and fineness).

Silica fume is a very active pozzolan which rarely provokes an increase in permeability. The use of 5 to 10% SF is usually sufficient to divide the permeability by a factor of 1.5 (Bonakdar et al., 2005; Shekarchi et al., 2010) or 3 (Nagataki and Ujike, 1986; Batrakov et al., 1992; Zadeh et al., 1998; Khan and Lynsdale, 2002). Larger replacement rates (17 to 30%), although they are not easy to achieve in practice, can lead to even larger reductions (Nagataki and Ujike, 1986; Batrakov et al., 1992). It has also been reported that silica fume concretes are less dependent on the curing time in water (Nagataki and Ujike, 1986).

8.3 Influence of supplementary cementitious materials (SCMs) on concrete deterioration

8.3.1 Internal attacks: endogenous swelling reactions

Endogenous reactions are defined here as reactions that are generated from the internal components of the initial concrete mixture, without the need for exterior aggressive agents. The reaction products are expansive compounds which induce stresses, and hence cracking, in concrete. The two principal types of endogenous reactions are the alkali–silica reaction (ASR) and delayed ettringite formation (DEF).

ASR

ASR occurs when the amorphous or poorly crystallized silica phase in an aggregate (e.g. chert, flint, chalcedony or opaline sandstone) is attacked and dissolved by the alkali hydroxides in the concrete pore solution. The formation of an alkali–silica gel leads to concrete swelling and cracking. ASR will develop only if the following conditions are found: sufficient amount of alkalis in concrete pore solution, sufficient moisture level in concrete and reactive aggregate.

To limit or suppress ASR expansion, it is necessary to act on one or more of the above factors. However, these precautions are often economically and technically insufficient to prevent disorders in concrete; the use of SCMs is regarded as one of the most effective methods for suppressing abnormal expansion due to ASR. Numerous papers have been published regarding the mitigation of ASR with pozzolans when used in sufficient proportion and correctly dispersed (Pepper and Mather, 1959; Stanton, 1959; Swamy, 1992; Massazza, 1998; Fournier and Bérubé, 2000; Duchesne and Bérubé, 2001; Carles-Gibergues and Hornain, 2008).

It should be noted that poor dispersion of the pozzolan may lead to deleterious effects. Many cases of ASR have been reported concerning abnormal expansions or pop-outs of concrete when large agglomerates of densified silica fume were incorrectly dispersed in concrete paste (Pettersson, 1992; Escadeillas et al., 2000; Marusin and Shotwell, 2000; Diamond et al., 2004; Juenger and Ostertag, 2004; Maas et al., 2007). Considering this, it has been suggested that the chemistry of ASR has common features with the pozzolanic reaction (Taylor, 1998).

Figure 8.5 gives a schematic representation of the effect of SCMs on the expansion of concrete due to ASR. The use of an insufficient quantity of pozzolan can sometimes lead to an increase of the expansion (curve Pz_b). A typical example is the pessimum effect of fly ash, especially when it has a high alkali content (Duchesne and Bérubé, 1994a; Carles-Gibergues and Hornain, 2008).

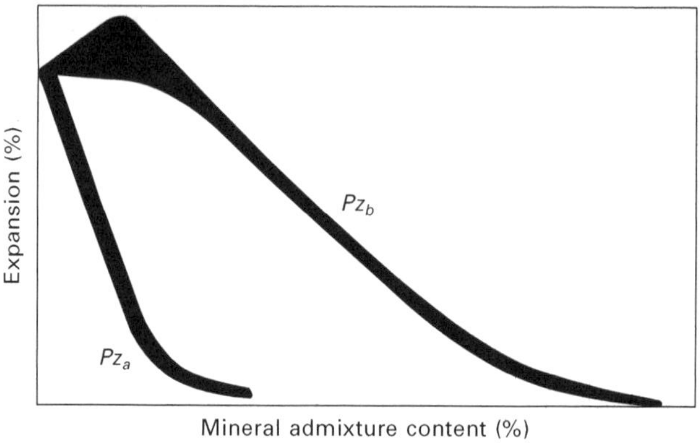

8.5 Schematic representation of the effect of pozzolan on ASR-expansion (inspired from Thomas and Folliard, 2007; Thomas 2011; Carles-Gibergues and Hornain, 2008).

The amount of SCMs required to control ASR-expansion in concrete increases for the following conditions (from Pz_a to Pz_b on Fig. 8.5):

- For a given pozzolan, when the amount of alkalis in the mixture (from cement or other internal or external sources) is high or becomes higher, the aggregates are more alkali-reactive, or the temperature of the test decreases (e.g. accelerated tests at 40 or 60 °C vs. *in situ* conservation).
- For a given mixture kept in fixed conditions, when the pozzolan is less reactive (less amorphous silica) or contains more alkalis or calcium.

In the case of very active pozzolans, the required content is lower (silica fume: 8–12%, metakaolin: 10–20%) than for less reactive SCMs (low-calcium fly ash: 20–30%; high-calcium fly ash: 40–60%; GGBS: 35–65%) (Thomas and Folliard, 2007).

Several mechanisms have been put forward to explain the effectiveness of SCMs against ASR. It is more than probable that many of them should be taken simultaneously to justify the role of SCMs.

- A densification of the hardened paste due to the physical and chemical actions of SCMs leads to a decrease in the porosity and the permeability (see Section 8.2.1) and the resulting reduction of the ionic mobility slows down the migration of alkalis towards the reactive aggregate. At the same time, these C-S-H lead to an improvement in the strength of the cement paste and, consequently, a higher resistance to the expansive stresses due to ASR gels (Turriziani, 1986; Bérubé and Duchesne, 1992; Glasser, 1992).
- A dilution of alkalis in concrete when low-alkali SCMs are used as cement replacement (Glasser, 1992). According to this hypothesis, pozzolans having high alkali contents, such as some silica fumes ($Na_2O_{eq} > 3\%$),

or fly ashes ($Na_2O_{eq} > 8\%$), are unable to reduce the ASR-expansion of concrete (Duchesne and Bérubé, 1994a, 1994b) but, on the other hand, this assertion cannot explain other results involving pozzolans such as glass powders, since glass brings a significant amount of alkalis (>12% Na_2O_{eq}), leading to higher alkali contents in the mixtures (Carles-Gibergues et al. 2008; Idir et al., 2010).

- A depletion of portlandite: it is considered that the consumption of portlandite plays a major role in the reduction or the suppression of the ASR expansion. According to some authors (Chatterji 1979; Chatterji et al., 1983; Bleszynski and Thomas, 1998), the depletion of portlandite leads to a decrease in pH and/or to the production of non-destructive gels. However, this view is contested by Duchesne and Bérubé (1994b). Other authors, e.g. Diamond (1989), affirm that calcium is involved in the formation of ASR gels (Ca is almost always found in various amounts in ASR-gels; Bérubé and Fournier, 1986). So the depletion of $Ca(OH)_2$ due to the pozzolanic reaction reduces the risk of producing harmful gels.
- A decrease of alkalis in the pore solution. Many authors (Bhatty and Greening, 1987; Taylor, 1987, Hong and Glasser, 1999; Lothenbach et al., 2011) have established that the ability of low Ca/Si C-S-H (produced by the reaction of pozzolans) to fix alkalis is enhanced compared to that of normal C-S-H. The depletion of free alkalis lowers the pH of the pore solution by decreasing the OH^- content and, consequently, reduces the attack of reactive aggregates (Uchikawa, 1986; Nixon and Page, 1987; Larbi et al., 1990; Glasser, 1992; Duchesne and Bérubé, 1994b). This last mechanism is the hypothesis most cited in the literature to explain the role of SCMs in counteracting the effect of ASR but cannot be used alone in all cases (Idir et al., 2010).

DEF

DEF can be defined as the formation of ettringite (AFt) several months or years after the setting of cement, without the need for new external sulphates. DEF is a consecutive consequence of a significant heating of the concrete (>65 °C) after its casting, either due to a thermal treatment (e.g. steam curing in the precast industry) or to the exothermic reaction of the cement (e.g. mass concrete). The consequences of DEF include swelling and cracking of the concrete, which can happen after several years of humid conservation.

It is believed that, since AFt is not a stable phase at high temperature, the initial heating of the concrete leads to the decomposition of primary ettringite. Some of the sulphate ions would then remain in the pore solution (Glasser et al., 1995), while others would be adsorbed by C-S-H (Brown and Bothe, 1993), which is a reversible process. Later with moist storage, their subsequent desorption would favour the reformation of ettringite in the hardened material, causing expansion and cracking in the concrete during its service life (Ramlochan et al., 2003, 2004).

General trends

Several parameters are necessary for DEF to be initiated and develop in concrete, and SCMs can affect most of them. Concrete must be subjected to a high temperature (65–90 °C), a temperature for which the stability of ettringite cannot be guaranteed, for a certain duration in the first few days of its hydration (Taylor et al., 2001; Carles-Gibergues and Hornain, 2008). It is recognized that fly ash and GGBS reduce the peak temperature of concrete when they are used in large quantities (Neville, 1995). These SCMs could thus help to limit the increase of heating by a cement dilution effect, especially in mass concretes.

The presence of water and humidity are essential to DEF since water is involved in the diffusion process and in the formation of the reaction products. DEF is essentially present in structures that are exposed to high humidity, in contact with water, or subjected to water ingress. The beneficial effect of pozzolan in the production of a denser and less permeable paste pore structure (Section 8.2.2) could help reduce the transport properties of the concrete.

The alkali content in the concrete has an effect on the solubility of ettringite, since high amounts of alkalis in the pore solution favour its dissolution and inhibit the precipitation of primary ettringite, making sulphate ions available for further delayed ettringite formation. Alkalis do not need to be at high concentrations in the concretes affected. For instance, in a study of eight bridges affected by DEF (LCPC, 2007), the alkali content in the concretes ranged from 2 to 4.6 kg/m^3. A decrease in the amount of alkalis increases the critical value of temperature causing DEF (LCPC, 2007).

As already stated for ASR, SCMs help to produce low Ca/Si C-S-H that fix more alkalis than normal C-S-H. All other things being equal, the decrease of alkali content in the pore solution reduces the initial sulphate uptake by C-S-H, thus limiting their future availability after desorption (Ramlochan et al., 2003).

The sulphate and aluminate contents have to be high enough to allow the formation of delayed ettringite and increase the risk of expansion (Odler and Chen 1995, 1996; Fu and Beaudoin, 1996). The critical SO_3 content is within the normal dosage in Portland cement (up to 3–4%) (Taylor, 1998). The amount of Al_2O_3 in Portland cement depends on the C_3A content, and damage due to DEF has been identified in bridges for quantities of C_3A as low as 7% (LCPC, 2007). However, DEF seems to remain insignificant for C_3A contents below 5% (Carles-Gibergues and Hornain, 2008).

According to Ramlochan et al. (2003, 2004), one of the major roles of pozzolan in controlling DEF expansion is related to the availability of reactive Al_2O_3, often found in SCMs such as fly ash, metakaolin or GGBS. The release of additional alumina from SCMs favours the precipitation of AFm instead of AFt (Taylor et al., 2001), as confirmed by the experimental work of Ramlochan et al. (2004), by decreasing the SO_3/Al_2O_3 ratio. This ratio is sometimes proposed to estimate the risk of DEF occurrence (Heinz and Ludwig, 1987; Grabowski et al., 1992; Odler and Chen, 1996). The presence of supplementary alumina moves the ratio away from the pessimum (around 0.8 in mass, according to Grabowski et al., 1992, Heinz et al., 1999, and Zhang et al., 2002), under the critical value for DEF to occur. Heinz and Ludwig (1987) and Heinz et al. (1999) affirm that blended cements with appropriate amounts of GGBS and siliceous fly ash must keep the SO_3/Al_2O_3 ratio under 0.45–0.55 to avoid DEF.

Particular effects of SCMs

Most SCMs show a beneficial effect in reducing DEF expansion when used in sufficient amounts (Fig. 8.6). Their efficiency could be related to several parameters, including their available aluminium content (MK, GGBS and class F fly ash) (Ramlochan et al., 2003, 2004). For *GGBS*, the minimum content varies between 25 and 35%, depending on the author (Fu, 1996; Kelham, 1999; Ramlochan et al., 2003), but it can reach 40% (Santos Silva et al., 2010) or higher levels when it is used with cements having very high sulphate or alkali contents (Ramlochan et al., 2003). Santos Silva et al.

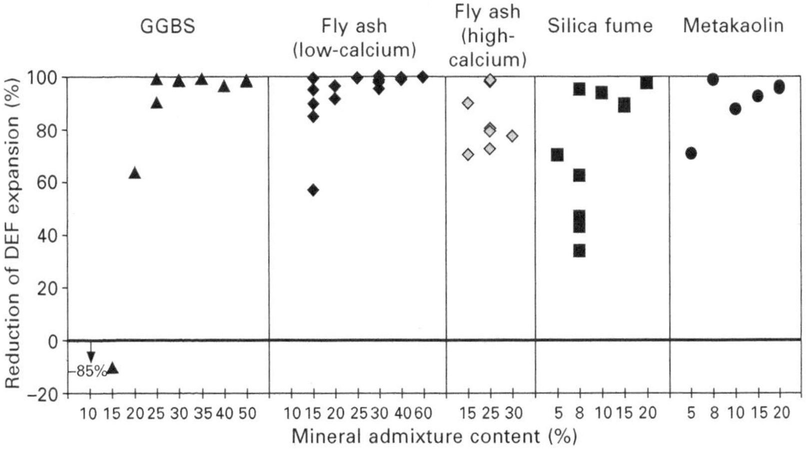

8.6 Reduction of DEF expansion (in %) due to the use of SCMs. Data from Ramlochan et al. (2003), Ghorab et al. (1980), Santos Silva et al. (2006, 2010), Fu (1996), Kelham (1999), Kurdowski and Duszak (2004).

(2010) have shown that low GGBS levels (e.g. 10 and 15%) can even lead to increased DEF expansions (Fig. 8.6).

In the case of *fly ash*, which performs well in existing structures (Thomas et al., 2008a), the distinction between low and high calcium must be made. In the former case, most of the results show that at least 15 to 25% is necessary to reduce the expansion significantly (Santos Silva, 2010: 15%; Ramlochan et al., 2003: 15–25%, Santos Silva et al., 2006: 20%, Heinz et al. 1999: 20%), and more is sometimes needed (Kurdowski and Duszak, 2004: 30%). In the latter case (high-calcium fly ash), 30% may not be enough to significantly reduce the expansion (Fu, 1996) (Fig. 8.6).

From the literature, it seems that 10 to 15% of *silica fume* is needed to counteract the harmful effect of DEF: Santos Silva et al. (2010) 10%, Fu (1996): 15%, Ramlochan et al. (2003): 15%. However, its efficiency is not always verified, probably due to its low-alumina content (Ramlochan et al., 2004).

Finally, 8% or more of *metakaolin* reduces the expansion significantly (Ramlochan et al., 2003: 8%, Santos Silva et al., 2010: 10%) but the LCPC's technical guide for the prevention of disorders due to DEF (LCPC, 2007) recommends using at least 20% of pozzolan. After 2 years of tests, Santos Silva et al. (2006) confirmed that the use of 20% of metakaolin was efficient in eliminating DEF expansion.

8.3.2 Aggressive chemical environments

Chemical attacks on concrete essentially concern the reactions of dissolution and precipitation that are produced when aggressive elements reach calcium hydrates of the cement paste by diffusion or permeation processes: dissolution of portlandite, leaching of calcium in C-S-H, and/or precipitation of more or less deleterious compounds. The principal effects of chemical attacks are as follows:

- Increase in the porosity of the cement paste due, for instance, to the dissolution and leaching of hydrates, leading to an augmentation of the permeability and the diffusivity of the concrete, and sometimes to a decrease in the mechanical characteristics (e.g. Young's modulus).
- Swelling and cracking of the concrete related, for instance, to the precipitation of secondary ettringite due to penetration of external sulphate.

These effects can cause deterioration of the concrete that may be simply unaesthetic but that can sometimes be harmful for the mechanical stability of the structure. SCMs are generally believed to be favourable to the chemical resistance of concrete since they consume portlandite and improve the transfer properties of the paste and ITZ. Among the multiple aggressive environments existing in a nature (soils, seawater, etc.) or occurring because of human activity (e.g. industry), only the two main ones are treated in this part: sulphate and acid attacks.

Sulphate attacks

Sulphate attack concerns the ingress of external sulphate producing mainly ettringite, which must be distinguished from primary AFt resulting from the reaction of C_3A with calcium sulphate, and DEF (Section 8.3.1). The sulphates can come from different sources, mainly soils, industrial or agricultural effluents or seawater, and they have different aggressiveness depending on the cation associated with the SO_4^{2-} anion: calcium, sodium or potassium, magnesium and ammonium are the main ones. The last two are known to be the most aggressive forms, leading to deleterious effects and rapid degradation of concrete (Massazza, 1998; Mehta and Monteiro, 2005; Escadeillas and Hornain, 2008).

Sulphate attack of hydrated cement is related to the reaction of sulphate ions with CH and C-A-H to form gypsum and ettringite. The sulphate also reacts with remaining C_3A in the cement to form ettringite. In the particular case of magnesium sulphate, this compound can react with C-S-H directly, resulting in the formation of magnesium silicate hydrate, which lacks cohesive properties.

Low permeability and diffusivity provide the best defence against sulphate attack by reducing sulphate penetration. This can partly be achieved by the use of SCMs. The dilution of C_3A content by the use of SCMs can also be beneficial (Escadeillas and Hornain, 2008).

In a general way, pozzolans or GGBS have a more beneficial effect against sodium sulphate than against magnesium sulphate (Fig. 8.7). For sodium sulphate, consumption of CH by SCMs reduces the amount of gypsum and ettringite produced. Numerous authors have reported beneficial effects of SCMs in counteracting sodium sulphate attack: see for example Frearson (1986), Frearson and Higgins (1992), Hogan (1981), Osborne (1991), Wee et al. (2000), Cao et al. (1997), Zeljkovic (2009), O'Connell et al. (2012) for GGBS; Kalousek et al. (1972), Cao et al. (1997), Torii and Kawamura (1994), Shashiprakash and Thomas (2001), Sideris et al. (2006) for fly ash; Wee et al. (2000), Moon et al. (2003), Cao et al. (1997), Torii and Kawamura (1994), Cohen and Bentur (1988), Al-Amoudi et al. (1995a), Shashiprakash and Thomas (2001) for silica fume; Zeljkovic (2009), Malolepszy and Pytel (2000), Khatib and Wild (1998), Courard et al. (2003), Al-Akhras (2006), Ramlochan and Thomas (2000) for metakaolin.

In the case of magnesium sulphate, a lower CH content in hydrated cement is undesirable because it encourages the reaction of sulphate with C-S-H (Al-Amoudi, 2002; Ganjian and Pouya, 2005; Lee et al., 2005). This could

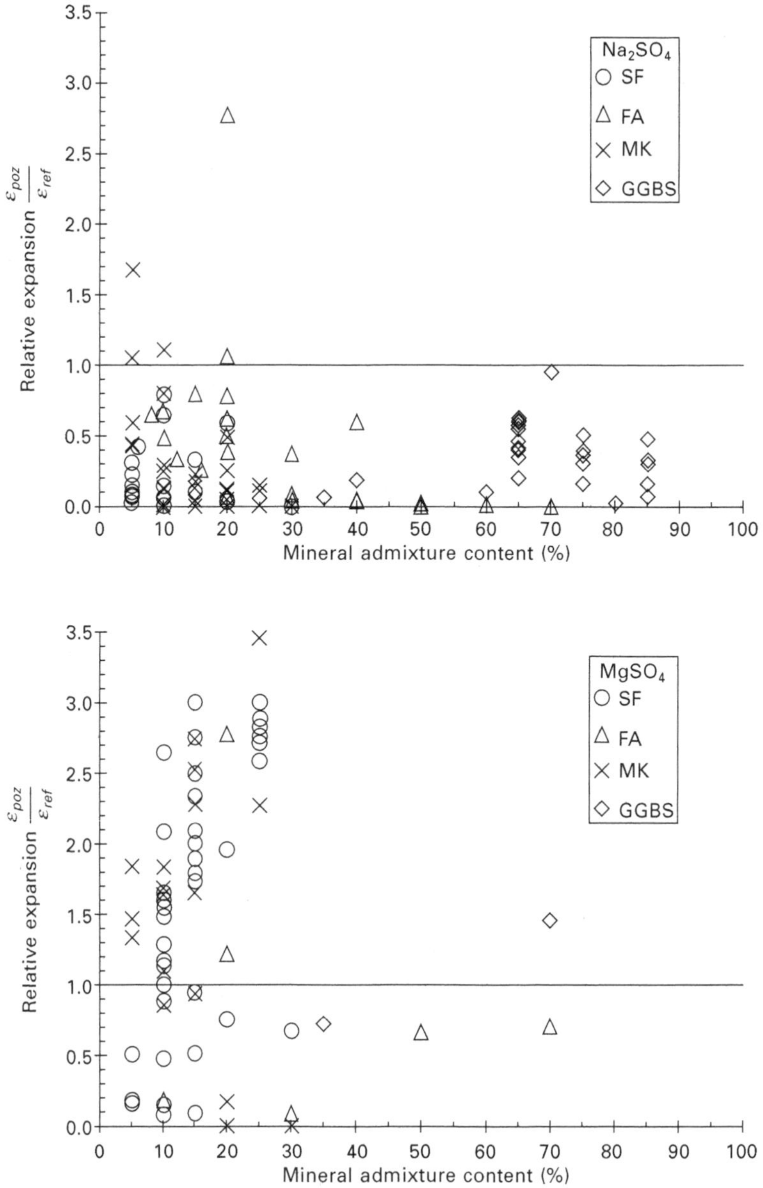

8.7 Relation between relative sodium and magnesium sulphate expansions (εpoz/εref) and mineral admixture contents for concretes containing GGBS, fly ash, silica fume and metakaolin. Data from Al-Akhras (2006), Al-Amoudi et al. (1995a, 1995b), Cao et al. (1997), Cohen and Bentur (1988), Courard et al. (2003), Diab et al. (2008a. 2008b), Hooton (1993), Kalousek et al. (1972), Khatib and Wild (1998), Lee et al. (2005), Malolepszy and Pytel (2000), Moon et al. (2003), Ramlochan and Thomas (2000), Shashiprakash and Thomas (2001), Sideris et al. (2006), Torii and Kawamura (1994), Wee et al. (2000), Zeljkovic (2009).

explain the inefficiency of pozzolans, especially silica fume and metakaolin (Fig. 8.7), to protect concrete from magnesium sulphate attack (Hekal et al., 2002; Moon et al., 2003; Lee et al., 2005; Diab et al., 2008a, 2008b; Al-Amoudi et al., 1995a).

Acid attacks

Any environment having a pH lower than 12.5 may be aggressive to concrete through leaching processes, but pH below 6 is much more critical and is responsible for most concrete damage, such as increases in porosity and permeability, decrease in strength, cracking and spalling (Mehta and Monteiro, 2005).

The parameters of concrete governing its quality (e.g. water–binder ratio, cure) are of major significance compared to the effect of SCMs (Hobbs and Matthews, 1998). Nevertheless, all other things being equal, including SCMs usually improves the concrete's resistance to mineral acids (hydrochloric, sulphuric, phosphoric, carbonic) or organic acids (acetic, lactic, citric) (Backes, 1986; Yamato and Emoto, 1989; Durning and Hicks, 1991; Torii and Kawamura, 1994; O'Donnell et al., 1995; Roy et al., 2001; Hill et al., 2003; Chang et al., 2005; Aydin et al., 2007; Larreur-Cayol et al., 2011; Gruyaert et al., 2012; O'Connell et al., 2012), as a result of a slower rate of acid attack due to lower permeability and reduced calcium hydroxide content. Enhanced resistance to silage effluent (lactic and acetic acids) or manure liquids (mixture of acetic, propionic, butyric, isobutyric and valeric acids) has also been reported (De Belie et al., 1996, 1997a, 1997b, 1997c, 2000; De Belie, 1997; Bertron et al., 2004, 2005a, 2005b, 2007; Pavia and Condren, 2008).

This improvement is illustrated in Fig. 8.8 on selected literature results for relative loss of mass ($\Delta M_{pz}/\Delta M_{ref}$) after short- and long-term immersion in different kinds of mineral and organic acids. It can be seen that all classic SCMs are able to reduce the degradation due to acid attacks (GGBS often performs better), although higher losses of mass can sometimes be found.

The short-term efficiency of SCMs is often higher than at later ages, meaning that SCMs retard the degradation due to acid attack without automatically preventing it (Fig 8.8). This has been confirmed by several authors, e.g. Durning and Hicks (1991), who obtained similar losses of mass, but over longer times when silica fume was used: 7.5% SF doubled the time for 25% mass loss for concrete in a 5% acetic acid solution, while 15% quadrupled this time.

8.3.3 Frost resistance

It is generally recognized that durability problems related to cold weather conditions can take two principal forms:

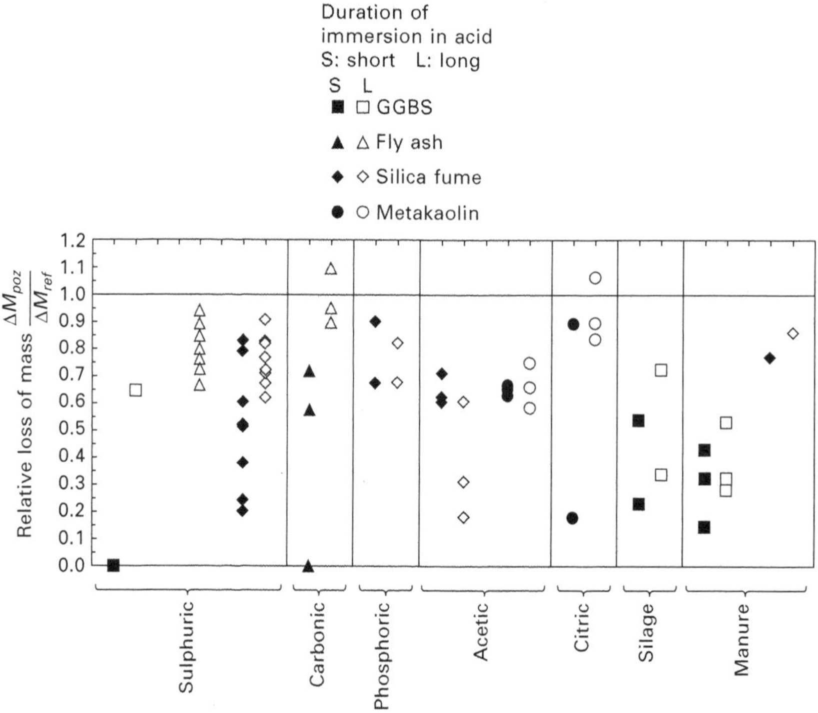

8.8 Effect of SCMs on the loss of mass due to different acid attacks, relative to reference. Data from Backes (1986), Yamato and Emoto (1989), Durning and Hicks (1991), Bertron et al. (2004, 2005b, 2007), Chang et al. (2005), Aydin et al. (2007), Pavia and Condren (2008), Larreur-Cayol et al. (2011).

- Internal cracking due to freezing and thawing cycles generally causes a loss of concrete strength and stiffness. It is often evaluated by the variation of the dynamic modulus of elasticity with freeze–thaw cycles (Fig. 8.9).
- Surface scaling due to freezing in the presence of de-icing salts, which leads to loss of concrete cover. The mass of scaling debris is usually the parameter measured to characterize this degradation (Fig. 8.9).

General trends

The mechanisms of frost damage have been attributed to several phenomena, partially due to the 9% expansion of water during freezing: hydraulic pressure build-up as water is forced away from the freezing front (Powers, 1954, 1975), osmotic pressure gradients driving water toward the freezing centres (Powers and Helmuth, 1956; Helmuth, 1960), vapour pressure potentials (Litvan,

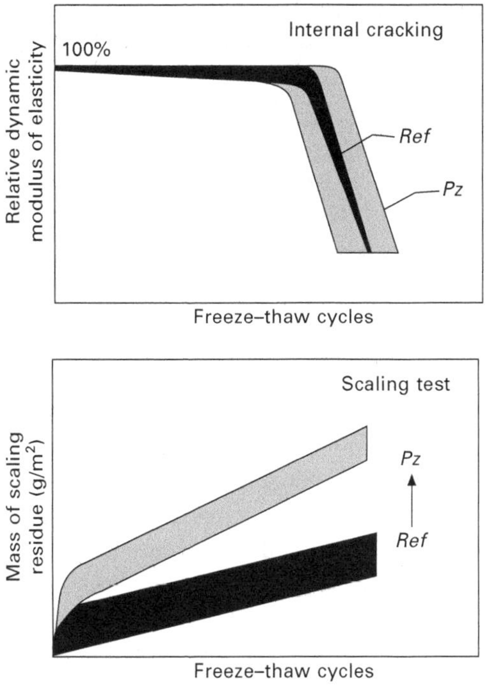

8.9 Schematic representation of frost resistance of SCMs concrete (*Pz*) compared to reference (*Ref*).

1972), and combinations of these processes. Most of these theories agree on the significant parameters responsible for frost damage in concrete:

- *Degree of saturation* – Concretes with or without pozzolan that have internal relative humidity below approximately 75 to 80% are normally not subject to internal damage from freezing, whatever the concrete composition (ACI Committee 201, 2008).
- *Air-void* – Frost damage is practically eliminated when a proper air-void system is achieved. Since air bubbles are practically never completely filled with water, ice can be formed without creating excessive internal pressure. In order to resist freeze–thaw cycles, it is generally recognized that concretes should contain more than 4% of air, with a bubble spacing factor of less than 200–250 μm. This is usually achieved by the use of air-entraining admixture.

It has been shown that the use of very active pozzolans such as silica fume in very high strength concretes (> 90 MPa) generally gives satisfactory behaviour in freeze–thaw cycles, without the need for a particular air void structure (Gagné et al., 1990; Pigeon et al., 1991). For other concretes with or without pozzolan, a correct air void structure is almost always necessary (Gagné and Linger, 2008).

SCMs can sometimes have a negative effect on entrained air voids, by increasing the demand for air-entraining agent (AEA) to attain a correct air-void spacing length. At equivalent AEA content, the use of certain SCMs can lead to larger spacing factors. This effect principally concerns fly ash (Gebler and Klieger, 1983; Zhang, 1996) but is also observed with GGBS (Fernandez and Malhotra, 1990; Saric-Coric and Aitcin, 2003) and silica fume (Khayat and Aitcin, 1992). This could be due either to a greater surface area or to high carbon content of the pozzolans. The carbon absorbs a portion of the air-entraining agent, thus limiting its availability for producing the required air bubble network. When the carbon content is low, pozzolans have practically no effect on air void structure (Pigeon et al., 1989).

- *Water–binder ratio* – Relatively high w/b (0.7–0.8) leads to mediocre freeze–thaw durability, even though the concrete contains air-entrained (Gagné and Linger, 2008). Water–binder ratios of less than 0.5 or 0.3 for air-entrained or non-air-entrained concretes, respectively, are usually needed to reach a correct level of durability. At these levels, there is less freezable water and concrete has a higher resistance to internal stresses due to the formation of ice. Moreover, a decrease of w/b tends to decrease the size of pores, leading to less freeze-water at a given temperature, since the ice needs a lower temperature to be formed in smaller pores (Gagné and Linger, 2008). In many cases, SCMs also tend to decrease the size of pores (see Section 8.2.1), which can be beneficial for frost resistance.
- *Maturity of the concrete* – As for air-void, many frost problems (especially for surface scaling) are eliminated when an adequate maturity of the concrete is achieved, since it helps decrease porosity and permeability. Concretes with SCMs are particularly vulnerable to short curing or to early freeze and thaw cycles. Numerous examples of SCMs having low reaction kinetics (e.g. GGBS and fly ash) are reported in the literature concerning short curing times leading afterward to insufficient frost resistance: see, for instance, Saric-Coric and Aitcin (2003).

Specific effects of SCMs

Internal cracking and surface scaling should be taken separately when analysing the effect of SCMs on the frost resistance of concrete conserved in a humid environment. *Internal cracking* is generally only slightly affected by SCMs (Fig. 8.9 left) for concretes having equivalent strength and adequate air void structure. Several authors have noted little or no significant influence of fly ash (Gebler and Klieger, 1986b; Langley et al., 1989; Naik et al., 1995a), GGBS (Malhotra, 1983; Sakai et al., 1992), silica fume (Bilodeau and Carette, 1989; Batrakov et al., 1992; Sabir and Kouyiali, 1991) and metakaolin (Zhang and Malhotra, 1995; Kim et al., 2007; Vejmelková et al., 2010), when pozzolans are used in limited amounts (e.g. 10% for silica fume). High-volume fly ash (> 40%) concrete can also be resistant to internal cracking, on condition that the w/b is lower than 0.35 (Langan et al., 1990; Langley et al., 1989).

Surface scaling is much more dependent on the type and content of pozzolan, and also on the curing time. The scaling resistance is usually reduced when SCMs are used (Fig. 8.9 right). All SCMs are concerned: fly ash (Gebler and Klieger, 1986b; Pigeon et al., 1996; Zhang et al., 1998; Boyd and Hooton, 2007), GGBS (Stark and Ludwig, 1997; Saric-Coric and Aitcin, 2003; Boyd and Hooton, 2007; Battaglia et al., 2010), silica fume (Sabir, 1997) and metakaolin (Zhang and Malhotra, 1995; Vejmelková et al., 2010). A few authors believe that the decrease in de-icer scaling resistance is related to carbonation of the surface layer of the concrete, which produces metastable calcium carbonates soluble in NaCl (Stark and Ludwig, 1997; Battaglia et al., 2010).

Nevertheless, a few exceptions have been found, for instance for fly ash up to 30–50% (Bilodeau et al., 1991; Naik et al., 1995a) or silica fume at 10% and low w/b (Bilodeau and Carette, 1989; Hammer and Sellevold, 1990). It should be noted that a difference is sometimes observed in field cured specimens since GGBS concrete, for instance, can perform relatively well up to 25–35% of GGBS (Boyd and Hooton, 2007; Bouzoubaa et al., 2008).

8.3.4 Mechanical stresses: wear of concrete surfaces

Mechanical damage such as wear of concrete can mainly be caused by impact, abrasion, erosion or cavitation. These effects concern the surface of the concrete and are due to attrition by sliding, rolling, scraping or percussion. Wear of concrete surfaces can occur on concrete pavements such as roads and industrial floors, for instance, when mechanical stresses locally exceed the strength of concrete. It can be initiated in the hydrated cement paste, in the aggregates or in the paste–aggregate interface by aggregate pull out. It has been pointed out that the wear resistance of concrete depends on the strength and hardness characteristics of the concrete constituents, including paste–aggregate bond strength. These characteristics are themselves influenced by several factors such as the water–cement ratio, the type and grading of aggregates, the presence of SCMs, air-entrainment, the finished surface condition and the duration of curing (Hilsdorf, 1995; Neville, 1995; Lawrence, 1998).

The behaviour of SCMs regarding abrasion resistance is strongly correlated to the compressive strength of the concrete, as shown in Fig. 8.10. This figure summarizes several relative results (81) taken from the literature for fly ash

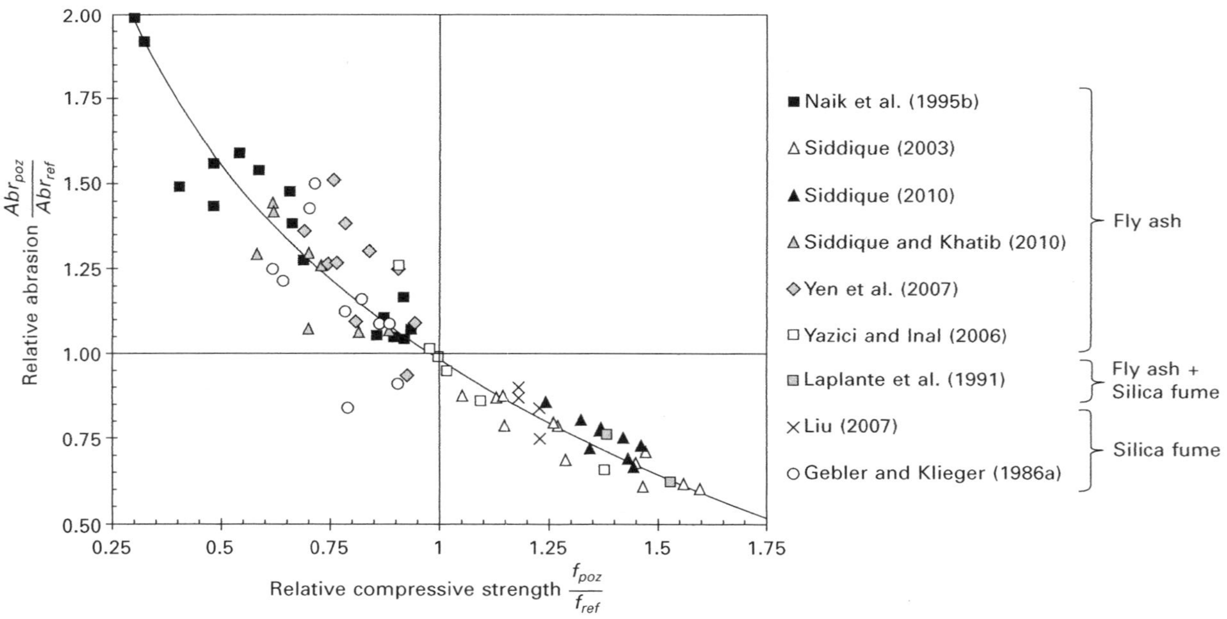

8.10 Effect of fly ash or silica fume on the relative abrasion of cement-based materials.

and silica fume incorporated at replacement rates from 10 to 70%, in all kinds of concrete having strengths in the range 16–80 MPa. It is noteworthy that the general trend is that SCMs improve the abrasion resistance compared to a reference concrete when they are accompanied by an increase in strength. Conversely, a decrease of strength almost systematically leads to worse abrasion resistance. The curing of the outer skin of the concrete is of a great importance, especially for pozzolan concretes.

This tendency is confirmed by several authors for all SCMs: Gebler and Klieger (1986a), Dhir et al. (1991), Laplante et al. (1991), Bilodeau and Malhotra (1992), Naik et al. (1995b), Siddique (2003, 2010a), Yazici and Inan (2006), Yen et al. (2007), Liu (2007), Siddique and Khatib (2010), Yetgin and Cavdar (2011).

8.4 Influence of supplementary cementitious materials (SCMs) on reinforced concrete deterioration

Most of the concrete used in the construction industry is intended to be included in reinforced concrete structures. The presence of steel compensates for the weakness of tensile strength of concrete but steel is very vulnerable to corrosion. Steel bars are naturally protected in sound concrete, since a shielding outer layer, called 'passivating layer', is formed on the steel surface, reducing the corrosion rate to a negligible level (passive state). The passivating layer is stable in alkali environments like the one provided by concrete under normal conditions ($pH > 13$).

The corrosion of steel in reinforced concrete is initiated when the passivating layer is partially or totally destroyed. Two principal phenomena are implicated in such destruction: the carbonation of concrete due to the penetration of CO_2 from the surface to the core of the concrete, and the penetration of chloride ions for concretes exposed to marine environments or de-icing salts. Both phenomena are significantly modified by the presence of SCMs.

8.4.1 Corrosion induced by carbonation

Carbon dioxide (CO_2) present in the air (~0.038%) enters the concrete by open porosity or cracks. In the presence of water, it reacts with the hydrates of cement (mainly portlandite – $Ca(OH)_2$, but also C-S-H, Dunster, 1989) to form calcium carbonate ($CaCO_3$) and silica gels (Dunster, 1989; Cowie and Glasser, 1991). This reaction, called 'carbonation' and for which the detailed mechanism can be found elsewhere (e.g. Baroghel-Bouny et al., 2008), has two main consequences:

- A reduction of global concrete porosity, since $CaCO_3$ and silica gels have a larger molar volume than the initial components ($Ca(OH)_2$ and C-S-H), and a modification of the pore size distribution (Pihlajavaara, 1968; Ngala and Page, 1997; Thiery et al., 2003; Villain and Thiery, 2005). The decrease of the porosity is not harmful and can lead to the improvement of some mechanical performances.

- The consumption of portlandite, which can be regarded as the 'basic reserve' of the concrete, causing a decrease in the pore solution pH, from 12.5–13.5 to around 9. At that level of pH, the passivating layer is not stable and dissolves, causing corrosion to start.

A schematic representation of the behaviour of concrete subjected to a carbonation test is given in Fig. 8.11. The carbonation depth usually follows a straight line versus the square root of the time, meaning that it is driven by the gas diffusion process (Baroghel-Bouny et al., 2008). At a given age of concrete, the carbonation depth is greater when the water–cement ratio is increased (Skjolsvold, 1986), the cement content is reduced (Venuat and Alexandre, 1968, 1969), or the curing time is shortened (Meyer, 1968) (A to B on Fig. 8.11). The carbonation is maximum at intermediate relative humidity (40–80%, depending on the author: Van Balen and Van Gemert, 1994; Saetta et al., 1995), while it is very low for completely dry or fully saturated concretes. In fact, the relative humidity must be sufficiently low to allow the permeation of CO_2 into the concrete but also sufficiently high for the carbonation reaction to be achieved in aqueous phase. Therefore, porosity and gas permeability are among the most important properties influencing the carbonation of concrete.

General trends

SCMs influence the carbonation of concrete by at least two antagonistic phenomena:

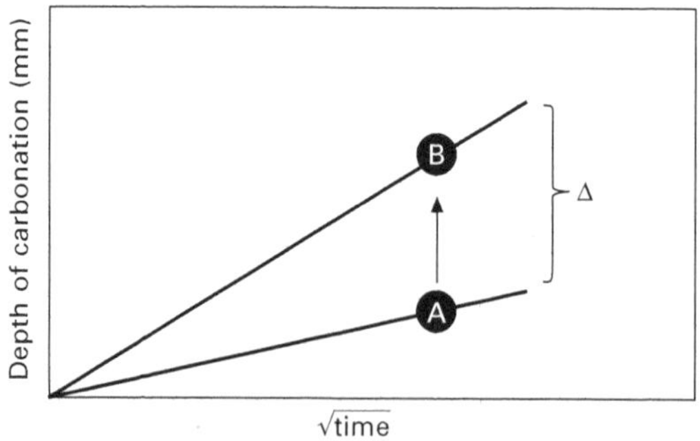

8.11 Schematic representation of the behaviour of concrete subjected to a carbonation test. A to B (Δ) is obtained by an increase in w/b, a decrease in curing time, or an increase in mineral admixture content.

- The consumption of portlandite due to the pozzolanic reaction, which implies that a smaller amount of CO_2 is required to carbonate the remaining hydrates. It has been shown that the depth of carbonation is greater when the amount of $Ca(OH)_2$ present is lower (Bier, 1986). Consequently the presence of SCMs could result in more rapid carbonation.
- Modification of the porosity and permeability of concrete (Sections 8.2.1 and 8.2.2), which could lead to an improvement in the transport properties of pozzolan concretes in certain conditions (active pozzolan, long curing time, etc.). The denser microstructure, refined and more tortuous pores, and reduced permeability encountered in pozzolan concretes tend to slow down the carbonation rate.

Depending on the dominant effect, pozzolan concretes can be more or less carbonated than control mixtures at a given age.

Although numerous papers comparing the carbonation of Portland cement with and without SCMs have been published, generalizations remain difficult. This could be partly due to different experimental conditions existing in the literature, e.g. accelerated test vs. natural carbonation, driven by the CO_2 content and the relative humidity. According to Massazza (1998), accelerated tests tend to overestimate the depth of carbonation of pozzolan concretes compared to long exposure on site. So care should be taken when interpreting the results of accelerated tests.

The overall trend found in the literature is that the depth of carbonation increases with the use of pozzolan in concretes (A to B on Fig. 8.11) (Massazza, 1998; Neville, 1995). However, some authors affirm that there is no significant difference in carbonation depth if the concretes with or without pozzolan have a similar strengths (Matthews, 1984; Hobbs, 1988; Kokubu and Nagataki, 1989). This affirmation is not always supported, as shown by Kim et al. (2007) with metakaolin and silica fume. All other things being equal, the carbonation of pozzolan concretes increases relative to control mixtures when the relative strength decreases (Shi et al., 2009). Compressive strength of concrete over 50 MPa can significantly reduce the depth of natural carbonation (indoor or outdoor), not only for ordinary Portland cement (OPC) but also for fly ash concretes (Sulapha et al., 2003). However, this is not always the case for accelerated conditions ($CO_2 > 4\%$), for which even high-strength concretes with very active pozzolans can remain vulnerable to carbonation (Kim et al., 2007, 75 MPa concretes). For relatively low pozzolan content, the differences might be negligible (Collepardi et al., 2004), but for high pozzolan content, it could be necessary to decrease the water–binder ratio to maintain the same carbonation performance (Cengiz Duran, 2003). One important parameter that greatly influences the result is the curing of concretes, which can be very problematic for pozzolan concrete.

Although concretes containing SCMs are sometimes considered as less resistant to carbonation (especially in accelerated tests) due to their low portlandite content, this property should not be considered alone regarding the risk of corrosion since the SCMs can sometimes reduce the permeability of the concrete, so that the high rate of carbonation might be confined to the first few centimetres, especially for high-strength concretes containing very active pozzolans. Consequently, further progression of the CO_2 can be slowed down, thereby providing appropriate protection to the steel (Bouikni et al., 2009). Moreover, it has been shown that many concretes with SCMs have very low probabilities of corrosion even after several tenths of years (due to lower permeability), often within the service life of the structure, meaning that carbonation effect may by insignificant in these cases (McNally and Sheils, 2012).

Specific effects of SCMs

The presence of *GGBS* often leads to an increase in the depth of carbonation for high percentages of GGBS (> 50%) (Bouikni et al., 2009; Gruyaert et al., 2010). An increase in the GGBS fineness has a beneficial effect on carbonation (Sulapha et al., 2003), as does long curing time. When the duration of the cure is prolonged beyond 7 days, the carbonation of concretes containing up to 50% of GGBS is not significantly higher than that of OPC concretes (Malhotra et al., 2000; Sulapha et al., 2003).

In the case of *fly ash*, the increase of carbonation is significant when the replacement rate is beyond around 30% (Thomas and Matthews, 1992b). This critical pozzolan content can sometimes be higher (e.g. 50%; Cengiz Duran, 2003).

The general tendency for *silica fume* concrete is an increase of carbonation depth, especially for high pozzolan content (>10%) (Khayat and Aitcin, 1992). Kim et al. (2007), using 5, 10, 15 and 20% of *metakaolin* in high strength concretes (w/b of 0.25, 75 MPa), obtained a continuous increase of carbonation depth (5% CO_2, 60% RH and 30 °C) with the increase of pozzolan, even for equivalent compressive strength. McPolin et al. (2007) found similar results on 58 MPa–10% metakaolin concrete (5% CO_2, 65% RH and 20 °C).

8.4.2 Corrosion induced by chlorides

The presence of chloride ions around the concrete rebar induces a local dissolution of the passivating layer, leading to localized corrosion pits due to the creation of electrochemical microcells. A minimum chloride concentration in the pore solution must be reached for the corrosion to be initiated. This critical concentration value is evaluated at around 0.4% of chlorides with respect to the cement mass (Baroghel-Bouny et al., 2008). The use of the ratio Cl^-/OH^- is often chosen as a criterion for corrosion initiation (Cl^- and OH^- are the concentrations of free chloride and hydroxyl ions, respectively, near the steel bar): the higher the ratio, the high the corrosion rate. Critical values are between 0.3 and 0.6 (Hausmann, 1967; Diamond, 1986).

Chlorides can be found in the components of the mixture themselves but most of the regulations in the world impose the use of chloride-free materials in concrete (e.g. chemical admixtures, or washed aggregates of marine origin). SCMs are also concerned by these restrictions and many standards limit the free chloride content to very low values: 0.10% for fly ash, GGBS and metakaolin, according to European standards EN 450-1 and EN 15167-1, and French standard NF P18-513, respectively, and 0.3% for silica fume (EN 13263-1).

Other possible sources of chlorides are seawater or de-icing salts. In the case of saturated conditions, the penetration of chlorides in the concrete is governed by a diffusion process following Fick's laws. When the concrete is subjected to wetting and drying cycles, chloride ions enter by capillary absorption and are transported into the concrete by convection of the liquid phase. Then chlorides can migrate by diffusion in the saturated zones. This mode of penetration allows rapid ingress of chlorides, explaining why concretes subjected to wetting and drying cycles are often degraded at higher rates than fully immersed concretes.

The two main properties responsible for the penetration of chlorides (diffusion and capillary absorption) are greatly influenced by the same parameters as govern the porosity and permeability of concrete (Section 8.2.2), i.e. the water–binder ratio, the cement content and the curing time. The diffusion of chlorides into the concrete depends strongly on the open porosity and pore tortuosity of the cement paste (Roy et al., 1986). A decrease in the former and an increase in the latter tend to reduce the coefficient of diffusion.

The presence of SCMs can play a major role in the modification of chloride diffusion, although it has been suggested that, in certain cases, the compactness of concrete (i.e. w/b) is more crucial than its pozzolan content. In other words, to improve the diffusion properties, it is better to reduce w/b than use pozzolan with a high water content (the best solution being obviously to decrease w/b and use SCMs).

The main effects of SCMs relative to the penetration of chlorides are the following:

- *Open porosity and pore tortuosity of the cement paste* – As stated earlier (Section 8.2.1), SCMs tend to refine and decrease the open porosity of cement-based materials, especially for low w/b and adequate curing times. This leads to a segmentation of the porosity and a creation of more tortuous paths in the paste, the main consequence being a decrease in the diffusivity of chloride ions. It has been shown that the chloride diffusion coefficient ratio between Portland cement and pozzolanic cement pastes can range between 3 and 10 (Page et al., 1981; Kumar and Roy, 1986; Byfors, 1987; Chatterji and Kawamura, 1992).

- *Reduction of pH* – The use of SCMs usually reduces the pH of the pore solution (Section 8.3.1). This reduction of OH^- concentration results in a lower threshold chloride value to initiate the corrosion, when the ratio Cl^-/OH^- is considered (Byfors, 1987). In other words, less chloride is needed to initiate corrosion when SCMs are used in concrete (due to a lower stability of the passivating layer when the pH is reduced). However, the ratio Cl^-/OH^- is not always increased by the use of SCMs since some Cl^- can be bound by C-S-H.
- *Chloride binding* – Chloride ions in the concrete exist in different forms (free, adsorbed and chemically bound) and not all of them are harmful regarding rebar corrosion. Of all the chloride ions entering the concrete, only the free chlorides are considered to be active in the process of depassivation and rebar corrosion (Baroghel-Bouny et al., 2008). Some of the chloride ions interact with the hydrated phases of Portland cement and the role of SCMs in the consumption of chloride ions is variable since the binding capacity differs considerably from one pozzolan to another:
  - Chlorides can be adsorbed on C-S-H, but to a lesser extent as the ratio Ca/Si decreases (Beaudoin et al., 1990). An increase in the proportion of active pozzolan, such as silica fume, causes a reduction in the amount of Cl^- adsorbed on C-S-H, due to the lower Ca/Si ratio of C-S-H. SCMs containing alumina are less affected by this phenomenon since there is evidence that calcium aluminate hydrates (C-A-H) produced by the pozzolanic reaction of fly ash (and probably GGBS and metakaolin) can bind the chlorides (Jensen and Pratt, 1989).
  - Chlorides can react with other compounds to give new hydrates such as calcium chloroaluminates (e.g. Friedel salts – $C_3A{\cdot}CaCl_2{\cdot}10H_2O$). In plain Portland cement, the C_3A (and C_4AF) content play an important role in the binding of chloride ions. It should be noted that the amount of C_3A decreases in pozzolan pastes, due to a dilution effect. On the one hand, for alumina-free pozzolans (e.g. silica fume), the amount of Friedel's salt decreases with increasing pozzolan content (Page and Vennesland, 1983), thus reducing the binding capacity. On the other hand, alumina-bearing SCMs such as fly ash, GGBS and metakaolin generally lead to an increase of chloride binding, due to the formation of more Friedel's salt (Dhir et al., 1996, 1997; Vejmelková et al., 2010).

Specific effects of SCMs

The decrease of pH caused by the introduction of SCMs causes a reduction in the amount of chlorides necessary to destroy the passive layer and initiate corrosion, but this negative effect is counterbalanced by the chloride binding of certain SCMs (GGBS, fly ash, metakaolin), and by the decrease of chloride diffusion due to the segmentation of the porosity (GGBS, fly ash, metakaolin, silica fume). The overall result is generally a significant decrease of chloride penetration, as shown on Fig. 8.12. All results show lower coefficients of diffusion, and the efficiency depends on the reactivity of the pozzolan. This could results in longer service life of structures, as reported by McNally and Sheils (2012) in their probabilistic service life prediction of concrete with GGBS.

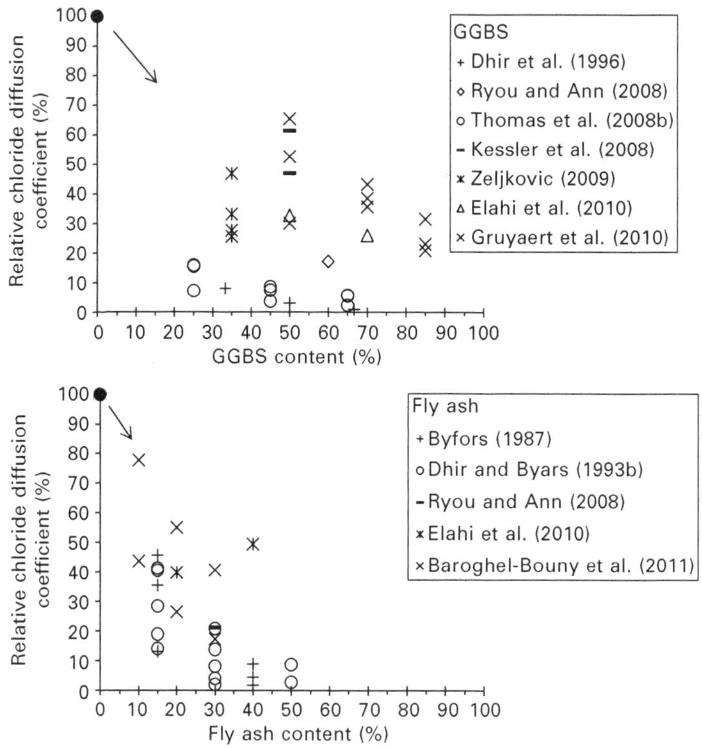

8.12 Effect of mineral admixture content on the coefficient of chloride diffusion, relative to the reference. Data from Gautefall (1986), Byfors (1987), Gautefall and Havdahl (1989), Luping and Nilsson (1992), Dhir and Byars (1993b), Dhir et al. (1996), Cabrera and Nwaubani (1998), Boddy et al. (2001), Oliveira et al. (2005), Kessler et al. (2008), Ryou and Ann (2008), Thomas et al. (2008b), Zeljkovic (2009), Elahi et al. (2010), Gruyaert et al. (2010), Baroghel-Bouny et al. (2011), San Nicolas (2011).

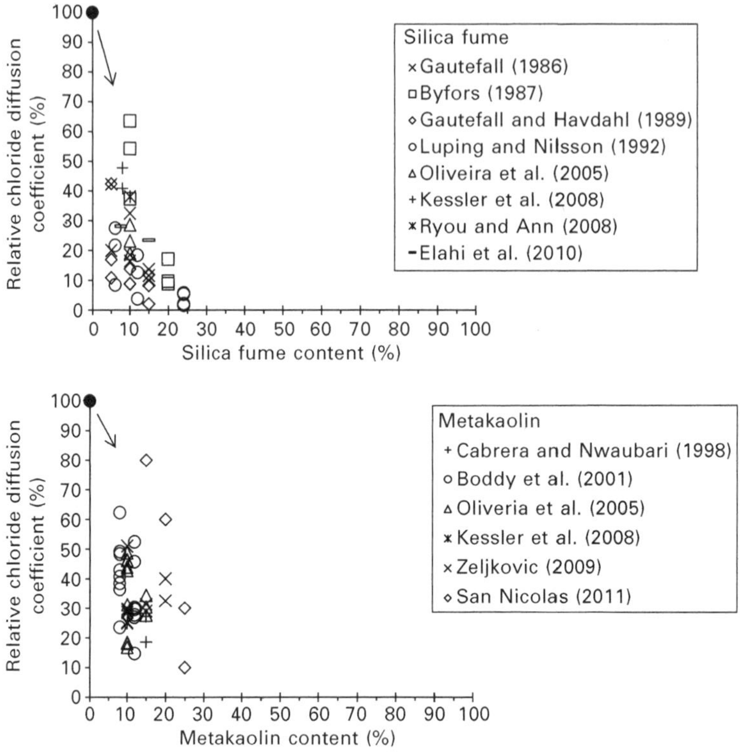

8.12 Continued

However, as stated by Angst et al. (2009) in their extensive review:

> the effect of some of these materials — namely SF, FA, and GGBS — has been studied several times, but the results were often contradictory and/or cannot be transferred to real structures due to unrealistic testing conditions. Regarding reinforcement corrosion, the behaviour of the mentioned pozzolanas, but also many other upcoming cementing materials, is completely unknown.

8.5 Future trends

Many important topics will need further research, for example:

- Since the interest in SCMs is continuing to increase, partly because of environmental, technological and economic concerns, other sources of SCMs still need to be characterized, including at nanoscale, in order to arrive at standards that will help their use in concrete. The use of ternary and quaternary binders is also a promising track.

- The development of mathematical models completing the existing ones (e.g. for ASR). These models could help the understanding and prediction of the behaviour of SCMs regarding the durability of concrete.

The comparison of the results of accelerated tests vs. *in situ* behaviour of concrete in real structures. It has been pointed out that accelerated tests are sometimes far from reality, so feedback of experience is still necessary to confirm the laboratory conclusions.

8.6 Sources of further information and advice

This chapter has presented a brief overview of some of the main effects of four artificial SCMs on major aspects of concrete durability. However it was not possible to include all interesting topics that would have merited further detailed discussion, such as shrinkage, pure water leaching, micro-organism attacks, high temperature effects (e.g. fire), etc. Further information on these subjects and on the ones treated in this chapter may be found in various media (books, conference proceedings, review papers, etc.) that deal specifically with the durability of concrete, for example:

- The effect of supplementary cementing materials on alkali–silica reaction: a review (Thomas, 2011).
- Resistance of silica fume concrete to de-icing salt scaling: a review (Zhang et al., 1999).
- Fire resistance of high strength/dense concrete with particular reference to the use of condensed silica fume: a review (Jahren, 1989).
- Biochemical attack on concrete in wastewater applications: a state of the art review (O'Connell et al., 2010).

8.7 References

ACI Committee 201 (2008), ACI 201.2 – Guide to Durable Concrete.

Al-Akhras N M (2006), Durability of metakaolin concrete to sulfate attack, *Cement and Concrete Research*, **36**, 1727–1734.

Al-Amoudi O S B (2002), Attack on plain and blended cements exposed to aggressive sulfate environments, *Cement and Concrete Composites*, **24**, 305–316.

Al-Amoudi O S B, Rasheeduzzafar, Maslehuddin M and Almana A I (1993), Prediction of long-term corrosion-resistance of plain and blended cement concretes, *ACI Materials Journal*, **90**, 564–570.

Al-Amoudi O S B, Maslehuddin M and Saadi M M (1995a), Effects of magnesium-sulfate and sodium-sulfate on the durability performance of plain and blended cements, *ACI Materials Journal*, **92**, 15–24.

Al-Amoudi O S B, Maslehuddin M and Abdul-Al Y A B (1995b), Role of chloride ions on expansion and strength reduction in plain and blended cements in sulfate environments, *Construction and Building Materials*, **9**, 25–33.

Angst U, Elsener B, Larsen C and Vennesland O (2009), Critical chloride content in reinforced concrete – A review, *Cement and Concrete Research*, **39**, 1122–1138.

Aydin S, Yazici H, Yigiter H and Baradan B (2007), Sulfuric acid resistance of high-volume fly ash concrete, *Building and Environment*, **42**, 717–721.

Backes H P (1986), Carbonic acid corrosion of mortars containing fly ash, in *SP-91: Fly Ash, Silica Fume, Slag, and Natural Pozzolans in Concrete*, ACI, 621–636.

Badogiannis E and Tsivilis S (2009), Exploitation of poor Greek kaolins: Durability of metakaolin concrete, *Cement and Concrete Composites*, **31**, 128–133.

Baroghel-Bouny V (2004), *Conception des bétons pour une durée de vie donnée des ouvrages – Maîtrise de la durabilité vis-à-vis de la corrosion des armatures et de l'alcali-réaction – Etat de l'art et guide pour la mise en oeuvre d'une approche performantielle et prédictive sur la base d'indicateurs de durabilité*, Bagneux.

Baroghel-Bouny V, Capra B and Laurens S (2008), La durabilité des armatures et du béton d'enrobage, *La durabilité des bétons – Bases scientifiques pour la formulation de bétons durables dans leur environnement*, 2nd ed. Paris, Presses de l'école nationale des Ponts et Chaussées (ENPC).

Baroghel-Bouny V, Kinomura K, Thiery M and Moscardelli S (2011), Easy assessment of durability indicators for service life prediction or quality control of concretes with high volumes of supplementary cementitious materials, *Cement and Concrete Composites*, **33**, 832–847.

Batrakov V G, Kaprielov S S and Sheinfeld A V (1992), Influence of different types of silica fume having varying silica content on the microstructure and properties of concrete, in *SP-132: Fly Ash, Silica Fume, Slag, & Natural Pozzolans & Natural Pozzolans in Conc: Proc 4th Intl Conf ACI*, 943–964.

Battaglia I K, Munoz J F and Cramer S M (2010), Proposed behavioral model for deicer scaling resistance of slag cement concrete, *Journal of Materials in Civil Engineering*, **22**, 361–368.

Beaudoin J J, Ramachandran V S and Feldman R F (1990), Interaction of chloride and C-S-H, *Cement and Concrete Research*, **20**, 875–883.

Bentur A and Cohen M D (1987), Effect of condensed silica fume on the microstructure of the interfacial zone in portland cement mortars, *Journal of the American Ceramic Society*, **70**, 738–743. 10.1111/j.1151–2916.1987.tb04873.x

Bertron A, Escadeillas G and Duchesne J (2004), Cement pastes alteration by liquid manure organic acids: Chemical and mineralogical characterization, *Cement and Concrete Research*, **34**, 1823–1835. 10.1016/j.cemconres.2004.01.002

Bertron A, Duchesne J and Escadeillas G (2005a), Attack of cement pastes exposed to organic acids in manure, *Cement and Concrete Composites*, **27**, 898–909. 10.1016/j.cemconcomp.2005.06.003

Bertron A, Duchesne J and Escadeillas G (2005b), Accelerated tests of hardened cement pastes alteration by organic acids: analysis of the pH effect, *Cement and Concrete Research*, **35**, 155–166. 10.1016/j.cemconres.2004.09.009

Bertron A, Duchesne J and Escadeillas G (2007), Degradation of cement pastes by organic acids, *Materials and Structures*, **40**, 341–354. 10.1617/s11527-006-9110-3

Bérubé M A and Duchesne J (1992), Does silica fume merely postpone expansion due to alkali-aggregate reactivity?, in *9th International Conference on Alkali-Aggregate Reaction in Concrete*, 71–80.

Bérubé M A and Fournier B (1986), Les produits de la réaction alcali-silice dans le béton: étude de cas de la région de Québec, *Canadian Mineralogist*, **24**, 271–288.

Bhatty M S Y and Greening N R (1987), Some long time studies of blended cements with emphasis on alkali–aggregate reaction, in *7th International Conference on Alkali–Aggregate Reaction in Concrete*, 85–92.

Bier T A (1986), Influence of type of cement and curing on carbonation progress and pore structure of hydrated cement pastes, in *Materials Research Society Symposium Proceedings*, **85**, 123–134.

Bilodeau A and Carette G G (1989), Resistance of condensed silica fume concrete to the combined action of freezing and thawing cycling and deicing salts, in *SP-114: Fly Ash, Silica Fume, Slag, & Natural Pozzolans in Conc: Proc 3rd Intl Conf, ACI*, 945–970.

Bilodeau A and Malhotra V M (1992), Concretes incorporating high volumes of ASTM Class F fly ashes: Mechanical properties and resistance to de-icing salt scaling and to chloride-ion Penetration, in *SP-132: Fly Ash, Silica Fume, Slag, & Natural Pozzolans & Natural Pozzolans in Conc: Proc 4th Intl Conf, ACI*, 319–350.

Bilodeau A, Carette G G and Malhotra V M (1991), Influence of curing and drying on salt scaling resistance of fly ash concrete, in *SP-126: Durability of Concrete: Second International Conference*, ACI, 201–228.

Bleszynski R F and Thomas M D A (1998), Microstructural studies of alkali-silica reaction in fly ash concrete immersed in alkaline solutions, *Advanced Cement Based Materials*, **7**, 66–78.

Boddy A, Hooton R D and Gruber K A (2001), Long-term testing of the chloride-penetration resistance of concrete containing high-reactivity metakaolin, *Cement and Concrete Research*, **31**, 759–765.

Bonakdar A, Bakhshi M and Ghalibafian M (2005), Properties of high-performance concrete containing high reactivity metakaolin, in *SP-228: 7th Intl Symposium on the Utilization of High-Strength/High-Performance Concrete*, ACI, 287–296.

Bouikni A, Swamy R N and Bali A (2009), Durability properties of concrete containing 50% and 65% slag, *Construction and Building Materials*, **23**, 2836–2845.

Bouzoubaa N, Bilodeau A, Fournier B, Hooton R D, Gagné R and Jolin M (2008), Deicing salt scaling resistance of concrete incorporating supplementary cementing materials: Laboratory and field test data, *Canadian Journal of Civil Engineering*, **35**, 1261–1275. 10.1139/l08-067

Boyd A J and Hooton R D (2007), Long-term scaling performance of concretes containing supplementary cementing materials, *Journal of Materials in Civil Engineering*, **19**, 820–825.

Brown P W and Bothe J V (1993), The stability of ettringite, *Advances in Cement Research*, **3**, 47–63.

Byfors K (1987), Influence of silica fume and flyash on chloride diffusion and pH values in cement paste, *Cement and Concrete Research*, **17**, 115–130.

Cabrera J G and Nwaubani S O (1998), The microstructure and chloride ion diffusion characteristics of cements containing metakaolin and fly ash, in *SP-178: Sixth CANMET/ACI/JCI Conference: Fly Ash, Silica Fume, Slag & Natural Pozzolans in Concrete*, ACI, 385–400.

Cao H T, Bucea L, Ray A and Yozghatlian S (1997), The effect of cement composition and pH of environment on sulfate resistance of Portland cements and blended cements, *Cement and Concrete Composites*, **19**, 161–171.

Caquot A (1937), Le rôle des matériaux dans le béton, *Mémoires de la Société des ingénieurs civils de France*, juillet–août, 562–582.

Carles-Gibergues A (1981), *Les ajouts dans les microbétons. Influence sur l'auréole de transition et sur les propriétés mécaniques*, Ph.D. thesis, Université Paul Sabatier, Toulouse.

Carles-Gibergues A and Hornain H (2008), La durabilité des bétons face aux réactions de gonflement endogènes, *La durabilité des bétons – Bases scientifiques pour la formulation de bétons durables dans leur environnement*. Paris, Presses de l'école nationale des Ponts et Chaussées (ENPC).

Carles-Gibergues A, Cyr M, Moisson M and Ringot E (2008), A simple way to mitigate alkali-silica reaction, *Materials and Structures*, **41**, 73–83. 10.1617/s11527-006-9220-y

Cengiz Duran A (2003), Accelerated carbonation and testing of concrete made with fly ash, *Construction and Building Materials*, **17**, 147–152.

Chang Z T, Song X J, Munn R and Marosszeky M (2005), Using limestone aggregates and different cements for enhancing resistance of concrete to sulphuric acid attack, *Cement and Concrete Research*, **35**, 1486–1494. 10.1016/j.cemconres.2005.03.006

Chatterji S (1979), The role of Ca(OH)2 in the breakdown of Portland cement concrete due to alkali-silica reaction, *Cement and Concrete Research*, **9**, 185–188.

Chatterji S and Kawamura M (1992), A critical reappraisal of ion diffusion through cement based materials. Part 1: Sample preparation, measurement technique and interpretation of results, *Cement and Concrete Research*, **22**, 525–530.

Chatterji S, Thaulow N, Christensen P and Jensen A D (1983), Studies of alkali–silica reaction with special reference to prevention of damage to concrete. A preliminary study, in *6th International Conference on Alkali–Aggregate Reaction in Concrete*, 253–257.

Cheng A, Huang R, Wu J-K and Chen C-H (2005), Influence of GGBS on durability and corrosion behavior of reinforced concrete, *Materials Chemistry and Physics*, **93**, 404–411.

Cohen M D and Bentur A (1988), Durability of Portland cement-silica fume pastes in magnesium and sodium sulfate solutions, *ACI Materials Journal*, **85**, 148–157.

Collepardi M, Collepardi S, Ogoumah Olagot J J, and Simonelli F (2004), The influence of slag and fly ash on the carbonation of concretes, in *SP-221: Eighth CANMET/ACI Intl Conf on Fly Ash, Silica Fume, Slag, and Natural Pozzolans in Concrete*, ACI, 483–494.

Courard L, Darimont A, Schouterden M, Ferauche F, Willem X and Degeimbre R (2003), Durability of mortars modified with metakaolin, *Cement and Concrete Research*, **33**, 1473–1479.

Cowie J and Glasser F P (1991), The reaction between cement and natural waters containing dissolved carbon dioxide, *Advances in Cement Research*, **4**, 119–134.

Cwirzen A and Penttala V (2005), Aggregate–cement paste transition zone properties affecting the salt-frost damage of high-performance concretes, *Cement and Concrete Research*, **35**, 671–679.

De Belie N (1997), On-farm trial to determine the durability of different concrete slats for fattening pigs, *Journal of Agricultural Engineering Research*, **68**, 311–316. 10.1006/jaer.1997.0209

De Belie N, Verselder H J, Deblaere B, Vannieuwenburg D and Verschoore R (1996), Influence of the cement type on the resistance of concrete to feed acids, *Cement and Concrete Research*, **26**, 1717–1725. 10.1016/s0008-8846(96)00155-x

De Belie N, De Coster V and Van Nieuwenburg D (1997a), Use of fly ash or silica fume to increase the resistance of concrete to feed acids, *Magazine of Concrete Research*, **49**, 337–344.

De Belie N, Debruyckere M, Van Nieuwenburg D and De Blaere B (1997b), Concrete attack by feed acids: Accelerated tests to compare different concrete compositions and technologies, *ACI Materials Journal*, **94**, 546–554.

De Belie N, Debruyckere M, Vannieuwenburg D and Deblaere B (1997c), Attack of concrete floors in pig houses by feed acids: Influence of fly ash addition and cement-bound surface layers, *Journal of Agricultural Engineering Research*, **68**, 101–108. 10.1006/jaer.1997.0185

De Belie N, Lenehan J J, Braam C R, Svennerstedt B, Richardson M and Sonck B (2000), Durability of building materials and components in the agricultural environment, Part III: Concrete structures, *Journal of Agricultural Engineering Research*, **76**, 3–16.

Dhir R K and Byars E A (1993a), Pulverized fuel-ash concrete – intrinsic permeability, *ACI Materials Journal*, **90**, 571–580.

Dhir R, Hewlett P and Chan Y (1991), Near-surface characteristics of concrete: Abrasion resistance, *Materials and Structures*, **24**, 122–128. 10.1007/bf02472473

Dhir R K, Hewlett P C and Chan Y N (1989), Near surface characteristics of concrete: Intrinsic permeability, *Magazine of Concrete Research*, **41**, 87–97.

Dhir R K and Byars E A (1993b), PFA concrete: chlorides diffusion rates, *Magazine of Concrete Research*, **45**, 1–9.

Dhir R K, El-Mohr M A K and Dyer T D (1996), Chloride binding in GGBS concrete, *Cement and Concrete Research*, **26**, 1767–1773.

Dhir R K, El-Mohr M A K and Dyer T D (1997), Developing chloride resisting concrete using PFA, *Cement and Concrete Research*, **27**, 1633–1639.

Diab A M, Awad A E M, Elyamany H E, Elmoty A and Elmoty M A (2008a), Blended cement is a bad recommendation for magnesium sulfate attack, in *Microstructure Related Durability of Cementitious Composites*, Vol. 1/2, RILEM Publications, 445–454.

Diab A M, Awad A E M, Elyamany H E and Elmoty A (2008b), Magnesium sulfate resistance of silica fume concrete specimens and R.C. columns, in *Microstructure Related Durability of Cementitious Composites*, Vol. 1/2, RILEM Publications, 859–868.

Diamond S (1986), Chloride concentrations in concrete pore solutions resulting from calcium and sodium chloride admixtures, *Cement, Concrete and Aggregates*, **8**, 97–102.

Diamond S (1989), ASR – Another look at mechanisms, in *8th International Conference on Alkali–Aggregate Reaction in Concrete*, 83–94.

Diamond S, Sahu S and Thaulow N (2004), Reaction products of densified silica fume agglomerates in concrete, *Cement and Concrete Research*, **34**, 1625–1632.

Duchesne J and Bérubé M A (1994a), The effectiveness of supplementary cementing materials in suppressing expansion due to ASR: Another look at the reaction mechanisms part 1: Concrete expansion and portlandite depletion, *Cement and Concrete Research*, **24**, 73–82.

Duchesne J and Bérubé M A (1994b), The effectiveness of supplementary cementing materials in suppressing expansion due to ASR: Another look at the reaction mechanisms part 2: Pore solution chemistry, *Cement and Concrete Research*, **24**, 221–230.

Duchesne J and Bérubé M A (2001), Long-term effectiveness of supplementary cementing materials against alkali-silica reaction, *Cement and Concrete Research*, **31**, 1057–1063.

Dunster A M (1989), An investigation of the carbonation of cement paste using trimethylsilylation, *Advances in Cement Research*, **2**, 99–106.

Durning T A and Hicks M C (1991), Using microsilica to increase concrete's resistance to aggressive chemicals, *Concrete International*, **13**, 42–48.

Elahi A, Basheer P A M, Nanukuttan S V and Khan Q U Z (2010), Mechanical and durability properties of high performance concretes containing supplementary cementitious materials, *Construction and Building Materials*, **24**, 292–299.

Escadeillas G and Hornain H (2008), La durabilité des bétons vis-a-vis des environnements chimiquement agressifs, *La durabilité des bétons – Bases scientifiques pour la formulation de bétons durables dans leur environnement*, 2nd ed., Presses de l'école nationale des Ponts et Chaussées (ENPC).

Escadeillas G, Massias, E. and Carles-Gibergues, A. (2000), Case study of alkali–silica reaction due to silica-fume pellets in a mortar repair, in *11th International Conference on Alkali-Aggregate Reaction in Concrete*, 841–849.

Feldman R F (1983), Significance of porosity measurements on blended cement performance, in *SP-79: Fly Ash, Silica Fume, Slag and Other Mineral By-Products in Concrete*, ACI, 415–434.

Feldman R F (1984), Pore structure damage in blended cements caused by mercury intrusion, *Journal of the American Ceramic Society*, **67**, 30–33. 10.1111/j.1151-2916.1984.tb19142.x

Fernandez L and Malhotra M V (1990), Mechanical properties, abrasian resistance, and chloride permeability of concrete incorporating granulated blast-furnace slag, *Cement, Concrete and Aggregates*, **12**, 87–100.

Fournier B and Bérubé M A (2000), Alkali–aggregate reaction in concrete: a review of basic concepts and engineering implications, *Canadian Journal of Civil Engineering*, **27**, 167–191. doi:10.1139/l99-072

Frearson J P H (1986), Sulfate resistance of combinations of Portland cement and ground granulated blast furnace slag, in *SP-91: Fly Ash, Silica Fume, Slag, and Natural Pozzolans in Concrete*, ACI, 1495–1524.

Frearson J P H and Higgins D D (1992), Sulfate resistance of mortars containing ground granulated blast-furnace slag with variable alumina content, in *SP-132: Fly Ash, Silica Fume, Slag, & Natural Pozzolans & Natural Pozzolans in Conc: Proc 4th Intl Conf*, ACI, 1525–1542.

Frias M and Cabrera J (2000), Pore size distribution and degree of hydration of metakaolin–cement pastes, *Cement and Concrete Research*, **30**, 561–569.

Fu Y (1996), *Delayed ettringite formation in Portland cement products*, Ph.D. thesis, University of Ottawa, Ottawa, 220p.

Fu Y and Beaudoin J J (1996), Microcracking as a precursor to delayed ettringite formation in cement systems, *Cement and Concrete Research*, **26**, 1493–1498.

Fuller W B and Thompson S E (1907), The laws of proportioning concrete, *Transactions of the American Society of Civil Engineers*, **59**, 67–143.

Gagné R and Linger L (2008), La durabilité des bétons en ambiance hivernale rigoureuse, *La durabilité des bétons – Bases scientifiques pour la formulation de bétons durables dans leur environnement*, 2nd ed. Paris, Presses de l'école nationale des Ponts et Chaussées (ENPC).

Gagné R, Pigeon M and Aïtcin P-C (1990), Durabilité au gel des bétons de hautes performances mécaniques, *Materials and Structures*, **23**, 103–109. 10.1007/bf02472569

Ganjian E and Pouya H S (2005), Effect of magnesium and sulfate ions on durability of silica fume blended mixes exposed to the seawater tidal zone, *Cement and Concrete Research*, **35**, 1332–1343.

Gao J M, Qian C X, Liu H F, Wang B and Li L (2005), ITZ microstructure of concrete containing GGBS, *Cement and Concrete Research*, **35**, 1299–1304.

Garboczi E J and Bentz D P (1996), Modelling of the microstructure and transport properties of concrete, *Construction and Building Materials*, **10**, 293–300.

Gautefall O (1986), Effect of condensed silica fume on the diffusion of chlorides through hardened cement paste, in *SP-91: Fly Ash, Silica Fume, Slag, and Natural Pozzolans in Concrete*, ACI, 991–998.

Gautefall O and Havdahl J (1989), Effect of condensed silica fume on the mechanism of chloride diffusion into hardened cement paste, in *SP-114: Fly Ash, Silica Fume, Slag, & Natural Pozzolans in Conc: Proc 3rd Intl Conf*, ACI, 849–860.

Gebler S and Klieger P (1983), Effect of fly-ash on the air-void stability of concrete, *Journal of the American Concrete Institute*, **80**, 341–341.

Gebler S H and Klieger P (1986a), Effect of fly ash on physical properties of concrete, in *SP-91: Fly Ash, Silica Fume, Slag, and Natural Pozzolans in Concrete*, ACI, 1–50.

Gebler S H and Klieger P (1986b), Effect of fly ash on the durability of air-entrained concrete, in *SP-91: Fly Ash, Silica Fume, Slag, and Natural Pozzolans in Concrete*, ACI, 483–520.

Ghorab H Y, Heinz D, Ludwig U, Meskendahl T and Welter A (1980), On the stability of calcium aluminate sulphate hydrates in pure systems and in cements, in *7th International Congress on the Chemistry of Cements*, IV, 496–503.

Glasser F P (1992), Chemistry of alkali–aggregate reaction, in Swamy, R. N. (Ed.) *The Alkali–Silica Reaction in Concrete*. London, Blackie and Son Ltd.

Glasser F P, Damidot D and Atkins M (1995), Phase development in cement in relation to the secondary ettringite problem, *Advances in Cement Research*, **7**, 57–68.

Gonen T and Yazicioglu S (2007), The influence of mineral admixtures on the short and long-term performance of concrete, *Building and Environment*, **42**, 3080–3085.

Grabowski E, Czarnecki B, Gillott J E, Duggan C R and Scott J F (1992), Rapid test of concrete expansivity due to internal sulfate attack, *ACI Materials Journal*, **89**, 469–480.

Gruyaert E, Van Den Heede P, Maes M and De Belie N (2010), A comparative study of the durability of ordinary Portland cement concrete and concrete containing (high) percentages of blast-furnace slag, in Brameshuber, W. (Ed.) *International RILEM Conference on Material Science*. Bagneux, RILEM Publications.

Gruyaert E, Van Den Heede P, Maes M and De Belie N (2012), Investigation of the influence of blast-furnace slag on the resistance of concrete against organic acid or sulphate attack by means of accelerated degradation tests, *Cement and Concrete Research*, **42**, 173–185.

Güneyisi E, Gesoğlu M and Mermerdaş K (2008), Improving strength, drying shrinkage, and pore structure of concrete using metakaolin, *Materials and Structures*, **41**, 937–949. 10.1617/s11527-007-9296-z

Hammer T A and Sellevold E J (1990), Frost resistance of high-strength concrete, in *SP-121: High-Strength Concrete: Second International Symposium*, ACI, 457–487.

Hausmann D A (1967), Steel corrosion in concrete – How does it occur?, *Materials Protection*, **6**, 19–23.

Heinz D and Ludwig U (1987), Mechanism of secondary ettringite formation in mortars and concretes subjected to heat treatment, in *SP-100: Concrete Durability: Proceedings of Katharine and Bryant Mather International Symposium*, ACI, 2059–2072.

Heinz D, Kalde M, Ludwig U and Ruediger I (1999), Present state of investigation on damaging late ettringite formation (DLEF) in Mortars and Concretes, in *SP-177: Ettringite, the Sometimes Host of Destruction*, ACI, 1–14.

Hekal E E, Kishar E and Mostafa H (2002), Magnesium sulfate attack on hardened blended cement pastes under different circumstances, *Cement and Concrete Research*, **32**, 1421–1427. 10.1016/s0008-8846(02)00801-3

Helmuth R H (1960), Capillary size restrictions on ice formation in hardened Portland cement pastes, in *4th International Symposium on the Chemistry of Cement*, 855–869.

Hill J, Byars E A, Sharp J H, Lynsdale C J, Cripps J C and Zhou Q (2003), An experimental study of combined acid and sulfate attack of concrete, *Cement and Concrete Composites*, **25**, 997–1003. 10.1016/s0958-9465(03)00123-9

Hilsdorf H K (1995), Concrete compressive strength, transport characteristics and durability, in Kropp, J. & Hilsdorf, H. K. (Eds.) *Performance Criteria for Concrete Durability – RILEM Report 12*. London, E & FN Spon.

Hobbs D W (1988), Carbonation of concrete containing PFA, *Magazine of Concrete Research*, **40**, 69–78.

Hobbs D W and Matthews J (1998), Minimum requirements for concrete to resist chemical attack, in Hobbs, D. W. (Ed.) *Minimum requirements for durable concrete Carbonation- and chloride-induced corrosion, freeze-thaw attack and chemical attack*. Crowthorne, United Kingdom, British Cement Association.

Hogan F (1981), Evaluation for durability and strength development of a ground granulated blast furnace slag, *Cement, Concrete and Aggregates*, **3**, 40–52.

Hong S-Y and Glasser F P (1999), Alkali binding in cement pastes: Part I. The C-S-H phase, *Cement and Concrete Research*, **29**, 1893–1903.

Hooton R D (1986), *Blended Cement*, American Society for Testing and Materials.

Hooton R D (1993), Influence of silica fume replacement of cement on physical properties and resistance to sulfate attack, freezing and thawing, and alkali–silica reactivity, *ACI Materials Journal*, **90**, 143–151.

Idir R, Cyr M and Tagnit-Hamou A (2010), Use of fine glass as ASR inhibitor in glass aggregate mortars, *Construction and Building Materials*, **24**, 1309–1312.

Jahren P A (1989), Fire resistance of high strength/dense concrete with particular reference to the use of condensed silica fume – A review, in *SP-114: Fly Ash, Silica Fume, Slag, & Natural Pozzolans in Conc: Proc 3rd Intl Conf*, ACI, 1013–1049.

Jensen H and Pratt P L (1989), The binding of chloride ions by pozzolanic product in fly ash cement blends, *Advances in Cement Research*, **2**, 121–129.

Juenger M C G and Ostertag C P (2004), Alkali–silica reactivity of large silica fume-derived particles, *Cement and Concrete Research*, **34**, 1389–1402.

Kalousek G L, Porter L C and Benton E J (1972), Concrete for long-time service in sulfate environment, *Cement and Concrete Research*, **2**, 79–89.

Kasai Y, Matsui I, Fukushima Y and Kamohara H (1983), Air permeability and carbonation of blended cement mortars, in *SP-79: Fly Ash, Silica Fume, Slag and Other Mineral By-Products in Concrete*, ACI, 435–452.

Kelham S (1999), Influence of cement composition on volume stability of mortar, in *SP-177: Ettringite, the Sometimes Host of Destruction*, ACI, 27–46.

Kessler R J, Powers R G, Vivas E, Paredes M A and Virmani Y P (2008), Surface resistivity as an indicator of concrete chloride penetration resistance, in *2008 Concrete Bridge Conference*, Portland Cement Association, 18p.

Khan M I (2003), Isoresponses for strength, permeability and porosity of high performance mortar, *Building and Environment*, **38**, 1051–1056.

Khan M I and Lynsdale C J (2002), Strength, permeability, and carbonation of high-performance concrete, *Cement and Concrete Research*, **32**, 123–131.

Khatib J M and Wild S (1996), Pore size distribution of metakaolin paste, *Cement and Concrete Research*, **26**, 1545–1553.

Khatib J M and Wild S (1998), Sulphate resistance of metakaolin mortar, *Cement and Concrete Research*, **28**, 83–92.

Khayat K H and Aitcin P C (1992), Silica fume in concrete – An overview, *SP-132: Fly Ash, Silica Fume, Slag, & Natural Pozzolans & Natural Pozzolans in Conc: Proc 4th Intl Conf*, ACI, 835–872.

Kim H-S, Lee S-H and Moon H-Y (2007), Strength properties and durability aspects of high strength concrete using Korean metakaolin, *Construction and Building Materials*, **21**, 1229–1237.

Kokubu M and Nagataki S (1989), Carbonation of concrete with fly ash and corrosion of reinforcements in 20 year tests, in *SP-114: Fly Ash, Silica Fume, Slag, & Natural Pozzolans in Conc: Proc 3rd Intl Conf*, ACI, 315–330.

Kumar A and Roy B D M (1986), Retardation of Cs^+ and Cl^- diffusion using blended cement admixtures, *Journal of the American Ceramic Society*, **69**, 356–360. 10.1111/j.1151-2916.1986.tb04747.x

Kurdowski W and Duszak S (2004), Can addition of limestone eliminate the expansion of the mortars due to DEF?, in *International RILEM Workshop on Internal Sulfate Attack and Delayed Ettringite Formation*, RILEM Publications SARL, 229–235.

Langan B W, Joshi R C and Ward M A (1990), Strength and durability of concretes containing 50% Portland cement replacement by fly ash and other materials, *Canadian Journal of Civil Engineering*, **17**, 19–27. doi:10.1139/l90–004

Langley W S, Carette G G and Malhotra V M (1989), Structural concrete incorporating high volumes of ASTM Class F fly ash, *ACI Materials Journal*, **86**, 507–514.

Laplante P, Aitcin P C and Vezina D (1991), Abrasion resistance of concrete, *Journal of Materials in Civil Engineering*, **3**, 19–28.

Larbi J A, Fraay A L A and Bijen J M J M (1990), The chemistry of the pore fluid of silica fume-blended cement systems, *Cement and Concrete Research*, **20**, 506–516.

Larreur-Cayol S, Bertron A, San Nicolas R and Escadeillas G (2011), Durability of different binders in synthetic agricultural effluents, in *7th International Symposium on Cement Based Materials for a Sustainable Agriculture*, 11p.

Lawrence C D (1998), Physicochemical and mechanical properties of Portland cements, in Hewlett, P. C. (Ed.) *Lea's Chemistry of Cement and Concrete*, 4th ed. London, Arnold Publishers.

LCPC (2007), *Recommandations pour la prévention des désordres dus à la réaction sulfatique interne des bétons*, Paris, LCPC.

Lee S T, Moon H Y, Hooton R D and Kim J P (2005), Effect of solution concentrations and replacement levels of metakaolin on the resistance of mortars exposed to magnesium sulfate solutions, *Cement and Concrete Research*, **35**, 1314–1323.

Litvan G G (1972), Phase transitions of adsorbates: IV, Mechanism of frost action in hardened cement paste, *Journal of the American Ceramic Society*, **55**, 38–42.

Liu Y-W (2007), Improving the abrasion resistance of hydraulic-concrete containing surface crack by adding silica fume, *Construction and Building Materials*, **21**, 972–977.

Lothenbach B, Scrivener K and Hooton R (2011), Supplementary cementitious materials, *Cement and Concrete Research*, **41**, 1244–1256.

Luping T and Nilsson L O (1992), Chloride diffusivity in high strength concrete at different ages, *Nordic Concrete Research*, **11**, 162–171.

Maas A J, Ideker J H and Juenger M C G (2007), Alkali silica reactivity of agglomerated silica fume, *Cement and Concrete Research*, **37**, 166–174.

Malhotra V M (1983), Strength and durability characteristics of concrete incorporating a pelletized blast furnace slag, in *SP-79: Fly Ash, Silica Fume, Slag and Other Mineral By-products in Concrete*, ACI, 891–922.

Malhotra V M, Zhang M H, Read P H and Ryell J (2000), Long-term mechanical properties and durability characteristics of high-strength/high-performance concrete incorporating supplementary cementing materials under outdoor exposure conditions, *ACI Materials Journal*, **97**, 518–525.

Malolepszy J and Pytel Z (2000), Effect of metakaolinite on strength and chemical resistance of cement mortars, in *SP-192: 2000 Canmet/ACI Conference on Durability of Concrete*, ACI, 189–204.

Manmohan D and Mehta P K (1981), Influence of pozzolanic, slag, and chemical admixtures on pore size distribution and permeability of hardened cement pastes, *Cement, Concrete and Aggregates*, **3**, 63–67.

Marusin S L and Shotwell L B (2000), Alkali–silica reaction in concrete caused by densified silica fume lumps: a case study, *Cement, Concrete and Aggregates*, **22**, 90–94.

Maso J-C (1980), The bond between aggregates and hydrated cement paste, Principal report Sub-theme VII-1, in *7th International Congress on the Chemistry of Cements*, Vol I, Theme VII-1/3.

Massazza F (1998), Pozzolana and pozzolanic cements, in Hewlett, P. C. (Ed.) *Lea's Chemistry of Cement and Concrete*, 4th ed. London, Arnold Publishers.

Matthews J D (1984), Carbonation of ten-year old concretes with and without pulverized-fuel ash, in *AshTech'84, 2nd International Conference on Ash Technology and Marketing*, 398A

McNally C and Sheils E (2012), Probability-based assessment of the durability characteristics of concretes manufactured using CEM II and GGBS binders, *Construction and Building Materials* **30**, 22–29.

McPolin D O, Basheer P A M, Long A E, Grattan K T V and Sun T (2007), New test method to obtain pH profiles due to carbonation of concretes containing supplementary cementitious materials, *Journal of Materials in Civil Engineering*, **19**, 936–946.

Mehta P K and Manmohan D (1980), Pore size distribution and permeability of hardened cement paste, in *7th International Congress on the Chemistry of Cements*, **3**, 7.1, 1–11.

Mehta P K and Monteiro P J M (2005), *Concrete – Microstructure, Properties, and Materials*,New York, McGraw-Hill Professional.

Meyer A (1968), Investigations on the carbonation of concrete, in *5th International Symposium on the Chemistry of Cement*, **3**, 394–401.

Moon H Y, Lee S T and Kim S S (2003), Sulphate resistance of silica fume blended mortars exposed to various sulphate solutions, *Canadian Journal of Civil Engineering*, **30**, 625–636. 10.1139/l03–024

Nagataki S and Ujike I (1986), Air permeability of concretes mixed with fly ash and condensed silica fume, in *SP-91: Fly Ash, Silica Fume, Slag, and Natural Pozzolans in Concrete*, ACI, 1049–1068.

Naik T R, Singh S S and Hossain M M (1994), Permeability of concrete containing large amounts of fly ash, *Cement and Concrete Research*, **24**, 913–922.

Naik T R, Singh S S and Hossain M M (1995a), Properties of high performance concrete systems incorporating large amounts of high-lime fly ash, *Construction and Building Materials*, **9**, 195–204.

Naik T R, Singh, S S and Hossain M M (1995b), Abrasion resistance of high-strength concrete made with Class C fly ash, *ACI Materials Journal*, **92**, 649–659.

Neville A M (1995), *Properties of concrete*, Harlow, England, Pearson Education Limited.

Ngala V T and Page C L (1997), Effects of carbonation on pore structure and diffusional properties of hydrated cement pastes, *Cement and Concrete Research*, **27**, 995–1007.

Nixon P J and Page C L (1987), Pore solution chemistry and alkali aggregate reaction, in *SP-100: Concrete Durability: Proceedings of Katharine and Bryant Mather International Symposium*, 1833–1862.

O'Connell M, McNally C and Richardson M G (2010), Biochemical attack on concrete in wastewater applications: A state of the art review, *Cement and Concrete Composites*, **32**, 479–485.

O'Connell M, McNally C and Richardson M G (2012), Performance of concrete incorporating GGBS in aggressive wastewater environments, *Construction and Building Materials*, **27**, 368–374.

Odler I and Chen Y (1995), Effect of cement composition on the expansion of heat-cured cement pastes, *Cement and Concrete Research*, **25**, 853–862.

Odler I and Chen Y (1996), On the delayed expansion of heat cured Portland cement pastes and concretes, *Cement and Concrete Composites*, **18**, 181–185.

O'Donnell C, Dodd V A, O'Kiely P and Richardson M (1995), A study of the effects of silage effluent on concrete: Part 1, Significance of concrete characteristics, *Journal of Agricultural Engineering Research*, **60**, 83–92.

Oliveira L, Jalali S, Fernandes J and Torres E (2005), L'emploi de métakaolin dans la production de béton écologiquement efficace, *Materials and Structures*, **38**, 403–410. 10.1007/bf02479308

Ollivier J P and Torrenti J M (2008), La structure poreuse des bétons et les propriétés de transfert, *La durabilité des bétons – Bases scientifiques pour la formulation de bétons durables dans leur environnement*, 2nd ed. Paris, Presses de l'école nationale des Ponts et Chaussées (ENPC).

Ollivier J P, Maso J C and Bourdette B (1995), Interfacial transition zone in concrete, *Advanced Cement Based Materials*, **2**, 30–38.

Osborne G J (1991), The sulphate resistance of Portland and blast furnace slag cement concretes, in *SP-126: Durability of Concrete: Second International Conference*, ACI, 1047–1072.

Page C and Vennesland Ø (1983), Pore solution composition and chloride binding capacity of silica-fume cement pastes, *Materials and Structures*, **16**, 19–25. 10.1007/bf02474863

Page C L, Short N R and El Tarras A (1981), Diffusion of chloride ions in hardened cement pastes, *Cement and Concrete Research*, **11**, 395–406.

Pavia S and Condren E (2008), Study of the durability of OPC versus GGBS concrete on exposure to silage effluent, *Journal of Materials in Civil Engineering*, **20**, 313–320. 10.1061/(asce)0899-1561(2008)20:4(313).

Pepper L and Mather B (1959), Effectiveness of mineral admixtures in preventing excessive expansion of concrete due to alkali silica reaction, proceedings of ASTM V, **59**, 1178–1203.

Pettersson K (1992), Effects of silica fume on alkali-silica expansion in mortar specimens, *Cement and Concrete Research*, **22**, 15–22.

Pigeon M, Plante P and Plante M (1989), Air-void stability. 1. Influence of silica fume and other parameters, *ACI Materials Journal*, **86**, 482–490.

Pigeon M, Gagné R, Aitcin P-C and Banthia N (1991), Freezing and thawing tests of high-strength concretes, *Cement and Concrete Research*, **21**, 844–852.

Pigeon M, Talbot C, Marchand J and Hornain H (1996), Surface microstructure and scaling resistance of concrete, *Cement and Concrete Research*, **26**, 1555–1566.

Pihlajavaara S (1968), Some results of the effect of carbonation on the porosity and pore size distribution of cement paste, *Materials and Structures*, **1**, 521–527. 10.1007/bf02473640

Poon C S, Lam L and Wong Y L (1999), Effects of fly ash and silica fume on interfacial porosity of concrete, *Journal of Materials in Civil Engineering*, **11**, 197–205. 10.1061/(asce)0899-1561(1999)11:3(197)

Poon C S, Lam L, Kou S C, Wong Y L and Wong R (2001), Rate of pozzolanic reaction of metakaolin in high-performance cement pastes, *Cement and Concrete Research*, **31**, 1301–1306.

Poon C S, Kou S C and Lam L (2006), Compressive strength, chloride diffusivity and pore structure of high performance metakaolin and silica fume concrete, *Construction and Building Materials*, **20**, 858–865.

Powers T C (1954), Void spacing as a basis for producing air-entrained concrete, *Journal of the American Concrete Institute*, **50**, 741–760.

Powers T C (1975), Freezing effects in concrete, in *SP-47: Durability of Concrete*, ACI, 1–11.

Powers T C and Helmuth R A (1956), Theory of volume changes in hardened Portland cement paste during freezing, *Proceedings of the Highway Research Board*, **32**, 285–297.

Ramlochan T and Thomas M (2000), Effect of metakaolin on external sulfate attack, in *SP-192: 2000 Canmet/ACI Conference on Durability of Concrete*, ACI, 239–252.

Ramlochan T, Zacarias P, Thomas M D A and Hooton R D (2003), The effect of pozzolans and slag on the expansion of mortars cured at elevated temperature: Part I: Expansive behaviour, *Cement and Concrete Research*, **33**, 807–814.

Ramlochan T, Thomas M D A and Hooton R D (2004), The effect of pozzolans and slag on the expansion of mortars cured at elevated temperature: Part II: Microstructural and microchemical investigations, *Cement and Concrete Research*, **34**, 1341–1356.

Roy D M, Kumar A and Rhodes J P (1986), Diffusion of chloride and cesium ions in Portland cement pastes and mortars containing blast furnace slag and fly ash, in *SP-91: Fly Ash, Silica Fume, Slag, and Natural Pozzolans in Concrete*, ACI, 1423–1444.

Roy D M, Arjunan P and Silsbee M R (2001), Effect of silica fume, metakaolin, and low-calcium fly ash on chemical resistance of concrete, *Cement and Concrete Research*, **31**, 1809–1813.

Ryou J S and Ann K Y (2008), Variation in the chloride threshold level for steel corrosion in concrete arising from different chloride sources, *Magazine of Concrete Research*, **60**, 177–187. 10.1680/macr.2008.60.3.177

Sabir B B (1997), Mechanical properties and frost resistance of silica fume concrete, *Cement and Concrete Composites*, **19**, 285–294. 10.1016/s0958-9465(97)00020-6

Sabir B B and Kouyiali K (1991), Freeze–thaw durability of air-entrained CSF concrete, *Cement and Concrete Composites*, **13**, 203–208.

Saetta A V, Schrefler B A and Vitaliani R V (1995), 2-D model for carbonation and moisture/heat flow in porous materials, *Cement and Concrete Research*, **25**, 1703–1712.

Sakai K, Watanabe H, Suzuki M, Hamazaki K (1992), Properties of granulated blast-furnace slag cement concrete, in *SP-132: Fly Ash, Silica Fume, Slag, & Natural Pozzolans & Natural Pozzolans in Conc: Proc 4th Intl Conf*, ACI, 1367–1384.

San Nicolas R (2011), *Performance-based approach for concrete containing metakaolin obtained by flash calcination* (in French), Ph.D. thesis, Université de Toulouse, Toulouse, 338p.

Santos Silva A, Bettencourt Ribeiro A, Jalali S and Divet L (2006), The use of fly ash and metakaolin for the prevention of alkali-silica reaction and delayed ettringite formation in concrete, in *International RILEM Workshop on Performance Based Evaluation and Indicators for Concrete Durability*.

Santos Silva A, Soares D, Matos L, Salta M M, Divet L, Pavoine A, Candeias A E and Mirao J (2010), Influence of mineral additions in the inhibition of delayed ettringite formation in cement based materials – A microstructural characterization, *Materials Science Forum, Advanced Materials Forum V*, 1272–1279.

Saric-Coric M and Aitcin P-C (2003), Bétons à haute performance à base de ciments composés contenant du laitier et de la fumée de silice, *Canadian Journal of Civil Engineering*, **30**, 414–428. doi:10.1139/l03-005

Scrivener K L, Bentur A and Pratt P L (1988), Quantitative characterization of the transition zone in high-strength concrete, *Advances in Cement Research*, **1**, 230–237.

Shafiq N and Cabrera J G (2004), Effects of initial curing condition on the fluid transport properties in OPC and fly ash blended cement concrete, *Cement and Concrete Composites*, **26**, 381–387.

Shashiprakash S G and Thomas M D A (2001), Sulfate resistance of mortars containing high-calcium fly ashes and combinations of highly reactive pozzolans and fly ash, in *SP-199: Seventh CANMET/ACI Intl Conf on Fly Ash, Silica Fume, Slag and Natural Pozzolans in Concrete*, ACI, 221–238.

Shekarchi M, Bonakdar A, Bakhshi M, Mirdamadi A and Mobasher B (2010), Transport properties in metakaolin blended concrete, *Construction and Building Materials*, **24**, 2217–2223.

Shi H-S, Xu B-W and Zhou X-C (2009), Influence of mineral admixtures on compressive strength, gas permeability and carbonation of high performance concrete, *Construction and Building Materials*, **23**, 1980–1985.

Siddique R (2003), Effect of fine aggregate replacement with Class F fly ash on the abrasion resistance of concrete, *Cement and Concrete Research*, **33**, 1877–1881.

Siddique R (2010), Wear resistance of high-volume fly ash concrete, *Leonardo Journal of Sciences*, **9**, 21–36.

Siddique R and Khatib J (2010), Abrasion resistance and mechanical properties of high-volume fly ash concrete, *Materials and Structures*, **43**, 709–718. 10.1617/s11527-009-9523-x

Sideris K K, Savva A E and Papayianni J (2006), Sulfate resistance and carbonation of plain and blended cements, *Cement and Concrete Composites*, **28**, 47–56.

Skjolsvold O (1986), Carbonation depths of concrete with and without condensed silica fume, in *SP-91: Fly Ash, Silica Fume, Slag, and Natural Pozzolans in Concrete*, ACI, 1031–1048.

Stanton T E (1959), Studies of use of pozzolanas for counteracting excessive concrete expansion resulting from reaction between aggregate and the alkalis in cement, in *Symposium on Pozzolanic Materials in Mortars and Concretes*, ASTM, STP 99, 178–203.

Stark J and Ludwig H-M (1997), Freeze–thaw and freeze-deicing salt resistance of concretes containing cement rich in granulated blast furnace slag, *ACI Materials Journal*, **94**, 47–55.

Sulapha P, Wong S F, Wee T H and Swaddiwudhipong S (2003), Carbonation of concrete containing mineral admixtures, *Journal of Materials in Civil Engineering*, **15**, 134–143. 10.1061/(asce)0899-1561(2003)15:2(134)

Swamy R N (1992), Role and effectiveness of mineral admixtures in relation to alkali–silica reaction, in Swamy, R. N. (Ed.) *The Alkali–Silica Reaction in Concrete*. London, Blackie and Son Ltd.

Taylor H F W (1987), A method for predicting alkali ion concentrations in cement pore solutions, *Advances in Cement Research*, **1**, 5–16.

Taylor H F W (1998), *Cement Chemistry*, Thomas Telford Publishing.

Taylor H F W, Famy C and Scrivener K L (2001), Delayed ettringite formation, *Cement and Concrete Research*, **31**, 683–693.

Thiery M, Villain G and Platret G (2003), Effect of carbonation on density, microstructure and liquid water saturation of concrete, in *Advances in Cement and Concrete IX*, 481–490.

Thomas M D A (2011), The effect of supplementary cementing materials on alkali–silica reaction: A review, *Cement and Concrete Research*, **41**, 1224–1231.

Thomas M D A and Folliard K (2007), Concrete aggregates and the durability of concrete, in Page, C. L. & Page, M. M. (Eds.) *Durability of concrete and cement composites*. Cambridge, Woodhead Publishing Limited and Maney Publishing Limited.

Thomas M D A and Matthews J D (1992a), The permeability of fly ash concrete, *Materials and Structures*, **25**, 388–396. 10.1007/bf02472254

Thomas M D A and Matthews J D (1992b), Carbonation of fly-ash concrete, *Magazine of Concrete Research*, **44**, 217–228.

Thomas M D A, Folliard K, Drimalas T and Ramlochan T (2008a), Diagnosing delayed ettringite formation in concrete structures, *Cement and Concrete Research*, **38**, 841–847.

Thomas M D A, Scott A, Bremner T, Bilodeau A and Day D (2008b), Performance of slag concrete in marine environment, *ACI Materials Journal*, **105**, 628–634.

Torii K and Kawamura M (1994), Effects of fly ash and silica fume on the resistance of mortar to sulfuric acid and sulfate attack, *Cement and Concrete Research*, **24**, 361–370.

Turriziani R (1986), Internal degradation of concrete: alkali–aggregate reaction, reinforcement steel corrosion, in *8th International Congress on the Chemistry of Cement*, 388–437.

Uchikawa H (1986), Blended cements – Effect of blending components on hydration and structure formation, in *8th International Congress on the Chemistry of Cement*, **1**, 250–280.

Van Balen K and Van Gemert D (1994), Modelling lime mortar carbonation, *Materials and Structures*, **27**, 393–398. 10.1007/bf02473442

Vejmelková E, Pavlíková M, Keppert M, Kersner Z, Rovnaníková P, Ondrácek M, Sedlmajer M and Cerný R (2010), High performance concrete with Czech metakaolin: Experimental analysis of strength, toughness and durability characteristics, *Construction and Building Materials*, **24**, 1404–1411.

Venuat M and Alexandre J (1968, 1969), De la carbonatation du béton, *Revue des Matériaux de Construction*, 421–427 and 5–15.

Verbeck G J and Helmuth R H (1968), Structures and physical properties of hardened cement paste, in *5th International Symposium on the Chemistry of Cement*, **3**, 7–1, 1–11.

Villain G and Thiery M (2005), Impact of carbonation on microstructure and transport properties of concrete, in *10DBMC International Conference on Durability of Building Materials and Components*, 8 p.

Wee T H, Suryavanshi A K, Wong S F and Rahman A (2000), Sulfate resistance of concrete containing mineral admixtures, *ACI Materials Journal*, **97**, 536–549.

Winslow D N, Cohen M D, Bentz D P, Snyder K A and Garboczi E J (1994), Percolation and pore structure in mortars and concrete, *Cement and Concrete Research*, **24**, 25–37.

Xu A, Sarkar S and Nilsson L (1993), Effect of fly ash on the microstructure of cement mortar, *Materials and Structures*, **26**, 414–424. 10.1007/bf02472942

Yamato T and Emoto Y (1989), Chemical resistance of concrete containing condensed silica fume, in *SP-114: Fly Ash, Silica Fume, Slag, & Natural Pozzolans in Conc: Proc 3rd Intl Conf*, ACI, 897–914.

Yazici S and Inan G (2006), An investigation on the wear resistance of high strength concretes, *Wear*, **260**, 615–618.

Yen T, Hsu T-H, Liu Y-W and Chen S-H (2007), Influence of class F fly ash on the abrasion-erosion resistance of high-strength concrete, *Construction and Building Materials*, **21**, 458–463.

Yetgin S and Cavdar, A. (2011), Abrasion resistance of cement mortar with different pozzolanic compositions and matrices, *Journal of Materials in Civil Engineering*, **23**, 138–145.

Yogendran V, Langan B W and Ward M A (1987), Utilization of silica fume in high strength concrete, Proceedings of the symposium of *Utilization of High Strength Concrete*, stavanger, Norway, 1987: 85–97.

Zadeh M S, Debicki G, Clastres P and Billard Y (1998), Influence of silica fume on permeability of concrete to oxygen for temperatures up to 500 °C, in *SP-178: Sixth CANMET/ACI/JCI Conference: Fly Ash, Silica Fume, Slag & Natural Pozzolans in Concrete*, ACI, 975–996.

Zeljkovic M (2009), *Metakaolin effects on concrete durability*, Ph.D. thesis, University of Toronto, Toronto, 172p.

Zhang D S (1996), Air entrainment in fresh concrete with PFA, *Cement and Concrete Composites*, **18**, 409–416.

Zhang M H, Bilodeau A, Shen G and Malhotra V H (1998), De-icing salt scaling of concrete incorporating different types and percentages of fly ashes, in *SP-178: Sixth CANMET/ACI/JCI Conference: Fly Ash, Silica Fume, Slag & Natural Pozzolans in Concrete*, ACI, 493–526.

Zhang M H, Bouzoubaa N and Malhotra V M (1999), Resistance of silica fume concrete to deicing salt scaling – A review, *in SP-172: High-Performance Concrete*, ACI, 67–102.

Zhang M H and Malhotra V M (1995), Characteristics of a thermally activated alumino-silicate pozzolanic material and its use in concrete, *Cement and Concrete Research*, **25**, 1713–1725.

Zhang Z, Olek J and Diamond S (2002), Studies on delayed ettringite formation in heat-cured mortars: II. Characteristics of cement that may be susceptible to DEF, *Cement and Concrete Research*, **32**, 1737–1742.

9
Performance of self-compacting concrete (SCC) with high-volume supplementary cementitious materials (SCMs)

E. GÜNEYISI, M. GESOĞLU and Z. ALGIN,
Gaziantep University, Turkey

DOI: 10.1533/9780857098993.2.198

Abstract: Since ordinary self-compacting concrete (SCC) is usually associated with high material costs due to its high binder and chemical admixture content, researchers are currently attempting to produce SCC with high volumes of supplementary cementitious materials (SCMs), to make it cost effective and more durable. The SCMs used in this type of SCC, namely fly ash (FA), ground-granulated blast furnace slag (GGBFS), silica fume (SF), metakaolin (MK) etc., are generally industrial by-products and waste materials. A subsidiary effect of this SCC composition is to conserve the environment and non-renewable natural material resources. This chapter reviews the current literature on fresh properties of SCC containing high-volume SCMs, and explains the effect of high-volume SCMs on the mechanical and durability-related properties of SCC.

Key words: durability, fresh properties, self-compacting concrete (SCC), supplementary cementitious materials (SCMs), strength.

9.1 Introduction

Self-compacting concrete (SCC), also known as self-consolidating concrete, is one of the most widely used concrete types, mainly because of its self-compacting characteristics and strength. SCC is a highly flowable, non-segregating, special concrete type that can settle into formworks, and encapsulates, heavily reinforced, narrow and deep sections by means of its own weight. Unlike conventional concrete, SCC does not require compaction using external force from mechanical equipment such as immersion vibrators. In addition to these attractive benefits, as a high-performance concrete SCC maintains all of concrete's common mechanical and durability characteristics. SCC was developed in Japan in 1980s in order to achieve high-performance durable concrete structures, and with advancements in concrete technology its use has become widespread all over the world (Ozawa et al., 1989; Okamura and Ouchi, 2003).

The advantages of SCC in its fresh and hardened states include economic efficiency (i.e. it shortens the construction time as well as it reducing the labour and equipment required), improvement in working and living environment (i.e. it may consume high amount of industrial by-products, it reduces construction noise and health hazards) and enhancement in automation of the construction process (Ozawa et al., 1995; Bartos and Cechura, 2001). Generally, SCC is used for constructing reinforced concrete elements with closely arranged reinforcement sections, construction elements with limited compaction possibilities, filigree construction elements, exposed concrete parts where high surface quality is required, texture surfaced concrete construction elements, and reinforced concrete parts in environmentally noise sensitive sites (Vejmelková et al., 2011).

Compared with ordinary concrete, SCC includes large amounts of binder, superplasticizer, and/or viscosity modifying admixtures (VMA) (Nehdi et al., 2004). The supplemented binder content is associated with SCMs such as fly ash (FA), ground-granulated blast furnace slag (GGBFS), silica fume (SF), metakaolin (MK), rice husk ash (RHA), etc. The incorporation of SCMs into cement or concrete mixes provides many benefits to fresh and hardened concrete, such as improvement in workability and ultimate strength values. It also reduces the construction cost (Dinakar et al., 2008). SCC with high-volume SCMs is defined by the replacement of large amount of SCMs (generally this ratio is more than 40–50%) with cement in SCC mixes.

9.2 Significance of using high-volume supplementary cementitious materials (SCMs) in self-compacting concrete (SCC)

The principal reason for using high-volume SCMs in SCC is to reduce cement consumption in SCC production using industrial by-products and waste materials, and subsequently to conserve the environment and non-renewable natural material resources. Moreover, the incorporation of SCMs not only enhances the fresh characteristics of SCC, but also contributes to the strength development of concrete and makes it more durable.

Cement manufacturing is an energy-intensive industry and nearly 40% of cement production cost is energy related (Marei, 1990). Companies worldwide are searching for new options for reducing energy consumption costs. The average energy consumption in a modern dry process cement plant is 850 000 kcal of fuel and 120 kWh of electricity per ton of cement. In a wet process plant, this amount becomes 1 500 000 kcal of fuel and 80 kWh of electricity. This suggests that on average, 100 kg of fuel oil (or 210 kg of coal) is required to produce 1 ton of clinker, which releases approximately 1 ton of carbon dioxide to the atmosphere (Marei, 1990; Kenai et al., 2004). The cement industry is responsible for 7% of the total carbon dioxide (CO_2) annual world emissions, assuming 1.6 billion tons of Portland cement production (Malhotra and Mehta, 2002; Mehta, 2002).

SCC with comparable compressive strength is usually associated with 20–50% higher material cost owing to its high binder and chemical admixture content. Replacing 50% of the Portland cement with SCM produces cost-effective SCC with good workability and higher 28- and 91-day strength values, compared with the results from a reference SCC containing 100% Portland cement (Nehdi et al., 2004).

The optimisation study on cost-effective SCC with high-volume SCMs indicates that cost-effective SCC can be achieved by replacing up to 50% of ordinary Portland cement with SCMs, such as FA and GGBFS (Bouzoubaa and Lachemi, 2001; Nehdi et al., 2003). The use of SCMs in the form of binary (two-component), ternary (three-component) or quaternary (four-component) blends provides excellent compressive strength values, even at early ages, plus it enhances the rheological behaviour of SCC and thereby reduces its material cost (Bouzoubaa and Lachemi, 2001).

9.3 Properties of fresh self-compacting concrete (SCC) with high-volume supplementary cementitious materials (SCMs)

Many researchers have stated the possibility of designing an SCC incorporating high volumes of SCMs. Table 9.1 summarises SCM compositions introduced and fresh properties obtained from laboratory tests conducted in some studies dealing with SCC containing high-volume SCMs. As shown in Table 9.1, these SCC compositions that satisfy the qualification norms of SCC demonstrate that the fresh properties are in agreement with the various consistency classes, providing high segregation resistance, good flowability and passing ability, which are defined by European guidelines (Bouzoubaa and Lachemi, 2001; Nehdi et al., 2004; Lachemi et al., 2007; Dinakar et al., 2008; Sukumar et al., 2008; Pathak and Siddique, 2012).

A high volume of FA replacements result in slump flow values between 500 and 700 mm, which indicate good deformability. Slump flow values are more sensitive to the dosage of superplasticiser than the percentage of FA or the water to binder ratio (Bouzoubaa and Lachemi, 2001). The justification for using ternary blended SCMs in SCC is that, for SCC mixes containing 60% FA, the slump flow time can be as low as 1 s due to the lack of cohesion of these mixtures. This does not comply with EFNARC (2005), whereas the combined use of SCMs such as FA and SF (ternary blend) increases slump flow time to 3 s which complies with EFNARC (2005). When the slump flow of SCC is considered, the use of ternary and quaternary blends of SCMs seems to provide better performance than binary blends (Gesoğlu et al., 2009).

The segregation index of SCC contains high-volume SCMs is low, especially with the use of SF, RHA or VMA (Nehdi et al., 2004). In addition, the

Table 9.1 The SCM compositions and the corresponding results of fresh properties obtained from some research on SCC containing high-volume SCMs

Sources	FA (%)	GGBFS (%)	SF (%)	MK (%)	RHA (%)	TBC (kg/ m^3)	w/b	SP (kg/m^3)	Slump flow (mm)	T_{50} (s)	V-funnel time (s)	L-box	Segregation index (%)
Güneyisi et al. (2011)	15–45	–	–	5–15	–	550	0.32	6.0–8.0	695–730	–	–	–	–
	–	15–45	–	5–15	–			8.0	705–730	–	–	–	–
	7.5–22.5	7.5–22.5	–	5–15	–			7.0–8.0	700–730	–	–	–	–
Vejmelková et al. (2011)	–	56	–	–	–	570	0.26	6.84	750	8	–	–	–
	–	–	–	40	–	607	0.32	9.105	730	9	–	–	–
Liu (2010)	40–100	–	–	–	–	439–495	0.35–0.37	1.98–3.71	705–730	–	6.1–9.1	–	5–13
Gesoğlu et al. (2009)	20–60	–	–	–	–	450	0.44	2.9–3.2	675–730	1–2	4–10.4	0.71–0.95	–
	–	20–60	–	–	–			2.8–3.7	670–710	3	10–14	0.70–0.73	–
	15–45	–	5–15	–	–			4.2–4.8	690–720	2–3	6–7.5	0.79–0.89	–
	–	15–45	4–15	–	–			4.0–5.8	680–715	3.0–4.0	5.2–11.2	0.82–0.93	–
	10–30	10–30	–	–	–			2.8–3.2	690–730	2.2–3.0	6.2–9.9	0.85–0.90	–
	7.5–22.5	7.5–22.5	5–15	–	–			4.2–5.0	675–700	2.8–3.4	4.2–6.0	0.85–0.87	–
Dinakar et al. (2008)	30–85	–	–	–	–	500–750	0.41–0.31	8.75–15	630–800	–	5–12	–	–
Sukumar et al. (2008)	30–52	–	–	–	–	525–570	0.32–0.34	2.1–2.85 (VMA=5.25–5.70)	773–793	1.0–1.5	3–4	0.96–1.0	3.0–5.5
Gesoğlu and Özbay (2007)	20–60	–	–	–	–	550	0.32	6.67–7.43	730	2.2–4.0	7.0–17.0	0.90–0.96	–
	–	20–60	–	–	–			8.89–10.43	700–730	1.2–3.3	11.0–15.0	0.88–0.95	–
	15–45	–	5–15	–	–			8.22–9.11	705–730	2.0–2.2	7.1–8.2	0.93–0.96	–
	–	15–45	5–15	–	–			9.78–10.78	710–730	2.3–4.0	8.0–13.6	0.94–0.95	–
	10–30	10–30	–	–	–			4.44–8.00	730	2.0	6.3–12.2	0.94–0.98	–
	7.5–22.5	7.5–22.5	5–15	–	–			6.40–8.00	670–700	3.0–5.0	8.3–13.1	0.83–0.88	–

Continued

Table 9.1 Continued

Sources	FA (%)	GGBFS (%)	SF (%)	MK (%)	RHA (%)	TBC (kg/ m^3)	w/b	SP (kg/m^3)	Slump flow (mm)	T_{50} (s)	V-funnel time (s)	L-box	Segregation index (%)
Nehdi et al. (2004)	50	–	–	–	–	425–430	0.38	1.5 (l/m^3)	635	–	–	0.84	8
	50	–	–	–	–			3.3 (l/m^3) (VMA=0.1)	640	–	–	0.80	6
	25	25	–	–	–			1.3 (l/m^3)	690	–	–	0.85	5
	25	25	–	–	–			4.9 (l/m^3) (VMA=0.2)	615	–	–	0.82	4
	24	20	6	–	–			3.0 (l/m^3)	620	–	–	0.79	5
	24	20	–	–	6			3.2 (l/m^3)	650	–	–	0.80	3
Bouzoubaa and Lachemi (2001)	40–60	–	–	–	–	387–423	0.45–0.35	0–3.8 (l/m^3)	450–650	–	3–7	–	–

segregation index of SCC with a water-to-binder ratio of 0.45 decreases as the FA replacement ratios increase. However, the segregation index increases with an increase in the dosage of superplasticiser, accompanied by a decrease in water-to-binder ratio for the SCC containing FA replacement around 40% (Bouzoubaa and Lachemi, 2001). Generally, the segregation resistance is improved by the use of high-volume FA replacements (80–100%) in SCC mixes (Liu, 2010).

The setting times for SCC with high-volume FA are generally longer than SCC containing only ordinary Portland cement (OPC), owing to the low cement content. The maximum temperature rise in SCC with high-volume FA replacement ratios is usually low due to its low cement content and the slow FA reaction process. This behaviour could enable SCC with high-volume FA to be used in large structural concrete members (Bouzoubaa and Lachemi, 2001). The binary use of FA or GGBFS significantly delays the initial and final setting times, which can be prolonged even further by increasing the FA or GGBFS replacement level. When ternary blends of SF, FA or GGBFS are used, the inclusion of SF reduces the setting time delay of SCC compared with the binary use of FA or GGBFS replacement values. The combined use of FA and GGBFS in SCC slightly reduces both the initial and final setting times compared to the binary blends of FA or GGBFS (Gesoğlu and Özbay, 2007).

Since some binary and ternary uses of blended SCMs (such as 40–60% GGBFS and 45% FA–15% SF) do not comply with the L-box test requirements of EFNARC (2005), ternary and quaternary blends of SCMs can be used (such as 30%FA–30%GGBFS, 22.5%FA–22.5%GGBFS–15%SF, 45%GGBFS–15%SF, etc.) to improve filling and passing. Using GGBFS in the binary blends causes high viscosity, whereas using FA around 40–60% replacement levels, or using ternary or quaternary blended high volume SCMs, reduces the viscosity and subsequently results in much lower V-funnel flow times (Gesoğlu et al., 2009). The increase in GGBFS content from 15 to 45% results in higher slump flow and passing ability, and lowers T_{50} slump flow and V-funnel flow time results (Güneyisi and Gesoğlu, 2011).

SCC containing 40% MK exhibits smaller slump flow diameter and longer T_{50} flow time values, as well as providing higher yield stress values compared to SCC containing 56% GGBFS. The loss of flowability over time is much faster for SCC containing MK than SCC with GGBFS. This could be related to higher reactivity due to the higher specific surface value of MK. The measured plastic viscosity of SCC with MK is around 100–136 Pa s. This could imply that this mixture composition is thixotropic and has non-Newtonian fluid properties. A non-linear relationship exists between shear stress and strain rate, and the apparent viscosity decreases with the duration of stress. SCC with GGBFS mixture exhibits Newtonian fluid properties with zero yield stress and significant plastic viscosity around 175–233 Pa s. SCC with MK requires more water and superplasticiser than SCC with GGBFS to comply with requirements (Vejmelková et al., 2011).

In order to achieve similar filling ability, and to maintain the constant slump flow and V-funnel time values compared with the control mix, a reduction in superplasticiser dosage and an increase in the amount of water might be required for SCC containing a high volume of FA (Liu, 2010). SCC with high-volume SCMs provides much lower slump flow time values, especially when FA is used (Gesoğlu and Özbay, 2007).

Figure 9.1 summarises some of the results in the current literature, which indicate that the superplasticiser dosage usually decreases with an increase in FA content (Nehdi et al., 2004; Sukumar et al., 2008; Gesoglu et al., 2009, 2011; Liu, 2010). The superplasticiser dosage increases as the water to binder ratio and the FA replacement level decrease (Bouzoubaa and Lachemi, 2001).

9.4 Mechanical properties of self-compacting concrete (SCC) with high-volume supplementary cementitious materials (SCMs)

Some researchers have explored the possibility of designing SCC with high-volume FA replacement values according to the required strength grades. The results show that the strength of SCC with an FA content of 50–70%, even at 28 days, is sufficient for use in reinforced concrete construction (Dinakar et al., 2007). SCC with lower strength grades of 20–30 MPa can be produced by using FA replacement ratios of 70–85%. Higher-strength grades

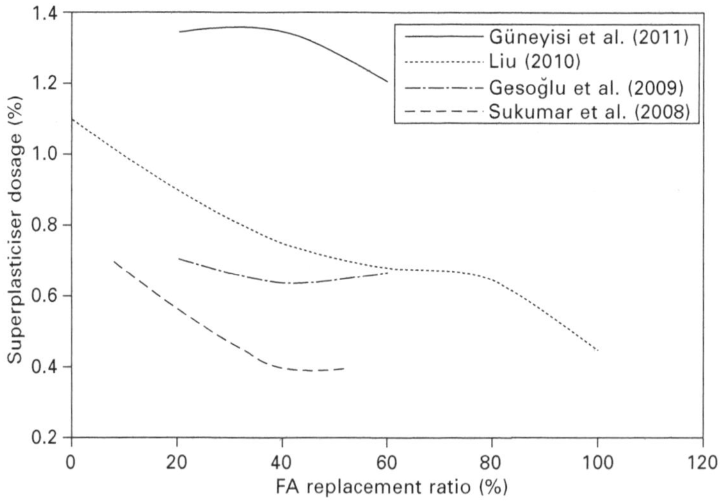

9.1 The influence of FA replacement ratio on superplasticiser dosage.

(about 60–90 MPa) can be produced with 30–50% FA replacement (Dinakar et al., 2008). SCC containing 40–60% FA and 0.35–0.45 water to binder ratio provides sufficient compressive strength, ranging from 25 to 50 MPa (Bouzoubaa and Lachemi, 2001). The long-term strength of SCC containing 60% high-volume FA is around 40 MPa (Khatib, 2008). This implies that high-volume FA can be used to produce SCC with adequate strength.

SCC containing more than 50% FA replacement produces compressive strength of 20–30 MPa at 3–7 days (Kumar et al., 2004). The early-age strength value of SCC containing 50% FA is low due to the slower reactivity of FA. This can be improved by using a ternary blend of 25% FA and 25% GGBFS; in this case the 28- and 91-day compressive strength values exceed the results from the reference mixtures (Nehdi et al., 2004). Similarly, the compressive strength decreases as the FA replacement ratio increases (Pathak and Siddique, 2012). In contrast to FA, SCC with GGBFS provides comparable strength values (Gesoğlu and Özbay, 2007; Güneyisi et al., 2011). The 28-day compressive strength of SCC incorporating 30% FA and 30% GGBFS with a suitable superplasticiser becomes around 70 MPa, and the resulting concrete mix provides adequate fluidity, cohesiveness and water-retaining properties (Fang et al., 1999).

As shown in Fig. 9.2, the compressive strength of SCC with binary blends

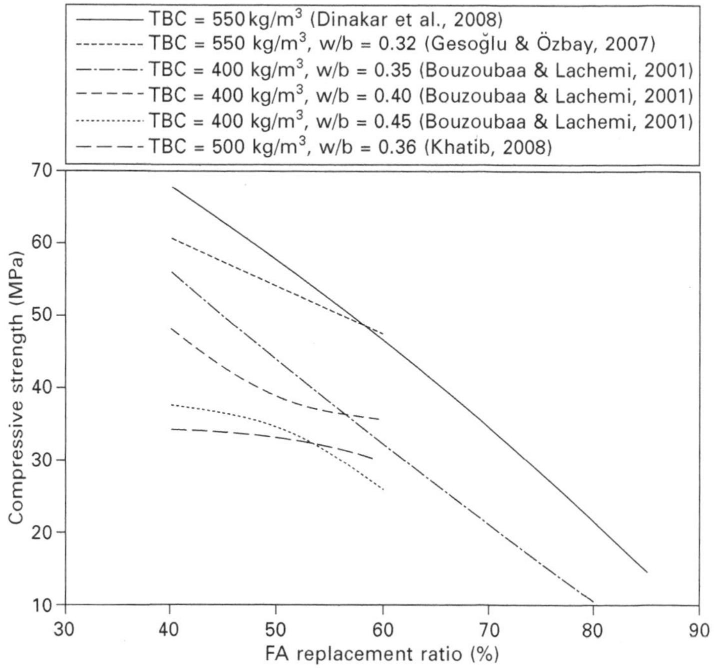

9.2 The 28-day compressive strength variation in terms of FA replacement ratio.

of FA falls as the FA replacement level rises. This could be attributed to the slower pozzolanic reaction of FA with $Ca(OH)_2$ in the hydrated cement. The higher the replacement level of FA in SCC, the lower the hardened properties, owing to the lower reaction rate between cement hydration products and FA (Liu, 2010). A 60% FA replacement ratio reduces the compressive strength by about 40%. However, this adverse effect of FA could be remedied by the combined use of mineral admixtures. SCC with GGBFS and/or SF generally provides comparable strength values to SCC containing OPC only (Gesoğlu et al., 2009).

In the case of ternary use of FA and SF, the compressive strength also gradually decreases in terms of the replacement ratio, but the rate of reduction is much less compared to the binary use of FA. However, as illustrated in Fig. 9.3, the ternary use of GGBFS and SF provides an increase in the compressive strength, especially at higher replacement levels (Gesoğlu and Özbay, 2007). For instance, as can be seen in Fig. 9.3, while the 90-day compressive strength of SCC containing 45% FA and 15% SF is about 45 MPa, SCC with 45% GGBFS and 15% SF provides a compressive strength of about 70 MPa.

Figures 9.4 and 9.5 show the compressive strength results obtained from ternary and quaternary use of MK, FA and/or GGBFS replacement combinations respectively. The MK and FA (or GGBFS) replacement combinations (ternary

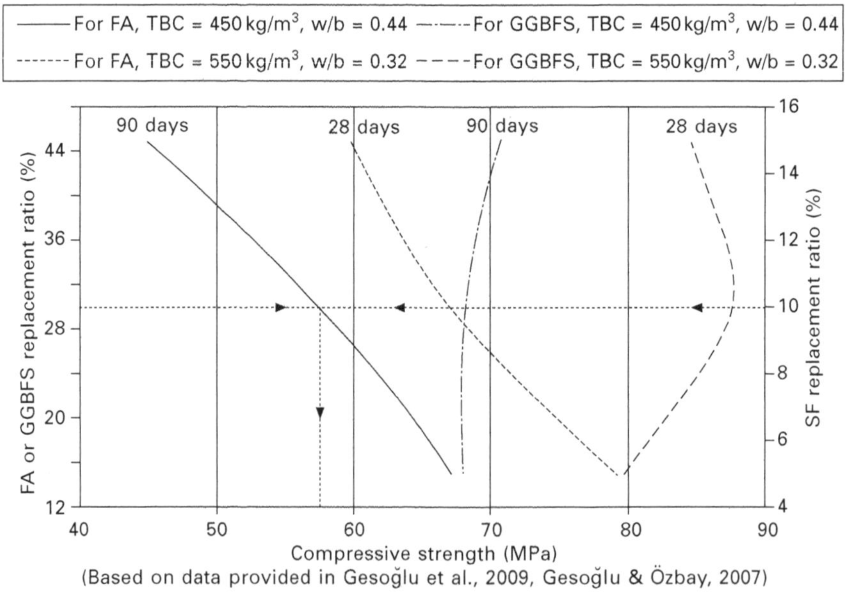

9.3 The influence of SF and FA or GGBFS replacement combinations (ternary blends) on the compressive strength values.

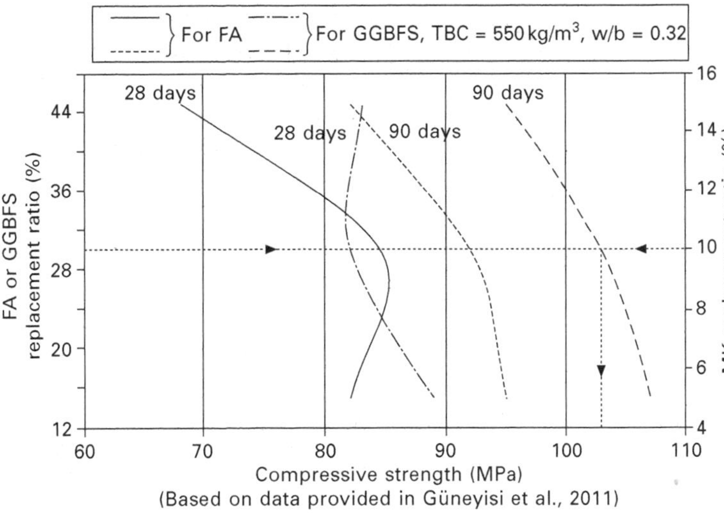

9.4 The influence of MK and FA or GGBFS replacement combinations (ternary blends) on the compressive strength values.

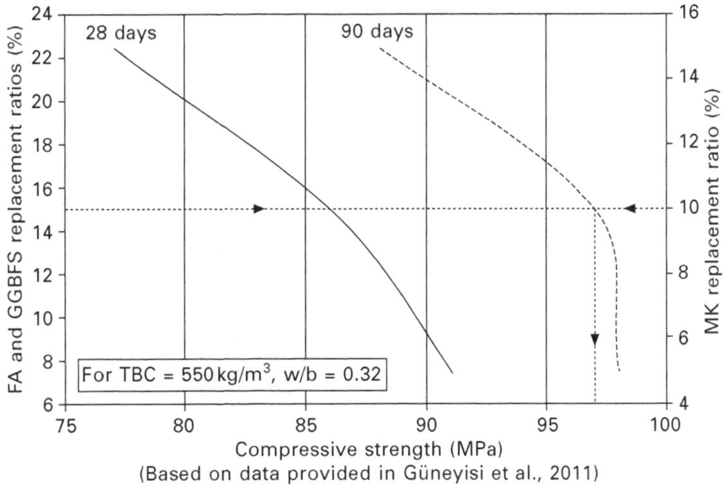

9.5 The influence of MK, FA and GGBFS replacement combinations (quaternary blends) on the compressive strength values.

blends) significantly improve the compressive strength compared to binary use of FA replacements. For example, while 60% FA replacement results in a compressive strength of 48 MPa at 28 days (see Fig. 9.2), the ternary use of 45% FA and 15% MK replacements provides a 42% compressive strength increase (see Fig. 9.4). The compressive strength obtained from the binary use of GGBFS slightly increases compared to its ternary combination with

MK (see Fig. 9.4). This increase at 90 days is about 6% for a 45% GGBFS and 15% MK replacement combination (Güneyisi et al., 2011).

Figure 9.5 indicates the effect of MK, FA and GGBFS replacement combinations (quaternary blends) on the compressive strength values. The quaternary blends provide significant increases compared to the binary use of FA replacements. While 60% FA replacement results in compressive strengths of 48 and 65 MPa at 28 days and 90 days, respectively, the quaternary use of 22.5% FA, 22.5% GGBFS and 15% MK replacements provides 60% and 35% compressive strength increases at 28 days and 90 days (see Fig. 9.5) (Güneyisi et al., 2011).

The compressive strength of SCC with MK increases much faster than SCC containing GGBFS during the initial setting time (up to 2 days). This could be related to higher reactivity caused by the binder blended with MK. After this period, the increase in relative strength is similar for both SCC mixes, but SCC with MK always provides higher strength values (Vejmelková et al., 2011).

Most SCC with FA combinations demonstrate higher splitting tensile strength values compared with the normal vibrated concretes, due to the high paste content, which exhibits slightly higher deformability. The splitting tensile strength values obtained from SCC with FA combinations are about 10% of the corresponding compressive strength (Dinakar et al., 2007). The splitting tensile strength of SCC increases with the decrease in percentage of FA content and the water-to-cementitious materials ratio (Pathak and Siddique, 2012).

The high-volume FA content in SCC reduces the static elastic modulus since it leads to higher paste volume and lower coarse aggregate content. The static elastic modulus of SCC is about 8% lower than that of normal concrete with similar compressive strength (Dinakar et al., 2007).

9.5 Durability of self-compacting concrete (SCC) with high-volume supplementary cementitious materials (SCMs)

The chloride diffusivity of SCC with high-volume FA replacements is much lower than in the corresponding normal vibrated concretes. Although SCC with high-volume FA demonstrates higher absorption and permeable voids, it shows better performance in terms of chloride permeability. This could be due to the chloride ion penetration, which depends on the chloride binding capacity of the constituent materials (Dinakar et al., 2008). SCC with high volume FA shows significantly lower chloride ion permeability than SCC without FA and normal concrete mixes (Amrutha et al., 2011; Pathak and Siddique, 2012).

SCC with high-volume FA admixed with GGBFS, SF, MK and RHA shows large improvements in terms of chloride penetration, that is, SCC with 60% FA and 12% MK has the lowest chloride penetration values compared to the other SCC mixes (Amrutha et al., 2011). The chloride ion permeability of SCC with GGBFS decreases when the replacement ratio of GGBFS is increased. A reduction of approximately 90% of the total chloride ion charge is obtained for SCC with 60% GGBFS compared with SCC with OPC only (Uysal et al., 2012).

Even though SCC containing high-volume FA decreases the rapid chloride penetrability compared with SCC containing OPC only, a larger reduction can be obtained by introducing high-volume multi-blended SCMs into SCC (Nehdi et al., 2004; Gesoğlu et al., 2009; Güneyisi et al., 2011). SCC containing a combination of GGBFS and SF replacement seems to be more effective, and the highest resistance to chloride permeability is obtained from the mixture containing 45% GGBFS and 15% SF (Gesoğlu et al., 2009).

The total charge passing through SCC containing OPC only is approximately 1009 coulombs, whereas, as shown in Figs 9.6 and 9.7, the total charges of SCC with 60% FA and 60% GGBFS replacements are nearly 715 and 264 coulombs, respectively (Güneyisi et al., 2011). Figure 9.6 indicates that the reduction in rapid chloride penetrability seems to be generally less effective

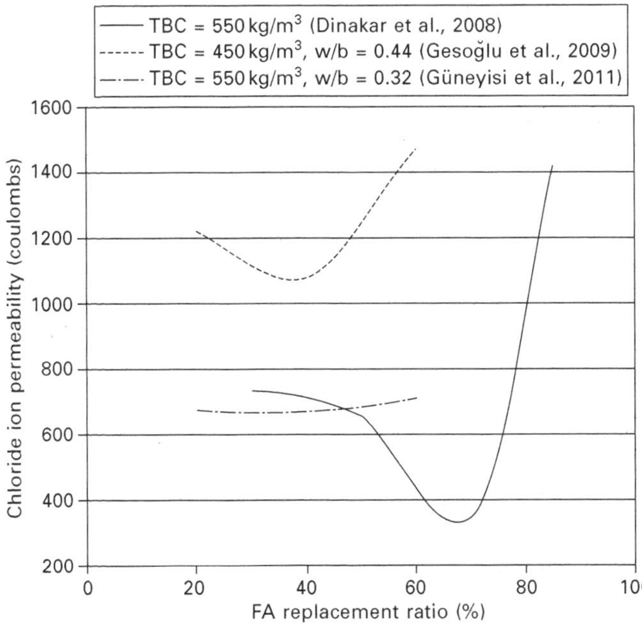

9.6 The variation of chloride ion permeability with FA replacement ratio.

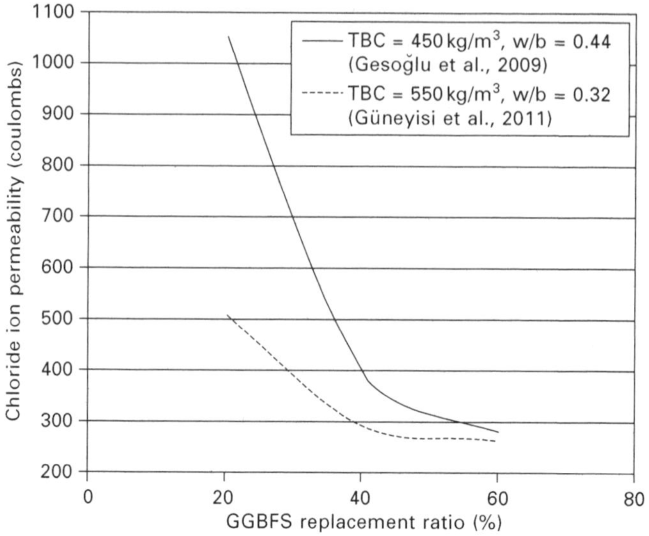

9.7 The variation of chloride ion permeability with GGBFS replacement ratio.

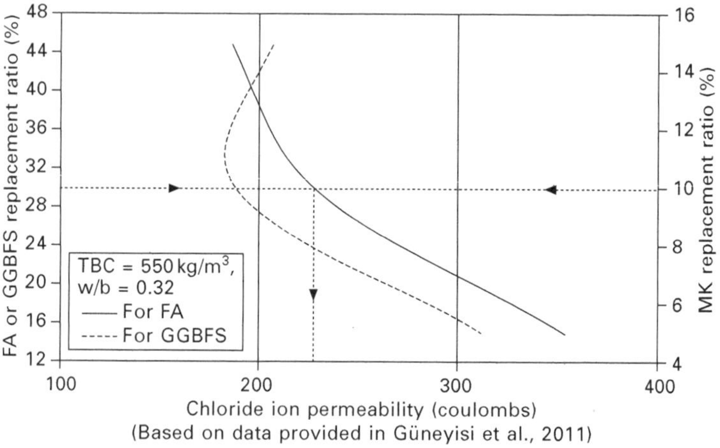

9.8 The influence of MK and FA or GGBFS replacement combinations (ternary blends) on the chloride ion permeability values.

with an increase in FA replacement ratio. However, as shown in Fig. 9.7, the rapid chloride penetrability clearly reduces when the GGBFS replacement ratio is increased.

Figure 9.8 indicates the effect of MK and FA or GGBFS ternary replacement combinations on the chloride ion permeability results. This ternary blend combination significantly reduces the rapid chloride penetration compared to the binary use of FA and GGBFS. This reduction becomes more apparent when high-volume ternary blends are used (see Fig. 9.8). For instance, while 60% FA and 60% GGBFS replacement ratios provide the rapid chloride penetration of 715 and 264 coulombs, respectively (see Figs 9.6 and 9.7), the ternary use of 45% FA–15% MK and 45% GGBFS–15% MK replacement combinations provide 74% and 22% reduction in the rapid chloride penetration, respectively (see Fig. 9.8) (Güneyisi et al., 2011).

The sorptivity index of SCC with high-volume SCMs is lower than for SCC with OPC only. The sorptivity values of SCC containing up to 40% FA slightly decrease with FA content, although the water to binder ratio increases from 0.33 (0% FA) to 0.35 (40% FA). SCC with 40% FA affords the lowest sorptivity since FA is finer than the cement, and so it fills the voids and leads to lower porosity. The sorptivity value of SCC with 60–80% FA replacement ratio significantly increases, since FA expands voids between the cement particles, resulting in higher porosity. According to the sorptivity results, it can be stated that the use of 60–80% FA in SCC may result in lower durability (Liu, 2010).

The effects of using ternary (SF and FA or GBBFS) and quaternary (FA, GBBFS and SF) blends of SCMs are to decrease the sorptivity index compared to SCC with OPC only (Gesoğlu et al., 2009). Figures 9.9 to 9.12 shows the variation of sorptivity index obtained from SCC containing multi-blended SCM combinations. When the sorptivity index values are compared between the results obtained from SCC with OPC only and from SCC containing high-volume ternary and quaternary blends of 45% FA–15% SF, 45% GGBFS–15% SF and 22.5% FA–22.5 GGBFS–15% SF, the reductions become 24%, 27%, and 33%, respectively (Gesoğlu et al., 2009). If SF is replaced with MK in these SCM replacement combinations, the reductions

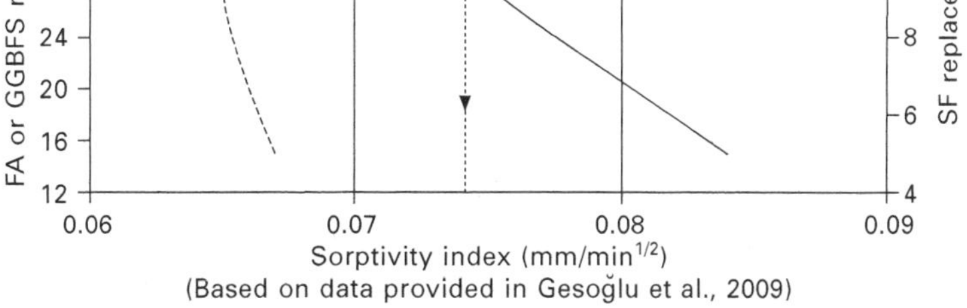

9.9 The influence of SF and FA (or GGBFS) replacement combinations (ternary blends) on the sorptivity index.

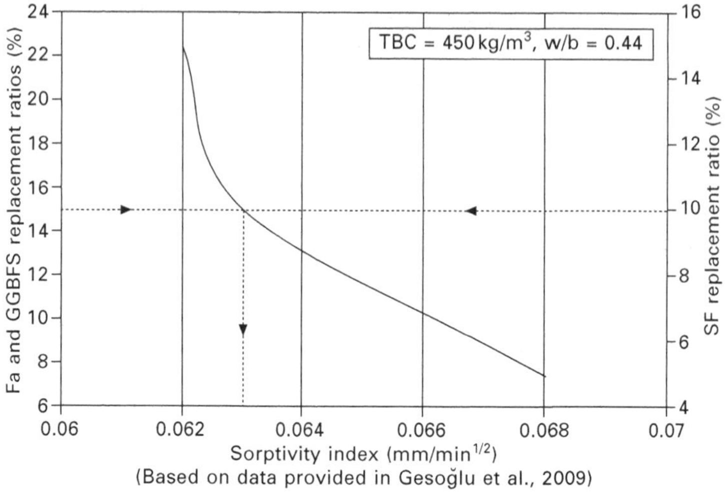

9.10 The influence of SF, FA and GGBFS replacement combinations (quaternary blends) on the sorptivity index.

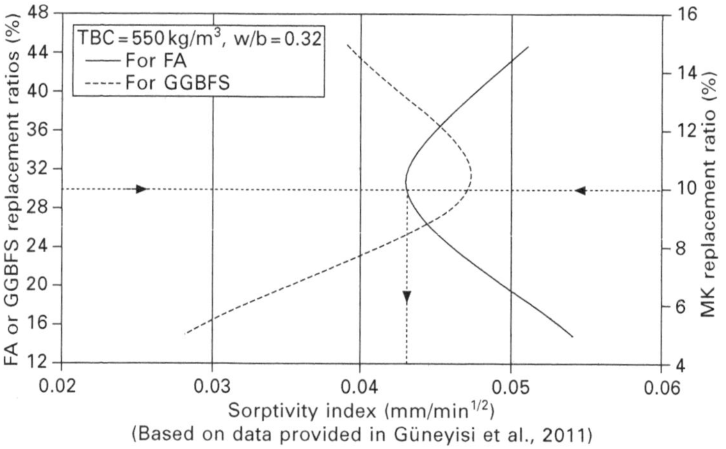

9.11 The influence of MK and FA (or GGBFS) replacement combinations (ternary blends) on the sorptivity index.

become 38%, 52%, and 45%, respectively (Güneyisi et al., 2011). As can be seen in Fig. 9.11, the lowest sorptivity index is obtained from SCC with the ternary blends of 15% MK and 45% GGBFS (see Figs 9.9 to 9.12). The use of MK seems to be more effective in reducing the sorptivity due to the reduced pore volume (Güneyisi et al., 2011).

The use of high-volume SCMs in SCC significantly reduces the water permeability of SCC depending on the type of SCMs and the replacement level used. When the water permeability values of SCC incorporating 40 and 60%

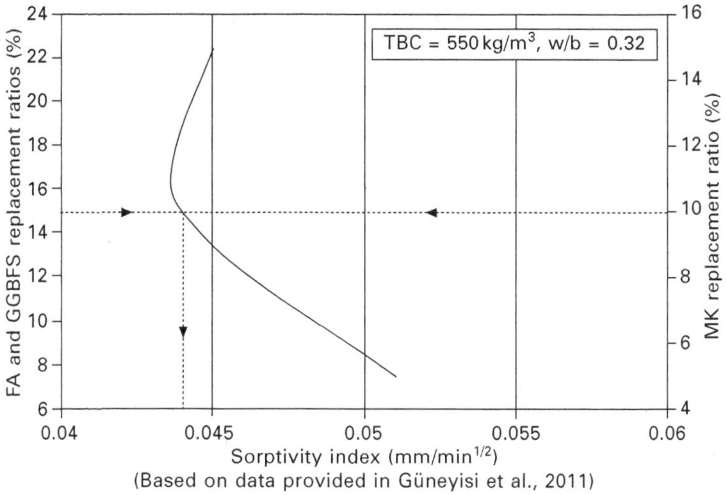

9.12 The influence of MK, FA and GGBFS replacement combinations (quaternary blends) on the sorptivity index.

FA (or GGBS) reduce to about 40–60% compared to SCC with OPC only, the ternary and quaternary blends of FA, GGBFS and MK cause a reduction of 70–80% (Güneyisi et al., 2011). The quaternary SCMs combination of 22.5% FA–22.5% GGBFS–15% SF provides about 45% water penetration depth reduction (Gesoğlu et al., 2009).

The water penetration depth of SCC containing 56% GGBFS replacement is about 40% higher than SCC containing 40% MK, after 28 days. However, after 90 days, the situation is reversed and SCC with GGBFS replacement exhibits only about 40% of the water penetration depth value obtained for the SCC with MK, which indicates a possible delay in the pozzolanic reaction compared to SCC containing MK (Vejmelková et al., 2011).

The initial absorption values of SCC with high-volume FA are slightly higher than the normal vibrated concretes and the absorption increases with an increase in FA replacement ratios. SCC with a high volume of 80–85% FA replacement shows the highest water absorption capacity; around 2–9% (Dinakar et al., 2008; Khatib, 2008). Similarly, the percentage permeable voids are also higher for SCC with high-volume FA than the normal vibrated concretes. This may be owing to the increasing porosity resulting from the high paste volumes, water contents and superplasticiser dosages used in SCC containing high volume FA replacement (Dinakar et al., 2008).

The carbonation depth of SCC with 20% FA is higher than SCC containing 15% FA replacement. If the FA content is increased to 25%, carbonation depth decreases. Carbonation depth becomes almost constant for SCC containing 25–35% FA at 90 days (Siddique, 2011). Research into the variation of carbonation depth of SCC in terms of the different initial water-curing period and curing condition has demonstrated that SCC with an initial water-curing period of 7 days has the lowest carbonation depths. SCC with 20% FA replacement under full room curing condition (28–32 °C, 75–85% RH) has a lower carbonation depth than that with full standard (19–21 °C, 95–100% RH) and full water curing condition (Zhao et al., 2012)

SCC containing 40% MK provides adequate freeze resistance, indicating no mass loss up to 56 freeze–thaw cycles. Additionally, the freeze resistance of SCC containing 56% GGBFS is also satisfactory (Vejmelková et al., 2011).

The deicing salt surface scaling following 50 freezing–thawing cycles was investigated, and the results indicate that SCC with a 50% FA replacement ratio provides the largest mass of scaled-off material and the worst visual scaling rating among all other SCC mixes containing the high-volume SCM replacements. Interestingly, as VMA is included in SCC with high-volume FA, the mass of scaled-off material considerably decreases and the visual rating increases from the rating of 4 to 2. SCC with 25% FA–25% GGBFS ternary blends along with the use of VMA results in sufficient resistance to deicing salt surface scaling compared to the results from SCC with OPC only (Nehdi et al., 2004).

The deterioration of SCC subjected to 3% H_2SO_4 solution demonstrates that weight loss significantly decreases with an increase in FA replacement ratio. This could be due to the low amount of reaction compounds like $Ca(OH)_2$ at lower levels of cement content in the deterioration process. Therefore, SCC with high-volume FA results in better concretes with higher resistance to acid attacks (Dinakar et al., 2008).

Most of the final drying shrinkage strains for SCC with high-volume FA are around 600×10^{-6}, which is the typical value for the final shrinkage strain of concrete in structures (see Fig. 9.13) (Bouzoubaa and Lachemi, 2001; Patel et al., 2004). SCC containing binary blends of FA or GGBFS provides lower shrinkage strain values. This beneficial effect appears to be more pronounced when the replacement levels of SCMs are increased (Khatib, 2008; Gesoğlu et al., 2009). Therefore, much lower cracking development is expected from SCC containing a high volume of SCMs.

9.6 Future trends

Although current studies into SCC with high-volume SCMs are attempting to explain fresh and hardened properties, the number of investigations into their characteristic behaviour and performance under the conditions likely to be applied in their structural uses is currently quite limited. In terms of current literature, the following list suggests probable future trends concerning SCC with high-volume SCMs:

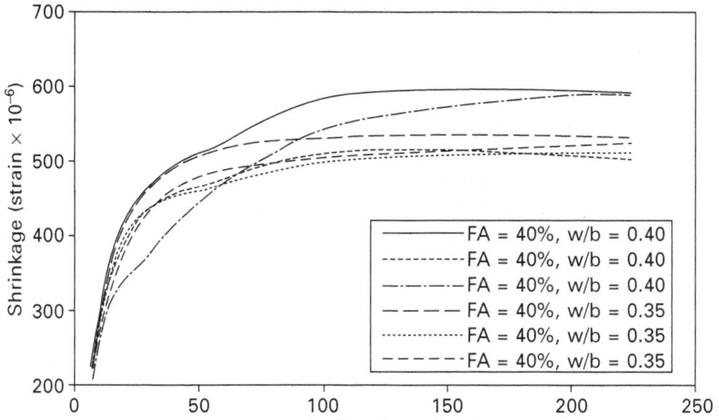

(Based on data provided in Bouzoubaa and Lachemi, 2001)

9.13 The drying shrinkage of SCC incorporating high-volume FA replacement values.

- The durability properties of SCC with high-volume SCMs such as carbonation, freezing and thawing, acid attack, and corrosion need to be investigated thoroughly for all possible combinations of SCMs.
- The effect of superplasticisers on the fresh and hardened properties of SCC containing various combinations of high volume SCMs need to be explored.
- A well-established mix design methodology needs to be developed for SCC with high-volume SCMs.
- The economic and environmental advantages of SCC with high-volume SCMs need to be embodied in more detail.
- The relationships between rheology, composition and hardened properties need to be investigated.
- The effect of formwork pressure, rheological problems related to thixotropy, the creep and bond to reinforcement behaviour during faster construction and application need to be clarified.
- Rheological parameters such as yield stress and plastic viscosity should be focused on, in order to better describe the effects of superplasticiser on types of SCMs in SCC with high-volume SCMs.
- Criteria relating to the aggregates (natural, manufactured or recycled, etc.) in terms of the shape, grading, amount and mineralogical characteristics need to be clarified.
- Since the multi-blended high volume SCMs (i.e. ternary and quaternary blends) seem to overcome many drawbacks, such as the early-age compressive strength reduction derived from the binary use of blends, the optimum rate of multi-component cementitious blends as well as

their effects on the fresh and hardened properties need to be investigated further.

9.7 References

Amrutha Nayak G, Narasimhan M C and Rajeeva S V (2011), 'Chloride-ion impermeability of self-compacting high-volume fly ash concrete mixes' *International Journal of Civil & Environmental Engineering*, **11**, 29–35.

Bartos P J M and Cechura J (2001), 'Improvement of working environment in concrete construction by the use of self-compacting concrete' *Structural Concrete*, **2**, 127–132.

Bouzoubaa N and Lachemi M (2001), 'Self-compacting concrete incorporating high volumes of class F fly ash: Preliminary results', *Cement and Concrete Research*, **31**, 413–420.

Dinakar P, Babu K G and Santhanam M (2007), 'Mechanical properties of high volume fly ash self compacting concretes', *5th International RILEM Symposium on Self-compacting Concrete*, Ghent, Belgium.

Dinakar P, Babu K G and Santhanam M (2008), 'Durability properties of high volume fly ash self compacting concretes', *Cement & Concrete Composites*, **30**, 880–886.

EFNARC (2005), Specification and Guidelines for Self-Compacting Concrete. (Available from: http://www.efnarc.org/pdf/SCCGuidelinesMay2005.pdf).

Fang W, Jianxiong C and Changhui Y (1999), 'Studies on self-compacting high performance concrete with high volume mineral additives', *First International RILEM Symposium on Self-compacting Concrete*, 569–578.

Gesoğlu M and Özbay E (2007), 'Effects of mineral admixtures on fresh and hardened properties of self-compacting concretes: binary, ternary and quaternary systems', *Materials and Structures*, **40**, 923–937.

Gesoğlu M, Güneyisi E and Özbay E (2009), 'Properties of self-compacting concretes made with binary, ternary, and quaternary cementitious blends of fly ash, blast furnace slag, and silica fume', *Construction and Building Materials*, **23**, 1847–1854.

Güneyisi E and Gesoğlu M (2011), 'Properties of self-compacting Portland pozzolana and limestone blended cement concretes containing different replacement levels of slag', *Materials and Structures*, **44**, 1399–1410.

Güneyisi E, Gesoğlu M and Özbay E (2011), 'Permeation properties of self-consolidating concretes with mineral admixtures', *ACI Materials Journal*, **108**, 150–158.

Kenai S, Soboyejo W and Soboyejo A (2004), 'Some engineering properties of limestone concrete', *Materials and Manufacturing Process*, **19**, 949–961.

Khatib J M (2008), 'Performance of self-compacting concrete containing fly ash', *Construction and Building Materials*, **22**, 1963–1971.

Kumar P, Haq Mohd A and Kaushik S K (2004), 'Early age strength of SCC with large volumes of fly ash', *Indian Concrete Journal*, **78**, 25–39.

Lachemi M, Hossain K M A, Patel R, Shehata M and Bouzoubaa N (2007), 'Influence of paste/mortar rheology on the flow characteristics of high-volume fly ash self-consolidating concrete', *Magazine of Concrete Research*, **59**, 517–528.

Liu M (2010), 'Self-compacting concrete with different levels of pulverized fuel ash', *Construction and Building Materials*, **24**, 1245–1252.

Malhotra V M and Mehta P K (2002), *High-performance, high volume fly ash concrete: materials, mixture proportioning, properties, construction practice, and case histories*, Supplementary Cementing Materials for Sustainable Development, Ottawa, Canada.

Marei A R (1990), *Energy and its impacts on cement industry with special reference to conversion and modernization of cement plants in Arab and developing countries*, Cement Industry in OIC Member Countries, Technical paper 1, Saudi Arabia.

Mehta P K (2002), 'Greening of the concrete industry for sustainable development', *Concrete International*, **24**, 23–28.

Nehdi M, El Chabib H and El Naggar M H (2003), 'Development of cost-effective self-consolidating concrete for deep foundation applications', *Concrete International*, **25**, 49–57.

Nehdi M, Pardhan M and Koshowski S (2004), 'Durability of self-consolidating concrete incorporating high-volume replacement composite cements', *Cement and Concrete Research*, **34**, 2103–2112.

Okamura H and Ouchi M (2003), 'Self-compacting concrete', *Journal of Advanced Concrete technology*, **1**, 5–15.

Ozawa K, Maekawa K, Kunishima M and Okamura H (1989), 'High-performance concrete based on the durability of concrete structures', In: *Proceedings of the second East Asia Pacific Conference on Structural Engineering and Construction* (EASEC-2), **1**, 445–450.

Ozawa K, Sakata N and Okamura H (1995), 'Evaluation of self-compactibility of fresh concrete by the use of self-compacting concrete', *Concrete Library International*, **25**, 59–75.

Patel R, Hossain K M A, Shehata M, Bouzoubaa N and Lachemi M (2004), 'Development of statistical models for mixture design of high-volume fly ash self-consolidating concrete', *ACI Materials Journal*, **101**, 294–302.

Pathak N and Siddique R (2012), 'Properties of self-compacting-concrete containing fly ash subjected to elevated temperatures', *Construction and Building Materials*, **30**, 274–280.

Siddique R (2011), 'Properties of self-compacting concrete containing class F fly ash', *Materials and Design*, **32**, 1501–1507.

Sukumar B, Nagamani K and Raghavan R S (2008), 'Evaluation of strength at early ages of self-compacting concrete with high volume fly ash', *Construction and Building Materials*, **22**, 1394–1401.

Uysal M, Yilmaz K and Ipek M (2012), 'The effect of mineral admixtures on mechanical properties, chloride ion permeability and impermeability of self-compacting concrete', *Construction and Building Materials*, **27**, 263–270.

Vejmelková E, Keppert M, Grzeszczyk S, Skalinski B and Robert C (2011), 'Properties of self-compacting concrete mixtures containing metakaolin and blast furnace slag', *Construction and Building Materials*, **25**, 1325–1331.

Zhao H, Sun W, Wu X and Gao B (2012), 'Effect of initial water-curing period and curing condition on the properties of self-compacting concrete', *Materials and Design*, **35**, 194–200.

10

High-volume ground granulated blast furnace slag (GGBFS) concrete

İ. B. TOPÇU, Eskişehir Osmangazi University, Turkey

DOI: 10.1533/9780857098993.2.218

Abstract: Ground granulated blast furnace slag (GGBFS) is a by-product of the iron-making process and because of its high calcium silicate content it has excellent cementious content and has been used in the construction industry for years as a replacement for ordinary Portland cement (OPC). GGBFS is also routinely used to limit the temperature rise in large concrete pours and is therefore a desirable material to use in mass concrete placements where control of temperatures is an issue. The more gradual hydration of GGBS cement generates both lower peak and less total overall heat than OPC. GGBFS was approved for use at a 70% replacement even though it is only allowed to use a maximum 25% instead of cement. It has a higher proportion of the strength-enhancing calcium silicate hydrates (CSH) than concrete made with only OPC, and a reduced content of free lime, which does not contribute to concrete strength.

Key words: ground granulated blast furnace slag, high replacement level, workability, strength, durability.

10.1 Introduction

Ground granulated blast furnace slag (GGBFS) has been used in the construction industry for years as a replacement for ordinary Portland cement (OPC). Ground granulated blast furnace slag also has a lower heat of hydration and, hence, generates less heat during concrete production and curing. GGBFS is a desirable material to use in mass concrete placements where control of temperatures is an issue. GGBFS can be used instead of cement with the ratios from 10% to 90%. However, because of its low heat generating characteristics, GGBFS was approved for use at a 70% replacement even though it is only allowed to use maximum 25% instead of cement (Richardson, 2006).

The general GGBFS literature indicates that the replacement of OPC by GGBFS typically results in lower early strengths (7 to 28 days), greater long-term strengths, lower chloride ion permeability, less creep, greater sulphate attack resistance, greater alkali silica reactivity (ASR) durability, enhanced workability, less bleeding, lower heat of hydration, and increased steel corrosion resistance. Results for drying shrinkage and freeze–thaw durability are somewhat mixed, although in general, the use of slag appears to be nondetrimental. Besides lower early strength, the downsides to the use of GGBFS include extended curing times, increased salt scaling, increased plastic shrinkage cracking, and increased air entrainment dosage required When Grade 120 (highest activity) slag is used, at least a 70% replacement may be needed to meet specification requirements. Most ready-mix concrete producers use 50% replacement with highly reactive slag during warm weather (Richardson, 2006).

10.2 The use of high-volume ground granulated blast furnace slag (GGBFS) concrete

Most cement plants consume great amounts of energy and have a negative effect on the environment because of the destructive greenhouse gases they emit to the atmosphere during the production process. Therefore, cement producers benefit from various mineral additives so as to reduce CO_2 emission and increase the production rate by decreasing energy consumption. One of those mineral additives is granulated blast furnace slag, which occurs during the production of pig iron in the iron and steel factories, a major part of which is ready to be used as an alternative binding material. The use of industrial by-products containing a high percentage of silica and alumina as an additive in cement or concrete, an area convenient for the utilization of large volumes of waste materials, will help the environmental problems to some extent. By preventing the rapid consumption of limited natural resources, the production of more qualified and more economic materials instead of traditional ones will also be possible.

The use of GGBFS, which has an amorph structure and shows pozzolan characteristics when finely granulated, as a replacement material within the systems of cement or concrete positively affects the properties of fresh and hardening concrete. The slag use provides an advantage especially in avoiding the thermal cracks in mass concrete and solving durability problems such as corrosion resistance, sulphate attack and ASR. On the other hand, when the slag is used in the concrete structures which will be exposed to hard conditions, there are some points, especially for cold climates, to be taken into consideration, such as preference of low rate of water/binder (w/b) in the mixture and, because the slag has a low speed of hydration, the need for a longer cure process compared with concrete with OPC.

10.3 Composition and properties of ground granulated blast furnace slag (GGBFS) concrete

When GGBFS is combined with water it shows a low rate of binding. Its speed of hydration is quite slow compared with OPC. It is used together with alkali salts or some amount of OPC so as to increase the speed of hydration. As a result of the combination of OPC and GGBFS, firstly calcium-silicate components such as C_3S and C_2S, which are structurally found in the OPC clinker, start the hydration and then $C_3S_2H_3$, gels of C-H-S together with $Ca(OH)_2$ (portlandite), which gives hydraulic binding property to cement, start to produce CH. The reaction between the additional CH and the pozzolan results in C-S-H. The emission of CH after the hydration of OPC gives no noticeable contribution to the resistance and is destructive in the sense of durability. Hence, it is aimed at reducing CH after its reaction with pozzolan. Moreover, owing to the fact that pozzolan provides economic advantages and positive environmental contributions, in *addition* to OPC, pozzolan mineral additives such as fly ash, blast furnace slag, silica fume and natural zeolite have started to be used. The necessary amount of CH for GGBFS to gain hydraulic binding property is provided by CH produced by the hydration of OPC. Similarly to the hydration of OPC, the reactions between GGBFS and CH engender the formation of new gels of C-H-S which has a binding property. The reactions of GGBFS and CH are quite similar to those of pozzolan. Under normal temperatures, the reaction between GGBFS and OPC occurs in two phases. When the hydration starts, the reaction of slag and alkali hydroxides stands in the forefront; then, however, the reactions of slag and CH become dominant (Erdoğan, 1995).

During the early hydration of the slag cement, the Portland cement releases calcium hydroxides (CH) with the ratio of 28% of its weight. In practice, however, this ratio does not exceed 20% of its weight. Hardened cement paste has a greater density and smaller pore sizes than an equivalent OPC paste, thus permeability and ionic diffusivity are reduced. Smaller pore size relates to lower permeability, although this does not necessarily mean lower total porosity. Slag–OPC and OPC mixes may result in similar total porosities, but the slag–OPC pore structures tend to be finer. Total porosity is important to mechanical properties such as compressive strength, but is less critical to properties that are associated with durability such as permeability. Durability seems to be related to larger pores. However, it has been reported that a 60% slag mortar mix not only had smaller pore sizes but also a somewhat smaller pore volume. Higher CH contents are associated with greater permeability and lower durability (Richardson, 2006).

The hydraulic binding property of GGBFS depends on the factors such as chemical characteristic of the slag, the concentration of alkali in the reactive system, the amount of vitreous structure in the slag, the fineness of the slag and the OPC used together with and the temperature during the hydration. Owing to the complexity of the factors which affect the hydraulic binding value of the slag, in previous years, to determine this value, a simple method was satisfactorily used. Provided that the slag is in accordance with the condition of $CaO + MgO + Al_2O_3/SiO_2 > 1$, its hydraulic binding value is accepted suitable. To assign the hydraulic binding value of the slag, the slag-activity index according to the ASTM C 989 standards is checked. The standard of ASTM C 989 constituted in 1982 divides GGBFS having hydraulic character and capability to be used as a mineral additive in concrete production into three groups as Grade 80, Grade 100 and Grade 120 regarding the slag-activity index. The standard states that the maximum amount of the slag remaining on the 45 μm sieve should be 20% and the maximum amount of air in the mortar prepared with this slag should be 12%. Furthermore, the amount of S in the slag is constrained to at most 2.5% and SO_3 at most 4% (Erdoğan, 1995). Because GGBFS cement reacts more slowly, especially at lower temperatures, it is typically ground finer than OPC. Lim and Wee (2000) found that more finely ground GGBFS had greater early strengths, but after 28 days the greater fineness caused little difference (Richardson, 2006). When the Blaine fineness increases at GGBFS, concrete shall have more brittle characteristics: this is an important factor at the mechanical properties of the concrete.

10.3.1 Fresh concrete properties

There are various studies on the unit weight and workability of fresh concrete having a high volume of GGBFS.

Fresh unit weight

The unit weight of fresh concrete when it is poured is between 2.35 and 2.55 kg/dm^3 and the theoretically calculated unit weight is between 2.27 and 2.50 kg/dm^3. The results are very close to each other (Bilim, 2006).

Need for water and workability

Concrete with GGBFS needs less water than those with OPC, because of the smooth surface texture of slag particles and the delay in the chemical reaction (Newman and Choo, 2003). Having less surface roughness and a lower density than clinker, GGBFS provides more volume of cement paste and positively affects the workability of slag concrete (Bilim, 2006). Where the workability of the concrete with GGBFS is concerned, the amount of cement, the rate of slag, the amount of aggregate, the type of slag and the interaction between them are very important. The reduction in the need for mixing water of GGBFS is related to its vitreous and non-absorbent surface structure (Mehta, 1989; Bilim, 2006). Owing to the fact that GGBFS have less surface roughness and a lower density compared to clinker, GGBFS provides more volume of cement paste and this renders the slag concrete more workable. In their study on workability of mortar with GGBFS, when low water/cement rate (0.20) and high rate of water-reducing admixture are used,

Ohama et al. (1995) determine that the fluency of the mortar is improved due to the partial replacement of slag with cement and the process is positively affected by the fineness of the slag. According to Güllü (1996), because of the increased volume of cement paste and smooth slip planes inside the paste, adding GGBFS to fresh concrete improves the workability of it and increases the set time 30–60 minutes below a temperature of 23 °C.

Because of its low density, GGBFS with high rates forms more paste, thus the workability of it develops. By means of good workability, the amount of rough aggregate is increased; hence the amount of the paste can be decreased. Nevertheless, Bush et al. (2000) obtained a mediocre reduction in sedimentation with GGBFS of 25%. Among the mixtures tested by Sanjayan and Sioulas (2000), for a constant w/b, the mixtures with slag showed 20–50% higher sedimentation. However, in the study of Duos and Eggers (1999), the sedimentation decreased by increasing the amount of slag below 23 °C.

10.3.2 Hardening concrete properties

The start and finish of setting time, hydration heat and bleeding issues are examined in a hardening concrete with a high ratio of GGBFS.

Setting time

Because GGBFS reacts with higher range water reducer (HRWR) than OPC, the slag use causes an increase in the setting time of the concrete (Bilim, 2006). This increase in drying time is much more at high-level replacements over 50% and at temperatures under 10 °C. The setting time of the concrete with GGBFS is longer than for OPC and the higher the amount of the slag, the longer the time. This characteristic is important in order to avoid cold joint formation during the pouring of large mass concrete. Hogan and Meusel (1981) found that both the initial and final setting time of the concrete with 40%, 50% and 65% slag is approximately one hour longer than the times of the concrete with OPC. Brooks et al. (2000) reported that the initial and final setting time of concrete with 40% or more GGBFS shows a noticeable delay respectively more than 11 and 17 hours.

Heat of hydration

The use of GGBFS in concrete decreases the maximum concrete temperature and prolongs the time to reach this maximum temperature by diminishing the hydration heat. Moreover, the rate of temperature change is reduced by the increased GGBFS rate. This feature is very beneficial in pouring large mass concrete since it avoids higher temperatures. The decrease in temperature during application depends on several factors such as the size of the section, the amount of the cement, the rate of the slag, the fineness of the binding components and their chemical composition (Bilim, 2006). The hydration speed of the slag and the temperature change is lower than those of the OPC. Therefore, because the partial replacement of the slag with OPC creates cement with lower hydration speed, the increase in the temperature of the concrete produced by such cement remains lower (Soroka, 1993). Alshamsi (1997) examined the effects of GGBFS and micro-silica on the hydration temperature of the concrete. In the conclusion of this study, he declared that the amount of the cement is significantly effective on the increase of hydration temperature and GGBFS reduces the temperature increase in cement paste more than micro-silica does. Additionally, it is reported that those materials substituted by cement certainly affect the time for reaching the peak temperature, and while GGBFS delays the time for reaching peak temperature, micro-silica accelerates it. Slag concrete both decreases the maximum temperature of the concrete and prolongs the time to reach this maximum temperature by reducing the hydration heat (Bilim, 2006).

Bleeding

Wainwright and Ait-Aider (1995) pointed out that the bleeding of fresh concrete is controlled by the fine materials in the mixture with respect to the amount and reactivity of the cement. While the replacement of GGBFS with OPC brings about a decrease in the speed of hydration, it causes an increase both in amount and speed of bleeding. The main reason for this increase is the delay in the hydration mechanism caused by the slag poured to cement paste; the decrease in the development speed of hydration products is a result of this. Olorunsogo (1998) examined the bleeding characteristics of the mortar produced by cement with slag and the distribution of the granule size of GGBFS. According to him, independently from the distribution of granule size and the w/b rate, for the mixture of both w/b rates of 0.35 and 0.45, the higher the slag rate, the higher the rate and capacity of bleeding. Whatever the rate of slag replacement and distribution of granule size were, for all mixtures, the increase in the w/b rate from 0.35 to 0.45 caused an increase at a percentage respectively of 86%, 83% and 71% in the bleeding, rate of mortar with 0%, 30% and 70% slag. This increase in bleeding, together with the increased rate of w/b, being a result of the larger average distance between the granules, is caused by weak cohesive attractive forces which occur because of the granule contact (Bilim, 2006).

10.3.3 Hardened concrete properties

Unit weight, compressive strength and strength development, tensile strength, splitting-tensile strength, modulus of elasticity, creep, shrinkage, permeability and chemical stability and capillary water absorption are examined in high-volume GGBFS.

Unit weight

The unit weight of hardened concrete with GGBFS is less than that of control concrete. Regarding the concrete age, the unit weight of the concrete with GGBFS increases more than normal concrete. While the unit weight in normal concrete does not change greatly for years, generally, it increases noticeably in concrete with GGBFS. The most significant reason for this is that the slag in the concrete with GGBFS produces new CSH, by combining with CH at the start of hydration. As is known, even if the concrete gets older, the hydration slowly continues. As a result of this the new C-S-H production goes on until the silicates finish deriving from slag and the unit weight value slightly increases. For further ages, however, since the density of the slag is less than that of the cement, the unit weight of the concrete with high-volume GGBFS is smaller than the control concrete. For instance, the dry unit weight of the 28-days-old concrete with 10%, 20%, 30% GGBFS are 5%, 10% and 14% less than the control concrete. The unit weight of the same concretes after 90 days are 2%, 5% and 6% less than the control concrete as well (Öz, 2006).

Compressive strength and strength development

When the equal amounts of cement and w/b rate are concerned, the slag concrete has less compressive strength at early ages and more compressive strength at late ages than has the concrete with OPC. Because GGBFS hydrates slower than OPC, the rate of strength development is lower at early ages for slag concrete. High-volume slag replacement instead results in a lower rate of strength development. However, when a suitable humidity condition is provided, the long-term strength of the slag concrete will probably be higher. This higher strength in late terms stems from the partial long-term strength of the slag and the micro-structure which is intensively hydrated due to the slower hydration reaction. When the temperature is increased, the rate of strength development of the slag concrete is also more than that of the concrete with OPC (Bilim, 2006). Yeau and Kim (2005) found that on 28th day the performance of the concrete with GGBFS is similar to that of control concrete; however, on 56th day the concrete with GGBFS has a superior performance to the concrete with OPC. According to the researchers, all of the concretes with up to 55% slag showed a lower compressive strength than the control concrete within 7 days. The results of the tests demonstrated that the hidden binding property of GGBFS, together with the increasing rate of slag replacement, slow down the strength development at early ages.

But at ages such as 28, 56 and 91 days, the strength development of the concrete with GGBFS is similar to greater than that of the slag-free control concrete with OPC.

Sakai et al. (1992), who studied the features of slag concrete which they produced mixing blast furnace slag of four different finenesses such as 300, 400, 500 and 600 m^2/kg into 300 m^2/kg cement with a replacement rate from 50% to 80%, found that the finer the slag, the higher the compressive strength according to the results of the experiments.

In research performed by Sanjayan and Sioulas (2000), for the compounds at 100 MPa pressure level, with slag of 70% when compared with the compounds with OPC, the strength percentage for the 7th, 28th, 56th and 91st days were respectively found to be 46%, 71%, 85% and 96%. The strength percentage on the 91st day, for the compounds with 50% slag at 40 MPa is 97%.

Güneyisi and Gesoğlu (2008) studied 0, 50, 60, 70 and 80% replacement rates of GGBFS in concretes and showed the changes of compressive strengths in accordance with the concrete's age and curing conditions. The compressive strength values of the control concretes are between 50.6 and 63.1 MPa and compressive strength values of the GGBFS concretes are between 27.9 and 66.2 MPa as shown in Fig. 10.1.

Flexural strength

The slag concrete, under a determined compressive strength, has a higher tensile strength than the concrete with OPC (Newman and Choo, 2003). Sivasundaram and Malhotra (1992) obtained nearly 8 MPa of strength in 14 days.

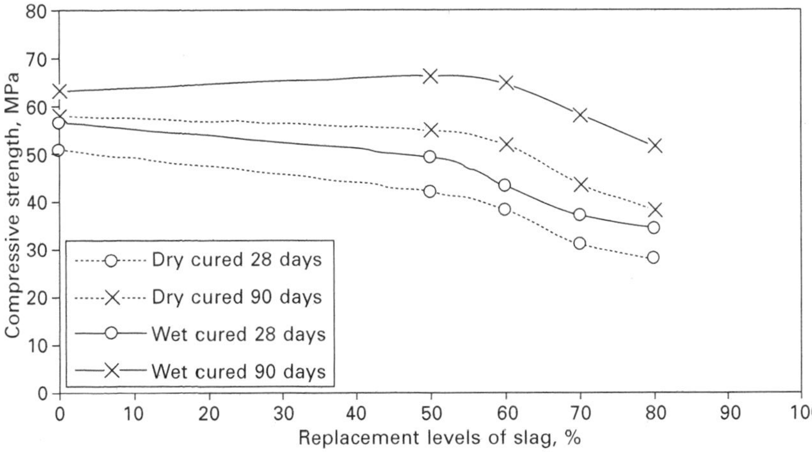

10.1 Effects of slag replacement and curing conditions on 28- and 90-day compressive strength of concretes (Güneyisi and Gesoğlu, 2008).

Splitting-tensile strength

The results of splitting experiments are similar to those for compressive strength. Increasing the slag amount, causes the splitting strength to decrease. Güneyisi and Gesoğlu (2008) stated that it is not valid for the samples containing 50% and 60% slag and cured in water for 90 days. Furthermore, as shown in Fig. 10.2, the decrease in splitting strength is less than the decrease in compressive strength. The results of splitting and compressive strength for the samples with 80% are lower.

Modulus of elasticity

Under a determined compressive strength, GGBFS has an improving impact on the modulus of elasticity of the concrete compared to concrete with OPC (Newman and Choo, 2003). At early ages, although the secant elasticity modulus of the concrete including 30–70% of slag and cured in water is similar to the concrete with OPC, it increases at later ages. Nevertheless, the modulus of elasticity of the concrete cured in air is lower than the concrete with high-volume slag cured in water at late ages (Brooks et al. 1992; Bilim, 2006). For equal strengths for 28 days, the modulus of elasticity of the concrete prepared with slag cement is a little bit higher.

Creep

Chern and Chan (1989) found that the creep behaviour of the concrete with OPC is similar to that with GGBFS. Under the cure conditions without

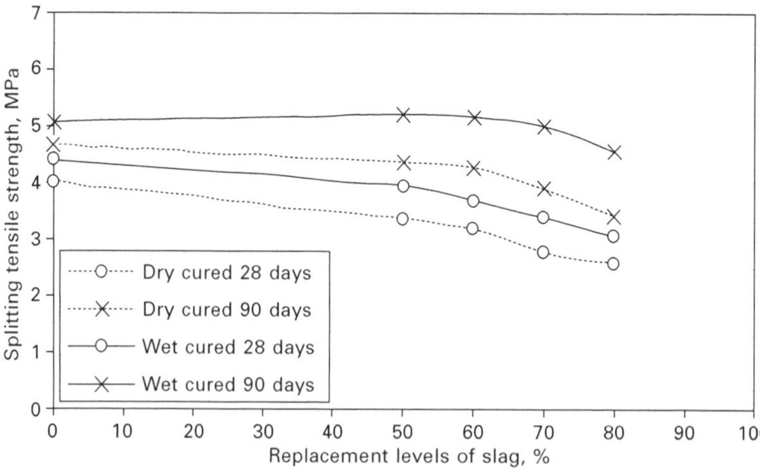

10.2 Effects of slag replacement and curing conditions on 28- and 90-day splitting tensile strength of concretes (Güneyisi and Gesoğlu, 2008).

humidity loss, along with the increasing slag replacement rate, the creep of the concrete decreases; on the other hand, under dry cure conditions, the concrete with GGBFS shows greater creep deformation than with OPC (Bilim, 2006). In their study on creep and drying shrinkage of high performance concrete, Jianyong and Yan (2001) substituted OPC with GGBFS and silica fume (SF) as a filling material and found that the highest creep value was obtained in the concrete in which SF and GGBFS were used together. Under the cure conditions without humidity loss, the more the slag ratio, the less the creep of concrete. With high replacement ratios such as 70%, the decrease in the creep of the concrete is about 50% (Newman and Choo, 2003). The decrease in creep is generally related to the higher strength that the concrete with slag gains at further ages. However, since there is less strength in later periods for the concrete under dry cure conditions, the differences observed in creep behaviour are less evident, and for most applications in which the drying shrinkage is not very much, the creep behaviour of the slag concrete is similar to of with OPC (Bilim, 2006).

Shrinkage

The results of research about the shrinkage of the slag concrete differ because of the experiment conditions and the various materials used, but those differences are not very significant. In general, the slag concrete affects shrinkage similarly to the concrete with OPC (Bilim, 2006).

It is known that GGBFS increases autogenous shrinkage. Lim and Wee (2000) obtained the highest shrinkage in 91 days for concrete with 50% GGBFS among other high-performance concretes with 30%, 50%, 65% and 50% GGBFS and 0.30 w/b. They showed that when the fineness of GGBFS is increased, autogenous shrinkage increases as well. A large amount of autogenous shrinkage causes deep cracks in concrete structures. Even though those cracks do not appear on surface, they create crucial tensile activity inside the concrete. Therefore, it is necessary to observe the autogenous shrinkage of the concrete with GGBFS. The autogenous shrinkage of the concrete with 0%, 10%, 30% and 50% GGBFS was examined. It was found that the autogenous shrinkage of the concrete with 10% GGBFS is higher, increasing autogenous shrinkage by a ratio of 6%.

Pore structure

The pozzolan mineral additives generally decrease the porosity of the concrete. As mentioned before, the slag in the concrete with GGBFS produces new C-S-H, by combining with CH in the first times of hydration. As is known, even if the concrete gets older, the hydration slowly continues.

Permeability

Slag is functional for long-term permeability in well-cured concretes, especially at high temperatures. The reasons for this are the continuation of hydration in slag concrete even after 28 days, in spite of no excessive change in porosity with the increasing slag amount, regarding the increased total finer gap distribution and curing temperature, while the gaps in the concrete with OPC is augmented the slag concrete is not affected by it (Bilim, 2006).

Sorptivity

The capillary water absorption values of the concrete formed by partial replacement of OPC with GGBFS decreases to an important extent, which is especially crucial for concretes which are exposed to moist and dry conditions. Taşdemir (2003), who found that the capillary water absorption value decreases by increasing compressive strength, also reports that the capillary water absorption value is vulnerable to the cure conditions, and the lowest coefficient of capillary water absorption is obtained from the samples which are kept under 20 °C in lime-saturated water cures during 28 days (Bilim, 2006).

Figure 10.3 shows the variation in water absorption rates with concrete age and curing condition for the control and slag concretes. It is clear that rate of water absorption decreases systematically with an increase in curing period (from 28 to 90 days), and the gradients of the water absorption tend to decrease with increase in the replacement level of slag. Generally, slag concrete performed better than the control concrete and marked improvements

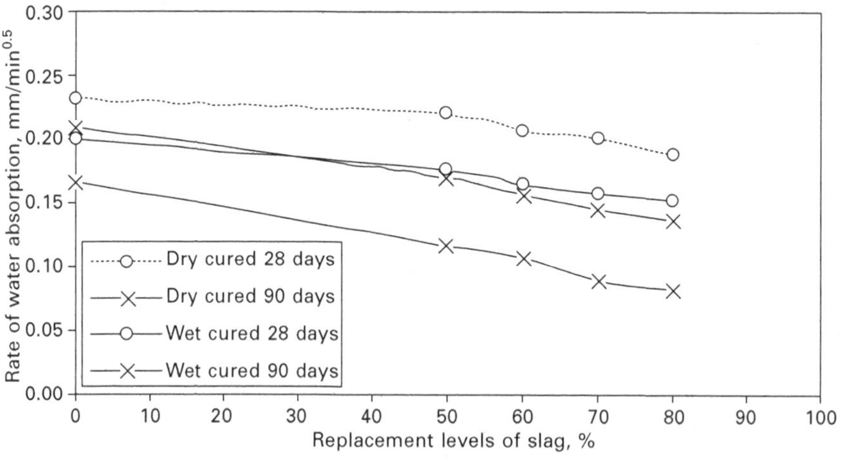

10.3 Effects of slag replacement and curing conditions on 28- and 90-day water absorption rate of concretes (Güneyisi and Gesoğlu, 2008).

in terms of lower rate of water penetration through capillary suction were apparent, particularly under wet curing condition.

Furthermore, the reduction in the water absorption with increasing test age was about 12% and 15% for air and wet-cured conventional concretes, respectively, while it was about 18–22% and 30–35% for air and wet-cured slag concretes, respectively. This reduction in the water absorption with age indicates better performance of slag blended cement concretes over conventional concrete. The improvement in the performance of the air-cured samples might be explained by the surface carbonation of the air-cured concrete resulted in enhanced reduction in sorptivity.

Güneyisi and Gesoğlu (2008) showed the correlation between water absorption with the curing conditions and the age of the concrete in Fig. 10.4. Similar to the rate of water absorption test results, water absorption characteristics of the concrete specimens decreased with increasing slag content, irrespective of curing condition and testing age. However, the differences in water absorption are more characteristic after 28 days of curing; water absorption of the GGBFS concretes are less than control concretes. At the age of 90 days, water absorption decreases with the ratio of between 12 and 15% at wet and dry cured control concretes. Decreasing ratio of water absorption at dry cured GGBFS concretes are between 18–22% and at wet cured GGBFS concretes are between 30 and 35%. This reduction in the water absorption with age indicates better performance of slag blended cement concretes over conventional concrete. The improvement in the performance of the air-cured samples might be explained by the surface carbonation of the air-cured concrete resulting in enhanced reduction in sorptivity.

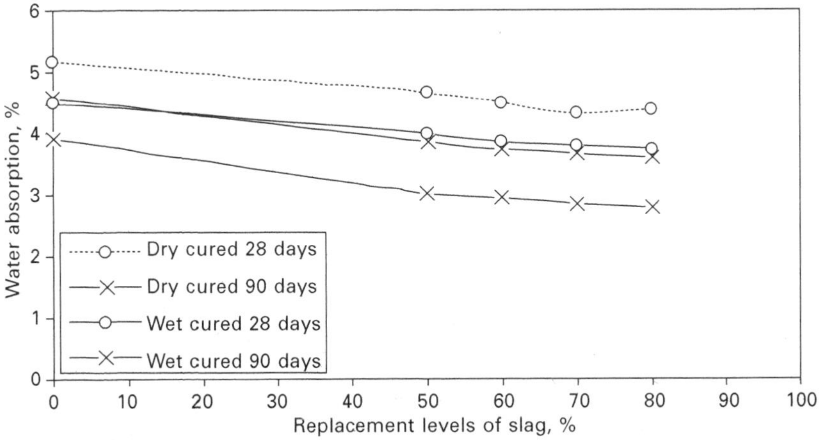

10.4 Effects of slag replacement and curing conditions on 28- and 90-day water absorption of concretes (Güneyisi and Gesoğlu, 2008).

10.4 Durability of ground granulated blast furnace slag (GGBFS) concrete

10.4.1 Sulphate resistance

In a study in which Hooton and Emery (1990) used solutions of $MgSO_4$, $CaSO_4$ and Na_2SO_4, 130 different cements were used. Proportionally with the amount of GGBFS, even though the concrete obtained by respectively 45% and 72% GGBFS replacement by sulphate OPC, did not show a deterioration inside 3000 mg/l SO_4 solution for 10.5 years, the concrete with C_3A of a ratio from 3.5% to 12.3% with OPC was badly affected. Frearson (1986) observed that the sulphate resistance of the concrete is affected by GGBFS content rather than water/cement content. The reason for this is related to the decrease in permeability caused by GGBFS. Furthermore, it is known that the increase in sulphate resistance, caused by GGBFS, is due to the decrease in C_3A content. Additionally, slag decreases $Ca(OH)_2$, increases the gels of C-S-H, so improving the strength, and weakens the permeability of the concrete (Abd El Aziz et al., 2005).

10.4.2 Resistance to chloride ion penetration and reinforcement corrosion

Resistance to the effects of chloride

Aldea et al. (2000), made a study on the effects of different cure conditions such as autoclave (175 °C, 0.5 MPa), steam cure (80 °C) and normal cure (20 °C, 100% RH) on the properties of concrete produced by four different slag replacements (0, 25, 50 and 75%). They reported that the cure at room conditions is the best way to improve durability, that it is best to choose steam cure instead of autoclave cure when early strength development is desired, and finally, since a reduction in chloride permeability occurs by increasing slag replacement rate, depending also on application, the use of different slag rates is possible to increase the strength and to decrease the chloride permeability (Bilim, 2006).

Chloride permeability

The slag cement or slag concrete is more resistant to penetration of chloride ions into the concrete than OPC. This positive effect stems from the chemical combination of chloride ions and slag hydration products which reduce the chloride stream in concrete and at the same time decreasing permeability of slag cement. Moreover, this increased strength has a potential to decrease the corrosion risk of steel in concrete structures. The important point in chloride permeability is the chloride binding capacity of the cement. Through studies, it is known that C_3A binds chloride. Even though OPC is supposed to bind more chloride ions and to decrease chloride penetration due to having C_3A more than slag cement, the studies prove the contrary. In the analysis done by cement with GGBFS, it is observed that the penetration of chloride ions in OPC is reduced, because the gels of C-S-H bind more chloride. Owing to the fact that the C-S-H occurring after the hydration of slag cement is more than OPC, due to binding chloride ions, the chloride permeability of slag concrete mainly decreases. Sato et al. (1998) examined the effects of different cure conditions (natural environment cure and wet cure) on chloride permeability of concrete with 70% GGBFS. At the end of their study, they observed that the use of GGBFS at a rate of 70% decreases the chloride ion permeability.

Güneyisi and Gesoğlu (2008) mentioned that the chloride permeability decreases considerably in concretes that include more than 50% of GGBFS, as shown in Fig. 10.5. For example, permeability value of 70 and 80% of GGBFS concretes at the age of 90 day is 1000 coulombs and this value is qualified as wonderful in accordance with ASTM 1202.

Corrosion resistance

Song et al. (2003) observed the free chloride permeability of the concrete with GGBFS and found that the higher the amount of GGBFS, the lower the values of free chloride permeability. Additionally, they found that the free chloride permeability of the concrete with 70% GGBFS is 43.8% less than of the control concrete. Yeau and Kim (2005) examined the corrosion

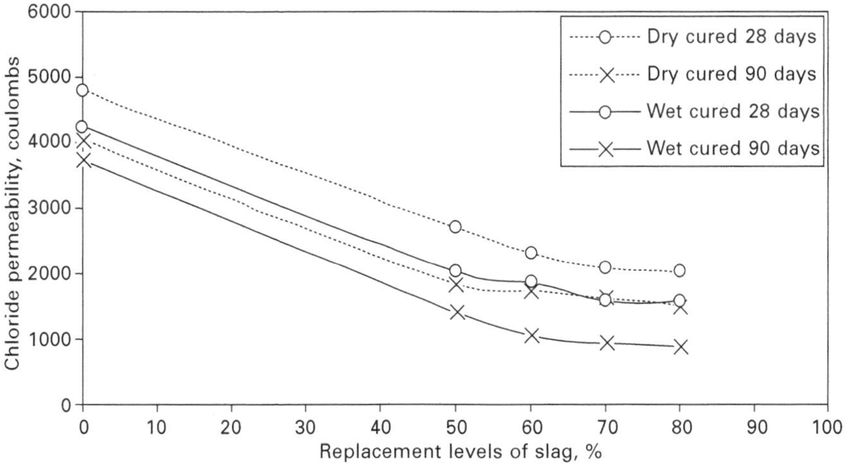

10.5 Effects of slag replacement and curing conditions on 28- and 90-day chloride permeability of concretes (Güneyisi and Gesoğlu, 2008).

resistance of the concrete with GGBFS and discovered that the steel corrosion potential of the concrete with GGBFS is less than 250 mV. This corrosion value is a very small value. In addition to that, they studied the surface of the steel exposed to corrosion according to the percentage of GGBFS and found that the surface under corrosion decreases, while the amount of GGBFS increases.

10.4.3 Durability to seawater

In his experimental study in which he worked with mortar samples completely sunk in seawater, Regourd (1986) examined the effect of the slag amount against seawater attack and pointed out that by increasing GGBFS amount in cement the expansion decreases. Osborne (1992), who a researched the performance of the concrete with slag and OPC of different C_3A content and seawater, reported that the strength development of the concrete with 70% slag that is completely sunk in seawater improves and it is exposed to fewer chemical attacks with a mediocre chloride permeability.

10.4.4 Resistance to carbonation

Carbonation occurs when CO_2 penetrates into the concrete. Both the porosity and the permeability of the concrete have an important role in the mechanism of carbonation. The carbonation of the concrete depends on the cure conditions, water/binding rate, the temperature at which the concrete is exposed to carbonation and relative humidity, and the properties of the mineral additives used. The decreased permeability delays the CO_2 penetration deeply through the concrete (Bilim, 2006). Pal et al. (2002) examined the carbonation depth of the concrete with 0%, 30%, 50% and 70% of GGBFS and they observed that the greater the slag rate, the less the carbonation depth.

Sulapha et al. (2003) found that for low Blaine fineness slag (4500 cm^2/g), although the slag concretes were denser, the carbonation rate increased with increasing amounts of slag. This was thought to be attributable to less CH being available due to the pozzolanic reaction, so the carbonation must progress deeper to get at the available CH. Less material is available that is prone to carbonation per unit area to react with CO_2. Initially the pozzolanic reaction is slow, thus porosity is higher and CO_2 diffusion is rapid. However, for higher fineness slags (6000–8000 cm^2/g), carbonation rates were lower than that for OPC, thus pore modification was more dominant than the change in CH content. So, slag mixes tend to have greater depths of carbonation and are more permeable (although not necessarily more porous) and therefore would tend to scale more, especially under severe environmental conditions (Richardson, 2006).

10.4.5 Freeze–thaw durability

Lane and Ozyildirim (1999) obtained perfect results in their study with the samples of 60% GGBFS and OPC. However, for all experiments, better results are gained for the mixtures with GGBFS rather than for with OPC. The highest resistance factor was obtained in the sample with 50% GGBFS. Pigeon and Regourd (1983) compared the mixtures with OPC and 66% slag. For both groups of samples, the same values are obtained for length changes, massive loss and changes in dynamic modulus of elasticity and compressive strength, air-gap systems and air-gap conditions.

10.4.6 Acid resistance

In their study of the long- and short-term performance of concrete with GGBS and FA, Li and Zhao (2003) prepared concretes with 40% FA and OPC. After 3 days of wet cure, among the samples left in 2% H_2SO_4 solution, under room temperature for times of 4, 8, 16 and 50 weeks, it was observed that concretes with 15% GGBFS and 25% FA are have superior protection against acid attack than concretes with OPC and 40% FA. The change in weight at the end of 50th week is approximately 8%. The weak resistance of concrete with OPC against H_2SO_4 attack is due to the presence of large pores and free $Ca(OH)_2$ in the concrete. The concrete with 40% FA is also poorly resistant to H_2SO_4 attack at an early age. The reason is that there are numerous un-hydrated FA particles in the concrete, and the matrix is significantly porous at an early age. Furthermore, the amount of C-S-H gel in the concrete with 40% FA formed by cement hydration is less than the amount of C-S-H in concretes with OPC or with FA and slag (Bilim, 2006).

10.4.7 Alkali–silica reaction

The replacement of GGBFS with some cement can reduce the amount of alkali. There are several alternative solutions to reduce the destruction deriving from the ASR risk such as constraining the alkali content of the concrete, use of GGBFS or FA, use of low-alkali cement and use of non-reactive aggregate. Kwon (2005), in his study with high-strength concrete (HSC) of 500 kg/m^3 cement, found that for ASR, the overdose alkali content in HSC is a disadvantage and that harmful volume expansions are not observed in concrete if aggregate is used which is non-reactive due to chemical methods and mortars. On the other hand, it is reported that the use of low-alkali cement or 30% GGBFS instead of OPC will avoid the expansions caused by ASR: an increase of approximately between 1.0 and 1.7 kg/m^3 is observed in the alkali content of the concrete with a replacement rate of

30%. Finally, increased slag fineness is effective in avoiding ASR (Bilim, 2006).

10.4.8 De-icing salt scaling resistance

Richardson (2006) has studied carbonation effects with regard to carbonation depth and rate on porosity, permeability, available calcium hydroxide and carbonation final products. It has been reported in many cases that slag concrete can have high frost resistance, if properly air entrained. However, the resistance to salt scaling of concrete with high (greater than 60%) slag contents cannot be improved by proper air entrainment alone.

10.4.9 Durability of high- and low-temperature effect

By use of GGBFS in concrete, the resistance of the concrete under high temperatures increases as a result of pozzolanic activity of the slag. Poon et al. (2001) exposed high-strength concretes including SF, FA and GGBFS to different temperature conditions maximally to 800 °C. According to experimental results, the concrete with FA and GGBFS, especially under 600 °C showed the best performance compared to control concrete. In previous studies, it was found that the resistance of the concrete with 40% GGBFS and exposed to 200 °C high temperature increased at a rate of 15% in comparison to those tested under room conditions. While the remaining resistance of the concrete with 40% under 400, 600 and 800 °C is 100%, 61 and 29%, the resistance of the controlled concrete is 89, 58 and 24%.

An earlier work on the performance of slag cement at elevated temperatures was done by Grainger (1980). He tested four cement pastes containing 0%, 50%, 70% and 90% replacement of slag by weight with OPC. The maximum tested temperature was 500 °C with an interval of 100 °C. All slag-cement paste specimens experienced an increase in strength between 100 and 250 °C. The 70% slag replacement showed the best results with a residual compressive strength of 190% of the original strength at 110 °C. Moreover, the residual strength of this paste was higher at all temperatures than the original strength. The other two slag-cement paste specimens also showed better residual strengths than the pure cement paste specimen. Sarshar and Khoury (1993) prepared cement paste and concrete specimens incorporating 65% slag by weight of cement and firebrick aggregates. The results were compared with pure OPC cement paste/concrete and 30% pulverized fuel ash (PFA) cement paste. The maximum temperature was 700 °C, while the residual properties were measured at every 100 °C interval. They found that the slag-cement paste and concrete gave the best results among all the specimens tested. The residual compressive strengths of slag concrete were 102% and 80% of the initial cold strength at 450 and 600 °C.

10.4.10 Abrasion durability

The abrasion durability of concrete with GGBFS increases by applying suitable and sufficient cure. While insufficient cure conditions decrease the abrasion durability to a large extent, slag concrete is affected by that more than concrete with OPC. In their study on abrasion durability and its mechanical features, Fernandez and Malhotra (1990) reported that the strength development of slag concrete indicates the partial usage of GGBFS instead of cement in concrete; however, the abrasion durability of slag concrete is lower than that of control concrete without slag. Provided that suitable and sufficient cure is applied, the use of GGBFS in concrete provides some advantage in abrasion durability (Newman and Choo, 2003).

For 28 days and above, to add 50–60% GGBFS to concrete results in higher resistance, whereas to add GGBFS over 80% results in lower resistance. It is understood that to obtain the maximum resistance, the optimum GGBFS rate is 40–60%. The more vitreous content, fineness and alkali content of GGBFS and greater fineness and alkali content of OPC used together will improve the strength development. Using GGBFS up to 80% can increase the resistance to 28 MPa. Therefore, it will be useful to prepare specifications for mixture design and regulations concentrated on the factory operating requirements for high-volume GGBFS and real expectations from the concrete. Hence, more qualified concrete will be obtained regarding workability, resistance and strength, as well as more economic concrete by waste reduction. At the same time air pollution could be reduced.

10.5 Future trends

The following are the types of concrete that are expected to be used in the concrete production applications of today and in the future, as the results of the studies described above:

- *Self-compacting concrete*: Self-compacting concrete (SCC), which is developed especially to be used in structures with dense reinforcement at the end of the 1980s in Japan, has a perfect workability and high resistance to segregation. SCC can easily furnish itself to the dense reinforced areas without using any vibration. The fresh SSC can be defined as a concrete which has the ability to fill the moulding and hold the reinforcement by means of its own weight, while maintaining homogeneity. There are positive effects of GGBFS used instead of cement on cost, and hydration will improve the strength feature of the concrete (Yalçınkaya and Yazıcı, 2010).
- *Roller compacted concrete*: Roller compacted concrete (RCC) is a low-viscosity concrete which is laid in plates before hardening. It is able to carry at least 5 tonnes of roller in order to be compacted. RCC is economic

because of its very low cement content. This small amount of cement in RCC brings about a decrease in hydration heat. Hence, RCC is not used for mass concrete constructions such as barrages. Karimpour (2010) examined the effects of the interval between mixing and compacting in RCC including high-volume GGBFS. For mixture with 0, 25, 50 and 75% of GGBFS, he observed that the interval between mixing and compacting mainly influences the compressive strength with increasing amounts of GGBFS. According to Karimpour, the increased distance between cement granules because of increased amount of GGBFS is the reason of this. For RCC, concrete without sedimentation is produced.

- *Non-slump concrete*: Togawka and Nakamoto (1992) noted that up to 30% GGBFS used instead of fine aggregate is useful for the compacting and compressive strength of RCC; the surface area of the GGBFS affects the development of strength. They added that larger surface area at early ages and smaller surface area at late ages show higher strength. They also mentioned that the GGBFS with small surface area is more effective at reducing the temperature of concrete.
- *Reactive powder concrete*: Yazıcı et al. (2008) mentioned that for reactive powder concretes with 20, 40 and 60% GGBFS even at 40 and 60%, they obtained compressive strength over 200 MPa. Yazıcı et al. (2010) used concrete with 20, 40 and 60% GGBFS and applied standard, steam and autoclave cure. They obtained compressive strength over 250 MPa in concrete that had autoclave cure with 20% GGBFS addition. In the presence of external pressure, they observed that the compressive strength up to 400 MPa. With the addition of 40 and 60% GGBFS they reached strengths of 210 and 254 MPa by standard cure in 90 days, obtaining 33.2 MPa of flexural strength, 6161 N/m of fractural energy and 17 210 N mm of toughness value and approximately the same results with autoclave cure.
- *Lightweight concrete*: Arreshvhina et al. (2006) observed that compressive strength in aerated concrete with GGBFS addition increases at a rate of 8–63% depending on the cure conditions. Adding a high volume of GGBFS improves compressive strength in aerated concretes and contributes to a better microstructure formation.
- *Ungrounded BFS*: Yüksel and Genç (2007) mentioned that, for concretes in which they use 10, 20, 30, 40, 50% of GBFS and BA (bottom ash) as aggregate, by increasing replacement rates, the resistance of the concrete decreases. BA gives rise to a greater reduction in resistance than GGBFS does and replacement over 40% is more effective. Yüksel et al. (2007) explored the resistance of concrete prepared by industrial waste such as polypropylene fibre, GBFS and BA as fine aggregate towards 800 °C high temperatures. They showed how it is possible to use these wastes with a ratio of 50% of aggregate which covers 30% volume of the concrete.

Topçu and Bilir (2010), in their studies on shrinkage done by GBFS as fine aggregate, determined that by increasing the GBFS rate, there is a decrease in shrinkage cracks. The reason of this is the gap form related to GBFS structure and, linked with the same reason, flexural strength, compressive strength and modulus of elasticity decrease.

- *Alkali-activated slags*: By means of the activation of GGBFS with alkalis such as NaOH, $NaCO_3$ and $NaSiO_3$ super-qualified concrete can be produced. Topçu and Canbaz (2008) found that alkali-activated mortar with GGBFS does not exceed the danger limit in ASR.

10.6 References and further reading

Abd El.Aziz M, Abd El.Aleem S, Heikal M and Didamony H El (2005), 'Hydration and durability of sulphate-resisting and slag cement blends in Caron's Lake water', *Cement and Concrete Research*, **35** (8), 1592–1600.

Aldea C-M, Young F, Wang K and Shah S P (2000), 'Effects of curing conditions on properties of concrete using slag replacement', *Cement and Concrete Research*, **30** (3), 465–472.

Alshamsi A M (1997), 'Microsilica and ground granulated blast furnacc slag effects on hydration temperature', *Cement and Concrete Research*, **27** (12), 1851–1859.

Arreshvhina N, Fadhadli Z, Warid H M, Zuhairy A H and Roswadi (2006), 'Microstructural behavior of aerated concrete containing high volume of GGBFS', *Proceedings of the 6th Asia-Pacific Structural Engineering and Construction Conference*, Kuala Lumpur, Malaysia.

Atis C D and Bilim C (2007), 'Wet and dry cured compressive strength of concrete containing ground granulated blast-furnace slag', *Building and Environment*, **42** (8), 3060–3065.

Bilim C (2006), '*The use of ground granulated blast furnace slag in cement based materials*', Ph. D Thesis, Department of Civil Engineering Institute of Natural and Applied Sciences, University of Çukurova, Adana, Turkey, 206p.

Bouikni A, Swamy R N and Bali A (2009), 'Durability properties of concrete containing 50% and 65% slag', *Construction and Building Materials*, **23** (8), 2836–2845.

Brooks J J, Wainwright P J and Boukendakji M (1992), 'Influence of slag type and replacement level on strength, elasticity, shrinkage and creep of concrete', *Proc. CANMET/ACI 4th Intern. Conf. on Fly Ash, Silica Fume, Slag and Natural Pozzolans in Concrete*, ACI SP-132, Vol. 2, Ed. V M Malhotra, ACI, Mich., Istanbul, pp. 1325–1341.

Brooks J J, Johari M A M and Mazloom M (2000), 'Effect of admixtures on the setting time of high-strength concrete', *Cement and Concrete Composites*, **22** (4), 293–301.

Bush Jr T D, Russell B W, Zaman M M, Hale M W and Ling T A (2000), 'Improving Concrete Performance Through the Use of Blast Furnace Slag', *Final Rpt*, FHWA-OK 00 (01), Oklahoma Dept. of Transportation, p.133.

Chern J C and Chan Y W (1989), 'Deformations of concretes made with blast-furnace slag cement and ordinary Portland cement' *ACI Materials Journal*, **86** (4) 372–382.

Duos C and Eggers J (1999), 'Evaluation of Ground Granulated Blast Furnace Slag in Concrete (Grade 120)', *Rpt. No. FHWA/LA-99/336*, Louisiana Trans. Res. Center, Baton Rouge, Louisiana, October, p.45.

Erdoğan T Y (1995), 'Ground Granulated Blast-Furnace Slag and Its Use', *Symposium Proceeding of the Use of Industrial Waste in The Construction Sector*, TMMOB – Chamber of Civil Engineer, Ankara, Turkey, pp. 1–13.

Fernandez L and Malhotra V M (1990), 'Mechanical properties, abrasion resistance and chloride permeability of concrete incorporating granulated blast-furnace slag', *Cement Concrete and Aggregates*, **12** (2), 87–100.

Frearson J P H (1986), 'Sulphate resistance of combination of Portland cement and ground granulated blast furnace slag', *Proc., 2nd Intern. Conf. on Fly Ash, Silica Fume, Slag and Natural Pozzolans in Concrete*, Madrid, Spain, ACI, SP-91, Detroit, USA, pp. 1495–1524.

Grainger B N (1980), *Concrete at High Temperature*, Central Electricity Research Laboratories, UK.

Güllü H (1996), '*Properties of fresh concrete containing fly ash and blast-furnace slag*', MSc Thesis, Gaziantep University, Gaziantep, Turkey, 153p.

Güneyisi E and Gesoğlu M (2008), 'A study on durability properties of high-performance concretes incorporating high replacement levels of slag', *Materials and Structures*, **41** (3), 479–493.

Hogan F J and Meusel J W (1981), 'Evaluation for durability and strength development of a ground granulated blast furnace slag', *Cement Concrete and Aggregates*, **3** (1), 40–52.

Hooton R D and Emery J J (1990), 'Sulphate resistance of a Canadian slag cement', *ACI Material Journal*, **87** (6), 547–555.

Jianyong L and Yan Y (2001), 'A study on creep and drying shrinkage of high performance concrete', *Cement and Concrete Research*, **31** (8), 1203–1206.

Karimpour A (2010), 'Effect of time span between mixing and compacting on roller compacted concrete (RCC) containing ground granulated blast furnace slag (GGBFS)', *Construction and Building Materials*, **24** (11), 2079–2083.

Kwon Y (2005), 'A study on the alkali–aggregate reaction in high-strength concrete with particular respect to the ground granulated blast-furnace slag effect', *Cement and Concrete Research*, **35** (7), 1305–1313.

Lane D S and Ozyildirim H C (1999), 'Combinations of Pozzolans and Ground, Granulated, Blast-Furnace Slag for Durable Hydraulic Cement Concrete', *Rpt. No. VTRC 00-R1*, Virginia Dept. of Transportation, p.189.

Li G and Zhao X (2003), 'Properties of concrete incorporating fly ash and ground granulated blast-furnace slag', *Cement and Concrete Composites*, **25** (3), 293–299.

Lim S N and Wee T H (2000), 'Autogenous shrinkage of ground granulated blast-furnace slag concrete', *ACI Materials Journal*, **97** (5), 587–593.

Mehta P K (1989), 'Pozzolanic and Cementitious By Products in Concrete- Another Look', *3rd Intern. Conf. on The Use of Fly Ash, Silica Fume, Slag and Other Mineral by-Products in Concrete*, Trondheim, Norway, ACI SP114-01, Vol. 114, pp. 1–44.

Newman J and Choo B S (2003), *Advanced Concrete Technology: Constituent Materials*, Butterworth-Heinemann, Oxford.

Ohama Y, Madej J and Demura K (1995), 'Efficiency of Finely Ground Blast Furnace Slags in High-Strength Mortars', Infront Outback – *Proc. 5th CANMET/ACI Int. Conf. Fly Ash, Silica Fume and Natural Pozzolans in Concrete*, Ed. V M Malhotra, Milwaukee, WI, ACI SP153-54, Vol. 153, pp. 1031–1050.

Olorunsogo F T (1998), 'Particle size distribution of GGBS and bleeding characteristics of slag cement mortars', *Cement and Concrete Research*, **28** (6), 907–919.

Osborne G J (1992), 'The Performance of Portland and Blast Furnace Slag Cement

Concretes in Marine Environments', *Proc. CANMET/ACI 4th Intern.Conf. on Fly Ash, Silica Fume, Slag and Natural Pozzolans in Concrete*, Ed. V M Malhotra, ACI SP132-70, Vol. 132, Istanbul, Turkey, pp. 1303–1324.

Öz A (2006), '*Thermo-mechanical properties of self compacting concrete containing natural zeolite and blast furnace slag*', MSc Thesis, Ataturk University, Institute of Natural and Applied Sciences, Erzurum, Turkey, p.90.

Pal S C, Mukherjee A and Pathak S R (2002), 'Corrosion behaviour of reinforcement in slag concrete', *ACI Materials Journal*, **99** (6), 521–527.

Pigeon M and Regourd M (1983), 'Freezing and thawing durability of three cements with various granulated blast furnace slag contents', *Fly Ash, Silica Fume, Slag and Other Mineral By-Products in Concrete*, ACI SP79-52, Ed. V M Malhotra, Vol. 79, pp. 979–998.

Poon C S, Azhar S, Anson M and Wong Y L (2001), 'Comparison of the strength and durability performance of normal- and high-strength pozzolanic concretes at elevated temperatures', *Cement and Concrete Research*, **31** (9), 1291–1300.

Reeves C M (1986), 'The use of ground granulated blast furnace slag to produce durable concrete', *Improvement of Concrete Durability*, Thomas Telford Limited, pp. 59–95.

Regourd M (1986), 'Slags and slag cements', *Cement Replacement Materials*, R.N. Swamy, Surrey University Press, pp. 73–79.

Richardson D N (2006), 'Strength and Durability Characteristics of a 70% Ground Granulated Blast Furnace Slag Concrete Mix', *Missouri Department of Transportation Organizational Results Final Report*, RI99-035/RI99-035B, p.85.

Sakai K, Watanabe H, Suzuki M and Hamazaki K (1992), 'Properties of Granulated Blast-furnace Slag Cement Concrete', *Proc. of CANMET/ACI 4th Intern. Conf. on Fly Ash, Silica Fume, Slag and Natural Pozzolans in Concrete*, Ed. V M Malhotra, İstanbul, Turkey, Vol. 132, pp. 1367–1383.

Sanjayan J G and Sioulas B (2000), 'Strength of slag cement concrete cured in place and in other conditions', *ACI Materials Journal*, **97** (5), 603–611.

Sarshar R and Khoury G A (1993), 'Material and environmental factors influencing the compressive strength of unsealed cement paste and concrete at high temperatures', *Magazine of Concrete Research*, **45** (162), 51–61.

Sato N M, Agopyan V and Quarcioni V A (1998), 'Permeability of Portland cement concrete and blast furnace slag cement concrete', *Proc. of the 2nd Intern. Conf. on Concrete Under Severe Conditions*, Eds. O E Sakai, K Sakai, N Banthia, CONSEC'98, Romsù, Norway, Vol. I, London and New York, pp. 2065–2073.

Sivasundaram V and Malhotra V M (1992), 'Properties of concrete incorporating low quantity of cement and high volumes of ground granulated slag', *ACI Materials Journal*, **89** (6), 554–563.

Song H W, Kwon S J, Lee S W and Byun K J (2003), A study of chloride ion penetration in ground granulated blast furnace slag concrete, *J. Korea Concr. Inst.*, **15** (3) 400–408.

Soroka I (1993), *Concrete in Hot Environments*, National Building Research Institute, Faculty of Civil Engineering, Technion-Israel Institute of Technology, Haifa, p.247.

Sönmez H T (2008), '*The strength properties of self-compacting concretes containing granulated blast-furnace slag*', MSc Thesis, Sakarya University, Institute of Natural and Applied Sciences, Sakarya, Turkey, 58p.

Sulapha P, Wong S F, Wee T H and Swaddiwudhipong S (2003), 'Carbonation of concrete containing mineral admixtures', *ASCE J of Materials in Civil Engineering*, **15** (2), 134–143.

Taşdemir C (2003), 'Combined effects of mineral admixtures and curing conditions on the sorptivity coefficient of concrete', *Cement and Concrete Research*, **33** (10), 1637–1642.

Togawka K and Nakamoto J (1992), 'Study of Effects of Blast-furnace Slag on Properties of No-slump Concrete Mixtures', *Proc. 4th Intern. Conf. of Fly Ash, Silica Fume, Slag, and Natural Pozzolans in Concrete*, ACI SP132-75, Istanbul, Turkey, Vol. 132, pp.1401–1412.

Tomisawa T and Fujll M (1995), 'Effects of high fineness and large amounts of ground granulated blast-furnace slag on properties and microstructure of slag cement', *Fly Ash, Silica Fume, Slag and Natural Pozzolans in Concrete*, Ed. V M Malhotra, ACI SP 153-50, Vol. 153, pp. 951–974.

Topçu İ B and Bilir T (2010), 'Effect of non-ground-granulated blast-furnace slag as fine aggregate on shrinkage cracking of mortars', *ACI Materials Journal*, **107** (6), 545–553.

Topçu İ B and Canbaz M (2008), Alkali-silica Reaction of Alkali Activated Slag Mortars, *Seminary of Structure Mechanics – 2008*, METU-ESOGU, Eskişehir, Turkey, pp. 117–124.

Wainwright P J and Ait-Aider H (1995), 'The influence of cement source and slag additions on the bleeding of concrete', *Cement and Concrete Research*, **25** (7), 1445–1456.

Yalçınkaya C and Yazıcı H (2010), 'The effect of high volume GGBFS replacement on mechanical performance of self-compacting steel fiber reinforced concrete', *9th Intern. Cong. on Advances in Civil Engineering*, Karadeniz Technical University, Turkey, pp.1–10.

Yazıcı H (2007), 'The effect of curing conditions on compressive strength of ultra high strength concrete with high volume mineral admixtures', *Building and Environment*, **42** (5), 2083–2089.

Yazıcı H, Yiğiter H, Karabulut A S and Baradan B (2008), 'Utilization of fly ash and ground granulated blast furnace slag as an alternative silica source in reactive powder concrete', *Fuel*, **87** (12), 2401–2407.

Yazıcı H, Yardımcı M Y, Yiğiter H, Aydın S and Türkel S (2010), 'Mechanical properties of reactive powder concrete containing high volumes of ground granulated blast furnace slag', *Cement and Concrete Composites*, **32** (8), 639–648.

Yeau K Y and Kim E K (2005), 'An experimental study on corrosion resistance of concrete with ground granulated blast-furnace slag', *Cement and Concrete Research*, **35** (7), 1391–1399.

Yüksel I and Genç A (2007), 'Properties of concrete containing nonground ash and slag as fine aggregate', ACI *Materials Journal*, **104** (4), 397–403.

Yüksel I, Bilir T and Özkan Ö (2007), 'Durability of concrete incorporating non-ground blast furnace slag and bottom ash as fine aggregate', *Building and Environment*, **42** (7), 2651–2659.

Zhang M, Bilodeau A, Malhotra V M, Kim K and Kim J (1999), 'Concrete incorporating supplementary cementing materials: effect on compressive strength and resistance to chloride-ion penetration', *ACI Materials Journal*, **96** (2), 181–189.

11
Recycled glass concrete

K. ZHENG, Central South University, China

DOI: 10.1533/9780857098993.2.241

Abstract: The chapter begins by introducing sources of waste glass and ways of recycling waste glass in concrete. It then summarizes fresh properties and mechanical properties of recycled glass concrete and discusses how recycled waste glass affects these properties. The chapter elaborates on durability of recycled glass concrete, especially on alkali–silica reactivity since this is the main concern for recycled glass concrete. Finally, the chapter presents suggestions for further studies on recycled glass concrete, and proposes future trends of using recycled glass in concrete in more economic and eco-efficient ways.

Key words: alkali–silica reactivity, concrete, mechanical properties, recycled glass.

11.1 Introduction

From his earliest origins, man has been making use of glass, and man-made glass dates back to 4000 BC. Glass as a material in its own right will be widely used in the future. It has a wide range of applications and is produced in many forms and types; these include packaging or container glass, flat glass, bulb glass, cathode ray tube glass (TV screens, monitors, etc.). All glass products have limited lifespan, though glass itself is an inert material. Hence, a lot of waste is generated and a huge amount of glass is produced annually. Even in the production of glass and/or glass products, there is the possibility of waste generation. The United Nations estimates the volume of yearly disposed solid waste to be 200 million tons, 7% of which is made up of glass over the world (Topcu and Canbaz, 2004).

The recycling of waste glass poses a major problem worldwide. Glass is a unique inert material that can be recycled many times without changing its chemical properties. Recycled glass has traditionally been used in glass manufacture. This practice brings about environmental advantages such as reducing energy and raw materials consumption, related air and water pollution. However, not all used glass can be recycled into new glass because of impurities, cost or mixed colours. In Europe, about 5 700 000 tons of container glass was not collected for recycling in 2010 (EU Container Glass Federation, 2012). In the United States, about 600 000 tons/year of the collected glass is not actually recycled into new glass (US Environmental Protection Agency, 2010). Large amounts of waste glass are still sent to landfill as residue. Since glass is not biodegradable, landfills do not provide an environmentally friendly solution. There is a need to develop new markets to recycle waste glass.

Use of recycled materials in construction is among the most attractive options because of the large quantity, low quality requirements and widespread sites of construction. The main applications include a partial replacement for aggregate in asphalt concrete, as fine aggregate in unbond base course, pipe bedding, landfill gas venting systems and gravel backfill for drains (Shi and Zheng, 2007). The use of recycled waste glass in Portland cement and concrete has also attracted a lot of interest worldwide due to increased disposal costs and environmental concerns. This is the background to the development of eco-efficient concrete with recycled waste glass, i.e. recycled glass concrete. Recently, many studies have focused on the use of waste glass as aggregate for cement concrete or as cement replacements. This chapter summarizes fresh properties, mechanical properties and durability of recycled glass concrete, and the influence of recycled glass on these properties.

11.1.1 Source of waste glass

Based on the most common compositions, glass can be classified into the following categories: vitreous silica, alkali silicates, soda-lime glass, borosilicate glass, lead glass, barium glass and alumino-silicate glass. Small amounts of additives are often added during the production of glass to give glass different colours or to improve specific properties. Soda-lime glass is most widely used to manufacture containers, float and sheets. Lead glass is mainly for crystal tableware, TV screens and display screen equipment. Borosilicate glass is for making glass-fibers, wool insulation, and ovenware and thermos flasks. The alumino-silicates glass is for scientific and optical apparatus. The typical compositions of different types for different applications are listed in Table 11.1.

Glass makes up a large component of household and industrial waste. In waste glass, container glass accounts for approximately 80% of waste glass. On a colour basis, 63% is clear, 25% is amber, 10% is green and 2% is blue or other colours. The main composition of these glasses is the same except for small amount of additives used for colour purpose. Soda-lime glass consists of approximately 73% SiO_2, 13–13% Na_2O and 10% CaO. Thus, based on their chemical composition, soda-lime glasses will be pozzolanic materials.

The second major type is lead glass, from colour TV funnel, neon tubing, electronic parts, etc. However, a serious concern for using this type of glass in cement and concrete is the high lead content in the glass, which can be potentially leached into the environment. Borosilicate (Pyrex) type glass is so expansive that under no circumstances should it be added to concrete.

Table 11.1 The typical compositions of different types glass for different applications

	SiO_2	Al_2O_3	B_2O_3	Na_2O	K_2O	MgO	CaO	BaO	PbO
Soda-lime glasses									
Containers	66–75	0.7–7		12–16	0.1–3	0.1–5	6–12		
Float	73–74			13.5–15	0.2	3.6–3.8	8.7–8.9		
Sheet	71–73	0.5–1.5		12–15		1.5–3.5	8–10		
Light bulbs	73	1		17		4	5		
Tempered ovenware	75	1.5		14			9.5		
Borosilicate									
Chemical apparatus	81	2	13	4					
Pharmaceutical	72	6	11	7	1				
Tungsten sealing	74	1	15	4					
Lead glasses									
Colour TV funnel	54	2		4	9				23
Neon tubing	63	1		8	6				22
Electronic parts	56	2		4	9				29
Optical dense flint	32			1	2				65

11.1.2 Use of recycled glass in concrete

There are two ways to use recycled waste glass in concrete, i.e. to use recycled waste glass as aggregate and finely ground waste glass powder as pozzolan. Waste glass can also be used as raw materials for cement production, and finally to produce Portland cement concrete, but this practice is beyond the scope of this chapter.

Research on the use of crushed glass as a partial replacement for aggregate dates back many decades (Pike, recycled glass et al., 1960; Schmidt and Saia, 1963; Phillips et al., 1972; Johnston, 1974). Because of the remarkable strength regression and excessive expansion, the use of recycled glass in concrete as part of the coarse aggregates is not satisfactory. Besides, the glass particles are likely to break down during the mixing process (Alexander and Mindess, 2005).

The use of crushed glass as aggregate for Portland cement concrete does have some negative effects on properties of the concrete; however, practical applicability can still be produced even using 100% crushed glass as aggregates (Meyer and Baxter, 1997, 1998). The main concerns for the use of crushed glasses as aggregate for Portland cement concrete are the expansion and cracking caused by the glass aggregates.

Being amorphous and containing relatively large quantities of silicon and calcium, glass is, in theory, pozzolanic or even cementitious in nature when it is finely ground. Thus, it can be used as a cement replacement in

Portland cement concrete. The use of finely ground glass as a pozzolanic material also started as early as 1970s (Pattengil and Shutt, 1973). There is an increasing research interest in this area recently because of the continual accumulation of waste glass and its consequent environmental issues (Jin et al., 2000; Shao et al., 2000; Shayan and Xu, 2004, 2006; Shi et al., 2004).

11.2 Properties of fresh recycled glass concrete

11.2.1 Workability

Both the content and particle properties of recycled glass can significantly affect the workability of recycled glass concrete. A smooth surface, sharp edges, harsh texture and low water absorption characterize the particles of recycled glass. On the one hand, the sharp edges and harsh texture of recycled glass increase the interlocking between particles, producing negative effects on the plastic properties of concrete. On the other hand, the weaker cohesion between the glass aggregates and the cement pastes due to their smooth surfaces can improve workability. Furthermore, due to hydrophobic and low water absorption, the presence of glass particles leads to w/c ratio increases relatively, i.e. the efficient water content in the concrete mix increases. This can also increase the flowing ability of fresh concrete. The outcome depends on the interaction of the size, the shape and texture of recycled glass particles and their content.

As shown in Fig. 11.1, the general trend observed indicates that the incorporation of recycled glass as a replacement for fine aggregate does not obviously reduce the workability until the replacement reaches a certain level (Mukesh, 2009; Wang, 2009a). With the increase in replacement level, once the negative effects of glass particles on workability outweigh the positive ones, concrete with recycled glass aggregate requires higher water content to reach the same workability. It was found that the consistency of fresh concrete was reduced; both segregation and bleeding became clearly noticeable as the content of recycled glass increased beyond 20% (Mukesh, 2009). Bleeding and segregation result from the inherent smooth surface and very low water absorption of recycled glass, both leading to a weaker adhesion between the paste and glass aggregates in the concrete mix. As the size is further reduced, improvements occur in the texture and shape properties of glass particle, and the incorporation of recycled glass can even improve the properties of fresh concrete, i.e. increase flow ability and segregation resistance.

For self-consolidating concrete, slump flow is used to determine flowability and the V-funnel test is adopted to evaluate cement paste viscosity in concrete and resistance to segregation. As shown in Fig. 11.2, the incorporation of recycled glass, which is finer than natural sand, increases both the slump flow and V-funnel time. The increase was ascribed to low water absorption,

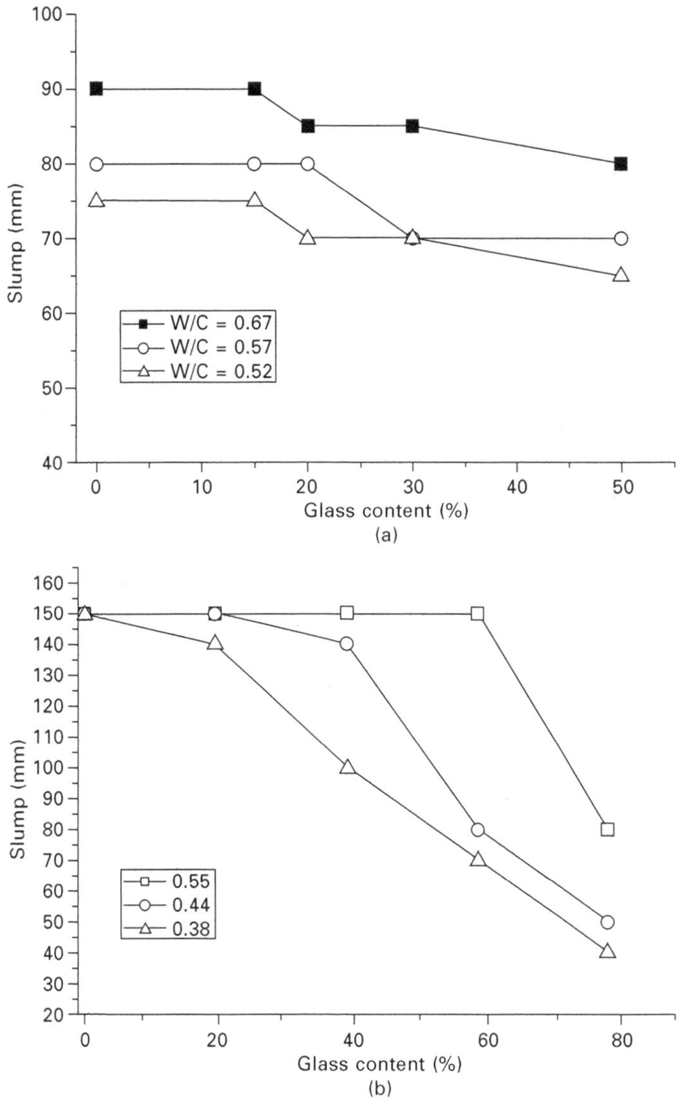

11.1 Slump of fresh recycled glass concrete with different content of (a) recycled mixed colour glass and (b) recycled LCD glass as fine aggregate. W/C = water/cement ratio. Adapted from Mukesh (2009) and Wang (2009a).

smooth surface and better fill-ability of glass powder (Wang and Huang, 2010a). Taha and Nounu (2008a) found that the presence of glass powder can also lead to increase in slump of fresh concrete whether it contains glass aggregate or not.

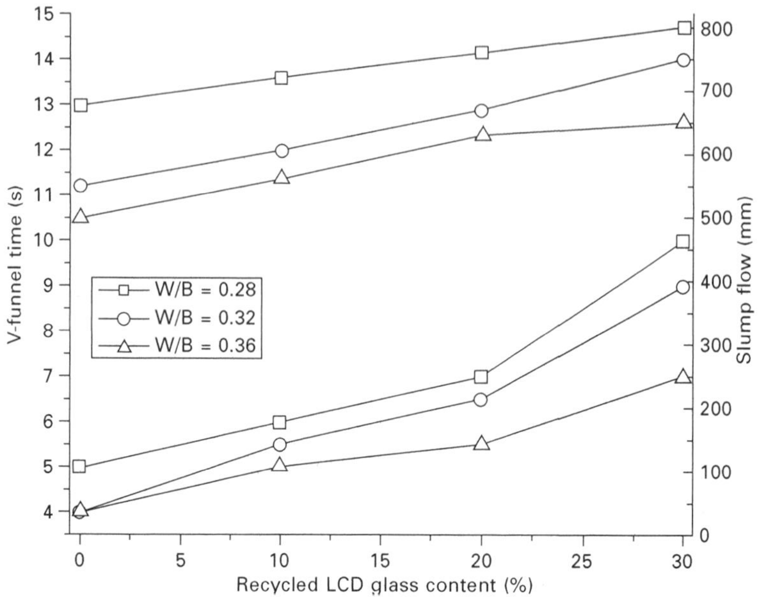

11.2 V-funnel time and slump flow of fresh self-consolidating concrete with different recycled LCD glass content. W/B = water/binder ratio. Adapted from Wang and Huang (2010a).

11.2.2 Air content and fresh unit weight

Compared with conventional concrete, fresh recycled glass concrete has a lower air content. There is a tendency that the air content decreases with the increase in glass content. This tendency can be attributed to the poor geometry and the smooth surface of recycled glass, which helps decrease porosity between recycled glass and cement paste.

In general, glass has a lower density than natural sand. As a result, unit weight of fresh recycled glass concrete is reduced as the glass aggregate content is increased with various water to binder ratios, though the incorporation of glass may decrease the air content (Topcu and Canbaz, 2004; Ismail and Al-Hashmi, 2009). Likewise, the use of recycled glass powder as a replacement of cement can also reduce the fresh concrete unit weight, as the density of glass is lower than that of cement.

11.2.3 Setting time

As discussed in Section 11.2.1, the presence of glass particles increases the efficient water content in the concrete mixture due to hydrophobic and lower water absorption of glass. As a result, concrete with recycled glass exhibits both prolonged initial and final setting times. Figure 11.3 indicates

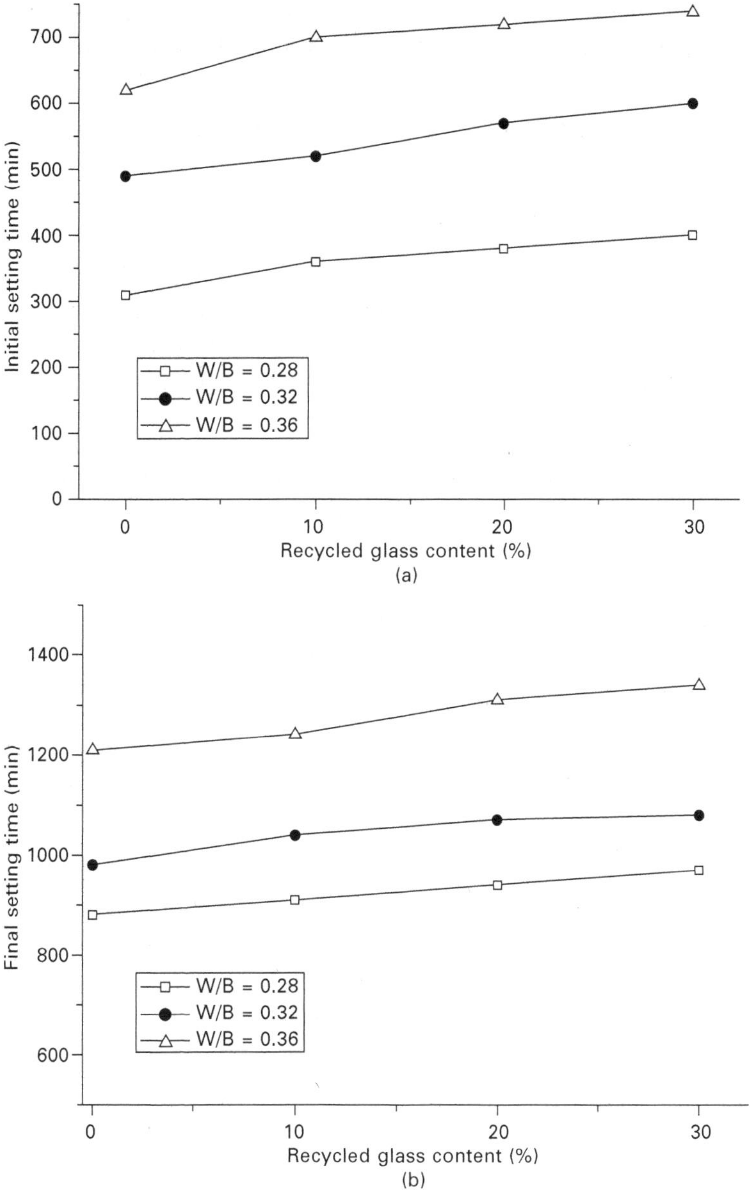

11.3 Effect of recycled glass on (a) initial setting time and final setting time of recycled glass concrete (Wang and Huang, 2010a).

that both initial and final setting times of recycled glass concrete increase almost linearly with the percentage of replacement with glass. As a result of delay in setting times, the slump loss of recycled glass concrete is also reduced compared to conventional concrete.

Additionally, if the recycled glass is not properly processed, the presence of impurities, such as sugar, can also prolong the setting time.

11.3 Properties of hardened recycled glass concrete

11.3.1 Compressive strength

Concrete with recycled glass aggregate

Effects of recycled glass on compressive strength of concrete can be attributed to the physical and mechanical properties of the glass particle itself and its influences on workability. In general, recycled glass concrete exhibits lower compressive strength than normal aggregate concrete with the same water to cement (binder) ratio or workability.

Properties of glass particles produce negative effects on compressive strength on several aspects, though the chemical bond, if any, between the aggregate and the cement paste might be greater for glass aggregate due to its amorphous surface potentially allowing pozzolanic behaviour. Firstly, it is difficult to achieve a homogeneous distribution of aggregates with the presence of recycled glass due to the poor geometry of recycled glass. Secondly, the high brittleness of recycled glass and inherent cracks in recycled glass particles existed due to the crushing process can be considered as a source of weakness. Thirdly, the smooth surface leads to relative weaken bond between the glass aggregates and cement paste.

Influences of glass aggregates on workability also produce negative effects on the compressive strength of recycled glass concrete. Glass aggregate requires more water for the same workability compared to natural aggregate when glass aggregate is coarser than the later. Satisfaction of this water demand increases the water to cement ratio of the concrete, thereby resulting in a lower strength (Polley, 1996). The increased bleeding and segregation will affect the handling and casting of fresh concrete, the meso-structure of concrete, and the final strength. Additionally, contamination and the organic content in recycled glass may pose negative effects on the development of strength and final results (Taha and Nounu, 2008a), if recycled glass is not processed properly.

Because of the strength regression and excessive expansion, the use of recycled glass in concrete as part of coarse aggregates is not satisfactory. Reduction in the size of recycled glass particles can relieve strength degrading. While recycled glass is incorporated in concrete as replacement for fine aggregate, as illustrated in Fig. 11.4, it exhibits slightly different effects on the strength of recycled glass concrete if the recycled glass content is lower than 20%. The strength decreases noticeably with an increase in recycled glass content as the recycled glass content exceeds 20%. This suggests that the optimal glass aggregate content is 20%. The later development of

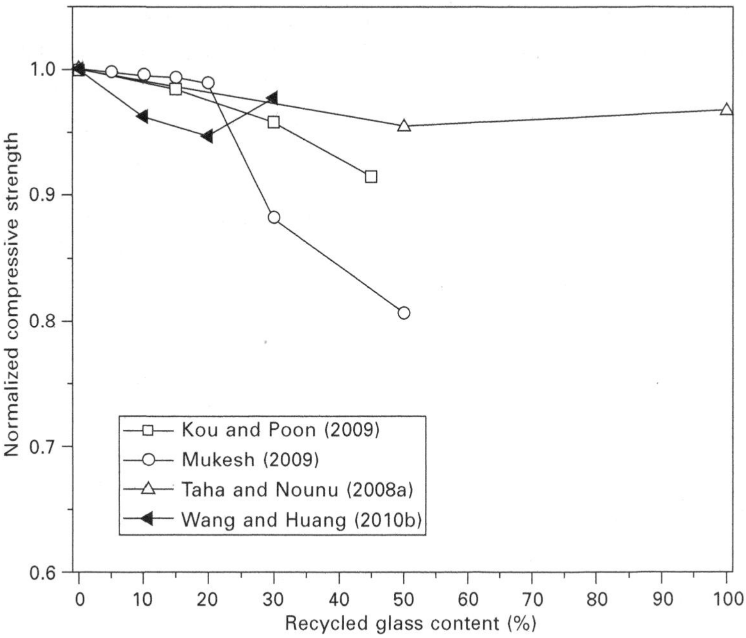

11.4 Normalized compressive strength at 28 days of recycled glass concrete with different recycled glass content.

compressive strength of recycled glass concrete shows a great improvement, though the early compressive strengths for concrete with different content of recycled glass content are not higher than that of the control group.

Concrete with glass powder

Being amorphous and containing relatively large quantities of silicon and calcium, glass is, in theory, pozzolanic or even cementitious in nature when it is finely ground. Moreover, the texture and shape properties of glass particle can be improved as the size is further reduced; thus, their negative effects on the strength decrease. Many investigations have confirmed that its pozzolanic reactivity increases as its fineness increass. Besides, a high content of Na_2O in glass is easy to leach out as the size of glass particle is reduced, hence influencing the strength of concrete, especially at early ages.

Based on observed compressive strength, the pozzolanic properties of glass are first noticeable at particle sizes below approximately 300 μm (Federico and Chidiac, 2009). Below 100 μm, glass can have a pozzolanic reactivity which is greater than that of fly ash at a low percentage of cement content and after 90 days of curing (Shi et al., 2005; Schwarz et al., 2008). Meyer et al. (1996) postulated that below 45 μm, glass may become pozzolanic.

Shi et al. (2005) compared the strength index of glass powder with different fineness to fly ash. The pozzolanic strength activity index of glass powder with Blaine fineness of 264 m^2/kg is only slightly lower at 1 to 7 days, but higher at 28 days than that of fly ash. Hence, by means of Blaine fineness, ground glass powders with Blaine specific surface higher than 250 m^2/kg exhibits very high pozzolanic activity.

Figure 11.5 shows the strength activity index of glass powder with different fineness and the comparison with that of fly ash (Shao et al., 2000). In Fig. 11.5 150 μm glass stands for ground glass having particles passing a #100 sieve (150 μm) and retained on a #200 sieve (75 μm); 75 μm glass for ground glass having particles passing a #200 sieve (75 μm) and retained on a #400 sieve (38 μm); and the 38 μm glass stands for ground glass powder having particles passing a #400 sieve (38 μm). As shown in Fig. 11.5, the strength reactivity index increases with the increase in powder fineness. The 38 μm glass concrete exhibited higher strength than the fly ash concrete at all ages. ASTM C618 recommends that a pozzolan has a minimum strength activity index of 75% for it to benefit concrete. According to ASTM C618, the activity index of 150 μm glass did not always satisfy the criteria, while 75 μm and 38 μm glasses are comparable to fly ash concrete at all ages. The relatively higher early strength index of 38 μm glass concrete may be attributed to the high content of Na_2O in glass.

The compressive strength of concrete with glass powder of different sizes

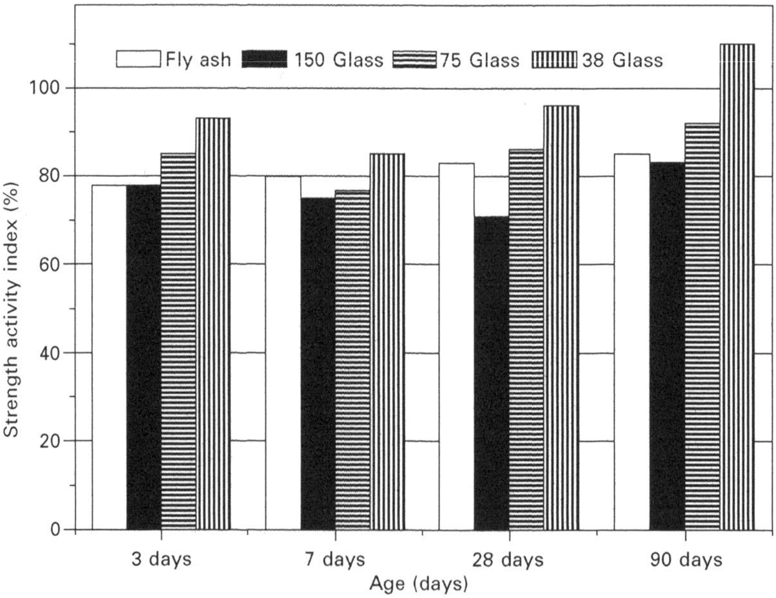

11.5 Comparison of strength activity index of different fineness glass powder with fly ash (Shao et al., 2000).

is given in Fig. 11.6. It is clear that concrete with finer glass powder exhibits higher strength, especially at later age, due to high-strength activity index.

Since finely ground glass powder exhibits pozzolanic activity, the presence of glass powder may reduce the strength at early ages due to the dilution effect; however it can improve the strength of concrete at later ages. As shown in Fig. 11.7, although the concrete mixtures containing glass powder have lower 28-day strength due to significantly lower cement content, they keep developing strength with time under moist curing conditions and approach the strength of the control mixture. Particularly when glass powder replaces sand, the strength is significantly greater than that of the control mixture. In Fig. 11.7, GP10, GP20, and GP30 stands for concrete mix with glass powder content of 10%, 20% and 30% respectively; GP30 RPL sand means the concrete mix with 30% glass powder replacement for sand.

Effects of glass colour on strength are not obvious. This suggests that influences of glass on mechanical properties of recycled glass concrete are more related to its physical characteristics than to the slight difference in chemical compositions.

11.3.2 Flexural strength

Factors that affect compressive strength also apply to flexural strength; recycled glass has similar effects on flexural strength of recycled glass concrete as on the compressive strength. As shown in Fig. 11.8, glass aggregate does not

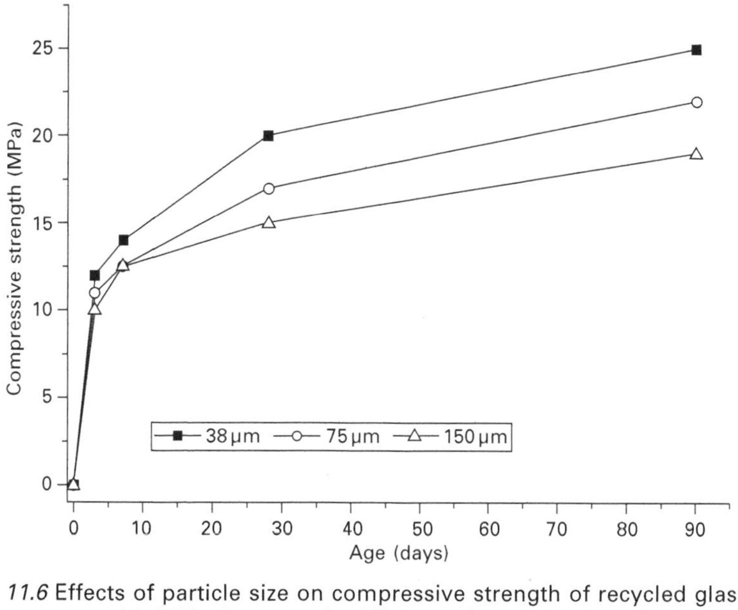

11.6 Effects of particle size on compressive strength of recycled glass concrete with 30% glass powder (Shao et al., 2000).

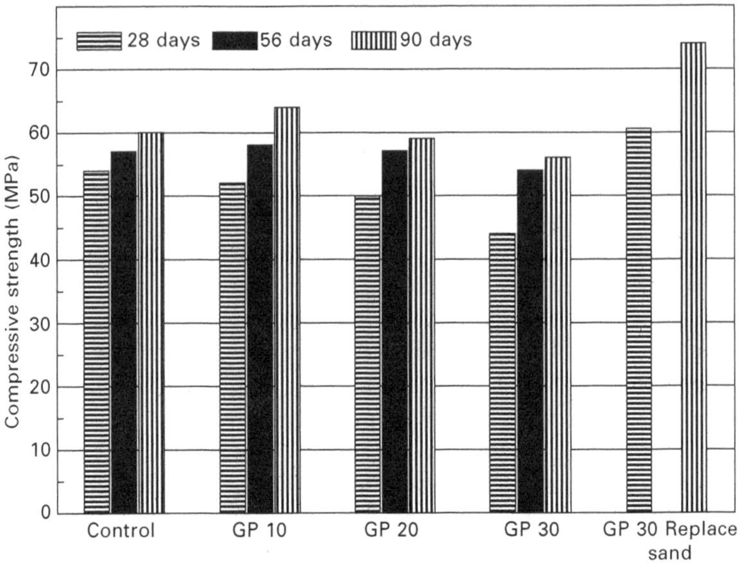

11.7 Compressive strength development of recycled glass concrete with glass powder. Adapted from Shayan and Xu (2004).

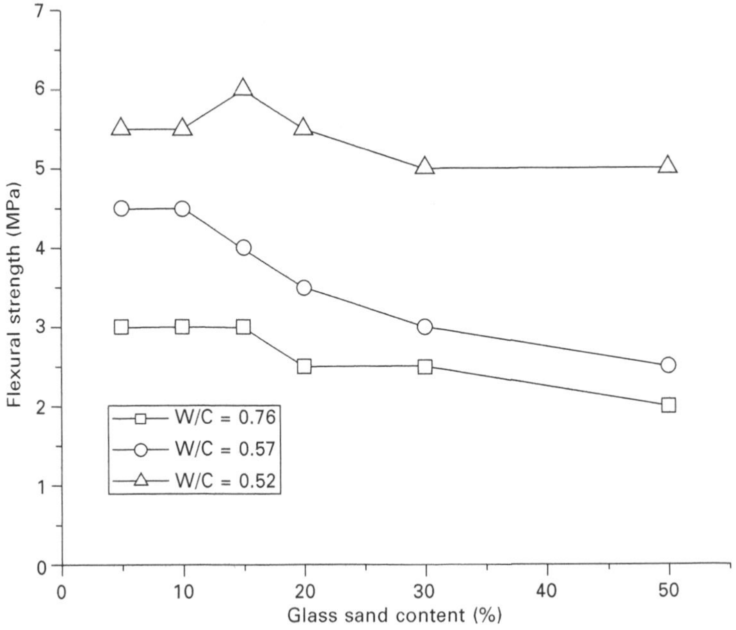

11.8 Flexural strength at 28 days of recycled glass concrete with various recycled glass content. Adapted from Mukesh (2009).

influence flexural strength when the replacement level is up to 20%. Then the increase in glass content leads to decrease in flexural strength. Though the flexural strength at later ages for concrete with different contents of glass aggregate may not be higher than that of the control group, the development of flexural strength of recycled concrete also shows a greater improvement.

11.3.3 Splitting tensile strength

The development of splitting tensile strength for recycled glass concrete is similar to that of flexural strength. It has been reported that the splitting tensile strength at later ages for the concrete with glass aggregate is higher than that of the normal group. According to Wang (2009a), concrete with 20% glass as replacement for fine aggregate exhibited the highest splitting strength. Furthermore, the specimens with glass aggregate developed a better splitting tensile strength at the later age.

11.3.4 Elastic modulus

The elastic modulus of recycled glass concrete is higher than that of normal concrete with comparable strength. The elastic modulus of concrete is generally considered to be the function of the elastic modulus of aggregate, hardened cement paste and their volume fraction. Since the elastic modulus of glass is higher than that of normal aggregate, it can be expected that the incorporation of recycled glass leads to an increase in elastic modulus of recycled waste glass concrete. Though the presence of glass may results in decreasing in strength of concrete, the elastic modulus can still be higher than concrete without recycled glass aggregate, when the ratios of water to cement are similar. As shown in Fig. 11.9, concrete with 20% of glass aggregate gave the largest elastic modulus values. Additionally, the elastic modulus decreased with increasing amounts of glass sand replacement, this may be caused by decrease in strength, but still be higher than that of the control.

11.3.5 Drying shrinkage

Drying shrinkage of concrete with glass aggregate is different from that of concrete with glass powder as replacement for cement. The presence of recycled glass aggregate reduces drying shrinkage. As shown in Fig. 11.10, the drying shrinkage values decreased with an increase in recycled glass content.

The main cause of drying shrinkage is the loss of water in concrete, and water content is the most important factor that affects drying shrinkage. There is a good relation between drying shrinkage and water absorption (or moisture loss) for recycled glass concrete. It was found that the amount of

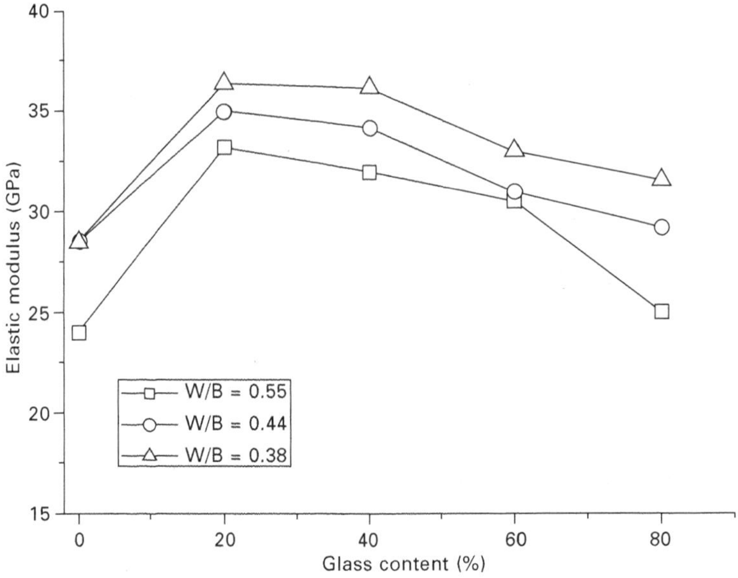

11.9 Elastic modulus of recycled glass concrete with different LCD glass. Adapted from Wang (2009a).

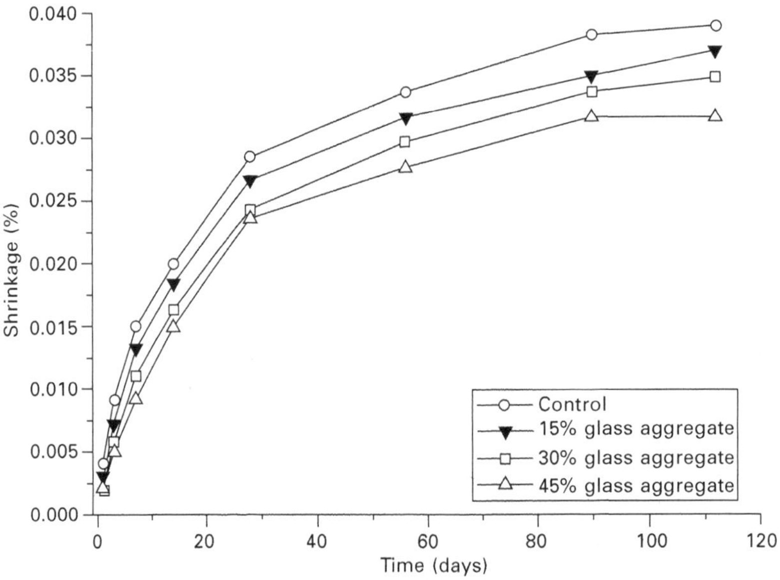

11.10 Typical drying shrinkage of recycled glass concrete with various glass contents (Kou and Poon, 2009).

the absorbed water in concrete was reduced as the content of recycled glass aggregate was increased in concrete (Taha and Nounu, 2008a). The effects of recycled glass aggregate on absorption of concrete can be attributed to the following factors: firstly, the presence of recycled glass in concrete will reduce the total demand for water absorption due to the negligible water absorption value of the glass; secondly, the presence of recycled glass aggregate in concrete can reduce the permeability of the concrete, for glass by nature is an impermeable material, and then restrict the migration of moisture inside the concrete. Reduction in water absorption of concrete with recycled glass aggregate, which means a relatively low water loss when concrete is exposed to drying, finally results in reduced drying shrinkage. Additionally, the increase in the elastic modulus of concrete with recycled glass aggregate, as discussed in Section 11.3.4, is also helpful to reduce drying shrinkage.

Adversely, the presence of glass powder results in an increase in dry shrinkage of concrete in general (Shayan and Xu, 2004, 2006; Shi and Wu, 2005). The influence of glass powder on drying shrinkage can still be related to its effects on water absorption. Unlike the recycled glass, the presence of glass powder as cement replacement in concrete increases the water absorption demand of concrete (Taha and Nounu, 2008a). This may be attributed to the differences in the nature of the microstructure of concrete mix and hydration product due to the presence of glass powder in concrete. However, the long-term drying shrinkage of recycled glass concrete is not excessive and easily meets the requirements of AS 3600, being values less than 0.075% at 56 days (Shayan and Xu, 2004, 2006).

It was also found that the drying shrinkage of concrete with glass powder decreased with an increase in the fineness of the glass (Shi and Wu, 2005).

11.4 Durability of recycled glass concrete

11.4.1 Resistance to sulphate attack

Recycled glass concrete exhibits good resistance to sulphate attack when compared with normal concrete. Figure 11.11 gives the weight loss of concrete with different recycled LCD (liquid crystal display) glass content under cycles of alternate drying and soaking in sulphate solution. As illustrated in Fig. 11.11, the presence of glass aggregate reduced the weight loss caused by the sulphate attack. It can also be seen that the weight loss decreased with the increase of glass content. This indicates that the addition of glass can improve the resistance to sulphate attack for concrete. The improvement in resistance to sulphate attack can be attributed to the following factors. Firstly, as an impermeable material, the presence of recycled glass can reduce

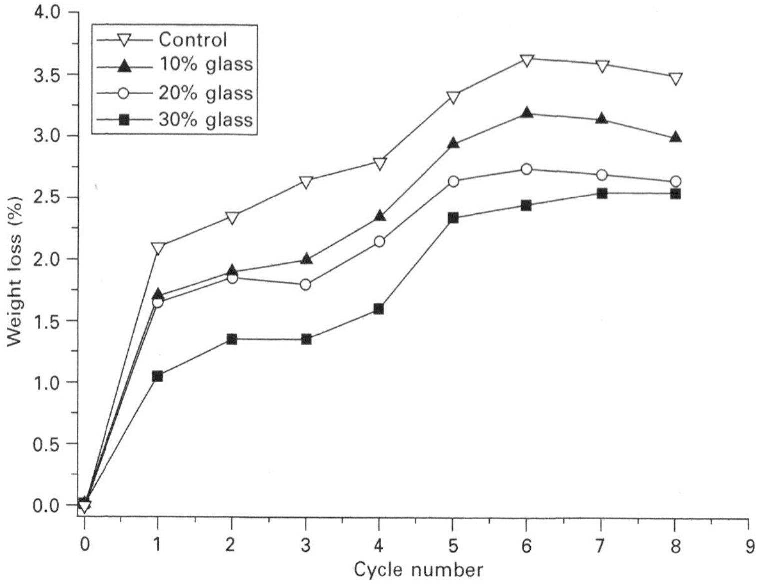

11.11 Weight loss of concrete with various recycled LCD glass content under cycling of drying and soaking in sulphate solution (Wang and Huang, 2010b).

permeability and absorption of the concrete, and then enhance the ability to restrict the migration of the water and ions. Secondly, glass particles have good acid and alkali-resistance ability.

11.4.2 Chloride penetration resistance

Chloride-induced corrosion of steel in concrete is the main cause responsible for durability of reinforced concrete structures which are exposed to chloride in the environment. Chloride penetration resistance of recycled glass concrete becomes a concern for its potential application in structures exposed to chloride in the environment.

Different chloride transport test methods have been used to evaluate chloride penetration resistance of recycled glass concrete, and recycled glass concrete exhibits improved chloride penetration resistance in recycled glass used either as fine aggregate or as cement replacement. Figure 11.12 shows the rapid chloride permeability (RCP) test results of different strength recycled glass concrete with recycled glass content ranges from 0 to 80% as replacement of fine aggregate. The penetration level of the chloride ion into concrete is obtained by measuring the difficulty and the amount of the current passing through the specimen, as regulated by ASTM C1202 (1997).

As shown in Fig. 11.12, the RCP values of specimens with glass sand

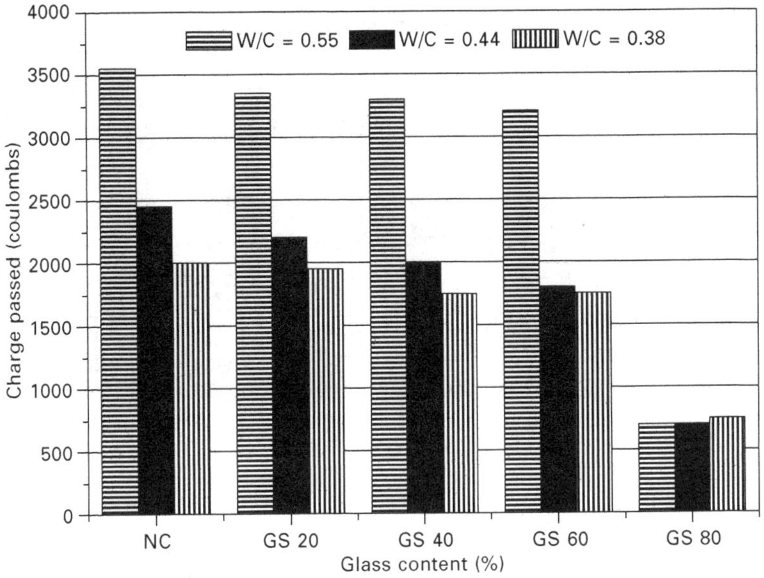

11.12 Chloride ion penetration of recycled glass concrete with different glass as replacement for sand (ASTM C1202 condition, age 28 days). Adapted from Wang (2009a).

replacement was less than that of the control groups. There is an obvious trend that RCP values decrease with the increase in glass aggregate content. The chloride ion penetrability decreased with the increase in recycled glass content which means the resistance to chloride ion penetration increased with an increase in recycled glass content. The RCP test basically measures the conductivity (or resistivity) of concrete, which depends on both the pore structure as well as the pore solution composition. It is clear that the test measures movement of charge related to all ionic species and not just chloride ions. The incorporation of glass aggregate may increase the content of alkalis in the pore solution, thus increasing the conductivity of the pore solution. Even with such an increase, the RCP values of recycled glass concretes are lower than that of plain concrete. It can be considered that the presence of glass particles in concrete can reduce the permeability of the concrete mix as discussed in Section 11.3.5, for the amount of the absorbed water in concrete was reduced as the content of recycled glass was increased in concrete.

Concrete with glass powder exhibits good chloride penetration resistance as well. Figure 11.13 gives the RCP value, non-steady-state migration (NSSM) coefficient and steady state conductivity (SSC) of concrete with different glass powder and fly ash content. As tested by different methods, recycled glass concrete exhibits higher chloride penetration resistance than normal concrete, especially at later ages.

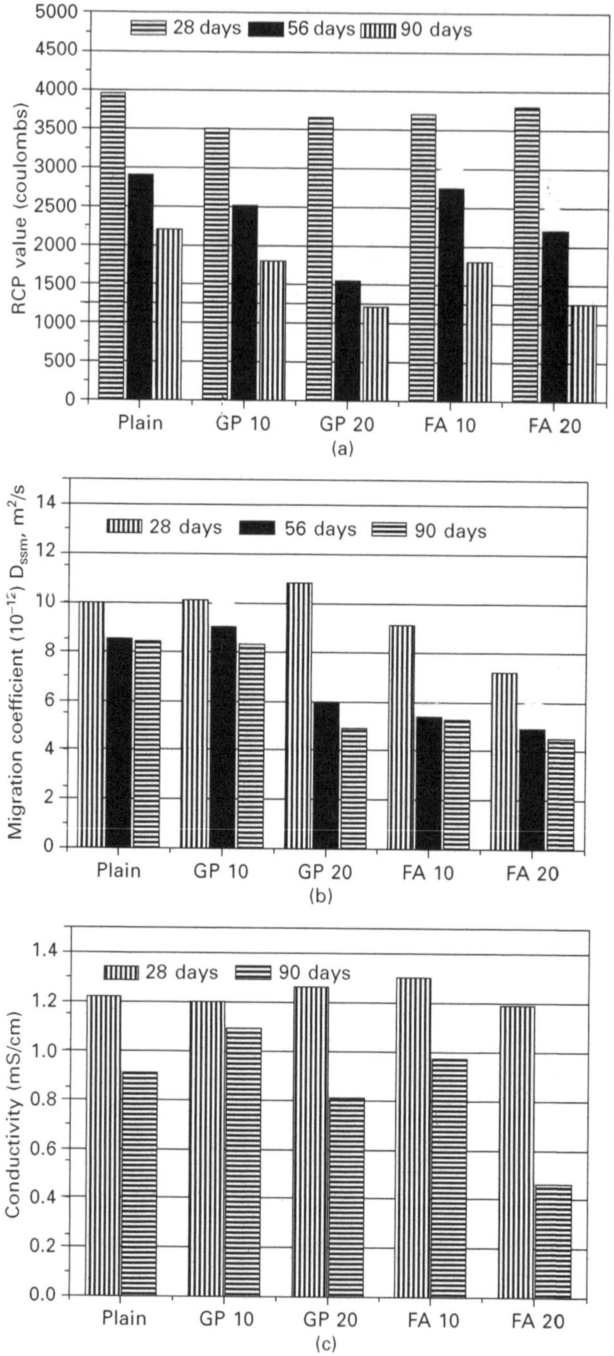

11.13 Effects of glass powder (GP) on chloride penetration resistance and comparison with fly ash (FA) (Jain and Neithalath, 2010).

Several aspects account for the improvement of chloride penetration resistance with the presence of glass aggregate or glass powder. Firstly, glass by nature is an impermeable material, the presence of glass particles in concrete can reduce the permeability of the concrete mix, and restrict the migration of ions inside the concrete. Secondly, the dense C-S-H gel hydrate was produced at the interface transition zone between the glass particle and the cement paste, which improves the structure of the interface transition zone. Furthermore, finely ground glass powders exhibited very high pozzolanic activity, and the presence of glass powder as replacement with cement can improve pore structure of hardened cement paste, hence increase the chloride penetration resistance. Additionally, the release of alkali ions results in an increase in the viscosity of the pore solution, and thereby, may hinder the transport of chloride ions, consequently reducing the RCP values.

11.4.3 Freezing–thawing

As far as resistance to freezing–thawing of recycled glass concrete is concerned, effects of glass on sorption, microstructure and strength should be taken into account. The outcome is controlled by the interaction of several aspects mentioned above. Hence, it is expected that there will be conflicting freezing–thawing test results on recycled glass concrete. For example, freezing–thawing tests following ASTM C666 indicate that concrete with glass aggregates exhibited a slightly poorer durability index than the control conventional concrete (Pollery 1996). However, according to Tuncan et al. (2001), the incorporation of recycled glass and fly ash into the concrete increases freezing–thawing resistance of concrete. Shi et al. (2005) also found that light aggregate concrete with glass powder exhibited good freezing–thawing resistance. The main concern for the use of waste glasses as concrete aggregates is expansion and cracking as discussed in detail in the following section.

11.4.4 Alkali–silica reactivity

The expansion and cracking of concrete containing glass aggregate have been known for decades. The main concerns for the use of crushed glasses as aggregates for concrete are the expansion and cracking caused by the recycled glass aggregates for they contain high content of both alkali and amorphous phase. Figure 11.14 shows the alkali–silica reactivity (ASR) affected glass particle with intra-particle gel formation in mortar in an accelerating test.

The ASR expansion of concrete with recycled glass aggregate mainly depends on glass colour and chemical composition, the size of glass particle and the content. These factors will be discussed in detail in the following. Above all, the expansion of recycled glass concrete is directly proportional

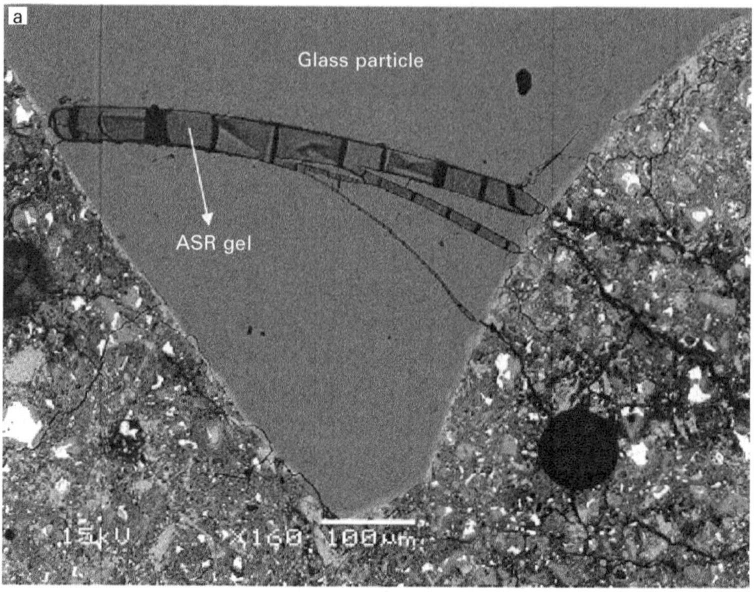

11.14 ASR affected glass particle (size range: 1.18–2.36 mm) showing intra-particle gel formation in mortar after 14 days in ASTM C1260 test (Maraghechi et al., 2012).

to the glass content. For some reactive siliceous aggregate, there is a pessimum proportion that causes the largest expansion of concrete, and that the expansion decreases when the content of the reactive aggregate in the concrete is increased or decreases from the pessimum proportion. Ichikawa (2009) explained the pessimum effects. However, it seems that the pessimum proportion doesn't apply to ASR of concrete with recycled glass. According to Jin et al. (2000), the expansion due to ASR increases with the mixed colour glass aggregate content at all tested ages. In the study of Jin et al. (2000), mixed colour recycled glass, which constituted 65–70% green, 25–30% clear and 5% brown by mass, was used as fine aggregate and the expansion of mortar bar was tested according to ASTM C126.

Secondly, glass colour also has a profound influence on ASR expansion; even the major chemistry is very similar. Figure 11.15 gives the ASR expansion of the green, blue, amber and flint glass with the similar chemical compositions (Zhu et al., 2009). As shown in Fig. 11.15, the ASR expansion is strongly dependent on the colour of the glass. Blue glass was most reactive, followed by flint glass, amber glass and green glass. The effect of different coloured glass on the concrete expansion can be attributed to the small amount of additives used for colour purpose. For an example, Cr_2O_3 in green glass could inhibit the expansion of concrete containing glass aggregate.

However, there are conflicting results according to different investigations.

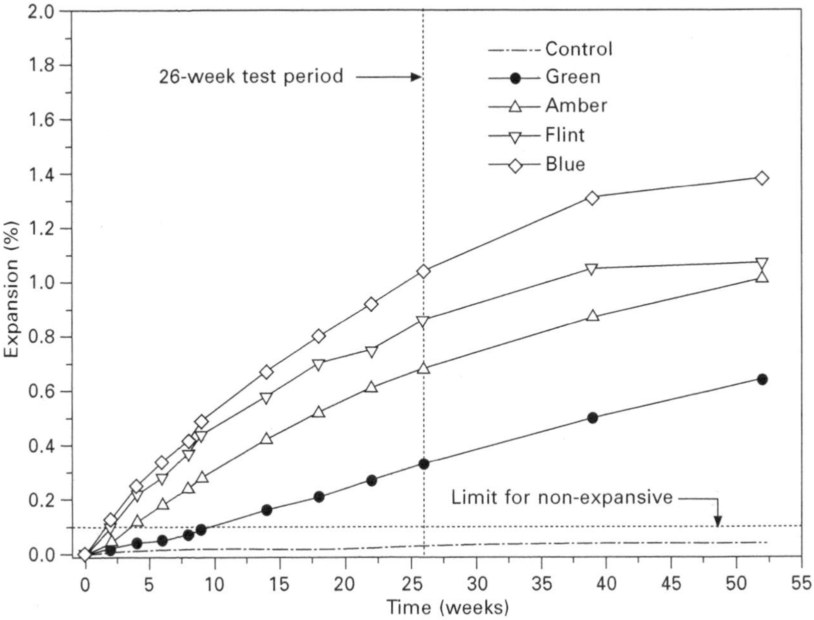

11.15 Effects of glass colour on ASR expansion (ASTM C 227 condition) (Zhu et al., 2009).

Idir et al. (2010) found that the magnitude of expansion increases in the sequence flint–amber–green, and the expansion behaviour of prisms containing flint glass was similar to that of the control. Jin et al. (2000) also observed concrete with flint glass aggregate producing the largest dimensional change. One possibility is that differences in the ASR behaviour of the different coloured glass samples reflect variations in manufacturing processes used for different glass products, which would influence levels of internal stress and consequently rates of leaching dissolution of the glass.

Last but not at least, the size of glass aggregate exerts evident influences on the expansion rate and value, with finer glass particles displaying considerably lower expansion. As shown in Fig. 11.16, the ASR expansion increased with the increase of glass aggregate size (Zhu et al., 2009). When the size of glass particle is smaller than one particular level, the glass particles themselves will not generate deleterious expansion (Meyer and Baxter, 1997; Shi et al., 2004). However, glass powder has a very high content of alkalis, and the alkalis in glass powders can be leached out, resulting in alkali–aggregate reaction (AAR) expansion when the aggregate is alkali-reactive. This particular size varies in a large range according to different observations, because it is also related to composition, colour and thermal history of glass. Therefore, it is necessary to find the maximum size that does not generate deleterious expansion by experiments for different sources of waste glass. There are also

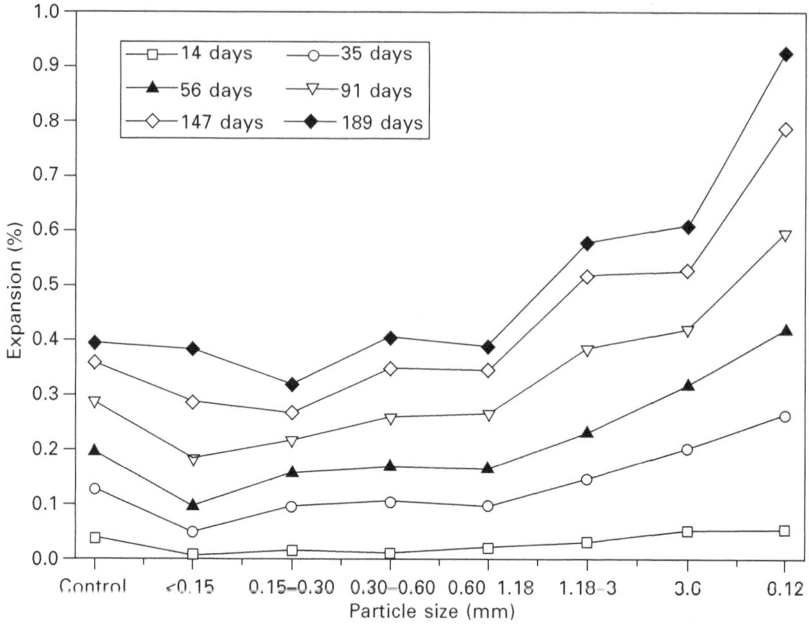

11.16 Effect of glass size on ASR reactivity (flint, ASTM C1260 Condition) (Zhu et al., 2009).

suggested a pessimium sizes (particle size that causes maximum expansion) (Jin et al., 2000), the pessimium size is related to composition, colour and thermal history of glass as well.

Different approaches have been made to reduce or eliminate the expansion of concrete containing glass aggregates. Practically, feasible concrete mixtures with expansion values lower than the limit that will not result in deleterious expansion, which may vary according to different test methods, can be produced in the laboratory or the field application. These approaches include: (1) reducing the size of glass particle, (2) restricting the glass content, (3)introducing air entrainment, (4) using porous lightweight aggregate or (5) supplementary cementing materials. Some admixtures such as lithium compounds were used as ASR suppressors as well. Even the fine ground glass powder itself can be used as ASR suppressors.

The introduction of air entrainment or use of porous lightweight aggregate is an effective method to reduce or eliminate the expansion. It is well known that porous aggregate can mitigate expansion resulted from AARs (Collins and Bareham, 1987). In that case, pores in porous aggregate permit the expansive reaction products to permeate into or relieve the expansive pressure, then reduce or eliminate the expansion. For example, when the volume of expanded shale is more than 60% of the total aggregate volume, the expansion of the specimens is greatly reduced and far below the deleterious expansion limit (Shi et al., 2005). It was found that the specimens do not expand at all when porous glass is used as concrete aggregates (Ducman et al., 2002). The introduction of air-entraining agent reduces the expansion by about 50% (Meyer and Baxter, 1998).

Supplementary cementing materials such as ground blast furnace slag, fly ash, silica fume and metakaolin are often used to reduce or eliminate the AARs. The use of these materials can also reduce the expansion of concrete containing glass aggregates. Figure 11.17 illustrates the effects of fly ash on expansion of mortar bars with different glass aggregate content. As shown in Fig. 11.17, serious ASR expansion was observed at 28 days when the mortar bars were prepared with 45% recycled glass, although it met the requirements prescribed in ASTM C1260 (<0.1% within 14 days). In comparison, the ASR expansion of all the specimens was significantly reduced by the use of fly ash in the mortar mixes.

The ASR expansion rate of the recycled glass concrete mixes was also measured using the concrete prisms prepared using the same exposure condition as the mortar bar test method (ASTM C1260). The results (Fig. 11.18) show that the concrete prisms displayed extremely small expansion even with a recycled glass content of 45% showing the expansion was suppressed by the use of fly ash in the concrete mixes. All 28-day measurements showed less than 0.1% expansion (25% fly ash content by mass of total binder).

With the use of supplementary cementing materials, concrete with 100% recycled glass did not generate deleterious expansion even at the age of 3 years (Dhir et al., 2009). The effectiveness will be dependent upon the chemical and physical characteristics. Metakaolin has been proven the most effective material (Meyer and Baxter, 1997; Jin et al., 2000; Zhu and Byars, 2004).

Furthermore, even the finely ground glass powder can be used to reduce the expansion. Figure 11.19 shows the effectiveness of finely ground glass powder to mitigate ASR expansion of the concrete-containing glass sand. With 20% cement replaced by glass powder, the expansion value of concrete prepared even with 100% glass sand is much lower than the 52-week expansion limit (0.2%) according to BS 812 part 123:1999. This indicates that fine glass powder acts as an effective ASR suppressor. Finely ground glass favours a relatively rapid pozzolanic reaction over the slower ASR reaction. The highly reactive glass powder reacts with lime and forms a calcium silicate hydrate (C-S-H) with a low C/S ratio, which retains the alkalis in the C-S-H, therefore reducing or even eliminating the ASR expansion.

11.5 Conclusion and future trends

As discussed in the above sections, the use of waste glass as concrete aggregates has a slight negative effect on the workability and strength of concrete.

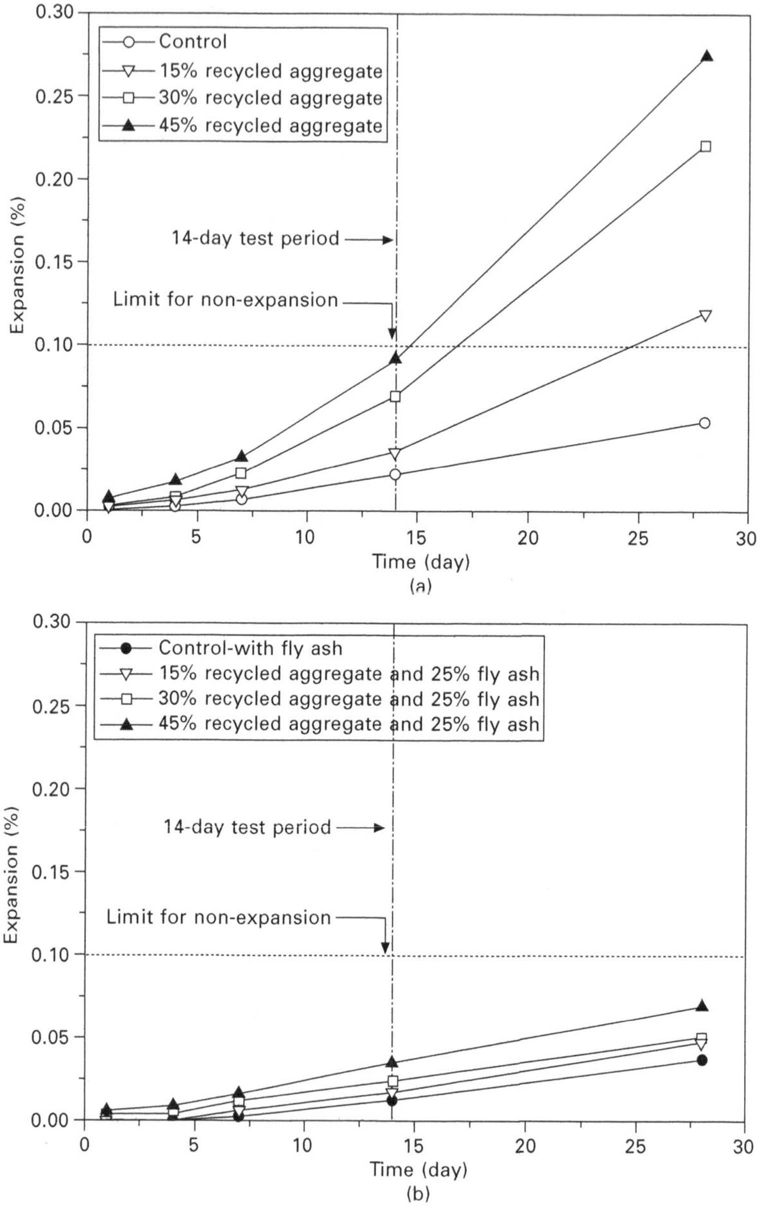

11.17 Comparison of effect of fly ash on expansion of mortar bar (a) without fly ash and (b) with fly ash containing different recycled glass content (Kou and Poon, 2009).

Nevertheless, effective measures can be taken to relieve these negative effects. Performance of concrete with glass powder is similar to those incorporated with supplementary cementing materials such as fly ash, and ground blast

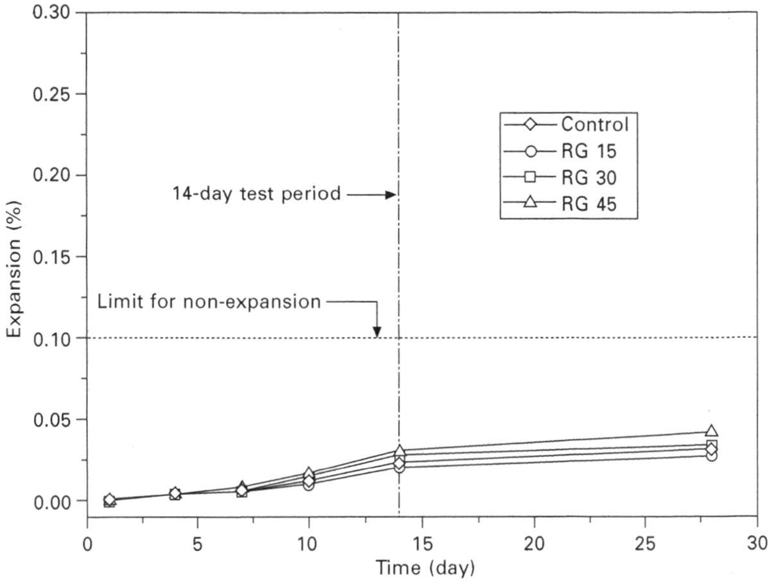

11.18 Expansion of concrete with fly ash and different glass aggregate content (Kou and Poon, 2009).

furnace slag, since it is pozzolanic or even cementitious in nature. Recycled glass aggregates can be safely used in concrete as long as the ASR potential is mitigated properly. However, the expansion phenomena of concrete with glass aggregate are different from the traditional ASR, the mechanism of ASR in recycled glass concrete; effects of glass chemical compositions and thermal history on ASR expansion are still to be understood fully. Owing to inherent relationships among constituents, microstructure and properties of recycled glass concrete, more efforts are needed to reveal microstructural features of hardened cement paste, with the presence of glass powder, the transition zone between cement paste and glass aggregate, and influences of glass powder on hydration and hydration products.

To use recycled glass safely and economically, there is a growing interest in using recycled glass as aggregate in concrete for non-structural applications, such as pre-cast concrete blocks, water-permeable concrete for seawater purification and controlled low-strength concrete. Recently, alternative approaches have also been made to use recycled glass in concrete in more eco-efficient and environmentally friendly ways. These approaches include using recycled glass and calcite for preparing construction materials by means of hydrothermal treatment, preparing concrete with 100% fly ash as cement replacement and recycled glass as aggregate. As for the process of waste glass for using in concrete, implosion technology is promising. Recycled waste glass particles crushed by imploding are free from sharp

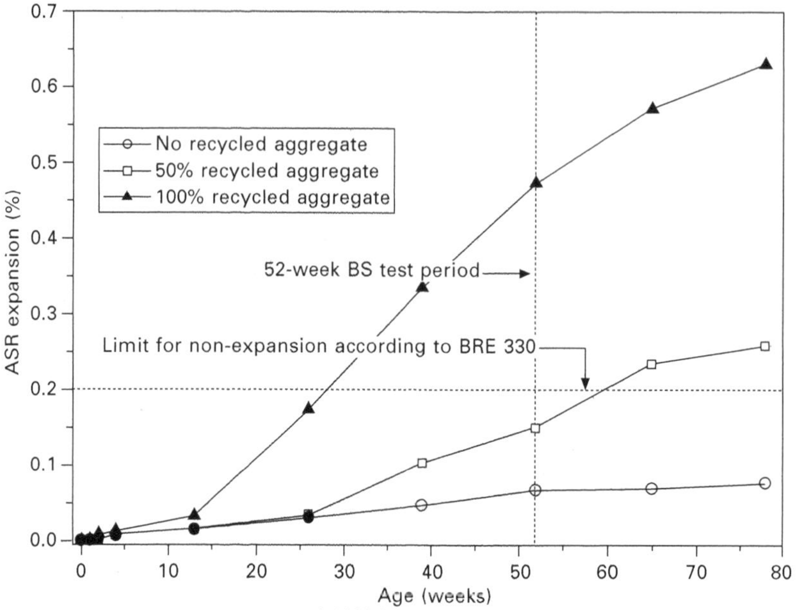

(a) Without glass powder

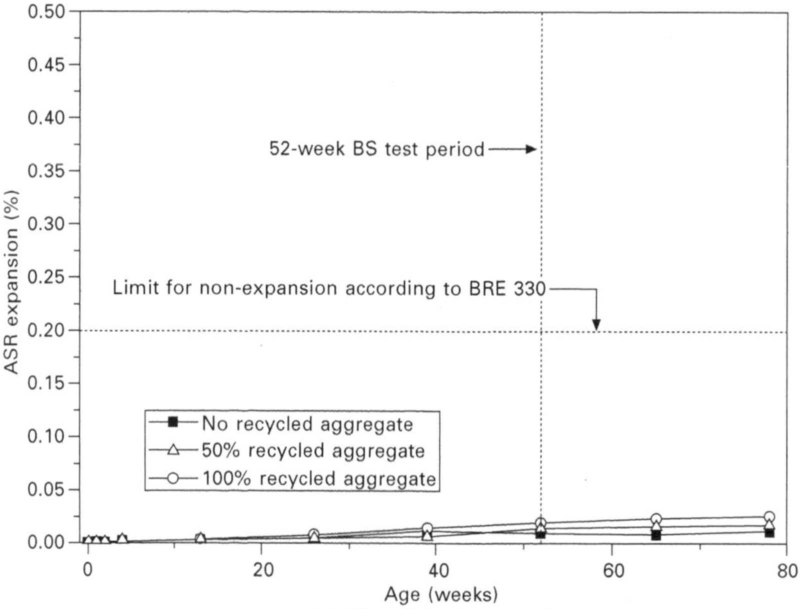

(b) With 20% glass powder

11.19 Mitigating effects of glass powder on ASR expansion of the concrete contained glass aggregate (BS 812 part 123:1999 test condition) (Taha and Nounu, 2008b).

edges. Furthermore, it is easy to separate other materials such as metal or cork tops, labels and straw from glass particles through sieving for these materials are left untouched during imploding process (Cassar and Camilleri, 2012). Obviously, glass culets obtained by implosion technology are more suitable for recycling in concrete than those crushed by abrasion or impact crushers.

11.6 Sources of further information and advice

The Use of Recycled Glass in Concrete, Available from: http://www.vitrominerals.com

The Waste & Resources Action Programme. Available from: http://www.wrap.org.uk

Berry M, Stephens J and Cross D (2011), 'Performance of 100% fly ash concrete with recycled glass aggregate', *ACI Materials Journal*, **108**, 378–384.

Lam C S, Poon C S and Chan D (2007), 'Enhancing the performance of pre-cast concrete blocks by incorporating waste glass-ASR consideration', *Cement & Concrete Composites,* **29**, 616–625.

Maeda H, Imaizumi H and Ishida E (2011), 'Utilization of calcite and waste glass for preparing construction materials with a low environmental load', *Journal of Environmental Management*, **92**, 2881–2885.

New York State Energy Research and Development Authority (1997), *Use of Recycled Glass for Concrete Masonry Blocks.* Final Report 97–15.

Park S B, Lee B J, Lee J and Jang Y (2010), 'A study on the seawater purification characteristics of water-permeable concrete using recycled aggregate', *Resources, Conservation and Recycling*, **54**, 658–665.

Sagoe-Crentsil K, Brown T, Taylor A (2001), *Guide for specification of recycled glass as sand replacement in premix concrete*, CSIRO, Building, Construction and Engineering.

Wang H (2009), 'A study of the engineering properties of waste LCD glass applied to controlled low strength materials concrete', *Construction and Building Materials*, **23**, 2127–2131.

11.7 References and further reading

Alexander M and Mindess S (2005), *Aggregates in concrete,* Oxon, Taylor & Francis.

Cassar J and Camilleri J (2012), 'Utilisation of imploded glass in structural concrete', *Construction and Building Materials,* **29**, 299–307.

Chen C, Huang R, Wu J, and Yang C (2006), 'Waste E-glass particles used in cementitious mixtures', *Cement and Concrete Research*, **36**, 449–456.

Chen G, Lee H, Young K, et al. (2002), 'Glass recycling in cement production – an innovative approach', *Waste Management*, **22**, 747–753.

Chen S, Chang C and Wang H (2011), 'Mixture design of high performance recycled liquid crystal glasses concrete (HPGC)', *Construction and Building Materials*, **25**, 3886–3892.

Collins R and Bareham P (1987), 'Alkali-silica reaction: suppression of expansion using porous aggregate', *Cement and Concrete Research*, **17**, 89–96.

Dhir R, Dyer T and M Tang (2009), 'Alkali–silica reaction in concrete containing glass', *Materials and Structures*, **42**, 1451–1462.

Ducman V, Mladenovic A and Suput J (2002), 'Lightweight aggregate based on waste glass and its alkali–silica reactivity', *Cement and Concrete Research*, **32**, 223–226.

EU Container Glass Federation (2012), Glass collection for recycling, 2010, Available from: http://www.feve.org/StatsFolder-2010/recycling-data-2010.html [Accessed April 4. 2012]

Federico L and Chidiac S (2009), 'Waste glass as a supplementary cementitious material in concrete – Critical review of treatment methods', *Cement & Concrete Composites*, **31**, 606–610.

Hendriks C and Janssen G (2003), 'Use of recycled materials in constructions', *Materials and Structures*, **36**, 604–608.

Ichikawa T. (2009), 'Alkali–silica reaction, pessimum effects and pozzolanic effect', *Cement and Concrete Research*, **39**, 716–726.

Idir R, Cyr M, Tagnit-Hamou A (2010), 'Use of fine glass as ASR inhibitor in glass aggregate mortars', *Construction and Building Materials,* **24**, 1309–1312.

Ismail Z and Al-Hashmi E (2009), 'Recycling of waste glass as a partial replacement for fine aggregate in concrete', *Waste Management,* **29**, 655–659.

Jain A and Neithalath N (2010), 'Chloride transport in fly ash and glass powder modified concretes – Influence of test methods on microstructure', *Cement & Concrete Composites*, **32**, 148–156.

Jin W, Meyer C and Baxter S (2000), '"Glascrete" – concrete with glass aggregate', *ACI Materials Journal*, **97**, 208–213.

Johnston C (1974), 'Waste glass as coarse aggregate for concrete', *Journal of Testing and Evaluation*, **2**, 344–350.

Kou S and Poon C (2009), 'Properties of self-compacting concrete prepared with recycled glass aggregate', *Cement & Concrete Composites*, **31**, 107–113.

Kozlova S, Millrath K, Meyer C and Shimanovich S (2004), 'A suggested screening test for ASR in cement-bound composites containing glass aggregate based on autoclaving', *Cement & Concrete Composites*, **26**, 827–835.

Lam C, Poon C and Chan D (2007), 'Enhancing the performance of pre-cast concrete blocks by incorporating waste glass – ASR consideration', *Cement & Concrete Composites*, **29**, 616–625.

Liu M (2011), 'Incorporating ground glass in self-compacting concrete', *Construction and Building Materials*, **25**, 919–925.

Maraghechi H, Shafaatian S, Fischer G, et al. (2012), The role of residual cracks on alkali silica reactivity of recycled glass aggregates, *Cement & Concrete Composites,* **34**, 41–47.

Meyer C and Baxter S (1997), *Use of Recycled Glass for Concrete Masonry Blocks*, Final Report 97-15, Albany, New York, New York State Energy Research and Development Authority.

Meyer C and Baxter S (1998), 'Use of recycled glass and fly ash for precast concrete', Rep. NYSERDA 98-18 (4292-IABR-IA-96) to New York State Energy Research and Development Authority, Dept. of Civil Engineering and Engineering Mechanics, Columbia University, New York.

Meyer C, Baxter S, Jin W (1996), 'Potential of waste glass for concrete masonry blocks'. In: *Proceedings of the fourth materials engineering conference*, Washington, 666–673.

Meyer C, Egosi N and Andela C (2001), 'Concrete with waste glass as aggregate' in *Recycling and Re-use of Glass Cullet*, Dhir, Dyer and Limbachiya, editors, *Proceedings of the International Symposium Concrete*, Technology Unit of ASCE and University of Dundee, 19–20.

Mukesh C (2009), 'Bulk engineering and durability properties of washed glass sand concrete', *Construction and Building Materials,* **23**, 1078–1083.

Park SB, Lee BC and Kim JH (2004), Studies on mechanical properties of concrete containing waste glass aggregate, *Cement and Concrete Research*, **34**, 2181–2189.

Pattengil M and Shutt TC (1973), 'Use of ground glass as a pozzolan', *Albuquerque Symp. on Utilisation of Waste Glass in Secondary Products*, Albuquerque, 137–153.

Phillips JC, Cahn DS and Keller GW (1972), 'Refuse glass aggregate in portland cement', *Proceedings of 5th MineralWaste Utilization Symposium.* Chicago, III: IIT Research Institute, 385–390.

Pike recycled glass, Hubbard D and Newman ES (1960), 'Binary silicate glasses in the study of alkali–aggregate reaction'. *High Research Board Bulletin*, **275**, 39–44.

Polley C (1996), 'The effects of waste glass aggregate on the strength and durability of Portland cement concrete', University of Wisconsin, Madison.

Poutos K, Alani A, Walden P and Sangha C (2008), 'Relative temperature changes within concrete made with recycled glass aggregate', *Construction and Building Materials,* **22**, 557–565.

Saccani A and Bignozzi M (2010), 'ASR expansion behaviour of recycled glass fine aggregates in concrete', *Cement and Concrete Research*, **40**, 531–536.

Schmidt A and Saia W (1963), 'Alkali–aggregate reaction tests on glass used for exposed aggregate wall panel work', *ACI Matererials Journal*, **60**, 1235–1236.

Schwarz N and Neithalath N (2008), 'Influence of a fine glass powder on cement hydration: Comparison to fly ash and modeling the degree of hydration', *Cement and Concrete Research*, **38**, 429–436.

Schwarz N, Cam H and Narayanan N (2008), 'Influence of a fine glass powder on the durability characteristics of concrete and its comparison to fly ash', *Cement & Concrete Composites*, **30**, 486–496.

Shao Y, Lefort T, Moras S et al. (2000), 'Studies on concrete containing ground waste glass', *Cement and Concrete Research*, **30**, 91–100.

Shayan A and Xu A (2004), 'Value-added utilisation of waste glass in concrete', *Cement and Concrete Research*, **34**, 81–89.

Shayan A and Xu A (2006), 'Performance of glass powder as a pozzolanic material in concrete: A field trial on concrete slabs', *Cement and Concrete Research*, **36**, 457–468.

Shi C and Wu Y (2005), 'Mixture proportioning and properties of self-consolidating lightweight concrete containing glass powder', *ACI Materials Journal*, **102**, 355–363.

Shi C and Zheng K (2007), 'A review on the use of waste glasses in the production of cement and concrete', *Resources, Conservation and Recycling*, **52**, 234–247.

Shi C, Wu Y, Shao Y and Riefler C (2004), 'AAR expansion of mortar bars containing ground glass powder', *Proc. 12th IAARC*, Beijing, 789–795.

Shi C, Wu Y, Riefler C and Wang H (2005), 'Characteristics and pozzolanic reactivity of glass powders', *Cement and Concrete Research*, **35**, 987– 993.

Taha B and Nounu G (2008a), 'Properties of concrete contains mixed colour waste recycled glass as sand and cement replacement', *Construction and Building Materials*, **22**, 713–720.

Taha B and Nounu G (2008b), 'Using lithium nitrate and pozzolanic glass powder in concrete as ASR suppressors', *Cement & Concrete Composites*, **30**, 497–505.

Terro M (2006), 'Properties of concrete made with recycled crushed glass at elevated temperatures', *Building and Environment*, **41**, 633–639.

Topcu IB and Canbaz M (2004), 'Properties of concrete containing waste glass', *Cement & Concrete Research*, **34**, 267–274.

Tuncan M, Karasu B and Yalc_ın M (2001), The suitability for using glass and fly ash in Portland cement concrete, In *The Proceedings of the Eleventh International Offshore and Polar Engineering Conference* (ISOPE 2001), Stavanger, Norway, **4**, 146–152.

US Environmental Protection Agency (2010), Municipal solid waste in the United States: 2009 Facts and figures, Washington, DC, Available from: http://www.epa.gov/wastes/nonhaz/municipal/msw99.htm [Accessed April 3. 2012]

Wang H (2009a), 'A study of the effects of LCD glass sand on the properties of concrete', *Waste Management*, **29**, 335–341.

Wang H (2009b), 'A study of the engineering properties of waste LCD glass applied to controlled low strength materials concrete', *Construction and Building Materials*, **23**, 2127–2131.

Wang H and Huang W (2010a), 'A study on the properties of fresh self-consolidating glass concrete (SCGC)', *Construction and Building Materials*, **24**, 619–624.

Wang H and Huang W (2010b), 'Durability of self-consolidating concrete using waste LCD glass', *Construction and Building Materials*, **24**, 1008–1013.

Wei Y, Lin C, Ko K et al. (2011), 'Preparation of low water-sorption lightweight aggregates from harbor sediment added with waste glass', *Marine Pollution Bulletin*, **63**, 135–140.

Zhu H and Byars E (2004), 'Alkali-silica reaction of recycled glass in concrete, alkali-aggregate reaction in concrete' in *Proceedings of the 12th International Conference on Alkali-Aggregate Reaction in Concrete*, 811–820.

Zhu H, Chen W, Zhou W and Byars E (2009), 'Expansion behaviour of glass aggregates in different testing for alkali-silica reactivity', *Materials and Structures*, **42**, 485–494.

Part III

Concrete with non-reactive wastes

12
Municipal solid waste incinerator (MSWI) concrete

M. TYRER, Mineral Industry Research Organisation, UK

DOI: 10.1533/9780857098993.3.273

Abstract: The potential to use municipal waste incineration products in concrete has been examined over many years, both as a source of aggregates and as supplementary cementitious materials (SCM). With an understanding of the chemical evolution of these materials in the cement environment, both applications provide routes to the re-use of these materials in construction. Fly ashes and air pollution control residues must be washed to remove their considerable chloride content before use, after which, the remaining solids (largely aluminosilicates) may replace some fraction of the cementitious binder. Care must be taken when using these materials as SCMs, as volatile heavy metals present in the original waste may be concentrated in the fly ash. The bottom ashes from municipal solid waste (MSW) incineration also show sufficient pozzolanic reactivity when finely ground, to replace some cement in the binder phase. They may also serve directly as aggregate in concrete, although the irregular and angular particle shape may reduce its workability somewhat. Most MSW bottom ashes contain sufficient free aluminium to disrupt the setting reactions of the binder through hydrolysis reactions at high pH. Consequently, this must be removed or reacted by pre-treatment, prior to use in concrete and means of doing so are described here. Slower detrimental reactions are also known, such as alkali silca reaction, especially of container-glass fragments and these are considered here. Lastly, the use of MSW incineration products in manufactured aggregates is described, comparing sintering, melting and recrystallization, and plasma processing, with low temperature methods such as rapid carbonation technology.

Key words: municipal solid waste (MSW), municipal solid waste incinerator (MSWI), fly ash, bottom ash, air pollution control (APC) residues, plasma processing, accelerated carbonation.

12.1 Introduction

The global trend towards incineration of municipal solid waste (MSW) reflects two important drivers. There is a widely recognized need to reduce the volumes of waste sent to landfills and this is especially true of highly industrialized and densely populated countries such as Japan and the United Kingdom, but is of wider global concern. Incineration of municipal waste offers two advantages over conventional landfill disposal; the first is volume reduction; typically around 90% of the original volume, equivalent to >70%

by mass (Ginés et al. 2009) and the second is a change in composition, in that its combustion products (largely oxides) are less likely to change their volumes due to decomposition after landfilling. The second driver for incineration is that of energy recovery. Although much of the potential energy released by combustion of these wastes is used in evaporation of the water they contain, there is a useful residue to be recovered and energy-from-waste (EfW) plants have been widely established in industrialized nations. In addition, one important benefit of waste incineration is the destruction of pathogens and organic compounds detrimental to health.

The residues of municipal solid waste incineration (MSWI) have some potential as construction materials such as unbound filler minerals, but it is their specific use in concrete which is discussed here. This chapter considers the composition of MSWI ashes, their hydration in concrete, their performance in service and the environmental implications of incorporating such residues.

12.2 Composition

Before looking in more detail at the composition of its combustion products, we should first look at municipal waste before it undergoes incineration. It is estimated (Chimenos et al. 1999) that in the year 2000, the rate of municipal waste incineration (MWI) residue produced in Europe, Japan and the United States was around 25 million tonnes per year. It is surprising, given the wide variety of materials disposed of as municipal waste throughout the world, that its bulk composition is relatively uniform. This largely reflects the modern practice of waste sorting – recovery of useful resources at source. Some countries are more exacting in their requirements than others, but the segregation of metals, paper and cardboard, glass, plastics and compostable materials are, to a greater of lesser extent, separated for individual collection by the municipal authorities. Combined with electromagnetic and eddy current separation for metal recovery, waste sorting has increased the value recovery from waste markedly. The impact on this over the last two decades is very far-reaching; for example some 80% of domestic paper and cardboard is recycled in Europe, where almost none was recovered 30 years ago. Conversely, the quantity of metal foil (principally aluminium) used in packaging has increased substantially over the same period and this has consequences for the use of MWI bottom ash in concrete – a subject expanded on later. A detailed discussion of waste composition, waste management practice and likely developments in waste generation lie outside the scope of this chapter. However, Parfitt (2002) and Burnley (2007) consider wastes in the United Kingdom, whilst Bartelings and Sterner (1999) discuss experiences in Sweden. For a more general reference, the book *Handbook of Solid Waste Management* (Tchobanoglous and Kreith (eds) 2002) gives a comprehensive view of the subject, particularly from a North American perspective.

Combustion technology has advanced from essentially a batch process to become dominated by continuous processes, of which the 'moving-grate' type is by far the most widely adopted (Fig. 12.1). Although furnaces of the fixed grate type are still in operation, they are eclipsed by moving grate incinerators, as these allow a continuous charge of municipal waste to be processed and their combustion products to be removed. Some installations use fluidized bed incinerators, in which the burning waste lies on top of a bed of mineral grains (commonly silica sand and dolomite) which is continuously agitated by an updraft of air flowing through it. Although some advantages are offered in terms of energy efficiency, these have been somewhat off-set by the longer burn times required and issues concerned with operational efficiency (Garg et al. 2009). In the interests of resource recovery and energy efficiency, the 'mass-burn' of unsorted waste is in its decline and the term 'refuse derived fuel' (RDF) has become widely adopted, to describe sorted waste streams which have a higher calorific value and lower volume of combustion products than unsorted waste. Critically, the energy content of unsorted waste is usually insufficient to maintain combustion (which requires 10–12 GJ/tonne) whilst sorted waste – RDF – burns readily above 800 °C with a forced air supply. The energy efficiency of incineration varies depending on both process efficiency and water content of the waste between around 14 and 28% in terms of electrical output (Blue Ridge Environmental Defense League 2009). Such efficiency calculations are notoriously difficult to perform, other than on a case-by-case basis, owing to the wide variations in the re-use of low-grade heat. Some modern incinerator designs incorporate

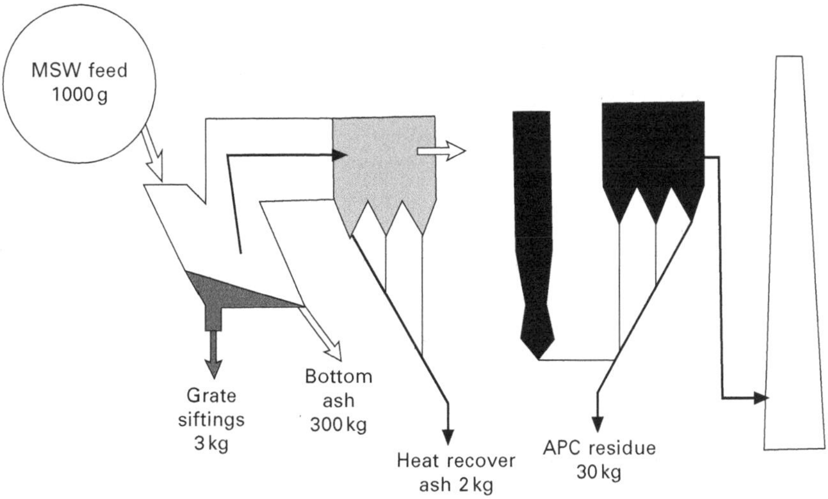

12.1 Proportions of waste generated during each stage of municipal waste incineration (APC = air pollution control). After Wiles (1996).

advanced heat integration technology, where the energy not used to generate electricity, is re-used by the plant, or by adjacent industries or municipal district housing schemes.

For more detailed discussions of waste incineration, readers are referred to the books by Brunner (1991), Wentz (1989), Williams (2005) and Dean (1988) which in combination, give a comprehensive view of the processes involved. In addition, The World Bank Technical Guidance Report (1998) on 'Municipal Solid Waste Incineration' provides an excellent overview of the issues influencing these activities.

12.3 Combustion products

Combustion of MSWI waste produces two principal combustion products: those which fall from the incinerator and those which rise from it during combustion. Both Sabbas et al. (2003) and Wiles (1996) describe in some detail the combustion products and distinguish between the following streams (note that mass fraction estimates are from Sabbas et al. (2003)):

- *Bottom ash*: Solids discharged from the furnace bottom, principally by falling through the grate, noting that the older term 'clinker' is in common usage in Europe and that 'grate siftings' or 'riddlings' (materials discharged during movement of the grate) are usually incorporated with the bottom ash. Together they represent 20–30% by mass of the original waste on a wet basis.
- *Heat recovery ash; 'economizer' ash*: Modern incinerators incorporate heat recovery devices and particulate ashes collected from this stage may be combined with either the bottom ash or air pollution control residues depending on the reactor design. They represent around 10% of the original wet mass of waste.
- *Fly ash*: Particles recovered from the flue gasses, prior to their chemical cleaning by electrostatic precipitation and/or injection of reactive solvents. Such ashes are enriched in volatile components (Pb, Zn, Cd, Sn, Sb, Na, K, Cl, S etc.) making their bulk composition distinct from the bottom ashes. In total, fly ashes account for 1–3% of the wet waste mass.
- *Air pollution control* (APC) residues: The products formed after flue gas cleaning, by reaction with injected solvents and sorbents. These contain precipitates from reactions with solutes in the spray in addition to fine ash particles and represent 2–5% of the original waste mass on a wet basis. These APC residues may be in solid, liquid or sludge form (Sabbas et al. 2003) depending on the scrubbing technology adopted and may be mixed with the fly ash fraction.

The different composition of these materials governs their potential re-use in construction and only certain fractions lend themselves readily to concrete

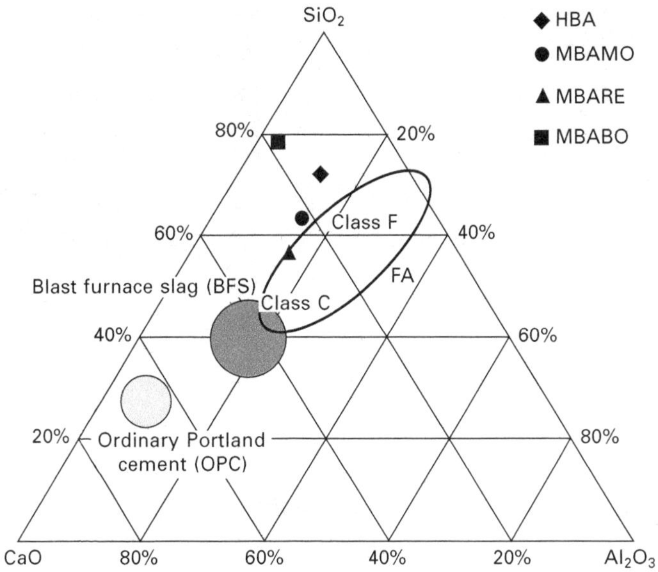

12.2 Comparison of MSWI fly ashes ('MBA...') and hospital incinerator waste ('HBA') with conventional cementitious materials. After Filliponi et al. (2003).

production. To understand why this is so, the composition of each of these materials should be considered in more detail.

12.3.1 Bottom ashes

These solids comprise around 80% of MWI combustion products (Chimenos et al. 1999) and consist of both glassy and crystalline particles, the latter containing diamagnetic and paramagnetic particles of metal. Approximately 10^9 tonnes of MSWI bottom ash are produced in Europe each year, but only 46% of this is re-used; the remainder poses a disposal problem. (Tiruta-Barna et al. 2007). This fraction is much coarser than the others; particles from 0.1 to 10 mm diameter being dominant (Ginés et al. 2009; Chimenos et al. 1999 – see Fig. 12.3). Johnson et al. (1995) compares bottom ashes to basic igneous rocks, as their bulk chemistry is dominated by Si, Ca, Al and Fe (see also Belevi et al. 1992) and describes them as 'composed of equal amounts of fine ash material and melted components, over half of which have crystallized, small quantities of metallic components, ceramics and stones', citing Lichtensteiger (1992) as a corroboration. She goes on to list some of the major crystalline components, such as calcite, ettringite, haematite, quartz, gypsum and silicate minerals (e.g. pyroxene, Fig. 12.4). Chimenos et al. (1999) add gypsum, feldspar (anorthite, $CaAl_2Si_2O_8$) and quick lime (CaO) to this list and describe the many residual ceramic materials such as

12.3 Municipal solid waste incinerator ash separated into size fractions After Rübner et al. (2008).

fired pottery fragments, cement and concrete, brickbat and porcelain. They further describe the magnetic components as dominantly magnetite (Fe_3O_4), haematite (Fe_2O_3) and wustite (FeO), noting that strictly only magnetite is ferromagnetic (see Fig. 12.5). Amongst the diamagnetic particles, copper and its alloys are noted and especially aluminium, which in its various forms represents over 90% by mass of the free metals. Numerous minor phases have been reported, amongst which are wollastonite, ($CaSiO_3$) Mayenite ('$C_{12}A_7$' – $Ca_{12}Al_{14}O_{33}$) and gehlenite ('C_2AS' – $Ca_2Al_2SiO_7$) (Qiao et al. 2008a).

The glassy particles are largely derived from container glass fragments incorporated in the waste, although of more variable composition after cooling, owing to incorporation of other elements whilst fused in the incinerator. Rübner et al. (2008) estimate the glass content comprises around 15% by mass of the bottom ash.

Also incorporated in the bottom ash is unburned carbonaceous material

12.4 Hedenbergite (iron-rich clinopyroxene; $CaFeSi_2O_6$) crystals in a MSWI bottom ash aggregate particle. Cross-polarized light. After Müller and Rübner (2006).

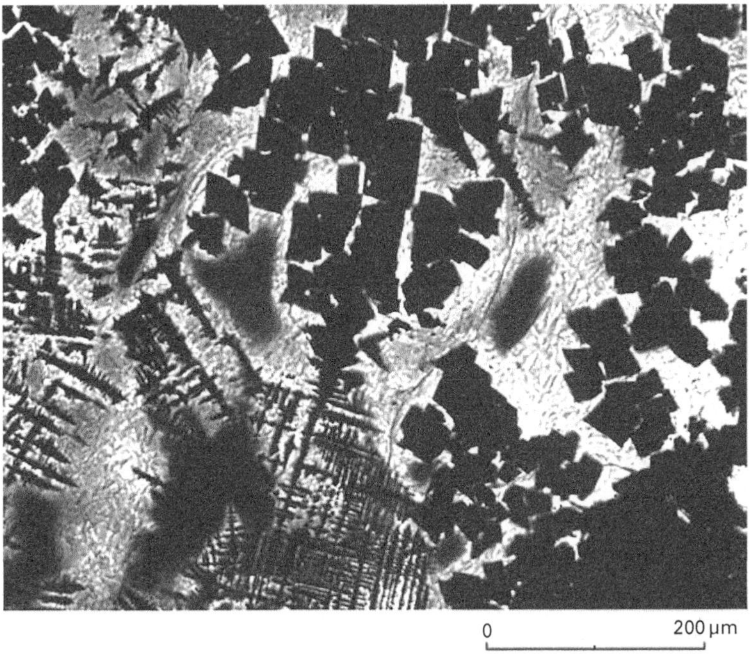

12.5 Magnetite crystal formation in silicate glass, showing 'graphic' texture (bottom left). After Müller and Rübner (2006).

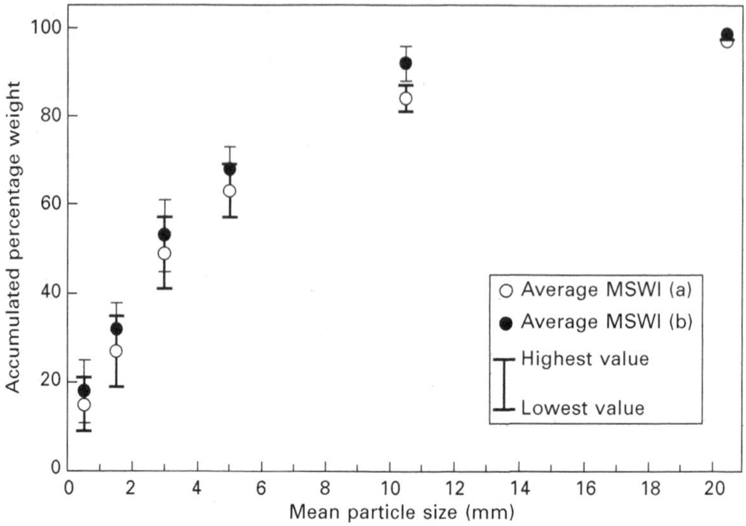

12.6 Particle size distribution in bottom ashes from Cataluña, Spain After Chimenos et al. (1999).

of varying composition, which represents a percent or so of the total mass. Carbon-rich, these particles have the potential to burn more fully, yet the residence time in the incinerator was insufficient to allow them to do so. Increasing the burn-time is an unattractive option as it reduces the overall energy efficiency of the plant, so the compromise is to allow a small (~1% of the original wet waste mass) to remain incompletely burnt. Bone fragments may also be present in the bottom ash, converted wholly or partially to apatite ($Ca_5(PO_4)_3(OH,F,Cl)$).

12.3.2 Fly ashes

The fine material (see Fig. 12.7) rising up the combustion stack comprises incombustible, inorganic solids such as molten glass fragments, and crystalline products from combustion. This fraction is collected by electrostatic precipitators and filters and represents 2–3% of the original wet mass of waste. Incorporated with these unburned fragments are condensates and reaction products from the gas phase and these distinguish the fly ash from the bottom ash.

Youcai et al. (2002) groups flue gas impurities into four groups:

- acid gases such as HCL, HF, SO_2, NO_x;
- products of incomplete combustion, such as hydrocarbons, CO, dioxins and furans;
- dust with heavy metals;
- volatile heavy metals.

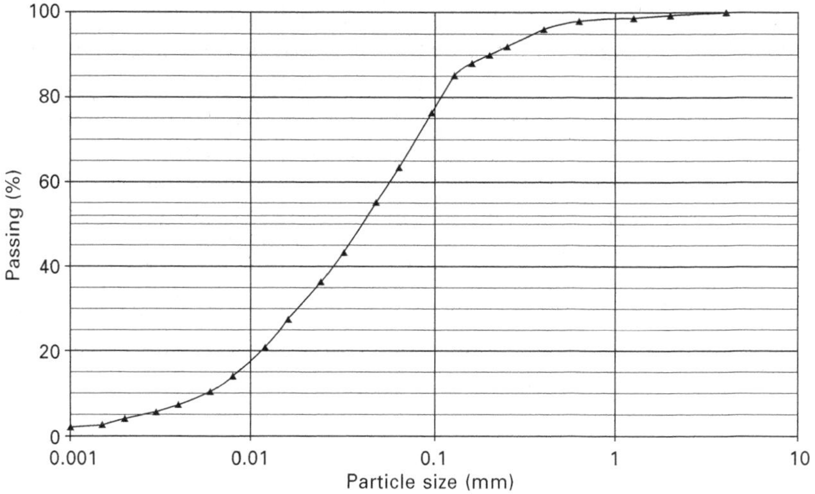

12.7 Particle size distribution of MSWI fly ash. After Aubert et al. (2004).

Volatile metals such as lead, zinc, tin cadmium and antimony, along with group I metal halides (especially sodium and potassium chlorides) and, to a lesser extent, sulphates, are concentrated in the fly ash. In terms of their elemental composition, the fly ashes are dominated by calcium, aluminium and silicon (mainly oxides) and by sodium and potassium halides (mainly chlorides) along with sulphur, largely present as sulphate. Youcai et al. (2002) confirm, as might be expected, that the heavy metal content of the ash is a function of the heavy metal content of the original waste. Much work has been done on the elemental composition of MSW fly ash, amongst which work by Fernandez et al. (1992), Greenberg et al. (1978) and Sawell and Constable (1993) are very informative. Readers are particularly referred to the work of the International Ash Working Group (IAWG) reports, especially 'Municipal Solid Waste Incinerator Residues' (Chandler et al. 1997). By comparison with the bulk elemental composition, relatively little has been done on the phase chemistry of fly ash. Eighmy et al. (1995) summarize much previous work and go on to identify two principal particle types:

- 'An extremely fine grained, polycrystalline aggregated platelet material containing many volatile elements such as Cl, K, Zn, Na S and Pb'; and
- 'spherical alumnosilicate particles'.

They describe the mineral phases associated with each; soluble salts such as NaCl, K_2ZnCl_4 and $KClO_4$ associated with the crystalline agglomorations and in the remainder, a wide range of calcium, sodium and potassium aluminosilicates, lead silicates and spinel phases. These observations are confirmed by Gong and Kirk (1994) who reported that these heavy metals exist largely as chlorides, primarily on the surfaces of larger particles (mostly silicates) a view supported by Greenberg and Zoller (1978). Importantly, they consider the relative solubility of these phases, using both leach testing methods and equilibrium thermodynamic calculations. Their findings that much of the fly ash (they estimate around 60%) is soluble and that heavy metals are mobilized from the ash following dissolution are confirmed and expanded by Johnson et al. (1996) and van der Sloot et al. (2001). The implication of this is far reaching; not only do the fly ashes contain heavy metal species, they are in a form which is soluble and therefore available for environmental dispersion, increasing the impact of their toxicity.

12.3.3 APC residues

This class of combustion product contains, to a greater or lesser extent, a portion of the fly ash which carries over from the previous stage, where treatment and collection of the two fractions are independent. APC residues are distinct from fly ash in that they are formed during the final stage of flue gas cleaning by injection of solvents and sorbents into the gas stream. To neutralize acid gases such as HCl, SO_2 and NO_x, alkaline solutions (typically calcium and/or sodium carbonate and/or hydroxides – finely ground limestone slurry is the most common) are sprayed into the gas in a 'scrubber'. The HCl and SO_2 are rapidly converted to halite and gypsum whilst some CO_2 can be captured as sodium carbonate where sodium hydroxide is present. Lundtorp et al. (2002) describe the use of $FeSO_4$ as a scrubbing reactant and show that it results in reaction products with significantly lower solubilities than do conventional methods. Organic compounds (especially dioxins and furans) are captured by sorption onto the surface of activated carbon particles also present in the scrubber fluids. Further oxidation of the gas stream may take place, through the use of regenerative oxidizers of either the thermal or catalytic types, but as these act solely in the gas phase (producing no solid waste) they need not concern us further. Alba et al. (1997) and Bethea (1978) review air pollution control technology and options for the treatment of its residues. More recent work by Quina et al. (2008) and Rani et al. (2008) completes the picture, discussing the production, properties, composition and treatment of MSW incinerator fly ashes. Both groups of authors advocate the thermal conversion of APC residues to glassy ceramics, feeling that the high solubility of some of the metal-bearing phases they contain precludes their direct use in concrete.

It should be noted that the term 'air pollution control residues' is used to describe more than one waste stream, which reflects the wide variation of technology and practice in the industry. APC residues may be the solids

Table 12.1 Elemental compositions of municipal waste combustion products

Element	Bottom ash Range (mg/kg)	Fly ash Range (mg/kg)	Dry/semi-dry APC residues Range (mg/kg)	Wet APC system residues without fly ash Range (mg/kg)
Ag	0.29–37	2.3–100	0.9–60	
Al	22 000–73 000	49 000–90 000	12 000–83 000	21 000–39 000
As	0.12–190	37–320	18–530	41–210
B	38–310			
Ba	400–3000	330–3100	51–14 000	55–1600
C	1 000–60 000			
Ca	37 000–120 000	74 000–130 000	110 000–350 000	87 000–200 000
Cd	0.3–71	50–450	140–300	150–1400
Cl	800–4200	29 000–210 000	62 000–380 000	17 000–51 000
Co	6–350	13–87	4–300	0.5–20
Cr	23–3200	140–1100	73–570	80–560
Cu	190–8200	600–3200	16–1700	440–2400
Fe	4100–150 000	12 000–44 000	2600–71 000	20 000–97 000
Hg	0.02–7.8	0.7–30	0.1–51	2.2223
K	750–16 000	22 000–62 000	5900–40 000	810–8600
Mg	400–26 000	11 000–19 000	5100–14 000	19 000–170 000
Mn	83–2400	800–1900	200–900	5000–12 000
Mo	2.5–280	15–150	9.3–29	1.844
N	110–900	1600		
Na	2 900–42 000	15 000–57 000	7600–29 000	720–3400
Ni	7–4 300	60–260	19–710	20–310
O	400 000–500 000			
P	1400–6400	4800–9600	1700–4600	
Pb	98–14 000	5300–26000	2500–10 000	3300–22 000
S	1000–5000	11 000–45 000	1400–25 000	2700–6000
Sb	10–430	260–1100	300–1100	80–200
Se	0.05–10	0.4–31	0.7–29	
Si	91 000–310 000	95 000– 210 000	36 000–120 000	78 000
Sn	2–380	550–2000	620–1400	340–450
Sr	85–1000	40–640	400–500	5–300
Ti	2600–9500	6800–14 000	700–5700	1400–4300
V	2–120	29–150	8–62	25–86
Zn	610–7800	9000–70 000	7000–20 000	8100–53 000

After Wiles (1996) modified after Chandler et al. (1997).

recovered from gas cleaning, or a mixture of these solids with some or all of the fly ash.

12.4 Hydration

Much work has been done on the leaching of MWI ashes, driven by two principal concerns. First, the potential to leach harmful substances from the solids governs their suitability for use in construction materials and second, determines the available disposal options, such as landfilling. In each case, partial dissolution of the solids results in an alkaline solution. For fly ashes this is both strongly buffered and saline, owing to the near complete dissolution of chlorides (Chandler et al. 1997).

12.4.1 Aluminium-induced expansion

Hydration in a cement pore solution, naturally provides an alkaline reserve buffered by both portlandite (calcium hydroxide) and CSH (calcium silicate hydrate) gel from the cement. Considering bottom ash in an alkaline solution, one reaction, above all others is dominant in determining the use of this material in concrete and that is the hydrolysis reaction associated with metallic aluminium (and similarly, though much less commonly, with zinc).

The amphoteric nature of aluminium allows it to react with both acids and alkalis. In acidic conditions, Al^{3+} is the dominant ion (strictly this is hydrated as $Al(H_2O)_6^{3+}$). As the pH of the solution rises, the aluminium speciates as the hydroxide $Al(OH)_3$ until its activity is sufficiently high to allow the solid hydroxide to precipitate. Under neutral conditions, gibbsite is the solubility limiting phase for aluminium at equilibrium; however, the related phases boehmite and diaspore (AlOOH) may both be formed initially, converting to the more stable gibbsite with time. At elevated pH, gibbsite dissolves in favour of aluminate ions; $[Al(OH)]_4]^-$ and in a cement pore solution (pH 12.5–13.5) solid hydroxide is absent at equilibrium. Equation (12.1) shows the alkaline hydrolysis at high pH:

12.8 Spalling of the edge of a concrete specimen containing MSWI bottom ash, caused by an aluminium grain (cross-section, scanner, reflected light). After Rübner et al. (2008).

$$2Al + 2OH^- + 6H_2O \rightarrow 2[Al(OH)_4]^- + 3H_2(gas) \quad [12.1]$$

The consequence of exposing bottom ash to a cement pore solution is that the aluminium will dissolve, releasing hydrogen gas bubbles. This is moderately rapid and will mechanically disrupt the formation of a continuous network of CSH, especially in the vicinity of the original aluminium grains, where voids may form during setting. If the aluminium is present largely as $Al(OH)_3$ (gibbsite) exposure to alkaline conditions will promote its dissolution, but without hydrogen liberation (Eqn 12.2):

$$Al(OH)_3 + OH^- \rightarrow [Al(OH)_4]^- \quad [12.2]$$

Müller and Rübner (2006) discuss the mechanism and consequence of aluminium-induced hydrolysis when MSWI bottom ash aggregates are used in concrete, showing the nature of the porosity induced and the form of the reaction products alongside other detrimental reactions. Aubert et al. (2004) presents a similar discussion for aluminium carried over into the fly ash and Bertolini et al. (2004) describe experiments on the swelling mechanisms of both fly ash and bottom ash. In each case, the authors note the potential for hydrogen gas generation and recommend that the metal is removed or reacted before use. Figures 12.9 to 12.12 illustrate the effects of expansion associated with aluminium-induced hydrolysis.

Looking at the other phases present in bottom ash which will undergo some reaction in cement pore solution, all appear to be the dissolution of solids which react to form solid hydrates and therefore contribute to the space-filling

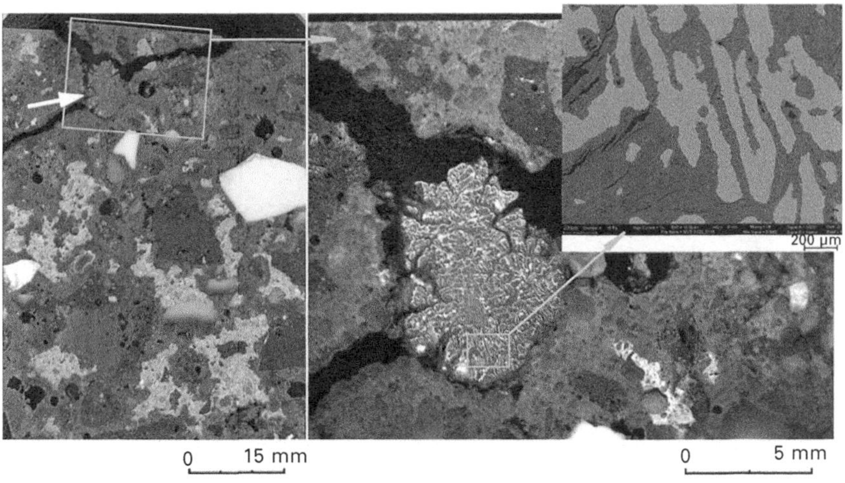

12.9 Aluminium grain (arrow), which caused spalling of the corner of specimen in Fig. 12.3 (polished cross-section, reflected light). The inset image shows a SEM image of the intergrowth of aluminum and aluminum hydroxide. After Müller and Rübner (2006).

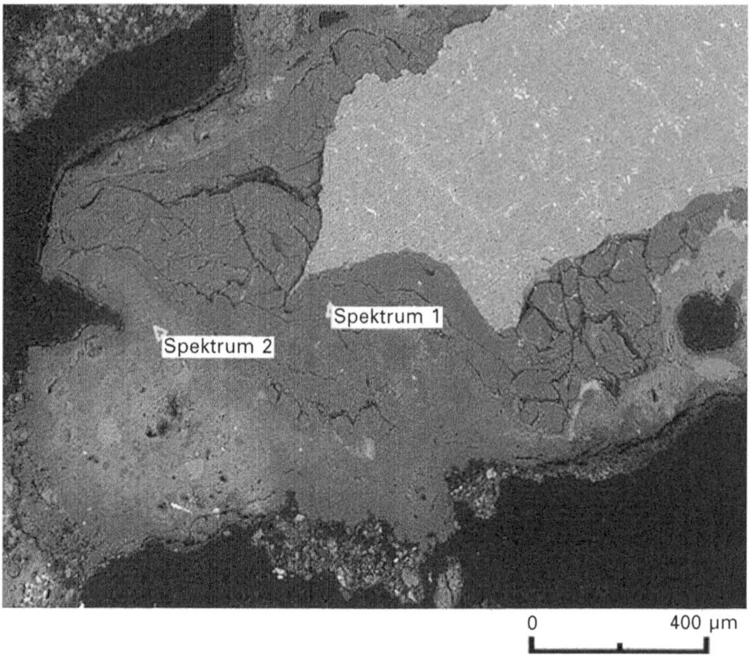

12.10 An amorphous Al-hydroxide layer (marked 'Spektrum 1') surrounding an aluminum grain (lightest phase; top right) in contact with a calcium aluminate hydrate (marked 'Spektrum 2'). Backscattered SEM. After Müller and Rübner (2006).

12.11 Expansion of 150 mm cubes after demoulding. Concrete samples containing MSWI bottom ash; wet ground (left) and dry ground (right). After Bertolini et al. (2004).

12.12 Example of entrapped voids on the fracture surface of the concrete made with dry ground MSWI bottom ash. After Bertolini et al. (2004).

of the concrete during setting. One consequence of this is that the volume of reaction products may exceed the space available to accommodate them, resulting in swelling and cracking of the cement or concrete as the reaction proceeds.

12.4.2 Glass-induced expansion

Considering next the glassy phases – a degree of reaction is expected in the silicate and aluminosilicate glasses; although as these exist as large grains, early-age dissolution before setting is relatively slight, being confined to surface reactions. The reaction products are dominantly alkali-bearing C(A)SH gels which contribute to the overall structure of the concrete. However, long-term reactions of container glass fragments are of much greater concern. Most container glass is sodium-rich and most of the glass present in MSW ashes is derived from it. As the glass is exposed to alkaline pore solutions, the potential for it to partially dissolve and re-precipitate as an expansive (Na-K-$SiO_2.nH_2O$) hydro-gel is a real threat to the long-term stability of cementitious materials containing it. Müller and Rübner (2006) show the nature of this expansive reaction product, forming in cracks within the glass particles, gradually forcing them apart. The morphology they describe is typical of alkali silica reaction seen in concrete made with other metastable silicates such as opaline silica and chert (Ingham 2011; St. John et al. 1998). Jin et al. (2000) and Xie et al. (2003) discuss this effect in detail

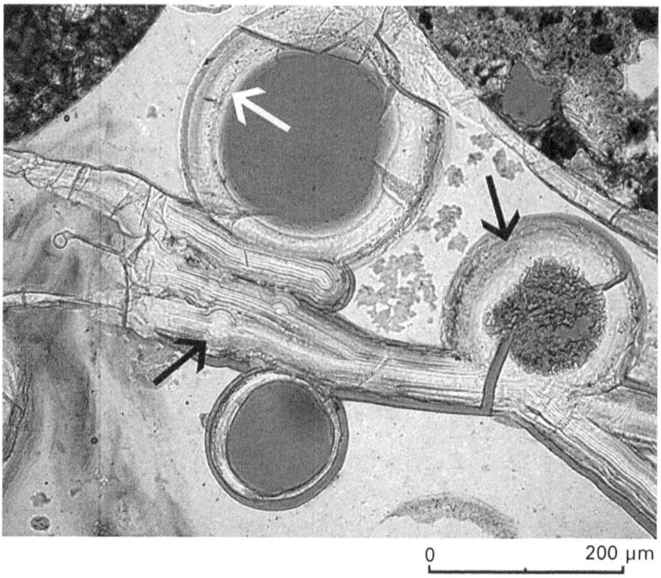

12.13 Silicate gel inside a bottle glass fragment (arrows). Eight-year old concrete, transmitted light. After Müller and Rübner (2006).

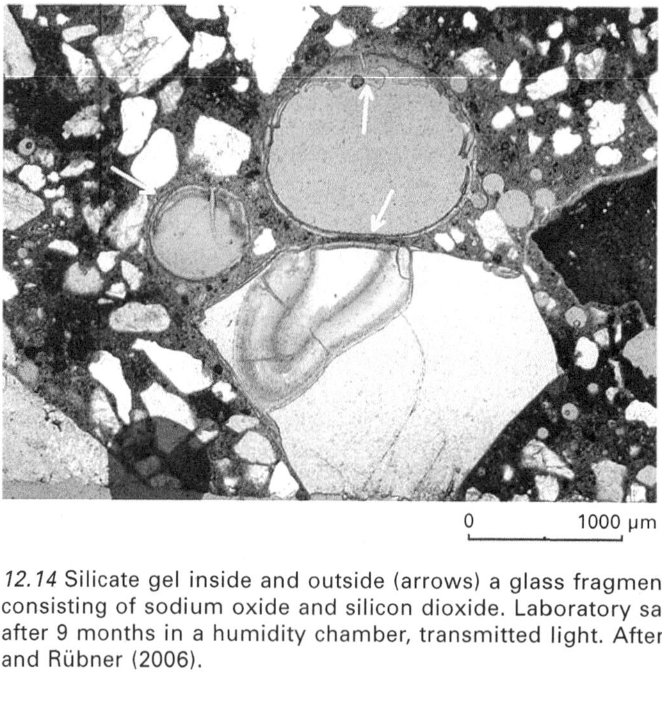

12.14 Silicate gel inside and outside (arrows) a glass fragment consisting of sodium oxide and silicon dioxide. Laboratory sample after 9 months in a humidity chamber, transmitted light. After Müller and Rübner (2006).

and excellent examples are shown in the micrographs of Müller and Rübner (2006).

12.4.3 Sulphate-induced expansion

In the case of weathered or oxidized bottom ash (where free aluminium metal is no longer present, after conversion to the oxide or hydroxide) an equilibrium will be established between calcite, ettringite and gypsum and the AFm and AFt phases, which will evolve with similar reactions in the hydrating cement. Practically, formation of both ettringite (the sulphate-AFt phase) and calcium monosulphoaluminate hydrate – or simply 'monosulphate' (the sulphate-AFm phase) will be governed by the availability of their constituents – in this case aluminate and sulphate ions – in solution. Competing with ettringite and monosulphate for the available aluminium are minor phases, such as the carbonate bearing AFm and hydrotalcite. These are very minor constituents of the phase assemblage and are noted here for completeness, although their occurrence is regularly reported in the literature (e.g. Qiao et al. 2008a). Iron, of course, is one of the four major constituents of MWI bottom ash, but the inherently low solubility of its salts at high pH, means it is of only minor interest. Certainly there may be some slight substitution of iron for aluminium in any of the Al-bearing hydrates, but that would reflect iron being present in (for example) the aluminosilicate glass phase of the waste. As most of the iron is present as oxides (haematite, magnetite and wüstite (FeO)) it is likely to remain there as these phases should not react appreciably in the cement environment.

In conclusion, there are numerous mechanisms by which MSWI combustion products may deteriorate when used in concrete. Some of these pose major concerns (aluminium-induced hydrolysis, alkali silica reaction and ettringite formation) and options for their mitigation are discussed below.

12.4.4 Minor reaction products

The formation of other phases may be considered minor reactions, but many have been noted by workers examining these complex materials as follows:

- *AFt formation.* This is dominated by ettringite precipitation but it is conceivable, for example where excess soluble carbonate is available in the system, that AFt-CO_3 could be formed (though not yet reported). AFt salts have an open channel structure and consequently high molar volume, so their potential to cause swelling is considerable.
- *AFm formation.* The AFm phases are denser than their AFt counterparts, so pose less of an expansion risk than AFt, but are widely reported minor

reaction products. Although there are many compounds in this mineral group, their impact on concrete durability is not expected to be great.

- AFm-SO_4 (calcium monosulphoaluminate hydrate; usually just 'Monosulphate' – $Ca_4(Al_6O_{12})(SO_4)$)). This phase is commonly seen in addition to ettringite at ambient temperatures, but will dominate above 50 °C where ettringite is no longer stable (Glasser et al. 1999).
- AFm-Cl (calcium monochloroaluminate hydrate. 'Friedel's salt' $Ca_2Al(OH)_6(Cl, OH) \cdot 2\ H_2O$). Reported by Zhu et al. (2010) in washed MSWI fly ash.
- AFm-CO_3 (calcium monocarboaluminate hydrate: $Ca_8Al_4O_{14}(CO_2.12H_2O)$)). Reported by Qiao et al. (2008a) – found in hydrated thermally treated bottom ash
- AFm-½CO_3 (calcium hemicarboaluminate hydrate: ($Ca_4Al_2O_6(CO_3)_{0.5}.12H_2O$). Reported by Qiao et al. (2008a) and found in hydrated, thermally treated bottom ash
- AFm-SO_4-CO_3 (calcium sulphoaluminate carbonate hydrate: $Ca_4Al_2O_6(CO_3)_{0.67}(SO_3)_{0.33}.11H_2O$). A similar phase to Kuzel's salt, reported by Qiao et al. (2008a) found in compression-moulded pastes containing hydrated thermally treated bottom ash (see Fig. 12.15).
- AFm-OH (Calcium aluminate hydroxide, 'C_4AH_X'). These phases are believed by Glasser et al. (1999) to be metastable with respect to hydrogarnet, gibbsite and in cements, other AFm phases.

- Additional phases in this AFm family are known as reported by Matschei et al. (2007) and a comprehensive treatment of their synthesis and stabilities was published by Glasser et al. (1999).
- Precipitation of hydrogarnet ($Ca_3[Al(OH)_6]_2$ 'hydrogrossularite', 'C_3AH_6') described by Rübner et al. (2008) as a phase which due to expansion, 'can cause damage to the concrete' according to the following reaction:

$$2[Al(OH)4]^- + 3Ca^{2+} + 4OH^- \rightarrow Ca_3[Al(OH)_6]_2 \text{ (hydrogarnet)} \qquad [12.3]$$

- *Hydration of oxides.* Thermal processing of MSWI ashes could convert calcite ($CaCO_3$) or dolomite ($MgCO_3$) to their respective oxides, which on subsequent hydration, would swell as they convert to the hydroxides $Ca(OH)_2$ and $Mg(OH)_2$ (see Qiao et al. 2008a). Note that the free oxides are not abundant in untreated MSWI ashes, but may form during thermal processing.

12.5 Use in concrete: assessment and pre-treatment

If MSWI ashes are to be used in concrete formulation, then a method must be found to minimize the reactive components they contain before

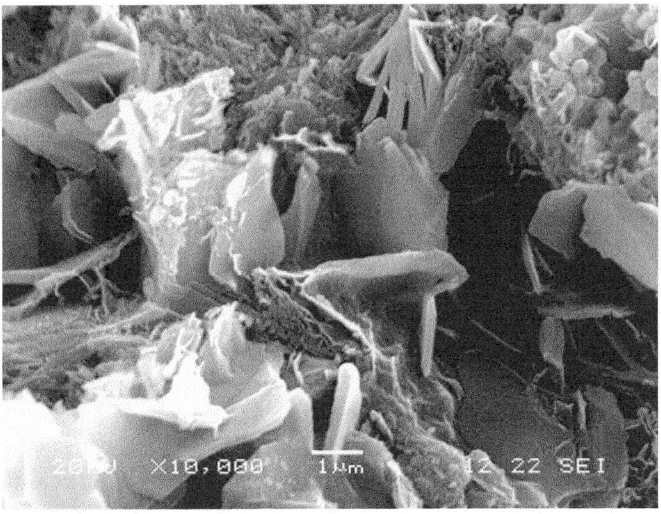

12.15 Tabular sulphate-carbonate AFm phase of intermediate composition ($Ca_4Al_2O_6(CO_3)_{0.67}(SO_3)_{0.33}.11H_2O$) formed in thermally treated bottom ash pastes hydrated under compression. After Qiao et al. (2008a).

mixing. Two means of reducing the aluminium metal in bottom ash have been proposed: direct oxidation of the metal by weathering or at elevated temperature in air; or a reactive pre-treatment which will convert the metal to its hydroxide. Qiao et al. (2008c) showed that heating MSWI bottom ash in a rotary tube furnace at 600–800 °C reduced the aluminium content and increased the content of hydraulic phases such as gehlenite ($Ca_2Al_2SiO_7$), and mayenite ($Ca_{12}Al_{14}O_{33}$). They showed that compacted samples did not suffer unduly from hydrogen evolution from the residual aluminium and produced hydrated materials of adequate compressive strength (~12–15 MPa) for re-use. Rübner et al. (2008) demonstrated the problems of using untreated MSWI bottom ash in concrete and showed how they may be overcome be reacting the metallic aluminium with sodium hydroxide solution before use. This 'lye' treatment (*sic.*) reduced the aluminium content to below 0.4% by mass with the advantage that the washing step removes both sulphates and chlorides by dissolution. The concretes they prepared in this way showed a moderately lower strength than the control samples (–15%) which suggests this pre-treatment is a suitable means of processing these solids. Pera et al. (1997) concur with Rübner et al. (2008) in the use of sodium hydroxide to react the aluminium (15 days' exposure) and found acceptable concrete could be made with 50% of the aggregate being processed MSWI bottom ash with a similar quantity of natural aggregate.

Glass removal was achieved by Zwahr (2005) using a combination of optical screening of the bottom ash particles and mechanical sorting, where the rejected (glass-rich) fraction was removed with an air jet. This process showed a ~50% reduction in the glass content, suggesting that after such treatment a reduction in alkali–silica reaction (ASR)-induced swelling would follow. In combination with up-stream washing of the bottom ash, to reduce the chloride, sulphate and lime content, opto-mechanical sorting increased greatly the value of processed bottom ash components in concrete. Boghetich et al. (2005) examined the effectiveness of chloride removal from MSWI bottom ash by washing, varying the water: solids ratio, contact time, temperature and particle size distribution. They conclude that temperature has the greatest effect on extraction efficiency, reporting only 15% chloride removal at 20 °C compared with 60% at 30 °C. They used the time to dissolve 20% of the initial chloride as a marker of extraction rate, noting that most of the dissolution takes place in the first 15 minutes. Their samples were ground to less than 180 μm and divided into three size fractions, which show surface area to volume effects as might be expected. The variations in chloride extraction between the size fractions was noticeable, but not great, suggesting that all the material was sufficiently fine to ensure adequate dissolution of the chloride salts. Stegemann et al. (1995) leached MSWI bottom ashes in lysimeters for two years, comparing the efficiency of distilled water with that of sodium hydroxide solution (as a potential means of pre-treating the ash). Subsequent examination of the ashes for the mobility of their heavy metals, showed little difference between the two, concluding 'Although alkaline washing clearly resulted in greater contaminant removal than did distilled water washing, the chemical properties of the alkaline-leached bottom ash were not significantly different from those of the water-leached ash.'

For pre-treatment of fly ashes and APC residues, Aubert et al. (2004) describe a commercial process developed in France: 'With a view of reducing the quantities to be landfilled, the SOLVAY Company, in collaboration with the Université Libre de Bruxelles, has been working on development of a new physicochemical treatment for these ashes, the REVASOL™ process.' This process was composed of three successive steps (Derie and Depaus 1999):

- Water dissolution of ash: during this washing, pH is kept below 10.5 to prevent the dissolution of heavy metals. In fact, the washing solution containing soluble salts (especially chlorides (KCl and NaCl), the alkaline sulphates being in an insoluble form) is treated to recover the sodium which is used in technical soda plants.
- Phosphation with phosphoric acid to stabilize heavy metals.
- Calcination to eliminate organic compounds (especially dioxins), at a temperature higher than 600 °C.

This step also enables the crystalline structure of the final product to be stabilized. The results of a micro-characterization study demonstrate that the process is able to stabilize heavy metals contained in MSWI fly ash (Piantone et al. 2004).

Although this approach was developed specifically as a pre-treatment prior to landfill disposal, it has implications for re-use of fly ashes in construction. The process may need further optimization for use in concrete production, however, to ensure no residual phosphates retard or even stop the hydration reactions.

There is an obvious research need to establish a means with which the potential for expansion may be predicted. The bulk oxide chemistry does not provide sufficient information on which to base operational guidance as in practice (even at very great age) much of the solid ash will remain unreacted. Rules based on the total sulphur, aluminium, calcium and silicon content of the ash could in principle be derived, assuming that all or a proportion of these elements would react to produce expansive phases. Given incomplete hydration, some empirical rule may be suggested as to what fraction may reasonably be associated with each reaction product. However, this approach seems to be fraught with uncertainties, as the overall composition of cements and concretes containing these ashes will vary widely in their composition. Moreover, the availability of reactive ions in the pore solution, assumes that water will be available to maintain a reactive pore solution, which may not always be the case. It seems inevitable that some discrepancy would arise between the need for conservative treatment of the possible phase chemistry in the interests of safety, and the need to use these material resources efficiently.

A thermodynamic treatment may offer more confidence, if a reasonable knowledge of the initial phase chemistry is available. The approach adopted by Glasser and co-workers is successful in the prediction of hydrate phase assemblages (e.g. Damidot et al. 1994; Lothenbach et al. 2008; Pöllmann 1989). This assumes that the system will approach thermodynamic equilibrium and that it is possible to predict the reaction products formed under these conditions. For this approach to be effective, certain assumptions must be made, such that a pore solution will persist and that reactants identified (for example) by leaching, diffraction and thermal methods will be available to form mineral hydrates but some will not. As with rules based on simple oxide chemistry, there will be a need for an empirical approach in deciding what fraction of the potentially reactive components are truly available to form the final assemblage of mineral hydrates.

Although such methods are attractive for prediction of AFt and AFm chemistry, they are unlikely to predict well the progress of alkali silica reaction, or of aluminium-induced hydrolysis. For these latter reactions, case by case analysis seems inevitable, irrespective of the pre-treatment of the ashes. Aubert et al. (2004) have shown a practical method (Figs 12.16 and 12.17) for the estimation of the residual aluminium content of ground

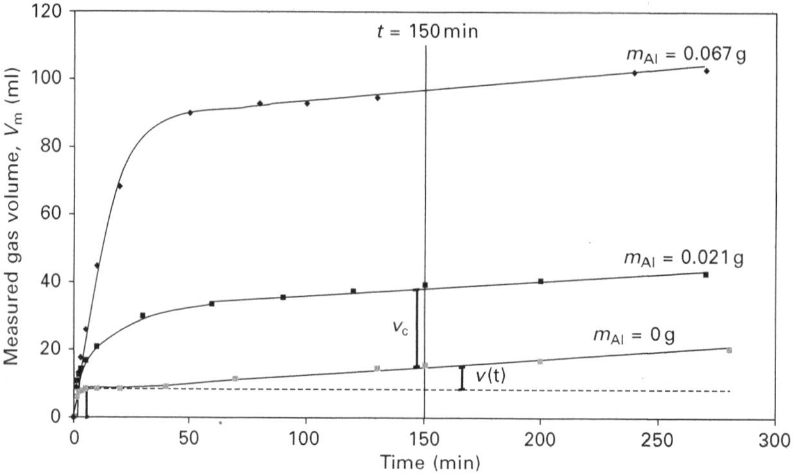

12.16 Rate of hydrogen liberation due to alkaline hydrolysis with aluminium in MSWI fly ash. After Aubert et al. (2004).

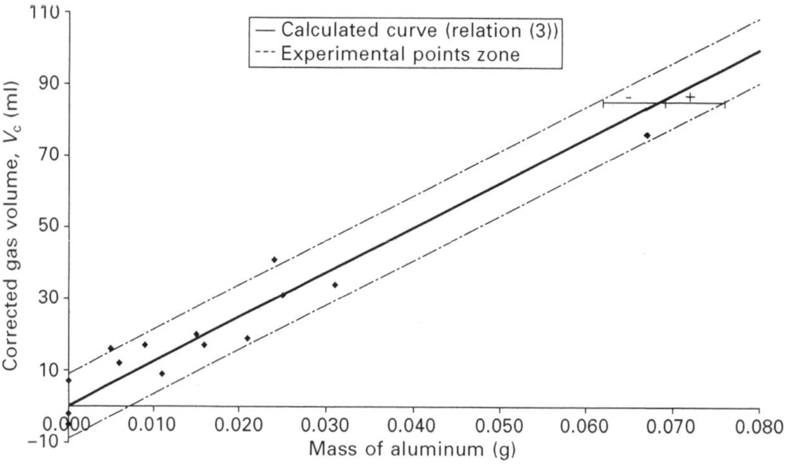

12.17 Comparison of predicted and measured quantities of hydrogen liberated during alkaline hydrolysis with aluminium in MSWI ash. After Aubert et al. (2004).

(< 125 μm) MSWI ashes, through reaction with sodium hydroxide and measurement of the volume of hydrogen liberated, which has been adopted elsewhere (Qiao et al. 2008b; see Fig. 12.18). They found rapid gas evolution in the first 50 minutes of reaction, after which the rate was reduced, the test yielding quantitive results after a few hours. Their final recommendation was to operate the test for 150 minutes and propose a means for estimation of the total free aluminium content of the ash. This pragmatic approach has

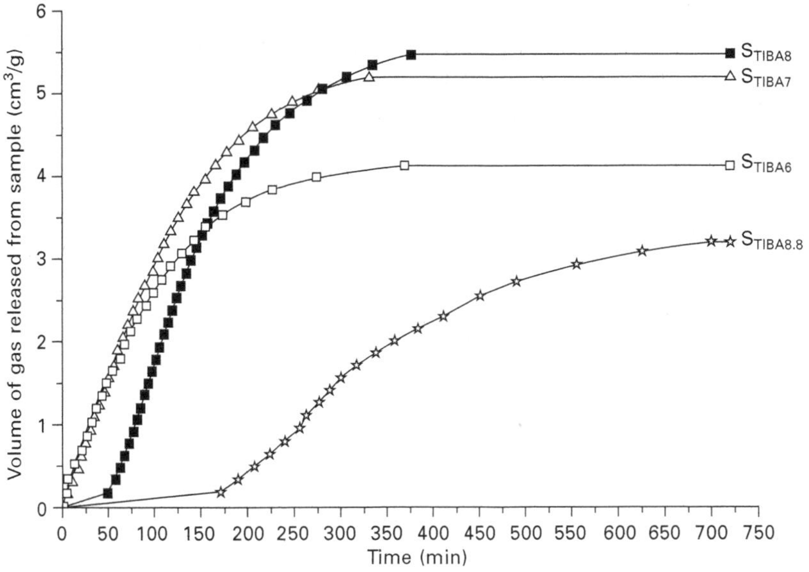

12.18 Comparison of the aluminium metal remaining in the system using Aubert's method following varying degrees of thermal pre-treatment of MSWI bottom ash After Qiao et al. (2008c).

much to recommend it and as it is a simple and robust technique, it may become widely adopted.

Quantifying the glass content and its reactivity is more challenging. Polarized light microscopy is well established and applicable to bottom ash characterization, where the particle size is relatively coarse, but is less attractive for very fine fly ash. The conventional method of point counting the optically isotropic particles (in extinction under crossed polarized light) would give a fair approximation of the glass content, but the results may be erroneous where the glass fragments are strained or devitrified. Despite this limitation, the method is robust and readily automated using image analysis. The alternative method of Reitveld refinement of X-ray diffraction (XRD) spectra in order to calculate the amorphous fraction seems too involved a process to be widely adopted in practice.

Glass reactivity poses further difficulty in assessment. Practically, the potential glass compositions may vary widely, depending on both the original glass type and its thermal history during incineration. Once fused at high temperature, the glasses are able to react with the materials near to them, selectively dissolving some components but not others. Analytical electron microscopy will quantify the range of glass chemistry within a sample, but currently no adequate correlation is available which links composition to reactivity. It seems likely that for the forseeable future, recourse must be made to accelerated tests of dimensional stability as a means of assessing the reactivity of the glass fraction. Aubert et al. (2004) compares methods for measuring expansion (total weight gain, X-ray diffraction (XRD) etc.) and ASTM methods C227 and C342 detail conventional expansion tests using mortar bars.

12.6 Use in concrete: examples

12.6.1 Conventional applications

Siddique (2010) reviews numerous examples of the use of MSWI ashes in concrete production and their re-use as other engineering materials such as ceramics and unbound fill. He concludes that the fly ashes have some potential as supplementary cementitious materials (SCM) but that leaching of heavy metals remains problematic. Bottom ash (he notes) has potential value as an aggregate, given the caveat that aluminium must first be removed.

It would seem, however, from the examination of the chemistry of the materials considered here, that no distinct division can be made between materials suited to aggregate use and those suited as ingredients of blended cements. Bottom ashes contain numerous reactive components which, if ground finely, hydrate either alongside the cement clinker minerals or through pozzolanic action. Where the bottom ash is coarse, this hydration is expected to be confined to its surface layers, presumably increasing the aggregate–paste bond strength and accounting for its behaviour as an acceptable aggregate (although somewhat weaker than many conventional aggregates). In any eventuality, the content of free sulphate and free (elemental) aluminium must be measured and reduced before use.

Numerous examples of the use of MSWI fly ash and bottom ash have been reported, where the combustion products are used as either cement replacement material or as aggregates. Hamernik and Frantz (1991) note the retarding effect of fly ash on the hydration of cement in concrete, as do Lin et al. (2004) who used the material as a partial replacement of commercial cements (types I, III and V). They found considerable variation in the rates of strength development at 28 days, but surprising similarities after 90 days. Their recommendation is that up to 20% by mass of cement may be replaced by fly ash without adverse effect on strength, confirming their earlier findings (Lin et al. 2003). Collivignarelli and Sorlini (2002) pre-treated their fly ashes by water washing and stabilisation with Portland cement, sodium silicate and slaked lime, milling the hydrated product to <1.5 mm. This processed fraction was used as a substitute for fine aggregate in concrete with compressive strengths of around 15 MPa. Pavlik et al. (2010) used untreated fly ash as a direct replacement for cement in concrete (2–15%) but concluded its performance was not sufficient to justify wider use. This group also use untreated bottom ash in a comparative study (*ibid.*) as a partial replacement for sand. In this case they conclude that satisfactory concrete could be made with up to 10% by mass of the aggregate being replaced by MSWI bottom ash. Berg and Neall (1992) describe MSWI bottom ashes as 'marginal aggregates' owing to both their relatively low strength and angular shapes, which reduce workability of fresh concrete.

Jaturapitakkul and Cheerarot (2003) examined finely ground bottom ash as a potential pozzolan in blended cement concrete. They also found the material influenced hydration kinetics in that at all blending ratios considered (0–30% by mass) the 28 day strengths were significantly lower than control samples made from conventional concrete. However, at low replacement levels (<20%) the late age strengths (60 days) were greater than those of the control samples. Using ground bottom ash, they produced concretes with strengths of ~30–40 MPa) at 28 days, rising to over 50 MPa at 90 days. Jurič et al. (2006) used ground bottom ash as a replacement for the binder in concrete and recommend that up to 15% by mass may be used as a cement replacement material, to produce acceptable strengths (~40 MPa) at 28 days.

12.6.2 Manufactured aggregates

One option for effective re-use of MSWI ashes in concrete is to process the material into a manufactured aggregate and this route has attracted considerable attention in recent times, through both low-temperature and high-temperature processing. Amongst the most promising approach for low temperature processing is the rapid carbonation of these materials either directly or during the early stages of hydration. Fernández Bertos et al. (2004) reviewed the potential of accelerated carbonation technology in the treatment of cement-based materials and sequestration of CO_2. The review followed their successful production of artificial aggregates using similar technology and this group went on to stabilise MSWI fly ashes using accelerated carbonation (Li et al. 2007). Previous work (e.g. van Gerven et al. 2005) has shown that the leachability of many metals was reduced during the rapid carbonation of MSWI bottom ash, but others (notably Cr) were enhanced. Gunning et al. (2010) reviewed the stability of many waste types (including MSWI) as a result of accelerated carbonation and presented additional results which confirm these findings. Arickx et al. (2006) found a marked reduction in leachability of rapidly carbonated bottom ash, but did not use it in concrete production. Latterly, the first commercial uptake of accelerated carbonation technology has been announced, in which artificial aggregates are being produced from APC residues (Gunning et al. 2012; Hills et al. 2012; Perella 2012). This application is claimed to operate with an overall consumption of CO_2 in a rapid process (minutes) sequestering

CO_2 into solid carbonates and greatly reducing the mobility of the metals contained in them. It is to be hoped that the product will be trialled as aggregate for concrete production in due course.

Re-melting of ashes has been carried out in Japan since the 1980s (Sakai and Hiraoka 2000) to greatly reduce the volume and leachability of this waste and to and provide a useful material for re-use, largely as engineering fills. This approach has been criticized for its high energy requirements and commercial uptake is rather limited; Ecke et al. (2000) describing applications in Japan, Sweden and Korea.

Wainwright et al. (1984) reviewed the options for using incinerator wastes in construction materials and subsequently developed sintering methods with which to produce synthetic aggregates. Wainwright and Cresswell (2001) used a novel trefoil kiln (Fig. 12.19) to sinter MSWI bottom ash with clays in order to produce lightweight aggregates with densities of around 1.8 g cm^{-3} and water absorption characteristics similar to commercial lightweight aggregates (~12% by mass at 24 hours –). The kiln was used for a range of synthetic aggregate types which incorporated a wide range of fine grained wastes and is shown in Fig. 12.19 (Laursen et al. 2006). The concrete produced by this method proved serviceable and the aggregates retained most of their heavy metals during leach testing – see Fig. 12.20.

Bethanis et al. (2002) milled and pelletized MSWI bottom ash before sintering them at a range of temperatures, showing that particle size distribution and sintering temperature control density and dimensional change, water absorption, mineralogy and microstructure (Fig. 12.21). Sintering at high temperature (>1080 °C) liberates gas and forms bubbles (which lower the density of the product) and suggest this is due to decomposition of sulphate, which presumably is lost as SO_2.

Cheeseman et al. (2005) report production and characterization of lightweight aggregate from finely ground bottom MSWI bottom ash. In this work, the

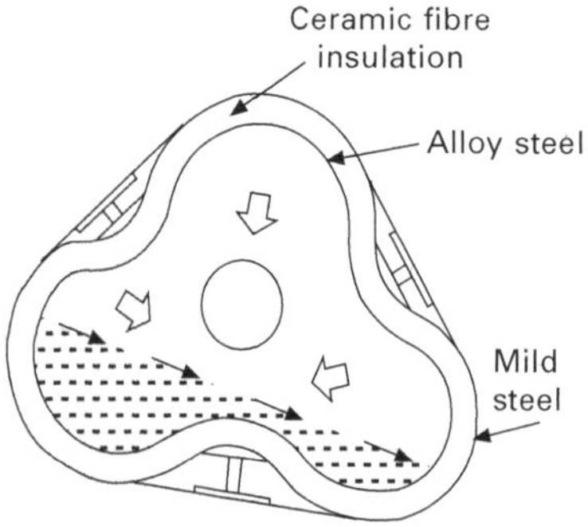

12.19 Trefoil kiln used by Wainwright and Cresswell (2001) for production of artificial aggregates from MSWI bottom ash sintered with other materials. After Laursen et al. (2006).

12.20 Artificial aggregates produced from MSWI bottom ash by sintering with other materials in a trefoil kiln. After Cresswell (2007).

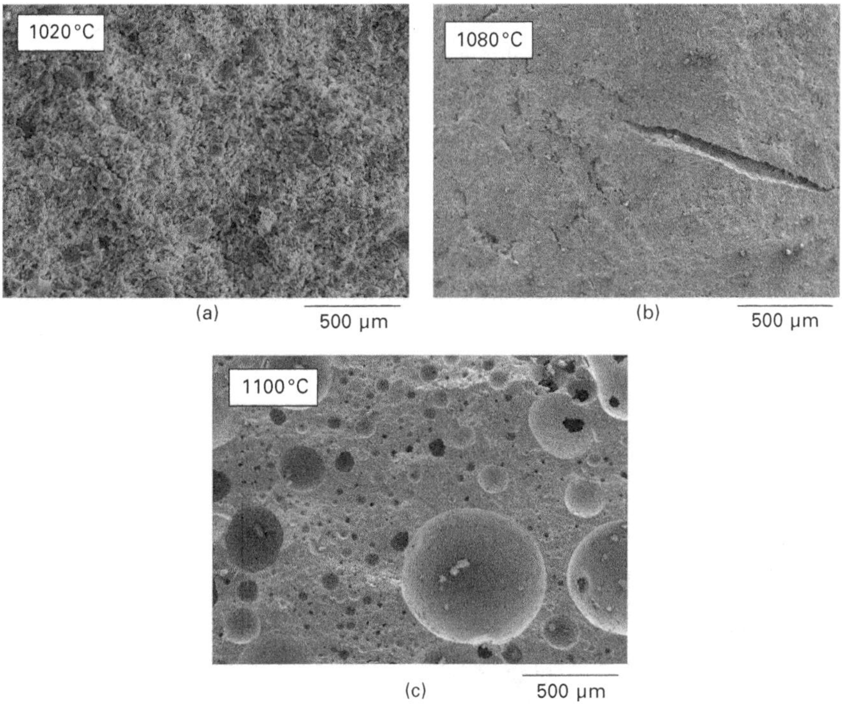

12.21 MSW incinerator bottom ash fines treated at increasing temperatures. The structure of the 1100 °C sample is attributed to decomposition of sulphate minerals and liberation of sulphur dioxide which foams the glassy matrix. After Bethanis et al. (2002).

ground ash was re-wetted (24% water by mass) and pelletized (8–10 mm) before oven drying and firing at 900–1080 °C. Examination of the product revealed considerable transformation to pyroxene (diopside: $CaMgSi_2O_6$ and clinoestatite: $Mg_2Si_2O_6$) in a glassy matrix containing minor albite feldspar ($NaAlSi_3O_8$) and wollastonite ($CaSiO_3$). Leach testing proved that some improvement (especially lead and zinc) was achieved but for other species, sintering made little long term difference in their mobility.

Appenido et al. (2004) melted MSWI bottom ash at higher temperatures (initially 1100 °C, rising to 1450 °C for 12 hours, then cooling to 1100 °C and final air quenching) finding a similar mineral assemblage to the above, containing corundum, feldspars (anorthite, $CaAl_2Si_2O_8$ and albite, $NaAlSi_3O_8$), pyroxenes (diopside, $Ca(Mg,Al)(Si,Al)_2O_6$, augite, $Ca(Fe,Mg)Si_2O_6$) and traces of metallic aluminium. These glassy ceramics were crushed and graded, being used as aggregates in cement and mortar samples (Ferraris et al. 2009; Fig. 12.22) which showed similar strength development curves to reference samples up to 180 days of curing where the particle size was coarse. Finer material (<5 mm diameter) showed a reduction in unconfined compressive strengths between around 10 and 30% in comparison with reference materials where the sintered aggregate replacement level varied between 25 and 75% by volume. In this case, where the manufactured aggregate is a direct replacement for sand in the concrete, they recommend a replacement level not exceeding 25% by volume and suggest the cause of the strength reduction is due to reduced integrity of the cement paste–aggregate bond, reflecting the smooth surface of the glass. Importantly, these authors used a finely ground fraction (<50 µm) as a filler material replacing part of the cement fraction. They found no evidence of alkali silica reaction after two years (using ASTM method C289), which suggests that the nature of

(a) (b)

12.22 Vitrified bottom ash (left) and a core of concrete containg vitrified bottom ash aggregate. After Ferraris et al. (2009).

the glass has been changed at high temperature. Moreover, this use suggests that no appreciable change in strength (with respect to reference mortars) was seen at loadings of up to 20% by volume.

This is an intriguing result, suggesting that although some pozzolanic reaction may be responsible for maintaining the strength, the glass is unaffected by alkali silica reaction. It is possible that on melting, a substantial quantity of alkali metal has been subject to volatile loss, depleting the system in an essential reactant necessary for ASR or that the modified glass is not very susceptible to ASR and this is worthy of further study. Notably, these authors find no evidence of ettringite-induced expansion. If the hypothesis of Bethanis et al. (2002) is correct, that sulphate is depleted from these materials at high temperature, this becomes a compelling explanation of Ferraris's findings: insufficient sulphate remains in these materials to promote ettringite formation. Conversely, it is possible that the gas evolution at high temperature is (at least in part) caused by some reduction of iron III in the melt, liberating oxygen as described by Sandrolini and Palmonari (1976) in their work on vitrified ceramics. Appenido et.al. (2004) allude to this in their discussion, but it would seem at the elevated temperatures they consider, the thermal decomposition of crystalline sulphates would occur (~1000–1150 °C) alongside this reduction.

Mangialardi (2002) treated MSWI fly ash in a similar way, following washing in water for 30 minutes. This step effectively removed the soluble salts (28–45% by mass of the original solids depending on source) which were shown to be sodium and potassium chlorides, calcium sulphate and and the double salt aphthitalite ($(Na,K)_3Na(SO_4)_2$). The remaining dried solids were re-wetted (8% water by mass) and compacted into cylindrical pellets prior to sintering at between 1090 and 1140 °C for an hour. Control samples of unwashed MSWI fly ash were treated similarly and all were subject to detailed thermal analysis (differential scanning calorimetry and thermogravimetry). Distinct mass losses due to thermal decomposition were seen at ~600 °C (siderite, $FeCO_3$) ~700 °C (calcite $CaCO_3$) and volatile loss of the alkaline sulphates around 1200 °C, this latter representing 6.8–12.5% of the initial mass. In the case of the unwashed fly ashes, volatilization of the alkali halides was seen between 800 and 1140 °C. Based on these findings, they suggest that the vast majority of sulphates and chlorides would be lost or decomposed by sintering at 1140 °C. Mechanical testing of the sintered products showed a marked increase in compressive strength between the unwashed and washed fly ashes, due to the loss of the mechanically weak alkali halides and sulphates. In addition, they note that washing relatively enriches the remaining solid in glass network formers (Si and Al) which increase greatly the integrity of the sintered product. Leach testing of the sintered products show levels of heavy metals released (Cd, Cr, Cu, Pb) to be well below the required Italian specifications and that the final compressive strengths (~28 MPa) and densities (2.17–2.50 g cm^{-3}) of the sintered aggregates were also acceptable.

In summary, it would appear that sintering or complete melting and recrystallization of bottom ash or washed fly ash derived from MSW is capable of producing satisfactory aggregates for use in concrete. There is further development work to be done before they enter the standards as routine construction materials, such as chemical optimization of the formulations and long term durability testing. Ultimately, the energy cost of processing these materials for re-use must be compared with the environmental cost of their disposal.

Air pollution control residues are, owing to their composition, the most problematic and difficult to treat products from municipal waste incineration. Rani et al. (2008) reviews the options for their treatment in the United Kingdom, following uncertainties in European Waste Acceptance Criteria, noting that alternatives to landfill disposal are widely sought. The persistence of dioxins and furans in the environment combined with their significant and cumulative toxicity, marks them as unattractive components when treated by conventional means. High-temperature plasma processing offers a novel route to their effective re-use and has been the subject of considerable interest in the research arena. The technology of plasma generation and its applications are beyond the scope of this discussion; however the subject was reviewed thoroughly by Gomez et al. (2009) and described recently by Kourti et al. (2010b). Using this method, high temperatures (~3000 °C) can be maintained and material passed under the plasma can be very rapidly heated. Such intense thermal treatment completely denatures organic species, volatilizes some material and condenses the remainder, which cool as glasses or glassy ceramics. It is these materials which offer some potential for use in construction, both as prospective aggregates in concrete and when finely ground, as components in blended cements.

Kourti et al. (2010a) blended APC residues with glass-forming additives and thermally processed them using a DC plasma to produce a high calcium aluminosilicate glass, which was ground to produce a wide particle size distribution. At a Si:Al ratio of 2.6, these powders hydrated with a 6M sodium hydroxide solution to produce a composite of coarse particles in a dense matrix of remarkably high compressive strength (~130 MPa). The authors describe the materials as 'geopolymers', but note that at high calcium content formation of amorphous CSH was also likely to form. Subsequent work (Kourti et al. 2011) confirms both geopolymer and CSH as reaction products and demonstrates these materials show low leachability.

Working at lower temperatures than those of a plasma, Dimech et al. (2008) washed APC residues to remove the soluble salts and blended the remaining solids with soda-lime glass and waste electrostatic precipitator dust containing boric oxide from the fibre glass industry. Pellet-pressing, then sintering at 900–100 °C produced dense glassy ceramics containing gehlenite and wollastonite of extremely high strengths (~4.5 GPa). The authors suggest that such materials may have applications as high-value construction products; however, it would seem a wasted resource to incorporate them in concrete as an aggregate.

Most of the work on APC residues described above requires the removal of chloride by washing as a first processing step. Singh et al. (2008) turned the chloride in these materials to their advantage, through imaginative thinking to produce alinite. Alinite cements were initially developed at the Scientific Research Institute of Building Materials in Tashkent, during the Soviet era. Stoichiometrically (and to an extent structurally) Alinite resembles alite (tricalcium silicate, 'C_3S') with some substitution of oxygen for chlorine. The phase was initially described by Ilyukhin et al. (1977) whose composition was revised by Noudelman et al. (1980) as Ca_{11} $(Si,Al)_4O_{18}Cl$ which better describes the composition, which varies considerably between limits and is subject to considerable substitution. Magnesium plays a particularly important role in stabilizing the structure (~3–4%) and in reducing the energy required for the synthesis and this was provided from MgO additions. The bulk oxide requirement for alite synthesis from both MSWI fly ash and bottom ash were adjusted in mixtures with bauxite and limestone. The mass fraction of chloride was 1% in the bottom as and 19% in the fly ash, so a stoichiometric chloride content was provided by additions of $CaCl_2$. After firing for four hours at 1100–1200 °C, the product was cooled, ground and characterized, then used to make cement pastes. Subsequent mechanical testing showed them to have similar (indeed slightly better) compressive strengths than the control samples made from Portland cement at around 32 kPa. It should be noted, however, that due to the chloride content of alinite cements, their applications must avoid steel-reinforced concrete, owing to the potential for chloride-induced corrosion.

12.7 Future trends

The combustion products of municipal waste incineration have some real potential to be used in concrete production, but this must be done with a clear and confident understanding of their composition. As these materials are not currently included in cement or concrete standards for structural use, their initial applications are likely to focus on mass-pours such as 'blinding' concrete, void fill, hard standing and similar uses. In this way, a body of knowledge will evolve which increases our understanding of their durability and environmental impact over the longer term. The area which seems most likely to see rapid growth in the future, is the area of manufactured aggregates. Both high temperature sintering or plasma processing and low temperature accelerated carbonation offer great potential in making efficient use of these materials and seem very close to commercial uptake. The final step will be governed in part by economics (environmental as much as financial) and in part by the availability of suitable standards and codes of practice for their commercial use. This is of critical importance and it falls on material producers, potential users and standard makers to consider how best this will be achieved. One thing is certain, that unless these materials are specifically included in relevant standards, they will join the long list of materials which never reach their commercial potential as constructors will choose alternatives for which standards do exist.

As time progresses, the constraints on waste disposal will become more restrictive and the availability of natural aggregates will reduce, such that manufactured aggregates become increasingly attractive materials. However, the technologies used in their production require energy (for heating or otherwise processing these materials, such as collection and compression of CO_2). It seems likely that establishment of aggregate manufacturing plants will be in locations where the lowest cost energy co-exists with materials suitable for aggregate manufacture. This may not necessarily be in close proximity to municipal waste incineration facilities; indeed other mineral feed stocks may fill this supply niche. Whilst the science and engineering communities advise political and economic planners, we do not always succeed in shaping their decisions in the best long-term interests of society. This particular example (manufacture of synthetic aggregates from waste) offers an appealing and resource-efficient use of MSWI combustion products, which is in the widest interests of society and deserving of thorough debate.

12.8 References and further reading

Alba, N., Gasso, S., Lacoret, T., Baldasano, J.M. (1997) Characterization of municipal solid waste incineration residues from facilities with different air pollution control systems. *Journal of the Air & Waste Management Association* **47**, 11, 1170–1179

Appenido, P.M., Ferraris, M., Matekovits, I., Salvo, M. (2004) Production of glass–ceramic bodies from the bottom ashes of municipal solid waste incinerators. *Journal of the European Ceramic Society* **24**, 803–810

Arickx, S., Van Gerven, T., Vandecasteele, C. (2006) Accelerated carbonation for treatment of MSWI bottom ash. *Journal of Hazardous Materials* **138**, 1, 201–204

ASTM International (American Society for Testing and Materials). ASTM C342-97 Standard Test Method for Potential Volume Change of Cement–Aggregate Combinations (withdrawn 2001, but without replacement)

ASTM International (American Society for Testing and Materials). ASTM E227 - 90 (1996) Standard Test Method for Optical Emission Spectrometric Analysis of Aluminum and Aluminum Alloys by the Point to Plane Technique (withdrawn 2002, but without replacement)

Aubert, J.E., Husson, B., Vaquier, A. (2004) Metallic aluminum in MSWI fly ash: quantification and influence on the properties of cement-based products. *Waste Management* **24**, 589–596

Bartelings, H., Sterner, T. (1999) Household waste management in a Swedish municipality: determinants of waste disposal, recycling and composting. *Journal Environmental and Resource Economics* **13** 4, 473–491

Belevi, H., Stämpfli, D.M., Baccini, P. (1992) Chemical behaviour of municipal solid waste incinerator bottom ash in monofills. *Waste Management & Research* **10**, 2, 153–167

Berg, E., Neal, J.A. (1992) Municipal waste combustor ash in concrete. *Proceedings of the Fifth International Conference on municipal solid waste combustor ash utilization.* Edited by Chesner W.H. and Roethel. F.J. Arlington, November 17–18, 1992, p. 155–170

Bertolini, L., Carsana, M., Cassago, D., Collepardi, M., Curzio Q. (2004) MSWI ashes as mineral additions in concrete. *Cement and Concrete Research* **34**, 1899–1906

Bethanis, S., Cheeseman, C.R., Sollars, C.J. (2002) Properties and microstructure of sintered incinerator bottom ash. *Ceramics International* **28**, 881–886

Bethea, R.M. (1978) *Air pollution control technology: An engineering analysis point of view*. New York: Van Nostrand Reinhold

Blue Ridge Environmental Defense League (2009) *Technical Report on: Waste gasification: Impacts on the Environment and Public Health*

Boghetich, G., Liberti, L., Notarnicola, M., Palma, M., Petruzzelli, D. (2005) Chloride extraction for quality improvement of municipal solid waste incinerator ash for the concrete industry. *Waste Management & Research* **23**, 57–61

Brunner, C.R. (1991) *Handbook of Incineration Systems*. New York: McGraw-Hill.

Burnley, S.J. (2007) A review of municipal solid waste composition in the United Kingdom. *Waste Management* **27**, 10, 1274–1285

Chandler, J.A., van der Sloot, H., Eighmy, T.T., Hartlén, J., Hjelmar, O., Kosson, D.S., Sawell, S.E., Vehlow, J. (1997) *Municipal Solid Waste Incinerator Residues.* International Ash Working Group Municipal Solid Waste Incinerator Residues Elsevier Studies in Environmental Science no. 67

Cheeseman, C.R., Makinde, A., Bethanis, S. (2005) Properties of lightweight aggregate produced by rapid sintering of incinerator bottom ash. *Resources, Conservation and Recycling*, **43**, 2, 147–162

Chimenos, J.M., Segarra, M., Fernandez, M.A., Espiell, F. (1999) Characterization of the bottom ash in municipal solid waste incinerator. *Journal of Hazardous Materials A* **64**, 211–222

Collivignarelli, C., Sorlini, S. (2002) Reuse of municipal solid wastes incineration fly ashes in concrete mixtures. *Waste Management* **22**, 909–912

Cresswell, D. (2007) Municipal waste incinerator ash in manufactured aggregate. *MIRO Characterisation of Mineral Wastes, Resources and Processing Technologies – Integrated waste management for the production of construction material*, WRT 177/WR0115

Damidot, D., Stronach, S., Kindness, A., Atkins, M., Glasser, F.P (1994) Thermodynamic investigation of the $CaO-Al_2O_3-CaCO_3-H_2O$ system at 25°C and the influence of Na_2O *Cement and Concrete Research* **24**, 563–572

Dean, R. (Ed) (1988) *Incineration of municipal waste*. Academic Press Inc

Derie, R., Depaus, C. (1999) Behavior of organic and inorganic pollutants during Neutrec process of stabilization of MSWI fly ash. In *Waste Stabilisation and Environment: towards the definition of objectives for stabilization of industrial wastes by taking into account the potential impact on health and the environment*, Société alpine de publications, Grenoble, France.

Dimech, C., Cheeseman, C.R., Cook, S., et al. (2008) Production of sintered materials from air pollution control residues from waste incineration. *Journal of Materials Science*, **43**, 4143–4151

Ecke, H., Sakanakura, H., Matsuto, T., Tanaka, N., Lagerkvist, A. (2000) State-of-the-art treatment processes for municipal solid waste incineration residues in Japan. *Waste Management and Research* **18**, 1, 41–51.

Eighmy, T.T., Dykstra Eusden, J., Krzanowski, J.E., Domingo, D.S., Stampfli, D., Martin, J.R., Erickson, P.M. (1995) Comprehensive approach toward understanding element speciation and leaching behavior in municipal solid waste incineration electrostatic precipitator ash. *Environmental Science and Technology*, **29**, 3, 629–646

Fernandez, M.A., Martinez, L., Segarra, M., Garcia, J.C., Espiell F. (1992) Behavior of heavy metals in the combustion gases of urban waste incinerators. *Environmental Science and Technology* **26**, 1040–1047

Fernández Bertos, M., Simons, S.J.R., Hills, C.D., Carey, P.J. (2004) A review of accelerated carbonation technology in the treatment of cement-based materials and sequestration of CO_2 *Journal of Hazardous Materials* **B112** 193–205

Ferraris, M., Salvo, M., Ventrella, A., Buzzi, L., Veglia, M. (2009) Use of vitrified MSWI bottom ashes for concrete production. *Waste Management* **29**, 1041–1047

Filipponia, P., Polettinia, A., Pomia, R., Sirinib, P. (2003) Physical and mechanical properties of cement-based products containing incineration bottom ash. *Waste Management* **23**, 145–156

Garg, A., Smith, R., Hill, D., Longhurst, P.J., Pollard, S.J.T., Simms N.J. (2009) An integrated appraisal of energy recovery options in the United Kingdom using solid recovered fuel derived from municipal solid waste. *Waste Management* **29**, 2289–2297

Germani, M.S., Zoller, W.H. (1994) Solubilities of elements on in-stack suspended particles from a municipal incinerator. *Atmospheric Environment* **28**, 8, 1393–1400

Ginés, O., Chimenosa, J.M., Vizcarro, A., Formosa, J., Rosell, J.R. (2009) Combined use of MSWI bottom ash and fly ash as aggregate in concrete formulation: Environmental and mechanical considerations. *Journal of Hazardous Materials* **169**, 643–650

Glasser, F.P., Kindness, A., Stronachach, S.A. (1999) Stability and solubility relationships in AFm phases: Part I. Chloride, sulfate and hydroxide. *Cement and Concrete Research* **29**, 861–866

Gomez, E., Rani, D.A., Cheeseman, C.R., et al. (2009) Thermal plasma technology for the treatment of wastes: A critical review., *Journal of Hazardous Materials* **161**, 614–626

Gong, Y., Kirk, D.W. (1994) Behaviour of municipal solid waste incinerator fly ash: I: General leaching study. *Journal of Hazardous Materials* **36**, 249–264

Greenberg, R.R., Zoller, W.H., Gordon, G.E. (1978) Composition and size distributions of particles released in refuse incineration *Environmental Science and Technology* **12**, 566–573

Greenberg, R.R., Zoller, W. H., Gordon, G.E. (1978) Composition and size distributions of particles released in refuse incineration. *Environmental Science & Technology* **12** 5, 566–573

Gunning, P.J., Hills, C.D., Carey, P.J. (2010) Accelerated carbonation treatment of industrial wastes. *Waste Management* **30**, 1081–1090

Gunning, P.J., Hills, C.D., Carey, P.J. (2012) Commercial production of accelerated carbonated aggregate from municipal solid waste air pollution control residues (APCR) Proc. WASCON 2012 – towards effective, durable and sustainable production and use of alternative materials in construction. 8th International conference on sustainable management of waste and recycled materials in construction, Gothenburg, Sweden,

30 May – 1 June, 2012. Proceedings. Ed by: Arm, M., Vandecasteele, C., Heynen, J., Suer, P. & Lind, B. In press

Hamernik, J.D., Frantz, G.C. (1991) Strength of concrete containing municipal solid waste fly ash (SP-105). *Materials Journal* **88** 5, 508–517

Hills, C.D., Gunning, P.J., Carey, P.J., Greig, S. (2012) Carbonation of MSWI APC residues for re-use as a lightweight aggregate. Proc. Conf. Ash Utilisation 2012 – Ashes in a Sustainable Society. Polhelmssalen. Citykonferensen, Malmskillnadsgatan, Stockholm, Sweden. In press

Ilyukhin, V.V., Nevsky, N.N., Bickbau, M.J., Howie, R.R. (1977) Crystal structure of alinite. *Nature* **269**, 397–398

Ingham, J. (2011) *Geomaterials under the microscope*. Manson Publishing

Jaturapitakkul, C. Cheerarot, R. (2003) Development of bottom ash as pozzolanic material. *Journal of Materials in Civil Engineering* **15** 1, 48–53

Jin, W., Meyer, C., Baxter, S. (2000) 'Glasscrete' – Concrete with glass aggregate. *ACI Materials Journal* **97**, 208–213

Johnson, C.A., Brandenberger, S., Baccini, P. (1995) Acid neutralizing capacity of municipal waste incinerator bottom ash. *Environmental Science and Technology* **29**, 142–147

Johnson, C.A., Kersten, M., Ziegler, F., Moor, H.C. (1996) Leaching behaviour and solubility controlling solid phases of heavy metals in municipal solid waste incinerator ash. *Waste Management* **16**, 3, 129–134

Johnson, C.A., Kaeppeli, M., Brandenberger, S., Ulrichs, A., Baumann, W. (1999) Hydrological and geochemical factors affecting leachate composition in municipal solid waste incinerator bottom ash Part II. The geochemistry of leachate from Landfill Lostorf, Switzerland. *Journal of Contaminant Hydrology* **40**, 239–259

Jurič, B., Hanžič, L., Ilić R., Samec, N. (2006) Utilization of municipal solid waste bottom ash and recycled aggregate in concrete. *Waste Management* **26**, 1436–1442

Kourti, I., Amutha Rani, D., Boccaccini, A.R., Cheeseman, C.R. (2010a) Geopolymers from DC plasma treated APC residues, metakaolin and GGBFS. Coventry University and The University of Wisconsin Milwaukee Centre for By-products Utilization, Second International Conference on Sustainable Construction Materials and Technologies, 28–30 June, Universita Politecnica delle Marche, Ancona, Italy

Kourti, I., Rani, D. A., Deegan, D., Cheeseman, C.R. and Boccaccini, A.R. (2010b) Development of geopolymers from plasma vitrified air pollution control residues from energy from waste plants. In *Design, Development, and Applications of Engineering Ceramics and Composites* (eds D. Singh, D. Zhu, Y. Zhou and M. Singh), John Wiley & Sons, Inc., Hoboken, NJ, USA

Kourti, I., Devaraj, A.R., Bustos, A.G., et al. (2011) Geopolymers prepared from DC plasma treated air pollution control (APC) residues glass: Properties and characterisation of the binder phase *Journal of Hazardous Materials*, **196**, 86–92

Laursen, K., White, T.J., Cresswell, D.J.F., Wainwright, P.J., Barton J.R.(2006) Recycling of an industrial sludge and marine clay as light-weight aggregates. *Journal of Environmental Management* **80**, 208–213

Li, X., Fernández Bertos, M., Hills, C.D., Carey, P.J., Simons, S.J.R. (2007) Accelerated carbonation of municipal solid waste incineration fly ashes. *Waste Management* **27**, 1200–1206

Lichtensteiger, T. (1992) VDI – Bildungswerk Verein Deutscher Ingenieure. Seminar 43–76, Schlackenaufbereitung, -verwertung und -entsorgung, (slag processing, recycling and disposal) 16 and 17 March, München

Lin, L.L., Wang, K.S., Tzeng, B.Y., Lin, C.Y. (2003) The reuse of municipal solid waste incinerator fly ash slag as a cement substitute. *Resources, Conservation and Recycling* **39**, 315–324

Lin, L.L., Wang, K.S., Lin, C.Y., Lin, C.H. (2004) The hydration properties of pastes containing municipal solid waste incinerator fly ash slag. *Journal of Hazardous Materials B* **109**, 173–181

Lothenbach, B., Matschei, T., Möschner, G., Glasser, F.P. (2008) Thermodynamic modelling of the effect of temperature on the hydration and porosity of Portland cement *Cement and Concrete Research* **38**, 1–18

Lundtorp, K., Jensen D.L., Sørensen, M.A., Christensen, T.H., Mogensen, E.P.B. (2002) Treatment of waste incinerator air-pollution-control residues with $FeSO_4$: Concept and product characterisation. *Waste Management & Research* **20**, 69–79

Mangialardi, T. (2001) Sintering of MSW fly ash for reuse as a concrete aggregate. *Journal of Hazardous Materials* **B87**, 225–239.

Mangialardi, T. (2002) Disposal of MSWI fly ash through a combined washing-immobilisation process *Journal of Hazardous Materials* **98**, 225–240

Matschei, T., Lothenbach, B., Glasser, F.P. (2007) The AFm phase in Portland cement. *Cement and Concrete Research* **37**, 118–130

Müller, U., Rübner, K. (2006) The microstructure of concrete made with municipal waste incinerator bottom ash as an aggregate component. *Cement and Concrete Research* **36**, 1434–1443

Noudelman, B., Bikbaou, M., Sventsitski, A., Ilukhine, V. (1980) Structure and properties of a finite and alinite cements. *Proc. 7th International Congress on the Chemistry of Cement* Vol 3 V-169–174

Parfitt, J. (2002) *Analysis of household waste composition and factors driving waste increases*. Rept. WRAP (Waste and Resources Action Plan, UK)

Pavlík, Z., Jerman, M., Keppert, M., Pavlíková, M., Reiterman, P., Černý, R. (2010) Use of municipal solid waste incineration waste materials as admixtures in concrete . Coventry University and The University of Wisconsin Milwaukee Centre for By-products Utilization, Second International Conference on Sustainable Construction Materials and Technologies, 28–30 June, Universita Politecnica delle Marche, Ancona, Italy

Pera, J., Coutaz, L., Ambroise, J., Chababbet, M. (1997) Use of incinerator bottom ash in concrete *Cement and Concrete Research* **27**, 1, 1–5.

Perella, M. (2012) *World's first APCr recycling and carbon capture plant nears completion*. Environmental Data Interactive Exchange (Faversham House Group, Jan. 2012)

Piantone, P., Bodenan, F., Chatelet-Snidarob, L. (2004) Mineralogical study of secondary mineral phases from weathered MSWI bottom ash: implications for the modelling and trapping of heavy metals *Applied Geochemistry* **19**, 1891–1904

Pöllmann, H. (1989) Solid solution in the system $3CaO\text{-}Al_2O_3\text{-}CaSO_4 \bullet aq\text{-}3CaO \bullet Al_2O_3 \bullet Ca(OH)_2 \bullet aq\text{-}H_2O$ at 25 °C, 45 °C, 60 °C, 80 °C. *Neues Jahrbuch Miner Abh*, **161**, 27–40

Qiao, X.C., Ng, B.R., Tyrer, M., Poon, C.S., Cheeseman, C.R. (2008a) Production of lightweight concrete using incinerator bottom ash. *Construction and Building Materials* **22** 4, 473–480

Qiao, X.C., Tyrer, M., Poon, C.S., Cheeseman, C.R. (2008b) Characterization of alkali-activated thermally treated incinerator bottom ash. *Waste Management* **28**, 1955–1962

Qiao, X.C., Tyrer, M., Poon, C.S., Cheeseman, C.R. (2008c) Novel cementitious materials produced from incinerator bottom ash. *Resources, Conservation and Recycling* **52**, 496–510

Quina, M.J., Bordado J.C., Quinta-Ferreira, R.M. (2008) Treatment and use of air pollution control residues from MSW incineration: An overview. *Waste Management* **28**, 2097–2121

Rani, D.A., Boccaccini, A.R., Deegan, D., Cheeseman, C.R. (2008) Air pollution control residues from waste incineration: Current UK situation and assessment of alternative technologies. *Waste Management* **28**, 2279–2292

Rani, D.A., Roether, J.A., Deegan, D.E., Cheeseman, C.R., Boccaccini, A.R. (2010) Castable glass and glass-ceramics from DC plasma treatment of air pollution control residues, in *Ceramics for Environmental and Energy Applications* (eds A. Boccaccini, J. Marra, F. Dogan, H.-T. Lin, T. Watanabe and M. Singh), Hoboken, NJ, USA: John Wiley & Sons, Inc.

Rübner, K., Haamkens, F., Linde, O. (2008) Use of municipal solid waste incinerator bottom ash as aggregate in concrete. *Quarterly Journal of Engineering Geology & Hydrogeology* **41**, 459–464

Sabbas, T., Polettini, A., Pomi, R., Astrup, T., Hjelmard, O., Mostbauer, P., Cappai, G., Magel, S., Speiserg, C., Heuss-Assbichler, S., Klein, R., Lechnera, P. (2003) Management of municipal solid waste incineration residues. *Waste Management* **23**, 61–88

Sakai, S., Hiraoka, M. (2000) Municipal solid waste incinerator residue recycling by thermal processes. *Waste Management* **20**, 249–258

Sandrolini, F., Palmonari, P. (1976) Role of iron oxides in the bloating of vitrified ceramic materials *Transactions of the British Ceramics Society* **75** 2, 25–32

Sawell, S.E., Constable, T.W. (1993) *The National Incinerator Testing and Evaluation Program: A Summary of the Characterization and Treatment Studies on Residues from Municipal Solid Waste Incineration*; Environment Canada Report EPS 3-UP-8; Environment Canada: Ottawa, Canada, Oct. 1993

Siddique, R. (2010) Use of municipal solid waste ash in concrete. *Resources, Conservation and Recycling* **55**, 83–91

Singh, M., Kapur, P.C., Pradip, A. (2008) Preparation of Alinite based cement from incinerator ash. *Waste Management* **28**, 1310–1316

St John, D., Poole, A., Sims, I. (1998) *Concrete petrography: A handbook of investigative techniques* Butterworth-Heinemann

Stegemann, J.A., Schneider, J., Baetz, B.W., Murphy, K.L. (1995) Lysimeter washing of MSW incinerator bottom ash. *Waste Management & Research* **13**, 149–165

Tchobanoglous, G., Kreith, F. (2002) *Handbook of Solid Waste Management*. New York: McGraw Hill

Tiruta-Barna, L., Benetto, E., Perrodin, Y. (2007) Environmental impact and risk assessment of mineral wastes reuse strategies: review and critical analysis of approaches and applications. *Resources, Conservation and Recycling* **50**, 351–379

van der Sloot, H.A. (2002) Characterization of the leaching behaviour of concrete mortars and of cement-stabilized wastes with different waste loading for long term environmental assessment. *Waste Management* **22**, 181–186

van der Sloot, H.A., Kosson, D.S., Hjelmar, O. (2001) Characteristics, treatment and utilization of residues from municipal waste incineration. *Waste Management* **21**, 753–765

van Gerven, T., Van Keer, E., Arickx, S., Jaspers, M., Wauters, G., Vandecasteele, C. (2005) Carbonation of MSWI-bottom ash to decrease heavy metal leaching, in view of recycling. *Waste Management* **25**, 291–300

Wainwright, P.J., Cresswell, D.J.F. (2001) Synthetic aggregates from combustion ashes using an innovative rotary kiln. *Waste Management* **21**, 241–246

Wainwright, P.J., Hadzinakos, I., Robery, P. (1984) A review of the methods of utilisation of incinerator residues as a construction material. In: *Proceedings of the International Conference on Low Cost and Energy Saving Construction Materials*. Rio de Janeiro, Brazil.

Wentz, C.A. (1989) *Hazardous waste management*. New York: McGraw-Hill.

Wiles, C.C. (1996) Municipal waste combustion ash: State of the knowledge. *Journal of Hazardous Materials* **47**, 325–344

Williams, P.T. (2005) *Waste treatment and disposal*, Second Edition, Chichester: John Wiley & Sons, Ltd UK.

World Bank Technical Guidance Report (1998) *Municipal Solid Waste Incineration*. The International Bank for Reconstruction and Development/World Bank, Washington, DC, USA

Xie, Z., Xiang, W., Xi, Y. (2003) ASR potentials of glass aggregates in water-glass activated fly ash and Portland cement mortars. *Journal of Materials in Civil Engineering* **15** 1, 67–74.

Youcai, Z., Lijie, S., Guojian, L. (2002) Chemical stabilization of MSW incinerator fly ashes *Journal of Hazardous Materials* **B95**, 47–63

Zhu, F., Takaoka, M., Oshitaa, K., Kitajima, Y., Inada, Y., Morisawa, S., Tsunoa, H. (2010) Chlorides behaviour in raw fly ash washing experiments. *Journal of Hazardous Materials* **178**, 547–552

Zwahr, H. (2004) Ash recycling: Just a dream. *North American Waste-to-Energy Conference NAWTEC 12*, Savannah, GA, USA

Zwahr, H. (2005) MV-Schlacke—Mehr als nur ein ungeliebter Baustoff. Pages 237–258 In: Bilitewski, B. Urban, A.I. and Faulstich, M. (Eds) *Proc. Symposium Thermal Waste Treatment*. Technical University of Dresden

13
Concrete with polymeric wastes

F. PACHECO TORGAL, University of Minho, Portugal
and Y. DING, Dalian University of Technology, China

DOI: 10.1533/9780857098993.3.311

Abstract: The volume of polymeric wastes such as tyre rubber and polyethylene terephthalate (PET) bottles is increasing at a fast rate. An estimated 1000 million tyres reach the end of their useful lives every year and 5000 million more are expected to be discarded in a regular basis by the year 2030. Only a small part is currently recycled and millions of tyres are just stockpiled, landfilled or buried. As for PET bottles annual consumption is over 300 000 million units. The majority is just landfilled. This chapter reviews research published on the performance of concrete containing tyre rubber and PET wastes. Furthermore it discusses the effect of waste treatments, the size of waste particles and the waste replacement volume on the fresh and hardened properties of concrete.

Key words: polymeric wastes, tyre rubber, polyethylene terephthalate (PET) bottles, concrete, properties, durability.

13.1 Introduction

Polymeric wastes, namely tyre rubber and polyethylene terephthalate (PET) represent a major environmental problem of increasing relevance. An estimated 1000 million tyres reach the end of their useful lives every year (WBCSD, 2010). At present enormous quantities of tyres are already stockpiled (whole tyre) or landfilled (shredded tyre), 3000 million inside the EU and 1000 millions in the US (Oikonomou and Mavridou, 2009). By the year 2030 the number of tyres to be discarded from motor vehicles on a regular basis is expected to reach almost 5000 million, representing 1200 million vehicles. Tyre landfill represents a serious ecological threat. Many waste tyre disposal areas contribute to the reduction in biodiversity; also tyres hold toxic and soluble components (Day et al., 1993). In addition, although waste tyres are difficult to ignite, this risk is always present. Once tyres start to burn due to accidental causes, high temperatures occur and toxic fumes are generated (Chen et al., 2007). The high temperature causes tyres to melt, thus producing an oil that will contaminate soil and water. In Wales a tyre dump with 10 million tyres has been burning continuously for nine years (Cairns et al., 2004).

The implementation of the Landfill Directive 1999/31/EC (European Commission, 1999) and the End of Life Vehicle Directive 2000/53/EC

(European Commission, 2000) banned the landfill disposal of waste tyres, creating the driving force behind the recycling of these wastes. However, millions of tyres are just being buried all over the world (www. pyreco.com). Although tyre rubber wastes are already used for paving purposes, only a part of these wastes can be recycled (Vieira et al., 2010). One alternative is the formation of artificial reefs but some investigations have already questioned the validity of this option (Hartwell et al., 1998). Tyre waste can also be used in cement kilns to provide energy (Siddique and Naik, 2004) and to produce carbon black by tyre pyrolysis (Farcasiu, 1993), a thermal decomposition of these wastes in the absence of oxygen in order to produce by-products that have low economic viability.

Some research has already been conducted on the used of waste tyres as aggregate replacement in concrete, showing that a concrete with enhanced toughness and sound insulation properties can be achieved. Rubber aggregates are obtained from waste tyres using two different technologies: mechanical grinding at ambient temperature or cryogenic grinding at a temperature below the glass transition temperature (Nadgi, 1993). The first method generates chipped rubber to replace coarse aggregates. As for the second method it usually produce crumb rubber (Eleazer et al., 1992) to replace fine aggregates. Since the cement market demand is expected to have a twofold increase (Fig. 13.1) this means that concrete volume is expect to increase

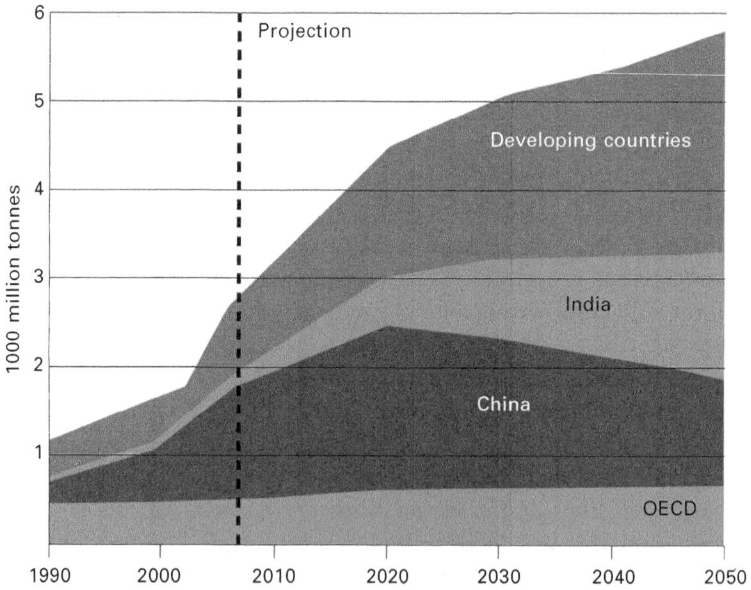

13.1 Global cement demand by region and country (Taylor and Gielen, 2006).

in a similar pattern representing an excellent way to reuse wastes like tyre rubber.

Similar considerations can be made for PET wastes. This polymeric waste represents one of the most common plastics in solid urban waste (Mello et al., 2009). In 2007 the world´s annual consumption represented 250 000 million terephthalate bottles (10 million tons of waste) with a growth increase of 15%. In the United States 50 000 million bottles are landfilled each year (Gore, 2009). Since PET waste is not biodegradable it can remain in the environment for hundreds of years. Previous investigations already confirmed the potential of PET waste in replacing aggregates in concrete which represents a better option than landfill. In this work the most relevant knowledge about the properties and the durability of concrete containing polymeric wastes (tyre rubber and PET wastes) will be reviewed.

13.2 Concrete with scrap-tyre wastes

13.2.1 Fresh concrete properties

Workability

Cairns et al. (2004) used long and angular coarse rubber aggregates with a maximum size of 20 mm, obtaining concretes with an acceptable workability for low rubber content. These authors reported a reduction in the workability for higher rubber content, since a rubber content of 50% led to a zero slump value. Other authors (Guneyisi et al., 2004) studied concretes containing silica fume, crumb rubber and tyre chips, reporting a decrease in slump with increasing rubber content, with a 50% rubber content leading to mixtures without any workability. The results obtained by those authors show that reducing the water/concrete (W/C) ratio is associated with a decrease in the slump values and that the silica fume worsens the workability performance.

Albano et al. (2005) replaced fine aggregates by 5% and 10% of scrap rubber waste (particle sizes of 0.29 and 0.59 mm), reporting a decrease of 88% in concrete slump. Bignozzi and Sandrolini (2006) used scrap tyre (0.5–2 mm) and crumb-tyre (0.05 to 0.7mm) to replace 22.2% and 33.3% of fine aggregates in self-compacting concretes, commenting that the introduction of the rubber particles does not influence the workability in a significant way if the superplasticizer also increases. Skripkiunas et al. (2007) used crumbed rubber to replace 23 kg of fine aggregates in concretes with 0.6% of a polycarboxylic superplasticizer by cement mass obtaining the same workability of the reference concrete. Other authors (Batayneh et al., 2008) used crumb rubber tyres (0.075–4.75 mm) in the concrete to replace sand in various percentages (20%, 40%, 60% and 100%). These authors stated that increasing rubber waste content decreases the concrete slump (Table 13.1).

Table 13.1 Slump performance according to crumb rubber content (Batayneh et al., 2008)

Rubber content (%)	Slump (mm)
0	75
20	61
40	36
60	18
80	10
100	5

Freitas et al. (2009) used scrap tyre (0.15–4.8 mm) in the replacement of sand, reporting a slump decrease along with the increase of scrap tyre content. However, these authors used 1% by cement mass of an unknown plasticizer in the mixtures with tyre wastes, so the workability reduction is probably related to the low performance of the plasticizer. Topçu and Bilir (2009) studied the influence of rubber waste with a maximum dimension of 4 mm in self-compacting concretes noticing that rubber replacing sand increase concrete workability which is due to the presence of viscosity agents even to a volume of 180 kg/m^3. Aiello and Leuzzi (2010) used tyre shreds (Fig. 13.2) to replace fine and coarse aggregates (10–25 mm) with 1% by cement mass of a plasticizer, observing increase workability with tyre shreds content (Table 13.2). Guneyisi (2010) used crumb rubber waste to replace sand in self-compacting concretes in different percentages (5%, 15% and 25%) and using also a polycarboxylic superplasticizer with different amounts. This author noticed that the mixture with 25% of rubber waste although that containing 4% by cement mass of the superplasticizer did not achieved the target slump flow of 750 ± 50 mm. He also reported that adding fly ash helps to lower the amount of superplasticizer in the mixtures with high rubber waste content.

Although the majority of investigations show that rubber aggregates lead to a decrease in concrete workability some authors reported no workability loss and others even observed the opposite behaviour. This means that workability is very dependent on the characteristics of the rubber aggregates. Future investigations should study what rubber wastes could be used to produce self-compacting concretes.

13.2.2 Hardened concrete properties

Compressive strength

Guneyisi et al. (2004) found that the strength of concretes containing silica fume, crumb rubber and tyre chips decreases with rubber content. These authors suggest that it is possible to produce a 40 MPa concrete replacing

13.2 Rubber particles: top as they come after the shredding process; bottom, during the mixture of concrete (Aiello and Leuzzi, 2010).

a volume of 15% of aggregates by rubber waste. Ghaly and Cahill (2005) studied the use of different percentages of rubber in concrete (5%, 10% and 15%) by volume, also noticing that as rubber content increase leads to a reduction of compressive strength. Valadares (2009) studied the performance of concretes with the same volume replacement of rubber wastes confirming the decrease of compressive strength. A waste rubber volume of 15% leads to a 50% compressive strength decrease. These authors mentioned that the rubber waste with low dimensions leads to lower strength loss and also that the rubber production (mechanical grinding or cryogenic process) does not influence the compressive strength.

Freitas el al. (2009) mentioned a 48.3% decrease in compressive strength for concretes with a waste rubber volume of 15%. Ganjian et al. (2009)

Table 13.2 Slump tests of fresh concrete with aggregates replaced by rubber particles (Aiello and Leuzzi, 2010)

Mixture	Slump (mm)
Reference concrete with rubber waste replacing coarse aggregates (W/B = 0.52)	180
25% rubber vol.	220
50% rubber vol.	215
75% rubber vol.	215
Reference concrete with rubber waste replacing fine aggregates (W/B = 0.60)	180
15% rubber vol.	220
30% rubber vol.	220
50% rubber vol.	215
75% rubber vol.	225

W/B = water/binder ratio.

also confirmed the decrease in compressive strength for increasing rubber content. However, these authors obtained a slight increase in compressive strength when 5% of chipped rubber replaced the coarse aggregates probably due to a better grading of the mixture. This finding had already caught the attention of other authors (Biel and Lee, 1996; Khatib and Bayomy, 1999). Snelson et al. (2009) used concretes with shredded tyre chips (15–20 mm) for aggregate replacement in several percentages (2.5%, 5% and 10%), reporting a loss in compressive strength. The results show that the rubber mixtures also containing pulverized fuel ash as partial cement replacement presented major compressive strength loss. This means that the low adhesion between the cement paste and the rubber waste becomes even lower if admixtures with low pozzolanic activity are used.

Aiello and Leuzzi (2010) used tyre shreds to replace fine and coarse aggregates, concluding that the size of the rubber particles has a major influence on the compressive strength. When coarse aggregates are replaced by tyre particles the compressive strength loss is much more profound than that of concretes in which fine aggregates were replaced by rubber particles (Table 13.3). These results contradict the those of Valadares (2009) which may be related to the origin of the wastes used in each case (car, truck or motorcycle): wastes of different origins may possess different chemical compositions, leading to different adhesion between the cement paste and the rubber waste. Vieira et al. (2010) studied three types of rubber waste (Fig. 13.3) and three volume percentages (2.5%, 5% and 7.5%), reporting that the best mechanical performance was obtained using just 2.5% of the tyre rubber with 2.4 mm.

Several authors mentioned the use of pretreatments of rubber waste to increase the adhesion between the cement paste such as the use of a 10%

Table 13.3 Compressive strength of concrete with aggregates replaced by rubber particles (Aiello and Leuzzi, 2010)

Mixture	Compressive strength (MPa)	Compressive strength decrease (%)
Reference concrete with rubber waste replacing coarse aggregates (W/B = 0.52)	45.8	–
25% rubber vol.	23.9	47.8
50% rubber vol.	20.9	54.4
75% rubber vol.	17.4	61.9
Reference concrete with rubber waste replacing fine aggregates (W/B = 0.60)	27.1	–
15% rubber vol.	24.0	11.6
30% rubber vol.	20.4	24.7
50% rubber vol.	19.5	28.3
75% rubber vol.	17.1	37.1

Size 1.2 mm Size 2.4 mm Size 4.8 mm

13.3 Size of tyre rubber (Vieira et al., 2010).

NaOH saturated solution to wash the rubber surface during 20 minutes (Naik and Singh, 1991; Naik et al., 1995). Raghavan et al. (1998) confirm that the immersion of rubber in NaOH aqueous solution could improve the adhesion leading to a high strength performance of concrete rubber composites. The NaOH removes zinc stearate from the rubber surface, an additive responsible for the poor adhesion characteristics, enhancing the surface homogeneity (Segre et al., 2002). Segre and Joekes (2000) mention several pretreatments to improve that the adhesion of rubber particles like acid etching, plasma and the use of coupling agents. Cairns et al. (2004) used rubber aggregates coated with a thin layer of cement paste (Fig. 13.4).

Albano et al. (2005) studied concrete composites containing scrap rubber previously treated with NaOH and silane in order to enhance the adhesion between the rubber and the cement paste without noticing significant changes,

(a) (b)

13.4 20 mm rubber aggregate particles: (a) plain; (b) coated with cement paste (Cairns et al., 2004).

when compared with the untreated rubber composites. Oikonomou et al. (2006) mentioned that the use of styrene-butadiene rubber (SBR) latex enhances the adherence between the rubber waste and the cement paste. Chou et al. (2010) suggested that the pretreatment of crumb rubber with organic sulphur modifies the rubber surface properties, so increasing the adhesion between the waste and the cement paste. Investigations about rubber waste concrete show a compressive strength loss with waste content increase. This behaviour is related to the low compressive strength of rubber aggregates and to the low adhesion between these wastes and the cement paste. Several treatments reveal interesting results in order to overcome this disadvantage. Further investigations are needed on this subject, especially to find out if different kinds of rubber behave in a similar manner to the same treatment.

Tensile strength

Guneyisi et al. (2004) analysed the tensile strength of concretes containing silica fume, crumb rubber and tyre chips, finding a decrease in tensile strength according to the rubber content and also that the presence of silica fume is beneficial because it is responsible for a higher filler effect. The results also confirm that the decrease of tensile strength reduction is less influenced by the rubber content than for the compressive strength reduction. This tendency was also observed by Pierce and Williams (2004). This result appears to be due to the fact that rubber particles prevent crack opening (Fig. 13.5). Valadares (2009) obtained a higher tensile strength for concretes with rubber waste particles with a higher dimension which agrees with the previous finding.

Ganjian et al. (2009) mentioned the opposite behaviour, reporting that the tensile strength of concrete with chipped rubber replacement for aggregates is considerably lower than for concrete containing powdered rubber. In the first case a reduction between 30% and 60% takes place for a replacement

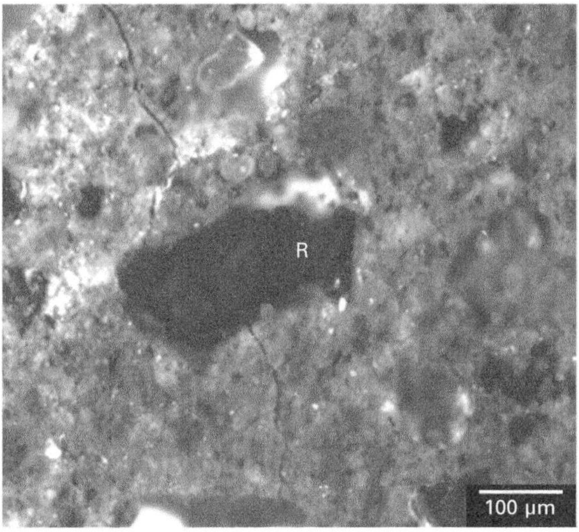

Note: R = rubber

13.5 Cracking bridging effect of rubber particles (Segre et al., 2006).

level of 5–10%, as for the latter case the reduction is between 15% and 30%. This behaviour maybe related to the very low adhesion between the chipped rubber and the cement paste. One reason for this is the fact that chipped rubber was prepared in laboratory with the help of scissors, a procedure quite different from the rubber waste particles shredded by a grinding process which favours a harsh surface. According to Aiello and Leuzzi (2010) when tyre shreds are used to replace fine aggregates a high tensile (flexural) strength is obtained. A replacement of the volume of fine aggregates (50% or 75%) leads to a strength reduction of only 5.8% and 7.30%. But if the same percentages were used to replace coarse aggregates a 28.2% strength reduction took place (Table 13.4). The tensile strength of rubber waste concrete is influenced by the characteristics of the rubber aggregate. Some are associated with a tensile strength loss but others present a high tensile strength. Future investigation should focus on the characteristics of the rubber aggregates that enhance tensile strength.

Toughness

Concrete composites containing tyre rubber waste are known for their high toughness (Li et al., 2004), having a high energy absorption capacity. ASTM C1018-97 defines several toughness indexes (I_5, I_{10} and I_{20}) as the area under load–deflection curve of a flexural specimen for different times of deflection after crack initiation related to area under the same curve up to the crack initiation. Some authors (Balaha et al., 2007) report a 63.2% increase in the

Table 13.4 Tensile strength (flexural) of concrete with aggregates replaced by rubber particles (Aiello and Leuzzi, 2010)

Mixture	Tensile strength (MPa)	Tensile strength decrease (%)
Reference concrete with rubber waste replacing coarse aggregates (W/B = 0.52)	3.51	–
25% rubber vol.	2.93	16.6
50% rubber vol.	2.52	28.2
75% rubber vol.	2.52	28.2
Reference concrete with rubber waste replacing fine aggregates (W/B = 0.60)	5.34	–
15% rubber vol.	5.10	4.49
50% rubber vol.	5.03	5.81
75% rubber vol.	4.95	7.30

damping ratio (self-capacity to decrease the amplitude of free vibrations) for concrete containing 20% rubber particles. Other authors (Zheng et al., 2008a, 2008b) confirmed the high damping potential of rubber waste concrete. They mentioned that concrete with ground rubber shows a 75.3% increase in the damping ratio and a 144% for crushed rubber concrete (Fig. 13.6).

Fioriti et al. (2007) found that concrete paving blocks containing 8% of tyre rubber waste have a resistant impact of almost 300% when compared to the reference concrete. Ling et al. (2009) also studied the performance of concrete paving blocks with crumb rubber, reporting a high toughness resistance due to the energy-absorbing capacity. These means that tyre waste concrete may be recommended for concrete structures located in areas of severe earthquake risk and also for the production of railway sleepers.

Modulus of elasticity

Since concrete with rubber waste has low compressive strength and a correlation exists between compressive strength and the modulus of elasticity, they should also possess lower modulus of elasticity. However, Skripkiunas et al. (2007) compared concretes with similar compressive strength (a reference one and another with 3.3% of crumb rubber) obtaining different static modulus of elasticity, 29.6 GPa versus 33.2 GPa for the reference concrete just 11% higher. The explanation for this behaviour is related to the low modulus of elasticity of rubber waste (Anison, 1964). Other authors (Turatsinze, 2007) report a decrease in the modulus of elasticity of 40% when the same percentage reduction takes place for compressive strength. Khaloo et al. (2008) confirmed that the inclusion of tyre rubber particles leads to a concrete with high ductility. Zheng et al. (2008a, 2008b) mentioned that crumb rubber (80% <2.62 mm) has less of an influence on the modulus of elasticity than crushed rubber (15–40 mm). Turatsinze and Garros (2008)

13.6 Tyre rubber waste: top ground; bottom crushed (Zheng et al., 2008a).

Table 13.5 Modulus of elasticity according to rubber content (Turatsinze and Garros, 2008)

Rubber content (%)	Compressive strength (MPa)	Modulus of elasticity (GPa)
0	43	35
10	30	23
15	20	19
20	15	15
25	12	10

state that the modulus of elasticity of self-compacting concrete increases with rubber (4–10 mm) waste content (Table 13.5). These authors also mentioned the risk of severe segregation with a high concentration of rubber waste at the top of the specimens which implies the needed for a proper combination between a viscosity agent and air-entraining agent to avoid segregation.

Son et al. (2010) studied the modulus of elasticity of concrete columns with two different sizes of crumb rubber (0.6 and 1 mm), reporting an increase in ductility performance of up to 90%. Those authors mentioned that crumb rubber concrete columns can undergo twice the lateral deformation before failure compared to the reference concrete columns. Mohammed (2010) confirmed that concrete slabs containing crumb rubber with a finesses modulus of 2.36 shows a higher ductility behaviour which fulfils the ductility requirements of EN1994 – Eurocode 4 (1994).

Thermal and sound properties

The replacement of fine aggregates by crumb rubber lowers the thermal conductivity of concrete (Sukontasukkul, 2009). The replacement of up to 30% reduces the thermal conductivity by more than 50% to a minimum of 0.241 W/mK. Crumb rubber concrete also show high noise reduction behaviour for high-frequency ranges (higher than 1000 Hz) when compared to the reference concrete. Crumb rubber concrete shows a noise reduction coefficient 36% higher. This reveals an ideal application for noise reduction barriers. However, further investigations are still needed regarding the aggregate characteristics and concrete mixes which could enhance soundproof performance and at the same time can keep a minimum compressive strength and durability.

Durability

Since rubber waste concrete has lower compressive strength and lower tensile strength than reference concrete it is expected that its behaviour under fast mechanical degradation actions would also be lower. Sukontasukkul and Chaikaew (2006) mentioned that crumb concrete blocks show less abrasion resistance and also that increasing the crumb rubber content leads to a reduction in the abrasion resistance. This result was confirmed by Ling et al. (2009). Freitas et al. (2009) studied the abrasion resistance by immersion of rubber waste concrete, reporting a lower degradation than the reference concrete when only 5% rubber per mass was used to replace the coarse aggregate. This result is quite interesting since the rubber addition leads to a 30% compressive strength decrease. However, since the tensile strength (Brazilian test) has been reduced by only 11% this helps to understand the high abrasion resistance. The authors used this mixture in the rehabilitation of a hydroelectric power plant.

Topçu and Demir (2007) mentioned that a high volume replacement of sand by rubber waste (1–4 mm) has lower durability performance assessed by freeze–thaw exposure, seawater immersion and high temperature cycles. According to them, the use of a 10% replacement is feasible for regions without harsh environmental conditions. The fact that these authors used a Portland cement II/B 32.5 which has a very low compressive strength may explain these results. Ganjian et al. (2009) studied the durability of concrete containing scrap-tyre wastes assessed by water absorption and water permeability revealing that a percentage replacement of just 5% is associated with a more permeable concrete (36% increase) but not a more porous one. Increasing the rubber percentage replacement to 10% doubles the concrete water permeability which means this kind of concrete cannot be used for applications where water pressure is present, such as underwater columns. Ling et al. (2010) tested 348 rubber waste paving blocks, reporting that an increase in the rubber waste decreases the abrasion resistance. Thus it is recommended that a 20% volume replacement should not be exceeded.

The durability of rubber waste concrete is a subject that needs further investigations. How different wastes influence durability parameters and most importantly how waste treatment can enhance the concrete durability are questions that must be addressed.

13.3 Concrete with recycled polyethylene terephthalate (PET) waste

13.3.1 Fresh concrete properties

Workability

Choi et al. (2005) found that increasing PET waste aggregates leads to an increase in workability. This result is related to the spherical and smooth shape of the aggregates used in the investigation which are made of PET waste and ground granulated blast furnace slag (GBFS). According to these authors these aggregates are made inside a mixer with a inner temperature of 250 ± 10 °C. At this temperature the PET particles start to melt and then mix with the GBFS particles, resulting in a composite aggregate with a PET core and a GBFS surface (Fig. 13.7). Batanyeh et al. (2007) reported a decrease of slump with increasing PET waste for aggregate replacement, a 20% replacement leading to a decrease of slump by 20% to 58 mm. The workability of PET concrete is influenced by the fact that PET wastes were previously submitted to a treatment. It remains to be seen if these treatments have an environmental impact that negates the ecological benefits of using PET wastes.

Shrinkage

Kim et al. (2008) studied the influence of three types of PET-based fibres (Fig. 13.8) on the control of plastic shrinkage. These fibres are obtained from melted PET waste to form a roll-type sheet. Then the sheet is cut into 0.5 mm long fibres and a deforming machine is used to change the fibre surface

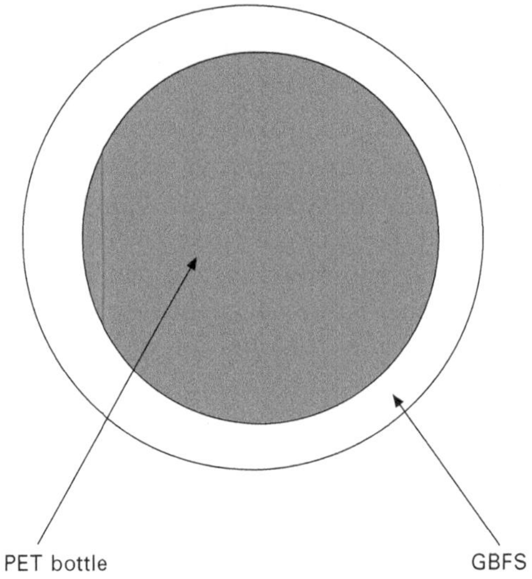

13.7 PET waste/GBFS aggregates (Choi et al., 2005).

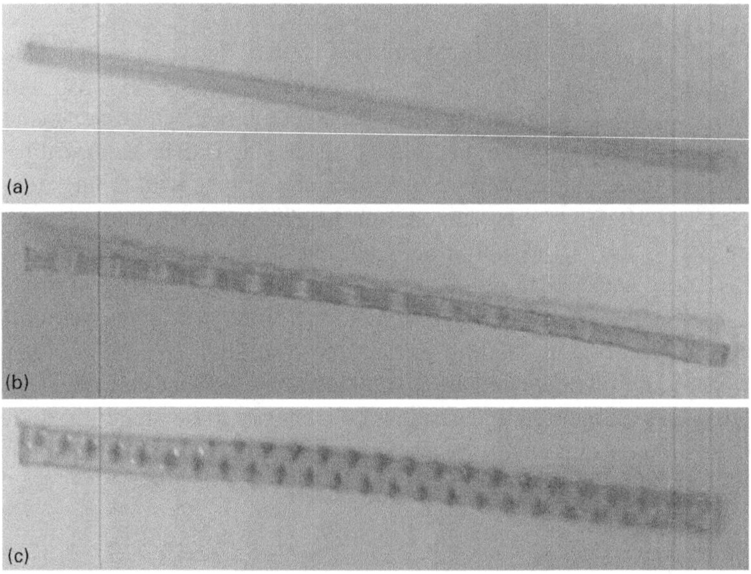

13.8 Geometry of recycled 50 mm long PET fibres: (a) Straight type (cross-section 0.5 × 1 mm^2); (b) crimped type (cross-section 0.3 × 1.2 mm^2); (c) embossed type (cross-section 0.2 × 1.3 mm^2) (Kim et al., 2008).

geometry. That the use of a volume of just 0.25% PET fibres was found to reduce the plastic shrinkage, while increasing PET fibre volume beyond 0.25% did very little to the shrinkage reduction. The results confirmed that the embossed-type fibre, the one that has the best mechanical resistance leads to the best shrinkage performance. Kim et al. (2010) confirmed the concrete crack control ability of PET fibre composites. Other authors compared the shrinkage performance of embossed type PET fibre previously submitted to a surface treatment to improve dispersion and bonding strength (Won et al., 2007) to the shrinkage performance of crimped polypropylene (PP) commercial fibres, reporting a slightly better beaviour for composites containing 0.5% of PP fibres. Nevertheless, composites containing 1% of PET fibres, future have a quite similar performance. Since investigations on concrete shrinkage performance used treated PET fibres, further investigations should study which treatment has the lowest environmental impact.

13.3.2 Hardened concrete properties

Compressive strength

Some authors suggest that PET waste can be used to produce an unsaturated polyester resin in the presence of glycols and dibasic acid. This material would serve as binder to produce polymer concrete with high mechanical performance (Rebeiz, 1994; Rebeiz and Fowler, 1996; Rebeiz et al., 1994a, 1994b, 1994c). Jo et al. (2006) confirm the high compressive strength of PET waste polymer concrete. Jo et al. (2008) also found that this type of concrete can also be used to incorporate recycled concrete aggregates with minor strength loss. Madhi et al. (2010) investigated the use of PET waste to produce a polyester resin, stating that the use of PET to glycol ratio of 2:1 has a positive effect on the compressive strength of polymer concrete. However, polymer concrete presents a significant strength decrease with increasing temperature. Rebeiz (1995) mentioned that polymer concrete can lose almost 45% of compressive strength for a temperature exposure of 60 °C.

Choi et al. (2005) mentioned that the replacement of fine aggregates for PET/GBFS aggregates (5–15 mm) leads to a decrease in the compressive strength. For a 25% replacement the mixtures with a W/C = 0.45 and 3 curing days lost just 6.4% in compressive strength. For 28 curing days the compressive strength loss reaches just 9.1%. Increasing the replacement percentage increases compressive strength loss but not in a proportional manner, for instance a 75% replacement the mixtures with a W/B = 0.45 and 3 curing days lost just 16.5% in compressive strength. This means that these treated PET aggregates perform in almost a similar way as natural aggregates.

Other authors (Batayneh et al., 2007) mentioned a severe compressive strength decrease of 72% for just 20% volume replacement of untreated PET waste. This behaviour is very different from the one reported by previous authors which means that using untreated PET waste implies the use of a very small volume in order to obtain an acceptable compressive strength concrete.

Marzouk et al. (2007) also studied the influence of crushed PET waste, reporting compressive strength decrease with increase replacement volume as long as the maximum dimension of the aggregates are below 2 mm. For plastic particles with a maximum dimension of 5 mm the compressive strength remains almost unchanged up to a volume replacement of 40%.

Ochi et al. (2007) mentioned that concrete mixtures with a low percentage of indented PET waste fibres obtained through a melting process (Fig. 13.9) have an acceptable mechanical performance. A mixture with 1.5% volume replacement show a minor compressive strength loss of 3.6% for W/B = 0.55. These authors mentioned that these fibres are already used in Japan at least since 2004 for tunnel support and also that they are as costly as steel fibres.

Modro et al. (2009) studied the influence of two types of PET waste (Fig. 13.10) on the compressive strength of concrete composites. The results differ according to the PET type. The concrete with sand PET presented a severe compressive strength loss with increasing waste content. As to the mixtures with the flake PET the strength loss was always less pronounced and in some conditions (replacement volume up to 10%) was negligible. According to these authors the strength loss is related to the porosity of the concrete specimens and also to the compressive strength of the PET waste.

Albano et al. (2009) studied concrete mixtures with two PET waste replacement percentages (10% and 20%) and with different PET dimensions (2.6 mm, 11.4 mm and a mix of the two). The results showed that concretes with a waste content of 20% and a higher waste dimensions (11.4 mm) have a higher compressive strength loss above 60%. Using just 10% PET waste replacement with a mix of the dimensions 2.6 mm and 11.4 mm (in equal parts) showed slight reductions in the strength loss between 15% and 20%. These authors also reported that mixtures with high dimension waste particles and a 20% waste volume should be avoided because they produce honeycomb formation (Fig. 13.11).

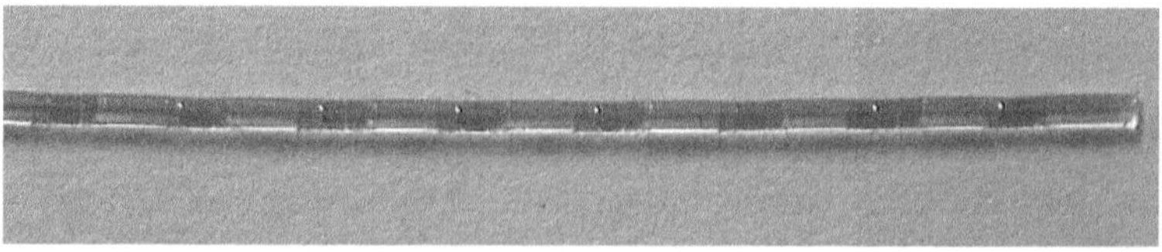

13.9 Indented PET fibre (Ochi et al., 2007).

13.10 PET waste types: (a) sand type; (b) flake type (Modro et al., 2009).

Choi et al. (2009) reported compressive strength losses of 6%, 16% and 30% for waste replacement percentages of 15%, 50% and 75%, stating that behaviour is independent of the W/C ratio. Such high mechanical performance can be explained by the treatment to which the PET waste was submitted (Fig. 13.12). Akcaozoglu et al. (2010) mentioned a slight reduction in compressive strength from 31.1 MPa to 28.8 MPa for 180 curing days specimens when the PET waste increase from 50% to 100% in mixtures of cement/GBFS (50%/50%). Frigione (2010) reported that the replacement of 5% PET waste (0.1–5 mm) by natural aggregates led to irrelevant compressive strength loss (0.4% to 1.9%) in specimens with 1 year curing. Other authors (Kim et al. 2010) have reported an irrelevant compressive strength loss for concretes with up to 1% embossed PET fibres waste. Hannawi et al. (2010) studied the performance of cementitious composites containing PET and polycarbonate wastes (Fig. 13.13), reporting a similar compressive strength behaviour for both wastes and revealing a serious strength loss with increasing waste content.

(a) (b)

13.11 Concrete specimens with honeycombs: (a) partial; (b) total (Albano et al., 2009).

Tensile strength

Jo et al. (2006) found that polymer concrete made with unsaturated polyester resin from PET waste can achieve a very high tensile strength (22.4 MPa in flexural and 7.85 MPa in splitting test). Considering the replacement of fine aggregates for PET/GBFS aggregates, Choi et al. (2005) mentioned a decrease in the splitting tensile strength of 16% for a 25% replacement volume. Since the loss percentage is higher than for the compressive strength of the same mixture, this would probably mean that artificial PET aggregates have a low adhesion to the cement paste. Ochi et al. (2007) found that the use of treated 30 mm long of indented PET fibres can lead to a tensile strength increase for volume replacements up to 1.5%. Marzouk et al. (2007) found that the tensile strength and the compressive strength of crushed PET concrete composites have a similar loss pattern. For Albano et al. (2009) the tensile strength loss is dependent on the volume of PET waste as to the PET particles dimensions, being that the mixtures with the smallest ones present lower tensile strength.

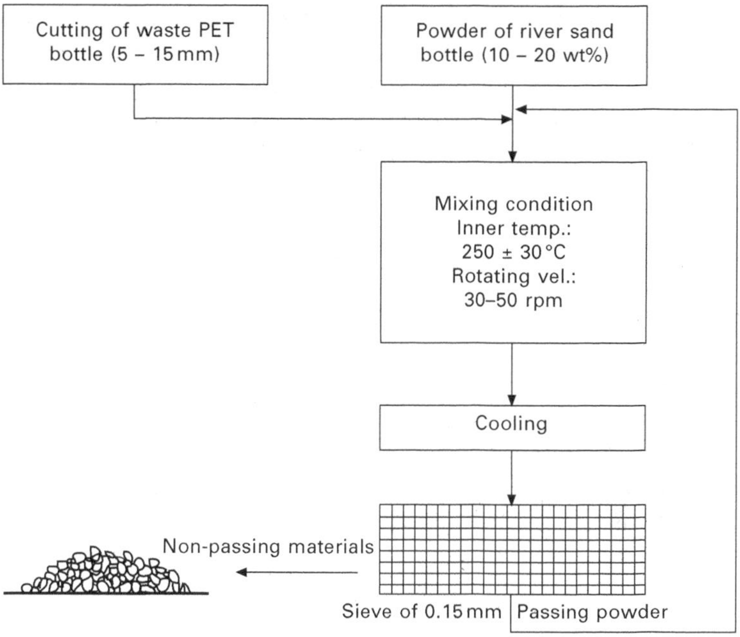

13.12 PET waste/sand production (Choi et al., 2009).

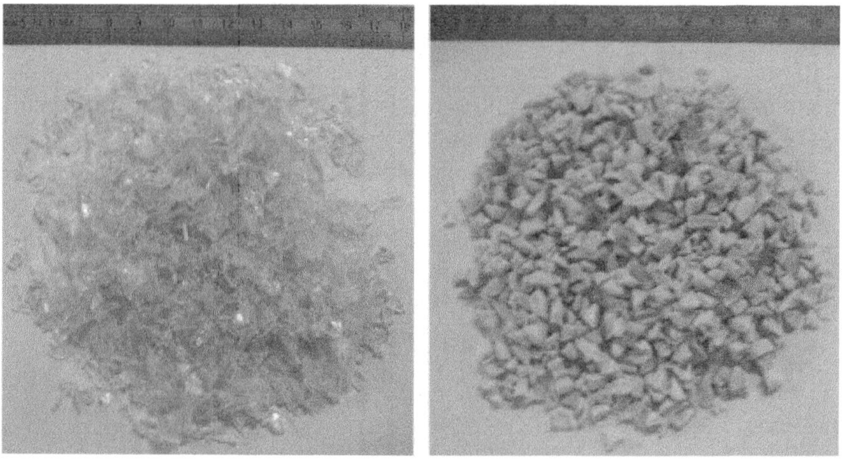

13.13 Polymeric wastes: left PET aggregates; right, polycarbonate aggregates (Hannawi et al., 2010).

Choi et al. (2009) reported that the ratio of flexural to tensile strength/compressive strength of PET waste mixtures was similar to that of current cementitious composites. Hannawi et al. (2010) mentioned that cementitious composites containing PET and polycarbonate wastes have similar tensile flexural strength performance. These authors also mentioned that a volume replacement up to 20% did not generate a relevant tensile strength loss. An increase in the replacement volume from 20% to 50% led to a 36% tensile strength loss for polycarbonate mixtures while the same replacement volume with PET waste was responsible for a tensile strength loss of 11%.

Toughness

Silva et al. (2005) mentioned that recycled monofilament PET fibres increased the toughness indexes of cementitious composites. Also Hannawi et al. (2010) mentioned that PET and polycarbonate wastes composites have a high energy-absorbing behaviour even with a high waste content. Kim et al. (2010) used embossed-type PET fibre, reporting that fibre concrete allows a mid-span deflection that is four times higher when compared with free fibre concrete. Investigations that clarify which treatment has the lowest environmental impact but maximizes the toughness characteristics of PET concrete are needed.

Modulus of elasticity

Marzouk et al. (2007) reported that the modulus of elasticity of PET-based composites decreased slightly with increasing waste content up to 20% (just 5%). Beyond this level the modulus of elasticity presented a severe decrease with waste content. A 40% replacement volume led to a decrease in the modulus of elasticity of 21.4%. Results also showed that mixtures with small PET particles have lower modulus of elasticity. Kim et al. (2010) confirmed the reduction in the elasticity modulus with increasing embossed PET waste fibres. However, since these authors used replacement volumes up just to 1.0%, they report irrelevant changes. The same happens with crimped PP fibre-based mixtures. For steel-reinforced concrete beams these authors found that using 0.5% embossed PET fibres led to maximum ductility. That makes it 10 times higher than that of concrete beams without fibres.

Thermal insulation

Marzouk (2005) mentioned that a 50% volume replacement of fine aggregates by PET wastes led to a reduction of the thermal conductivity by 46% from 1.28 (W/m K) to 0.69 (W/m K). Yesilata et al. (2009) obtained a 10% reduction in the thermal insulation of concrete specimens with square PET particles with a volume of 0.9% related to the volume of the concrete specimens. The results also showed that when strip or irregular PET particles were used the thermal insulation of the specimens increased from 10% to 17%. The results showed that square-type PET particles have low adhesion to cement, resulting in a composite with high thermal conductivity. PET concrete composites show an interesting thermal insulation performance. However, a high PET waste volume is needed to ensure a low thermal conductivity coefficient and that implies a high mechanical strength loss.

Durability

Silva et al. (2005) found that, with time, PET fibres degraded in the alkaline environment of the cement paste. The mechanism of PET degradation involves a depolymerization reaction that breaks the polymer chain, splitting it in two groups (the aromatic ring and the aliphatic ester). The infrared spectrum of PET fibre after immersion in alkaline solutions that reproduce the alkaline conditions of pore solution shows the presence of bands assigned to the aromatic ring. As a result of PET fibre degradation the toughness performance of cementitious composites decreases with time. These authors mentioned a 20% loss between specimens with 42 curing days and 104 curing days.

Ochi et al. (2007) mentioned that treated indented PET fibres showed high resistance to chemical degradation in an alkaline medium. These authors compared the alkali resistance of treated PET fibres with the resistance of PP and polyvinyl acetate (PVA) fibres reporting that the former retains 99% of their tensile strength while PP only retains 86% and PVA shows a severe degradation retaining only 56% of its tensile strength. Benosman et al. (2008) reported that the partial replacement of cement by PET wastes contributed to the reduction of the chloride ion diffusion coefficient.

Won et al. (2010) studied the durability of embossed PET fibres composites. Although the results of chloride permeability were similar to the control concrete and freeze–thaw resistance was better than the control concrete, the results related to exposure to an alkaline environment and to sulphuric acid revealed a performance that is not acceptable when it comes to real applications.

Galvão et al. (2010) compared the performance of PET-based concrete composites versus the performance of concrete with low-density polyethylene fibres and tyre waste fibres. The results showed that PET composites had the highest compressive strength with almost 30 MPa for the worst case scenario (7.5% volume replacement). Even for the tensile strength PET composites outperformed the other wastes. As to the erosion-abrasion under water test the best results were obtained for mixtures with 5% PET and 5% low-density polyethylene fibres. When compared with the reference concrete erosion-abrasion resistance, the 5% PET mixture presented less than 23% of mass loss and and 5% low-density polyethylene fibres mixture presented less than 40% of mass loss.

13.4 Other polymeric wastes

Several other polymeric wastes have been investigated regarding their potential for use as aggregate replacement in cementitious composites. Laukaitis et al. (2005) studied the development of lightweight thermo-insulating cementitious composites containing crumbled polystyrene waste and spherical blown polystyrene waste. The authors reported the need to use a 0.2% sulphonol and 0.03% glue hydrosolution to increase the adhesion between the polystyrene granules and the cement paste. The results showed that it is possible to produce a composite with 150–170 kg/m^3 and a thermal conductivity of coefficient between 0.06 and 0.0.64 W/m K. Ismail and Al-Hashmi (2008) mentioned that the use of polymeric wastes composed of 80% of polyethylene and 20% polystyrene as fine aggregate replacement increased the toughness of concrete with minor compressive and tensile strength decrease. However, the same authors reported workability issues that need to be addressed. For instance mixtures with a 15% replacement volume have an almost null slump.

Panyakapo and Panyakapo (2008) studied concretes with ground thermosetting polymer (melamine) waste reporting a reduction in compressive strength with waste, content increase related to the poor adhesion between waste plastic and cement paste. Yadav (2008) also confirmed the strength reduction associated with concrete polymeric waste composites. Nevertheless, the results showed that in spite of compressive strength reduction it is possible to produce non-load-bearing lightweight concrete.

Dweik et al. (2008) also used concrete with a thermosetting polymer (ground melamine-formaldehyde) waste as sand replacement. The authors mentioned a strength (both compressive and tensile) increase with waste content increase and also an increase in the thermal insulation. Mounanga et al. (2008) analysed concrete composites containing polyurethane wastes from insulation panels and found that although they have low thermal conductivity they present high compressive strength loss due to the weak and porous polyurethane aggregates. Results also showed an increase in high drying shrinkage with increasing polymeric waste content.

Fraj et al. (2010) also studied polyurethane waste-based concrete, reporting that waste replacement is responsible for a high compressive strength reduction. According to these authors the use of pre-soaked wastes together with a high W/C > 0.5 was responsible for the high porosity, so explaining the compressive strength reduction. They suggested that using a lower W/C ratio can increase the mechanical performance. Since the present conditions also led to a high drying shrinkage and an increase in both gas permeability and chloride diffusion using a lower W/C will probably improved the durability of these composites.

Alhozaimy and Shannag (2009) used minor amounts (up to 3%) of low-density polyethylene fibres and reported an improvement of the toughness characteristics of concrete composites of up to seven times. Kou et al. (2009) studied the influence of crush PVC wastes (Fig. 13.14) on the properties of concrete composites, reporting a relevant compressive strength decrease with the increase in waste content. Since PVC aggregates have a compressive strength of 65 MPa the compressive strength reduction of the concrete mixtures must be related to a low adhesion between PVC and the cement paste. The PVC aggregates also contribute to the reduction of the modulus of elasticity because of the low modulus of PVC. The results suggested that concrete with increased PVC volume show a decrease in drying shrinkage. For instance the use of just 15% PVC aggregates led to a 50% reduction in the drying shrinkage. Studies about the durability of other polymeric wastes are scarce but some authors have already reported that polyester and acrylic fibres can suffer from chemical degradation when immersed in the alkaline environment of a cement paste (Wang et al., 1987; Jelidi, 1991; Houget, 1992).

13.5 Conclusion

Tyre rubber and PET wastes represent a serious environmental issue that needs to be addressed with urgency by the scientific community. Investigations carried out so far reveal that tyre waste concrete is specially recommended for concrete structures located in areas of severe earthquake risk and also for applications subjected to severe dynamic actions such as railway sleepers. This material can also be used for non-load-bearing purposes such as noise reduction barriers. Investigations of rubber waste concrete show that concrete performance is very dependent on the waste aggregates. Further investigations are needed to clarify, for instance, which characteristics maximize concrete performance. As for PET-based concrete the investigations show that this

13.14 PVC waste: left, before crushing; right, after crushing (Alhozaimy and Shannag, 2009).

material is very dependent on the treatment of these wastes. At present PET fibres are used to replace steel fibres and some authors even report the use of PET concrete mixtures for repairing concrete structures submitted to high underwater erosion. Nevertheless, future investigations should clarify which treatments can maximize concrete performance while having the lowest environmental impact. Further investigations should also be carried on about the use of other polymeric wastes in concrete.

13.6 References

Aiello, M.; Leuzzi, F. (2010) Waste tyre rubberized concrete: Properties at fresh and hardened state. *Waste Management*, Vol.30, 1696–1704.

Akcaozoglu, S.; Atis, C.; Akcaozoglu, S.K. (2010) An investigation on the use of shredded waste PET bottles as aggregates in lightweight concrete. *Waste Management*, Vol.30, 285–290.

Albano, C.; Camacho, N.; Reyes, J.; Feliu, J.; Hernández, M. (2005) Influence of scrap rubber addition to Portland concrete composites: destructive and non-destructive testing. *Composite Structure*, Vol.71, 439–446.

Albano, C.; Camacho, N.; Hernandez, M.; Gutierrez, A. (2009) Influence of content and particle size of waste PET bottles on concrete behaviour at different W/C ratios. *Waste Management*, Vol.29, 2707–2716.

Alhozaimy, A.; Shannag, M. (2009) Performance of concretes reinforced with recycled plastic fibres. *Magazine of Concrete Research*, Vol.61, 293–298.

Anison, M. (1964) An investigation into a hypothetical deformation and failure mechanism for concrete. *Magazine of Concrete Research,* Vol.47, 73–82.

ASTM C1018-97 (1997) Standard test method for flexural toughness and first-crack strength of fibre-reinforced concrete. Pennsylvania.

Balaha, M.; Badawy, A.; Hashish, M. (2007) Effect of using ground waste tire rubber as fine aggregate on the behaviour of concrete mixes. *Indian Journal of Engineering and Materials Sciences*, Vol.14, 427–435.

Batayneh, M.; Marie, I.; Asi, I. (2007) Use of selected waste materials in concrete mixes. *Waste Management*, Vol.27, 1870–1876.

Batayneh, M.; Marie, I.; Asi, I. (2008) Promoting the use of crumb rubber concrete in developing countries. *Journal of Waste Management*, Vol.28, 2171–2176.

Benosman, A.; Taibi, H.; Mouli, M.; Belbachir, M.; Senhadji, Y. (2008) Diffusion of chloride ions in polymer–mortar composites. *Journal of Applied Polymer Science*, Vol.110, 1600–1605.

Biel, T.; Lee, H. (1996) Magnesium oxychloride cement concrete with recycled tire rubber. Transportation Research Board, Report No. 1561. Washington, DC: Transportation Research Board, 6–12.

Bignozzi, M.; Sandrolini, F. (2006) Tyre rubber waste recycling in self-compacting concrete. *Cement and Concrete Research*, Vol.36, 735–739.

Cairns, R.; Kew, H.; Kenny, M. (2004) The use of recycled rubber tyres in concrete construction. Final Report, The Onyx Environmental Trust, University of Strathclyde, Glasgow.

Chen, S.; Su, H.; Chang, J.; Lee, W.; Huang, K. Hsieh, L.; Huang, Y.; Lin, W.; Lin, C. (2007) Emissions of polycyclic aromatic hydrocarbons (PAHs) from the pyrolysis of scrap tyres. *Atmospheric Environment*, Vol.41, 1209–1220.

Choi, Y.; Moon, D.; Chung, J.; Cho, S. (2005) Effects of waste PET bottles aggregate on the properties of concrete. *Cement and Concrete Research*, Vol.35, 776–781.

Choi, Y.; Moon, D.; Kim, Y.; Lachemi, M. (2009) Characteristics of mortar and concrete containing fine aggregate manufactured from recycled waste polyethylene terephthalate bottles. *Construction and Building Materials*, Vol.23, 2829–2835.

Chou, L.; Lin, C.; Lu, C.; Lee, C.; Lee, M. (2010) Improving rubber concrete by waste organic sulfur compounds. *Waste Management and Research*, Vol.28, 29–35.

Day, K.; Holtze, K.; Metcalfe, J.; Bishop, C.; Dutka, B. (1993) Toxicity of leachate from automobile tyres to aquatic biota. *Chemosphere*, Vol.27, 665–675.

Dweik, H.; Ziara, M.; Hadidoun, M. (2008) Enhancing concrete strength and thermal insulation using thermoset plastic waste. *International Journal of Polymeric Materials*, Vol.57, 635–656.

Eleazer, W.; Barlaz, M.; Whittle, D. (1992) Resource recovery alternatives for waste tires in North Carolina. School of Engineering, Civil Engineering Department, NCSU, US.

EN1994 – Eurocode 4 (1994) Design of composite steel and concrete structure. British Standards Institution London.

European Commission (1999) Council Directive 1999/31/EC of 26 April 1999 on the landfill of waste. *Official Journal of the European Communities*, L182, 1–19.

European Commission (2000) Directive 2000/76/EC of the European Parliament and of the Council of 4 December 2000 on incineration of waste. *Official Journal of the European Communities*, L332, 91–111.

Farcasiu, M. (1993) Another use for old tyres. *Chemtech*, 22–24.

Fioriti, C.; Ino, A.; Akasaki, J. (2007) Concrete paving blocks with tyre wastes. *Revista Internacional Construlink*, No. 15, Vol.5, 56–67.

Fraj, A.; Kismi, M.; Mounanga, P. (2010) Valorization of coarse rigid polyurethane foam waste in lightweight aggregate concrete. *Construction and Building Materials*, Vol.24, 1069–1077.

Freitas, C.; Galvão, J.; Portella, K.; Joukoski, A.; Filho, C. (2009) Desempenho fisico-químico e mecânico de concreto de cimento Portland com borracha de estireno-butadieno reciclada de pneus. *Química Nova*, Vol.32, 913–918.

Frigione, M. (2010) Recycling of PET bottles as fine aggregates in concrete. *Waste Management*, Vol.30, 1101–1106.

Galvão, J.; Portella, K.; Joukoski, A.; Mendes, R.; Ferreira, E. (2010) Use of waste polymers in concrete repair for dam hydraulic surfaces. *Construction and Building Materials*, Vol.25, 1049–1055.

Ganjian, E.; Khorami, M.; Maghsoudi, A. (2009) Scrap-tyre-rubber replacement for aggregate and filler in concrete. *Construction and Building Materials*, Vol.23, 1828–1836.

Ghaly, A.; Cahill, J. (2005) Correlation of strength, rubber content, and water to cement ratio in rubberized concrete. *Canadian Journal of Civil engineering*, Vol.32, 1075–1081.

Gore, A. (2009) *Our choice. A plan to solve the climatic crisis.* Penguin.

Guneyisi, E. (2010) Fresh properties of self-compacting rubberized concrete incorporated with fly ash. *Materials and Structures*, Vol.43, 1037–1048.

Guneyisi, E.; Gesoglu, M.; Ozturan, T. (2004) Properties of rubberized concretes containing silica fume. *Journal of Cement and Concrete Research*, Vol.34, 2309–2317.

Hannawi, K.; Kamali-Bernard, S.; Prince, W. (2010) Physical and mechanical properties of mortars containing PET and PC waste aggregates. *Waste Management*, Vol.30, 2312–2320.

Hartwell, S.; Jordahl, D.; Dawson, C.; Ives, A. (1998) Toxicity of scrap tyre leachates in estuarine salinities: are tyres acceptable for artificial reefs? *Transactions of the American Fisheries Society*, Vol.127, 796–806.

Houget, V. (1992) Etude dês caracteristiques mecaniques et physico-chimiques de composites ciments-fibres organiques. PhD dissertation, Inst. Nat. Sci. Appl., Lyon, France.

Ismail, Z.; Al-Hashmi, E. (2008) Use of waste plastic in concrete mixture as aggregate replacement. *Waste Management*, Vol.28, 2041–2047.

Jelidi, A. (1991) Conception d'un materiau composite a matrice cimentaire reinforcee par des fibres de polyester. PhD dissertation, Inst. Nat. Sci. Appl., Lyon, France.

Jo, B.; Seung-Kook, P.; Cheol-Hwan, K. (2006) Mechanical properties of polyester polymer concrete using recycled polyethylene terephthalate. *ACI Structural Journal*, Vol.103, 219–225.

Jo, B.; Park, S.; Park, J. (2008) Mechanical properties of polymer concrete made with recycled PET and recycled concrete aggregates. *Construction and Building Materials*, Vol.22, 2281–2291.

Khaloo, A.; Dehestani, M.; Rahmatabadi, P. (2008) Mechanical properties of concrete containing a high volume of tire-rubber particles. *Waste Management*, Vol.28, 2472–2482.

Khatib, Z.; Bayomy, F. (1999) Rubberized Portland cement concrete. *ASCE Journal of Materials in Civil Engineering*, Vol.11, 206–213.

Kim, J.; Park, C.; Lee, S.; Lee, S.; Won, J. (2008) Effects of the geometry of recycled PET fibre reinforcement on shrinkage cracking of cement-based composites. *Composites: Part B*, Vol.39, 442–450.

Kim, S.; Yi, N.; Kim, H.; Kim, J.; Song, Y. (2010) Material and structural performance evaluation of recycled PET fiber reinforced concrete. *Cement and Concrete Composites*, Vol.32, 232–240.

Kou, S.; Lee, G.; Poon, C.; Lai, W. (2009) Properties of lightweght aggregate concrete prepared with PVC granules derived from scraped PVC pipes. *Waste Management*, Vol.29, 621–628.

Laukaitis, A.; Zurauskas, R.; Keriené, J. (2005) The effect of foam polystyrene granules on cement composites properties. *Cement and Concrete Composites*, Vol.27, 41–47.

Li, G.; Garrick, G.; Eggers, J.; Abadie, C.; Stubblefield, M.; Pang, S. (2004) Waste tire fiber modified concrete. *Composites: Part B*, Vol.35, 305–312.

Ling, T.; Nor, H.; Hainin, M.; Chik, A. (2009) Laboratory performance of crumb rubber concrete block pavement. *International Journal of Pavement Engineering*, Vol.10, 361–374.

Ling, T.; Nor, H.; Lim, S. (2010) Using recycled waste tyres in concrete paving blocks. *Proceedings of Institution of Civil Engineers: Waste and Resource Management*, Vol. 163, 37–45.

Mahdi, F.; Abbas, H.; Khan, A. (2010) Strength characteristics of polymer mortar and concrete using different compositions of resins derived from post-consumer PET bottles. *Construction and Building Materials*, Vol.24, 25–36.

Marzouk, O. (2005) Valorization of plastic waste: thermal conductivity of concrete formulated with PET. In 1st International Conference on Engineering for Waste Treatment, École de Mines d'Albi-Carmaux, France.

Marzouk, O.; Dheilly, R.; Queneudec, M. (2007) Valorization of post-consumer waste plastic in cementitious concrete composites. *Waste Management*, Vol.27, 310–318.

Mello, D.; Pezzin, S.; Amico, S. (2009) The effect of post consumer PET particles on the performance of flexible polyurethane foams. *Polymer Testing*, Vol.28, 702–708.

Modro, N.L.; Modro, N.; Modro, N.R.; Oliveira, A. (2009) Avaliação de concreto de cimento Portland contendo resíduos de PET. *Revista Matéria*, Vol.14, 725–736.

Mohammed, B.S. (2010) Structural behavior and m-k value of composite slab utilizing concrete containing crumb rubber. *Construction and Building Materials*, Vol.24, pp. 1214–1221.

Mounanga, P.; Gbongbon, W.; Poullain, P.; Turcry, P. (2008) Proportioning and characterization of lightweight concrete mixtures made with rigid polyurethane foam wastes. *Cement and Concrete Composites*, Vol.30, 806–814.

Nagdi, K. (1993) *Rubber as an engineering material: Guidelines for user*. Hanser Publication.

Naik, T.; Singh, S. (1991) Utilization of discarded tyres as construction materials for transportation facilities. Report No. CBU – 1991–02, UWM Center for by-products utilization. University of Wisconsin, Milwaukee, 16.

Naik, T.; Singh, S.; Wendorf, R. (1995) Applications of scrap tire rubber in asphaltic materials: state of the art assessment. Report No. CBU-1995-02, UWM Center for by-products utilization. University of Wiscosin, Milwaukee, 49.

Ochi, T.; Okubo, S.; Fukui, K. (2007) Development of recycled PET fibre and its application as concrete-reinforcing fibre. *Cement and Concrete Composites*, Vol.29, 448–455.

Oikonomou, N.; Mavridou, S. (2009) The use of waste tyre rubber in civil engineering works. In *Sustainability of construction materials*, Ed. J. Khatib, Woodhead Publishing Limited.

Oiknomou, N.; Stefanidou, M.; Mavridou, S.(2006) Improvement of the bonding between rubber tire particles and cement paste in cement products. 15th Conference of the Technical Chamber of Greece, Alexandroupoli, Greece, 234–242.

Panyakapo, P.; Panyakapo, M. (2008) Reuse of thermosetting plastic waste for lightweight concrete. *Waste Management*, Vol.28, 1581–1588.

Pierce, C.; Williams, R. (2004) Scrap tire rubber modified concrete: Past, present and future. In Used/Post-consumer Tires, *Proceedings of the International Conference on Sustainable Waste Management and recycling, Research Group*, Eds M., Limbachiya & J., Roberts, Thomas Telford, 1–16.

Raghavan, D.; Huynh, H.; Ferraris, C. (1998) Workability, mechanical properties, and chemical stability of a recycled tyre rubber filled cementitious composite. *Journal of Materials Science*, Vol.33, 1745–1752.

Rebeiz, K. (1994) Precast use of polymer concrete using unsaturated polyester resin based on recycled PET waste. *Construction and Building Materials*, Vol.10, 215–220.

Rebeiz, K. (1995) Time temperature properties of polymer concrete using recycled PET. *Cement and Concrete Composites*, Vol.17, 119–124.

Rebeiz, K.; Fowler, D. (1996) Flexural strength of reinforced polymer concrete made with recycled plastic waste. *ACI Structural Journal*, Vol.93, 524–530.

Rebeiz, K.; Fowler, D.; Paul, D. (1994a) Mechanical properties of polymer concrete systems made with recycled plastic, *ACI Materials Journal*, Vol.91, 40–45.

Rebeiz, K.; Serhal, S.; Fowler, D. (1994b) Structural behaviour of polymer concrete beams using recycled plastics, *ASCE Journal Materials Civil Engineering*, Vol.6, 150–165.

Rebeiz, K.; Yang, S.; David, W. (1994c) Polymer mortar composites made with recycled plastics. *ACI Materials Journal*, Vol.91, 313 319.

Segre, N.; Joekes, I. (2000) Use of tire rubber particles as addition to cement paste. *Cement and Concrete Research*, Vol.30, 1421–1425.

Segre, N.; Monteiro, P.; Sposito, G. (2002) Surface characterization of recycled tire rubber to be used in cement paste matrix. *Journal of Colloid and Interface Science*, Vol.248, 521–523.

Segre, N.; Ostertag, C.; Monteiro, P. (2006) Effect of tire rubber particles on crack propagation in cement paste. *Materials Research*, Vol.9, 311–320.

Siddique, R.; Naik, T. (2004) Properies of concrete containing scrap-tyre rubber – an overview. *Waste Management*, Vol.24, 563–569.

Silva, D.; Betioli, A.; Gleize, P.; Roman, H.; Gomez, L.; Ribeiro, J. (2005) Degradation of recycled PET fibers in Portland cement-based materials. *Cement and Concrete Research*, Vol.35, 1741–1746.

Skripkiunas, G.; Grinys, A.; Cernius, B. (2007) Deformation properties of concrete with rubber waste additives. *Materials Science*, Vol.13, 219–223.

Snelson, D.; Kinuthia, J.M.; Davies, P.; Chang, S. (2009) Sustainable construction: Composite use of tyres and ash in concrete. *Waste Management*, Vol.29, 360–367.

Son, K.; Hajirasouliha, I.; Pilakoutas, K. (2010) Strength and deformability of waste tyre rubber-filled reinforced concrete columns. *Construction and Building Materials*, Vol.25, 218–226.

Sukontasukkul, P.; Chaikaew, C. (2006) Properties of concrete pedestrian block mixed with crumb rubber. *Construction and Building Materials*, Vol.20, 450–457.

Sukontasukkul, P. (2009) Use of crumb rubber to improve thermal and sound properties of pré-cast concrete panel. *Construction and Building Materials*, Vol.23, 1084–1092.

Taylor, M.; Gielen, D. (2006) *Energy efficiency and CO_2 emissions from the global cement industry*. International Energy Agency.

Topçu, I.; Bilir, T. (2009) Experimental investigation of some fresh and hardened properties of rubberized self-compacting concrete. *Materials and Design*, Vol.30, 3056–3065.

Topçu, I.; Demir, A. (2007) Durability of rubberized mortar and concrete. *Journal of Materials in Civil Engineering*, Vol.19, 173–178.

Turatsinze, A.; Garros, M. (2008) On the modulus of elasticity and strain capacity of self-compacting concrete incorporating rubber aggregates. *Resources, Conservation and Recycling*, Vol.52, 1209–1215.

Turatsinze, A.; Bonnet,S.; Granju, J. (2007) Potential of rubber aggregates to modify properties of cement based-mortars: Improvement in cracking shrinkage resistance. *Construction and Building Materials*, Vol.21, 176–181.

Valadares, F. (2009) Mechanical performance of structural concretes containing rubber waste from waste tires. Master Dissertation in Civil Engineering, IST-UTL, Lisbon.

Vieira, R.; Soares, R.; Pinheiro, S.; Paiva, O.; Eleutério, J.; Vasconcelos, R. (2010) Completely random experimental design with mixture and process variables for optimization of rubberized concrete. *Construction and Building Materials*, Vol.24, 1754–1760.

Wang, Y.: Backer, S.; Li, V. (1987) An experimental study of synthetic fibre reinforced cementitious composites. *Journal of Materials Science*, Vol.22, 4281–4291.

Won., J.; Park, C.; Kim, H.; Lee, S. (2007) Effect of hydrophilic treatments of recycled PET fibre on the control of plastic shrinkage cracking of cement-based composites. *Journal of Korean Society of Civil Engineers*, Vol. 27, 413–419.

Won, J.; Jang, C.; Lee, S.; Lee, S.; Kim, H. (2010) Long-term performance of recycled PET fibre-reinforced cement composites. *Construction and Building Materials*, Vol.24, 660–665.

WBCSD (2010) *End-of-life tyres: A framework for effective management systems*. World Business Council for Sustainable Development.

Yadav, I. (2008) Laboratory investigations of the properties of the concrete containing recycled plastic aggregates. Master of Engineering in Structural Engineering, Thapar University, Patiala, India.

Yesilata, B.; Isiker, Y.; Turgut, P. (2009) Thermal insulation enhancement in concretes by assign waste PET and rubber pieces. *Construction and Building Materials*, Vol.23, 1878–1882.

Zheng, L.; Huo, X.; Yuan, Y. (2008a) Strength, modulus of elasticity, and brittleness index of rubberized concrete. *Journal of Materials in Civil Engineering*, Vol.20, 692–699.

Zheng, L.; Huo, S.; Yuan, Y. (2008b) Experimental investigation on dynamic properties of rubberized concrete. *Construction and Building Materials*, Vol.22, 939–947.

14
Concrete with construction and demolition wastes (CDW)

A. E. B. CABRAL, Federal University of Ceará, Brazil

DOI: 10.1533/9780857098993.3.340

Abstract: This chapter discusses the recycling of construction and demolition wastes (CDW) and the use of recycled aggregates in concrete. Classification and characteristics of recycled aggregates, physical and mechanical properties, and durability of recycled aggregate concrete are also discussed.

Key words: construction and demolition waste (CDW), recycled aggregate, recycled aggregate concrete, mechanical and durability properties.

14.1 Introduction: use of construction and demolition wastes (CDW) in concrete

As in any industrial process, the construction industry generates waste on a large scale, which needs to be managed. According to John (2001), the macro complex of the construction industry accounts for 40% of waste generated in the economy.

Given the important role of the construction industry in developing nations, it is advisable to seek and adopt urgent measures in order to achieve sustainable development. For Terry (2004), legislation is the biggest factor affecting waste management in the construction industry. Without effective legislation and strong enforcement, as well as an effective collection system, it will not be possible to resolve the complex waste management issue. This can be defined as construction waste

> from the buildings, renovations, repairs and demolition of civil works, as well as the resulting preparation and excavation of land, such as bricks, ceramic blocks, concrete in general, soils, rocks, metals, resins, glues, paints, wood, plywood, liners, mortar, plaster, tiles, asphalt pavement, glass, plastics, pipes, electrical wiring, among others, which are commonly called debris works, glass or shrapnel (MMA, 2007).

It is estimated that most of the waste is from construction and demolition work, as well as construction sites or construction services, known as construction and demolition waste (CDW).

There is a great tendency to sort out CDW as inert waste, due to the large amount of mineral components that are clean and chemically inert. However, this trend is already viewed with some trepidation, since such waste may be contaminated with paint, surface treatment of substances or even heavy metals that can leach out and contaminate water and soil.

Oliveira (2002) concluded in his research that building wastes consisting solely of concrete waste, mineralize the water and change the soil, i.e., are non-inert, suggesting that these wastes are classified as non-hazardous and non-inert. Hansen (1992) also mentions that the CDW may contain components, that are considered as toxic from an environmental point of view. So, it seems that such waste can be classified as inert or non-inert, depending only on its origin and constitution.

One particular point that demonstrates the importance of CDW is its increasing share of total municipal solid waste (MSW). Several surveys indicate that CDW currently represents around 50% of the total MSW produced in Brazilian cities, with an average rate of generation around 0.52 tonne/habitant/year (Cabral, 2007). In the European Union, there is no consensus about CDW generation but it represents approximately 22–49% of the total waste generation, representing 450–970 million tonnes of CDW generated per year, which corresponds to 0.9–2.0 tonne/habitant/year (Solis-Guzman et al., 2009; Sonigo et al., 2010; Sáez et al., 2011).

According to Sáez et al. (2011), France and Luxembourg generate 5.5 and 15 tonnes/habitant/year, respectively. Germany and Ireland generate between 2 and 4 tonnes/habitant/year, whereas the rest of the European countries generate between 0.2 tonnes/habitant/year (Norway) and 1.9 tonnes/habitant/year (United Kingdom). In the United States, the generation of CDW during 90 years was 0.43 tonne/habitant/year (Hansen, 1992) and in 2002 it was estimated 2.0–2.57 tonnes/habitant/year (Cochran and Townsend, 2010).

The generation of construction and demolition waste is influenced by many different factors including the construction and demolition practices adopted; the economic and market factors such as market size, availability and cost of natural aggregates compared to the costs of delivery for recycled aggregates; the regulatory framework that provides incentives to minimize waste generation in construction sites and disincentives for waste disposal in landfills; perceptions regarding the quality of recycled materials and the absence of use in codes of practice, specifications and quality assurance mechanisms (Bakoss and Ravindrarajah, 1999).

The composition of construction and demolition waste is also variable, depending on geographic region, time of the year and type of work, among other factors. When coming from construction, the composition is dependent on the stage of the work, since the stage of concrete structure is a higher incidence of fragments from concrete, steel, wooden forms, among others, while in the finishing stage, predominance of residual mortar, bricks, tiles, ceramics, among others (Poon et al., 2001). If construction reforms, there will be a higher incidence of ceramics, wood, natural stone, glass, metals and plastics (Esin and Cosgun, 2007).

In Brazil, it is estimated that an average 65% of the waste materials is of mineral origin, 13% wood, 8% plastics and 14% other materials. Construction companies are responsible for generating 20% to 25% of that debris, while the remainder comes from reforms and works of self.

For demolition work, the characteristics of waste also vary with the type of structure to be demolished and the technique used. However, in general, demolition waste consists of a high percentage of inert material such as bricks, sand and concrete. Metals, wood, paper, glass, plastics and other materials also appear; however, to a lesser extent (Poon et al., 2001).

In general, the vast majority of CDW has great potential to be recycled. In Europe, Henrichsen (2000) states that over 90% of the CDW can be recycled although on average 50% is recycled (Sonigo et al., 2010; Sáez et al., 2011). Of course there are countries with a high recycling rate, as Belgium (Flanders), with over than 90%, Denmark, Estonia, Germany, Ireland and Netherlands with over 70%, and countries with intermediate recycling rate, as Austria, Belgium, France, Lithuania and UK with rates between 60 and 70%, Luxembourg, Latvia and Slovenia between 40 and 60%, and countries with low recycling rates, as Cyprus, Czech Republic, Finland, Greece, Hungary, Poland, Portugal and Spain (below 40%). There are no available data from Bulgaria, Italy, Malta, Romania, Slovakia and Sweden (Sonigo et al., 2010).

14.2 Management of construction waste

As already discussed, the generation of construction waste is significant within the context of solid waste. Some Brazilian cities already have laws that are specific to the management of CDW (Marques Neto, 2005). Nevertheless, for such waste to be recycled and thus reused, as raw material, the characteristics of these recycled products must be compatible with the intended purpose. The recycling of CDW contaminated with non-inert materials produces low-quality recycled materials. Therefore, it is essential to separate the various types of waste produced, where inert phase is the one with greatest potential for recycling, in order to produce good quality recycled material to be reused in construction itself.

Sim and Park (2011) present in Fig. 14.1 the manufacturing process of recycled aggregate from CDW. From the purely economic point of view, recycling of CDW is attractive only when the recycled products are competitive with the natural materials, in terms of cost and quantity. Therefore, recycled materials will be more competitive in areas where there is scarcity of natural materials and landfill areas (Tam and Tam, 2006).

14.1 Recycled concrete aggregate manufacturing process (Sim and Park, 2011).

Hendriks and Janssen (2001) discussed several ways to reuse the various constituents of the CDW, some of which are listed below:

- Concrete debris reused without any improvement in road construction or as fill material for low-lying areas, among other applications. After the concrete waste has been crushed and the aggregates separated into various sizes, the waste can be used as aggregate for asphalt concrete production, sub-bases for roads and concrete with recycled aggregates.
- Wood, when the undamaged part can be reused in the construction itself and the non-reusable can be reduced to small sizes, in order to be processed, making paper and cardboard. Alternatively, wood can be burned as energy to be used, or decomposed by pyrolysis or gasification, which can be used in the chemical industry after hydrolysis.
- The waste of asphalt can be reused in road construction, both for the processing of new asphalt and the manufacturing of sub-bases, such as granular material.
- Metals can be reused to produce new metal.
- Glasses, which would not be allowed to be processed and recycled aggregates, are part of a function of alkali–silica reaction, although there are studies that indicate its use as microphylous in the production of concrete. These can also be recycled into new glass, fiber glass, tiles and paving blocks or as additive in the manufacture of asphalt.
- Masonry waste, including brick, stone and ceramics, can be used in the production of concrete, although there is a reduction in compressive strength and special concretes, such as lightweight concrete with high thermal insulation. The aforesaid can be used as aggregate in the manufacture of bricks, even with the use of his fine piece, as filling material, which can be burned and turned into ashes with the reuse in their own construction.
- Paper and cardboard, as well separated and collected, can be recycled, generally as a packaging material.
- Waste plastic from polystyrene (PS), polypropylene (PP), polyethylene (PE) and polyvinyl chloride (PVC) can be recycled, although other resins are difficult to be reprocessed.
- Hazardous waste must be incinerated or grounded with specific procedures. Some waste such as oil, paints and solvents, abrasives and batteries, can be recycled.

In order to ensure that recycled materials are permanently incorporated into the market as raw material to be used in construction, it is necessary to convince designers and builders, as well as the final consumer, that their use has some competitive advantage and low technical and environmental risks. This is necessary to overcome the prejudice against the use of material which is considered to be a second rate, and rather to explore the ecological side of recycling.

14.3 Recycled aggregates

14.3.1 Classification of recycled aggregates from CDW

In order to be suitable for certain applications, the recycled aggregate from CDW must meet certain requirements for particle size and a minimal presence of contaminants, in line with other requirements of stability and durability. In the case of use for the production of concrete, it must be tough enough for the class of concrete to be produced in addition to being dimensionally stable for changes in the content of moisture. The recycled aggregate should also have no deleterious reaction with cement or armor, and thereafter, it must have a satisfactory shape and particle size to produce an acceptable workability concrete (Hansen, 1992).

Taking into account the considerations duly mentioned above, some institutions classify the aggregates, as shown below:

- Kawano (2000) cites a classification proposed by the Ministry of Construction in Japan, dividing the recycled aggregates into classes, according to Table 14.1.
- Rilem (1994) classifies as Type I aggregates those that originate from masonry waste (Fig. 14.2); Type II are those that originate from waste concrete (Fig. 14.3) and Type III are those that are a mixture of natural aggregates and recycled aggregates, being the ratio of at least 80% from natural aggregate and a maximum 10% of households Type I.
- The European Standard EN 12620 sorts the recycled coarse aggregate in more detail, as the content of their constituents, as can be seen on Tables 14.2 and 14.3 (Lay, 2006). In order to specify what is recycled aggregate according to the recommendations of EN 12620, one example can cite as following: the aggregate naming RCU90, RB10, RA5, $FL_{NS}0$ and XRG1, means an aggregate of more than 90% by mass of concrete and clean aggregate, less than 10% masonry, less than 5% of bituminous material and so on.
- The Brazilian Association of Technical Standards (ABNT) published in 2004 the first edition of the Standards series that relate to construction

Table 14.1 Classification of recycled aggregates proposed by the Ministry of Construction in Japan (Kawano, 2000)

Coarse aggregate				Fine aggregate		
Class	Water absorption		Loss in weight	Class	Water absorption	Loss in weight
I	<3%		<12%	I	<5%	<10%
II	<3%	and	<40%	II	<10%	–
		or				
	<5%	and	<12%			
III	<7%		–			

14.2 Recycled aggregates from masonry waste.

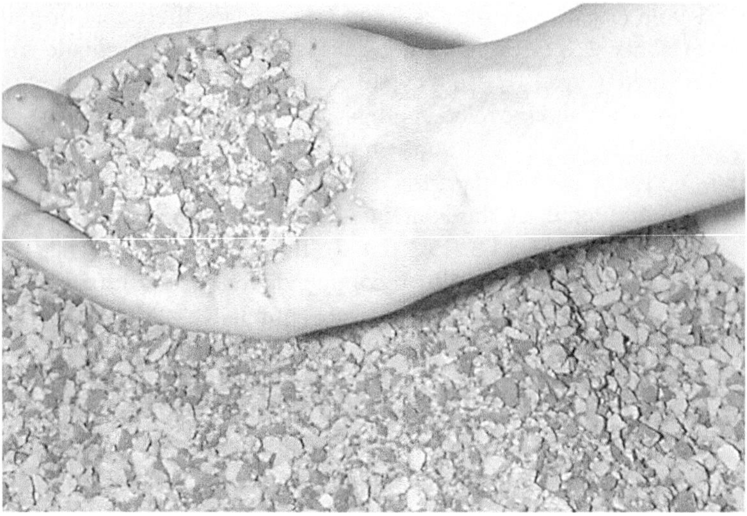

14.3 Recycled aggregates from concrete waste.

waste, these being the NBR 15112 to 15116. However, only NBR 15116 is about the requirements of the recycled aggregates for use in concrete pavement and in nonstructural function. According to the specifications of this last provision, in order to produce aggregates from construction waste, they ought to be classified as Class A, according to CONAMA Resolution 307. Once produced, recycled aggregates can be classified as recycled concrete aggregate (RCA), if its coarse fraction is composed

Table 14.2 Categories according to the constituents of the recycled coarse aggregates (Lay, 2006)

Constituent	Content (wt %)	Category
R_C	≥90	R_C 90
	≥70	R_C 70
	<70	R_C D
	no requirement	R_C SR
$R_C + R_U$	≥90	R_{CU} 90
	≥70	R_{CU} 70
	≥50	R_{CU} 50
	<50	R_{CU} D
	no requirement	R_{CU} SR
R_B	<10	R_B 10
	<30	R_B 30
	<50	R_B 50
	>50	R_B D
	no requirement	R_B SR
R_A	<1	R_A 1–
	<5	R_A 5–
	<10	R_A 10–
FL_S+FL_{NS}	<1	$FL_{total}1$
	<3	$FL_{total}3$
FL_{NS}	<0.01	$FL_{NS}0.01$
	<0.05	$FL_{NS}0.05$
	<0.1	$FL_{NS}0.1$
$X+R_G$	<0.2	$XR_G0.2$
	<0.5	$XR_G0.5$
	<1	XR_G1

Table 14.3 Constituents of recycled coarse aggregates (Lay, 2006)

Constituent	Description
R_C	Concrete, concrete products, mortar, concrete brick
R_U	Natural stone, recycled aggregate clean (without mortar)
R_B	Bricks, tiles, masonry units, calcium silicate, non-floating aerated concrete
R_A	Bituminous material
R_G	Glass
FL_S	Floating stone material (<1 mg/m^3)
FL_{NS}	Floating non-stone material (<1 mg/m^3)
X	Others: cohesive materials (soils and clays), metals, non-floating wood, plastic, rubber

of at least 90% by mass of fragments, based on Portland cement and stones. If its coarse fraction is composed of less than 90% by mass of fragments based on Portland cement and stones, it is named recycled mix aggregate (RMA)

14.3.2 Characteristics of recycled aggregates from CDW

To be satisfactory in use and in certain applications, the recycled material must meet certain requirements of maximum grain size, presence of contaminants and other requirements of stability and durability. Usually recycled aggregates have a more elongated, irregular and a rougher surface texture, as well as being more porous than natural aggregates, and therefore often presenting a fissured surface (Ravindrarajah and Tam, 1987b; Carneiro et al., 2001; Zaharieva et al., 2003; Tu et al., 2006).

These characteristics are reflected directly in the water absorption of recycled aggregates, often with values much higher than those of natural aggregates. This increased absorption is assigned to the mortar adhered to the natural aggregate concrete duly recycled (Ravindrarajah and Tam, 1985; Hansen, 1986; Katz, 2003; Rakshvir and Barai, 2006). However, the presence of ceramic materials in the composition of recycled aggregate can lead to a high absorption recycled aggregate. Agrela et al. (2011) obtained average values of absorption for recycled concrete aggregate (more then 90% of concrete) between 3.7% and 7.1% and for mix recycled aggregate (ceramic content <30% and concrete content <90% and >70%) between 5.3% and 7.2%. However, for ceramic recycled aggregate (ceramic content >30% and concrete content <70%) the absorption values are very superior to those obtained for the other two groups, ranging from 9.9% to 13.5%.

According to Tam et al. (2005), aggregates with high absorption rates usually lead to concrete with lower performances, thus affecting properties such as strength, durability, creep and shrinkage. As for the red ceramic recycled aggregates, it appears that the porosity thereof is directly proportional to the porosity of the ceramic artifacts created and, consequently, the mechanical strength of the same (Schulz and Hendricks, 1992).

The specifications of Rilem (1994) set a ceiling for the water absorption of recycled concrete and ceramic coarse aggregate, 20% and 10%, respectively. As for Japanese Standards, the maximum absorption of recycled concrete coarse and fine aggregate should be 7% and 13%, respectively (Hansen, 1992). According to NBR 15116, the maximum absorption for coarse and fine RCA are 7% and 12%, respectively. For aggregates of mixed waste, these rates rise to 12% and 17% for the coarse and fine aggregates, respectively.

Another striking feature of these recycled aggregates is the speed with which water is absorbed. According to Schulz and Hendricks (1992), within 30 minutes of submersion in water, recycled ceramic aggregates absorb 98% of all the water absorbed in 24 hours of immersion. It is not different for the recycled aggregate concrete: Bairagi et al. (1993) state that for this type of aggregate, absorption also occurs very quickly in the first 30 minutes of submersion, and in their experiment it absorbed 76% of all the water absorbed in 24 hours, while for 4 hours of submersion, the stated figure rises to 94%. Figure 14.4 shows the behavior of water absorption by time of recycled concrete coarse aggregate (RCCA), recycled brick coarse aggregate (RBCA), recycled mortar coarse aggregate (RMCA), recycled concrete fine aggregate (RCFA), recycled brick fine aggregate (RBFA) and recycled mortar fine aggregate (RMFA) (Cabral, 2007).

Generally, recycled aggregates have a bulk density and a specific gravity lower than those of natural aggregates. For the specific gravity, this reduction is a function of the characteristics for the raw materials of the same type, since they are less dense than the natural coarse aggregate. Regarding the bulk density, the reduced density of the material itself and the characteristic of high porosity aggregates, the irregular shape of the aggregates in particles contributes to the reduction of the same. However, these stated reductions are also dependent on the particle size of recycled aggregates and on the composition of recycled aggregate. Agrela et al. (2011) noticed an increment of 2.5% and 8.3% in a specific gravity of recycled concrete aggregate when compared with mix recycled aggregate and ceramic recycled aggregate specific gravities, respectively.

In general, some of the other characteristics of recycled aggregates, such as resistance to impact, to abrasion, to crushing and mass loss, among others, are also considerably lower than those of natural aggregates (Ravindrarajah and Tam, 1985, 1987b; Hansen, 1992; Zakaria and Cabrera, 1996; Nagataki et al., 2000; Senthamarai and Manoharan, 2005).

14.4 Characteristics of concrete with recycled aggregates

Since the performance of concrete with the replacement of natural aggregates by recycled aggregates is modified, it is necessary to understand the behavior

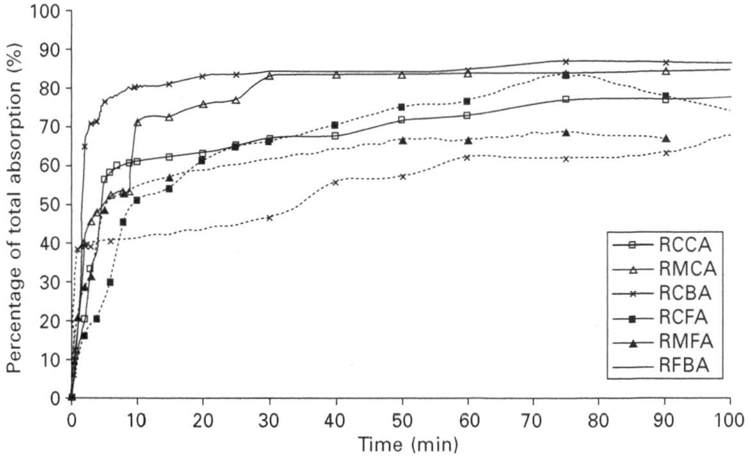

14.4 Water absorption of recycled aggregates by time (Cabral, 2007).

of concrete with respect to mechanical properties and with respect to its durability.

14.4.1 Specific gravity

As seen above, recycled aggregates generally have a specific density lower than that of natural aggregates. As a result, the concrete produced by these aggregates generally has a specific gravity lower than that of concrete produced with natural aggregates, both in fresh and hardened states. Some studies indicated that the level of entrained air in concrete with recycled aggregates is higher than in conventional concrete (Hansen, 1986; Schulz and Hendricks, 1992; Katz, 2003).

There seems to be a linear relationship between the specific gravity of the particle of recycled aggregate and of the concrete produced with this aggregate (Schulz and Hendricks, 1992; Brito and Alves, 2010). Such behavior is demonstrated in studies by Poon et al. (2002) and Bairagi et al. (1993). The fact that the concrete produced with recycled aggregates has a lower density than conventional concrete suggests that it may be used in situations where the weight of the structure is an issue beyond what the structural parts of sections can be reduced to, hence, representing considerable financial savings.

14.4.2 Workability

In general, concrete with recycled aggregates has a lower workability than concrete with natural aggregates for the same ratio of dry materials/paste. This is possibly due to the greater absorption of recycled aggregates, making the mixture drier and therefore less workable. Another reason for this behavior is due to the crushing and grinding processes, once recycled aggregates become more angular with a ratio surface/volume greater than the natural aggregates, which are more spherical and smoother on the surface. As a result, the internal friction of recycled aggregate concretes is high, therefore requiring much more paste to have the same workability of natural aggregate concretes (Hansen and Narud, 1983; Rashwan and Abourizk, 1997; Rakshvir and Barai, 2006; Corinaldesi and Moriconi, 2010).

For concrete made with recycled aggregates that have mortar, such as recycled aggregates of concrete and mortar, there is also the possibility of generation of fillers during the mixing process, which is due to the wear of the old mortar in recycled aggregate, thereby increasing the concrete's cohesion and decreasing the workability (Hansen and Narud, 1983; Hansen, 1986). However, this decrease in workability is usually solved by adding an additional amount of mixing water, ranging between 5% and 8% (Ravindrarajah and Tam, 1985), or an additive. Obviously, mechanical properties tend to decrease.

Mukai et al. (1978; cited in Hansen, 1992), are more finicky. According to these authors, concretes produced with the replacement of recycled coarse aggregate demanded about 10 l/m³ or 5% more water than concrete control, in order to achieve the same stated workability. When replacing both coarse and fine aggregates by their recycled approximately 25 l/m³ or 15% more water will be required. In order not to change the water/cement ratio, a corresponding amount of cement must also to be added to the mixing. Note, however, that the above values cited depend on the composition of the recycled aggregates and the type of crushing process: they cannot be taken as absolute.

14.4.3 Porosity, water absorption and permeability

Generally, the concrete made with recycled aggregates are characterized by a high percentage of meso- and macro-pores, hence suggesting a greater tendency to porosity with water absorption and leaching, than those prepared with natural aggregates (Sani et al., 2005). According to studies by Gómez-Soberón (2002), the distribution of pores in concretes with recycled aggregates replacing natural aggregates is modified, while being the most sensitive for high levels of replacement.

For concrete made with recycled concrete and/or mortar aggregate, this porosity increases as the level of replacement of natural aggregate by recycled aggregate rises. This gradually increases the content of paste in the concrete, consequently increasing the pore volume, since the mortar is clearly more porous than the natural aggregate (Gómez-Soberón, 2003; Etxeberria et al., 2006; Evangelista and Brito, 2010; Brito and Alves, 2010).

For concrete with recycled red ceramic aggregates, Zakaria and Cabrera (1996) comment that such concrete had 53% higher porosity than the reference concrete. Also according to these said authors, since the recycled red ceramic aggregate has a higher porosity than that of natural aggregates, then concrete made with these aggregates possibly also has a high porosity. The total porosity of these concretes is also affected by the high angularity of recycled ceramic aggregates, and this generally increases the mixture pore volume.

This increased porosity of the recycled aggregate concrete is then reversed in higher water absorption. According to Hansen (1992), it is no surprise, since the recycled aggregate concrete contains a large volume fraction with the porous recycled aggregates distributed in the matrix, while the conventional concrete has natural aggregates (less porous) distributed in the same matrix. Lovato et al. (2012) state that the concrete's water absorption is more negatively affected by the fine aggregate replacement then the coarse aggregate replacement and according to López-Gayarre et al. (2011), up to 50% of replacement of natural aggregates by recycled ones, water absorption increases up to 34%.

The permeability of recycled aggregates concrete is dependent mainly on the quality of the matrix, since a matrix with low permeability will not allow water to penetrate. However, since this matrix, for the vast majority of concrete made routinely, is not of good quality, the quality of recycled aggregate starts to be of great importance.

According Rasheeduzzafar and Khan (1984; cited in Hansen, 1992), it seems that the high water absorption of concrete with recycled aggregates can be compensated by the production of concrete with a water/cement ratio from 0.05 to 0.10 lower than that of conventional concrete. Therefore, based on the experiments above, the performance of concrete produced with recycled aggregates, the porosity, water absorption and permeability is dependent on the quality of recycled aggregate and the cement matrix of the new concrete.

14.4.4 Compressive strength

Several studies (Ravindrarajah and Tam, 1985, 1987a; Hansen, 1992; Bairagi et al., 1993; Ajdukiewicz and Kliszczewicz, 2002; Gómez-Soberón, 2002, 2003; Katz, 2003; Zaharieva et al., 2003; Topçu and Sengel, 2004; Xiao et al., 2005; Rakshvir and Barai, 2006; Tu et al. 2006; Rahal, 2007; Xiao and Falkner, 2007; Brito and Alves, 2010; Corinaldesi, 2010; Lovato et al., 2012) have shown that the compressive strength of concrete produced with recycled aggregates is usually smaller than the concrete produced with natural aggregates for the same consumption of cement. According to these authors' data, these reductions can reach up to about 45% of the reference concrete strength.

However, some authors (Hansen, 1992; Leite, 2001; Ajdukiewicz and Kliszczewicz, 2002; Khatib, 2005; Evangelista and Brito, 2007) showed increases in concrete strength up to 33%, when natural aggregate was replaced by the recycled ones. According to López-Gayarre et al. (2011), the percentage of replacement does not affect the compressive strength of concrete, being affected only by the quality of the recycled aggregates employed. This discrepancy is due to the different variables involved, such as the type of crusher used in the production of recycled aggregates, which influence the form of it and consequently the content of voids in the concrete, the type of cement used, the recycled aggregate composition, as well as the methodology used, among other factors.

When the cement matrix of concrete produced with recycled aggregates is less resistant than the recycled aggregate, this aggregate does not exert a great influence on the concrete mechanical strength, since the cement matrix will be the concrete's weakest link; therefore, possibly, the concrete will break in the matrix. Furthermore, when the concrete's cement matrix is more resistant than the recycled aggregate, this aggregate will have substantial influence on the concrete's strength as it may break in the concrete aggregate.

Generally, recycled aggregate is less resistant than natural aggregate, due to its physical characteristics, with a high porosity, high water absorption and a low bulk density (see Section 14.3.2). In the specific case for concrete made with recycled concrete aggregate, it appears that the strength of concrete that originated from recycled aggregate does not affect the compressive strength of concrete produced with the aforesaid when the water/cement ratio is high. However, as the strength of recycled aggregate concrete increases, the strength of the concrete that originated the recycled aggregate becomes more important (Hansen and Bøegh, 1985; Hansen and Narud, 1983; Otsuka and Miyazato, 2000; Kokubu et al., 2000).

Therefore, it is possible to produce concrete with recycled aggregates with the same or better strength than the original concrete, for the same water/cement ratio and identical control. For this, it is necessary to use a recycled aggregate and a cement matrix of excellent qualities. There are authors (Nagataki et al., 2000; Limbachiya et al., 2000; Shayan and Xu, 2003; Dhir et al., 2004b; Fonseca et al., 2011) that produced recycled aggregate concretes with strength higher than 50 MPa, at 28 days.

It seems that the interfacial transition zone (ITZ) formed between the recycled aggregate and the cement matrix is better for recycled aggregates concrete than for conventional concrete, due to the greater angularity and rougher texture of the recycled aggregates, collaborating an increase in the adhesion between the paste and the aggregate. Additionally, the recycled material produces a higher absorption of cement paste, causing precipitation of hydration crystals in the pores between the aggregate and the paste, thus, providing a greater closure in the ITZ (Leite, 2001).

Specifically for recycled concrete made with recycled concrete aggregate, it appears that there are two ITZs between the aggregate and the matrix, one new and one old (Fig. 14.5). Some researchers (Otsuka and Miyazato, 2000; Kokubu et al. 2000; Nagataki et al. 2000; Ryu, 2002; Masce et al., 2003) state that for concrete produced with low water/cement ratio, both ITZs have influence on the strength of new concrete. This corroborates the statement made earlier, that the quality of recycled concrete aggregate only affects the concrete's compressive strength when the recycled concrete's water/cement ratio is low, since only in this case would the former ITZ be required. When the quality of the new ITZ is better than the old, the strength of recycled aggregate concrete will depend on the quality of the old ITZ, i.e., depends on the quality of recycled aggregate.

It seems that the replacement of coarse and fine aggregates have different degrees of influence on the compressive strength in the concrete.

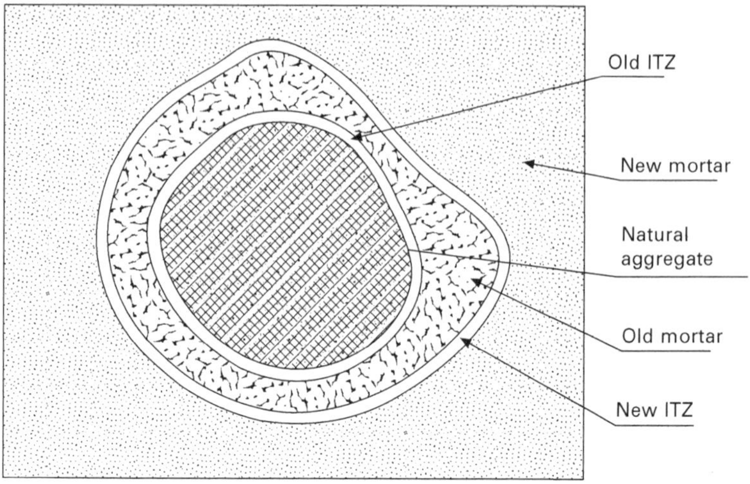

14.5 Interfacial transition zones (ITZ) of recycled aggregate concrete.

Ravindrarajah and Tam (1987a), Cabral (2007) and Lovato et al. (2012) claim that the recycled concrete coarse aggregate exerts a greater influence on the compressive strength than the recycled fine aggregate. Nevertheless, Ujike (2000) and Kokubu et al. (2000) found that fine recycled aggregates starkly affected the compressive strength of recycled aggregate concrete rather than the recycled coarse aggregate.

Replacing the coarse and fine natural aggregates with their respective recycled aggregates, it seems that the compressive strength, declining trend holds, and in some cases, this particular reduction is even enhanced (Hansen and Marga, 1989, cited in Hansen, 1992; Kokubu et al., 2000, Sani et al., 2005). However, Lovato et al. (2012) state that a 50% fine recycled aggregate and 50% coarse recycled aggregate content in concretes with water/cement ratios up to 0.60 showed satisfactory results for compressive strength.

For ceramic recycled aggregates, some studies point out an increase in the compressive strength of concrete made with these materials, due to possible pozzolanic reactions caused by the fines. However, Brito and Alves (2010) state that the concrete compressive strength falls as the replacement rate increases due to the lower mechanical characteristics of the ceramics and of the mortar adhering to the natural aggregates (in recycled concrete).

According to the recommendations of Rilem (1994), for the use of recycled coarse aggregate in concrete production, aggregates originated from masonry waste can only be used to produce concrete with maximum 16–20 MPa of strength, while recycled concrete aggregates are suitable for concretes up to 50–60 MPa. However, there are records of recycled aggregates used in structural concrete, although in such particular cases the amount of aggregate used is generally limited to a small amount, being not more than 30–40% (Poon et al., 2002; Evangelista and Brito, 2007).

Some techniques can be used to increase the recycled aggregate concrete's strength up to levels equal to or greater than the conventional concrete's strength, such as the combined use of mineral admixtures (silica fume, metakaolin, slag furnace, blast furnace or copper) and super-plasticizer additives. Another way to offset the loss of strength is simply forcing a decrease in recycled aggregate concrete water/cement ratio, which implies a higher consumption of cement.

Another technique is the method of mixing the constituents of concrete called double-mixing, where the difference to the traditional method is that water is added in two steps. The goal of this stated method is to make the recycled aggregates into contact firstly with a low water/cement ratio mortar, being surrounded by a layer of this mortar, hence, obtaining best properties in the new ITZ. As a consequence, concrete properties like compressive strength, tensile strength, depth of carbonation and chloride penetration improve (Otsuki and Miyazato, 2000).

Tsuji et al. (2000) describe a technique that consists of submerging the recycled aggregates in a colloidal silica solution for 30 minutes before making the concrete. According to these authors, the pozzolanic solution is absorbed by aggregates, and later on, fills the aggregate's micro-cracks with a gel from pozzolanic reaction, therefore improving recycled concrete performance. Using this technique, these aforementioned authors produced concretes with a compressive strength very similar to the reference concrete's strength.

The concrete's curing condition also does not influence the compressive strength of recycled aggregate concrete. Fonseca et al. (2011) tested four types of curing (laboratory curing; outer environment curing; wet chamber and water immersion curing) in natural and recycled concretes and found similar behavior for all concretes.

In general, recycled aggregate concretes usually still have enough strength to be useful, except in structural applications. However, depending on the substitution methodology of natural aggregates by recycled aggregates, the recycled aggregates composition, among other factors, it is possible to produce recycled concretes with a high strength, make them able to be used in structural applications.

14.4.5 Module of deformation

Knowledge of a concrete's module of deformation is extremely important, since it is used in structural calculations, in order to predict the maximum admissible strain, consequently, a degree of cracking in concrete pieces. For concrete made with recycled aggregates, several authors (Akhtaruzzaman and Hasnat, 1983; Hansen and Bøegh, 1985; Ravindrarajah and Tam, 1985, 1987a, 1987b; Hansen, 1992; Bairagi et al., 1993; Leite, 2001; Levy, 2001; Ajdukiewicz and Kliszczewicz, 2002; Gómez-Soberón, 2002, 2003; Dhir et al., 2004a, 2004b; Xiao et al., 2005, 2006; Rakshvir and Barai, 2006; Rahal, 2007; Evangelista and Brito, 2007; Brito and Alves, 2010; Corinaldesi, 2010; López-Gayarre et al., 2011; Fonseca et al., 2011; Lovato et al., 2012) indicate that its module of deformation usually is smaller than conventional concrete's module of deformation. According to Ujike (2000), this reduction in modulus of deformation is mostly felt, when producing recycled aggregate concrete with low water/cement ratio.

However, the shape of the stress–strain curve for recycled aggregate concrete is very similar to that of conventional concrete, regardless of the replacement of natural aggregate by recycled, which suggests that structures made with recycled aggregate concrete can be designed according to the plasticity theory, like conventional concrete structures (Xiao et al., 2005).

For concrete with recycled concrete aggregate, module of deformation reductions are usually attributed to the cement matrix that remains bound to the particles of natural aggregate in recycled aggregates after crushing (Hansen and Bøegh, 1985; Hansen 1986). Kokubu et al. (2000) say that concrete's modulus of deformation decreases with the increasing content of adhered mortar in recycled aggregates. Brito and Alves (2010) state that this occurs due to the lower mechanical characteristics, mainly the stiffness, of recycled aggregates. For Lovato et al. (2012), since the recycled aggregates, both coarse and fine, are more porous and less dense than natural aggregates, they can be said to diminish the concrete deformation module, as confirmed in the experimental results obtained by them. For these aforesaid authors, a content of 50% fine recycled aggregates and 50% coarse recycled aggregate presents satisfactory results as to the module of deformation, although Corinaldesi (2010) points to 30% as the best replacement content.

Some studies, such as Frondeurs-Yann (1977; cited in Khalaf and DeVeeny, 2004), Ravindrarajah and Tam (1985), Hansen and Bøegh, (1985), Bairagi et al. (1993), Katz (2003), Khatib (2005), Xiao et al. (2005, 2006), Evangelista and Brito (2007) and Lovato et al. (2012), showed modules of deformation of concretes made with recycled concrete aggregate are 15–45% smaller than the conventional concrete modules.

However, it seems that it is not only the recycled aggregate concrete that modifies the module of deformation. Akhtaruzzaman and Hasnat (1983) found in a concrete with recycled ceramic coarse aggregates a modulus of deformation around 30% lower than a module of conventional concrete. According to Khatib (2005), replacing only the natural fine aggregate by recycled ceramic fine aggregate, the average reduction would be 20%. Schulz and Hendricks (1992) as well as Rilem (1994) state that recycled ceramic concretes have modules of deformation between a half and two-thirds of the module of deformation of a similar strength conventional concrete.

It seems that the aggregate governs the behavior of concrete's module of deformation. Thus, as recycled aggregate is more deformable than the natural aggregate, the concrete produced with this aggregate is more deformable than the concrete produced with natural aggregates.

14.4.6 Tensile strength

It seems that the replacement of natural aggregate by the recycled aggregate also causes a reduction in the tensile strength, nevertheless, this appears to be less intense than those that cause reductions in compressive strength. Several authors (Dhir et al. 2004b; Ajdukiewicz and Kliszcczewicz, 2002; Gómez-Soberón, 2002, 2003; Topçu and Sengel, 2004; Sagoe-Crentsil et al., 2001; Evangelista and Brito, 2007; Fonseca et al., 2011; Lovato et al., 2012) state reductions in tensile strength from 6% to 30% for concrete made with the recycled aggregate concrete.

According to Hansen (1986), the loss of tensile strength in concrete with recycled aggregate concrete is less acute when only the coarse aggregates are replaced. When both aggregates (coarse and fine) are replaced, usually the loss is greater, up to 20%. Consistent with this, Ravindrarajah and Tam (1987a) observed a loss of 10% in the replacement of coarse aggregate, which increased to 15% when both aggregates were replaced. However, Lovato et al. (2012) state that coarse aggregates have a greater negative influence on this property.

The replacement of natural aggregates by the recycled red ceramic aggregates also causes changes in the tensile strength in concrete. Brito et al. (2005) state that replacing natural coarse aggregate by recycled red ceramic coarse aggregate in contens of 33%, 66% and 100% achieves a reduction in tensile strength of about 8.6%, 15.7% and 25.7%, respectively.

Therefore, it seems the reduction caused by the recycled aggregates on tensile strength is not as strong as the reduction in compressive strength (Brito and Alves, 2010). The tensile strength appears to take into account the physical mechanisms of adhesion between particles. Once recycled aggregates promote a good adhesion between the paste and the aggregate due to its most irregular and rough form, the ITZ of recycled concrete is very good (Leite, 2001). Thus, based on the good performance of ITZ of recycled concrete, the tensile strength of recycled concrete is not as affected as the compressive strength.

14.4.7 Abrasion resistance

It seems that the abrasion resistance of recycled aggregate concrete is also lower than that of conventional concrete although some authors (Evangelista and Brito, 2007; Brito and Alves, 2010) state an improvement. This reduction is attributed to reductions in physical and mechanical properties from the recycled aggregate itself, since it generally has abrasion resistance values lower than those of conventional aggregates.

The high abrasion of recycled concrete aggregate partly reflects the large amount of mortar adhered to the natural aggregate (Tavakoli and Soroushian, 1996). This behaviour also occurs to brick ceramic recycled aggregate (Brito et al., 2005).

The poor performance by recycled concrete may be avoided when preparing high compressive strength concrete. Limbachiya et al. (2000) found similar abrasion resistance of 50, 60 and 70 MPa concrete produced with 100% recycled aggregate concrete and Poon et al. (2002) found a 60 MPa recycled concrete abrasion around 12% higher than conventional concrete abrasion. This probably occurs due to the high abrasion resistance of the high-strength mortar.

Brito and Alves (2010) state the abrasion resistance of concrete with recycled aggregates tends to increase (less loss of mass) due to the greater adherence between particles provided by the more porous surface of recycled aggregates.

It may be concluded that even with a possible decrease in the abrasion resistance for recycled aggregate concrete for medium strength, it is possible to produce recycled concrete with a satisfactory abrasion resistance. For this, it is necessary to increase the concrete mortar abrasion resistance.

14.4.8 Drying shrinkage

According to Mehta and Monteiro (2008), the particle size distribution, the maximum size, and the shape and texture of the aggregates are factors that influence the concrete drying shrinkage. However, the aggregate's modulus of deformation is considered the most important factor.

Corinaldesi (2010) points out that when the finer coarse recycled aggregate fraction is used to produce concrete, low values of shrinkage are detected at an early age (7 days). Corinaldesi and Moriconi (2010) found significantly lower drying shrinkages in recycled aggregate concretes. In both cases, the authors suggest the 'curing effect' as the reason for this behavior, propitiated by the pre-soaking of water into porous of recycled aggregate particles. However, the great majority of authors, such as Ravindrarajah and Tam (1985, 1987b) Tavakoli and Soroushian (1996), Sagoe-Crentsil et al. (2001), Poon et al. (2002), Ajdukiewicz and Kliszczewicz (2002), Shayan and Xu (2003) and Dhir et al. (2004b), had drying shrinkage 12% to 61% higher in recycled aggregate concrete than conventional concrete. Besides, there are studies that show increases up to 100% (Ravindrarajah and Tam, 1987a; Katz, 2003).

Therefore, when much of the natural aggregates are replaced by recycled aggregates, the concrete has a higher drying shrinkage, since recycled aggregates have a lower modulus of deformation than natural aggregates, being more deformable. They also have high water absorption, requiring higher water content that leads to increase the drying shrinkage (Poon et al., 2002). Brito and Alves (2010) blame the higher shrinkage of hardened concrete due to the higher porosity and lower stiffness of the recycled aggregates.

Recycled concrete aggregate generally consists of 60–70% of its volume by natural aggregate and 30–40% by mortar, being the latter much more porous than the previous one (Hansen and Narud, 1983). This high content of mortar causes the appearance of some undesirable effects on the recycled concrete, such as an increase of drying shrinkage, thereby enhancing the appearance of cracks. According to Tavakoli and Soroushian (1996), the greater the amount of adhered mortar in recycled concrete aggregate, the greater the possibility of having high drying shrinkage in recycled concrete.

Therefore, it is common to try to release the mortar from the recycled concrete aggregate. However, the procedure that is often used to accomplish this is to put recycled aggregates in a concrete mixer and let them mix alone. The friction between the concrete aggregates and between the recycled concrete aggregate's mortar and the mixer walls makes the mortar come off (Tavakoli and Soroushian, 1996). According Hansen (1992), this process also improves the aggregate's shape, contributing to a better mix.

Drying shrinkage also increases as the content of natural aggregate are replaced by recycled aggregate (Limbachiya et al., 2000; Gómez-Soberón, 2002, 2003; Khatib, 2005). It is also dependent on the quality of the concrete from which the recycled aggregates came. The greater the resistance of the original concrete, the greater the shrinkage of concrete made with their recycled aggregates. This is probably occurs due to the high content of mortar in higher-strength concretes (Ravindrarajah and Tam, 1985).

It is generally accepted that recycled concrete made with the recycled concrete coarse aggregate performs better on the drying shrinkage than concrete made with recycled ceramic coarse aggregate (Khalaf and DeVeeny, 2004). The biggest drying shrinkage presented by recycled ceramic concrete can be explained due to the lower resistance offered by these recycled aggregates to deformations of the concrete's cement paste, once these aggregates have a lower modulus of deformation than recycled concrete aggregates (Schulz and Hendricks, 1992).

One may associate the drying shrinkage of recycled aggregate concrete with the type of recycled aggregate, since the more porous the recycled aggregate is, the less restricted is the shrinkage, thus allowing the concrete to retract further. Hence, concrete produced with recycled aggregates containing a high content of cement paste, will possibly suffer major drying shrinkages.

14.4.9 Resistance to fire

When concrete is subjected to high temperatures such as fires, major changes occur in its components, leading to reductions in the compressive strength and modulus of deformation. The above modifications come from the loss of free water and water gel, changes in the structure of hydrated cement as well as great aggregate expansions, giving rise to internal tensions that may even disrupt the concrete (Cánovas, 1998).

Surveys show that recycled ceramic aggregate concretes have a superior performance to conventional concretes in relation to the loss of the compressive strength after both being exposed to high temperatures (Khoury, 1996, and Newman, 1946, both cited in Khalaf and DeVenny, 2004; Schulz and Hendricks, 1992). This may be explained because the recycled ceramic aggregate is thermostable, in contrast to natural aggregates, such as limestone, which are not. It occurs due to the ceramic's raw material characteristics: a fairly high ability to retain heat, not spreading it, non-flammable and having refractory properties, which means that it maintains its integrity and resistance to high temperatures, in some cases approaching 1000 °C. As a consequence of this behavior, recycled ceramic aggregate concretes have more protection against heat, which means that they can maintain their structural integrity against fire for a much longer period than conventional concrete (Khalaf and DeVenny, 2004). Therefore, recycled aggregate concrete that contain a certain amount of recycled ceramic aggregate possibly will have a better performance than conventional concrete when submitted to high temperatures.

14.4.10 Depth of carbonation and chloride penetration

Depth of carbonation and chloride penetration are two of the most vulnerable aspects of introducing recycled aggregates into concrete production and are explained, as for the other durability-related characteristics (e.g. water absorption), by the greater porosity of recycled aggregates compared with that of natural aggregates (Brito and Alves, 2010). Tests conducted by Katz (2003) and Evangelista and Brito (2010) show that the depth of carbonation in recycled aggregate concrete is 1.1 to 2.5 times the reference concrete carbonation depth. Figures 14.6 and 14.7 compare the carbonation depth in a reference concrete and a recycled aggregate concrete, respectively, according to Lovato et al. (2012).

Ryu (2002) and Tu et al. (2006), investigating the penetration depth of chloride ions in recycled concrete aggregate concrete, found that the penetration depth of these concretes was higher than for concrete with natural

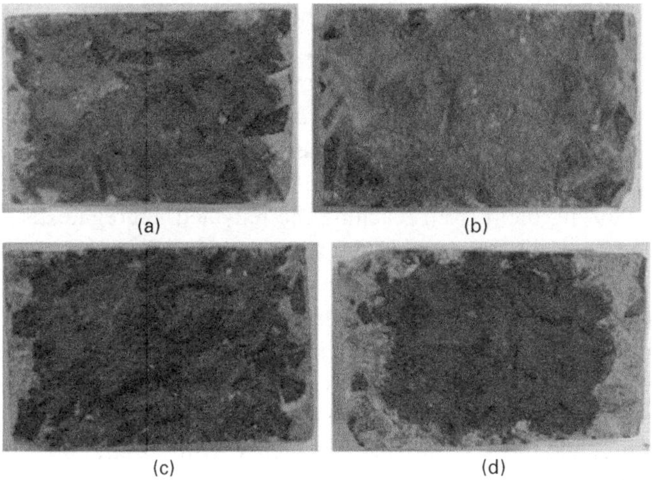

14.6 Evolution of carbonation in a reference concrete: (a) 7 days; (b) 14 days; (c) 28 days; (d) 56 days (Lovato et al., 2012).

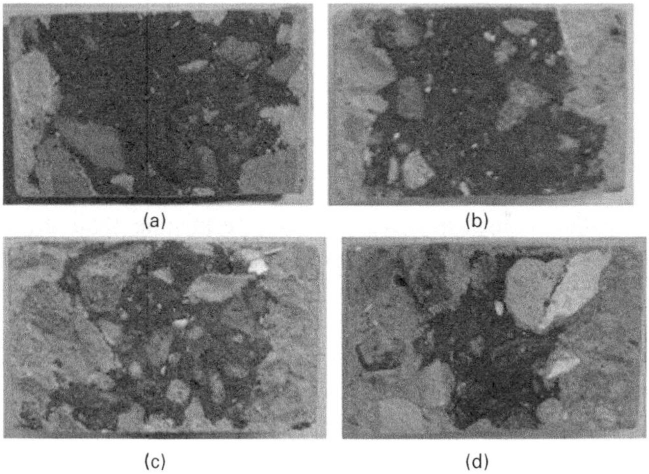

14.7 Evolution of carbonation in a recycled aggregate concrete: (a) 7 days; (b) 14 days; (c) 28 days; (d) 56 days (Lovato et al., 2012).

aggregates, especially for a great water/cement ratio. In experiments similar to those of the previous authors, Masce et al. (2003), Otsuki and Miyazato (2000) and Evangelista and Brito (2010) reached the same conclusions.

However, according to these aforesaid authors, this occurs due to the former ITZ and adhered mortar in the recycled aggregate concrete that makes recycled concretes more permeable than conventional concretes. Once the quality of the new transition zone is improved, carbonation depth and chloride penetration decrease.

According to Olorunsogo and Padayachee (2002), the reduction in durability and performance presented by recycled concrete occurs due to the cracks in the aggregates created during the recycling process, which becomes an easy way for the passage of fluids and aggressive components besides the presence of a more porous structure. Nevertheless, it is known that if recycled concretes have a good new mortar this will prevent the passage of deleterious agents, preventing them from reaching the recycled aggregates.

If we compare concrete within the same range of compressive strength, it appears that recycled concretes are more resistant to the penetration of chlorides and CO_2. This behavior can be explained by the reduction of water/cement ratio by the recycled concrete in order to achieve the same strength of conventional concrete. This particular reduction causes a substantial improvement in the cement matrix, making it less permeable, thereby impeding the penetration of aggressive agents (Dhir et al., 2004a).

Levy (2001) concluded that the carbonation and chloride penetration in recycled aggregate concrete are much more connected to the water/cement ratio and cement content, than to the type of recycled aggregate duly used. Consistent with this, Otsuki and Miyazato (2000) state that these two properties are more related to the quality of the ITZ in concrete, regardless of whether the aggregate used is recycled or not.

Therefore, comparing low resistance and the same water/cement ratio conventional concrete with recycled aggregate concrete, the recycled concrete would be expected to have a worse performance, since both the matrix and the recycled aggregate will be porous, allowing the penetration of aggressive agents, while the natural aggregate is less permeable than the recycled, thus hindering the penetration of aggressive agents. On the flip side, when comparing conventional concrete with recycled aggregate concrete in high strength with the same water/cement ratio, we expect the behavior of both to be similar, since the matrix will be so little permeable that it will not allow the penetration of aggressive agents.

14.5 Future trends

Further investigations about recycled aggregate concrete are being made worldwide, researching about:

- self-compacted concrete with recycled aggregates – the content of ultra-fines (<0.075 mm) produced in the CDW recycling is high and it can be used in SCC;
- structural lightweight recycled aggregate concrete – recycled concrete usually has an specific gravity lower than conventional concrete but it is laborious to reach structural strength in lightweight concretes;
- reinforced fiber recycled aggregate concrete – the use of fibers in

recycled aggregate concrete can minimize undesired effects such as cracks by drying shrinkage and also improve tensile and flexural strength;
- high performance recycled aggregate concrete – structural concrete is the most worldwide construction material used and high consumption of recycled aggregate in it is still a challenge.

14.6 References and further reading

Agrela, F.; Sánchez de Juan, M.; Ayuso, J.; Geraldes, V. L.; Jiménez, J. R. Limiting properties in the characterization of mixed recycled aggregates for use in the manufacture of concrete. *Construction and Building Materials*, Vol. 25, p. 3950–3955, 2011.

Ajdukiewicz, A.; Kliszczewicz, A. Influence of recycled aggregates on mechanical properties of HS/HPC. *Cement and Concrete Composites*, Vol. 24, p. 269–279, 2002.

Akhtaruzzaman, A. A.; Hasnat, A. Properties of concrete using crushed brick as aggregate. *Concrete International*, Febuary, p. 58–63, 1983.

Bairagi, N. K.; Ravande, K.; Pareek, V. K. Behaviour of concrete with different proportions of natural and recycled aggregates. *Resources, Conservation and Recycling*, Vol. 9, p. 109–126, 1993.

Bakoss, S. L.; Ravindrarajah, R. S. Recycled construction and demolition materials for use in roadworks and other local government activities. *Scoping Report*. Sydney, 1999. 72 p. Centre for Built Infrastructure Research. University of Technology, Sydney.

Brito, J. de; Alves, F. Concrete with recycled aggregates: the Portuguese experimental research. *Materials and Structures*, Vol. 43 p. 35–51, 2010.

Brito, J. de; Pereira, A. S.; Correia, J. R. Mechanical behavior of non-structural concrete made with recycled aggregates. *Cement & Concrete Composites*, Vol. 27, No. 4, p. 429–433, 2005.

Cabral, A. .E. B. Mechanical properties and durability modeling of recycled aggregates concrete, considering the construction and demolition waste variability. São Carlos-SP, 2007. 280p. *Thesis (Doctoral)*. São Carlos School of Engineering, University of São Paulo (in Portuguese).

Cánovas, M. F. *Pathology and therapy of reinforced concrete*. 399p. São Paulo: PINI, 1998.

Carneiro, A. P.; Quadros, B. E. C., Oliveira, A. M. V.; Brum, I. A. S., Sampaio, T. S., Alberto, E. P. V.; Costa, D. B. Features of the rubble and recycled aggregate. In: *Recycling of rubble for the production of building materials*, organized by Pires, A. C.; Brum, I. S. A. and Cassa, J. S. C. Salvador: EDUFBA, p. 312, 2001, p. 144–187.

Cochran, K. M., Townsend, T. G. Estimating construction and demolition debris generation using a materials flow analysis approach. *Waste Management*, 30, vol. p. 2247–2254, 2010.

Corinaldesi, V. Mechanical and elastic behaviour of concretes made of recycled-concrete coarse aggregates. *Construction and Building Materials*, Vol. 24, p. 1616–1620, 2010.

Corinaldesi, V., Moriconi, G. Recycling of rubble from building demolition for low-shrinkage concretes. *Waste Management*, vol. 30, p. 655–659, 2010.

Dhir, R.; Paine, K.; Dyer, T. Recycling construction and demolition wastes in concrete. *Concrete*, March, p. 25–28, 2004a.

Dhir, R.; Paine, K.; Dyer, T. Tang, A. Value-added recycling of domestic, industrial and construction arisings as concrete aggregate. *Concrete Engineering International*, Spring, p. 43–48, 2004b.

Esin, T.; Cosgun, N. A study conducted to reduce construction waste generation in Turkey. *Building and Environment*, Vol. 42, p. 1667–1674, 2007.

Etxeberria, M.; Vázquez, E.; Marí, A. Microestructure analysis of hardened recycled aggregate concrete. *Magazine of Concrete Research*, Vol. 58, p. 683–690, 2006.

European Thematic Network on Recycling in Construction (ETNRC). *An EC report on construction and demolition waste*. Combined Vol. 1, No. 1/2, March/September, p. 9. 1999.

Evangelista, L.; Brito, J. Mechanical behaviour of concrete made with fine recycled concrete aggregates. *Cement & Concrete Composites*, No. 29, p. 397–401, 2007.

Evangelista, L.; Brito, J. Durability performance of concrete made with fine recycled concrete aggregates. *Cement & Concrete Composites*, No. 32, p. 9–14, 2010.

Fonseca, F.; de Brito, J.; Evangelista, L. The influence of curing conditions on the mechanical performance of concrete made with recycled concrete waste. *Cement & Concrete Composites,* Vol. 33, p. 637–643, 2011.

Gómez-Soberón, J. M. V. Porosity of recycled concrete with substitution of recycled concrete aggregate: an experimental study. *Cement and Concrete Research*, Vol. 32, p. 1301–1311, 2002.

Gómez Soberón, J. M. V. Relationship between gas absorption and the shrinkage and creep of recycled aggregate concrete. *Cement, Concrete and Aggregates*, Vol. 25, No. 2, p. 42–48, 2003.

Hansen, T. C. Recycled aggregates and recycled aggregate concrete: second state-of-art report developments 1945–1985. *Matériaux et Constructions*, Vol. 19, No. 111, 1986.

Hansen, T. C. Recycled aggregates and recycled aggregate concrete: third state-of-the-art report 1945–1989. In: *Recycling of Demolished Concrete and Masonry*, RILEM Technical Committee Report N. 6, Editor: T. C. Hansen, E & FN Spon, London, p. 1–163, 1992.

Hansen, T. C.; Bøegh, E. Elasticity and drying shrinkage of recycled-aggregate concrete. *ACI Journal*, Vol. 82, No. 5, p. 648–652, 1985.

Hansen, T. C.; Narud, H. Strenght of recycled concrete made from crushed concrete coarse aggregate. *Concrete International*, Vol. 5, No. 1, 1983.

Hendriks, Ch. F. e Janssen, G. M. T. Application of construction and demolition waste. *Heron*, Vol. 46, No. 2, p. 95–108, 2001.

Henrichsen, A. Use of recycled aggregate. In: International Workshop on Recycled Aggregate. *Proceedings*. Niigata, Japan. p. 1–8, 2000.

John, V. M. Use of waste as construction materials. In: *Recycling of rubble for the production of building materials*. Salvador: EDUFBA; 312 p.; p. 27–45, 2001 (in Portuguese).

Katz, A. Properties of concrete made with recycled aggregate from partially hydrated old concrete. *Cement and Concrete Research*, Vol. 33, p. 703–711, 2003.

Kawano, H. Outline of JIS/TR on recycled concrete using recycled aggregate. International Workshop on Recycled Aggregate. *Proceedings*. Niigata, Japan. p. 43–48, 2000.

Khalaf, F. M.; DeVenny, A. Recycling of demolished masonry rublle as coarse aggregate in concrete: review. *Journal of Materials in Civil Engineering*, Vol. 16, No. 04, p. 331–340, 2004.

Khatib, J. M. Properties of concrete incorporating fine recycled aggregate. *Cement and Concrete Research*, Vol. 35, p. 763–769, 2005.

Kokubu, K.; Shimizu, T.; Ueno, A. Effects of recycled aggregate qualities on the mechanical properties of concrete. International Workshop on Recycled Aggregate. *Proceedings*. Niigata, Japan. pp. 107–115, 2000.

Lamond, J. F.; Campbell, R. L.; Campbell, T. R., Cazares, J. A.; Giraldi, A.; Halczak, W.; Hale, H. C.; Jenkins, N. J.; Miller, R.; Seabrook, P. T. Removal and reuse of hardened concrete. *ACI Materials Journal*, May–June, p. 300–325, 2002.

Lay, J. European standardization of recycled aggregates. *Concrete Engineering International*. Autumn, p. 62–63, 2006.

Leite, M. B. Evaluation of mechanical properties of concrete produced with recycled aggregates from construction and demolition waste. Porto Alegre-RS, 2001. 270 p. *Thesis (Ph.D.)*. Post-Graduate Program in Civil Engineering, Federal University of Rio Grande do Sul (in Portuguese).

Levy, S. M. Contribution to the study of the durability of concrete produced with waste concrete and masonry. Sao Paulo-SP, 194 p 2001. *Thesis (Ph.D.)*. Polytechnic University of Sao Paulo (in Portuguese).

Limbachiya, M. C.; Leelawat, T.; Dhir, R. K. Use of recycled concrete aggregate in high-strength concrete. *Materials and Structures*, Vol. 33, November, p. 574–580, 2000.

López-Gayarre, F., López-Colina, C., Serrano-López, M. A., García Taengua, E., López Martínez, A. Assessment of properties of recycled concrete by means of a highly fractioned factorial design of experiment. *Construction and Building Materials*, Vol. 25, p. 3802–3809, 2011.

Lovato, P.; Possan, E.; Dal Molin, D.; Masuero, A.; Ribeiro, J. Modeling of mechanical properties and durability of recycled aggregate concretes. *Construction and Building Materials*, Vol. 26, pg. 437–447, 2012.

Marques Neto, J. C. M. *Management of construction and demolition waste in Brazil*. São Carlos: Rima, 162 p. 2005. (in Portuguese).

Masce, N. O.; Miyazato, S.; Yodsudjai, W. Influence of recycled aggregate on interfacial transition zone, strength, chloride penetration and carbonation of concrete. *Journal of Materials in Civil Enginnering*, Vol. 15, No. 5, p. 443–451, 2003.

Mehta P. K., Monteiro, P. J. M. *Concrete: structure, properties and materials*. São Paulo, Ed. PINI, 2008.

MMA Ministry of Environment, Water Resources and Legal Amazon. Available at http://www.mma.gov.br. Accessed in 19 January 2007 (in Portuguese).

Nagataki, S.; Iida, K.; Saeki, T.; Hisada, M. Properties of recycled aggregate and recycled aggregate concrete. International Workshop on Recycled Aggregate. *Proceedings*. Niigata, Japan. p. 53–68, 2000.

Oliveira, M. J. E. Materials discarded by construction work: studies of concrete waste for recycling. Rio Claro, SP, 191 p 2002. *Thesis (Ph.D.)*. Institute of Geosciences and Exact Sciences, Paulista University (in Portuguese).

Olorunsogo, F. T.; Padayachee, N. Performance of recycled aggregate concrete monitored by durability indexes. *Cement and Concrete Research*, Vol. 32, p. 179–185, 2002.

Otsuki, N.; Miyazato, S. The influence of recycled aggregate on ITZ, permeability and strength of concrete. International Workshop on Recycled Aggregate. *Proceedings*. Niigata, Japan. pp. 77–93, 2000.

Poon, C. S.; Ann, T. W. Yu; Ng L. H. On site sorting of construction and demolition waste in Hong Kong. *Resources, Conservation and Recycling*, Vol. 32, p. 157–172, 2001.

Poon, C. S.; Kou, S. C.; Lam, L. Use of recycled aggregates in molded concrete bricks and blocks. *Construction and Building Materials*, Vol. 16, p. 281–289, 2002.

Rahal, K. Mechanical properties of concrete with recycled coarse aggregate. *Building and Environmental*, Vol. 42, p. 407–415, 2007.

Rakshvir, M.; Barai, S. V. Studies on recycled aggregates-based concrete. *Waste Management and Research*, Vol. 24, p. 225–233, 2006.

Rashwan, M. S.; Abourizk, S. The properties of recycled concrete: factors affecting strength and workability. *Concrete International*, Vol. 19, No. 07, 1997.

Ravindrarajah, S. R.; Tam, C. T. Properties of concrete made with crushed concrete as coarse aggregate. *Magazine of Concrete Research*, Vol. 37, No. 130, 1985.

Ravindrarajah, S. R.; Tam, C. T. Recycled concrete as fine and coarse aggregates in concrete. *Magazine of Concrete Research*, Vol. 39, No. 141, 1987a.

Ravindrarajah, S. R.; Tam, C. T. Recycling concrete as fine aggregates in concrete. *The International Journal of Cement Composites and Lightweight Concrete*, Vol. 9, No. 4, 1987b.

Rilem Recommendation. Specifications for concrete with recycled aggregates. 121- DRG guidance for demolition and reuse of concrete and masonry. *Materials and Structures*, Vol. 27, p. 557–559, 1994.

Ryu, J. S. Improvement on strength and impermeability of recycled concrete made from crushed concrete coarse aggregate. *Journal of Materials Science Letters*, Vol. 21, p. 1565–1567, 2002.

Sáez, P. V.; Merino, M. R.; Amores, C. P.; González, A. S. A. European legislation and implementation measures in the management construction and demolition waste. *The Open Construction and Building Technology Journal*, No. 5, p. 156–161, 2011.

Sagoe-Crentsil, K. K.; Brown, T.; Taylor, A. H. Performance of concrete made with commercially produced coarse recycled concrete aggregate. *Cement and Concrete Research*, Vol. 31, p. 707–712, 2001.

Sani, D.; Moriconi, G.; Fava, G. Corinaldesi, V. Leaching and mechanical behavior of concrete manufactured with recycled aggregates. *Waste Management*, Vol. 25, p. 177–182, 2005.

Schulz, R. R.; Hendricks, Ch. F. Recycling of masonry rubble. In: *Recycling of Demolished Concrete and Masonry*, RILEM Technical Committee Report N. 6, Editor: T. C. Hansen, E & FN Spon, London, p. 164–255, 1992.

Senthamarai, R. M.; Manoharan, P. D. Concrete with ceramic waste aggregate. *Cement and Concrete Composites*. Vol. 27, p. 910–913, 2005.

Shayan A.; Xu, A. Performance and properties of structural concrete made with recycled concrete aggregate. *ACI Materials Journal*, Vol. 100, No. 5, p. 371–380, 2003.

Sim, J.; Park, C. Compressive strength and resistance to chloride ion penetration and carbonation of recycled aggregate concrete with varying amount of fly ash and fine recycled aggregate. *Waste Manag*ement, Vol. 31, No. 11, p. 2352-2360, 2011.

Solis-Guzman, J.; Marrero, M.; Montes-Delgado M.; Ramirez-De-Arellano, A. A Spanish model for quantification and management of construction waste. *Waste Management*, Vol. 29, p. 2542–2548, 2009.

Sonigo, P.; Hestin, M.; Mimid, S. Management of construction and demolition waste in Europe. Stakeholders Workshop, Brussels, 2010.

Tam, V. W. Y.; Tam, C. M. A review on the viable technology for construction waste recycling. *Resources, Conservation and Recycling*, Vol. 47, p. 209–221, 2006.

Tam, V. W. Y.; Gao, X. F.; Tam, C. M. Microstructural analysis of recycled aggregate concrete produced from two-stage mixing approach. *Cement and Concrete Research*, Vol. 35, p. 1195–1203, 2005.

Tavakoli, M.; Soroushian, P. Drying shrinkage behavior of recycled aggregate concrete. *Concrete International*, Vol. 18, No. 11, p. 58–61, 1996.

Terry, M. Waste minimization in the construction and demolition industry. Sydney, 78p,

2004. Capstone Project (*Thesis of Bachelor in Civil & Environmental Engineering*) – Faculty of Engineering, University of Technology, Sydney.

Topçu, I. B.; Sengel, S. Properties of concretes produced with waste concrete aggregate. *Cement and Concrete Research*, Vol. 34, p. 1307–1312, 2004.

Tsuji, M.; Sawamoto, T.; Kimachi, Y. Technical method to improve properties of recycled aggregate concrete. International Workshop on Recycled Aggregate. *Proceedings*. Niigata, Japan. pp. 105–176, 2000.

Tu, T.; Chen Y.; Hwang, C. Properties of HPC with recycled aggregates. *Cement and Concrete Research*, Vol. 36, p. 943–950, 2006.

Ujike, I. Air and water permeability of concrete with recycled aggregate. International Workshop on Recycled Aggregate. *Proceedings*. Niigata, Japan. p. 95–106, 2000.

Xiao, J.; Falkner, H. Bond behavior between recycled aggregate concrete and stell rebars. *Construction and Building Materials*, Vol. 21, p. 395–401, 2007.

Xiao, J.; Li, J.; Zhang, Ch. Mechanical properties of recycled aggregate concrete under uniaxial loading. *Cement and Concrete Research*, Vol. 35, p. 1187–1194, 2005.

Xiao, J.; Sun, Y.; Falkner, H. Seismic performance of frame structures with recycled aggregate concrete. *Engineering Structures*, Vol. 28, p. 1–8, 2006.

Zaharieva, R.; Buyle-Bodin, F.; Skoczylas, F. Wirquin, E. Assessment of the surface permeation properties of recycled aggregate concrete. *Cement and Concrete Composites*, Vol. 25, p. 223–232, 2003.

Zakaria, M.; Cabrera, J. G. Performance and durability of concrete made with demolition waste and artificial fly ash-clay aggregates. *Waste Management*, Vol. 16, No. 1–3, p. 151–158, 1996.

15
An eco-efficient approach to concrete carbonation

F. PACHECO-TORGAL, University of Minho, Portugal, S. MIRALDO and J. A. LABRINCHA, University of Aveiro, Portugal and J. DE BRITO, Technical University of Lisbon, Portugal

DOI: 10.1533/9780857098993.3.368

Abstract: Carbonation is a major cause of concrete structure deterioration, leading to expensive maintenance and conservation operations. The eco-efficient construction agenda favors the increase of the use of supplementary cementitious materials (SCMs) to reduce Portland cement's consumption and also the use of recycled aggregates concrete (RAC) in order to reduce the consumption of primary aggregates and to avoid landfill disposal of concrete waste. A wide range of literature has been published in the field of concrete carbonation related to the use of SCMs and/or RCA. However, the different conditions used limit comparison, and in some cases contradictory findings are noted. Besides, since most investigations are based on the use of the phenolphthalein indicator, which provides a poor estimate of the real concrete carbonation depth, there is a high probability that past research could have underestimated the corrosion potential associated with concrete carbonation. This chapter reviews current knowledge on concrete carbonation addressing carbonation depth measurement and the use of SCMs and RAC.

Key words: concrete carbonation, eco-efficient construction, supplementary cementitious materials (SCMs), recycled aggregates concrete (RAC)

15.1 Introduction

Carbonation is a major cause of concrete structure deterioration. Concrete carbonation is a process by which atmospheric carbon dioxide reacts with the cement hydration products to form calcium carbonate. The importance of this phenomenon is related to the fact that it reduces the alkalinity of the concrete to a pH near 8. Since the steel passivation layer, an iron oxide layer that protects the steel from corrosion, needs a pH between 12 and 14 (Hobbs, 1998, suggested that 9.5 is the pH threshold value for depassivation), the carbonation phenomenon can be responsible for the steel depassivation thus leading to corrosion. After entering concrete, CO_2 will first react with calcium hydroxide available in the pore solution and then with calcium silicate hydrates (CSH) after the calcium hydroxide has been depleted (Peter et al., 2008). The carbonation rate is controlled by the ingress of CO_2 into the concrete pore system by diffusion which in turn is influenced by the relative humidity of concrete. The diffusion of CO_2 is actually 10^4 higher in air than in water (Younsi et al., 2011). For a low relative humidity (RH less than 50%) the diffusion of CO_2 into concrete is high but there is not enough water in the pores to generate carbonation. For a high HR the diffusion of CO_2 is very low also, reducing the carbonation rate (Papadakis et al., 1991, 1992). That is why the majority of the research on concrete carbonation use RH values between 50% and 70%.

Previous investigations have shown that concrete carbonation is influenced by several parameters. For instance in a concrete with a water binder (W/B) ratio = 0.6 a carbonation depth of 15 mm can be achieved after 15 years, but if the concrete has a lower W/B = 0.45, the same carbonation depth will take 100 years to reach (Wiering, 1984). Wasserman et al. (2009) found that for a given W/B the carbonation was independent of binder content (160–200 kg/m^3), and this was explained in terms of two competing processes, of reduction in penetration and reduction in CO_2 binding at lower cement contents. Other authors tried to correlate this mechanism with different concrete properties. Tam et al. (2008) cited an extensive survey by Brown (1991) which found that carbonation depths correlated well with concrete quality. She argues that the factors that increase concrete permeability can increase the carbonation rate. Atis (2003) reported the existence of a strong correlation (R^2 = 0.9) between carbonation depth and compressive strength for fly-ash (FA) concrete. Muntean and Böhm (2009) stated that carbonation is strongly dependent on the degree of porosity, which is 'the path for carbon dioxide and water to transport in concrete'. However, this is in contradiction with the findings of Schutter and Audenaert (2004) who found no correlation between carbonation rate and porosity.

Roziere et al. (2009) studied the possible correlations between porosity, chloride diffusivity, gas permeability and carbonation rate, confirming there is no correlation between them. Those authors mention that porosity, chloride diffusivity and gas permeability deal with properties of the porous net of concrete, but they do not take into account chemical reactivity of binder and carbonatable content. The same authors confirmed the findings of Assie et al. (2007) about a strong correlation between carbonation rate and chloride diffusivity, for concrete mixes with the same initial CaO content. These findings highlight the need of review efforts that try to address the gaps and the contradictions already detected in this field, so they may help to focus future investigations.

15.2 Carbonation evaluation

15.2.1 Preconditioning of concrete specimens

Table 15.1 shows that different researchers used different preconditioning factors in their tests. The CO_2 concentration seems to be the most concerning parameter. Very few authors used outdoor atmosphere exposure and very few used the so-called normal CO_2 concentration (0.03%), while the majority of the authors used high and very high CO_2 concentration. Dhir et al. (2007) mentioned that a comparison between a 0.035% CO_2 concentration and accelerated test (4% concentration) gave good agreement thus suggesting that the accelerated test method can provide an indication of likely long-term concrete carbonation resistance. Limbachiya et al. (2012) used accelerated carbonation tests (3.5% CO_2), stating that '1 week-time exposure of concrete specimen in the carbonation chamber is somewhat equivalent to 12 months exposition under natural environment'. These statements seem to forget that non-uniform CO_2 concentration exists and that it is not stable over time.

According to Tam et al. (2008) small CO_2 concentrations are associated with rural air where CO_2 content is about 0.03%. In a non-ventilated laboratory, the concentration of CO_2 may rise to above 0.1%. In large cities, it is about 0.3% and in some exceptional cases, it can increase to 1%. Conciatori et al. (2010) refer to the following carbon dioxide concentrations, land (0.015%), town center (0.036%) and industrial area (0.045%). However, Yoon et al. (2007) stated that CO_2 concentration in the atmosphere is increasing by 0.5% per year, and that the conditions over metropolitan areas are even more critical. Furthermore, it remains to be investigated how high CO_2 concentrations influence microstructure of the cement hydration products.

15.2.2 Carbonation depth assessment

The majority of research works on concrete carbonation use a phenolphthalein indicator to assess carbonation depth. This involves spraying concrete broken faces after flexural strength tests with 1% phenolphthalein in 70% ethyl alcohol (Rilem, 1988). When the pH of the pore solution is less than 7.5, the degree of carbonation of the specimen is 100%. When the pH value of the pore solution is between 7.5 and 9.0, the degree of carbonation is 50–100%. When the pH of the pore solution is 9.0–11.5, the degree of carbonation is 0–50%. However, when the pH of the pore solution exceeds 11.5, the specimen is not carbonated (Siddique, 2011). In the uncarbonated part of the specimen where the concrete is still highly alkaline, a purple-red colour is obtained. In the carbonated part where the alkalinity of concrete is reduced, no coloration occurs. The average depth of the colourless phenolphthalein

Table 15.1 Preconditioning of concrete specimens

Reference	Initial curing conditions	Carbonation chamber		
		Temp. (°C)	RH (%)	CO_2 concentration (%)
Papadakis (2000)	1 month laboratory conditions	25	61	3
Buyle-Bodin and Hadjieva-Zaharieva (2002)		20	65	0.03
Otsuki et al. (2003)		40	70	10
Atis (2003)		20	65	5
Katz (2003)		30	60	5
Bai et al. (2002)		20	65	4
Chang and Chen (2006)	28 days water cured	23	70	20
Khunthongkeaw et al. (2006)	28 days water cured	40	55	4
Villain et al. (2007)	3 month water cured	20	53	45
Dinakar et al. (2007)	28 days water cured	exposed to the outdoor atmosphere during 12 months		
Gonen and Yazicioglu (2007)	28 days in a moisture chamber		55	40
Dhir et al. (2007)		20	65	0.035
		20	65	4
Kulakowski et al. (2009)	28 days in a moisture chamber			≥50
Wassermann et al. (2009)		30	50	5
Abbas et al. (2009)	28-day curing period	23	60	3
Roziere et al. (2009)	28 days water cured	20	65	50
Shi et al. (2009)	90 days curing	20	70	20
Chatveera and Lertwattanaruk (2011)	28-day curing period	23	70	0.03
Sim and Park (2011)	100% RH during 7 days plus 14 days in laboratory conditions	30	60	10
Zega and Di Maio (2011)		exposed to an urban industrial natural environment		
Werle et al. (2011)	cured in water for 63 days		70	1
Limbachiya et al. (2012)	28 days water curing plus 14 days cured in dry air	20	65	3
Lovato et al. (2012)	63 days plus 16 days at 25 °C and of 65% RH	21	65	6
Zhao et al. (2012)	26 days curing	20	70	20

region has been measured in three points, perpendicular to the two edges of the split face, immediately after spraying the indicator and 24 h later.

Bouikni et al. (2009) used a more time-consuming procedure based on 40 measurements on the eight faces of each broken prism, and with three prisms, the reported values of the depth of carbonation being the mean of 120 readings. In the meantime new methods have been developed to assess the carbonation depth in a more accurate way. Lo and Lee (2002) mention that the carbonation rate constant determined by infrared (IR) spectrum analysis was 23.9% higher than that obtained by using the phenolphthalein indicator. Other authors (Chang and Chen, 2006) compared the carbonation depths determined from the thermo-gravimetric analysis (TGA), X-ray diffraction analysis (XRDA) and Fourier transform infrared spectroscopy (FTIR) methods, with the results obtained using the phenolphthalein indicator and found that the TGA, XRDA and FTIR results showed the depth of carbonation front was on average twice that determined with the phenolphthalein indicator (Table 15.2).

Villain et al. (2007) compared thermo-gravimetry, chemical analysis (CA) and gamma-densimetry to assess the carbonation profiles. They mention that TGA has to be supplemented with CA to give accurate quantitative profiles. Chemical analysis allows obtaining the cement content in a part of the powder sample taken in concrete specimens and tested also by TGA. TGA-CA can be used to determine carbonation profile either in structure cores or in laboratory carbonated specimens. Gamma-densimetry cannot easily and accurately quantify the CO_2 content in a core of an aged concrete structure. Gamma-densimetry is recommended to monitor laboratory accelerated tests. Tam et al. (2008) used FTIR spectroscopy to compare the carbonation depth among the samples. Five layers of cement paste around an aggregate were studied. The content of carbon monoxide (CO) and carbon dioxide (CO_2) in each layer were examined.

Muntean et al. (2011) proposed a moving-interface model to forecast the maximum penetration depth of gaseous CO_2 in the porous concrete matrix. Bouchaala et al. (2011) used nonlinear resonant ultrasound spectroscopy (NRUS), finding that a nonlinear parameter is significantly affected by the

Table 15.2 Relationship between phenolphthalein colorless depths and carbonation front depths (Chang and Chen, 2006)

Accelerated carbonation time (weeks)	Phenolphthalein colourless depth (X_p, mm)	Carbonation front depth (X_C, mm)			X_C/X_p		
		TGA	X-ray	FTIR	TGA	X-ray	FTIR
8	12	25	25	24	2.1	2.1	2.0
16	17	35	35	35	2.1	2.1	2.1

presence of carbonation. Research carried out to assess the depth of concrete carbonation show that the phenolphthalein indicator provides a poor estimate of the real concrete carbonation. This means that much of what has been previously said and done about potential corrosion safety due to concrete carbonation must now be reassessed. As a more direct consequence many concrete structures in which the uncarbonated depth was higher than the 10 mm safety threshold (Yoon et al., 2007) could already have initiated reinforcement corrosion.

15.3 Supplementary cementitious materials (SCMs)

According to Schubert (1987) the action of supplementary cementitious materials (SCMs) is twofold, since they are associated with the consumption of $Ca(OH)_2$ in the pozzolanic reaction which reduces the pH and increases the rate of carbonation, while at the same time the formation of new CSH blocks capillary pores decreasing carbonation. Park (1995) found that the greater the amount of pozzolanic materials, the deeper the carbonation depth becomes. This researcher stated that this phenomenon is primarily due to the reduction in the alkali content in the cementitious materials and the calcium silicate hydrate formed from the pozzolanic reaction absorbs more alkali ions, hence lowering the pH level in concrete.

15.3.1 Silica fume

Skjolsvold (1986) investigated the influence of silica fume (SF) on the carbonation depth reporting that higher carbonation depths are associated with the use of SF. These results were confirmed by Grimaldi et al. (1989). Khan and Lynsdale (2002) reported that SF slightly increases carbonation. Kulakowski et al. (2009) report the existence of a 'critical threshold' in the carbonation behavior of concrete with SF, which is delimited by an interval of W/B ratios (0.45 and 0.50). Below the lower W/B ratio limit, carbonation is determined mainly by the porosity of the cementitious matrix while the concentration of $Ca(OH)_2$ and pH have little influence on carbonation depths at this ratio. For values above the upper W/B ratio limit, chemical characteristics start to play a more significant role in carbonation depth and the consumption of $Ca(OH)_2$ in the pozzolanic reactions caused by silica fume starts to have a detrimental effect on carbonation. According to these authors the effect of silica fume, in practice, is detrimental only for W/B ratios above the 'critical carbonation threshold'.

15.3.2 Fly-ashes

Ho and Lewis (1983) reported an increase in concrete carbonation when FA is used. Kokubu and Nagataki (1989) mentioned that the carbonation depths of FA concrete decreased significantly with increasing strength grades of concrete. Ogha and Nagataki (1989) reported that the carbonation increases with the replacement ratio of cement by FA. These results were not confirmed by those obtained by Atis (2003), who found that FA concrete made with a 70% replacement ratio showed higher carbonation than that of concrete with 50% replacement and of the reference concrete for both moist and dry curing conditions. However, FA concrete made with a 50% replacement ratio showed lower or comparable carbonation than that of reference concrete for both curing conditions. The reason may be W/B related, because the reference concrete has a W/B = 0.55 and the 50% FA mix has a W/B = 0.33.

Khunthongkeaw et al. (2006) found that the carbonation coefficient of concrete increases with the FA content (above 30%) and W/B. However, they also mention that the carbonation coefficient changes very little for less than 30% FA and does not change when a small amount (10%) of FA is used. Roziere et al. (2009) found that the curing conditions of concrete with and without FA influences the carbonation depth.

Siddique (2011) studied the carbonation of self-compacting concrete (SCC), mentioning that increasing FA from 15 to 20% leads to an increase in the carbonation depth. However, a further increase in FA content to 25%, decreases carbonation depth. This author mentioned that carbonation depth was almost constant for mixes with 25% and 35% FA and that the overall results of carbonation depth were very low, as already found by Assie et al. (2007). Zhao et al. (2012) studied SCC with FA report the importance of initial curing on the carbonation depth. The initial water-curing period of 7-day plus 21-day room curing at 75–85% RH leads to the lowest carbonation depth and 28-day water curing leads to the highest carbonation depth.

According to Younsi et al. (2011) when FA concrete mixes are stored for 28 days under water before testing (water-curing), similar carbonation depths were measured for the two mixes (30% and 50% FA), also whatever the prior oven-drying period. Water-curing has a beneficial effect on all the mixes tested. The carbonation depths are about 20–50% lower in the case of water-curing than in the case of air-curing. Chatveera and Lertwattanaruk (2011) mention that the depths of carbonation of the concretes mixed with black rice husk ash (BRHA) are higher than the ordinary Portland cement (OPC) concrete. Increasing the BRHA replacement ratio (from 20% to 40% by weight of binder) tends to increase the depth of carbonation. The ratio of paste volume to void volume content of the compacted aggregates (*g*) is an important factor for the damage of concrete. Increasing the *g* value tends to have an adverse effect of carbonation on concrete with the higher BRHA replacement ratio due to the higher volume of cement paste. In addition, increasing the W/B ratio tends to increase the porosity and volume of capillary pores in concrete, and significantly affect the carbonation depth.

15.3.3 Slags

Tori and Kawamura (1992) noted that the depth of carbonation of concrete with mineral additions was much higher than that of the corresponding OPC concrete both at dry and wet curing conditions. In particular, 50% slag concrete displayed higher carbonation depth than OPC concrete. A typical value of carbonation depth for slag concrete cured initially in water and exposed to dry environment for one year was 1.3 mm, and the corresponding value for OPC concrete was 0.2 mm. Dinakar et al. (2007) mention that the blast furnace slag concrete (BFSC) showed higher carbonation depths than the corresponding OPC in all grades. For these authors the reason lies on the fact that blending of cement with mineral admixtures leads to a lowering of the $Ca(OH)_2$ content in the hardened cement paste so that a smaller amount of CO_2 is required to remove all the $Ca(OH)_2$ by producing $CaCO_3$. Bouikni et al. (2009) mention that concrete with 65% slag replacement always showed higher carbonation penetration than concrete with 50% slag, and that water curing is clearly a major factor in reducing carbonation.

15.3.4 Blended SCMs

Byfors (1985) investigated the carbonation of SF and pulverized fuel ash (PFA) blended cement concrete and found that the incorporation of SF has no effect on carbonation, while PFA exhibited higher rate of carbonation. Tori and Kawamura (1992) noted that the carbonation depth of blended concrete increased with the (W/C) ratio and decreased with the period of initial curing. Khan and Lynsdale (2002) studied high-performance concrete (HPC) with a W/B = 0.27 and the use of binary and ternary blended cementitious systems based on ordinary Portland cement (PC), PFA and SF. They observed that there was an increase in carbonation with PFA content. These authors mention that SF did not exhibit significant influence on carbonation depth.

Papadakis (2000) studied several SCMs (SF, low- and high-calcium FA), reporting that the carbonation depth decreases as aggregate replacement by SCM's increases, and increases as cement replacement by SCMs increases. Bai et al. (2002) studied PC-PFA-metakaolin (MK) concrete mixes, mentioning two main trends: (i) increasing replacement of PC with PFA increases carbonation depth and (ii) systematically substituting the PFA with increasing levels of MK reduces carbonation depth. They also reported a strong correlation between carbonation depth decrease and sorptivity decrease. Dhir et al. (2007) studied concrete mixes produced using combinations of

PC and limestone powder, finding that for 15% limestone powder there is no influence on carbonation resistance. Increasing the limestone content leads to a reduction of carbonation resistance which is higher for higher W/B ratios.

Gonen and Yazicioglu (2007) reported that the carbonation depth of concrete mixes containing FA was slightly higher than that of reference concrete. In concrete mixes containing silica fume and fly-ash (SFAC) at the same time, the carbonation depth was lower than that of other concrete mixes, where silica fume had little effect on carbonation. These authors attribute the lower depth of carbonation in SFAC to their lower porosity; because they noticed that the porosity of the FA mix was twice that of the SFAC concrete. It is worth noting that a similar trend also occurs for capillary water absorption.

Shi et al. (2009) studied HPC with different replacement levels (0–60%) of FA and ground granulated blast furnace slag (GGBFS). They mention that for a W/B = 0.30 the carbonation depth significantly increases with FA replacement ratio (almost 50 mm for W/B = 0.30 and 60%). However, for W/B = 0.25 an optimum effect occurs for 30% FA. They also mention that concrete mixes with GGBFS show significantly lower carbonation depth (below 5.5 mm) than that of HPC with FA.

According to the final report of Rilem TC 205-DSC (De Schuter and Audenaert, 2008) on the durability of SCC, this material sometimes displays a larger carbonation depth and at other times a smaller one in comparison with conventional concrete with the same amount of water and cement content, although the differences are small. If properly cured, the pore structure of SCC could be denser and less permeable. Nevertheless it seems that a slightly increased vulnerability is noticed concerning carbonation of SCC with limestone filler. Valcuende and Parra (2010) studied the natural carbonation of SCC containing limestone fines, noting that this material has lower carbonation depths than normal vibrated concrete due to a refinement in the pore structure provided by the limestone fines. They also report that carbonation decreases when W/B is reduced and that for the same W/B carbonation is lower when 45 R cement is used instead of 32.5N cement.

15.4 Recycled aggregates concrete (RAC)

Larbi and Steijaert (1994) mention that the depth of carbonation is greater for porous aggregates (with interconnected pores) concrete than for normal (dense) aggregate concrete. Sagoe-Crentsil et al. (2003) mention a 10% increase in the carbonation rate of concrete with coarse recycled aggregates. The coarse recycled aggregates had 5.6% water absorption and were pre-saturated for 10 minutes. Buyle-Bodin and Hadjieva-Zaharieva (2002) found that the carbonation rate of concrete with both coarse and fine recycled aggregates is 3.2 times higher than for primary aggregate concrete. The concrete mix containing natural sand and coarse recycled aggregates showed a carbonation rate 1.8 times higher than for primary aggregates concrete. Fine and coarse recycled aggregates had a water absorption value respectively of 12% and 6% and were used pre-soaked. Katz (2003) found that the depth of carbonation of the recycled concrete aggregates was 1.3–2.5 times greater than that of the reference concrete. Coarse, medium and fine recycled aggregates were used with water absorptions equal to (3%, 9% and 11%). According to the author the properties of aggregates made from crushed concrete and the effect of the aggregates on the new concrete (e.g. strength, modulus of elasticity) resemble those of lightweight aggregate concrete, and similar considerations apply when dealing with this type of aggregates.

Otsuki et al. (2003) mention that carbonation the rate of concrete (W/B = 0.4) with coarse recycled aggregates is just slightly higher than for primary aggregates concrete. They used coarse recycled aggregates with a water absorption in a range of 3–5%. According to Rao et al. (2007) the increase in the carbonation depth of recycled aggregates concrete (RAC) could be attributed to the higher permeability of the recycled aggregates on account of the presence of old mortar adhering to the original aggregate, and the old interfacial transition zone (ITZ) between them. Tam et al. (2008) found several acceptable correlations between carbonation depth and absorption (R^2 = 0.76); particle density (R^2 = 0.73) and porosity (R^2 = 0.77). Werle et al. (2011) reported that the carbonation depth is influenced by the porosity of the recycled aggregates. Aggregates with a porosity lower than the matrix porosity led to a reduction of the carbonation depth.

15.4.1 Curing

Balayssac et al. (1995) found that carbonated depth decreases rapidly when the curing period increases from 1 to 3 days. After 18 months, for a concrete with a cement content of 350 kg/m^3, increasing the curing period from 1 to 28 days halves the carbonation depth. For concrete stored for 18 months, increasing the curing period from 1 to 3 days increases the durability performance by a value of 10% for a concrete with a cement content of 300 kg/m^3 and 50% for a concrete with a cement content of 420 kg/m^3; increasing the curing period from 3 to 28 days still improves the durability performance by a value of 30% the concrete with the lowest cement content, but only by a value of 10% the concrete with the highest cement content.

Bai et al. (2002) mention that water-curing reduces sorptivity, which reflects a finer pore structure that will, inhibit ingress of aggressive elements into the pore system reducing carbonation. Buyle-Bodin and Hadjieva-Zaharieva (2002) found that the depth of carbonation is cut by half when concrete is cured in water. The decrease in the depth of carbonation might be partially due to the higher internal humidity content of this concrete. However, this influence would be less pronounced for concrete with recycled aggregates because of its higher porosity allowing faster water evaporation after curing. Lo and Lee (2002) found that large differences in carbonation depth were recorded between water-cured and air-cured samples but the difference decreased with time. Water-cured concrete was found to have carbonated to 72% of the level reached by air-cured samples after 3 months of accelerated curing. Atis (2003) showed that the longer initial curing period resulted in lower carbonation depth. The effect is more marked with moist curing.

Haque et al. (2007) studied lightweight aggregates, mentioning that the greater the extent of initial water curing, the lesser the depth of carbonation. Lo et al. (2008) found that mixes under hot water curing exhibited higher carbonation than mixes under normal curing. Lo et al. (2009) stated that the carbonation depth under accelerated curing was higher than that of concrete under normal curing. The trend was more prominent for mixes with a higher W/C ratio, than for the mixes, with a lower W/C ratio. One of the reasons was that the samples were cured in hot water for 3 days and thereafter stored in water for normal curing for 24 days. This indicates that the initial curing period of PFA-incorporated concrete in hot water for 3 days resulted in larger inter-pores in the cement paste.

Limbachiya et al. (2012) mentioned that carbonation increase with W/B and also that the water stored in the pore system of the recycled aggregates and released throughout hydration process may contribute to the carbonation process of the RAC mixes compared to the control mixes. Moreover, it is well recognized that carbonation of concrete occurs at an RH from about 40% to 70%. Owing to the high water absorption of the coarse recycled aggregates, the control mixes have generally lower moisture content than the corresponding RAC. This may also explain the low resistance to carbonation of concretes containing coarse recycled aggregates.

15.4.2 RAC content

The carbonation depth increases with recycled aggregates content (Levy and Helene, 2004). Li (2008) cites research published in Chinese journals confirming that the recycled aggregates content has some influence on the carbonation resistance of concrete. As the recycled aggregates content increases, so does the carbonation depth. When the recycled aggregates content is 60%, the carbonation depth increases as much as 62% compared with that of the reference concrete. Gomes and de Brito (2009) mentioned that recycled aggregates concrete shows just a slight increase in the carbonation depth when compared with primary aggregates concrete. However, when more than 50% of the volume is replaced by recycled aggregates the carbonation depth increases about 10%. Lovato et al. (2012) mentioned that the increased content of recycled aggregates, both fine and coarse, did not have much influence on the carbonation depth of concrete. However, concrete produced with recycled aggregates tends to present slightly higher rates of carbonation depth (about 5 mm) than the reference concrete. Concrete mixes with 100% coarse and fine recycled aggregates present carbonation depth rates that are about 43% higher than those produced with 100% primary aggregates. The higher the *m* (recycled aggregate/cement) ratio and the W/C ratio the greater the concrete carbonation depths are. Limbachiya et al. (2012) mentioned that adding various proportions of coarse recycled aggregates as a partial replacement of primary aggregates has resulted in a lower resistance to carbonation, especially for the C30 and C35 concrete grades. The carbonation depth and the rate of carbonation of concrete increase when the content of recycled aggregates enhances.

Evangelista and Brito (2010) used only two RAC with 30% and 100% replacement of fine primary aggregates with fine recycled aggregates (FRA) and obtained the following results concerning carbonation depth after 90 days in the carbonation chamber: the 30% RAC had a performance 27% better than the reference concrete's and the 100% RAC 35% worse. The authors considered the first result anomalous and possibly due to the limited number of specimens tested. However, they mentioned that results were compatible with the ones from the compression strength test where all the RAC mixes achieved values very similar to the one of the reference concrete. Zega and Di Maio (2011) studied natural carbonation during 310 and 620 days of concrete mixes with a partial replacement (20 and 30%) of FRA. These authors reported a similar carbonation depth for mixes with different FRA content, which they attribute to the low W/B ratio (0.41–0.43).

15.4.3 Combined effect of RAC and SCMs

Corinaldesi and Moriconi (2009) mention that the carbonation depth after 1 year was 8.6, 5.9 and 6.5 mm, respectively, for the reference mix (REF), recycled aggregates mix (REC) and recycled aggregates plus fly-ash mix (REC + FA). They concluded that, for mixes prepared with lower W/C ratio and due to the refinement of the pore system, carbonation did not present evidence of risks for reinforcement corrosion. This is due to the very low permeability of these mixes, even if porous aggregates, such as recycled aggregates, were used. Abbas et al. (2009) found that mixes with FA had the greatest carbonation depth throughout the 140 days of exposure. This behavior can be attributed to the pozzolanic action of the SCMs, which consume $Ca(OH)_2$ and consequently lower the alkalinity of concrete. The carbonation coefficients of specimens with FA were the largest and were almost twice the values of specimens without SCMs. The carbonation depths of RAC with and without SCMs fall in the expected range for structural-grade conventional concrete. RAC specimens without SCMs showed the lowest level of carbonation, followed by specimens containing blast furnace slag and FA, respectively. Specimens with high cement content were found to have high resistance to carbonation.

Yang et al. (2011) mentioned that the incorporation of coarse recycled aggregates and FA significantly cut down the concrete's carbonation resistance, which is related with the replacement rate; the content of $Ca(OH)_2$ in the RAC decreased and there are also obvious interface transition zones between the coarse recycled aggregate and the new paste. There are obvious cracks and large voids before the RAC is loaded, which leads directly to lower carbonation resistance.

According to Tian et al. (2011) the use of FA as a substitute for cement decreased the carbonation depth of the RAC. It is observed that the largest influence on the depth of carbonation of the RAC comes from W/B ratio, which means that the carbonation age and FA content have a lower effect. Zhu et al. (2011) studied the carbonation resistance in concrete mixes with several replacement levels of primary aggregates by both fine and coarse recycled aggregates. Several replacement levels of FA and slag were also used. They found that SCMs mitigate carbonation increase caused by recycled aggregates.

Sim and Park (2011) studied the combined effect of FA addition and the recycled aggregates incorporation by volume replacement. The overall carbonation depths were lower at 60% and 100% replacement ratio than at 0% and 30%. They mentioned that this unexpected result may in part be due to the error in the experiments or the nature of the recycled aggregates that is largely dependent on the properties of the original concrete. It can be concluded that the RAC even with some addition of FA can provide sufficient resistance to carbonation based on the measured carbonation depths which were mostly below 10 mm. Limbachiya et al. (2012) mentioned that for a given design strength, the carbonation depth of all the FA concrete mixes was quite greater than the one of the ordinary Portland cement (OPC) concrete mixes. They also mention that combining FA with coarse recycled aggregates in concrete may enhance the long-term resistance to carbonation. A reliable linear relationship between the carbonation coefficient and time of exposure as well as between the carbonation coefficient and compressive strength was observed. As expected, the higher the compressive strength of concrete, the lower is the carbonation coefficient.

15.4.4 Possible remedial actions

Several authors suggested some remedial actions to minimize concrete carbonation when recycled aggregates are used. Shayan and Xu (2003) studied four treatments that were applied to the RAC as follows: 1(a) concentrated

Table 15.3 Types of surface improving agents (Tsujino et al. 2007)

Type	Oil (O)		Silane (S)	
Application	Release agent used in wooden form		Water-repellent agent with permeability to the concrete surface	
Main constituent	Mineral oil (paraffin)	85–95%	Silicon analogue	28–32%
	Emulsifying agent	1–5%	Emulsifying agent	Minute quantity
	Lanolin fatty acid salt	1–5%	Water	68–72%
State	Emulsion		Emulsion	

sodium silicate solution, designated N42 solution; 1(b) diluted (50%) N42 solution; 2 diluted N42 solution and lime; 3 diluted N42 solution and silica fume; and 4 diluted N42 solution, silica fume and lime. These treatments were initially applied to coarse RCA for the assessment of their effect on the RCA surface features.

They mentioned that the mixes with no chemical treatment had lower carbonation depths. They also mentioned that since carbonation depth is a function of the amount of water in concrete, lower carbonation depths could result from high humidity or wet concrete. Tsujino et al. (2007) used an oil and silane-type surface improving agents (Table 15.3) that increases the resistance to carbonation both for W/B = 0.4 and 0.6. Otsuki et al. (2003) suggested a double mixing method to improve carbonation resistance. In the double mixing method, the addition of water is divided into two stages. The first portion of water is added into the mixer for 30 s after fine and coarse aggregates have been mixed for 30 s. The mixing is then stopped while the cement is being added into the mixer. This is followed by 60 s of mixing by machine and 60 s of mixing by hand. After that, the final portion of water is added for 30 s and the mixing continues for the last 90 s. This method is devised to coat RCA with mortar of a lower W/B ratio than the rest of mortar matrix. Xiao et al. (2010) suggested the use of a maximum volume of coarse recycled aggregates and the use of recycled aggregates from concrete with minimum strength class as a way to increase resistance to carbonation.

15.5 References

Abbas, A.; Fathifazl, G.; Isgor, O.; Razaqpur, A.; Fournier, B.; Foo, S. (2009) Durability of recycled aggregate concrete designed with equivalent mortar volume method. *Cement and Concrete Composites* **31**, 555–563.

Assie, S.; Escadeillas, G.; Waller, V. (2007) Estimates of self-compacting concrete 'potential' durability. *Construction and Building Materials* **21**, 1909–1917.

Atis, C. (2003) Accelerated carbonation and testing of concrete made with fly ash. *Construction and Building Materials* **17**, 147–152.

Bai, J.; Wild, S.; Sabir, B. (2002) Sorptivity and strength of air-cured and water-cured

PC–PFA–MK concrete and the influence of binder composition on carbonation depth. *Cement and Concrete Research* **32**, 1813–1821.

Balayssac, J.; Detriche, H.; Grandet, J. (1995) Effects of curing upon carbonation of concrete. *Construction and Building Materials* **9**, 91–95.

Bouchaala, F.; Payan, C.; Garnier, V.; Balayssac, J. (2011) Carbonation assessment in concrete by nonlinear ultrasound. *Cement and Concrete Research* **41**, 557–559.

Bouikni, A.; Swamy, R.; Bali, A. (2009) Durability properties of concrete containing 50% and 65% slag. *Construction and Building Materials* **23**, 2836–2845.

Brown, J.H. (1991) The effect of exposure and concrete quality: field survey results from some 400 structures. The 5th International Conference on the Durability of Materials and Components, Bringhton, 249–259.

Buyle-Bodin, F.; Hadjieva-Zaharieva, R. (2002) Influence of industrially produced recycled aggregates on flow properties of concrete. *Materials and Structures* **35**, 504–509.

Byfors, K. (1985) Carbonation of concrete with silica fume and fly ash. *Journal of Nordic Concrete Research* **4**, 26–35.

Chang, C.-F.; Chen, J.-W. (2006) The experimental investigation of concrete carbonation depth. *Cement and Concrete Research* **36**, 1760–1767.

Chatveera, B.; Lertwattanaruk, P. (2011) Durability of conventional concretes containing black rice husk ash. *Journal of Environmental Management* **92**, 59–66.

Conciatori, D.; Laferrière, F.; Brühwiler, E. (2010) Comprehensive modeling of chloride ion and water ingress into concrete considering thermal and carbonation state for real climate. *Cement and Concrete Research* **40**, 109–118.

Corinaldesi, V.; Moriconi, G. (2009) Influence of mineral additions on the performance of 100% recycled aggregate concrete. *Construction and Building Materials*, **23**, 2869–2876.

De Schuter, G.; Audenaert, K. (2008) Final report of RILEM TC 205-DSC: durability of self-compacting concrete. *Materials and Structures* **41**, 225–233.

Dhir, R.; Limbachiya, M.; McCarthy, M.; Chaipanich, A. (2007) Evaluation of Portland limestone cements for use in concrete construction. *Materials and Structures* **40**, 459–473.

Dinakar, P.; Babu, K.; Santhanam, M. (2007) Corrosion behaviour of blended cements in low and medium strength concretes. *Cement and Concrete Composites* **29**, 136–145.

Evangelista, L.; de Brito, J. (2010) Durability performance of concrete made with fine recycled concrete aggregates. *Cement and Concrete Composites* **32**, 9–14.

Gomes, M.; de Brito, J. (2009) Structural concrete with incorporation of coarse recycled concrete and ceramic aggregates: durability performance. *Materials and Structures* **42**, 663–675.

Gonen, T.; Yazicioglu, S. (2007) The influence of mineral admixtures on the short and long-term performance of concrete. *Building and Environment* **42**, 3080–3085.

Grimaldi, G.; Carpio, J.; Raharinaivo, A. (1989) Effect of silica fume on carbonation and chloride penetration in mortars. In: Mohammed Alasali, editor. *Third CANMET/ACI International Conference on Fly Ash, Silica Fume, Slag and Natural Pozzolans in Concrete*; 320–334.

Haque, M.; Al-Khait, H.; Kayali, O. (2007) Long-term strength and durability parameters of lightweight concrete in hot regime: importance of initial curing. *Building and Environment* **42**, 3086–3092.

Ho, D.; Lewis, R. (1983) Carbonation of concrete incorporating fly ash or a chemical admixture. *First International Conference on The Use of Fly Ash, Silica Fume, Slag and other Mineral By-Products in Concrete. Montebello, Canada*: SP-79, 333–346.

Hobbs. D.W. (1988) Carbonation of concrete in PFA. *Magazine of Concrete Research* **40**, 69–78.

Katz, A. (2003) Properties of concrete made with recycled aggregate from partially hydrated old concrete. *Cement and Concrete Research* **33**, 703– 711.

Khan, M.; Lynsdale, C. (2002) Strength, permeability, and carbonation of high-performance concrete. *Cement and Concrete Research* **32**, 123–131.

Khunthongkeaw, K.; Tangtermsirikul; S.; Leelawat, T. (2006) A study on carbonation depth prediction for fly ash concrete. *Construction and Building Materials* **20**, 744–753.

Kokubu, M.; Nagataki, S. (1989) Carbonation of concrete with fly ash and corrosion of reinforcements in 20 years tests. *Third international Conference on Fly Ash, Silica Fume, Slag and Natural Pozzolans in Concrete, Trondheim, Norway*. SP-114. ACI; 315–329.

Kulakowski, M.; Pereira, F.; Dal Molin, D. (2009) Carbonation-induced reinforcement corrosion in silica fume concrete. *Construction and Building Materials* **23**, 1189–1195.

Larbi, J.; Steijaert, P. (1994) Microstructure of concretes containing artificial aggregates and recycled aggregates. In *Environmental Aspects of Construction with Waste Materials*, J. Gourmans, H. Van der Sloot and T. Aalbers (Ed), 877–888. Elsevier.

Levy, S.M.; Helene, P. (2004) Durability of recycled aggregates concrete: a safe way to sustainable development. *Cement and Concrete Research* **34**, 1975–1980.

Li, X. (2008) Recycling and reuse of waste concrete in China: Part I. Material behaviour of recycled aggregate concrete. *Resources, Conservation and Recycling* **53**, 36–44.

Limbachiya, M., Meddah, M.S., Ouchagour, Y. (2012) Use of recycled concrete aggregate in fly-ash concrete. *Construction and Building Materials* **27**, 439–449.

Lo, T.; Tang, W.; Nadeem, A. (2008) Comparison of carbonation of lightweight concrete with normal weight concrete at similar strength levels. *Construction and Building Materials* **22**, 1648–1655.

Lo, T.; Nadeem, A.; Tang, W.; Yu, P. (2009) The effect of high temperature curing on the strength and carbonation of pozzolanic structural lightweight concretes. *Construction and Building Materials* **23**, 1306–1310.

Lo, Y.; Lee, H. (2002) Curing effects on carbonation of concrete using a phenolphthalein indicator and Fourier-transform infrared spectroscopy. *Building and Environment* **37**, 507–514.

Lovato, P.S., Possan, E., Molin, D.C.C.D., Masuero, Â.B., Ribeiro, J.L.D. (2012) Modeling of mechanical properties and durability of recycled aggregate concretes. *Construction and Building Materials* **26**, 437–447.

Mindess, S.; Young, J.F.; Darwin, D. (2002) *Concrete*, 2nd ed. Prentice Hall, New Jersey, USA.

Muntean, A; Böhm, M. A. (2009) Moving-boundary problem for concrete carbonation: global existence and uniqueness of weak solutions. *Journal of Mathematical Analytical Applications* **350**, 234–251.

Muntean, A.; Boohm, M.; Kropp, J. (2011) Moving carbonation fronts in concrete: A moving-sharp-interface approach. *Chemical Engineering Science* **66**, 538–547.

Ogha, H.; Nagataki, S. (1989) Prediction of carbonation depth of concrete with fly ash. *Third International Conference on Fly Ash, Silica Fume, Slag and Natural Pozzolans in Concrete. Trodheim, Norway*; SP-114, 275–294.

Otsuki, N.; Miyazato, S.; Yodsudjai, W. (2003) Influence of recycled aggregate on interfacial transition zone, strength, chloride penetration and carbonation of concrete. *Journal of Materials in Civil Engineering* **15**, 443–451.

Papadakis, V. (2000) Effect of supplementary cementing materials on concrete resistance against carbonation and chloride ingress. *Cement and Concrete Research* **30**, 291–299.

Papadakis, V.G.; Vayenas, C.G., Fardis, M.N. (1991) Experimental investigation and mathematical modeling of the concrete carbonation problem. *Chemical Engineering Science* **46**, 1333–1338.

Papadakis, V.G.; Vayenas C.G.; Fardis, M.N. (1992) Hydration and carbonation of pozzolanic cements. *ACI Materials Journal* **89**, 119–130.

Park, G.K. (1995) Durability and carbonation of concrete. *Magazine of Korean Concrete Institute* **7**, 74–81.

Peter, M.; Munteen, A.; Meier, S.; Bohm, M. (2008) Competition of several carbonation reactions in concrete: A parametric study. *Cement and Concrete Research* **38**, 1385–1393.

Rao, A.; Jha, K.N.; Misra, S. (2007) Use of aggregates from recycled construction and demolition waste in concrete. *Resources, Conservation and Recycling* **50**, 71–81.

Rilem (1988) CPC-18 measurement of hardened concrete carbonation depth. *Materials and Structures* **21** (6), 453–455.

Roziere, E.; Loukili, A.; Cussigh, F. (2009) A performance based approach for durability of concrete exposed to carbonation. *Construction and Building Materials* **23**, 190–199.

Sagoe-Crentsil, K.; Brown, T.; Taylor, A. (2003) Performance of concrete made with commercially produced coarse recycled concrete aggregate. *Cement and Concrete Research* **31**, 707–712.

Schubert, P. (1987) Carbonation behaviour of mortars and concrete made with fly ash. *ACI Special Publications SP-100* 1945–1962.

Schutter, G.; Audenaert, K. (2004) Evaluation of water absorption of concrete as a measure for resistance against carbonation and chloride migration. *Materials and Structures* **37**, 591–596.

Shayan, A.; Xu, A. (2003) Performance and properties of structural concrete made with recycled concrete aggregate. *ACI Materials Journal* **100**(5), 371–380.

Shi, H.-S.; Xu, B.-W.; Zhou, X.-C. (2009) Influence of mineral admixtures on compressive strength, gas permeability and carbonation of high performance concrete. *Construction and Building Materials* **23**, 1980–1985.

Siddique, R. (2011) Properties of self-compacting concrete containing class F fly ash. *Materials and Design* **32**, 1501–1507.

Sim, J.; Park, C. (2011) Compressive strength and resistance to chloride ion penetration and carbonation of recycled aggregate concrete with varying amount of fly ash and fine recycled aggregate. *Waste Management* **31**, 2352–2360.

Skjolsvold, O. (1986) *Carbonation depths of concrete with and with out condensed silica fume*. ACI Special Publications SP-91, 1031–1048.

Tam, V.; Wang, K.; Tam, C. (2008) Assessing relationships among properties of demolished concrete, recycled aggregate and recycled aggregate concrete using regression analysis. *Journal of Hazardous Materials* **152**, 703–714.

Tian, F.; Hu, W.; Cheng, H.; Sun, Y. (2011) Carbonation depth of recycled aggregate concrete incorporating fly ash. *Advanced Materials Research*_261–263, 217–222.

Tori, K.; Kawamura, M. (1992) Pore structure and chloride permeability of concretes containing fly ash, blast furnace slag and silica fume. In *Fly ash, silica fume, slag and natural pozzolans in concrete*, Detroit: American Concrete Institute; SP-132, vol. 1., 135–150.

Tsujino, M.; Noguchi, T.; Tamura, M.; Kanematsu, M.; Maruyama, I. (2007) Application of conventionally recycled coarse aggregate to concrete structure by surface modification treatment. *Journal of Advanced Concrete Technology* **5**, 13–25.

Valcuende, M.; Parra, C. (2010) Natural carbonation of self-compacting concretes. *Construction and Building Materials* **24**, 848–853.

Villain, J.; Thiery, M.; Platret, G. (2007) Measurement methods of carbonation profiles in concrete: Thermogravimetry, chemical analysis and gammadensimetry. *Cement and Concrete Research* **37**, 1182–1192.

Wassermann, R.; Katz, A.; Bentur, A. (2009) Minimum cement content requirements: a must or a myth? *Materials and Structures* **42**, 973–982.

Werle, A.; Kazmierczak, C.; Kulakowski, M. (2011) Carbonation of concrete with recycled concrete aggregates. *Ambiente Construído, Porto Alegre*, **11** (2), 213–228 (in Portuguese).

Wiering, P. (1984) Long time studies on the carbonation of concrete under normal outdoor exposure. RILEM Symposium on Durability of Concrete under Normal Outdoor Exposure. Hannover.

Xiao, J.-Z.; Lei, B.; Zhang, C.-Z. (2010) Effects of recycled coarse aggregates on the carbonation evolution of concrete. *Key Engineering Materials* **417–418**, 697–700.

Yang, H.; Deng, Z.; Li, X. (2011) Microscopic mechanism study on carbonation resistance of recycled aggregate concrete. *Advanced Materials Research* **194–196**, 1001–1006.

Yoon, I.-S.; Copuroglu, O.; Park, K.-B. (2007) Effect of global climatic change on carbonation progress of concrete. *Atmospheric Environment* **41**, 7274–7285.

Younsi, A.; Turcry, P.; Rozière, E.; Aït-Mokhtar, A.; Loukili, A. (2011) Performance-based design and carbonation of concrete with high fly ash content. *Cement and Concrete Composites* **33**, 993–1000.

Zega, C.J.; Di Maio, A. (2011) Use of recycled fine aggregate in concretes with durable requirements. *Waste Management* **31**, 2336–2340.

Zhao, H.; Sun, W.; Wu, X.; Gao, B. (2012) Effect of initial water-curing period and curing condition on the properties of self-compacting concrete. *Materials and Design* **35**, 194–200.

Zhu, P.; Wang, X.; Feng, J. (2011) Carbonation behavior of recycled aggregate concrete under loading. *Advanced Materials Research* **250–253**, 779–782.

16
Concrete with polymers

M. FRIGIONE, University of Salento, Italy

DOI: 10.1533/9780857098993.3.386

Abstract: This chapter describes the use of different polymers that can be added to fresh or hardened hydraulic cement, or used to replace the cement. The introduction of polymers can modify the characteristics and properties of concrete, and protect or repair concrete elements.

Key words: polymer admixtures, superplasticizers, polymer-modified concrete (PMC) polymer-impregnated concrete (PIC), polymer concrete, (PC), coatings, adhesives.

16.1 Introduction

In Western Europe polymers represent only 1% in weight of all construction materials (2007 data), but in financial turnover polymers represent more than 10% of the construction industry. The available data indicates a steady increase in the use of polymers in construction (EuPC 2006). The amount of polymers used in concrete–polymer composites is only a minor part of the total used for concrete modification (total consumption of 486 000 tons in 2000), which also includes water-reducing agents (121 500 tons of super-plasticizer in EU in 1998). The use of polymer admixtures in concrete will further increase. New environmental regulations and higher-quality demands will require admixtures for nearly all ready-mixed concretes. This evolution opens a yearly market of several million tons of admixtures (Van Gemert and Knapen 2010).

Polymers used in concrete have been developed within the past 60–70 years. Four main kinds of polymers can be distinguished as increasing function of the polymer content with respect to cement:

- In Portland cement concrete (PCC) the properties and performance of concrete are largely determined by the properties of the hydraulic cement paste that is the active constituent. Polymer admixtures are ingredients that, added in small quantities (usually below 3% wt) to the concrete batch, immediately before or during mixing, interact in the hydrated-cementitious system by physical, chemical and physical-chemical action modifying one or more properties allowing wider concrete applications, conferring certain beneficial effects to concrete, including enhanced durability, improved workability, increasing strengths.

- Polymer-modified concrete (PMC) is prepared by mixing a small amount of polymer (up to 25% wt) in the fresh concrete mixture. The resulting composite has an excellent adhesion to steel reinforcement and to old concrete and good durability.
- In polymer-impregnated concrete (PIC), the capillary pores of hydrated PCC are filled with a low-viscosity monomer that subsequently polymerizes. As a consequence, the permeability is lowered, while the compressive and tensile strengths are enhanced.
- Finally, in polymer concrete (PC) the polymers are used in total replacement of the hydraulic cement that, consequently, contains no Portland cement. Although the cost of the monomer can be quite high, PC can be useful for maintenance and repairs of highway and aircraft runways and compares well on a strength-to-cost basis.

Polymers are also used to repair or strengthen damaged concrete, as adhesives, or to protect it, as coatings.

16.2 Water-reducing admixtures for Portland cement concrete

Corrosion of reinforcing steel, frost and physico-chemical effects in aggressive environments are the major causes of concrete deterioration. There is a general agreement that the permeability of concrete, rather than normal variations in the composition of Portland cement, is the key to all durability problems. Among the recent advancements, most noteworthy is the development of superplasticized concrete mixtures which give very high fluidity at relatively low water contents and, consequently, lower porosity. Hardened concretes possessing low porosity are generally characterized by high durability (Mehta, 1991).

The addition of some water-soluble organic polymers dramatically improves workability of fresh concrete batches, giving at the same time strengths as high as those of non-admixtured concretes with the same water/cement (W/C) ratio, and offers many advantages over casting of concrete.

Water-reducing admixtures are classified broadly into two categories: normal water reducers (WR), also called 'plasticizers', and high-range water reducers (HRWR), called 'superplasticizers'. While the normal WRs (generally based on lignosulfonates, obtained as a waste liquor during the process of paper-making pulp from wood) can reduce the water demand by 5–10%, the HRWRs (based on polymers obtained from chemical synthesis) can cause a reduction of 15–40%. In general terms, a decrease of workability from 15 cm slump to 7–8 cm is achievable by producing a concrete with superplasticizers (Papayianni et al., 2005).

These chemical admixtures, always water-soluble, can be used in several ways to influence the properties of a concrete mix:

- to increase workability so as to ease placing in inaccessible locations;
- to reduce water content for increased strength and reduced permeability and improve durability;
- to achieve the same workability by decreasing the cement content (Oetiker, 2009).

The so-called 'second generation' superplasticizers are based on polynaphthalene sulfonate (sulfonated naphthalene formaldehyde, SNF) or polymelamine sulfonate (sulfonated melamine formaldehyde, SMF), substituting those based on lignosulfonate.

16.2.1 Mechanism of action of SNF and SMF

When the dry particles of cement are mixed with a small amount of water, the electric charges upon the solid particles tend to cause their aggregation, avoiding a good distribution of the water between solid particles. This prevents an optimal repartition of the hydrates formed between the particles. The action of the dispersants is, therefore, to prevent the flocculation of fine particles of cement. The grinding of cement results in ground particles with a surface charge. These charges derive from the non-compensation of positive and negative valences of the lattice atoms, which are brought to the surface owing to the grinding of the original grain. These charges are localized in different points and therefore a particle can simultaneously have positive and negative charges. The opposite sign charges present in different particles cause aggregation and form a mass of flocculate. The dispersants (such as SNF and SMF) are surface-active chemicals consisting of long-chain organic molecules, having several anionic polar groups [sulfonic ($-SO_3H$) and hydroxyl ($-OH$)] in the hydrocarbon chain. When they are added to the cement–water system, the polar chain gets adsorbed on the surface of cement particle, giving them a negative charge, which leads to electrostatic repulsion between the particles and results in stabilizing their dispersion. In addition, the negative charge causes the development of a sheath of oriented water molecules around each particle, thus separating the particles. Hence, there is a greater particle mobility, and water, freed from the restraining influence of the flocculated system, becomes available to lubricate the mix so that workability is increased (Malhotra, 1997; Ramachandran et al., 1998).

The use of superplasticizers allows stiff fresh concrete to be changed into a very flowable concrete which is easy and quick to place and has the same properties, in the hardened state, as a stiff concrete without admixture (Morin et al., 2001). Additional mechanisms of superplasticizer action include dispersion of cement particles by reduction in surface tension of mixing water and a decrease in frictional resistance because of the orientation of polymers along the concrete flow direction and lubrication properties produced by low molecular weight polymers (Uchikawa et al., 1995).

Fresh concrete is well known to lose its workability with time. This phenomenon is called 'slump loss'. The slump loss reduces the beneficial effect of high workability at the time of placement. When a concrete mix must be transported for a long time, particularly in hot weather, it should keep as far as possible its initial slump level to avoid the practice of re-dosing the concrete with water above that required in the mix design. Factors that determine slump loss include initial slump value, type and amount of cement, type, amount and time of addition of the superplasticizer, temperature, mixing criteria and the presence of other admixtures in the mix.

The chemical composition of the cement, in particular the C_3A content, is critical to ensure good compatibility between cement and superplasticizer (Sugiyama et al., 2003; Nkinamubanzi and Aitcin 2004; Zimmermann et al., 2009). According to Bundyra-Oracz and Kurdowski (2011) the amount of ettringite arising from the C_3A is responsible for the slump loss. Bensted (2002) reported, on the other hand, that all the aluminatic phases present in cement, i.e. $C_3A + C_4AF$, are responsible for slump loss. The slump loss increased significantly also with the increase in cement fineness (Aydin et al., 2009).

Several methods have been adopted to slow the rate of slump loss. The first is to add the superplasticizer at the point of discharge; however, there are some practical problems associated with this approach. Other methods include the addition of a higher dosage of superplasticizer or the use of some type of retarding admixture in the formulation (Uchikawa et al., 1992; Aiad et al., 2002). However, there are some limits, mainly economic, even in this approach. Slump retention characteristics are also improved by blending SNF with lignosulfonates (Heikal and Aiad, 2008). Another common problem with SMF and SNF admixtures is the excessive set retardation. The low molecular weight fractions, commonly present in the polymers, cause excessive retardation by covering reactive sites on the cement surface and inhibiting reactions.

Pei et al. (2000) synthesized a sodium sulfanilate–phenol–formaldehyde (SSPF) condensate. The latter has the advantage over the SNF and SMF admixtures of maintaining around 80% of initial slump values for about 120 min, so it is more suitable to prepare pumping concrete (Fig. 16.1). The smallest slump-loss rate of SSPF was attributed to the higher number of functional groups in its molecule, which results in a strong adsorption to cement particles and an increase in the density of electric charges on the surfaces of particles.

Compressive strengths of concrete at 7 and 28 days (cement = 300 kg/cm^3, slump 8 ± 1 cm) with SSPF were compared with those of concretes with either SNF or SMF, as shown in Fig. 16.2. The compressive strengths

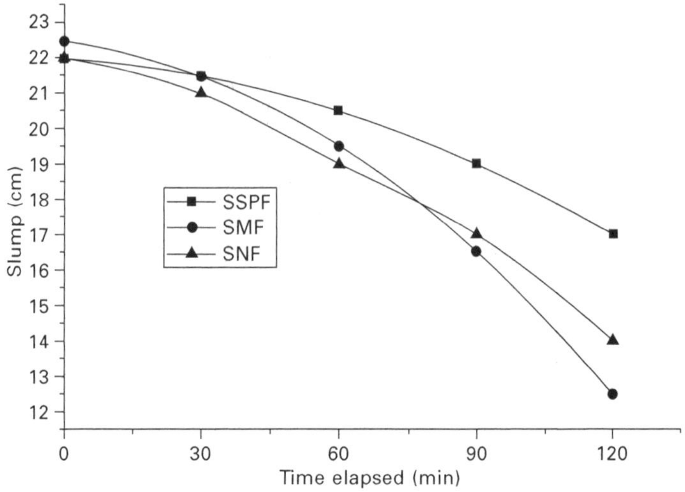

16.1 Effect of admixture type on the slump loss of concrete. Amount of each dispersant added = Cement × 0.65 wt%. SSPF W/C = 0.55; SNF W/C = 0.58; SMF W/C = 0.60 (Pei et al., 2000).

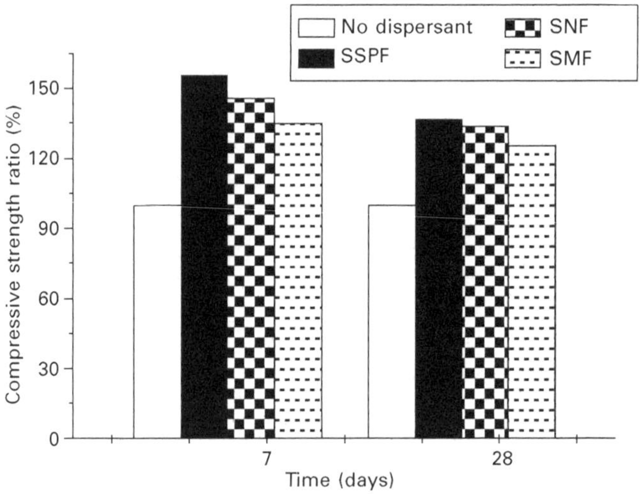

16.2 Effect of admixture type on the compressive strength of concrete (Pei et al., 2000).

of concretes containing SNF or SMF superplasticizers are clearly higher than those of untreated concrete, and that the compressive strength of concrete containing SSPF is higher than those of concretes with either SNF or SMF. This was explained in terms of SSPF's higher water reduction.

16.2.2 Polycarboxylate superplasticizers

In the late twentieth century, new formulations based on the family of polycarboxylate polymers (called 'third generation superplasticizers') were developed. By changing the polymeric structure in these products, for instance the length and the kind of the chain, it was possible to confer a wide range of properties to the concretes.

Polycarboxylates are copolymers whose chemical structures can be easily modified. Unlike SNF they have high dispersion property. Their 'comb-type' molecule consists of one main linear chain with lateral carboxylate and ether groups (Hommer, 2009), shown in Fig. 16.3. Different superplasticizers based on polycarboxylate polymers with a specific function have been synthesized in relation to their final employment: PA (polyacrylate), PE (polyether) and SLCA (slump loss controlling agent).

Referring to the mechanism of action of polycarboxylates, their groups are active in the adsorption of these admixtures to cement particles. Dispersion is due to electrostatic repulsion (as in SNF and SMF admixtures) owing to the carboxylate groups, but primarily to the steric repulsion effect associated with the long lateral ether chains rather than to the presence of negatively charged anionic groups (COO^-) (Collepardi, 1998; Puertas et al., 2005b).

The high degree of the fluidity and duration of this state that this admixture offers to concrete is related to structural factors; hence, the shorter the main chain and the longer and more numerous the lateral chains, the greater and more long lasting is the fluidity induced. The molecular weight of these admixtures likewise has a substantial effect on their performance (Uchikawa et al., 1997; Liao et al., 2006; Felekoğlu and Sarikahya, 2008; Ran et al., 2009a; Zingg et al., 2009).

Besides a lower slump loss, polycarboxylate-based superplasticizers perform better than the traditional sulfonated polymers even in terms of either higher reduction in the W/C at a given workability, or higher slump level at a given mixture composition, although the performance of the SNF products seems

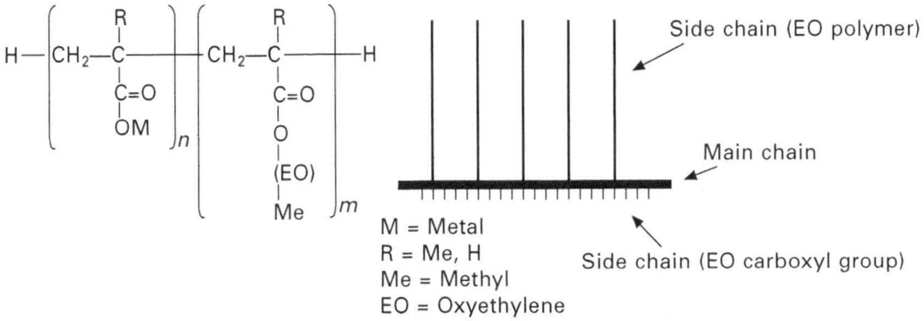

16.3 Chemical structure of polycarboxylate admixtures (adapted from Puertas et al., 2005a).

to be less dependent on the cement type (Coppola et al., 2010). Used at low dosage, polycarboxylate admixtures still can reduce water as much as high dosage of conventional admixtures based on lignosulfonates.

Tanaka et al. (1996) studied the effect of a partially cross-linked superplasticizer based on copolymer of acrylic acid and polyethylene glycol mono-alkyl ether (CLAP) on the slump-loss of concrete mixture. According to the authors, the cross-linked polymer is hydrolyzed by the alkaline water phase of the cement paste and, than, converted into a polycarboxylate-based polymer. The negative carboxylic groups, due to the alkaline hydrolysis, would be adsorbed on cement surface of cement particles and then would be responsible for the dispersion of cement particles and the fluidizing action of the admixture. The low slump-loss effect of this superplasticizer was attributed to the increasing number of the extension side chains of the acrylic polymer which would prolong the dispersion of hydrated cement particles through a steric hindrance effect.

Other polycarboxylic acid-based copolymers with block and graft groups of polyethylene oxide (PEO) chains were synthesized by Li et al. (2005). The effects of the different PEO chains on the fluidity and adsorption in cement paste and on performance of the copolymer in concrete were studied. The properties of the copolymer were found to be affected by the length and density of PEO graft and block chains; copolymers with some block PEO chains at a certain length had good performances in the water-reducing capability and fluid-retaining ability. A PEO-based superplasticizer with much longer side chains of ethylene oxide (EO) – 130 moles of EO instead of 10–25 moles as in traditional acrylic polymer-based superplasticizer – produces a lower adsorption speed and reduces the typical retarding effect related to the early adsorption (Hamada et al., 2003; Qiu et al., 2011). Longer PEO side chains gave more fluidity at the same dosage, along with a shorter setting time. Concrete cured at low temperature when added with a polycarboxlate-based superplasticizer with longer graft chain possess good early strength development at and better durability (Dan et al., 2001). Polycarboxylate-type superplasticizers find, therefore, large applications in the precast industry as an alternative of the steam curing, for quick turnaround of forms and casting beds. With the use of these new generation superplasticizers it is possible, in fact, to overcome the negative effects of steam curing on strength loss, permeability, shrinkage, creep and frost resistance (Khurana and Torresan, 1997).

A modified PE-based superplasticizer, where a great number of carboxylic groups are replaced by a slump-loss controlling agent, is able to achieve a still higher slump retention with minimal setting retardation. Subsequently to the hydrolytic effect related with the OH^- presence in the aqueous phase of the cement paste, the slump can still increase by prolonging the mixing time, owing to the increasing adsorption of the polymer on the surface of the cement particles. In conclusion, a shorter mean backbone chain length gave more fluidity at the same dosage, along with a slightly longer setting time; while it did not affect the loss of workability with time. The concrete produced with this kind of superplasticizer, even displaying a slow initial development of mechanical resistance due to the delay of the cement hydration as the effect of the high number of adsorbed molecules on the cement surface (Hamada et al., 2009), is able to achieve the typical value of strength in a few (two to three) days. This concrete mix can be transported for long time, even in hot weather, keeping its initial slump level, avoiding the practice of re-dosing the concrete with water (Collepardi, 2005). However, its applications in preparing flowing concrete and high-performance concrete are still limited due to economic reasons. In other words, SNF is still the main superplasticizer used because of its relatively low cost.

16.2.3 Influence on durability

The influence of a superplasticizer on the pore structure of hardened cement paste mainly depends on its chemical nature. When polycarboxylate superplasticizer types are added to concrete, the amount of pores greater than 0.1 μm diameter decreases, while the amount of those < 0.1 μm increases (Sakai et al., 2006). It is commonly known that capillary pores of diameters $>$ 0.1 μm adversely influence paste tightness and permeability (Deja, 2002).

The general effect of SNF and polycarboxylate-type superplasticizers is to retard the initial cracking time of mortars and decrease their cracking sensitivity (Ma et al., 2007). The maximum crack width of mortars with polycarboxylate-type is reduced with respect to reference ones.

16.2.4 Applications

A brief overview of the application fields of superplasticizers is given. The most important innovative products that can be manufactured by the use of superplasticizers, due to the remarkable reduction of W/C ratio, are: high-performance concrete (HPC); ultra-high-performance concrete (UHPC); self-compacting concrete (SCC); and reactive powder concrete (RPC).

The term high-performance concrete refers to cement mixtures with a water–binder (W/B) ratio as low as 0.30–0.35, so that a 28-day compressive strength as high as 70–100 MPa and even 1-day compressive strength as high as 45–55 MPa can be obtained (Ran et al., 2009b). UHPCs exhibit exceptional tensile and compressive strengths and high durability. They possess a very low W/C ratio (< 0.25). Over the last years UHPCs have become a vanguard product in industrial and structural applications thanks to outstanding properties, such as compressive strength of 150–200 MPa, tensile strength of 8–15 MPa, with significant remaining post-cracking bearing capacity, and remarkable fracture energy of 20–30 kJ/m^2. The superior performance of these products have been achieved by tailoring their microstructures, that is by maximizing the packing density with very fine minerals, quartz powder and silica fume, and by enhancing the matrix toughness with an optimal fiber reinforcement (Sorelli et al., 2008; Plank et al., 2009; Schröfl et al., 2010).

SCC is considered to be a concrete that, once poured, will compact under its self-weight with no requirement for vibration. Moreover, it remains cohesive enough to be handled without segregation or bleeding (Selvamony et al., 2010). These characteristics allow SCC to completely fill formwork even when congested with reinforcement. A more complete description of SCC was given in Chapter 9.

By replacing a part of Portland cement with fly ash, slag or pozzolans in the concrete, it is possible to obtain the required features of the concrete mixture and, at the same time, contribute to the environment protection by reducing the consumption of raw materials, lowering the energy consumption and reducing the emissions of greenhouse gases (Malhotra, 2006; Damtoft et al., 2008; Mehta and Walters, 2008).

Uzal et al. (2007) indicated that concretes with high-volume natural pozzolan, in the presence of high range water reducers, are suitable for structural concrete applications with 12–14 MPa and 29–38 MPa compressive strengths at 3 and 28 days, respectively. SCCs made with a high volume of mineral admixtures achieve good workability, high long-term strength, good de-icing salt surface scaling resistance, low sulfate expansion and very low chloride ion penetrability (Swamy, 1997; Uysal and Summer, 2011).

16.2.5 Underwater concreting

Placing concrete under water can be particularly challenging because of the potential for washout of the cement and fines from the fresh mixture, with relative reduction in strength and integrity of the in-place concrete. Although several techniques have been used successfully to place concrete under water, other situations, requiring enhanced cohesiveness of the concrete mixture, necessitate the use of an anti-washout or viscosity-modifying admixtures (AWA). As a long-chain polymer 'bridge', AWA enables mixture to form a stable network structure with a flexible space, improves the cohesion of fresh concrete, minimizes the dispersion and segregation of fresh concrete and, in presence of superplasticizers, makes it possible to form high-quality, uniform, self-compacting concrete. Hu et al. (2011) compared the compatibilities between AWA and different superplasticizer, finding that compatibility of AWA with PC is better than that with SNF.

16.2.6 Air entrainment admixtures

Almost one-third of the concrete volume cast in the Nordic countries must possess adequate freeze–thaw durability. Since the damage of freezing and thawing involves expansion of water on freezing, air-entraining admixtures are used to introduce and stabilize microscopic air bubbles in concrete. These bubbles should be distinguished from accidentally entrapped air, which is in the form of larger bubbles left behind during the compaction of fresh concrete. The use air-entraining admixtures improve the durability of concrete exposed to cycles of freezing and thawing.

Air-entraining agents belong to a class of chemicals called 'surfactants'. A surfactant is a material whose molecules are strongly adsorbed at air–water or solid–water interfaces. Such substances have a dual nature, one portion of their molecules being polar and the other one being markedly non-polar.

For more than 50 years, neutralized Vinsol natural resin has been effectively used for air entrainment. Other synthetically manufactured surfactant agents have been introduced in recent years, based on blends of alkylbenzene sulfonates or salts of sulfonated hydrocarbons. Their use increased due to the high cost and limited availability of Vinsol resin.

Nagi et al. (2007) compared the properties of concrete containing a variety of air-entraining admixtures: two based on synthetic surfactants (sodium olefin sulfonate and alpha olefin sulfonate, indicated with ES and EA, respectively) and two based on Vinsol resin (indicated with EN and EV). Admixtures ES, EA and EN are commercially available while EV is an admixture accepted and used by most highway agencies. The air content of the concretes prepared with the air-entraining admixtures was set at 6 ±1.0%. The adequacy of an air-entraining admixture in a given concrete can be estimated by a spacing factor. The latter is an index of the maximum distance of any point in the cement paste from the periphery of nearby air void. A maximum spacing factor of 0.2 μm is required for satisfactory frost protection. The spacing factor for the four mixes was within this reference limit. Compressive, flexural and split tensile strength values of concrete containing EV and EA admixtures, at both 7 and 28 days, were relatively high compared to a non-air- entraining concrete. The compressive strengths of the concretes containing the other two admixtures, EN and ES, were lower (i.e., less than 70% of the reference concrete).

The difficulty with the use of any air-entraining admixture is that the spacing factor cannot be controlled in advance. This difficulty is avoided if rigid-foam particles of suitable size are used, such as compressible hollow plastic microspheres. They have a diameter of 10–60 μm which is a narrower range of sizes than in the case of air-entraining admixtures.

16.2.7 Self-curing agents

The purpose of curing, i.e. the procedure used for promoting the hydration of cement, at ambient temperature is to keep concrete saturated, or as nearly saturated as possible, until the originally water-filled space in the fresh cement paste has been occupied to the desired extent by the products of hydration of cement. Poor curing can reduce significantly the performance expected from the specified W/C ratio and cement content. When the W/C ratio of concrete is lower than a critical value, a self-desiccation may occur, leading to autogenous shrinkage. This phenomenon is of particular interest in high-strength concretes with a low W/C ratio. In the latter, the autogenous shrinkage can induce stress, causing micro-cracks if the stress exceeds the local tensile strength of the concrete (Bentur et al., 2001).

Concrete self-curing agents are able to reduce the water evaporation from concrete, hence increasing the water retention capacity of the concrete compared to standard concrete. According to ACI Committee 308, Curing Concrete, 'Internal curing refers to the process by which the hydration of cement occurs because of the availability of additional internal water that is not part of the mixing water'. The additional internal water is typically supplied by adding small amounts of saturated lightweight fine aggregates or water-soluble superabsorbent polymers (SAP). The benefits of internal curing include increased hydration and strength development, reduced autogenous shrinkage and cracking, reduced permeability and increased durability (Bentz et al., 2005). SAP polymeric materials have the ability to absorb a significant amount of liquid from the surroundings and to retain the liquid within their structure without dissolving. Many different types of SAPs are known; they can be produced by either solution or suspension polymerization. The commercially available SAPs are covalently cross-linked polyacrylates and copolymerized polyacrylamides/polyacrylates. They can absorb large quantities of water due to their ionic nature and interconnected structure (Jensen and Hansen, 2001).

El-Dieb (2007) measured the water permeability coefficient on self-curing and conventional concretes, at two cement contents and at different ages up to 56 days. Figure 16.4 shows the permeability coefficients as a function of time. Self-curing concrete shows higher water permeability than moist-cured conventional mixes, irrespective to the cement content.

16.3 Polymer-modified concrete (PMC)

16.3.1 Technology and properties

PMC, also known as latex-concrete modified (LCM), is developed by mixing a polymer dispersion (latex) with Portland cement type concrete mix with the aim of enhancing some of the characteristics of the hardened concrete.

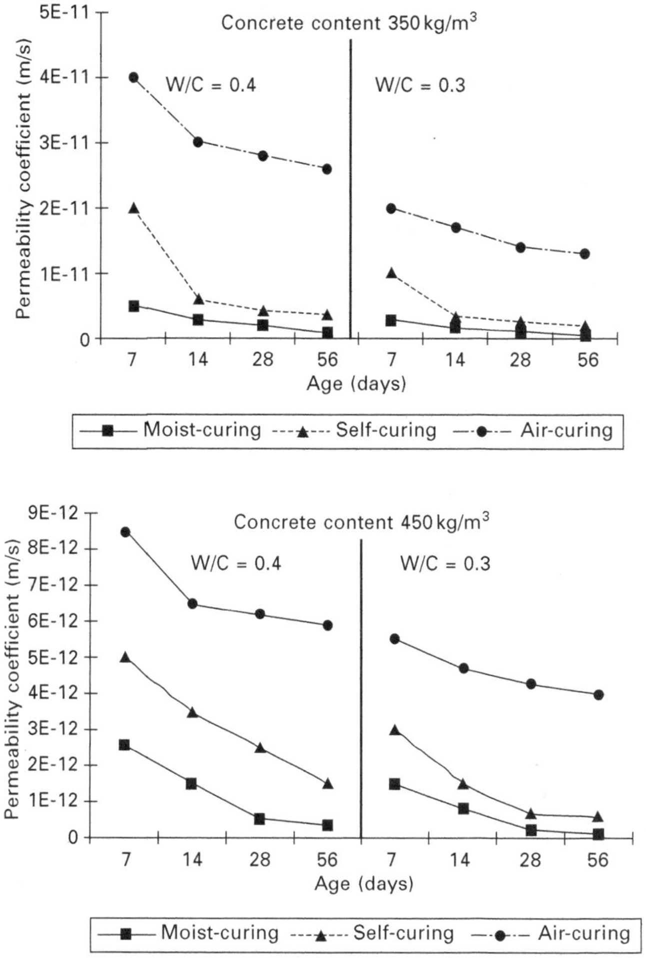

16.4 Water permeability coefficient vs. time for self-curing and traditional mixes (El-Dieb, 2007).

Standard cement concrete is a brittle material: the addition of polymer in the mix improves to some extent its flexibility (Wang et al., 2012).

Polymeric latex consists of very small polymer particles (0.05–5 μm) dispersed in water and are generally formed by polymerization in emulsion. Different polymers (i.e. thermoplastics, thermosettings and elastomers) can be incorporated also in other forms than latex: re-dispersible powder, water-soluble polymers or resins (Chung, 2004; Cestari et al., 2008). The content of polymer (solid basis) used in typical PMC formulations is 10–25% wt of cement.

The polymeric particles with a high molecular weight are held in dispersion with the aid of surface active agents (such as polyvinyl alcohol or alkyl phenols reacted with ethylene) (Mehta and Monteiro, 2006). The presence of such agents leads to the incorporation of air in concrete. Since an excess of air inclusion is detrimental for the mechanical properties of PMC as well as for its durability, air detraining agents are, then, usually added to commercial latexes (Wang et al., 2005).

PMCs are mixed with water and directly added to the cement/aggregate mixes, using the same equipment and tools used for mortar or concrete. They display a better workability over standard concrete. The slump tends to increase with polymer/cement ratio. This was attributed to the anti-friction action of polymeric particles and to the presence of air bubbles. The resistance to bleeding and segregation of latex-modified mortar or concrete is excellent in spite of their larger flowability characteristics (Paiva et al., 2006).

At the same level of polymer/cement ratio, better performance, especially resistance to freeze–thaw cycles, is obtained by pre-mixing polymer with sand (Wu et al., 2002). Curing under wet conditions, such as water immersion or moist curing, is detrimental to PMC. The PMCs require, in fact, a different curing method: the best properties are achieved by a combined wet and dry curing (Knapen and Van Gemert, 2009a). The setting and hardening of PMC are delayed to some extent in comparison with ordinary cement concrete, depending on the polymer type and polymer/cement ratio (Ohama, 1998; Puterman and Malorny, 1998).

According to Gajbhive et al. (2010), the polymers are adsorbed on the surface of cement hydration products and a weak interaction takes place between these products and the polymer. PMC hardening evolution is influenced by both cement hydration and polymer film formation processes in their binder phase. The cement hydration generally precedes the polymeric film formation, formed by coalescence of polymeric particles in latexes (Beeldens et al., 2001). Chemical reactions may also take place between the polymeric film and cement hydrates and aggregates, improving the properties of hardened latex-modified mortar and concrete (Ohama, 1998).

The tendency of certain water-soluble polymers to retard the flocculation of the cement particles minimizes the formation of a water-rich layer around the aggregate surfaces. They also provide a more uniform distribution of un-hydrated cement particles in the matrix. Both effects enable a reduction in the interfacial transition zone (Knapen and Van Gemert, 2011).

The presence of polymers influences the morphology of the $Ca(OH)_2$ crystals developed by the Portland cement hydration. Polymer bridges are formed between the layered $Ca(OH)_2$ crystals, increasing the interphase bonding and producing a better structure. Water-soluble polymers are found to improve further the internal cohesion of the cement paste. The strengthening of the hardened paste results in a significantly lower crack formation (Knapen and Van Gemert, 2009b). PMC has an excellent adhesion strength and durability and it is, therefore, widely used as a repairing material, enabling conservation of damaged or obsolete structures (Van Gemert, 2011). In addition to the increase in adhesion strength, PMCs are also characterized by an increase in tensile strength, resistance to dynamic loads, impact toughness and durability, the latter due to the reduction in permeability. On the other hand, the elastic modulus of concrete is reduced, even though no changes in the compressive strength are observed. A reduced elastic modulus might be particularly helpful when PMC is applied as a bridge deck overlay or repair surface, since it implies a reduction of the stress due to different shrinkage and thermal coefficients. PMC can also improve the resistance to corrosion, chemical attack and severe environments (acid rain and freeze–thaw cycles) (Su et al., 1996; Kardon, 1997; Silva et al., 2001; Barluenga and Olivares, 2004; Morlat et al., 2007; Bode and Dimmig-Osburg, 2010).

The properties of the fresh and hardened PMC are affected by several factors that tend to interact each other: the nature of materials used as latex, cement and aggregates; the mix proportions (W/C, polymer/cement ratio, etc.); air content; type and amount of surfactants and anti-foaming; and curing methods (de S. Almeida and Sichieri, 2006). The influence of any parameter depends on the type of polymeric backbone and, for the same backbone chain, on its molecular weight (Nair and Thachil, 2011).

16.3.2 Types of polymeric latex

A great variety of latexes, of a single or combinations of polymers, is available for use in PMC, such as: polyvinyl acetate, copolymers of vinyl acetate–ethylene, styrene–butadiene, styrene–acrylic and acrylic and styrene–butadiene rubber (SBR) emulsions (Beeldens et al., 2005; Wong et al., 2003). As already underlined, the chemical nature of the polymer has a profound effect on the properties of the relative PMC.

The molecular structure of SBR comprises both the flexible butadiene chains and the rigid styrene chains; their combination offers to the SBR-modified mortar and concrete many desirable characteristics, such as good mechanical properties, water tightness and abrasion resistance, especially when an appropriate polymer/cement ratio is used (Yang et al., 2009). SBR-modified lightweight aggregate concrete is an ideal material for precast components due to its low weight, high strength and high performance under severe service conditions (Rossignolo and Agnesini, 2004). SBR–cement composites give better strength than acrylonitrile butadiene rubber (NBR) (Nair and Thachil, 2011). However, the chemical resistance of NBR-cement composites is slightly better than SBR-ones. Styrene–acrylic ester copolymer (SAE) latex decreases the elastic modulus and increases the toughness of cement mortar to a larger extent.

The addition of methylcellulose to cement concrete increases tensile strength and tensile ductility, and compressive modulus; on the other hand, tensile modulus, compressive strength and compressive ductility are decreased. All these effects increase with increasing the methylcellulose content (Fu and Chung, 1996). The combined use of silica fume (15% wt of cement) and methylcellulose (0.4% wt of cement) as admixtures was found to give concrete exhibiting high bond strength to steel rebar (Fu and Chung, 1998).

Ethylene-vinyl acetate (EVA) copolymer, as latex or re-dispersible powder, is usually added to concrete to improve the fracture toughness, impermeability and bond strength to various substrates. EVA copolymer reduces the calcium hydroxide in the cement paste due to chemical interactions between Ca^{2+} ions and anions acetate released by EVA alkaline hydrolysis. These interactions decrease the EVA flexibility, thus promoting the increase of elasticity modulus of cement-based materials (Betioli et al., 2009).

Compared with polyvinyl alcohol (PVAl), polyvinyl acetate (PVAc) yields concretes with better strength and chemical resistance properties due to the formation of a strong chemical bond between the two phases, i.e. cement and polymer. Re-dispersible polymer powder of vinyl acetate-versatic vinylester copolymer (VA-VeoVa) exhibits excellent water-reduction and water-retention effects in cement concrete and improves concrete properties effectively. Moreover, the addition of VA-VeoVA copolymer strongly reduces the portlandite formation, being this result attributed principally to lower Portland cement hydration (Gomes and Ferreira, 2005).

The introduction of an epoxy resin, even at low ratios, increases the corrosion resistance of the resulting PMC, due to the reduction of the concrete permeability (El-Hawary and Abdul-Jaleel, 2010). The epoxy emulsion has a greater effect on improvement of properties of the mortar than the acrylic one, at the same amount of polymer/cement ratio. Moreover, epoxy emulsion is considered non-re-emulsifiable latex (Type II ASTM 1059) and, therefore, it is not destabilized under humid and alkaline environment, but acrylic-based mortar can be. Thus, epoxy emulsion-based mortar can be used for repair works in humid and industrial environments (Aggarwal et al., 2007).

The results obtained by Morin et al. (2011) showed that the presence of thin polymer layers at the paste aggregate interfaces in 'micro-concretes', with a maximum aggregate diameter of 10 mm, can have a significant effect on cracking behavior at much lower polymer dosages than those commonly used in conventional polymer-modified mortars.

16.3.3 Durability

The durability of PMC to freeze–thaw is generally very high due to the introduction of air in the latex concrete. On the other hand, the durability to strong acids is not always appreciable, due to the alkalinity of concrete.

Concretes with the styrene–acrylic ester polymer or with vinyl-copolymer, with a content of polymer of 7.5% wt. of cement, showed a higher resistance in a solution of 0.5% sulfuric acid (pH around 1.0) than the reference concrete without polymer (Monteny et al., 2001). Furthermore, according to Xiong et al. (2001) a mortar modified with soluble glass (Na_2SiO_3) and PVAc shows an appreciable high sulfuric acid resistance. According to Li et al. (2009), after a 5-year sulfuric acid attack, the compressive strength loss of PMC with Na_2SiO_3–PVAc was 29%, whereas the control mortar had a much higher strength loss of 50%.

The fire-protecting performance and thermal resistance of polymer-modified mortars are generally limited, since they contain polymers which are thermally unstable and combustible. Shirai et al. (2010) found that PMC based on SBR, EVA and PAE latexes showed an improvement in fire-protecting performance by adding appropriate flame retardants: magnesium hydroxide for EVA-modified concrete; calumite for SBR-modified concrete; and aluminum hydroxide and nitrite type hydrocalumite for PAE-modified concrete.

16.4 Polymer-impregnated concrete (PIC)

16.4.1 Technology and properties

PIC is a hardened hydraulic cement concrete impregnated with a monomer that is subsequently polymerized *in situ*. Concrete is characterized by channels and voids, partially full of water. The basic concept of PIC is that, if the voids are responsible for poor durability and strength of concrete, then filling the voids with a polymer should improve the characteristics of the concrete (Angle et al., 1999). To this aim, the voids must be cleared by water by drying the concrete in some manner. PIC provides advantages over conventional concretes in terms of strength and durability; wear resistance appreciably improves with respect the traditional concrete (Sebök and Stránêl, 2004).

The porosity and pore-size distribution assessed by scanning electron microscopy (SEM) observations and mercury intrusion porosimetry (MIP) tests indicate that the polymer impregnation reduce the pore volume and maximum pore size, which are the key factors affecting the permeability of concrete. The maximum pore size of PIC can be reduced to less than 300 nm, which is very small compared with 1200 nm in traditional concrete, thus inhibiting permeation of water and aggressive solutions (Chen et al., 2006).

The degree to which the available voids in the concrete are filled with monomer determines whether the concrete is partially or fully impregnated. In the partially impregnated concrete (sometimes called 'surface impregnated concrete') the monomer is introduced into the concrete by soaking at atmospheric pressure; on the other hand, fully impregnated concrete is impregnated to the full depth of the section. Actually, full impregnation implies that about 85% of the available void space after drying is filled. The different methods used for full and partial impregnation produce concrete of differing physical characteristics.

The simplest and cheapest approach for impregnation is to immerse the concrete in a low-viscosity monomer; this technique is applicable to precast concrete elements and is based upon the easiness of penetration of a low-viscosity monomer into a limited depth. Full impregnation is obtained by first thoroughly removing free water from the concrete in order to provide the maximum amount of previously water-filled pore space for monomer filling. This is followed by complete monomer saturation, usually under pressure, and subsequently its polymerization. The techniques employed for free water removal and monomer saturation needed to fully impregnate the concrete depend on the size and configuration of concrete elements. Therefore, this treatment is restricted to concretes that can be made and handled in precast plants.

A simplified process description to obtain a PIC is shown in Fig. 16.5. A vacuum pump is used to remove water and air incorporated inside the pores of concrete specimen. After evacuation, the monomer is allowed to pass into the impregnation unit. After penetration, the monomer polymerizes *in situ*.

The polymerization can take place in three different ways: (i) polymerization catalyzed by various catalysts at room temperature; this process is slow and scarcely controllable; (ii) polymerization at room temperature induced by gamma radiation; and (iii) polymerization in the presence of a catalyst at moderate temperature (about 70° C), by heating the concrete with steam, hot water or infrared (IR) heaters. The polymerization reactions consist of molecules of the monomer that chemically link together to form a long repeated chain-like structure, with high molecular weight, into the concrete voids.

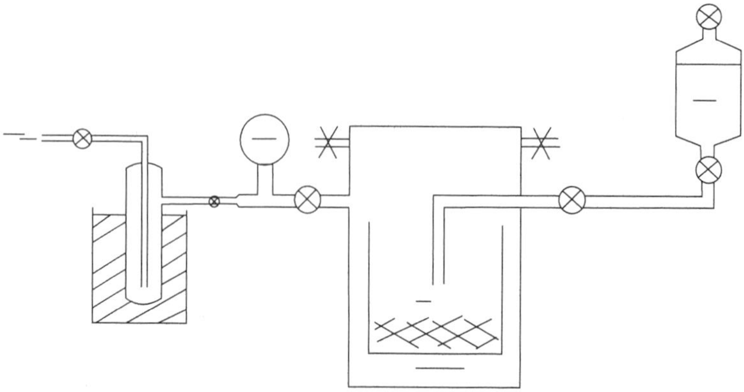

16.5 Schematic impregnation unit (Khattab et al., 2011).

The resulting composite material consists essentially of two interpenetrating networks: one is the original network of hydrated cement concrete and the other is a continuous network of polymer that fills most of the voids in the concrete.

Monomer evaporation is a problem when high-vapor pressure monomers, such as methyl-methacrylate, are used. Several techniques have been used to minimize monomer evaporation, i.e.:

- Wrapping monomer-saturated concrete element in polyethylene sheet or aluminum foil.
- Encapsulating the concrete element in a tight form during impregnation and polymerization.
- Impregnating with monomer, followed by dipping the impregnated concrete in high-viscosity monomer prior to polymerization.
- Polymerizing monomer-saturated concrete element underwater.

Underwater polymerization appears to be the most feasible method for large-scale applications, for instance pipe, beams and panels. The method has been successfully used in conjunction with radiation and thermal-catalytic polymerization, observing a very little surface depletion. Underwater polymerization does not have any detrimental effects on the PIC properties and can produce elements with highly reproducible characteristics.

Any types of aggregates, cements and admixtures used to produce concrete can be impregnated. Similarly, curing procedures and curing age for strength development prior to impregnation are not critical. On the other hand, the kind of materials and curing conditions used influence the final properties of PIC. The highest strengths have been obtained with high-pressure steam cured concrete. A good quality dense concrete requires a lower polymer amount for full impregnation than a more porous, poorer-quality concrete. Fractured concretes can also be repaired to some extent using impregnation techniques.

In principle, any monomer capable of undergoing polymerization in voids of a hardened concrete can be used to produce PIC. The selection of a suitable monomer, however, is usually based on the impregnation and polymerization characteristics, the availability and cost and the achievable properties of the polymer and the PIC. Plasticizers, such as dibutyl phthalate, may be added to monomers to improve the flexibility of inherently brittle polymers, such as polymethyl methacrylate and polystyrene.

Monomers such as methylmethacrylate and styrene are commonly used for penetration because of relatively low viscosity, high boiling point (small loss due to volatilization) and low cost (Ohama, 1997; Fowler 1999). Methylmethacrylate is an acrylic-type polymer, whose characteristics include lightweight, high toughness, high hardness and abrasion resistance. Concrete impregnated with styrene display the advantages of not altering the visual aspects of the concrete and of developing a new technology involving recycled polymers. Compared with the process used for conventional PIC, concrete impregnated with recycled polymer is simpler, involves little technological resources and consumes less energy. For the production of concrete impregnated with styrene, the impregnation monomer is prepared by the dissolution of recycling waste expanded polystyrene in a solution of acetone (70%) and cyclohexane (30%) (Amianti and Botaro, 2008).

16.4.2 Characteristics and durability

The increases in strength are function of the depth of impregnation of concrete. Fully impregnated PIC typically develops compressive strengths three to four times greater than the concrete from which it was made; increased tensile and flexural strength; excellent durability, particularly to freeze and thaw cycles, and acid resistance, due to its extremely low permeability. Inexplicably, the modulus of elasticity is 50–100% higher than traditional concrete, even though the modulus of the polymer is no more 10% higher than that of the concrete (Fowler, 1999). PIC exhibits almost zero creep properties. Table 16.1 gives the main properties of PIC containing radiation-polymerized methylmethacrylate.

A new impregnation process involving evacuation by ultrasonic vibration

Table 16.1 Properties of polymer-impregnated concrete containing radiation-polymerized methylmethacrylate (adapted from Naidu, 1992)

Property	Unit	Concrete without polymer	Concrete polymer – impregnated
Polymer loading	% weight	0	5–7
Compressive strength	N/mm^2	36	140
Tensile strength	N/mm^2	2.8	11.2
Modulus of elasticity	N/mm^2	24 000	43 000
Modulus of rupture	N/mm^2	5.1	18.1
Flexural modulus of elasticity	N/mm^2	30 000	43 000
Water permeability	m/yr	1.9×10^{-4}	0
Water absorption	%	5.3	0.29
Freeze–thaw durability	Nr cycles	590	2420
Hardness-impact ('L' Hammer)		32.0	55.3
Corrosion by 15% HCl (84-day exposure)	% Wt loss	10.4	3.6
Corrosion by sulfates (300-day exposure)	% expansion	0.144	0
Corrosion by distilled water		Severe attack	No attack

and polymerization process of monomers using microwaves has been experimented with for the manufacture of PIC. In particular, a modified microwave reactor using magnetron was used for polymerization of styrene/methylmethacrylate (1:1). The degree of polymerization increased up to 30% and more homogeneous PIC is achieved as compared to the conventional thermal method. The mechanical strengths also increased more than two or three times (800–1200 MPa) and the resistance to acids improved up to 25% (Ku et al., 2008). The improved durability was attributed to the sealing of pores and cracks by the polymer, as confirmed by SEM and porosimetry data (Nair et al., 2010a, 2010b). The microstructural changes in the mortar, on exposure of the specimens to hydrochloric acid and seawater for 7 and 28 days, indicated that the polymer addition decreased the voids in the mortar, thereby preventing leaching of water-soluble salts present in the ordinary cement concrete (Nair et al., 2010c).

Specimens of Portland cement paste with 5% of rice husk ash were impregnated with unsaturated polyester resin and irradiated at a dose of 30 kGy. The durability was analyzed after immersion in seawater and in magnesium sulfate solutions at different percentages, up to 6 months. The impregnated blended cement paste irradiated displayed a good resistance towards sulfate and seawater attack compared with the neat blended cement paste (Khattab et al., 2011).

In conclusion, the use of polymer impregnation process enhances the chemical resistance of hardened concrete. This solution is economically viable, especially for pipes of small diameters. However, if sulfate resistance is the aim, the use of pozzolanic cements are mostly recommended since no increase in costs or procedures of fabrication is involved (Pacheco-Torgal and Jalali, 2009). The stress–strain curve reported in Fig. 16.6 indicates that

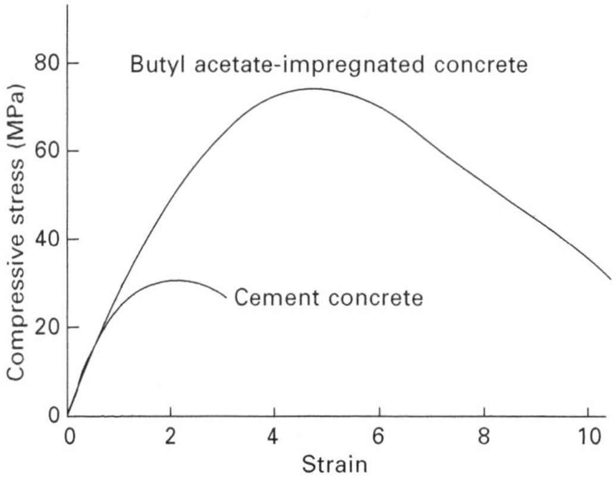

16.6 Stress-strain properties of polymer-impregnated concrete compared with cement concrete.

the impregnation of a concrete with polymer transforms the traditional brittle concrete into a ductile material.

PIC has a wide range of application areas due to its improved durability and structural properties compared to conventional concrete. However, its commercial utilization is restricted for several reasons, the main one being its complicated technology (Naidu, 1992). PIC has been used for the production of high-strength precast products and to improve the durability of bridge deck surfaces. Most of the work on partially impregnated concrete has been done in developing a technique to protect concrete bridge decks and spillways from damage caused by de-icing salts and freeze–thaw deterioration. The process has also been applied to concrete stilling basins, curbstones, concrete pipes and mortar linings, deteriorated buildings and floor constructions.

16.5 Polymer concrete (PC)

16.5.1 Components

PC is a composite material in which the aggregate is bound together in a polymeric matrix. The composite, therefore, does not contain hydrated cement phase. Because the use of a polymer instead of Portland cement represents a substantial increase in cost, PC is employed only in applications in which the higher cost can be justified by superior properties (Rebeiz, 1996).

In PC, thermosetting resins are generally used as the principal polymeric component due to their high thermal stability; thermoplastic polymers are also used to a minor extent. Among the thermosetting, epoxy resins are mainly employed to produce PC due to the strong adhesion exerted by these resins to most building materials, their low shrinkage, good creep and fatigue resistance and low water sorption; however, they are relatively expensive. Unsaturated polyester resins are also used, due to their superior chemical and mechanical properties, combined with their lower cost (Rebeiz and Fowler, 1996). Other materials that are used for PC include: methylmethacrylate, styrene, vinyl esters, furfuryl alcohol and furan resins, and their copolymers. Curing agents or initiators are used to promote the curing reactions. For any PC, the working and curing times depend on the concentration of any component, the curing temperature, the mass volume.

Referring to the aggregates, from a petrological standpoint, limestone, basalt, silica, quartz, granite, and other high-quality materials are employed, whether crushed or naturally reduced in size. Aggregates can, however, also be manufactured from industrial by-products. The use of fly ash as a filler in polymer concrete is very appealing since it improves the physical properties of PC, in particular its compressive and flexural strengths (Harja et al., 2009; Bărbuță et al., 2010). The effect of $CaCO_3$ as a filler on PC was found to be even more effective than that of fly ash. This was attributed to the larger surface area of $CaCO_3$ particles and the higher adhesion between the resin binder and the aggregate (Jo et al., 2007).

The aggregates must be dry and free from dust and organic materials. Moisture and/or dust on the aggregates, in fact, reduce the bond strength between the polymeric phase and the aggregate. A proper grading of aggregates would assure a minimum void volume for packed aggregate. This minimizes the amount of monomer required to guarantee proper bonding of all the aggregate particles and results in a more economical PC. The aggregate/resin ratio varies with the monomer formulation; it has been used in ratios from 1:1 to 15:1 by weight, depending on the aggregate gradation (Muthukumar and Mohan, 2004). Generally speaking, the optimum polymer content varies from 12% to 14%; the very fine aggregates (filler), in particular, form the major component of the polymer concrete.

Immediately after the addition of curing agents or initiators, the liquid monomer is mixed with coarse and fine aggregates. Formwork for production casting PC must be solvent resistant, must possess a low coefficient of thermal expansion, have smooth, cleanable surfaces and preferably should be a good heat conductor.

The process of curing can be controlled through temperature and curing agent content and type. The process can take from minutes to hours. Some formulations benefit from heating during the curing period, whereas others simply require time at ambient temperatures. Curing can take place at temperatures varying from about –15 °C to 60 °C (Rebeiz, 1996).

16.5.2 Characteristics and applications

Performance of PC is generally determined by the nature of the polymer binder to a greater extent than by the type, the amount and the grading curve of filler. The properties of the matrix polymer are highly dependent on time and the temperature to which it is exposed. The content of filler plays an important role in restricting the deformation of polymer concrete (Aniskevich and Hristova 2002).

The typical age effect on the compressive and flexural strength of a PC based on unsaturated polyester is shown in Fig. 16.7. PC achieves more than 80% of its 28-day strength value in the first day. Conversely, hydraulic cement concrete usually achieves about 20% of its 28-day strength in one day. The early strength gain is important in precast applications.

The compression stress–strain curve for the same PC is shown in Fig. 16.8. The ultimate compressive strength and strain of PC are typically about two to three times greater than those of a typical Portland cement concrete. The relatively high modulus of the PC confirms that the material is mostly suitable in precast applications. However, different resins can be chosen in order to produce PC with low stiffness, thus making the material suitable

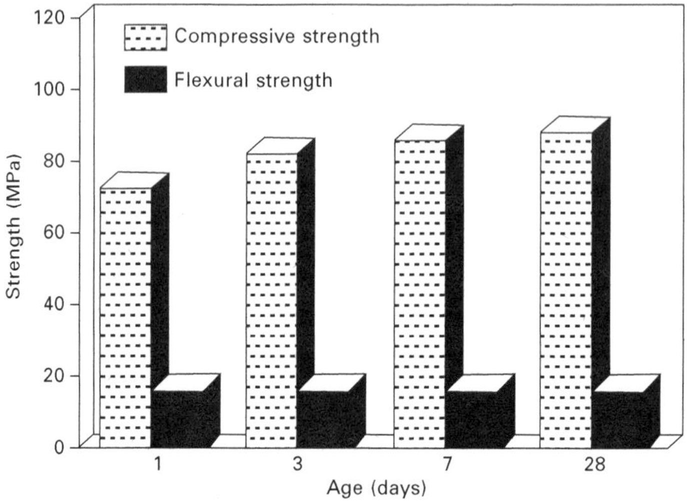

16.7 Age effect on compressive and flexural strength (Rebeiz, 1996).

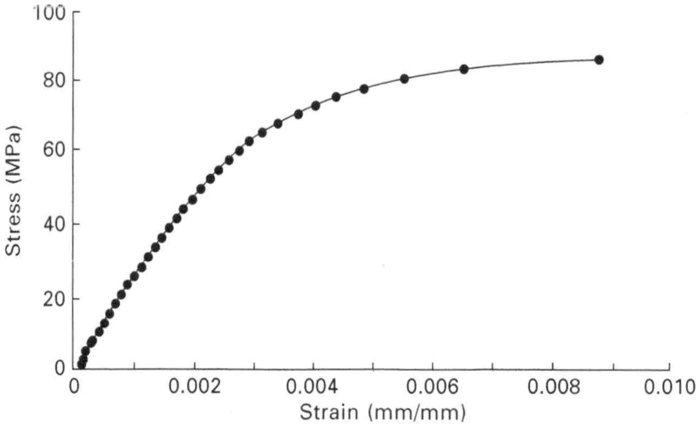

16.8 Typical stress–strain curve in compression (Rebeiz, 1996).

for overlay and repair applications on Portland cement concrete, especially when large thermal and mechanical movements occur (Rebeiz, 1996).

The viscoelastic properties of the polymer binder give rise to high creep values for PCs (Blaga, 1973). The elastic modulus depends on the formulation and may range from 20 to about 50 GPa; the tensile failure strain is usually 1%. Shrinkage strains vary with the polymer used (high for polyester and low for epoxy-based binder).

Abdel-Fattah and El-Hawary (1999) studied the influence of the type of resin (epoxy and polyester) on the flexural and ultimate compressive strength, strain at failure and ductility of PC. The results show that the maximum flexural strength for polymer concrete can be three times higher than that of an ordinary Portland cement concrete with the same ultimate compressive strength. The highest values were generally found for epoxy-based PCs. To achieve the full potential of PC for certain applications, various fiber reinforcements are used: glass fibers and fabrics, carbon fibers, metal fibers, nano-particles.

Vipulanandan and Mantrala (1996) found a maximum compressive stress of about 55 MPa for polyester PC. Moreover, they showed that the compressive strength increase up to 16% with the addition of 6% wt of glass fiber. Ribeiro et al. (2003) determined the influence of the fibers on thermal expansion coefficient of epoxy polymer mortars reinforced with both chopped carbon and glass fibers. They concluded that the presence of chopped glass fibers (1% wt) has no significant effect on thermal expansion coefficient of epoxy polymer mortar, while the inclusion of carbon fibers (2% wt) on the same mortar formulation reduces the thermal expansion coefficient of the PC at temperatures above the room temperature.

Nano-sized particles, acting as mechanical, optical or electrical defects, can provide multifunctionality to the polymer concrete (Baur and Silverman, 2007; Maity et al., 2006). The addition of even a small volume fraction of nano-scale fillers introduces a large region of interfacial surface, possessing different properties with respect the polymeric matrix. Winey and Vaia (2007) estimated that for a 1% volume fraction of 2 nm particles dispersed in a matrix, as much as 63% of the total volume could be occupied by the interphase region. The use of nano-particles led to a significant increase in flexural and compressive elastic modulus of the tested polymer mortars (Reis et al., 2011). No flexural or compressive strength enhancement was, however, observed with the introduction of nano-particles into the polymer. Improved ductility and toughness in brittle thermoset polymers were generally reported due to the addition of silica nano-particles (Zhao et al., 2008). However, further research and development is recommended before these materials can gain significant position in the market (Móczó and Pukánszky, 2008).

There is a growing interest aimed at using post-consumer polyethylene-terephthalate (PET) and polypropylene wastes for the production of PC due to environmental and economic reasons (Meran et al., 2008; Hugo et al., 2011), as was illustrated in Chapter 13. Different authors (Chen, 2003; Tawfik and Eskander, 2006; Mahdi et al., 2007; Jurumenha and Dos Reis, 2010) achieved the chemical transformation of recycled PET through degrading glycolysis reaction with different glycols, namely: propylene glycol, diethylene glycol, triethylene glycol and a mixture of diethylene glycol with propylene glycol or triethylene glycol; the obtained monomers are, then, used to prepare PC. The glycolized products are converted into unsaturated polyester after the reaction with maleic anhydride. Choi and Ohama (2004) reported that the obtained polystyrene mortars possess similar physical properties and durability than polymer mortars based on commercial unsaturated polyester resin.

The compressive strength of polymer mortars and PCs produced starting from recycled PET was found to vary from 15 to 28 MPa and from 20 to 42 MPa, respectively. The tensile strength of PC was even higher than that of equivalent grade cement concrete (Mahdi et al., 2010).

The properties of PC based on unsaturated polyester resins, from recycled PET plastic waste, and recycled concrete aggregates were analyzed by varying the coarse and fine aggregate ratio and the resin content (Jo et al., 2008). The strength of the PC was found to increase with increasing the resin content; beyond a certain resin content, however, the strength did not change appreciably. By comparing the stress–strain curves of PCs with 100% natural aggregate and 100% recycled aggregate, different failure mechanisms under compression were found. Referring to the resistance to acid, the polymer concrete with 100% recycled aggregate showed poor acid resistance; on the other hand, it was not attacked by alkali compounds, as observed from the weight change and the compressive strength.

The exceptional mechanical resistance of PC enables production of lightweight elements; it is also widely used in electrical applications (Pratap, 2002). Furthermore, its outstanding durability and its minimum wear from abrasion are other characteristics that make PC a high quality material. In particular, the inherent very low permeability of the material to liquids and gases (unlike Portland cement concrete, a highly porous material) and its exceptional corrosion resistance (unlike metallic materials) allow PC to be used as safe disposal of toxic and acid wastes over long periods of time. PC is widely used as an alternative to cast iron and steel for the machine tool manufacturers, especially for the production of machine tools beds. The advantages of using PC in such applications are: the ease of manufacturing; resistance to corrosion; high strength-to-weight ratio; low thermal conductivity; and the most important, vibration damping (Orak, 2000). PCs are also frequently used as coatings used for protecting concrete structures, as described in the next section.

16.5.3 Durability

PCs display an exceptional stability toward the freeze–thaw cycles, good resistance to to chemicals and corrosive agents, low permeability to water. Polyester-based PC is more acid-resistant than the epoxy one; it is, however, less resistant to alkalis than epoxy-based PC. However, strength and stiffness of PCs are typically influenced by the service environment, in particular temperature and thermal cycles.

The effect of temperature on the compressive and flexural strength of unsaturated polyester resin PC is shown in Fig. 16.9. The influence of the temperature on the mechanical strengths of three different PCs (epoxy, polyester and acrylic) was also analyzed by Pardo et al. (1995). PC specimens were

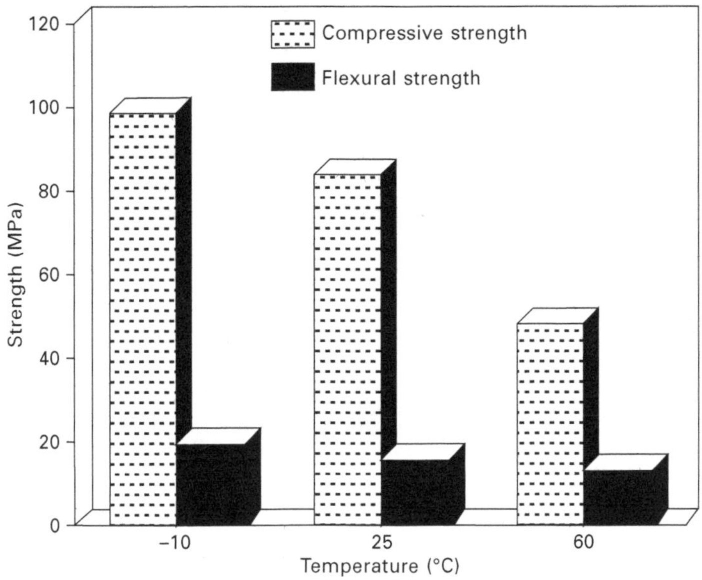

16.9 Temperature effect on compressive and flexural strength (Rebeiz, 1996).

tested in bending and compression, after conditioning at different temperatures (from 20 to 200 °C). The results showed that a significant decrease in both flexural and compressive strength took place by testing the specimens at temperatures greater than the ambient temperature.

Ribeiro et al. (2004) investigated the influence on the flexural behavior of two different binders, i.e. unsaturated polyester and epoxy, of several thermal treatments: permanence at a constant temperature (between –20 and +100 °C); repeated positive thermal cycles (+20 to +100 °C); and repeated freeze–thaw cycles (–10 to +10 °C). The positive thermal cycles have negative effect on the flexural strength of epoxy mortars: a drop of 14% and of 75% with respect to its initial value was measured on the flexural strength after 50 and 100 thermal cycles, respectively. Exposure to freeze–thaw cycles resulted in no relevant influence on the flexural strength of both polymer mortars.

PCs produced using orthophthalic or isophthalic polyester and fly ash as filler displayed increased compressive strengths by increasing the concentration of fly ash. High values of modulus of elasticity were also obtained (up to 29 GPa), comparable to those observed in high-strength Portland concrete (Gorninski et al., 2004). The resistance to acid environments, representative of those often encountered in corrosive processes of industrial environments, was also investigated: polymer mortars made with both types of unsaturated polyester resins did not show any evidence of physical surface changes or weight loss. However, a decrease in the flexural strength of the specimens exposed to corrosive agents was registered and this effect was more pronounced at lower filler concentrations (Gorninski et al., 2007; Reis, 2009).

In most of their applications, PCs are exposed to outdoor environment. Polyester and epoxy-based PCs are to some extent sensitive to degradation effects due to natural weathering, with epoxy-based mortars being in general more weather resistant than polyester-based ones. One-year outdoor exposure to corrosive maritime environment has a moderate harsh effect on mechanical flexural strength of both types of PCs (Ribeiro et al., 2008).

16.6 Coatings

16.6.1 Components and properties

Materials to protect the surface, also called coatings, of cement concrete can be classified into three groups: pore liner (makes the concrete water repellent), pore blocker (reacts with certain soluble cement constituents and forms insoluble products) and coatings (form continuous film on concrete surface) (Medeiros and Helene, 2009). The main function of a coating is to prohibit water and any soluble salts from penetrating the concrete, causing corrosion, leaking and other problems. Coatings must be also very effective in minimizing the rate of corrosion once it has been initiated by preventing further access of moisture and oxygen to the steel surface.

The coatings used for protecting concrete structures can be: cement-based, polymeric resins and silane/siloxane products. They vary in terms of method of application, durability performance and price. Most surface treatments must be applied to a clean, dry and sound substrate at moderate temperature and humidity conditions in a well-ventilated space. A relatively smooth surface is needed for liquid applied membranes (Fowler and Whitney, 2011). Since these conditions do not always prevail, the difficulty and cost in achieving the appropriate installation conditions may influence the choice of a system.

The base polymers of polymeric coatings include, but are not limited to, acrylics, epoxies, polyesters, polyurethanes, styrene–butadienes, polyvinyl acetates, chlorinated rubbers. Silanes and siloxanes are also frequently used. Thin polymer concrete overlays (TPCOs), with a thickness of 25 mm or less, are widely used as coatings. Their advantages include the rapid development of high strength and the ability to cure or harden under a wide range of environmental conditions, very low permeability, low weight due to the small thicknesses, wear and abrasion resistance and resistance to a wide range of aggressive chemicals, excellent bond to concrete, especially in the case of polyester–styrene and epoxy resins (Fowler, 1998; Tabor, 2004).

The bond between a coating material and concrete substrate must remain intact: even with proper surface preparation, this bond can fail. For instance, if the coating is highly impermeable to water vapor, water may condense at the concrete/coating interface and destroy the bond. Generally speaking, the adhesion strength of a resin-based coating to concrete is somehow higher that that measured for cement-based coatings (Al-Dulaijan et al., 2000a). The epoxy-based coatings display a better adhesion to concrete substrate than the acrylic-based ones (Al-Dulaijan et al., 2000b). In addition, the conditions at which the cure of the resin takes place have a noticeable influence on the adhesion exerted to concrete: the adhesion decreases when the substrate is wet and is promoted by moderate (at least 23 °C) curing temperatures (Littmann and Schwamborn, 1999).

Surface treatments should be as permeable to water vapor as the untreated concrete, in order to ensure that concrete, especially in freezing climates, has sufficient permeability to allow the substrate (and any reinforcement it contains) to remain substantially dry. It is reported that silicon-based treatments, in particular silanes and siloxanes, possess these characteristic (Basheer et al., 1997). Epoxy resins, especially if applied in multiple coats, and acrylic coatings decrease to some extent the breathability of concrete. In the case of acrylic coatings, however, in the longer term there is no difference in permeability between untreated and surface-treated concretes (Basheer et al., 1990).

A high resistance to liquid water penetration must be first assured by coatings: epoxies, silane-based coatings, methylmethacrylate, alkyl alkoxysilane and oligomeric siloxane have been found to be very effective in reducing the water intake of concrete. Silane-siloxane was found to be slightly inferior to pure silane in reducing the sorptivity of concrete. When an acrylic top coat was applied on to the base coating of silane, superior performance is obtained (Basheer et al., 1997). However, the protective properties of a coating may degrade in the long term when the permeability of the coating is increased by deterioration.

The surface treatments should also improve the resistance of concrete to carbonation, chloride ingress, sulfate ingress, freeze–thaw, salt scaling and, above all, corrosion of embedded steel (Basheer et al., 1997). Referring to the performance of the different coatings against carbon dioxide penetration, epoxy resins and some acrylic coatings have been found effective as anticarbonation coatings. Alky-alkoxy silanes do not improve the resistance of concrete to carbonation, while siloxane with large alkyl group may marginally reduce the carbon dioxide diffusion.

Although many polymeric coatings, such as those based on epoxies, methylmethacrylates and polyurethanes, provide an excellent barrier against the ingress of chloride ions into concrete (Pfeiffer and Scali, 1981), with a consequent appreciable increase of the time to initiation of corrosion of steel bars (Cleland and Basheer, 1995), the best performance is usually obtained with silanes. It is reported that the application of a silane coating is able to reduce the corrosion current compared with the untreated specimens, bringing about a reduction of corrosion by-product (Pfeifer, 1987). Good performance has also been reported for silane-siloxane treatments: these sealants with a topcoat and an acrylic coating were among the most effective in reducing reinforcement corrosion.

Vassie (1991) reported that the most effective coatings to slow down the corrosion rate of active steel in concrete (0.5 W/C ratio) were: solvent-free urethane, alkyl alkoxy silane, silane/acrylic and liquid plastic dressing. The alkyl alkoxy silane, in particular, was able to reduce the ongoing corrosion by around 35%; however, it failed to stop the corrosion completely.

Medeiros and Helene (2009) compared different surface treatments (silane/siloxane dispersed in solvent, acrylic coating, polyurethane coating) able to inhibit chloride penetration in concrete. They concluded that the polyurethane coating presented the best efficiency in reducing the chloride penetration (up to about 85%). The authors, however, pointed out that their result cannot be universally considered since the efficiency of the treatment does not depend only on the type of resin that composes it.

Different authors reported the success of methylmethacrylate impregnation on US bridges to prevent the onset of corrosion (Smoak, 1990; Weyers et al., 1990). A silane coating was also found to be effective in wet zones of a bridge (Van Es, 1992), even though the effectiveness of silane in a freeze–thaw environment is somehow questionable: silane does not improve the freeze–thaw resistance of concrete (Perenchio, 1988). However, other researchers reported that the freeze–thaw resistance of concrete treated with silanes and siloxanes, with or without an upper coat of acrylic, was much better than that of untreated concrete (Basheer et al., 1990; McGill and Humpage, 1990). This discrepancy can be explained taking onto account two different effects: if the concrete is initially dry, the application of silane will ensure a low degree of water saturation during freezing and thawing. As a result, the resistance offered by the surface-treated concrete to freeze–thaw deterioration may be significant. On the other hand, if the concrete is saturated, the application of silane may not improve the resistance to freeze–thaw deterioration because dilatation of water-filled capillary pores will take place as a result of expansion of the water already contained in concrete during freezing.

16.6.2 Durability

The durability of surface treatments depends on the type of polymer on which the coating is based and, very importantly, on the depth of penetration. A significant depth of penetration has been generally observed for silane and siloxanes (Basheer et al., 1997). Other polymeric treatments penetrate to a few millimeters or less and, in fact, many form a film on the surface.

In normal environments a coating for concrete must be resistant to oxidation as well as to ultraviolet (UV) and IR radiation exposure. On floors, resistance to abrasion and punctures as well as resistance to mild chemicals (salts, grease and oil, battery acid and detergents) are also important. In order to assess the durability of coatings, they are typically subjected to accelerated aging by exposure to UV light (alone or in combination with other agents), IR radiation, water, thermal cycling, with and without the influence of salt, freeze and thaw. Epoxy, methylmethacrylate and alkyl alkoxysilane coatings were found highly resistant to the artificial weathering tests (Basheer et al., 1997). The performance of a silane, with or without an upper coat of acrylic, under an artificial weathering was fairly good. The effects of solar radiation were found appreciable for acrylic rubber emulsion coatings (around 1 mm thickness), whose elongation appreciably decreased (Swamy and Tanikawa, 1990).

After accelerated AASHTO (American Association of State Highway and Transportation Officials) diffusion tests, epoxy coating demonstrated a very high chloride resistance (Cabrera and Hassan, 1994). Structural members protected with an acrylic rubber coating showed an almost total absence of penetration of chlorides throughout the exposure period and the coating maintains long-term durability of reinforced concrete structures under marine environment (Yamada et al., 1997). Acrylic-based coatings were found also able to prevent the penetration of carbon dioxide into the concrete (Swamy et al., 1998).

A broad range of coatings for reinforced concrete, including epoxy, silicone, urethane, acrylic rubber, acrylic resin and polyester, along with polymer-modified mortar, were studied in relation to the resistance to simulated marine exposure (Umoto et al., 1994). After 18 months, no evidence of corrosion was observed in any of the coated beams while uncoated specimens exhibited corrosion at the interface and in adjacent concrete substrate.

Accelerated testing indicates that epoxy coatings provides excellent protection to simulated marine and bridge deck environments and are able to extend the service life of reinforced concrete significantly (Erdoğdu et al., 2001). Almost no loss of resistance was recorded in coatings based on epoxy formulations as a consequence of the immersion in alkaline solution (Wolff et al., 2006). Epoxy coatings are also able to limit the chemical sulfate-attack and improve the sulfate resisting performance of concrete partially immersed in sodium sulfate solution (Liu et al., 2010). Polyurethane coatings are known to be extremely durable if exposed to a wide range of environmental conditions; they are chemical resistant and UV stable (Vipulanandan and Liu, 2005). Results from accelerated corrosion tests showed that coatings for concrete based on waterproofing elastomeric polyurethane performed better than those based on other waterproofing materials (Al-Zahrani et al., 2002). Fairly good behavior was found for the specimens coated with epoxy-based coatings. Moreover, the two polymer-based coatings showed better performance than a cement-based coating in terms of water absorption, water permeability, chloride permeability and adhesion.

Good durability to biogenic sulfuric acid has been reported by De Muynck et al. (2009) for epoxy and polyurea coatings for sewer pipes. Epoxy coatings were found particularly effective in protecting concrete from microbiological attack taking place in cooling towers, typically caused by sulfur oxidizing or other acid-producing bacteria (Berndt, 2011).

16.7 Adhesives

16.7.1 Components and properties

The cement concrete deterioration manifests itself in the form of spalling and cracking of concrete, caused by reinforcement corrosion and softening and/or cracking. The formation of cracks and micro-cracks can be the consequence of several phenomena, such as bleeding, shrinkage, dynamic loads, thermal gradients, freeze–thaw and alkali–aggregate reaction. These unsealed cracks and micro-cracks may form potential flow channels, providing easy access to aggressive agents, independently of the porosity of the cement matrix. The cracks have, therefore, to be repaired to achieve a reduction in permeability and corresponding increase in durability.

The most frequently used method of repair is by conventional concrete placement, consisting of the replacement of defective concrete with new concrete. This technique is applicable to a wide range of situations and it is usually also the most economical. However, it should not be used in situations where an aggressive factor has caused the deterioration of the concrete being replaced. For example, if the deterioration noted has been caused by acid attack or abrasion-erosion, it may be expected that a repair made with conventional concrete will deteriorate again for the same reasons.

Among other traditional techniques, especially for local interventions, those involving the injections of mortar or reinforced grouted perforations are employed to repair damaged concrete or upgrading structures under both gravity and seismic loads. This traditional technique, however, presents severe drawbacks, such as the scarce adhesion between two solids, the new mortar and the old concrete (Burlamacchi, 1994). Other disadvantage are related to esthetics and functionality because of the possible obstruction of new areas, long time of interventions and therefore of activities interruption, conservation when referring to historical buildings and monuments.

Polymeric adhesives are frequently employed in substitution of cement grouts. Thermosetting resins are generally used, being able to harden without adversely affecting any metals or the concrete boundaries of the opening or void into which it has been injected. They may contain various inert fillers to modify physical properties, such as viscosity and heat generation, and to increase volume. Issa and Debs (2007) reported that, while cracks cause a reduction in compressive strength of concrete of about 40%, the polymeric adhesives, when properly applied, are able to restore the compressive strength by decreasing the reduction down to less than 10%.

The selection of a polymeric adhesive must be done according to whether it will harden to a rigid condition or to a flexible one. High stiff polymeric adhesives of high shear and tensile strength (up to 30 MPa), such as epoxy and acrylate, are usually preferred in bonding and repair applications. They bond exceedingly well to dry substrate and some will bond to wet concrete, can prevent all movement and restore the full strength of a cracked concrete member. However, these materials have low ultimate strain (under 4%), thus are inconvenient in bonding concrete when high deformability or dynamic action are expected; moreover, if tensile or shear stresses exceed the capability of the concrete, new cracks will appear in the concrete near, but generally not at the adhesion interface. The liquid adhesives can penetrate cracks somewhat finer than 0.05 mm, the penetration being dependent on viscosity, injection pressure, temperature and set time.

Flexible polymers of elastomeric behavior, such as acrylamides and polyurethanes, on the other hand, have ultimate strengths of 1–20 MPa but much higher ultimate strains (up to 1000%) (Kwiecień et al., 2010). With respect to rigid adhesives, they display higher values of deformation energy and ductility, i.e. higher toughness. Flexible repair joints allow for local slight deformation, absorbing or dissipating strain energy in this way, and can tolerate repeated cyclic deformations (Kwiecień, 2010). Flexible-type polymeric adhesives are mainly used to shut off or greatly reduce water movement. They will not restore strength to a structure, but they generally will maintain water tightness despite minimal movement across a crack. Most flexible adhesives are water solutions and will, therefore, exhibit shrinkage if allowed to dry, but they will recover when rewetted. Some of them can be formulated at very low viscosity in order to be injected into any hole through which water will flow. Others can be made to yield a foam that can be used in cavities down to 100 mm wide.

One of the main characteristic of adhesives is the wettability, i.e. they are able to wet the surface of structural elements that have to be joined. The application of the adhesive, moreover, must be done to a properly prepared substrate, in order to guarantee a satisfactory degree of adhesion. To this aim, surface pre-treatment is often required to remove contaminations and weak surface layers and to enhance the roughness of the surface. It is well recognized, in fact, that a substantial roughness of the cementitious substrate promotes mechanical anchoring of the polymeric adhesive and tends to increase the contact surface area, both resulting in a higher bond strength (Abu-Tair et al., 2000; Toutanji and Ortiz, 2001).

Structural adhesives are characterized by high strength in order to support loads, even higher by that of the concrete, although they are less stiff. Since they are employed to bond rigid structural elements, they should provide uniform stress distribution through continuous bonding but, at the same time, sufficient flexibility to counter the effects of displacement, impact or vibrations. Polymeric adhesives posses a low density and usually only a thin layer is applied, thus avoiding a weight penalty compared with other repairing materials. Polymeric adhesives, finally, must be: not toxic, compatible with the original substratum, adequately resistant to high/low-temperature cycles and aggressive environments and, as far as possible, removable.

Among other thermosetting resins, epoxy and unsaturated resins, such as polyester and mainly vinyl ester, are widely employed as adhesives, assuring good performance in terms of flexibility and effectiveness of repairing technique and moderate costs. In the field of adhesives and resins for structural applications epoxy resins, in particular, are known to exert a strong interfacial bond between cement concrete and a broad range of reinforcing materials. These resins are traditionally employed for making extensive repairs to concrete structures, as adhesives to fill concrete cracks, to join precast concrete elements, to bond fresh concrete to hardened one, for anchoring reinforcing bars to concrete, to bond concrete to other materials and as matrices for FRP (fiber reinforced polymer) composites, recently successfully used for strengthening damaged structures.

With an opportune choice of the base components and the curing cycle, epoxy and vinyl ester resins display low viscosity in the un-cured state, high reactivity on curing, no release, or very small, of volatile products and excellent adhesion properties. The addition of suitable fillers to the original formulation allows one to obtain products with a modulated viscosity, adjusted for both repair and gap-filling applications. Moreover, these adhesives provide a high level of mechanical properties, excellent resistance to corrosion and to chemicals for both the concrete and its steel reinforcement. The irreversibility of such resins represents one their main drawbacks: once formed, they do not melt or soften upon reheating and do not dissolve in any solvent (Hollaway, 2010).

For economic and practical reasons, the epoxy resins used in civil engineering repairing applications are 'cold-curing' types, normally based on bisphenol epoxy resins, cured with the addition of aliphatic amines at ambient temperatures on site (May, 1989; Hollaway, 2010). Providing any kind of heat sources, over the large areas required for the described applications, in fact, is very difficult and prohibitively expensive.

Vinyl ester resins can sometimes be preferred to epoxies as adhesives, due to advantages in terms of processing conditions, reduced curing cycles and lower costs. Like the epoxies, moreover, they are often used in applications which require high acid resistance and/or high toughness, strength and stiffness; its resistance against corrosion is reported to be even higher than that of epoxies (Riffle et al., 1998). In addition, they possess relatively high ultimate failure strains and damage tolerance. The presence of styrene, used as reactive diluent, however, can result in an excessive shrinkage of vinyl ester during curing, with the possible formation of micro-cracks or micro-gels in the bulk, resulting in micro-inhomogeneities and incomplete polymerization.

The use of unsaturated polyester resins in the field of construction is rather limited, due to the high shrinkage occurred during curing (up to 10% by volume) that severely restricts the volume of material that can be employed (Mays and Hutchinson, 1992). Reservations have also been expressed regarding the suitability of polyesters in such applications because the poor creep resistance under sustained load, low glass transition temperatures (T_g), a high susceptibility to moisture and their scarce bonding efficiency in damp or wet conditions and in particular to alkaline substrates. They are, finally, easily susceptible to moisture/aqueous attack.

16.7.2 Durability

The final performance and durability of polymeric adhesives (such as adhesive strength, heat and chemical resistance) depend mainly on the chemical nature of the base materials, on the conditions in which the resins set and harden, and, mainly, on the environmental conditions in which they 'operate' (Frigione, 2007).

The main consequences of a cure at ambient (often uncontrolled) temperature of epoxy adhesives are: long curing times (in the order of weeks) are necessary to achieve sufficient mechanical properties, the lower the curing temperature, the longer the curing time (Moussa et al., 2012); the curing (cross-linking) reactions taking place in epoxy resins at ambient temperatures are often not completed because of kinetic restraints; a moderate glass transition temperature (T_g, as explained below), in practice never greater than 65–70 °C, is attainable by these systems (Frigione et al., 2001, 2006a, 2006b; Sciolti et al., 2010). In addition, the absorption of external water (for example, as atmospheric moisture or rain), produces a decrease in the initial T_g of the resin, which in turn affects the mechanical and adhesive properties. The mean value of the temperature range below which the typical properties of a cured adhesive vary in a manner similar to that of a solid (glassy) phase and above which it behaves in a manner similar to that of rubber is known as the glass transition temperature, T_g. Approaching the T_g, the behavior of a resin drastically changes from that of a solid adhesive, able to effectively bond two different materials, to that of a soft material, unable to guarantee the stress transfer between the same materials. When the range of the T_g is approached, a dramatic decrease (up to 70%) in stiffness and strength is measured on the adhesive (Frigione et al., 2000; Shin et al., 2011). Above T_g both properties become equal to zero. The detrimental effects of moderate temperatures on the performance of a resin are reproduced also in its behavior as adhesive. The service temperature is able to reduce appreciably the adhesion strength to concrete, i.e. by over 80% at 50 °C, and the fatigue resistance (Aiello et al., 2002; Shin et al., 2011). However, the effects of these agents on the adhesive strength depend on the strength of the concrete in relation to that of the adhesive and on the thickness of the adhesive layer. As temperature increases, the mechanism of failure occurring in the samples changed, from predominantly concrete failure, to mixed failure of epoxy and at the interface. In conclusion, due to their moderate T_g, attention must be given to the site temperature when using the cold-cured resins: the environmental temperature under working conditions should be some 20 °C below the glass transition temperature (Hollaway, 2010).

The presence of humidity in the air, either in the form of moisture or actual water through rain, is probably the most harmful environment that can be encountered by adhesives for civil engineering applications. Epoxy resins are able to absorb substantial amounts of water due to the presence of polar groups able to attract water molecules. The ingress of moisture over time is particularly significant if the polymer is permanently immersed in water, or salt solution or is exposed to de-icing salt solutions. An excessive penetration of water is generally considered harmful, since it leads to a reduction in stiffness and strength with a consequent marked unsuitable decrease of load-bearing capacity through plasticization effects (Mays and Hutchinson, 1992; Shaw, 1994; Frigione et al., 2006a, 2006b). The T_g of the adhesive is also appreciably reduced, as already underlined.

The presence of water can be particularly dangerous when the adhesive is used to join two dissimilar adherents. Water may easily penetrate through a permeable adherent like the concrete, which possesses from 10% to 40% of volumetric fraction of voids and capillary pores, and it can diffuse or be transmitted along the interfaces through capillary action. After having accessed the joint, water may cause deterioration of the bond by reducing mechanical properties and adhesive displacement at the interface. The displacement is even augmented by pre-existing micro-cracks or debonded areas at the interface, which originate from poor wetting by the adhesives (Sung, 1990). Several authors reported that substantial decreases in bond strength over aging time (up to 40–50% loss in some cases) are experienced under wet and moist environments (Tu and Kruger, 1996; Toutanji and Gomez, 1997; Malvar et al., 2003; Grace and Singh, 2005; Frigione et al., 2006a; Marouani et al., 2008; Silva and Biscaia, 2008; Karbhari and Gosh, 2009; Benzarti et al., 2011).

Other environmental factors have been found to have a deleterious effect on the durability of polymeric adhesives, in particular epoxy-based, such as temperature cyclical variations, freeze–thaw, alkaline environment, salt solutions or chemicals, and UV radiation. Some of them can act together, possibly giving rise to synergistic effects.

The typical coefficient of thermal expansion (CTE) of a thermosetting adhesive is of an order of magnitude higher than those of the conventional civil engineering materials (Hollaway, 2010). The deterioration of the bond strength, resulting after repeated freeze–thaw or temperature cycles, is often attributed to the different CTEs (Bisby and Green, 2002). This property must, therefore, be considered in structural design when critical temperature variations are expected.

Exposure to saline and alkaline environments may also cause chemical degradation of the resin. Environmental pollution, which includes ozone, nitrogen and sulfur oxides, fuels, oils, and other harmful chemicals, interacts with the adhesives, affecting their properties. Polymeric adhesives have, in fact, a poor resistance to concentrated sulfuric and nitric acids. Furthermore, the attack of aqueous solutions occurs through hydrolysis in which moisture degrades the molecular bonds (Hollaway, 2010).

The UV light from sun radiation is strong enough to cleave the covalent bonds in polymers, causing embrittlement and yellowing, weight loss and a general reduction of the performance of the materials (Roylance and Roylance, 1978; Hollaway, 2010). Both vinyl ester and polyester styreneated resins are particularly sensitive to UV radiation and should be stabilized and protected from this type of radiation (Lesko et al., 2001). However, the adhesives used to repair cracks are hardly subjected to UV radiation, since the effects of such radiation are usually confined to the top few microns of the surface; however, this effect can be amplified by the action of moisture, temperature and other environmental agents.

The polymeric adhesives are organic materials composed of carbon, hydrogen and nitrogen atoms. They are, therefore, flammable to a certain degree. Consequently, a major concern for their use resides in problems associated with fire. Large variations in the resin's mechanical properties occur in fires, which in turn influence the load-bearing capacity of the repaired concrete (Pinoteau et al., 2011). Moreover, great health hazards derived from polymers in a fire accident are generated from the toxic combustion products produced during burning of the material. In order to limit the fire hazard, the introduction of proper additives (halogens or phosphorus compounds) is usually employed.

The typical features displayed by cold-cured adhesives produce great concerns about the efficiency and durability of such materials especially when thinking of their typical outdoor applications. As few examples of experimental adhesives recently proposed by academic and industrial research, organic–inorganic hybrid materials, based on silane functionalized epoxy resins containing interpenetrating inorganic, highly hydroxylated silica, co-

continuous domains, proved to be an attractive alternative to the traditional cold-cured epoxy (Lettieri et al., 2010, 2011). These epoxy–silica hybrid systems displayed, in fact, much larger increases in the T_g over the control (unmodified) epoxy, with the formation of a denser siloxane network. When exposed to humidity environments, moreover, a reduction of plasticization effects of the absorbed water was experienced, attributed to the rigidity of the siloxane domains. Such systems, finally, displayed also improvements in the adhesion to concrete surfaces (Lionetto et al., 2011).

Epoxy layered silicate nanocomposites introduced into the polymer at the time of manufacture has the potential to lower its permeability, thus improving its barrier properties and its mechanical strengths (Hackman and Hollaway, 2006). Thus, by improving the barrier property, a reduction of the ingress of moisture, aqueous and salt solutions is achieved in the concrete. These nanocomposites are also able to reduce the fire hazards in polymers. However, the utilization of nanocomposites is still too expensive for the construction industry and can be currently justified only under very special circumstances.

16.8 Future trends

The concrete for the future is supposed to be: stronger, more workable, extremely durable, preserving environmental, social and economic aspects. In particular, there is expected to be an increase in concrete stability with regard to mechanical and temperature factors and particularly resistance to corrosion and degradation caused by atmospheric and gaseous media containing vapors and aggressive media. Most such requirements will be achievable with proper use of polymer admixtures.

In addition, an effective advantageous employment of polymer admixtures in traditional and latex concrete would be found in new applications, such as: foundations and structures subjected to dynamic loads; marine and offshore structures; mass production of large anti-corrosive structural elements and weather-proofed roof decks.

Research devoted to concretes possessing a very high polymer content, will probably be focused on the development of monomers that are environmentally sustainable, economical, realizable with a low energy consumption and, possibly, by waste polymer. According to Ohama (2010), the sustainable concrete–polymer composites which are expected to become the promising and challenging construction materials in the next future will be:

- Concrete–polymer composites for the longevity of reinforced concrete structures. For instance: polymer-modified mortar used as repair materials for deteriorated reinforced concrete structures, possibly containing compound with chloride-induced corrosion-auto-inhibiting function for path repair work; durability-improving materials for reinforced concrete structures, such as microcrack-autohealing hardener-free epoxy-modified cementitious systems or water repellency coatings based on silane or silane-siloxane.

- Resource-saving concrete–polymer composites. As an example, artificial marble would be produced using recycled aggregates and waste polymer.
- Ecologically safe concrete–polymer composites. In eco-friendly systems, low-volatility liquid resins would be used as binders for polymer concrete and mortar. Similarly, waterproofing systems, based on liquid-applied membranes, not containing toxic organic solvents, would be possibly realized. The use of polymer-modified cement paste containing a photo-catalyst, anatase type titanium dioxide, would prevent air pollution due to the decomposition of car exhausted gases with nitrogen oxides on road surfaces by the action of the photo-catalyst.

In the field of adhesives and coatings for concrete, the use of nano-structured polymers, such as nano-composites or organic–inorganic hybrid materials, is expected to become a realistic alternative to traditional polymeric products due to their superior properties, especially in terms of higher durability against weathering.

The actual efforts of academic and industrial research are mainly devoted, therefore, to a reduction of costs production of polymeric material in order to employ them effectively as properties-aid/substitute in cement and concrete.

16.9 References

Abdel-Fattah H, El-Hawary M M, (1999), 'Flexural behavior of polymer concrete', *Constr Build Mater*, **13**, 253–262. http://dx.doi.org/10.1016/S0950-0618(99)00030-6.

Abu-Tair A I, Lavery D, Nadjai A, Rigden S R, Ahmed T M A, (2000), 'A new method for evaluating the surface roughness of concrete cut for repairing or strengthening', *Constr Build Mater*, **14**, 171–176. http://dx.doi.org/10.1016/S0950-0618(00)00016-7.

Aggarwal L K, Thapliyal P C, Karade S R, (2007), 'Properties of polymer-modified mortars using epoxy and acrylic emulsions', *Constr Build Mater*, **21**, 379–383. http://dx.doi.org/10.1016/j.conbuildmat.2005.08.007.

Aiad I, Abd El-Aleem S, El-Didamony H, (2002), 'Effect of delaying addition of some concrete admixtures on the rheological properties of cement pastes', *Cem Concr Res*, **32**, 1839–1843. http://dx.doi.org/10.1016/S0008-8846(02)00886-4.

Aiello M A, Frigione M, Acierno D, (2002), 'Effects of environmental conditions on performance of polymeric adhesives for restoration of concrete structures', *ASCE J Mater Civil Eng*, **14**, 185–189. DOI: 10.1061/(ASCE)0899-1561(2002)14:2(185).

Al-Dulaijan S U, Maslehuddin M, Al-Zahrani M M, Al-Juraifani E A, Al-Idi S H, Al-Mehthel M, (2000a), 'Performance evaluation of cement-based surface coatings', In: *Repair, Rehabilitation, and Maintenance of Concrete Structures, and Innovations in Design and Construction,* Proc 4th ACI Inter Conf, SP-193-20, Seoul, Korea, 321–335.

Al-Dulaijan S U, Maslehuddin M, Al-Zahrani M M, Sharif A M, Al-Juraifani E A, Al-Idi S H, (2000b), 'Performance evaluation of resin-based surface coatings', In: *Deterioration and Repair of Reinforced Concrete in the Arabian Gulf*, Proc. 6th ACI Inter Conf Bahrain, 345–362.

Al-Zahrani M M, Al-Dulaijan S U, Ibrahim M, Saricimen H, Sharif F M, (2002), 'Effect of waterproofing coatings on steel reinforcements corrosion and physical properties of concrete', *Cem Concr Compos*, **24**, 127–137. http://dx.doi.org/10.1016/S0958-9465(01)00033-6.

Amianti M, Botaro V R, (2008), 'Recycling of EPS: A new methodology for production of concrete impregnated with polystyrene (CIP), *Cem Concr Compos*, **30**, 23–28. http://dx.doi.org/10.1016/j.cemconcomp.2007.05.014.

Angle P M, Antonio A, Alejandro J, (1999), 'Fatigue behavior of polymer modified porous concretes', *Cem Concr Res*, **29**, 1077–1083. http://dx.doi.org/10.1016/S0008-8846(99)00095-2.

Aniskevich K, Hristova J, (2002), 'Aging and filler effects on the creep model parameters of thermoset composites', *Compos Sci Technol*, **62**, 1097–1103. http://dx.doi.org/10.1016/S0266-3538(02)00055-6.

Aydin S D, Aytac A H, Kambiz Ramyar K, (2009), 'Effects of fineness of cement on polynaphthalene sulfonate based superplasticizer cement interaction', *Constr Build Mater*, **23**, 2402–2408. http://dx.doi.org/10.1016/j.conbuildmat.2008.10.004.

Bărbuță M, Harja M, Baran I, (2010), 'Comparison of mechanical properties for polymer concrete with different types of filler', *ASCE J Mater Civil Eng*, **22**, 696–701.
http://dx.doi.org/10.1061/(ASCE)MT.1943-5533.0000069.

Barluenga G, Olivares F H, (2004), 'SBR latex modified mortar rheology and mechanical behaviour', *Cem Concr Res*, **34**, 527–535. http://dx.doi.org/10.1016/j.cemconres.2003.09.006.

Basheer P A M, Montgomery F R, Long A E., Batayneh M, (1990), 'Durability of surface treated concrete', In: *Protection of Concrete*, eds. Dhir R K and Green J W, Publ E and FN Spon, 212–221.

Basheer P A M, Basheer L, Cleland D J, Long A E, (1997), 'Surface treatments for concrete: assessment methods and reported performance', *Constr Build Mater*, **11**, 413–429. http://dx.doi.org/10.1016/S0950-0618(97)00019-6.

Baur J, Silverman E, (2007), 'Challenges and opportunities in multifunctional nanocomposite structures for aerospace applications', *MRS Bull*, **32**, 328–334.

Beeldens A, Monteny J, Vincke E, De Belie N , Van Gemert D, Taerwe L, Verstraete W, (2001), 'Resistance to biogenic sulphuric acid corrosion of polymer-modified mortars', *Cem Concr Compos*, **23**, 47–56. http://dx.doi.org/10.1016/S0958-9465(00)00039-1.

Beeldens A, Van Gemert D, Schorn H, Ohama Y, Czarnecki L, (2005), 'From microstructure to macrostructure: an integrated model of structure formation in polymer-modified concrete', *Mater Struct*, **38**, 601–607. DOI: 10.1007/BF02481591

Bensted J (2002), 'Hydration of portland cement', In: *Advances in Cement Technology: Chemistry, Manufacture and Tersting*, Second Edition, Ed. Ghosh S N, Publ TBi, Tech Books International, New Delhi, India, 37.

Bentur A, Igarashi S, Kovler K, (2001), 'Prevention of autogenous shrinkage in high-strength concrete by internal curing using wet lightweight aggregates', *Cem Concr Res*, **31**, 1587–1591. http://dx.doi.org/10.1016/S0008-8846(01)00608-1.

Bentz D P, Lura P, Roberts J W, (2005), 'Mixture proportioning for internal curing', *Concr Int*, **27**, 35–40.

Benzarti K, Chataigner S, Quiertant M, Marty C, Aubagnac C, (2011), 'Accelerated ageing behavior of the adhesive bond between concrete specimens and CFRP overlays', *Constr Build Mater*, **25**, 523–538. http://dx.doi.org/10.1016/j.conbuildmat.2010.08.003.

Berndt M L, (2011), 'Evaluation of coatings, mortars and mix design for protection of concrete against sulphur oxidising bacteria', *Constr Build Mater*, **25**, 3893–3902. http://dx.doi.org/10.1016/j.conbuildmat.2011.04.014.

Betioli A M, Filho J H, Cincotto M A, Gleize P J P, Pileggi R G, (2009), 'Chemical interaction between EVA and Portland cement hydration at early-age', *Constr Build Mater*, **23**, 3332–3336. http://dx.doi.org/10.1016/j.conbuildmat.2009.06.033.

Bisby L A, Green M F, (2002), 'Resistance to freezing and thawing of fiber-reinforced polymer–concrete bond', *ACI Struct J*, **99**, 215–223.

Blaga A, (1973), *Properties and behaviour of plastics*, Division of Building Research, National Research Council Canada, Canadian Building Digest, no. 157, Ottawa.

Bode, K A, Dimmig-Osburg A, (2010), 'Shrinkage properties of polymer-modified cement mortars (PCM)', *13th International Congress on Polymers in Concrete, ICPIC 2010*, 10–12 February, Madeira Portugal, Eds Aguiar J B, Jalali S, Camoes A, Ferreira R M, 89–94.

Bundyra-Oracz G, Kurdowski W, (2011), 'Effect of the cement type on compatibility with carboxylate superplasticisers', *Mater Construct*, **61**, 302, 227–237. DOI: 10.3989/mc.2011.54309.

Burlamacchi L (1994), '*Capire il calcestruzzo*', Hoepli, Milano, Italy, 228.

Cabrera J G, Hassan K G, (1994), 'Assessment of the effectiveness of surface treatments against the ingress of chlorides into mortar and concrete', In: *Corrosion and corrosion protection of steel in concrete*, Ed Swamy R N, Sheffield University Press, 1028–1043.

Cestari A R, Vieira E F S, Pinto A A, Rocha F C, (2008), 'Synthesis and characterization of epoxy-modified cement slurries – Kinetic data at hardened slurries/HCl interfaces', *J Colloid Interface Sci*, **327**, 267–274. http://dx.doi.org/10.1016/j.jcis.2008.08.008.

Chen C-H, (2003), 'Study of glycolysis of poly(ethylene terephthalate) recycled from post consumer soft drink bottles. III. Further investigations' *J Appl Polym Sci*, **87**, 2004–2010. DOI: 10.1002/app.11694

Chen C-Hs, Huang R, Wu J K, Chen C-Hu, (2006), 'Influence of soaking and polymerization conditions on the properties of polymer concrete', *Constr Build Mater*, **20**, 706–712. http://dx.doi.org/10.1016/j.conbuildmat.2005.02.003.

Choi N W, Ohama Y, (2004), 'Basic properties of new polymer mortars using waste expanded polystyrene solution-based binders', *J Polym Eng*, **24**, 369–383.

Chung D D L, (2004), 'Use of polymers for cement-based structural materials', *J Mater Sci*, **39**, 2973–2978.

Cleland D J, Basheer L, (1995), 'Surface treatments for concrete – Quantifying the improvement'. *Extending the life span of structures, IABSE Symp*, San Francisco, USA.

Collepardi M, (1998), 'Admixtures used to enhance placing characteristics of concrete', *Cem Concr Compos*, **20**, 103–112. http://dx.doi.org/10.1016/S0958-9465(98)00071-7.

Collepardi M, (2005), 'Admixture-enhancing concrete performance', In *Proceedings of the 6th International Congress: Repair and Renovation of Concrete Structures. Global Construction: Ultimate Concrete Opportunities*, July, Dundee, Scotland, Eds Dhir R K, Jones M R, Zheng L, Thomas Telford, London, 469–476.

Coppola L, Buoso A, Lorenzi S, (2010), 'Compatibility issues of NSF-PCE superplasticizers with several lots of different cement types (long-term results)', *J China Ceram Soc*, **38**, 1631–1637.

Damtoft J S, Lukasik J, Herfort D, Sorrentino D, Gartner E M, (2008), 'Sustainable development and climate change initiatives', *Cem Concr Res*, **38**, 115–127. http://dx.doi.org/10.1016/j.cemconres.2007.09.008.

Dan N, Toyoharu N, Noboru Y, Hiroshi O, (2001), 'Effect of chemical structure of superplasticizer on strength development and frost resistance of high strength concrete', *JCA Proc Cement & Concrete (Japan Cement Association)*, **55**, 500–506.

De Muynck W, De Belie N, Verstaete W, (2009), 'Effectiveness of admixtures, surface treatments and antimicrobial compounds against biogenic sulphuric acid corrosion of concrete', *Cem Concr Compos*, **31**, 163–170. http://dx.doi.org/10.1016/j.cemconcomp.2008.12.004.

de S. Almeida A E F, Sichieri E P, (2006), 'Mineralogical study of polymer modified mortar with silica fume', *Constr Build Mater*, **20**, 882–887. http://dx.doi.org/10.1016/j.conbuildmat.2005.06.029

Deja J, (2002), 'Immobilization of Cr^{6+}, Cd^{2+}, Zn^{2+} and Pb^{2+} in alkali-activated slag binders', *Cem Concr Res*, **32**, 1971–1979. http://dx.doi.org/10.1016/S0008-8846(02)00904-3.

El-Dieb A S, (2007), 'Self-curing concrete: Water retention, hydration and moisture transport', *Constr Build Mater*, **21**, 1282–1287. http://dx.doi.org/10.1016/j.conbuildmat.2006.02.007.

El-Hawary M M, Abdul-Jaleel A, (2010), 'Durability assessment of epoxy modified concrete', *Constr Build Mater*, **24**, 1523–1528. http://dx.doi.org/10.1016/j.conbuildmat.2010.02.004.

Erdoğdu S, Bremner T W, Kondratova I L, (2001), 'Accelerated testing of plain and epoxy-coated reinforcement in simulated seawater and chloride solutions', *Cem Concr Res*, **31**, 861–867. http://dx.doi.org/10.1016/S0008-8846(01)00487.

EuPC, (2006), 'The European markets for plastic building products. European Plastics Convertors–Building'. http://www.plasticsconverters.eu

Felekoğlu B, Sarikahya H, (2008), Effect of chemical structure of polycarboxylate-based superplasticizers on workability retention of self-compacting concrete', *Constr Build Mater*, **22**, 1972–1980. http://dx.doi.org/10.1016/j.conbuildmat.2007.07.005.

Fowler D W, (1998), 'Current status of polymer concrete in the United States', *IXth International Congress on Polymers in Concrete, ICPIC'98*, 14–18 September, Bologna, Italy, Ed Sandrolini F, 37–44.

Fowler D W, (1999), 'Polymers in concrete: A vision for the 21st century', *Cem Concr Compos*, **21**, 449–452. http://dx.doi.org/10.1016/S0958-9465(99)00032-3.

Fowler D W, Whitney D P, (2011), 'State of practice for polymer concrete overlays', *American Concrete Institute*, ACI 278 SP, 1–22.

Frigione M, (2007), 'Durability aspects of polymer composites used for restoration and rehabilitation of structures', *Leading-edge composite material research*, Ed Wouters T G, Publ Nova Science Publishers, Inc., New York, Chap. 1, 23–69.

Frigione M, Naddeo C, Acierno D, (2000), 'Epoxy resins employed in civil engineering applications: Effects of exposure to mild temperatures', *Mat Eng*, **11**, 59–80.

Frigione M, Naddeo C, Acierno D, (2001), 'Cold-curing epoxy resins: Aging and environmental effects. I – Thermal properties', *J Polym Eng*, **21**, 23–51.

Frigione M, Aiello M A, Naddeo C, (2006a), 'Water effects on the bond strength of concrete/concrete adhesive joints', *Constr Build Mater*, **20**, 957–970. http://dx.doi.org/10.1016/j.conbuildmat.2005.06.015.

Frigione M, Lettieri M, Mecchi A M, (2006b), 'Environmental effects on epoxy adhesives employed for restoration of historical buildings', *ASCE J. Mater Civil Eng*, **18**, 715–722. DOI: 10.1061/(ASCE)0899-1561(2006)18:5(715)

Fu X, Chung D D L, (1996), 'Effect of methylcellulose admixture on the mechanical properties of cement', *Cem Concr Res*, **26**, 535–538. http://dx.doi.org/10.1016/0008-8846(96)00028-2.

Fu X, Chung D D L, (1998), 'Combined use of silica fume and methylcellulose as admixtures in concrete for increasing the bond strength between concrete and steel rebar', *Cem Concr Res*, **28**, 487–492. http://dx.doi.org/10.1016/S0008-8846(98)00016-7.

Gajbhiye N S, Sing N B, Middendorf B, (2010), 'Interaction of sodium salt of polyacrylate polymers during Portland cement hydration', *13th International Congress on Polymers in Concrete, ICPIC 2010*, 10–12 February, Madeira, Portugal, Eds Aguiar J B, Jalali S, Camoes A and Ferreira R M, 65–72. ISBN: 978-972-99179-4-3.

Gomes C E M, Ferreira O P, (2005), 'Analyses of microstructural properties of VA/VeoVA copolymer modified cement pastes', *Polímeros*, **15**, 3 (*Print version*).

Gorninski J P, Dal Molin D C, Kazmierczak C S, (2004), 'Study of the modulus of elasticity of polymer concrete compounds and comparative assessment of polymer concrete and Portland cement concrete', *Cem Concr Res*, **34**, 2091–2095. http://dx.doi.org/10.1016/j.cemconres.2004.03.012.

Gorninski J P, Dal Molin D C, Kazmierczak C S, (2007), 'Strength degradation of polymer concrete in acidic environments', *Cem Concr Compos*, **29**, 637–645. http://dx.doi.org/10.1016/j.cemconcomp.2007.04.001

Grace N F, Singh S B, (2005), 'Durability evaluation of carbon fiber-reinforced polymer strengthened concrete beams: Experimental study and design', *ACI Struct J*, **102**, 40–51.

Hackman I, Hollaway L C, (2006), 'Epoxy-layered silicate nanocomposites in civil engineering', *Compos Part A Appl Sci Manuf*, **37**, 1161–1170. http://dx.doi.org/10.1016/j.compositesa.2005.05.027.

Hamada D, Sato H, Yamamuro H, Izumi T, Mizunuma T, (2003), Development of Slump-Loss Controlling Agent with Minimal Setting Retardation', Proc 7th CANMET/ACI Inter Conf on superplasticizers and other chemical admixtures in concrete, *American Concrete Institute*, ACI 217 SP, 127–142.

Hamada D, Sagawa K, Tanisho Y, Yamamuro H, (2009), 'Pursuance of workability retention by new superplasticizer', *American Concrete Institute,* ACI 262 SP, 297–307.

Harja M, Barbuta M, Rusu L, (2009), 'Obtaining and characterization of the polymer concrete with fly ash', *J Appl Sci*, **9**, 88–96. DOI: 10.3923/jas.2009.88.96.

Heikal M, Aiad I, (2008), 'Influence of delaying addition time of superplasticizers on chemical process and properties of cement pastes', *Ceramics – Silikaty*, **52**, 8–15.

Hollaway L C, (2010), 'A review of the present and future utilization of FRP composites in the civil infrastructure with reference to their important in-service properties', *Constr Build Mater*, **24**, 2419–2445. http://dx.doi.org/10.1016/j.conbuildmat.2010.04.062.

Hommer H, (2009), 'Interaction of polycarboxylate ether with silica fume', *J Eur Ceram Soc*, **29**, 1847–1853. http://dx.doi.org/10.1016/j.jeurceramsoc.2008.12.017.

Hu H M, Yao Z X, Zeng SY, Yu B B, (2011), 'Experimental research on compatibility between underwater anti-washout admixture and superplasticizer', *Key Eng Mater*, **477**, 190–199. DOI: 10.4028/www.scientific.net/KEM.477.190.

Hugo A-M, Scelsi L, Hodzic A, Jones F R, Dwyer-Joyce R, (2011), 'Development of recycled polymer composites for structural applications', *Plast Rubber Compos*, **40**, 317–323. http://dx.doi.org/10.1179/1743289810Y.0000000008.

Issa C A, Debs P, (2007), 'Experimental study of epoxy repairing of cracks in concrete', *Constr Build Mater*, **21**, 157–163. http://dx.doi.org/10.1016/j.conbuildmat.2005.06.030.

Jensen O M, Hansen P F, (2001), 'Water-entrained cement-based materials: I. Principles and theoretical background', *Cem Concr Res*, **31**, 647–654. http://dx.doi.org/10.1016/S0008-8846(01)00463-X.

Jo B W, Tae G H, Kim C H, (2007), 'Uniaxial creep behavior and prediction of recycled-PET polymer concrete', *Constr Build Mater*, **21**, 1552–1559. http://dx.doi.org/10.1016/j.conbuildmat.2005.10.003.

Jo B W, Park S K, Park J C, (2008), 'Mechanical properties of polymer concrete made with recycled PET and recycled concrete aggregates', *Constr Build Mater*, **22**, 2281–2291. http://dx.doi.org/10.1016/j.conbuildmat.2007.10.009.

Jurumenha M A G, Dos Reis J M L, (2010), 'Fracture mechanics of polymer mortar made with recycled raw materials', *Mat Res*, **13**, 475–478.

Karbhari V M, Gosh K, (2009), 'Comparative durability evaluation of ambient temperature cured externally bonded CFRP and GFRP composites for repair bridges', *Compos Part A Appl Sci Manuf*, **40**, 1353–1363. http://dx.doi.org/10.1016/j.compositesa.2009.01.011

Kardon J B, (1997), 'Polymer-modified concrete: review', *ASCE J Mater Civ Eng*, **9**, 85–93. http://dx.doi.org/10.1061/(ASCE)0899-1561(1997)9:2(85).

Khattab M M, Abdel-Rahman H A, Younes M M, (2011), 'Durability of gamma irradiated polymer-impregnated blended cement pastes', *Constr Build Mater*, **25**, 651–657. http://dx.doi.org/10.1016/j.conbuildmat.2010.07.026.

Khurana R, Torresan I, (1997), 'New admixtures for eliminating steam curing and its negative effects on durability', *International Concrete Research & Information Portal*, ACI 173 SP, 83–104.

Knapen E, Van Gemert D, (2009a), 'Effect of under water storage on bridge formation by water-soluble polymers in cement mortars', *Constr Build Mater*, **23**, 3420–3425. http://dx.doi.org/10.1016/j.conbuildmat.2009.06.007.

Knapen E, Van Gemert D, (2009b), 'Cement hydration and microstructure formation in the presence of water-soluble polymers', *Cem Concr Res*, **39**, 6–13. http://dx.doi.org/10.1016/j.cemconres.2008.10.003.

Knapen E, Van Gemert D, (2011), 'Microstructural analysis of paste and interfacial transition zone in cement mortars modified with water-soluble polymers', *Key Eng Mat*, **466**, 21–28. DOI: 10.4028/www.scientific.net/KEM.466.21

Ku D H, Park J S, Park H Y, Hur M J, Lee W M, (2008), 'Physical properties of polymer impregnated concrete prepared using microwave radiation', *J Kor Ind Eng Chem*, **19**, 345–350.

Kwiecień A, (2010), 'New repair method of cracked concrete airfield surface using polymer joints', *13th International Congress on Polymers in Concrete, ICPIC 2010*, 10–12 February, Madeira, Portugal, Eds Aguiar J B, Jalali S, Camoes A, Ferreira R M, 657–664.

Kwiecień A, Gruszczyński M, Zając B, (2010), 'Tests of flexible polymer joints repairing of polymer modified concretes', *13th International Congress on Polymers in Concrete, ICPIC 2010*, 10–12 February, Madeira, Portugal, Eds Aguiar J B, Jalali S, Camoes A, Ferreira R M, 559–566.

Lesko J J, Hayes M D, Schniepp T J, Case S W, (2001), 'Characterization and durability of FRP structural shapes and materials', *Int. Workshop Composites in Construction: A Reality,* Capri, Italy, Eds Cosenza E, Manfredi G, Nanni A, ASCE, American Society of Civil Engineering, Reston, Virgina, 110–119.

Lettieri M, Frigione M, Prezzi L, Mascia L, (2010), 'Novel cold-cured epoxy-silica hybrids as potential adhesives for concrete/masonry rehabilitation: environmental aging', *Restor Build Monum*, **16**, 353–365. DOI 10.1002/pen.21817,

Lettieri M, Lionetto F, Frigione M, Prezzi L, Mascia L, (2011), 'Cold-cured epoxy-silica hybrids: effects of large variation in specimen thickness on the evolution of the T_g and related properties', *Polym Eng Sci*, **51**, 358–368. DOI 10.1002/pen.21817.

Li C-Z, Fengb N-Q, Lib Y-D, Chen R-J, (2005), 'Effects of polyethylene oxide chains on the performance of polycarboxylate-type water-reducers', *Cem Concr Res*, **35**, 867–873. http://dx.doi.org/10.1016/j.cemconres.2004.04.031.

Li G, Xiong G, Lü Y, Yin Y, (2009), 'The physical and chemical effects of long-term sulphuric acid exposure on hybrid modified cement mortar', *Cem Concr Compos*, **31**, 325–330. http://dx.doi.org/10.1016/j.cemconcomp.2009.02.014

Liao T-S, Hwang C-L, Ye Y-S, Hsu K-C, (2006), 'Effects of a carboxylic acid/sulfonic acid copolymer on the material properties of cementitious materials', *Cem Concr Res*, **36**, 650–655. http://dx.doi.org/10.1016/j.cemconres.2005.10.005.

Lionetto F, Frigione M, Mascia L, (2011), 'Structural epoxy-silica hybrid adhesive and adhesion to concrete', *International Conference on Structural Adhesive Bonding, AB 2011*, 7–8 July, Porto, Portugal, Eds. da Silva F M, Öchsner A, Adams R D, 49.

Littmann K, Schwamborn B, (1999), 'Influences on the adhesion quality of epoxy coatings on concrete', *2nd International RILEM Symposium on Adhesion between Polymers and Concrete, ISAP '99*, 14–17 September, Dresden, Germany, Ed Ohama Y, Puterman M, RILEM, Cachan, Cedex, France, 193–203.

Liu Z, De Schutter G, Deng D, (2010), 'The effect of epoxy coating attack on cement paste partially immersed in sulfate environment', *13th International Congress on Polymers in Concrete, ICPIC 2010*, 10–12 February, Madeira, Portugal, Ed Aguiar J B, Jalali S, Camoes A, Ferreira R M, 577–583.

Ma B, Wang X, Liang W, Li X, He Z, (2007), 'Study on early-age cracking of cement-based materials with superplasticizers', *Constr Build Mater*, **21**, 2017–2022. http://dx.doi.org/10.1016/j.conbuildmat.2006.04.012.

Mahdi F, Khan A A, Abbas H, (2007), 'Physiochemical properties of polymer mortar composites using resins derived from post-consumer PET bottles', *Cem Concr Compos*, **29**, 241–248. http://dx.doi.org/10.1016/j.cemconcomp.2006.11.009.

Mahdi F, Abbas H, Khan A A, (2010), 'Strength characteristics of polymer mortar and concrete using different compositions of resins derived from post-consumer PET bottles', *Constr Build Mater*, **24**, 25–36. http://dx.doi.org/10.1016/j.conbuildmat.2009.08.006.

Maity P, Basu S, Parameshwaran V, Gupta N, (2006), 'Surface degradation studies in polymer dielectrics with nano-sized fillers', *IEEE 8th International Conference on Properties and Applications of Dielectric Materials (ICPADM-f)*, 26–30 June, Bali, Indonesia, 171–174. DOI: 10.1109/ICPADM.2006.284145.

Malhotra V M, (1997), 'Innovative applications of superplasticizers in concrete – A review', *Proc Mario Collepardi Symp Advances in Concrete Science and Technology*, 7 October, Rome, Italy, 271–314.

Malhotra V M, (2006), 'Reducing CO_2 emissions – Role of fly ash', *ACI Concr Int*, **28**, 42–45.

Malvar L J, Joshi N R, Bearn J A, Novinson T, (2003), 'Environmental effects on the short term bond of carbon fiber reinforced polymer (CFRP) composites', *ASCE J Compos Constr*, **7**, 58–63. http://dx.doi.org/10.1061/(ASCE)1090-0268(2003)7:1(58).

Marouani S, Curtil L, Hamelin P, (2008), 'Composites realized by hand lay-up process in a civil environment: initial properties and durability', *Mater Struct*, **41**, 831–851. DOI: 10.1617/s11527-007-9288-z.

May C A, (1989), 'Epoxy resin', *ASM International, ASM Engineered Materials Reference Book,* Ohio.

Mays G C, Hutchinson A R, (1992), *Adhesives in civil engineering*, Cambridge: Cambridge University Press.

McGill L P, Humpage M, (1990), 'Prolonging the life of reinforced concrete structures by surface treatments', In: *Protection of concrete*, Eds Dhir R K, Green J W, Publ E and FN Spon, 191–200.

Medeiros M H F, Helene P, (2009), 'Surface treatment of reinforced concrete in marine environment: influence on chloride diffusion coefficient and capillary water absorption', *Constr Build Mater*, **23**, 1476–1484. http://dx.doi.org/10.1016/j.conbuildmat.2008.06.013.

Mehta P K, (1991), 'Durability of concrete – Fifty years of progress?', In: *SP 126 CANMET/ACI. Durability of Concrete. Second International Conference*, Montreal, Canada, Ed Malhotra V M, Detroit, Michigan, 1–31.

Mehta P K, Monteiro P J M, (2006), *Concrete: Microstructure, properties and materials*, Third Edition, McGraw-Hill, NY, 522–525.

Mehta P K, Walters M, (2008), 'Roadmap to a sustainable concrete construction industry', *The Construction Specifier*, **61**, 48–57.

Meran C, Ozturk O, Yuksel M, (2008), 'Examination of the possibility of recycling and utilizing recycled polyethylene and polypropylene', *Mater Des*, **29**, 701–705. http://dx.doi.org/10.1016/j.matdes.2007.02.007

Móczó J, Pukánszky B, (2008), 'Polymer micro and nanocomposites: Structure, interactions, properties', *J Ind Eng Chem*, **14**, 535–563. http://dx.doi.org/10.1016/j.jiec.2008.06.011.

Monteny J, De Belie N, Vincke E, Verstraete W, Taerwe L, (2001), 'Chemical and microbiological tests to simulate sulfuric acid corrosion of polymer-modified concrete', *Cem Concr Res*, **31**, 1359–1365. http://dx.doi.org/10.1016/S0008-8846(01)00565-8.

Morin V, Cohen Tenoudji F, Feylessoufi A, Richard P, (2001), 'Superplasticizer effects on setting and structuration mechanisms of ultrahigh-performance concrete', *Cem Concr Res*, **31**, 63–71. http://dx.doi.org/10.1016/S0008-8846(00)00428-2.

Morin V, Moevus M, Dubois-Brugger I, Gartner E, (2011), 'Effect of polymer modification of the paste-aggregate interface on the mechanical properties of concretes', *Cem Concr Res*, **41**, 459–466. http://dx.doi.org/10.1016/j.cemconres.2011.01.006.

Morlat R, Orange G, Bomal Y, Godard P, (2007), 'Reinforcement of hydrated Portland cement with high molecular mass water-soluble polymers', *J Mater Sci*, **42**, 4858–4869. DOI: 10.1007/s10853-006-0645-z.

Moussa O, Vassilopoulos A P, Keller T, (2012), 'Effects of low-temperature curing on physical behavior of cold-curing epoxy adhesives in bridge construction', *Int J Adhes Adhes*, **32**, 15–22. http://dx.doi.org/10.1016/j.ijadhadh.2011.09.001

Muthukumar M, Mohan D, (2004), 'Studies on polymer concretes based on optimized aggregate mix proportion', *Eur Polym J*, **40**, 2167–2177. http://dx.doi.org/10.1016/j.eurpolymj.2004.05.004.

Nagi M, Okamoto P A, Kozikowski R L, Hover K, (2007), *Evaluating air-entraining admixtures for highway concrete*', National Cooperative HIGHWAY Research Program. Transportation Research Board of the National Academies USA NCHRP 2007, Report 578, Washington, DC.

Naidu Y C, (1992), 'Properties and applications of polymer concretes', In: *Cement and concrete science & technology*, Vol 1, Part II, Ed Ghosh S N, ABI Books Ltd, New Delhi, India, 267–287.

Nair P S, Thachil E T, (2011), 'Effect of structure of polymer on properties of polymer cement composite', *13th International Congress on Polymers in Concrete, ICPIC*

2010, 10–12 February, Madeira, Portugal, Eds Aguiar J B, Jalali S, Camoes A, Ferreira R M, 64–72.

Nair P, Ku D H, Lee C W, Park J S, Park H Y, Lee W M, (2010a), 'Mechanical properties and durability of PMMA impregnated mortar', *Korean J Chem Eng*, **27**, 334–339. DOI: 10.1007/s11814-009-0346-9.

Nair P, Ku D H, Lee C W, Park H Y, Song H Y, Lee S S, Lee W M, (2010b), 'Microstructural studies of PMMA impregnated mortars', *J Appl Polym Sci*, **116**, 3534–3540. DOI: 10.1002/app.31887.

Nair P, Park J S, Lee C W, Park H Y, Lee W M, (2010c), 'Physical-chemical changes in polymer impregnated mortars on exposure to sea water', *Korean J Chem Eng*, **27**, 1323–1327. DOI: 10.1007/s11814-010-0192-9.

Nkinamubanzi P-C, Aitcin P-C, (2004), 'Cement and superplasticizer combinations: Compatibility and robustness', *Cement, Concrete and Aggregates*, **26**, 102–109.

Oetiker D, (2009), *Polycarboxylate ether polymers and their influence on sustainability in concrete production*, American Concrete Institute, ACI 262 SP, 297–307.

Ohama Y, (1997), 'Recent progress in concrete–polymer composites', *Adv Cem Bas Mater*, **5**, 31–40. http://dx.doi.org/10.1016/S1065-7355(96)00005-3.

Ohama Y, (1998), 'Polymer-based admixtures', *Cem Concr Compos*, **20**, 189–212. http://dx.doi.org/10.1016/S0958-9465(97)00065-6.

Ohama Y, (2010), 'Concrete–polymer composites – The past, present and future', *13th International Congress on Polymers in Concrete, ICPIC 2010*, 10–12 February, Madeira, Portugal, Eds Aguiar J B, Jalali S, Camoes A, Ferreira R M, 1–13.

Orak S, (2000), 'Investigation of vibration damping on polymer concrete with polyester resin', *Cem Concr Res*, **30**, 171–174. http://dx.doi.org/10.1016/S0008-8846(99)00225-2.

Pacheco-Torgal F, Jalali S, (2009), 'Sulphuric acid resistance of plain, polymer modified and fly ash cement concretes', *Constr Build Mater*, **23**, 3485–3491. http://dx.doi.org/10.1016/j.conbuildmat.2009.08.001.

Paiva H, Silva L, Labrincha J, Ferreira V, (2006), 'Effects of a water-retaining agent on the rheological behaviour of a single-coat render mortar', *Cem Concr Res*, **36**, 1257–1262. http://dx.doi.org/10.1016/j.cemconres.2006.02.018.

Papayianni I, Tsohos G, Oikonomou N, Mavria P, (2005), 'Influence of superplasticizer type and mix design parameters on the performance of them in concrete mixture', *Cem Concr Compos*, **27**, 217–222. http://dx.doi.org/10.1016/j.cemconcomp.2004.02.010.

Pardo A, Maribona I R Z, Urreta J, San José J T, Muguerza A, (1995), 'Influence of dosage and temperature on mechanical properties of polymer concrete', In: *Proc 8th Inter Congr Polymers in Concrete, ICPIC VIII*, Oostende, Belgium.

Pei M, Wang D, Hu X, Xu D, (2000), 'Synthesis of sodium sulfanilate-phenol-formaldehyde condensate and its application as a superplasticizer in concrete', *Cem Concr Res*, **30**, 1841–1845. http://dx.doi.org/10.1016/S0008-8846(00)00389-6.

Perenchio W F, (1988), 'Durability of concrete treated with silanes', *Concr Int*, *Nov*, 34–40.

Pfeifer D W, (1987), *Protective systems for new prestressed and substructure concrete*. Report No. FHWA/RD-86/193, Federal Highway Administration, Washington, DC, USA.

Pfeiffer D W, Scali M J, (1981), *Concrete sealing for protection of bridge structures*. National Cooperative Highway Research Program (NCHRP 244), Transportation Research board, Washington, DC, USA.

Pinoteau N, Pimienta P, Guillet T, Rivillon P, Rémond S, (2011), 'Effect of heating

rate on bond failure of rebars into concrete using polymer adhesives to simulate exposure to fire', *Int J Adhes Adhes*, **31**, 851–861. http://dx.doi.org/10.1016/j.ijadhadh.2011.08.005.
Plank J, Schröfl C, Gruber M, (2009), *Use of a supplemental agent to improve flowability of ultra-high-performance concrete*, American Concrete Institute, ACI 262 SP, 1–16.
Pratap A, (2002), 'Vinyl-ester and acrylic based polymer concrete for electrical applications', *Prog Cryst Growth Charact Mater*, **45**, 117–125.
Puertas, F, Alonso, M M, Vázquez, T, (2005a), 'Influencia de aditivos basados en policarboxilatos sobre el fraguado y el comportamiento reológico de pastas de cemento Portland', *Mater Construc*, **55**, 61–73. DOI: 10.3989/mc.2005.v55.i277.180.
Puertas F, Santos H, Palacios M, Martínez-Ramírez S, (2005b), 'Polycarboxylate superplasticiser admixtures: effect on hydration, microstructure and rheological behaviour in cement pastes', *Adv Cem Res*, **17**, 77–89. DOI: 10.1680/adcr.2005.17.2.77.
Puterman M, Malorny W, (1998), 'Some doubts and ideas on the microstructure formation of PCC', *IXth International Congress on Polymers in Concrete, ICPIC'98*, 14–18 September, Bologna, Italy, Ed Sandrolini F, 165–178.
Qiu X, Peng X, Yi C, Deng Y, (2011), 'Effect of side chains and sulfonic groups on the performance of polycarboxylate-type superplasticizers in concentrated cement suspensions', *J Disp Sci Technol*, **32**, 203–212. DOI: 10.1080/01932691003656888.
Ramachandran V M, Malhotra V M, Jolicoeur C, Spirattos N, (1998), *Superplasticizers: Properties and applications in concrete*, Natural Resources Canada, CANMET, Ottawa, Canada, Ontario, 404.
Ran Q, Somasundaran P, Miao C, Liu J, Wuc S, Shen J, (2009a), 'Effect of the length of the side chains of comb-like copolymer dispersants on dispersion and rheological properties of concentrated cement suspensions', *J Colloid Interface Sci*, **336**, 624–633. http://dx.doi.org/10.1016/j.jcis.2009.04.057.
Ran Q, Liu J, Miao C, Mao Y, Shang Y, Sha J, (2009b), *Development of new ultra-high-early-strength superplasticizer and its application*, American Concrete Institute, ACI 262 SP, 187–199.
Rebeiz K S, (1996), 'Precast use of polymer concrete using unsaturated polyester resin based on recycled PET waste', *Constr Build Mater*, **10**, 215–220. http://dx.doi.org/10.1016/0950-0618(95)00088-7.
Rebeiz K S, Fowler D W, (1996), 'Shear and flexure behavior of reinforced polymer concrete made with recycled plastic wastes', In: *Properties and uses of polymers in concrete*, American Concrete Institute, Farmington Hills, 62–77.
Reis J M L, (2009), 'Mechanical characterization of polymer mortars exposed to degradation solutions', *Constr Build Mater*, **23**, 3328–3331. http://dx.doi.org/10.1016/j.conbuildmat.2009.06.047.
Reis J M L, Moreira D C, Nunes L C S, Sphaier L A, (2011), 'Experimental investigation of the mechanical properties of polymer mortars with nanoparticles', *Mater Sci Eng A*, **528**, 6083–6085. http://dx.doi.org/10.1016/j.msea.2011.04.054.
Ribeiro M C S, Reis J M L, Ferreira A J M, Marques A T, (2003), 'Thermal expansion of epoxy and polyester polymer mortars-plain mortars and fibre-reinforced mortars', *Polym Test*, **22**, 849–857. http://dx.doi.org/10.1016/S0142-9418(03)00021-7.
Ribeiro M C S, Nóvoa P R, Ferreira A J M, Marques A T, (2004), 'Flexural performance of polyester and epoxy polymer mortars under severe thermal conditions', *Cem Concr Compos*, **26**, 803–809. http://dx.doi.org/10.1016/S0958-9465(03)00162-8.
Ribeiro M C S, Ferreira A J M, Marques A T, (2008), 'Weatherability of epoxy and polyester polymer mortars under maritime environments' *Challenges for Civil Construction,* Porto, Portugal, Eds Torres Marques et al., 1–12.

Riffle J S, Lesko J J, Puckett P M, (1998), 'Chemistry of polymer matrix resins for infrastructure', *2nd Int. Conf. on Composites in Infrastructures*, Vol. 1, University of Arizona, Tucson, AZ, USA, Eds Saadatmanesh H, Ehsani M R, 23–34.

Rossignolo J A, Agnesini M V C, (2004), 'Durability of polymer-modified lightweight aggregate concrete', *Cem Concr Compos*, **26**, 375–380. http://dx.doi.org/10.1016/S0958-9465(03)00022-2.

Roylance D, Roylance M, (1978), 'Weathering of fiber-reinforced epoxy composites', *Polym Eng Sci*, **18**, 249–254. DOI: 10.1002/pen.760180402.

Sakai E, Kasuga T, Sugiyama T, Asaga K, Masaki Daimon M, (2006), 'Influence of superplasticizers on the hydration of cement and the pore structure of hardened cement', *Cem Concr Res*, **36**, 2049–2053. http://dx.doi.org/10.1016/j.cemconres.2006.08.003.

Schröfl C, Gruber M, Plank J, (2010), 'Interactions between polycarboxylate superplasticizers and components present in ultra-high strength concrete', *J China Ceram Soc*, **38**, 1605–1612.

Sciolti M S, Frigione M, Aiello M A, (2010), 'Wet lay-up manufactured FRP's for concrete and masonry repair. Influence of water on the properties of composites and of their epoxy components', *ASCE J Compos Constr*, **14**, 823–833. http://dx.doi.org/10.1061/(ASCE)CC.1943-5614.0000132.

Sebök T, Stránêl O, (2004), 'Wear resistance of polymer-impregnated mortars and concrete', *Cem Concr Res*, **34**, 1853–1858. http://dx.doi.org/10.1016/j.cemconres.2004.01.026.

Selvamony C, Ravikumar M S, Kannan S U, Basil Gnanappa S, (2010), 'Investigations on self-compacted self-curing concrete using limestone powder and clinkers', *ARPN J Eng Appl Sci*, **5**, 1–6.

Shaw S J, (1994), 'Epoxy resin adhesives', In: *Chemistry and technology of epoxy resins*, Ed Blackie Academic & Professional, Chapman and Hall, London, UK, 206–255.

Shin H-C, Miyauchi H, Tanaka K, (2011), 'An experimental study of fatigue resistance in epoxy injection for cracked mortar and concrete considering the temperature effect', *Constr Build Mater*, **25**, 1316–1324. http://dx.doi.org/10.1016/j.conbuildmat.2010.09.013.

Shirai A, Ohama Y, Kokubun Y, (2010), 'Effects of flame retardants on fire-protecting performance of polymer-modified mortars', *13th International Congress on Polymers in Concrete, ICPIC 2010*, 10–12 February, Madeira, Portugal, Eds Aguiar J B, Jalali S, Camoes A, Ferreira R M, 119–126.

Silva D A, John V M, Ribeiro J L D, Roman H R, (2001), 'Pore size distribution of hydrated cement pastes modified with polymer', *Cem Concr Res*, **31**, 1177–1184. http://dx.doi.org/10.1016/S0008-8846(01)00549-X.

Silva M A G, Biscaia H, (2008), 'Degradation of bond between FRC and RC beams', *Compos Struct*, **85**, 164–174. http://dx.doi.org/10.1016/j.compstruct.2007.10.014.

Smoak W G, (1990), 'Field application of concrete surface impregnation', In *Protection of Concrete*, Eds. Dhir R K, Green J W, E and FN Spon, 451–467.

Sorelli L, Constantinides G, Ulm F-J, Toutlemonde F, (2008), 'The nano-mechanical signature of ultra high performance concrete by statistical nanoindentation techniques', *Cem Concr Res*, **38**, 1447–1456. http://dx.doi.org/10.1016/j.cemconres.2008.09.002.

Su Z, Sujata K, Bijen J M, Jennings H M, Fraaij A L A, (1996), 'The evolution of the microstructure in styrene acrylate polymer-modified cement pastes at the early stage of cement hydration', *Adv Cem Bas Mater*, **3**, 87–93. http://dx.doi.org/10.1016/S1065-7355(96)90041-3.

Sugiyama T, Sugamata T, Ohta A, (2003), 'The effects of high range water reducing agent on the improvement of rheological properties', *Proc Seventh CANMET/ACI Inter Conf on superplasticizers and other chemical admixtures in concrete*, American Concrete Institute, ACI 217 SP, 343–360.

Sung N-H, (1990), 'Moisture effects on adhesive joints', In: *Adhesives and sealants. Engineered materials handbook*, vol. 3, USA, ASM International, 622–627.

Swamy R N, (1997), 'Design for durability and strength through the use of fly ash and slag', *Proc Mario Collepardi Symp on Advances in Concrete Science and Technology*, 7 October, Rome, Italy, 127–194.

Swamy R N, Tanikawa S, (1990), 'Surface coatings to preserve concrete durability', In: *Protection of concrete*, Eds Dhir R K, J W Green, E and FN Spon, 149–165.

Swamy R N, Suryavanshi A K, Tanikawa S, (1998), 'Protective ability of an acrylic based surface coating system against chloride and carbonation penetration into concrete', *ACI Mater J*, **95**, 101–112.

Tabor L J, (2004), 'Repair materials and techniques'. In: *Durability of concrete structures. Investigation, repair, protection*, Ed Mays G, E and FN Spon, 82–129.

Tanaka M, Matsuo S, Ohta A, Veda M, (1996), 'A new admixture for high performance concrete', In *Proc. Concrete in The Service of Mankind,* Eds Dhir R K, McCarthy M J, 291–300.

Tawfik M E, Eskander S B, (2006), 'Polymer concrete from marble wastes and recycled poly(ethylene terephthalate)', *J Elastomers Plast*, **38**, 65–79. DOI: 10.1177/0095244306055569.

Toutanji H, Gomez W, (1997), 'Durability characteristics of concrete beams externally bonded with FRP composite sheets', *Cem Concr Compos*, **19**, 351–358. http://dx.doi.org/10.1016/S0958-9465(97)00028-0.

Toutanji H, Ortiz G, (2001), 'The effect of surface preparation on the bond interface between FRP sheets and concrete members', *Compos Struct*, **53**, 457–462. http://dx.doi.org/10.1016/S0263-8223(01)00057-5.

Tu L, Kruger D, (1996), 'Engineering properties of epoxy resins used as concrete adhesives', *ACI Mater J*, **93**, 26–35.

Uchikawa H, Hanehara S, Shirasaka T, Sawaki D, (1992), 'Effect of admixture of hydration of cement, adsorptive behaviour of admixture and fluidity and setting of fresh cement paste', *Cem Concr Res*, **22**, 1115–1129. http://dx.doi.org/10.1016/0008-8846(92)90041-S.

Uchikawa H, Sawaki D, Hanehara S, (1995), 'Influence of kind and added timing organic admixture type and addition time on the composition, structure and property of fresh cement paste', *Cem Concr Res*, **25**, 353–364. http://dx.doi.org/10.1016/0008-8846(95)00021-6.

Uchikawa H, Hanehara S, Sawaki D, (1997), 'The role of steric repulsive force in the dispersion of cement particles in fresh paste prepared with organic admixture', *Cem Concr Res*, **27**, 37–50. http://dx.doi.org/10.1016/S0008-8846(96)00207-4.

Umoto T, Ohga H, Yorezawa T, Ibe H, (1994), 'Durability of repaired reinforced concrete in marine environment', In: *Durability of Concrete*, Proc 3rd Inter Conf, Nice, France, ACI 145 SP, 445–468.

Uysal M, Sumer M, (2011), 'Performance of self-compacting concrete containing different mineral admixtures', *Constr Build Mater*, **25**, 4112–4120. http://dx.doi.org/10.1016/j.conbuildmat.2011.04.032.

Uzal B, Turanli L, Mehta P K, (2007), 'High-volume natural pozzolan concrete for structural applications', *ACI Mater J*, **104**, 535–538.

Van Es R, (1992), 'Bridge 425 – a coated bridge', *Construction Repair*, **Sept/Oct**, 25–27.

Van Gemert D, (2011), 'Contribution of concrete–polymer composites to sustainable construction and conservation procedures', *European Symposium on Polymers in Sustainable Construction, ESPSC 2011, Czarnecki Symposium,* 6–7 September, Warsaw, Poland, 44–52.

Van Gemert D, Knapen E, (2010), 'Contribution of C-PC to sustainable construction procedures', *13th International Congress on Polymers in Concrete, ICPIC 2010*, 10–12 February, Madeira Portugal, Eds Aguiar J B, Jalali S, Camoes A, Ferreira R M, 27–36.

Vassie P R, (1991), 'Concrete coatings, do they reduce on-going corrosion of reinforcing steel', In: *Protection of concrete*, Proc Inter Conf, Eds Dhir R K, Green J W, E and FN Spon, London, 281–291.

Vipulanandan C, Liu J, (2005), 'Performance of polyurethane-coated concrete in severe environment', *Cem Concr Res*, **35**, 1754–1763. http://dx.doi.org/10.1016/j.cemconres.2004.10.033.

Vipulanandan C, Mantrala S K, (1996), 'Behavior of fiber reinforced polymer concrete'. In: *Proceedings of Fourth Materials Engineering Conference*, 10–14 November, Washington, DC, 1160–1169.

Wang J, Zhang, S, Yu H, Kong X, Wang X, Gu Z, (2005), 'Study of cement mortars modified by emulsifier-free latexes', *Cem Concr Compos*, **27**, 920–925. http://dx.doi.org/10.1016/j.cemconcomp.2005.05.005.

Wang R, Wang P-M, Yao L-J, (2012), 'Effect of redispersible vinyl acetate and versatate copolymer powder on flexibility of cement mortar', *Constr Build Mater*, **27**, 259–262. http://dx.doi.org/10.1016/j.conbuildmat.2011.07.050.

Weyers R E, Cady P D, Hentry M, (1990), 'Field service life performance of deep polymer impregnation as a bridge deck corrosion protection-method'. In: *Protection of concrete*, Eds Dhir R K, Green J W, E and FN Spon, 397–412.

Winey K I, Vaia R A, (2007), 'Polymer nanocomposites', *MRS Bull*, **32**, 314–319.

Wolff L, Hailu K, Raupach M, (2006), 'Mechanisms of blistering of coatings on concrete', *Int. Symp. Polymers in Concrete, ISPIC 2006*, University of Minho, Guimarães, Portugal, Eds Aguiar J B, Jalali S, A Camões, R M Ferreira, 213–223.

Wong W G, Fang P, Pan J K, (2003), 'Dynamic properties impact toughness and abrasiveness of polymer-modified pastes by using nondestructive tests', *Cem Concr Res*, **33**, 1371–1375. http://dx.doi.org/10.1016/S0008-8846(03)00069-3.

Wu K R, Zhang D, Jun-Mei Song J M, (2002), 'Properties of polymer-modified cement mortar using pre-enveloping method', *Cem Concr Res*, **32**, 425–429. http://dx.doi.org/10.1016/S0008-8846(01)00697-4.

Xiong G, Chen X, Li G, Chen L, (2001), 'Sulphuric acid resistance of soluble soda glass–polyvinyl acetate latex–modified cement mortar', *Cem Concr Res*, **31**, 83–86. http://dx.doi.org/10.1016/S0008-8846(00)00426-9.

Yamada Y, Oshiro T, Tanikawa S, Swamy R N, (1997), *Field evaluation of an acrylic rubber protective coating system for reinforced concrete structures*, ACI Inter. Concrete Research & Information Portal, Special Publication, no. **170**, 23–40.

Yang Z, Shi X, Creighton A T, Peterson M M, (2009), 'Effect of styrene–butadiene rubber latex on the chloride permeability and microstructure of Portland cement mortar', *Constr Build Mater*, **23**, 2283–2290. http://dx.doi.org/10.1016/j.conbuildmat.2008.11.011.

Zhao S, Schadler L S, Duncan R, Hillborg H, Auletta T, (2008), 'Mechanisms leading to improved mechanical performance in nanoscale alumina filled epoxy', *Compos Sci Technol*, **68**, 2965–2975. http://dx.doi.org/10.1016/j.compscitech.2008.01.009.

Zimmermann J, Hampel C, Kurz C, Frunz L, Flatt R J (2009), *Effect of polymer structure on the sulfate–polycarboxylate competition*, American Concrete Institute, ACI 262 SP, 165–175.

Zingg A, Winnefeld F, Holzer, L, Pakusch J, Becker S, Figi R, Gauckler L, (2009), 'Interaction of polycarboxylate-based superplasticizers with cements containing different C_3A amounts', *Cem Concr Compos*, **31**, 153–162. http://dx.doi.org/10.1016/j.cemconcomp.2009.01.005.

Part IV

Future alternative binders and use of nano and biotech

17
Alkali-activated based concrete

I. GARCÍA-LODEIRO, A. FERNÁNDEZ-JIMÉNEZ and A. PALOMO, Eduardo Torroja Institute for Construction Science, Spain

DOI: 10.1533/9780857098993.4.439

Abstract: Portland cement-based products (primarily concretes) are the world's most commonly used building materials. Due to its huge production worldwide, the Portland cement industry poses economic, energy and environmental problems (7% of total world-wide CO_2 emissions). International concern over how to reduce CO_2 emissions has given rise within the scientific community to a growing interest in the development of materials and technologies able to reduce the impact of Portland cement and make construction a more sustainable industry. One of the possible alternative materials whose study and use has intensified in recent years is so-called *alkaline cement*, produced by alkali-activating aluminosilicates, whether of natural or industrial (blast furnace slag, fly ash, etc.) origin. Based on the nature of their cementitious components (the $CaO–SiO_2–Al_2O_3$ system), alkaline cements may be grouped under two main categories: high-calcium and low-calcium cements. The reaction products governing the characteristics of the two types of end product differ. While the main reaction product precipitating in calcium-rich systems is a C-A-S-H type gel, in low calcium systems the reaction product is a N-A-S-H gel. A third type of alkaline cement, blended or hybrid cements, is a combination of the above. In this group, the reaction products comprise complex mixes of C-A-S-H + N-A-S-H gels. This chapter stresses the interest that has arisen around the study of these new cementitious systems.

Key words: alkali activation, alkaline cements, N-A-S-H gel, hybrid cement.

17.1 Introduction: alkaline cements

Portland cement concrete is today's construction material par excellence. It owes this pre-eminence to its mechanical strength, high value for money and generally good performance. Nonetheless, Portland cement manufacture raises certain energy and environmental issues, since it calls for temperatures of up to 1500 °C and raw materials whose quarrying mars the landscape, while emitting gases such as CO_2 and NOx. Moreover, concrete poses certain durability problems, such as the aggregate-alkali reaction and attendant expansion, chloride-induced corrosion in reinforcing steel, retarded ettringite formation and the generation of thaumasite (Metha, 1991; Taylor, 1997).

One of the options widely accepted in the industry today to reduce this impact while contributing to solve other environmental problems (finding a use for industrial by-products or waste that must otherwise be stockpiled, a costly and pollution-prone procedure) is to include active (mineral or industrial by-product) additions in Portland cement clinker. This has given rise to different types of ordinary cements, presently listed in Spanish and European standard UNE-EN 197-1:2000, which specifies both the type and maximum amount of additions allowed.

Another more innovative option consists of developing alternative, less expensive and less environmentally damaging cements (involving lower CO_2 emissions or the re-use of industrial by-products), that exhibit characteristics or performance comparable to or even better than ordinary Portland cements (OPC). One such category of materials includes a series of binders generically known as alkaline cements (Wu et al., 1983; Glukhovsky 1994; Wang and Scrivener, 1995; Wang *et al.* 1995; Fernández-Jiménez, 2000; Fernández-Jiménez et al., 2005; Shi et al., 2006, 2011; Duxson et al., 2007a; Provis et al., 2009; Li et al., 2010; Pacheco-Torgal et al., 2008a, 2008b; Temuujin et al., 2010; Lemougna et al., 2011; Juenger et al., 2011).

Alkaline cements are cementitious materials formed as the result of the dissolution of natural or industrial waste materials (with amorphous or vitreous structures) in an alkaline medium. When mixed with alkaline activators, these materials set and harden, yielding a material with good binding properties. Alkaline cements are characterised by:

- Low Portland clinker content (from 0 to 30%);
- The use of solid or liquid alkaline activators.

A wide variety of alkali-activated cements has been developed in the last few decades. Based on the nature of the cementitious components ($CaO–SiO_2–Al_2O_3$ system), alkaline cements may be grouped under two main categories: (1) high-calcium and (2) low-calcium cements. The activation pattern differs in each category.

- **Model 1:** activation of calcium- and silicon-rich materials ($(Na, K)_2O–CaO–Al_2O_3–SiO_2–H_2O$ system); an example of this first model is the activation of *blast furnace slag* (SiO_2 + CaO > 70%) under relatively moderate alkaline conditions (Schilling et al., 1994; Bakharev et al., 2000; Fernández-Jiménez 2000; Fernández-Jiménez et al., 2003; Shi et al., 2006; Bernal et al., 2011a; Puertas et al., 2011; Ben Haha et al., 2012). In this case the main reaction product is a *C-S-H* (calcium silicate hydrate) gel, similar to the gel obtained during Portland cement hydration, which takes up a small percentage of Al in its structure (C-(A)-S-H gel).
- **Model 2:** activation of materials comprising primarily *aluminium and silicon* $(Na,K)_2O–Al_2O_3–SiO_2–H_2O$ system); this second major alkali activation model involves materials with low CaO contents such as metakaolin or type F fly ash (from coal-fired steam power plants). In

this case more intense working conditions are required to kick-start the reactions (highly alkaline media and curing temperatures of 60–200 °C). The main reaction product formed in this case is a three-dimensional inorganic alkaline polymer, a *N-A-S-H* (or alkaline aluminosilicate) gel that can be regarded as a zeolite precursor (Palomo et al., 1999, 2004; Duxson et al., 2007a; Provis and van Deventer, 2009). This gel is also known as a *geopolymer* or *inorganic polymer*.

Today, a *third* alkaline activation model, a combination of the preceding two, can also be described. The product in this case is a new type of binder known as a *blended* or *hybrid alkaline cement*, formed as the result of the alkaline activation of materials with CaO, SiO_2 and Al_2O_3 contents around 20%. These materials can be divided into two groups. *Group A* covers the alkaline activation of materials having low portland cement clinker and a high proportion (over 70%) of mineral additions (Alonso and Palomo, 2001; Yip et al., 2005; Palomo et al. 2007a). Examples are to be found in cement + slag, cement + fly ash, cement + slag + fly ash, or cement + slag + metakaolin blends. *Group B* comprises blends containing no Portland cement: blast furnace slag + fly ash, phosphorus slag + blast furnace slag + fly ash and similar. The cementitious gels forming in this group are very complex and are in fact mixed, (C,N)-A-S-H or N-(C)-A-S-H-type gels. Each of these groups can be subdivided into sub-systems whose development and applications differ substantially (Shi et al., 2006, 2011; Palomo et al., 2007a).

The aforementioned models are described in some detail in the following discussion, which focuses particularly on the low calcium and hybrid systems, since the first group of calcium-rich systems has been amply addressed in the literature (Fernández-Jiménez, 2000; Shi et al., 2006).

17.2 Alkali activation of calcium-rich systems

17.2.1 Composition of starting materials

The material most commonly used in the manufacture of calcium-rich alkaline cement and concrete is blast furnace slag. As Fig. 17.1 shows, this material comprises primarily calcium, silicon and aluminium oxide.

Vitreous blast furnace slag is a steel industry waste, formed when the acid clay gangue in iron ore and the sulphur ash from coke combine with the lime and magnesium in the limestone or dolomite used as fluxes. The high-temperature (1600 °C) fusion of acid (SiO_2 and Al_2O_3) with basic (CaO and MgO) oxides and their subsequent abrupt cooling to 800 °C yields vitreous blast furnace slag which, after grinding, is stored in silos for subsequent removal. The vitreous phase, which contains network-forming anions $(SiO_4)^{4-}$, $(AlO_4)^{5-}$ and $(MgO_4)^{6-}$ and network-modifying cations, Ca^{2+}, Al^{3+} and Mg^{2+}, accounts for 90–95% of this slag, on average. The percentage composition of

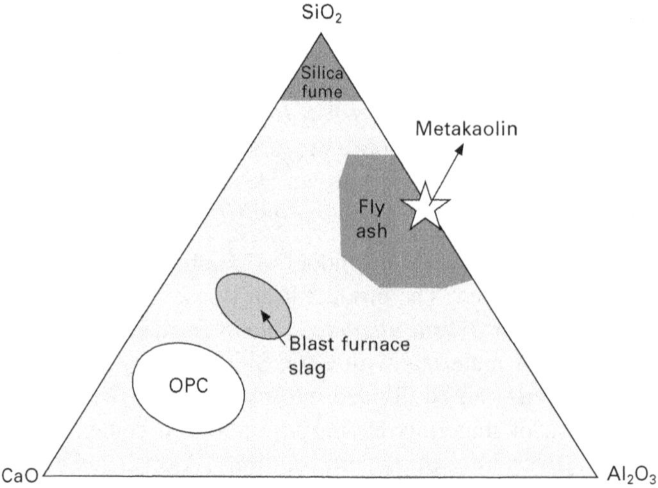

17.1 Standard composition of the prime materials used to manufacture alkaline cements on a $CaO–SiO_2–Al_2O_3$ diagram.

this material is shown in Fig. 17.1. The minority crystalline phases include a solid solution of two melilite family minerals, gehlenite and akermanite, which crystallise in the tetragonal system (Fernández-Jiménez, 2000; Shi et al., 2006).

The main requirements to be met by such slag for use in activated slag cement are listed below (Fernández-Jiménez 2000; Shi et al., 2006):

- It must be granulated or pelletised and have a vitreous phase content of 85–95%.
- It must exhibit structural disorder, since the lower the degree of polymerisation in the glass, the higher is its hydraulic activity: the degree of polymerisation depends on the (SiO_4) tetrahedral and the Al and Mg coordination in the vitreous phase of the slag.
- It must be pH-basic, i.e., have a CaO + MgO/SiO_2 ratio of greater than 1. Basic slag has a higher hydraulic potential, since the lime content in the slag controls its activation. That notwithstanding, acid slag may also be alkali-activated.
- It must be ground to a specific surface of 400–600 m^2/kg. Specific surface plays an important role in the rate and intensity of the activation reaction.

17.2.2 Reaction mechanisms

The hydration of the calcium silicates in Portland cement (C_3S and C_2S) yields portlandite ($Ca(OH)_2$) and a non-crystalline calcium silicate hydrate, generically known as C-S-H, the compound primarily responsible for the binding properties exhibited by the material (Cong and Kirkpatrick, 1996; Richardson and Groves, 1997; Taylor, 1997; García-Lodeiro et al., 2012a). The reaction mechanisms in alkaline cements differ from the mechanisms observed in OPC, as discussed below.

Glukhovsky (1994) and Krivenko (1992, 1994) proposed a model for the alkaline activation of SiO_2- and CaO-rich materials (such as blast furnace slag) to describe the reaction sequence in which the alkaline cation (R^+) acts as a mere catalyser in the initial phases of hydration, via cationic exchange with the Ca^{2+} ions. The mechanism proposed by these authors is summarised below:

$$=\text{Si}-\text{O}^- + \text{R}^+ = =\text{Si}-\text{O}-\text{R}$$

$$=\text{Si}-\text{O}-\text{R} + \text{OH}^- = =\text{Si}-\text{O}-\text{R}-\text{OH}^-$$

$$=\text{Si}-\text{O}-\text{R}-\text{OH}^- + \text{Ca}^{2+} = =\text{Si}-\text{O}-\text{Ca}-\text{OH} + \text{R}^+$$

These same authors believed that as the reactions advance, the alkaline cations are taken up into the structure.

Fernández-Jiménez (2000) and Fernández-Jiménez et al. (2003) reported that the nature of the anion in the solution also plays an instrumental role in activation, especially at early ages and in particular with regard to setting. The model that describes the reaction mechanisms (based on a model proposed by Glasser, 1990) is depicted in Fig. 17.2.

The slag particles first undergo chemical attack due to the high pH in the liquid medium. The insoluble hydration products resulting from that attack settle on the surface of the grains, forming a semi-protective layer or barrier that prevents the reactions from proceeding at the desired pace. These hydration products are amorphous and consequently undetected by X-ray

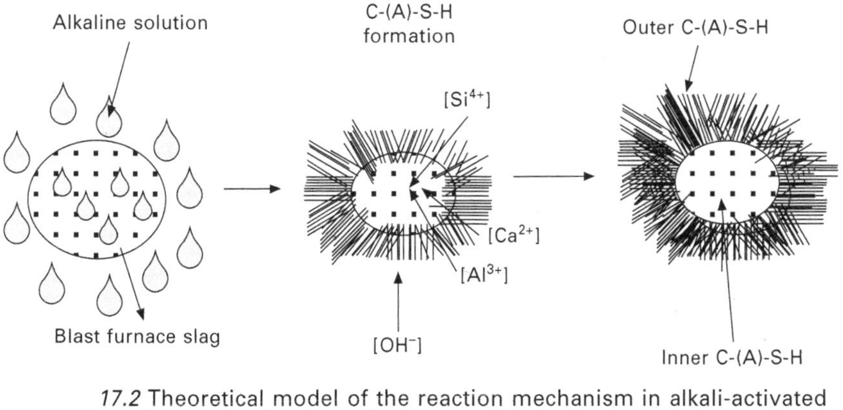

17.2 Theoretical model of the reaction mechanism in alkali-activated slag.

diffraction (XRD) or electron diffraction. Since these C-S-H-type gels, with lower C/S ratios than in the product present in Portland cement pastes, also contain Al in their structure, they are actually C-(A)-S-H gels. This layer of gel partially isolates the most anhydrous part of the slag particles. Total slag hydration may, however, lead to the formation of pores that would favour the mobility of dissolved ions, in turn inducing the formation of hydrotalcite-like solids ($Mg_6Al_2CO_3(OH)_{16}·4H_2O$). Nonetheless, hydration also redistributes paste porosity, rendering the matrix denser and porosity more concentrated to thereby reduce permeability.

17.2.3 Reaction products

The major reaction product in Portland cement hydration, a C-S-H-type gel, is the substance primarily responsible for the mechanical properties of the end material. Portlandite, ettringite and calcium monosulphoaluminate, among others, form as secondary products (Taylor, 1997). The process is similar in alkaline cements. The main reaction product is a C-(A)-S-H gel (whose composition and structure vary with respect to the standard C-S-H generated in OPC hydration), while a series of secondary products also form. The type of secondary product generated depends on starting material composition, activator type and concentration and curing conditions (Cheng et al., 1992; LaRosa et al., 1992; Roy et al., 1994; Fernández-Jiménez, 2000; Fernández-Jiménez et al., 2003; Shi et al., 2006; Rodriguez et al., 2008; Bernal et al., 2011a). Table 17.1 compares the most common reaction products forming in normal Portland cement hydration to the products of (calcium-rich and low calcium) alkaline cement activation (Wang and Scrivener, 1995; Fernández-Jiménez et al., 1996, 2003; Taylor, 1997; Palomo et al., 1999; Fernández-Jiménez, 2000; Shi et al., 2006; Duxson et al., 2007a; Provis and van Deventer, 2009; Winnefeld et al., 2010; Puertas et al., 2011).

Broadly speaking, the nature and composition of the reaction products forming during the alkaline activation of materials are among the most controversial aspects of this field of research and the areas most in need of

Table 17.1 Products precipitating in different types of binders

Binder type		OPC	Alkaline cement	
			$(Na,K)_2O$-CaO-Al_2O_3-SiO_2-H_2O	$(Na,K)_2O$-Al_2O_3-SiO_2-H_2O
Reaction product	Primary	C-S-H	C-A-S-H	N-A-S-H
	Secondary	$Ca(OH)_2$ AF_m AF_t	Hydrotalcite $[Mg_6Al_2CO_3(OH)_{16}·4H_2O]$ C_4AH_{13} $CASH_8$ C_4AcH_{11} $C_8A_2cH_{24}$	Zeolites: hydroxysodalite, zeolite P, Na-chabazite, zeolite Y, faujasite
	C = CaO	S = SiO_2	A = Al_2O_3 N = Na_2O H = H_2O	C = CO_2

continued study. The discrepancies are favoured by the essentially amorphous nature of the reaction products, although authors tend to differ most over the minority phases formed: and indeed, no consensus has yet been reached for a number of these phases. Some authors (Cheng et al., 1992; LaRosa et al., 1992; Roy et al., 1994; Wang and Scrivener, 1995) report that compounds such as $Ca(OH)_2$, C-S-H (II)-type gel and a $(M,C)_4AH_{13}$ solid solution may appear in alkali-activated slag. Glukhovsky (1994), however, contends that alkali-activated slag cement contains no $Ca(OH)_2$ or any of the calcium aluminate hydrates or calcium sulpho-aluminate hydrates typical of Portland cement pastes. By contrast, this author identifies compounds such as hydronephelines ($R_2O{\cdot}Al_2O_3{\cdot}2SiO_2{\cdot}nH_2O$), natroline ($R_2O{\cdot}Al_2O_3{\cdot}3SiO_2{\cdot}nH_2O$), analcime ($R_2O.Al_2O_3.4SiO_2{\cdot}nH_2O$), muscovite and paragonite ($Na_2O{\cdot}3Al_2O_3{\cdot}6SiO_2{\cdot}nH_2O$), gismondine ($CaO{\cdot}Al_2O_3{\cdot}2SiO_2{\cdot}nH_2O$), hydrogarnets and a mix of potassium-calcium and sodium-calcium hydroaluminosilicates.

In another vein, LaRosa et al. (1992) report that the conditions that exert the greatest effect on possible zeolite formation are curing temperature, chemical composition of the initial slag and the concentration of the alkaline solution. Although some zeolites may co-exist with C-S-H gel, their presence would require a system with a relatively high Al/Si ratio and a sufficiently low Ca/Si ratio. Zeolite formation is quite improbable in alkali-activated slag cement systems, since the calcium content in slag is normally high enough to consume all the silicon comprising the C-S-H gel. With respect to the $Ca(OH)_2$ formation, in turn, the slag would have to have a high Ca/Si ratio, inasmuch as $Ca(OH)_2$ would not be likely to co-exist with C-S-H gel on a long-term basis where the Ca/Si ratio is ≈ 1.

The reactions leading to the formation of most of the aforementioned products take place because alkaline aluminosilicate solubility is considerably lower than the solubility of the hydrated compounds in Portland cement paste, as shown in Table 17.2 (Puertas, 1995). That lower product solubility ensures higher chemical stability in alkali-activated than in Portland cement pastes.

The authors consulted do appear to agree, however, on the points listed below.

- The main hydration product is a C-S-H gel-type calcium silicate hydrate. This gel differs slightly from the gel form in portland cement paste and has a lower C/S ratio (C/S under 1.0).
- The structure and composition of the C-S-H gel and the presence of other secondary phases or compounds depend on the type and amount of activator used, slag structure and composition and the curing conditions in which the material hardens.

The minority phases whose formation during the alkaline activation of slag has marshalled the widest consensus are listed below:

Table 17.2 Solubility of alkali-activated slag and Portland cement reaction products (Puertas, 1995)

Alkali-activated slag cements			Portland cement		
Mineral	Stoichiometric formula	Solubility (kg/m^3)	Mineral	Stoichiometric formula	Solubility (kg/m^3)
C-S-H (B)	$5CaO.6SiO_2{\cdot}nH_2O$	0.050	Calcium hydroxide$_{<0\}}$	$Ca(OH)_2$	1.300
Xonotlite	$6CaO.6SiO_2{\cdot}H_2O$	0.035	C_2SH_2	$2CaO{\cdot}SiO_2{\cdot}nH_2O$	1.400
Riversidite	$5CaO.6SiO_2{\cdot}3H_2O$	0.050	CSH (B)	$5CaO{\cdot}SiO_2{\cdot}nH_2O$	0.050
Plombierite	$5CaO.6SiO_2{\cdot}10.5H_2O$	0.050	Tetracalcium hydroaluminate	$4CaO{\cdot}Al_2O_3{\cdot}13H_2O$	1.080
Gyrolite	$2CaO.3SiO_2{\cdot}2.5H_2O$	0.051	Tricalcium hydroaluminate	$3CaO{\cdot}Al_2O_3{\cdot}6H_2O$	0.560
Calcite	$CaCO_3$	0.014			
Hydrogarnet	$3CaO.Al_2O_3{\cdot}1.5SiO_2{\cdot}3H_2O$	0.020	Ettringite	$3CaO{\cdot}Al_2O_3{\cdot}3CaSO_4$ $31H_2O$	1.754
Na and Ca hydrosilicate	$(Na,Ca)SiO_2{\cdot}nH_2O$	0.050			
Thomsonite	$(Na,Ca)Al_2O_3Si_2O_5{\cdot}6H_2O$	0.050			
Hydronepheline	$Na_2O{\cdot}Al_2O_3{\cdot}2SiO_2{\cdot}2H_2O$	0.020			
Natrolite	$Na_2O{\cdot}Al_2O_3{\cdot}3SiO_2{\cdot}2H_2O$	0.020			
Analcime	$Na_2O{\cdot}Al_2O_3{\cdot}4SiO_2{\cdot}2H_2O$	0.020			

- A phase known as hydrotalcite ($Mg_6Al_2CO_3(OH)_{16}·4HO$) has been detected in slag activated with NaOH (Fernández-Jiménez, 2000; Ben Haha et al., 2011) and waterglass (Cheng et al., 1992). Hydrotalcite is a natural mineral whose structure consists of layers of brucite ($Mg(OH)_2$) with interstitial water molecules and CO_3^{2-} ions. It forms sub-microscopically, making it difficult to identify with scanning electron microscopy (SEM). These tiny hydrotalcite crystals are dispersed throughout the C-S-H gel. Phases of this type have also been found in cement–slag blends.
- C_4AH_{13}-type phases have been detected in slag activated with NaOH. These phases can be identified as platelets 0.1–0.2 μm thick and approximately 1.5 μm in diameter (LaRosa et al., 1992; Wang et al., 1995). Other authors (Cheng et al., 1992) have observed carbonated phases such as C_4AcH_{11} and $C_8A_2cH_{24}$ in slag pastes activated with NaOH and $Ca(OH)_2$.

17.2.4 Cementitious gel structure

A detailed comparison between the C-S-H gel forming in alkaline cements and the gel found in OPC follows. A full understanding of gel structure and composition is vital, since this is the compound primarily responsible for the physical-mechanical properties of the material and hence its performance during its service life.

Cementitious gels are not readily characterised due to their amorphous nature. Nonetheless, the information reported in recent years as a result of the application of a number of characterisation techniques, most prominently nuclear magnetic resonance, has proved to be very helpful for establishing models able to explain and describe the structure of the various gels formed.

Taylor based his well-known model for the C-S-H formed during OPC hydration on the structure of defective tobermorite, also known as dreirketten type chains (Cong and Kirkpatrick, 1996; Taylor, 1997; Richardson, 2008; Garcia-Lodiero et al., 2012a) (see Fig. 17.3(a)). Perfect tobermorite consists of two linear chains of silica tetrahedra arranged on either side of a central sheet of CaO. Every fourth silicate in the linear chain is repeated; of the three tetrahedra in each group, two are connected to the centre sheet of CaO by two oxygen bridges while the third, known as the bridging tetrahedron, is not connected to the CaO. On the nanostructural scale, then, tobermorite consists of infinite linear chains of silica tetrahedra (Q^2 units), whereas in C-S-H gel many of the bridging tetrahedra are missing, giving rise to finite two-, three- or five-link chains (see Fig. 17.3(b)).

Schilling et al. (1994), in turn, proposed a model for the C-S-H gel generated by the alkaline activation of CaO-rich materials (such as blast furnace slag), in which Al_T is taken up into the structure, replacing a Si tetrahedron in a bridging tetrahedron position (see Fig. 17.4). This model was subsequently

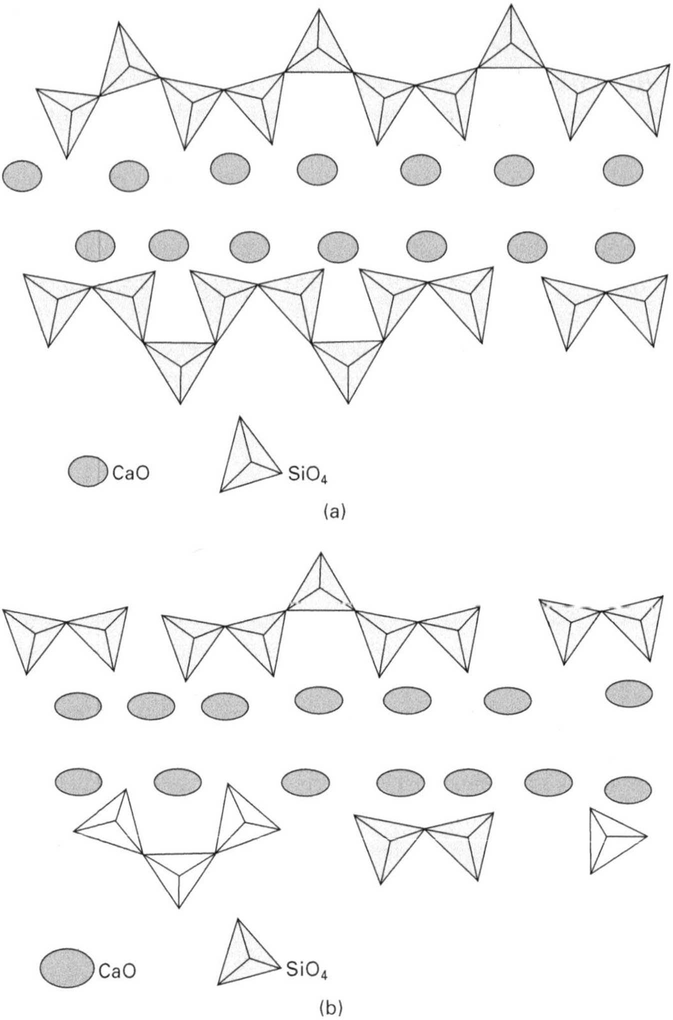

17.3 Structural model proposed for a C-S-H gel with a dreirketten-type chain.

ratified by Fernández-Jiménez (2000). Based on ^{29}Si and ^{27}Al magic angle spinning nuclear magnetic resonance (MAS NMR) studies (Fernández-Jiménez; 2000; Fernández-Jiménez et al., 2003; Fernández-Jiménez and Palomo, 2010; Puertas et al., 2011), the author showed that the presence of Al gives rise to gels with longer linear chains as well as to the possible existence of sporadic inter-chain, Si—O—Al bonds and consequently two-dimensional structures (Q^3(nAl) structures). Under those conditions, the C-S-H gels would become C-(A)-S-H gels (see Fig. 17.4(b)).

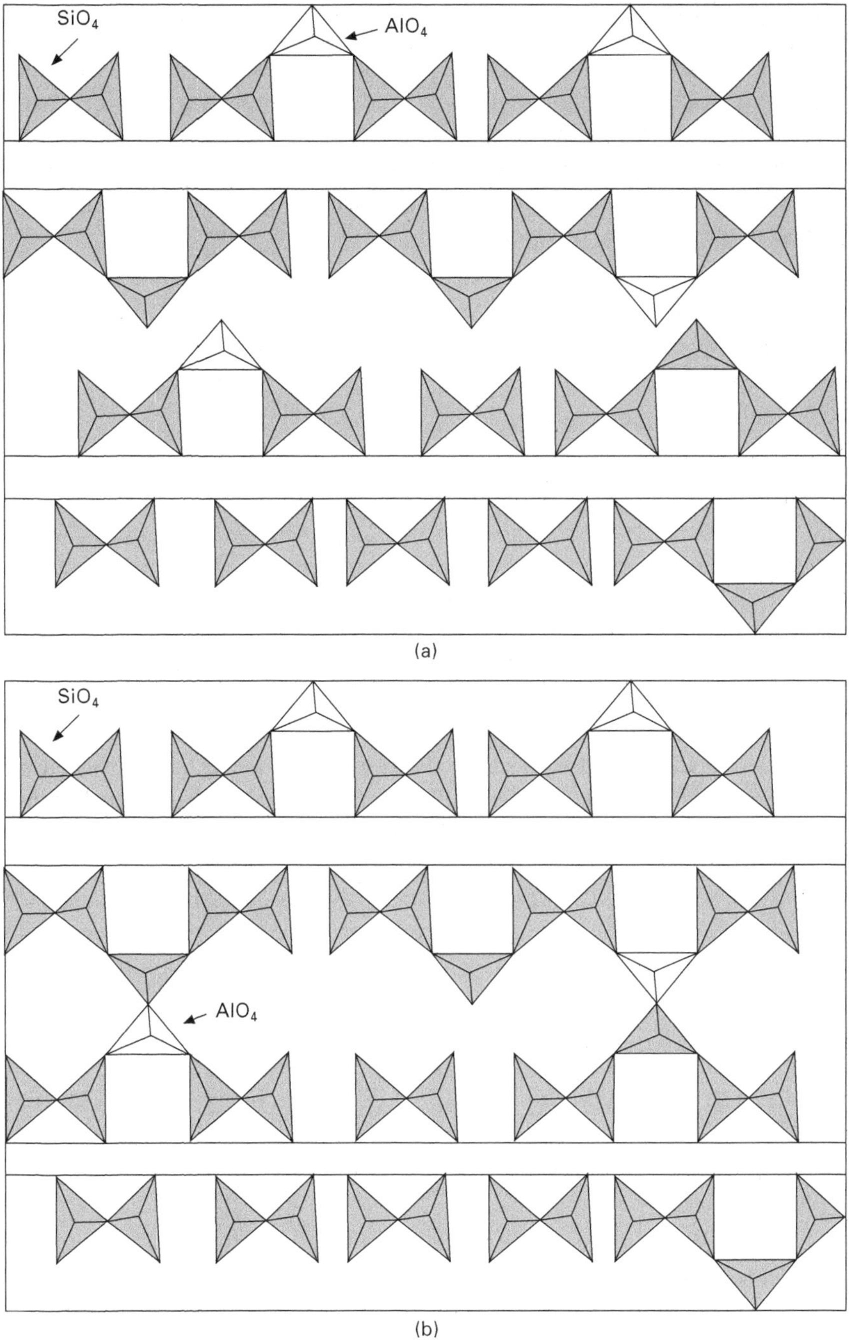

17.4 Structural model for a C-S-H gel with Al (a) linear chains (Schilling et al., 1994); (b) linear chains with sporadic bonds, forming planes (Fernández-Jiménez, 2000; Fernández-Jiménez et al., 2003).

17.2.5 Industrial applications

From the outset when alkali-activated cements and concretes were discovered, they were produced commercially and used for different purposes in a wide variety of structures in the former Soviet Union, and more specifically in Ukraine. In parallel with the substantial advances in the scientific understanding of these materials in the last 40 years, considerable experience has been gained in their design, production and applications, which include:

- structural concretes;
- pavements;
- pipes for cableways;
- railway sleepers;
- small bridge piers;
- refractory concretes.

Between 1990 and 2000, a group of Ukrainian scientists examined buildings and structures built with alkali-activated slag concrete and cement: a 15-storey building erected in 1960, special concrete pavements designed for heavy (50–60 t) vehicles laid in 1984; a 24 storey residential building erected in 1994; prestressed concrete railway sleepers manufactured in 1988. All these the alkali-activated slag concretes and cements exhibited excellent service conditions and even improved on the performance of Portland cement concretes used in similar situations. The tests conducted to observe the performance of these systems in different media and the microstructural analysis of some of the samples showed that concrete properties depend primarily on the prime materials used, service conditions and time (Shi et al., 2006; Xu et al., 2008).

Brief mention should also be made of other examples of structures in service that were erected with activated slag concrete. These include a storehouse built in Krakow, Poland, in 1974, using steel-reinforced alkali-carbonate-activated BFS precast concrete in floor slabs and wall panels (Malolepszy, 1989; Deja, 2002); two alkali-activated concrete roads built in the City of Magnitogorsk, Russia in 1984; and underground and trench structures, such as a drainage collector built in 1966 in Odessa, Ukraine, and silage trenches built in 1982 in Zaporohye, Ukraine. All these structures exhibited good service performance in inspections conducted many years later (Shi et al., 2006; Xu et al., 2008).

17.3 Alkali activation of low calcium systems

17.3.1 Composition of starting materials

Fly ash and metakaolin are the most commonly used low calcium materials in alkaline cement and concrete manufacture, although for reasons of cost metakaolin is adopted more sparingly. The most representative compositions of these materials are shown in Fig. 17.1.

Fly ash is an industrial by-product generated in coal-fired steam power plants. Before its use as a fuel in these plants, coal is ground to a very fine powder. Coal combustion gives rise to heavy ash (known as bottom ash) and other much finer particles that are carried by smoke and trapped in precipitators to prevent their release into the air. Known as fly ash, this material is characterised by its peculiar morphology: hollow spheres that may or may not house other smaller spheres. It consists essentially of a vitreous phase and a few minority crystalline phases such as quartz (5–13%), mullite (8–14%) and magnetite (3–10%) (Fernández-Jiménez and Palomo, 2003).

Fly ash composition may nonetheless vary depending on the type of coal used and the incineration process in place. After an exhaustive study of a large number of types of ash, Fernández-Jiménez and Palomo (2003) concluded that the characteristics required of type F fly ash for successful use in the manufacture of alkaline cements are (Fernández-Jiménez and Palomo, 2003, 2005b):

- unburnt percentage < 5%;
- $Fe_2O_3 \leq 10\%$;
- $CaO \leq 10\%$;
- reactive $SiO_2 > 40\%$
- 80–90% particles < 45 μm
- vitreous phase content > 50%;
- $[SiO_2]_{reactive}/[Al_2O_3]_{reactive}$ ratio > 1.5

17.3.2 Reaction mechanisms

Since the alkaline activation of SiO_2- and Al_2O_3-rich materials is a relatively recent line of research, knowledge is less advanced than in the case of slag activation. Glukhovsky (1994) proposed a general mechanism for the activation reactions in these materials, consisting of three stages: (a) destruction-coagulation; (b) coagulation-condensation; and (c) condensation-crystallisation. More recently, Palomo and Fernández-Jiménez (Palomo et al., 2004) proposed a model for the specific activation of fly ash based on zeolite synthesis. Under this model, the process consists of two stages: (a) *nucleation*, with the dissolution of the aluminates present in the ash and the formation of complex ionic species (this stage would combine the first two stages proposed by Glukhovsky and is heavily dependent upon thermodynamic and kinetic parameters); (b) *growth*, in which the nuclei reach a critical size and the crystal begins to grow. This stage is very slow due to experimental conditions. The final result of the alkaline activation of fly ash is an amorphous matrix with cementitious properties whose main component is an aluminosilicate gel known as N-A-S-H gel or 'zeolite precursor'. Hypothetically, this gel would eventually evolve into a zeolite.

Based on SEM analysis, these same authors proposed a conceptual model for the alkaline activation of fly ash (see Fig. 17.5) (Fernández-Jiménez et al., 2005; Fernández-Jiménez and Palomo, 2005b). The process would begin with a chemical attack on the ash surface (Fig. 17.5(a)). The outcome would be the formation of small cavities in the wall of the ash particles, exposing the tiny particles on the inside to the action of the alkalis. In this stage of the reaction the alkalis would attack from inside and outside the particles (Fig. 17.5(b)). The ash would continue to dissolve and the reaction products generated inside and outside the ash crust would precipitate (Fig. 17.5(c)), covering the smaller unreacted spheres and hindering their contact with the alkaline solution (Fig. 17.5(e)). Finally, alkaline activation continues slowly since the ash particles covered by the reaction products can only be attacked by alkalis through diffusion mechanisms. This leads to a final situation in which different morphologies may co-exist in the same paste (unreacted ash particles, particles under alkaline attack and reaction products (N-A-S-H gel, zeolites, etc.) (see Fig. 17.5(d)).

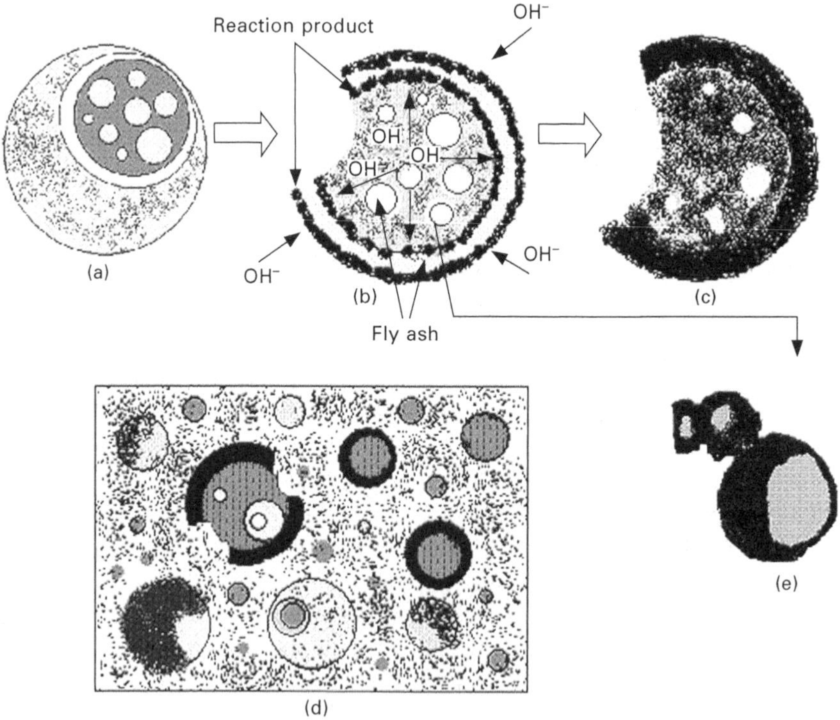

17.5 Conceptual model for the alkaline activation reaction mechanism in fly ash (Fernandez-Jimenez et al., 2005; Fernández-Jiménez and Palomo, 2005b).

17.3.3 Reaction products

The alkaline activation of silica- and alumina-rich (and low-calcium) materials such as fly ash and metakaolin leads to the precipitation of the main reaction product, an amorphous alkaline aluminosilicate hydrate (M_n-(SiO_2)-(AlO_2)$_n$ $\cdot wH_2O$) known as N-A-S-H gel (Fernández-Jiménez et al., 2006a). This product also goes by the name alkaline inorganic polymer. Its silicon and aluminium tetrahedra are distributed at random, forming a three-dimensional skeleton (see Fig. 17.6) (Palomo et al., 2004; Fernández-Jiménez and Palomo, 2005b). The secondary reaction products in this type of systems are zeolites such as hydroxysodalite, zeolite P, Na-chabazite, zeolite Y and faujasite (Criado, 2007).

17.3.4 Cementitious gel structure

The structure of the gels forming in the alkaline activation of low-calcium, silicoaluminous materials differs substantially from the gels formed in the activation of calcium-rich cements. These N-A-S-H gels are characterised by a three-dimensional structure in which the Si is found in a variety of environments, with a predominance of Q^4(3Al) and Q^4(2Al) units. The Si^{4+} and Al^{3+} cations are tetrahedrally coordinated and joined by oxygen bonds. The negative charge on the AlO_4^- group is offset by the presence of alkaline cations (typically Na^+ and K^+).

N-A-S-H gel formation entails a series of stages that can be summarised

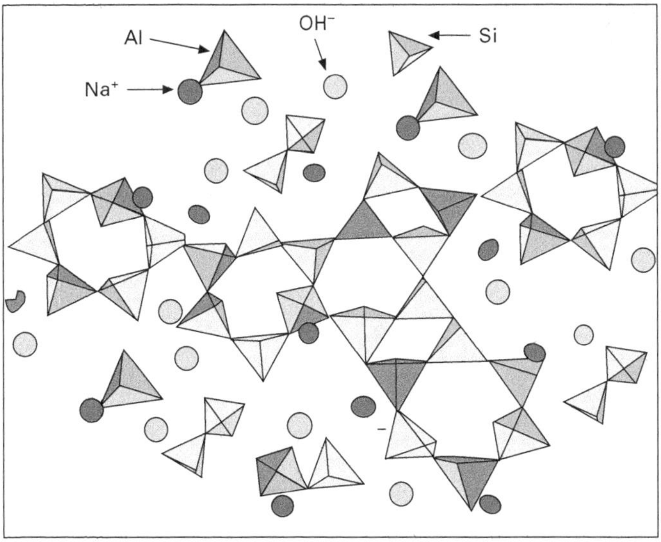

17.6 Plan view projection of the three-dimensional structure of a N-A-S-H gel (Criado, 2007).

as follows (see Fig. 17.7): when the source of aluminosilicate comes into contact with the alkaline solution it dissolves into several species, primarily silica and alumina monomers. These monomers interact to form dimers, which in turn react with other monomers to form trimers, tetramers and so on. When the solution reaches saturation an aluminosilicate gel, N-A-S-H gel, precipitates. This gel, known as *Gel 1*, is initially Al-rich (Si/Al≈1) and constitutes a metastable, intermediate reaction product (Fernández-Jiménez et al., 2006a). Its formation can be explained by the high Al^{3+} ion content in the alkaline medium in the early stages of the reaction (from the first few minutes through the first few hours), since the reactive aluminium dissolves more quickly than silicon because Al—O bonds are weaker than Si—O bonds. As the reaction progresses, more Si—O groups in the original source of aluminosilicate dissolve, raising the silicon concentration in the reaction medium and its uptake in the gel. The N-A-S-H gel so formed, *Gel 2*, is richer in silica (Si/Al ≈ 2) and is primarily responsible for the high mechanical strength observed in (OPC-free) alkali-activated fly ash cement (Fernández-Jiménez et al., 2006a).

These structural reorganisation processes determine the final composition of the polymer as well as pore microstructure and distribution in the material, which are critical factors in the development of many physical properties of the resulting cement (White et al., 2011).

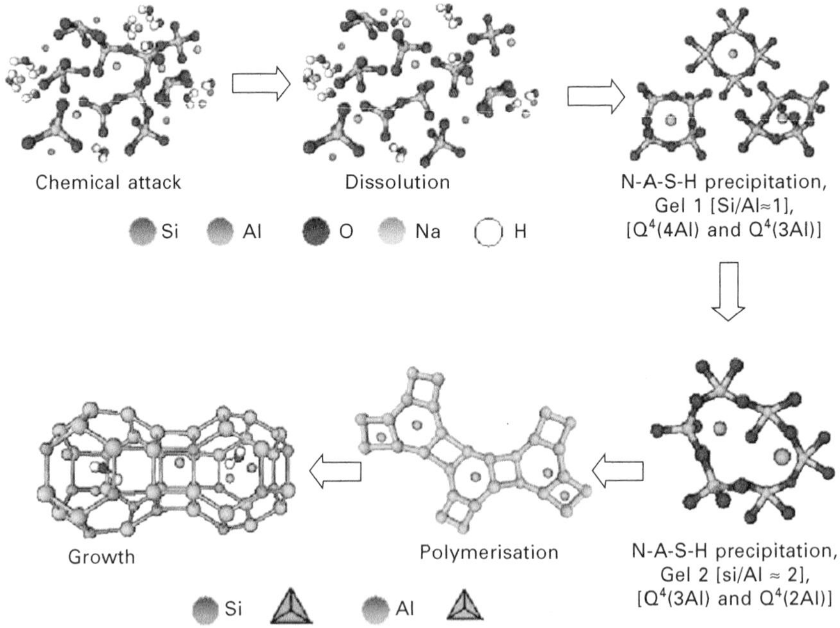

17.7 Structural model proposed for N-A-S-H gel formation (Shi et al., 2011).

17.3.5 Engineering properties of concretes made with aluminosilicate cements

The study of the macroscopic properties of the cementitious materials obtained by alkali-activating aluminosilicates, and specifically the mortars and concretes made with (Portland cement-free) alkali-activated fly ash, is a highly topical question for the scientific community, given the impact of these properties on concrete technology and hence the possible applications of such materials. Depending on the type of alkali activator used, the thermally cured material exhibits a promising list of properties and characteristics, including: high early bending and compressive strength, low drying shrinkage, good matrix–steel bonding, and excellent resistance to acid attack and fire. Given their desirable technological characteristics and durability, together with the ease with which they can be adapted to existing facilities in precast and bulk concrete plants, these new cements can be readily used in a number of applications. Some of their more promising properties are discussed below.

Mechanical strength

Many papers provide data on mechanical strength development in alkali-activated fly ash pastes and mortars (Palomo et al., 1999; Škvára et al., 2005; Shi et al., 2006, 2011; Duxson et al., 2007a; Fernández-Jiménez and Palomo, 2011). Very few studies have reported on Portland cement-free concretes in which the binder is alkali-activated fly ash, however. The earliest papers found in the literature were published by Fernández-Jiménez and Palomo (2005a), Fernández-Jiménez et al. (2006a) and Hardjito et al. (2002). These studies generally showed that the properties of alkali-activated fly ash concretes, like the properties of conventional concretes, depend on a number of factors, including blend dosage and curing conditions. In fact, many of the mix design factors (blend proportions, mixing procedure, curing time and temperature, nature and concentration of the alkali activator and solution/ash ratio) affect setting and hardening and consequently subsequent strength development (Van Jaarsveld et al., 2002; Fernández-Jiménez and Palomo, 2009).

Research conducted by Fernández-Jiménez and Palomo (2009) on alkali-activated concretes showed that they develop high early age compressive strength, with 1-day values on the order of 45 MPa, which is higher than observed for conventional OPC concretes (see Fig. 17.8). Moreover, strength continues to gradually rise over time. The use of different types of activators also plays an important role in strength performance; the inclusion of soluble silica with the activator, for instance, visibly enhances mechanical strength.

The effects of changing the aggregate/binder and liquid/binder ratios on the properties of activated fly ash concrete are fairly similar to the effects

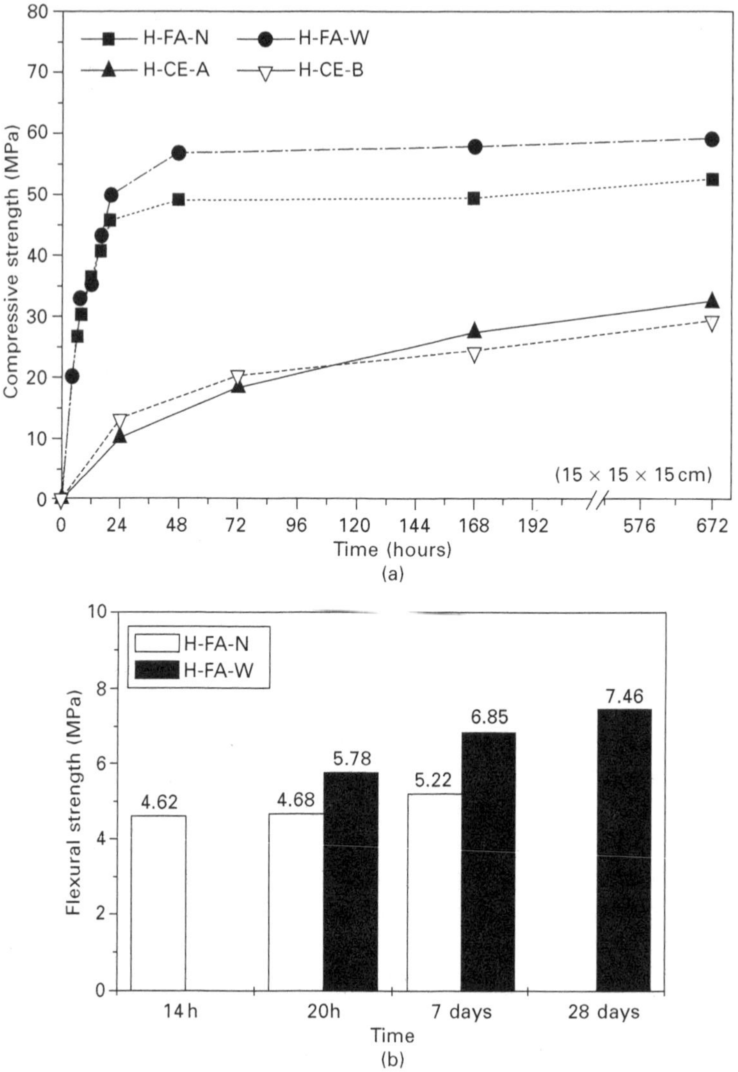

17.8 Mechanical strength in concretes alkali-activated with NaOH (H-FA-N) and NaOH+Na_2SiO_3 (H-FA-W, compared to the values obtained for traditional Portland cement (H-CE-A and H-CE-B)): (a) compressive strength in cubic (15 × 15 × 15 cm) specimens; (b) bending strength in (15 × 10 × 70 cm) prismatic specimens (Fernández-Jiménez and Palomo, 2009).

of such changes on traditional Portland cement concretes. Increasing the amount of ash raises compressive strength (Fernández-Jiménez and Palomo, 2009). The findings show that the most favourable results are obtained for liquid/binder ratios of 0.4–0.5 (Fernández-Jiménez and Palomo, 2009). With lower ratios, plasticity declines substantially, leading to poorly compacted concrete, which has an adverse impact on subsequent strength development. Conversely, higher liquid/binder ratios, while improving fresh material workability, also generate more porous concretes, with a greater risk of segregation and lower compressive strength.

Temperature and curing time also have a considerable impact on the strength of these materials. This type of binder, unlike traditional Portland cement mortars and concretes, for which the curing temperature is generally limited to 50 °C, can bear curing temperatures up to 120 °C while maintaining their strength and durability intact (see Fig. 17.9 (Fernández-Jiménez and Palomo, 2009)). Moreover, the findings show that raising the curing temperature up to

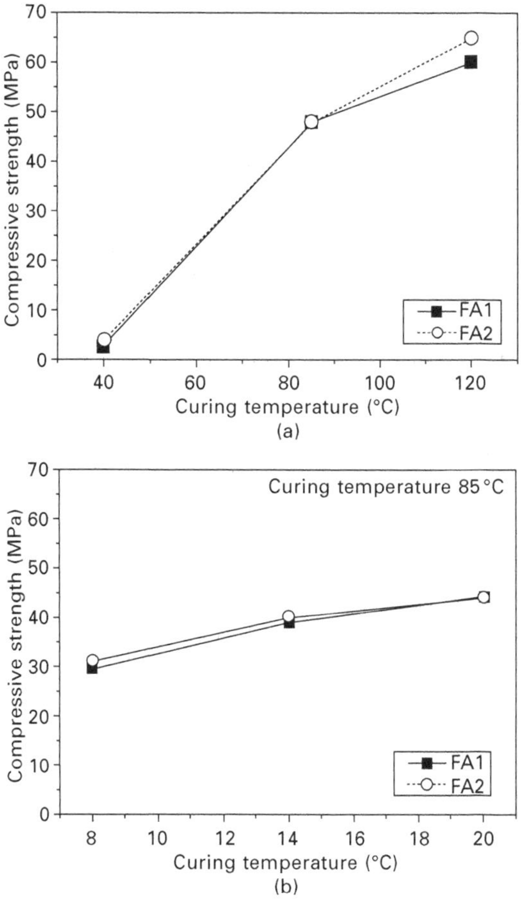

17.9 Effect of (a) curing temperature and (b) curing time on mechanical strength development in alkali-activated fly ash concrete (FA) (Fernández-Jiménez and Palomo, 2009).

a given value (Criado et al., 2010) has a beneficial effect on the compressive strength of these concretes, regardless of the type of ash used.

Drying shrinkage

Mortars made entirely of alkali-activated fly ash exhibit promising shrinkage properties. In 2009, Fernández-Jiménez and Palomo (2009) studied shrinkage in (OPC-free) alkali-activated fly ash mortar and compared the results with the findings for a series of traditional Portland cement mortars. The fly ash mortars had very small drying shrinkage values, clearly smaller than found for the Portland cement mortars (see Fig. 17.10), an indication of their dimensional stability.

Pull-out test

Bonding provides for the transfer of stress between the steel and the concrete. This characteristic is the key to the advantage in combining concrete, which has a high compressive strength, and steel, which has a high tensile strength, in reinforced concrete structures. When a steel bar embedded in a concrete matrix is subjected to tensile force, the stress is transferred from the steel to the concrete via slanted compressive forces originating at and following the

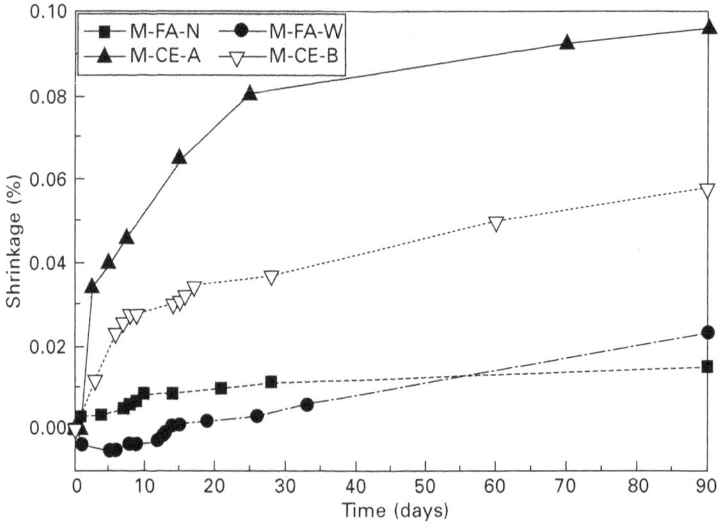

17.10 Drying shrinkage over time in fly ash mortars (cured at 85 °C) alkali-activated with two types of activators: NaOH (M-FA-N) and NaOH +Na_2SiO_3 (M-FA-W) compared with the values obtained for traditional Portland cement mortars cured at ambient temperature (M-CE-A) or at 50 °C (M-CE-B) (Fernández-Jiménez and Palomo, 2009).

same angle α, as the ribs on the bar. The radial component of this compressive force is offset by a tensile ring that appears in the concrete surrounding the bar, generating internal longitudinal cracks. In the absence of transverse reinforcement, these cracks grow outward through the entire cover, causing brittle or splitting failure at the concrete surface. If the concrete is well confined by the transverse reinforcement, however, failure occurs because the bar breaks away from the surrounding concrete. The result is pull-out failure (see Fig. 17.11).

Fernández-Jiménez and Palomo (2009) also compared bonding in alkali-activated fly ash concrete to bonding in portland cement concretes (see Fig. 17.11). Some of the findings are given in Table 17.3. In the alkali-activated fly ash concrete (H-FA-N and H-FA-W), failure occurred in the matrix, whereas in OPC concrete (H-CE-A), pull-out failure was observed.

17.11 (a) Apparatus used in the pull-out test; (b) fly ash concrete activated with NaOH + Na_2SiO_3, in which the matrix failed; and (c) fly ash concrete activated with NaOH, in which the 16-mm diameter steel bar failed (Fernández-Jiménez and Palomo, 2009).

Table 17.3 Pull-out test findings (20×20×20 cm moulds) (Fernández-Jiménez and Palomo, 2009)

Sample denomination	Bar diameter (mm)	Maximum load (Q = kN)	Maximum load τ (MPa)	Cracked area
H-CE	16	95.49	11.88	Pull-out
H-CE	16	122.6	15.24	Pull-out
H-FA-N	16	136.6	17.0	Specimen fails
H-FA-N	16	142.15	17.7	Specimen fails
H-FA-N	16	124.59	17.5	Bar fails
H-FA-W	16	102.94	12.8	Specimen fails
H-FA-W	16	101.09	11.6	Specimen fails
H-FA-W	16	99.95	12.4	Specimen fails

H-CE, Portland cement concretes; H-FA-N, fly ash concretes alkali-activated with NaO; H-FA-WN fly ash concretes alkali-activated with NaOH + Na_2SiO_3.

Alkali-activated fly ash concrete bonding properties differ depending on the activator used, but in all cases the materials tested by these authors exhibited higher than the minimum 9.70 N/mm^2 beam stress requirement for 16 mm diameter bars laid down in the Spanish structural concrete code (EHE). Concretes activated with a NaOH solution reached higher maximum stress values than the materials containing soluble silica in the alkaline activator. In this latter case, in one of the three specimens tested the steel bar failed before the matrix cracked (Fig. 17.11(c)). This is an indication of the excellent matrix/steel bonding attained in these systems.

Rheology

The rheological performance of portland cement pastes, mortars and concretes is, even today, an area difficult to analyse, very likely because of the many factors involved in cement blending, mixing and hydrating (presence of admixtures, batching, etc.). Several papers have shown the utility of rheology for monitoring fly ash or metakaolin alkaline cement paste setting (Palomo et al., 2005; Criado et al., 2009; Poulesquen et al., 2011; van Deventer et al., 2011). By contrast, practically no data can be found on mortar or concrete rheology in the literature.

In a study of rheology in alkali activated fly ash pastes, Palomo et al. (2005) and Fernández-Jiménez and Palomo (2011) found that, like hydrated Portland cement pastes, the Bingham visco-plastic flow model fits their behaviour fairly well. They also observed that the type of activator is the most significant parameter in determining paste rheology.

Criado et al. (2009) studied the effect of commercial admixtures commonly used in Portland cement concrete manufacture, such as lignosulphonates, melamines (first and second generation products) and polycarboxylates (latest generation) on alkali-activated fly ash pastes. According to their findings, these chemical admixtures do not have the same effect on alkali-activated fly ash systems as on Portland cement pastes. As most of these admixtures decompose in the highly alkaline media needed to activate fly ash, they are ineffective.

17.3.6 Durability

Alkali-activated fly ash cements perform extremely well not only in terms of strength, but also of durability. Material durability is closely related to mineralogical and microstructural composition. The chief durability problems in OPC pastes, mortars and concretes involve the presence of calcium. In alkali-activated fly ash cements, by contrast, the main reaction product formed is a three-dimensional aluminosilicate hydrate (non-calcium N-A-S-H gel), regarded as a pre-zeolite (Palomo et al., 2004), which is clearly different from the C-S-H gel formed in OPC pastes. Durability, then, must necessarily differ in OPC and alkaline cements.

Sulphate and seawater resistance

The aggressive chemicals to which portland cement mortars and concretes are most commonly exposed are sulphates or ions found in sea water (Johansen et al., 1995; Allahverdi and Škvára, 2001; Rostami and Brendley, 2003; Bakharev, 2005; Fernández-Jiménez et al., 2007; Pacheco-Torgal et al., 2012). Fernández-Jiménez et al. (2007) monitored the resistance to these attacks in fly ash mortars activated with different alkalis (N: NaOH, W: Na_2SiO_3 + NaOH). Generally speaking, they observed no significant decay in these materials as a result of the action of these agents. In fact, mechanical strength continued to develop in the mortars studied regardless of the medium in which they were immersed (see Fig. 17.12). After immersion in aggressive media for one year the mortars showed no visible signs of decay. Nonetheless, microscopic analysis revealed a gel richer in silicon and with some magnesium in the specimens exposed to sea spray, while the samples immersed in a sodium sulphate solution were found to contain sodium sulphate. In both cases, however, these decay products were detected only sporadically.

Bakharev (2005) also studied the performance of alkali-activated ash mortars when immersed in sodium sulphate and magnesium sulphate solutions. This author found that mortar stability depended largely on the intrinsic order of the aluminosilicate gel (N-A-S-H gel) constituents, observing that the greater the degree of crystallinity in the gel, the greater was its stability in aggressive media. Like Fernández-Jiménez et al. (2007), Bakharev found that N-A-S-H gels prepared with NaOH performed better than the ones prepared with sodium silicate as the activator.

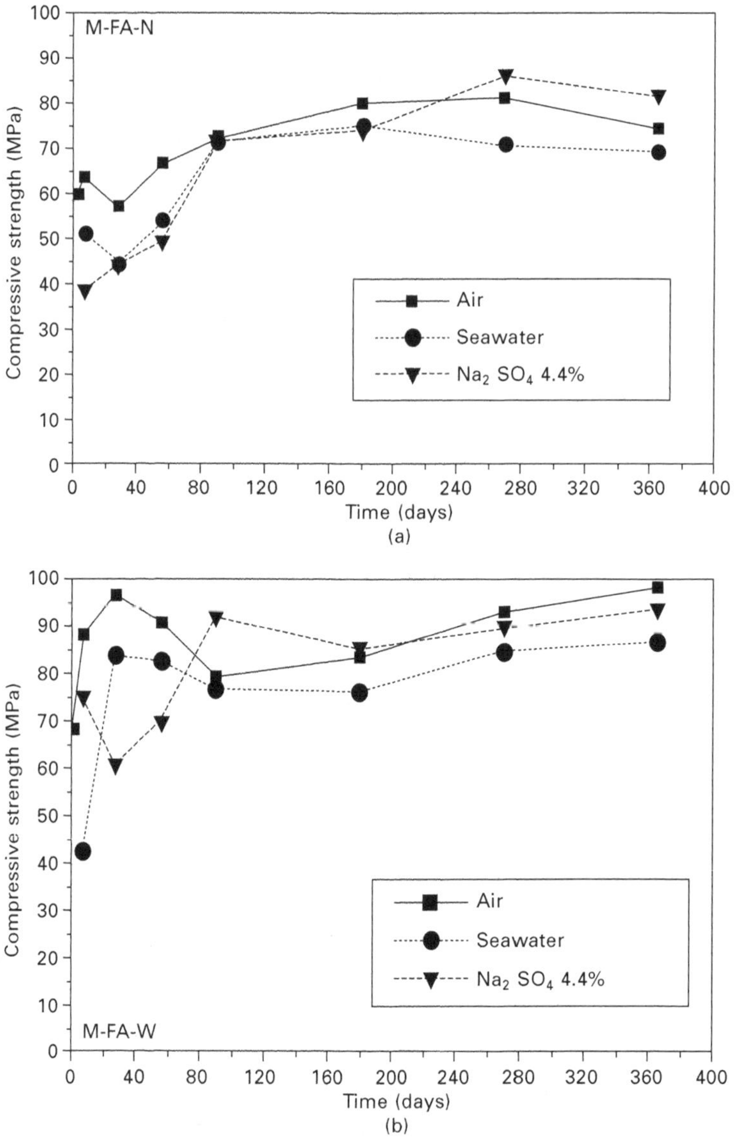

17.12 Compressive strength in fly ash mortars activated with (a) NaOH (M-FA-N) or (b) NaOH + Na_2SiO_3 (M-FA-W) (Fernández-Jiménez et al. 2007).

Resistance to acid attack

Resistance to acid attack is another test applied to assess material durability. Several authors (Allahverdi and Škvára, 2001; Rostami and Brendley, 2003; Pacheco-Torgal et al., 2012) have reported that alkali-activated metakaolin or fly ash perform well when exposed to solutions such as nitric or hydrochloric acid. In 2001 Allahverdi and Škvára (2001) proposed that the decay mechanism in hardened alkaline-activated cement paste in contact with acid solutions can be divided into two stages. The first involves leaching, in which alkali cations present in the aluminosilicate framework that offset the charge imbalance in the tetrahedral aluminium are exchanged for H^+ or H_3O^+ ions in the acid solution. In the second, the voids in the skeleton are re-occupied primarily by silicon atoms, resulting in the formation of a highly imperfect siliceous framework. The aluminium eliminated in the process, known as de-alumination (Allahverdi and Škvára, 2001), becomes octahedrally coordinated and accumulates mainly in the interstitial space.

Fernández-Jiménez and Palomo (2009) also compared stability in activated fly ash mortar and traditional Portland cement when exposed to HCl and observed higher performance in the former than the latter. While compressive strength declined in both types of mortar after extended contact with the (0.1 M) HCl solution, the 90-day downturn in strength in the ash systems (M-FA) was on the order of 23–25%, compared with nearly double that figure, 47%, in the cement mortars (M-CE-A) (see Fig. 17.13(a)). Moreover, after 90 days of exposure to the acid solutions, the ash specimens appeared to be generally healthy. The cement samples, by contrast, exhibited severe decay after 56 days (with a clear change of colour and mass loss on the arris (see Fig. 17.13(b)).

Alkali–silica reaction (ASR)

The alkali–silica reaction (ASR) is another durability-related development in mortars and concretes of great interest to the scientific community (Swamy, 1992; García-Lodeiro et al., 2007; Pacheco-Torgal et al., 2012). García-Lodeiro et al. (2007) studied the performance of mortars prepared with 100% alkali-activated fly ash and the so-called *pessimum proportion* of reactive aggregate (Ozol, 1975) when exposed to an accelerated alkaline attack by immersion in 1 M NaOH at 85 °C. The main conclusion reached was that alkaline mortars performed better than traditional Portland cement mortars (see Fig. 17.14), supporting the decisive role of calcium in the expansive nature of the gels formed.

Efflorescence

As a rule, efflorescence in alkaline cements is the result of the combination of an excess of alkalis and inappropriate curing (humidity). These cementitious materials must be cured in relatively high humidity (preferably over 90%). If sufficient water is available to ensure the transport of ionic species as the material cures, most of the alkalis are fixed in the form of N-A-S-H gel

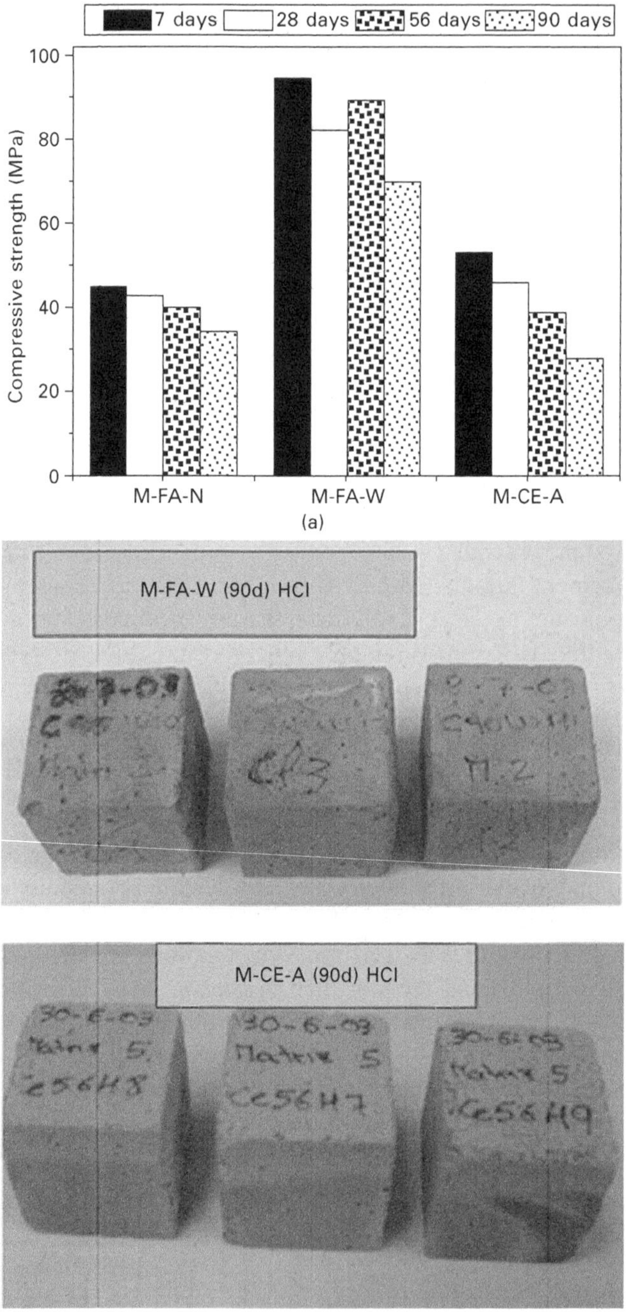

(b)

17.13 (a) Mechanical strength of fly ash mortars activated with NaOH (M-FA-N) or NaOH + Na_2SiO_3 (M-FA-W) and cement (M-CE) immersed in a 0.1 N HCl solution (b) specimens after immersion for 90 days in the 0.1 N HCl solution (Fernández-Jiménez and Palomo, 2009).

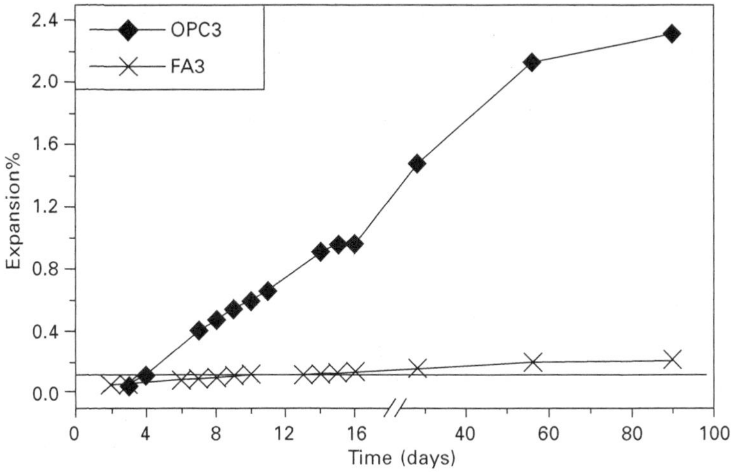

17.14 Activated fly ash (FA3) and traditional OPC (OPC3) mortar expansion (%) over time (García-Lodeiro et al., 2007).

(main reaction product), while the rest remain in the liquid housed in the pore system (Criado et al., 2005; Kovalchuk et al., 2007). When the curing temperature and humidity are unsuitable, the excess alkalis spread across the pore system, precipitating as salts on the surface of the material (Temuujin et al., 2009; van Deventer et al., 2010). Recent studies conducted by Pacheco-Torgal and Jalali (2010) show that sodium efflorescence is especially prevalent in alkali-activated aluminosilicate binders burnt at a temperature below the dehydroxylation temperature and containing sodium carbonate additions as a source of sodium cations.

Kani et al. (2011) recently showed that efflorescence can be reduced either by adding alumina-rich admixtures or by hydrothermal curing at temperatures of 65 °C or higher.

Frost resistance

The excellent resistance to extreme environments, and in particular to frost, exhibited by traditional Portland cement concretes and mortars is based on the degree of saturation and pore system of the hardened paste. If the concrete never becomes saturated, the risk of freeze/thaw-induced deterioration is nil. If the concrete expands beyond its stress resistance as a result of the presence of ice, however, it will be damaged. The information available on geopolymer expansion in response to frost is contradictory. According to Zhang and Wei (2006) alkali-activated fly ash concrete can withstand 2.2 times more freeze-thaw cycles than concrete made from OPC with the same compressive strength. Škvára (2007) reported that alkali activated fly ash-based materials exhibit excellent frost resistance. The mass of alkali-activated fly ash concrete samples containing sodium silicate remained essentially unchanged during freeze–thaw cycles conducted in an aqueous environment. No visible defects or deformation were detected after 150 cycles, although strength dipped to about 70% of the strength measured in unexposed samples stored under ambient conditions for the same period of time.

Protection of steel reinforcement against corrosion

Another significant aspect of durability is system corrosion performance. In Portland cement concretes, the primary reason for premature failure of reinforced concrete structures is reinforcement corrosion. The capacity of activated fly ash mortars and concretes to passivate steel reinforcement contributes largely to guaranteeing the durability of corrosion resistant steel (CRS). According to the few studies addressing reinforcement corrosion in alkaline concrete available in the scientific literature (Miranda et al., 2005; Bastidas et al., 2008; Fernández-Jiménez et al., 2010), fly ash mortars passivate steel reinforcement as rapidly and effectively as OPC mortar. The stability of the passive state in changing environmental conditions depends heavily on the activating solution used, however: carbonation is retarded substantially in mortars activated with waterglass and caustic soda due to the lower permeability of these materials. Moreover, the presence of chlorides in activated fly ash mortars in concentrations higher than the critical depassivation threshold multiplies corrosion rates by approximately 100, i.e., practically the same factor as observed in traditional OPC mortars.

Yodmunee and Yodsudjai (2006) report similar results after conducting accelerated corrosion testing of reinforcement in alkali-activated fly ash concrete. Alkaline concrete protects steel bars from corrosion more effectively than conventional concrete and the higher the compressive strength of the geopolymer concrete, the more effective is the protection afforded.

Resistance to high temperatures

The behaviour of alkali-activated aluminosilicate systems at high temperatures is a matter of particular interest. Research by a number of authors has shown that these systems perform very well at high temperatures (Barbosa and Mackenzie, 2003; Bakharev, 2006), especially compared with Portland concretes and mortars (Davidovits et al., 1999; Davidovits, 2008).

Fernández-Jiménez and Palomo (2009) compared the behaviour of fly ash pastes activated with two alkalis, NaOH and a mix of NaOH + WG, at temperatures of 200, 400, 600, 800 and 1000 °C to the performance of traditional Portland cement pastes under the same conditions. The post-treatment specimens are depicted in Fig. 17.15. These authors reported

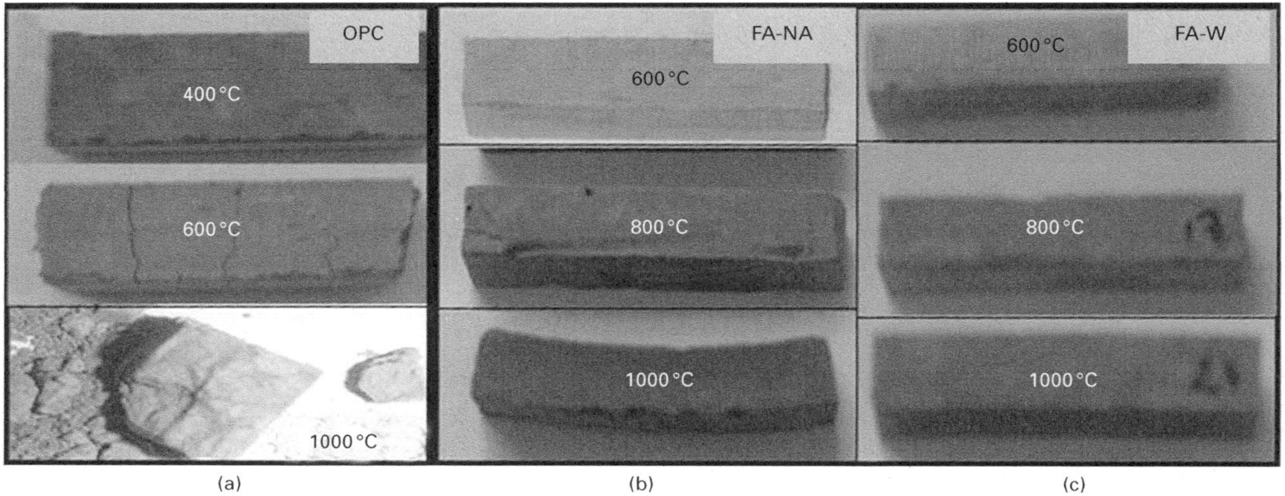

17.15 Physical aspect of the mortars after the thermal treatment (a) Portland cement (OPC); (b) fly ash activated with NaOH (FA-N) and (c) fly ash activated with NaOH + waterglass mixture (Fernández-Jiménez and Palomo, 2009).

that the cement specimens cracked at 600 °C and were wholly deteriorated at 1000 °C, whereas no cracks appeared in the ash specimens at 600 °C, although the material did undergo plastic deformation at that and higher temperatures. Such deformation increased with exposure time and proved to be more intense for the NaOH-activated materials.

In the post-thermal treatment trials conducted in that study, the residual bending and compressive strength values for Portland cement declined steadily: up to 400 °C, the flexural strength values fell by 33% and by nearly 100% after 600 °C. The pastes prepared with alkali-activated aluminosilicates (with no OPC) performed better at all temperatures. The compressive strength of Portland cement proved to be practically nil after 600–800 °C, at which temperatures it even exploded inside the kiln. The ash-based materials, by contrast, exhibited nearly constant residual compressive strength, which actually rose slightly at 800 and 1000 °C.

17.3.7 Industrial applications

Given the good mechanical properties and excellent matrix–steel bonding and dimensional stability in alkali-activated fly ash mortars and concretes, the first industrial application for these materials tested was concrete precasting (Palomo et al., 2007b; van Deventer et al., 2011).

The idea of using concretes made with alkali-activated fly ash as the binder in the manufacture of precast construction elements was based on the following proven facts.

- Many concrete precast elements are of 'manageable' size but extraordinarily complex from the technological standpoint. Their design and production require materials that ensure strength, durability, good bonding to steel reinforcement and volume stability. Activated fly ash concrete meets all these requirements.
- Accelerated thermal curing of concrete is advisable and at times required in conventional precasting processes. Optimal alkaline activation of fly ash is attained under similar thermal curing conditions. Consequently, production processes need not be substantially altered to accommodate the change from one prime material (Portland cement) to others (fly ash and alkaline activators). The difficulties encountered in attaining acceptable quality and durability in conventional concrete end products cannot be readily surmounted today. The problem lies in the fact that the measures recommended in codes of good practice (such as limiting curing temperatures to 60 °C or the use of aggregate that remains inert at high temperatures) are normally insufficient and nearly always incompatible or scantly reconcilable with the methods involved in mass production. The use of activated ash could change the present scenario radically.

- In prestressed elements, the steel reinforcement is prestressed before the concrete is cast in the mould. Given that these alkali-activated fly ash concretes can develop very high early age strength (in the first 12–20 hours) and bond well to the reinforcement, the time the concrete must remain in the mould can be shortened. That, in turn, would lead to a significant rise in plant productivity.
- Finally, the economic and ecological issues involved in the alkaline activation of fly ash also merit attention. The material proposed as a substitute for Portland cement in concrete manufacture is an industrial by-product widely available on all six inhabited continents, generally in stockpiles occupying broad expanses of land because this waste is non-consumable.

The photographs in Fig. 17.16 depict the industrial-scale manufacture of monoblock railway sleepers (one of the many applications for these concretes) (Fernández-Jiménez et al., 2006b; Fernández-Jiménez and Palomo, 2009). The sole significant change that had to be introduced in the manufacturing process used for portland cement sleepers was a rise in the curing temperature from 50 to 85–90 °C, which entailed building a pressure steam tunnel (see Fig. 17.16). When tested for the static and dynamic performance specified in the respective standards (see Table 17.4), these sleepers were found to be amply compliant.

Sumajouw and Rangan (2006) have studied slender beams and columns made with (OPC-free) reinforced fly ash concretes (with low Ca contents) alkali activated with a NaOH and sodium silicate solution. These concretes were cured in a high humidity environment at 60 °C for 24 hours. The strength values and performance and failure mode in the reinforced geopolymer beams and columns were compliant with Australian legislation. The values observed were similar to or better than the results obtained with portland cement reinforced concrete beams.

17.4 Blended alkaline cements: hybrid cements

17.4.1 Introduction

This group of materials may well attract growing attention in the context of the cements of the immediate future, since one of the possibilities being considered to address the environmental issues discussed in the introduction to this chapter is to dilute Portland cement with large volumes of supplementary cementitious materials. In fact, SCMs such as blast furnace slag, phosphorus slag, coal fly ash and natural pozzolans are widely used in the manufacture of blended cement (Buchwald et al., 2007; Bernal et al., 2011b) or as a cement replacement in concrete (Palomo et al., 2007b), (generally speaking, the use of these materials tends to lengthen setting times and lower early

17.16 Photographs of the industrial-scale manufacture of monoblock railway sleepers with (OPC-free) alkali-activated fly ash concrete (Fernández-Jiménez and Palomo, 2005a, 2009; Fernández-Jiménez et al., 2006b).

Table 17.4 Results for monoblock railway sleepers made with Portland cement-free, alkali-activated fly ash concretes (Palomo et al., 2007b; Fernández-Jiménez and Palomo, 2009)

		Minimum load (kN)		^{3}H-FA-W sleeper (kN)
		[1]RENFE	[2]UIC	
Static test*	1st crack	127.53	147.15	417.62
	Failure	280.56	323.73	583.40
Dynamic test	1st crack > 0.05 mm	195	228	350
	Failure	286	334	390

[1]RENFE: Red Nacional de Ferrocarriles Españoles (Spanish national railway system).
[2]UIC: International Union of Railways.
[3]H-FA-W: sleeper made with fly ash alkali-activated with NaOH + Na_2SiO_3.
*Further to European standard EN 13230 (parts 1 and 2).

age strength in cement and concrete). Many researchers have shown that the addition of alkaline activators could enhance the potential pozzolanicity of such supplementary materials and improve the properties of the respective cementitious systems, especially at early ages.

Dilution of the PC content reduces the effectiveness of the activation of these materials unless an alkaline activator can be added, however. This is the solution adopted in *hybrid Portland cement – alkali activated aluminosilicate* systems (Fernández-Jiménez et al., 2011; Macphee and García-Lodeiro, 2011). These systems are complex cementitious blends in which the type of product formed depends largely on the reaction conditions, including the chemical composition of the prime materials, alkaline activator type and concentration and curing temperature. In the systems containing portland clinker (clinker + blast furnace, phosphorous or steel mill slag, clinker + fly ash, clinker + slag + fly ash), C-S-H gel normally prevails as the main reaction product in slightly alkaline media (i.e., 2 M NaOH), whereas N-A-S-H gel prevails in highly basic environments (10 M NaOH) (Alonso and Palomo, 2001).

Consequently, the compatibility of the two cementitious gels, C-S-H (the main reaction product of ordinary Portland cement hydration) and N-A-S-H (the main product of the alkali activation of aluminosilicate materials), may have important technological implications for future cementitious systems in which both products might be expected to precipitate (Alonso and Palomo, 2001; Yip et al., 2005; Palomo et al., 2007a; García-Lodeiro et al., 2012b).

The hybrid cementitious systems most frequently studied include:

- Portland cement–blast furnace slag blends;
- Portland cement–phosphorus slag blends;
- Portland cement–fly ash blends;
- Portland cement–steel mill and blast furnace slag blends;

- Portland cement–fly ash–blast furnace slag blends;
- multi-constituent cement blends.

17.4.2 Co-precipitation of cementitious gels: C-S-H + N-A-S-H

Hybrid alkaline cements are complex cementitious blends with initial CaO, SiO_2 and Al_2O_3 contents around 20% (Alonso and Palomo, 2001; Yip et al., 2005; Palomo et al., 2007a), whose reaction products are intricate mixes of different gels (the type of product formed depends largely on the reaction conditions). The reaction products forming during the alkaline activation of cement and ash blends is an area of keen scientific and technological interest and the compatibility between the two main cementitious gels, N-A-S-H and C-S-H, is the object of considerable research today (Shi et al., 2011; García-Lodiero et al., 2012b).

Prior studies have shown that the co-precipitation of these two gels in hybrid cements is possible (Alonso and Palomo, 2001; Yip et al., 2005; Palomo et al., 2007a), although recent research has revealed that the two products do not develop separately, as two separate gels, but that they interact, undergoing structural and compositional change in the process (García-Lodeiro et al., 2011).

Research conducted on synthetic gels to determine the effect of the constituents of each on the other has led to the conclusion that high pH and the presence of aqueous aluminate impact C-S-H composition and structure (García-Lodeiro et al., 2009, 2010a). Similarly, aqueous Ca modifies N-A-S-H gels, in which part of the sodium is replaced by calcium to form (N,C)-A-S-H gels (García-Lodeiro et al., 2010b). Recent studies conducted on synthetic samples to analyse C-S-H/N-A-S-H compatibility in greater depth showed that the stability of the N-A-S-H structure in the presence of calcium depends heavily on the pH in the medium (Garcia-Lodeiro et al., 2011). In the presence of sufficient calcium, pH values of over 12 favour the formation of a C-A-S-H rather than N-A-S-H gel. The experiments yielding these findings were conducted in equilibrium conditions, however, which are not normally present during binder hydration, particularly in the early stages of the reaction.

Palomo et al. (2007a) reported the results of research on hybrid cements containing 30% Portland cement clinker and 70% fly ash. Their characterisation studies of the hardened matrices showed that in all cases the reaction products consisted of a mixture of amorphous gels (C-S-H + N-A-S-H). The mechanical strength developed by this blended cement differed significantly depending on the hydrating solution used (see Table 12.5). While strength development in the water- and NaOH-hydrated systems was relatively poor, the blends activated with waterglass + NaOH were compliant with the

Table 17.5 Mechanical strength development in hybrid cements (Palomo et al., 2007b; Shi et al., 2011)

Material			Paste	
Blended cement	Liquid	Liquid/solid ratio	Compressive strength (MPa)	
			2 days	28 days
S=70% FA	L1	0.325	11.23	28.91
+	L2	0.407	4.83	24.72
30% clinker	L3	0.487	12.91	36.94

L1 = deionised water; L2 = NaOH solution; L3 = sodium silicate solution (waterglass) + NaOH solution with SiO_2/Na_2O = 1.5.

strength requirements laid down in European standard EN 197-1:2000 and comparable to the values for standard 32.5 cement.

These authors concluded that Portland cement hydration follows different pathways depending on the OH^- concentration and the presence of soluble silica in the medium, while fly ash activation at ambient temperatures is accelerated by the presence of Portland cement. The favourable effect of the presence of Portland cement on ash activation at ambient temperature may be explained by the heat released during the hydration reaction; the energy from this heat would in turn activate the chemical reactions that originate fly ash setting and hardening in highly alkaline media.

^{29}Si NMR analysis (Palomo et al., 2007a) provides further insight into the microstructural evolution of these cementitious matrices and therefore into the nature of the gels making up the skeleton responsible for the strength of these materials (see Fig. 17.17(a)). Two clearly differentiated areas can be distinguished on the ^{29}Si spectrum for the non-hydrated powder (Fig. 17.17(a), spectrum A). The fly ash component has a wide asymmetric signal consisting of a series of peaks at around –89, –98, –104, –108 and –119 ppm. The clinker component, in turn, contains a more clearly defined signal at –72 ppm attributed to the Q^0 units in the anhydrous calcium silicates. When pure water (L1) is added to the cement blend, the clinker component is the primary object of attack (see spectrum B in Fig. 17.17). As a result, the peak at –72 ppm (generated by the anhydrous silicates in the clinker) nearly disappears, while a new peak begins to emerge at –86 ppm. This new peak can be attributed to the Q^2 units in the C-S-H gel. Spectrum B also shows that the fly ash component of the solid barely reacts even after 28 days of hydration: the pozzolanic reaction between the fly ash and the portlandite released during clinker hydration takes place slowly if at all. The situation depicted in spectrum C is quite different, however. According to this technique, the signal located at –86 ppm may be either C-S-H gel (Q^2 units) or N-A-S-H gel (Q^4(4Al) units). The presence of anhydrous silicates

(–72 ppm) supports the premise that C-S-H is not the only binder forming in systems hydrated with NaOH or waterglass + NaOH. The gel does, however, form a part of both these binders (compare this signal in spectra B and C to

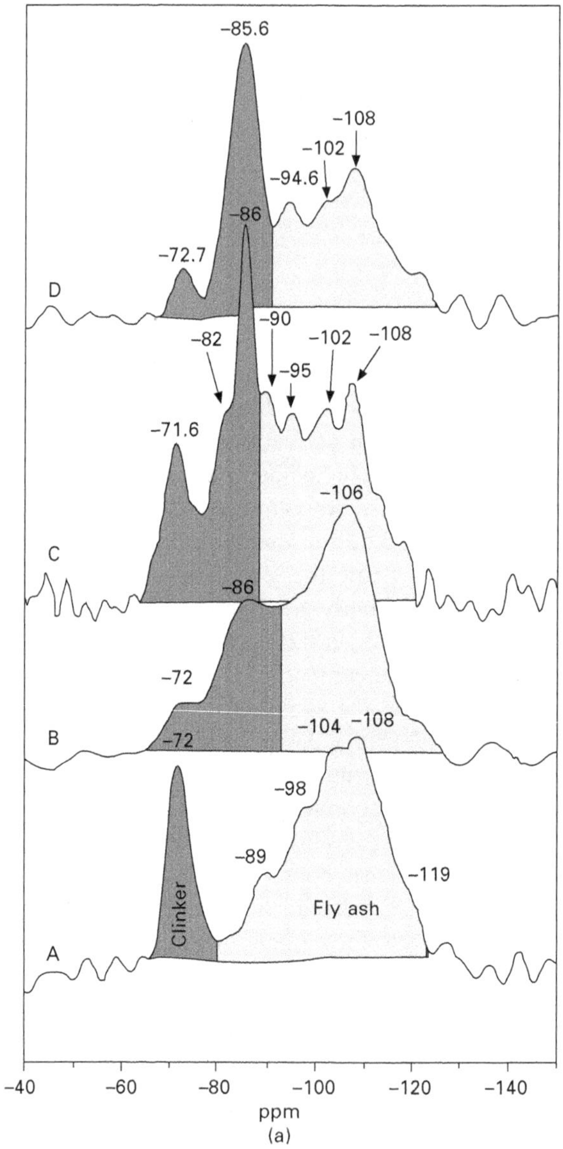

17.17 (a) ^{29}Si MAS NMR spectra: A, initial raw mix (S = solid); B, C and D, specimens hydrated for 28 days with water, NaOH and NaOH + Wg, respectively (Palomo et al., 2007a); (b) cementitious matrix (70% fly ash + 30% clinker) hydrated with caustic soda + Wg (Palomo et al., 2007a).

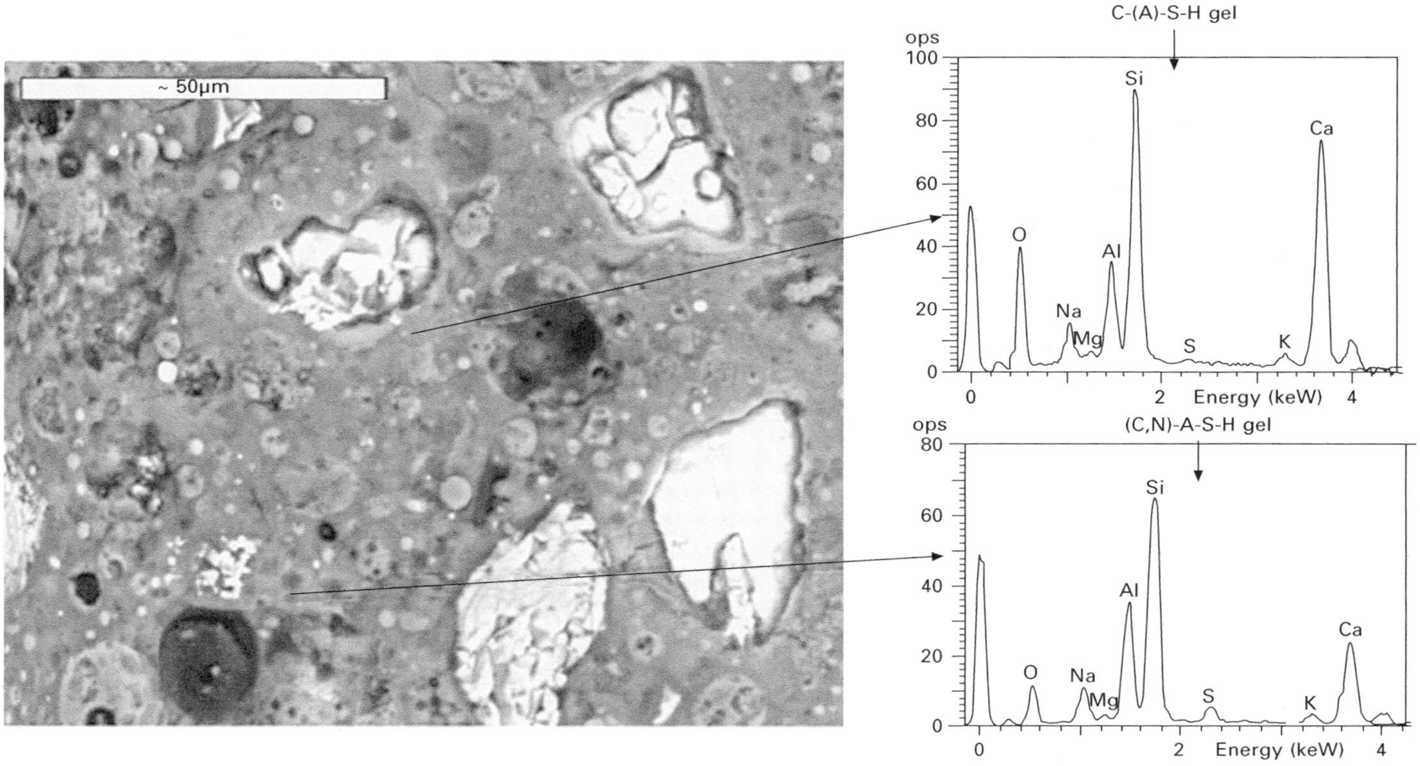

17.17 Continued

the same signal in spectrum A). The presence in these pastes of compounds previously identified as pertaining to N-A-S-H gel (Engelhardth and Michel, 1987; Fernández-Jiménez and Palomo, 2003; Criado et al., 2008) (–90, –95, –102 ppm) attributed to Q^4 units with three, two or one aluminium atoms) also supports the co-existence of the two reaction products.

Finally, the presence of phases with different chemical compositions has also been confirmed with SEM (light and dark phases in Fig. 17.17(b)). Microanalysis showed that the light phase is attributed to a mix of C-S-H and C-(A)-S-H gels and the dark phase to a mix of N-A-S-H and ((C,N)-A-S-H) gels.

This accumulation of evidence confirms that the precipitation of a mix of (C-S-H and N-A-S-H) gels is responsible for the setting and hardening of this type of fly ash-rich super-blended cement when highly alkaline hydration solutions are used. These findings are wholly consistent with data reported by Krivenko (1992) who studied blast furnace slag/aluminosilicate blends and van Deventer and coworkers (Yip and van Deventer, 2002; van Deventer et al., 2011), who conducted in-depth studies of the reactivity of systems containing metakaolin and blast furnace slag.

The fact that all the foregoing refers to early age findings only (28 days at most) must not be overlooked, however (Alonso and Palomo, 2001; Yip et al., 2005; Palomo et al., 2007a). Recent research conducted by García-Lodeiro et al. (2012a) on alkali-activated blends of 70% fly ash and 30% OPC over long reaction times (1 year), revealed the presence of different, mostly C-A-S-H-type gels (see Fig. 17.18). These authors also observed that (N,C)-A-S-H gels evolved into compositions with a higher calcium and a lower aluminium content. These findings, which would appear to support the hypothesis that the formation of C-A-S-H-type gels is favoured over time, are consistent with reports in the literature on the compatibility between N-A-S-H and C-A-S-H gels in equilibrium conditions (García-Lodeiro et al., 2011).

17.4.3 Applications

The alkaline activation of Portland cement and aluminosilicate blends is being studied with growing interest as a solution for substantially reducing the clinker content in cement while maintaining the mechanical properties and durability of the products generated (Palomo et al., 2004). Hybrid cements are a technologically viable alternative to the traditional cement industry based exclusively on Portland cement.

This technology is relatively new and while it is still under development and optimisation, the precast industry has already begun to adopt the knowledge transferred for certain promising applications. The photographs in Figs 17.19 and 17.20 depict the hybrid concretes (which look very much like concrete

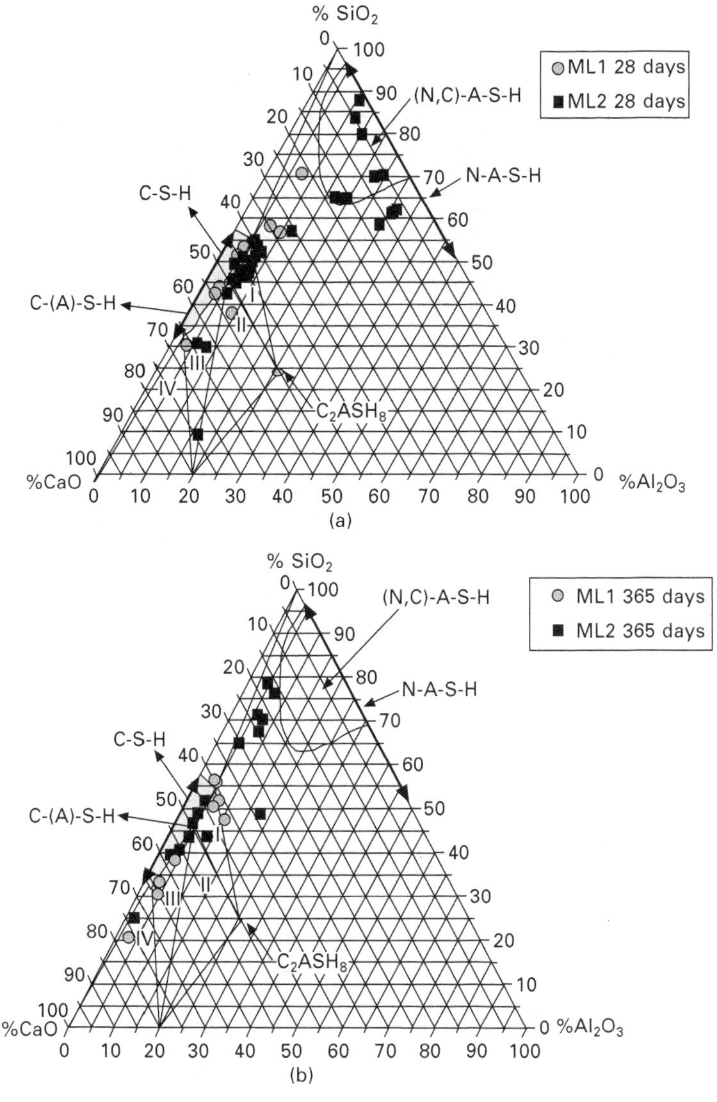

17.18 Transmission electron microscopy/energy dispersive X-ray (EM/EDX) analyses of (a) 28-day and (b) 365-day gels precipitating in water-hydrated (ML1) and alkali-activated (ML2), 70% FA + 30% OPC systems, projected onto the $CaO-Al_2O_3-SiO_2$ plane (García-Lodiero et al., 2012a).

manufactured with OPC), at several stages in the process of precast element manufacture and the end products.

The concrete in Fig. 17.19 was prepared with binder consisting of blast furnace slag + fly ash + OPC clinker (<20%) + solid alkaline activator. Concrete blocks such as are commonly used in building construction and

17.19 Low-cost concrete block manufacture using slag-fly ash-clinker-solid alkaline activator blended cements.

17.20 Concrete blocks for building construction and pavers, both made with slag-clinker hybrid cements.

blocks laid on the slopes of river banks for stabilisation purposes were made from this material. No changes had to be made in the standard aggregate dosage (fines/coarse) used at the plant to manufacture these blocks with OPC-containing concrete to manufacture these new elements.

The binder used for the material in Fig. 17.20 was a blend of clinker (<20%) + solid alkaline activator + different types of slag with different Ca, Al and Si contents. In this case, the binder/aggregate dosage was initially the same as normally applied at the plant (that ratio is higher in pavers than in blocks, given the different strength requirements for the two materials). In light of the good initial strength afforded by these hybrid cements or concretes, however, the binder/aggregate ratio could subsequently be lowered.

Cements of different strength categories can be simulated by modifying the ash, slag and alkaline component dosages (with clinker contents of consistently under 30%). This is extraordinarily important, since it means that this new technology is at least as versatile as Portland cement.

In all the specific construction elements discussed above, the manufacturing process entailed the use of extrusion facilities, i.e., pressure compacting the products in the moulds. The rheology of the fresh concrete plays an essential role in this process, since the products must neither collapse (due to excess water) nor be under-compacted (lack of sufficient water). Despite the substantial variation in the nature of the binder used to make the concrete (hybrid cement instead of the standard Portland cement), the rheological properties of the product were wholly unaltered. The use of solid instead of liquid activators provides for better control of concrete rheology.

By way of summary, the entire procedure, which involves blending the dry materials (alkaline binder, aggregate and so on), their mixing with water, carrying the fresh concrete on a conveyor belt to the extrusion machine, pressure compacting the fresh concrete and storing the fresh blocks in a curing chamber, poses no problems worthy of mentions and can be carried out with no need for even the slightest modification of a plant's *modus operandi*.

17.5 Future trends and technical challenges

This chapter aims to provide a certain amount of elementary information on alkaline activation technology and alkaline cements, while addressing the environmental issues posed by the binders used in construction. More specifically, the message is that such technology may shortly reach a stage of development in which it will serve as a link in the necessary transition from Portland cement to the cements of the future.

The environmental drawbacks to Portland cement production are obvious: it accounts for 5–8% of total CO_2 emissions, primarily because of the very high temperature required to decompose the enormous amounts of limestone involved in generating reactive calcium silicates. Gartner (2004) calculated that for a typical CEM I cement, thermal decomposition of the prime materials (mainly carbonates) alone releases around 0.5 t of CO_2 into the air per tonne of clinker produced. One of the most notorious advantages of alkaline over traditional Portland cements is the significant reduction of CO_2, primarily thanks to the elimination of the high temperatures required to burn the prime material (Habert et al. 2011). Alkaline activators do, however, have a certain carbon footprint. Duxson et al. (2007b) calculated the CO_2 emissions due to a number of alkali activated fly ash/metakaolin binders in terms of the dissolved solids (Na_2O + SiO_2) content in the activating solution. Their findings showed CO_2 savings on the order of 80% over OPC on a binder-to-binder basis.

Nonetheless, gaps can still be detected in the knowledge base in areas that will be essential to progress in the technological development of alkaline binders and to the aforementioned transition from traditional Portland to new, more sustainable cements. One significant example of the scientific and technical gap in alkali activation procedures is the lack of a systematic and orderly study of the mechanisms governing the effect of known alkali activators (sodium and potassium hydroxides, silicates, carbonates and sulphates) on aluminosilicate materials. The relationship between reaction mechanisms, the chemistry of alkaline activating solutions and end product properties needs to be explored, along with the decisive effect of calcium (with enormous technological implications) on such mechanisms. While alkaline silicate solutions have likewise been widely used in alkaline activation, essential aspects of the reactive process are still poorly understood, such as the effect of the various chemical species present in the solution on reaction kinetics or the composition of the end product.

Working with solid activators instead of alkaline solutions would afford an enormous technological advantage, for the former would emulate one of the most estimable properties of Portland cement: its conversion from a dehydrated solid state into an effective binder by mere mixing with water. In pursuit of such activators, some authors have proposed using cementitious formulas (with aluminosilicate materials) that contain sodium and/or potassium carbonates or even sodium and/or potassium sulphates. In any event, the literature on alkaline activation with these products is scant, and primarily geared to obtaining cementitious products. Nothing, or barely anything, has been published that would relate the interaction of aluminosilicate materials containing concentrated sodium or potassium carbonate or sulphate solutions to the formation of N-A-S-H or K-A-S-H type cementitious gels. Nor has any research been forthcoming to date that would indicate (in formulas with carbonates or sulphates) the structural destination of carbonate or sulphate anions if they are taken up into the three-dimensional cementitious skeleton of Si/Al binders. These questions would need to be explored.

Another understudied area is the formation of other phases not generally found in these cementitious systems. Prime materials also constitute a further sizeable gap in the understanding of these cements. The institution of a universal, standardised, economical and sustainable system for processing raw materials (such as is in place for Portland cement) and activating the resulting products would be extremely beneficial.

The rheology of these systems is yet another area poorly studied despite its enormous importance. The existing range of commercially available superplasticisers, developed specifically to suit the complex series of chemical reactions that take place in the OPC system, are generally ineffective in alkali-activated aluminosilicate systems. Moreover, most of the admixtures used to control slump, air dispersion, water retention and other properties of the OPC system are less effective when used with alkali-activated aluminosilicates. Consequently, a whole set of new admixtures needs to be developed for the alkali-activated aluminosilicates system, a significant challenge for an emerging industry that has yet to acquire scale but that must compete with the well-established OPC industry.

Lastly, a word is in order on hybrid cements. Work is needed on hybrids with a low Portland clinker content and high proportion of aluminosilicates, especially on their behaviour in alkaline environments. Setting, rheology, mechanical strength development and durability in this type of hybrids must also be studied. Work is needed on hybrids with a low Portland clinker content and high proportion of aluminosilicates, especially on their behaviour in alkaline environments. Setting, rheology, mechanical strength development and durability in this type of hybrids must also be studied. This latter issue is of special interest since, as noted above, many of the durability problems posed by Portland cement pastes, mortars and concretes are associated with the presence of calcium. Portland cement-free, aluminosilicate-based alkaline cements, by contrast, perform better in this regard. Since a mix of C-A-S-H/N-A-S-H gels precipitate in these hybrid cements (Alonso and Palomo, 2001; Yip et al., 2005; Palomo et al., 2007a) and all the evidence indicates that calcium replaces sodium in cementitious gels (García-Lodeiro et al., 2010b, 2011), behaviour somewhere between these two extremes might be expected. The durability of hybrid cements has yet to be researched, however.

17.6 Acknowledgement

This research project was funded by the Spanish Ministry of Science and Innovation under project BIA2010-17530.

17.7 References

Allahverdi A., Škvára F. (2001), 'Nitric acid attack on hardened paste of geopolymeric cements' *Ceramics-Silikaty* **45** (4) pp. 143–149

Alonso S., Palomo A. (2001), 'Alkaline activation of metakaolin-calcium hydroxide solid mixtures: Influence of temperature, activator concentration and metakaolin /$Ca(OH)_2$ ratio' *Mater. Lett.* **47** pp. 55–62

Bakharev T. (2005), 'Durability of geopolymer materials in sodium and magnesium sulphate solutions' *Cem. Con. Res.* **35** pp. 1233–1246

Bakharev T. (2006), 'Thermal behaviour of geopolymers prepared using class F fly ash and elevated temperature curing' *Cem. Con. Res.* **36** pp. 1134–1147

Bakharev T., Sanjayan J.G. Cheng Y.B. (2000), 'Effect of admixtures on properties of alkali-activated slag concrete' *Cem. Con. Res.* **30** pp 1367–1374

Barbosa V.F.F., Mackenzie K.J.D. (2003), 'Thermal behaviour of inorganic geopolymers and composites derived from sodium polysialate' *Mater. Res. Bull.* **38** pp. 319–331

Bastidas D.M., Fernández-Jiménez A., Palomo A., Gonzalez J.A. (2008), 'A study on the passive state stability of steel embedded in activated fly ash mortars' *Corrosion Sci.* **50** (4) pp. 1058–1065

Ben Haha M., Lothenbach B., Le Saout G., Winnefeld F. (2011), 'Influence of slag chemistry on the hydration of alkali-activated blast furcace slag-part I. Effect of MgO' *Cem. Con. Res.* **41** pp. 995–963

Ben Haha M., Lothenbach B., Le Saout G., Winnefeld F. (2012), 'Influence of slag chemistry on the hydration of alkali-activated blast furnace slag–part II. Effect of Al_2O_3' *Cem. Con. Res.* **42** pp. 74–83

Bernal S.A., Mejía de Gutierrez R., Pedraza A.L., Provis J.L., Rodriguez E.D., Delvasto S. (2011a), 'Effect of binder content the performance of alkali-activated slag concretes' *Cem. Con. Res.* **41** pp. 1–8

Bernal S.A., Provis J.L., Rose V., Mejia de Gutierrez R. (2011b), 'Evolution of binder structure in sodium silicate-activated slag-metakaolin blends' *Cem. Con. Comp.* **33** pp. 46–54

Buchwald A., Hilbig H., Kaps C. (2007), 'Alkali-activated metakaolin-slag blends – performance and structure in dependence on their composition' *J. Mat. Sci.* **42** pp. 3024–3032

Cheng Q.H., Tagnit-Hamou A., Sarkar S.L. (1992), 'Strength and microstructural properties of waterglass activated slag' *Mater. Res. Soc. Symp. Proc.* **245** pp. 49–54

Cong X.D., Kirkpatrick R.J. (1996), '^{29}NMR study of the structure of the calcium silicate hydrate' *Adv. Cem. Bas. Mat.* **3** pp. 144–156

Criado M. (2007), 'Nuevos materiales cementantes basados en la activación alcalina de cenizas volantes. Caracterización de geles N-A-S-H en función del contenido de sílice soluble. Efecto del Na_2SO_4' Ph.D. Tesis, Universidad Autónoma de Madrid

Criado M., Palomo A., Fernández-Jiménez A. (2005), 'Alkali activation of fly ashes. Part 1: Effect of curing conditions on the carbonation of the reaction products' *Fuel* **84** pp. 2048–2054

Criado M., Fernández-Jiménez A., Palomo A., Sobrados I. (2008), 'Effect of the SiO_2/Na_2O ratio on the alkali-activation of fly ash. Part II: ^{29}Si MAS-NMR survey', *Micr. Mes. Mater.* **109** pp. 525–534.

Criado M., Palomo A., Fernández-Jiménez A., Banfill P.B.G. (2009), 'Alkali activated fly ash: Effect of admixtures on paste rheology' *Rheol. Acta* **48** (4), pp. 447–455

Criado M., Palomo A., Fernández-Jiménez A. (2010), 'Alkali activation of fly ashes. Part III: Effect of curing conditions on reaction and its graphical description' *Fuel* **89** pp. 3185–3192

Davidovits J. (2008), *Geopolymer. Chemistry and Applications* Institut Géopolymere, Saint-Quentin, France

Davidovits J., Buzzi L., Rocher P., Gimeno D., Marini C., Tocco S. (1999), 'Geopolymeric Cement Based on Low Cost Geopolymer Material', Results from European Research Project GEOCISTEM, Geopolymer 99 Proceeding pp. 83–96

Deja J. (2002), 'Carbonation aspects of alkali activated slag mortars and concretes' *Silic. Industr.* **67** (1) pp. 37–42

Duxson P., Fernández-Jiménez A., Provis J.L., Lukey G.C., Palomo A., van Deventer J.S.J. (2007a) 'Geopolymer technology: The current state of the art' *J. Mat. Sci.* **42** pp. 2917–2933

Duxson P., Provis J.L., Lukey G.C., van Deventer J.S.J. (2007b), 'The role of inorganic polymer technology in the development of 'Green concrete' *Cem. Con. Res.* **37** pp. 1590–1597

Engelhardth G., Michel D. (1987), *High-resolution Solid-state NMR of Silicates and Zeolites*. John Wiley and Sons (1987)

Fernández-Jiménez A. (2000), 'Cementos de escorias activadas alcalinamante: influencia de las variables y modelización del proceso' Ph.D. Thesis, Universidad Autónoma de Madrid.

Fernández-Jiménez A., Palomo A. (2003) 'Characterization of fly ashes. Potential reactivity as alkaline cements', *Fuel*, **82** (18) pp. 2259–2265

Fernández-Jiménez A., Palomo A. (2005a), 'Composition and microstructure of alkali activated fly ash mortars. Effect of the activator' *Cem. Concr. Res.* **35** pp. 1984–1992

Fernández-Jiménez A., Palomo A. (2005b), 'Some main factors influencing the alkali activation of fly ashes factors' 2nd Inter. Sym. Non-Tradicional Cem. Con. Czech Republic

Fernández-Jiménez A., Palomo A. (2009), 'Propiedades y aplicaciones de los cementos alcalinos', X Congreso Latinoamericano de Patología y XII Congreso de Calidad en la Construcción. CONPAT, Valparaíso-Chile

Fernández-Jiménez A., Palomo A. (2010), 'Si and Al MAS NMR characterization of cement – State of art', *Cemento y Hormigón*, pp. 936

Fernández-Jiménez A., Palomo A. (2011), 'Factors affecting early compressive strength of alkali activated fly ash (OPC-free) concrete' *Materiales de la Construcción*, **57** (287) pp. 5–20

Fernández-Jiménez A., Puertas F., Fernández-Carrasco L. (1996). 'Procesos de activación alcalino-sulfaticos de una escoria Española de Alto Horno' *Materiales de Construcción*. **46** (241), pp. 23–37

Fernández-Jiménez A., Puertas F., Sobrados I., Sanz J. (2003), 'Structure of calcium silicate hydrate formed in alkaline activated slag. Influence of the type of alkaline activator' *J. Am. Cer. Soc.* **86** (8) pp. 1389–1394

Fernández-Jiménez A., Palomo A., Criado M. (2005), 'Microstructure development of alkali-activated fly ash cement: a descriptive model' *Cem. Con. Res.* **35** (6), pp. 1204–1209

Fernández-Jiménez A., Palomo A., Sobrados I., Sanz J. (2006a) 'The role played by the reactive alumina content in the alkaline activation of fly ashes' *Micr. Mes. Mat.* **91** pp. 111–119

Fernández-Jiménez A., Palomo A., López-Hombrados C. (2006b), 'Engineering properties of alkali activated fly ash concrete' *ACI Mater. J.*, **103** (2) pp. 106–112

Fernández-Jiménez A., García-Lodeiro I., Palomo A. (2007), 'Durable characteristics of alkali activated fly ashes' *J. Mater. Sci.* **42** pp. 3055–3065

Fernández-Jiménez A., Miranda J.M., Gonzales J.A., Palomo A. (2010) 'Steel passive state stability in activated fly ash mortars' *Materiales de Construcción* **60** pp. 561–565

Fernández-Jiménez A., Sobrados I., Sanz J., Palomo A. (2011), 'Hybrid cements with very low OPC content' Proceeding of 13th International Congress on the chemistry of Cement (XIII ICCC), 'Cementing a Sustainable Future'

García-Lodeiro I., Palomo A., Fernández-Jiménez A. (2007), 'Alkali–aggregate reaction in activated fly ash systems' *Cem. Con. Res.* **37** pp. 175–183

García-Lodeiro I., Macphee D.E., Palomo A., Fernández-Jiménez A. (2009), 'Effect of alkalis on fresh C-S-H gels. FTIR analysis', *Cem. Con. Res.* **39** pp. 147–153

García-Lodeiro I., Fernández-Jiménez A., Palomo A., Macphee D.E. (2010a), 'Effect on fresh C-S-H gels of the simultaneous addition of alkali and aluminium' *Cem. Con. Res.* **40** pp. 27–32

García-Lodeiro I., Fernández-Jiménez A., Palomo A., Macphee D.E. (2010b), 'Effect of calcium additions on N–A–S–H cementitious gels', *J. Am. Ceram. Soc.* **93** pp. 1934–1940

García-Lodeiro I., Palomo A., Fernández-Jiménez A., Macphee D.E. (2011), 'Compatibility studies between N-A-S-H and C-A-S-H gels. Study in the ternary diagram Na_2O-CaO-Al_2O_3-SiO_2-H_2O' *Cem. Con. Res.* **41** pp. 923–931

Garcia-Lodiero I., Fernández-Jiménez A., Palomo A. (2012a), 'Variation in hybrid cements over time. Alkaline activation of fly ash-portland cement blends' *Cem. Con. Res.* (submitted)

García-Lodeiro I., Fernández-Jiménez A., Sobrados I., Sanz J., Palomo A. (2012b), 'C-S-H gel. Interpretation of ^{29}Si MAS-NMR spectra' *J. Am. Ceram. Soc.* **95** (4) pp. 1440–1446

Gartner E. (2004), 'Industrially interesting approaches to 'low CO_2' cements' *Cem. Con. Res.* **34** pp. 1489–1498

Glasser F.P. (1990), 'Cements from micro to macrostructure' *Br. Ceram. Trans. J.* **89** (6) pp. 192–202

Glukhovsky V. (1994), 'Ancient, modern and future concretes' First Inter. Conf. Alkaline Cements and Concretes. Vol. 1, pp. 1–8. Kiev, Ukraine.

Habert G., d'Espinose de Lacaillerie J.B., Roussel N. (2011), 'An environmental evaluation of geopolymer based concrete production: reviewing current research trends' *J Cleaner Production* **19** pp. 1229–1238

Hardjito D., Wallah S.E., Rangan B.V. (2002), 'Research into engineering properties of geopolymer concrete', International Conference 'Geopolymer 2002 – turn potential into profit', Melbourne, Australia, October

Johansen V., Thaulow N., Skalny J. (1995), 'Chemical degradation of concrete', presentation at the 1995 TRB meeting, Washington, DC, January 1995.

Juenger M.C.G., Winnefeld F., Provis J.L., Ideker J.H. (2011), 'Advances in alternative cementitious binders' *Cem. Con. Res.* **41** pp. 1232–1243

Kani E., Allahverdi A., Provis J. (2011), 'Efflorescence control in geopolymer binders based on natural pozzolan' *Cem. Con. Comp.* **34** pp. 25–33

Kovalchuk G., Fernández-Jiménez A., Palomo A. (2007), 'Alkali-activated fly ash: effect of thermal curing conditions on mechanical and microstructural development – part II' *Fuel* **86** pp. 315–322

Krivenko P.V. (1992), *Special slag alkaline cements*, Budivelnyk, Kiev.

Krivenko P.V. (1994), '*Theory, Chemistry, Structure and Properties of Alkaline Cementitious Systems*', First Inter. Conf. Alkaline Cements and Concretes, Vol. 1, pp. 12–129, Kiev, Ukraine

LaRosa J.L., Kwan S., Grutzeck M.W. (1992), 'Zeolite formation in Class F fly ash blended cement pastes' *J. Am. Ceram. Soc.* **75** (6) pp. 1574–1578.

Lemougna P.N., MacKenzie K.J.D., Chinje Melo U.F. (2011), 'Synthesis and thermal properties of inorganic polymers (geopolymers) for structural and refractory applications from volcanic ash' *Ceram. Int.* **37** pp. 3011–3018

Li C., Sun H., Li L. (2010), 'A review. The comparison between alkali activated (Si+Ca) and meatakaolin (Si+Al) cements', *Cem. Con. Res.* **40** pp. 1341–1349

Macphee D.E., García-Lodeiro L. (2011), 'Activation of aluminosilicates-Some Chemical considerations', 2nd International Slag Valorisation Symposium, Leuven, Belgium

Malolepszy J. (1989), 'The hydration and the properties of alkali activated slag cementitious materials' *Ceramika* **53** pp. 7–125

Metha P.K. (1991), 'Durability of concrete: Fifty years of progress?, *American Concrete Institute On-line Journals* **126**, pp. 1–32

Miranda J.M., Fernández-Jimebea A., Gonzalez J.A., Palomo A. (2005) 'Corrosion resistance in activated fly-ash mortars' *Cem. Con. Res.* **35** pp. 1210–1217

Ozol M.A. (1975), 'The pessimum proportion as a reference point in modulating alkali–silica reaction', Proceedings of a Symposium on Alkali–Aggregate Reaction, Preventive Measures, Reykjavik, Iceland, pp. 113–130

Pacheco-Torgal F., Jalali S. (2010), 'Influence of sodium carbonate addition on the thermal reactivity of tungsten mine waste mud based binders' *Construction Building Mater.* **24** pp. 56–60

Pacheco-Torgal F., Castro-Gomes J., Jalali S. (2008a), 'Alkali-activated binders: a review. Part 1. Historical backround, terminology, reaction mechanims and hydration products' *Construction Building Mater.* **23** pp. 1305–1314

Pacheco-Torgal F., Castro-Gomes J., Jalali S. (2008b), 'Alkali-activated binders: a review. Part 2: About materials and binders manufacture' *Construction Building Mater.* **23** pp. 1315–1322

Pacheco-Torgal F., Abdollahnejad Z., Camões A.F., Jamshidi M., Ding Y. (2012),

'Durability of alkali-activated binders: A clear advantage over portland cement or an unproven issue?' *Construction Building Mater.* **30** pp. 400–405

Palomo A., Grutzeck M.W., Blanco M.T. (1999), 'Alkali-activated fly ashes- A cement for the future' *Cem. Con. Res.* **29** (8) pp. 1323–1329

Palomo A., Alonso S., Fernández-Jiménez A., Sobrados I., Sanz J. (2004), 'Alkaline activation of fly ashes. A NMR study of the reaction products' *J. Am. Ceramic. Soc.* **87** (6) pp. 1141–1145

Palomo A., Banfill P.F.G., Fernández-Jiménez A., Swift D.S. (2005), 'Properties of alkali-activated fly ashes determined from rheological measurements' *Adv. Cem. Res.* **17** pp. 143–151

Palomo A., Fernández-Jiménez A., Kovalchuk G., Ordoñez L.M., Naranjo M.C. (2007a), 'OPC-fly ash cementitious system. Study of the gel binders produced during alkaline hydration' *J. Mater. Sci.*, **42** pp. 2958–2966

Palomo A., Fernández-Jiménez A., Lopez-Hombrados C., Lleyda J.L. (2007b), 'Railway sleepers made of alkali activated fly ash concrete' *Revista Ingeniería de Construcción* **22** (2) pp. 75–80

Palomo A., Fernández-Jiménez A. (2011), 'Alkaline activation' as a procedure for the transformation of fly ashes into new materials. Part I: Applications', World of Coal Ash (Woca) Conference 'Science, applications and sustainability', Denver, Co.

Poulesquen A., Frizon F., Lambertin D. (2011), 'Rheological behavior of alkali-activated metakaolin during geopolymerization' *J. Non-Crystalline Sol.* **357** pp. 3565–3571

Provis J., van Deventer J.S.J. (eds) (2009), *Geopolymers, structure, processing, properties and industrial applications*, Woodhead Publishing Limited

Puertas F. (1995), 'Cementos de escorias activadas alcalinamente: Situación actual y perspectivas de futuro' *Materiales de la Construcción* **45** (239) pp. 53–64

Puertas F., Palacios M., Manzano H., Dolado J.S., Rico A., Rodriguez J. (2011), 'A model for the CASH gel formed in alkali-activated slag cements' *J. Eur. Cer. Soc.* **31** pp. 2043–2056

Richardson I., Groves G.W. (1997), 'The structure of the calcium silicate hydrate phases present in hardened pastes of white portland cement/blast-furnace slag blends' *J. Mater. Sci.*, **32** pp. 4793–4802

Richardson I (2008), 'The calcium silicate hydrates' *Cem. Con. Res.* **38** pp. 137–158

Rodriguez E., Bernal S., Mejia de Gutierez R., Puertas F. (2008), 'Alternative concrete based on alklai-activated slag' *Materiales de Construcción* **58** pp. 53–67

Rostami H., Brendley W. (2003), 'Alkali ash materials: A novel fly ash-based cements' *Environ. Sci. Technol.* **37** p. 3454

Roy A., Schilling P.J., Eaton H.C. (1994), 'Activation of ground blast-furnace slag by alkali-metal and alkaline-earth hydroxides' *J. Am. Ceram. Soc.* **75** (12), pp. 3233–3240

Schilling P.J., Butler L.G., Roy A., Eaton H.C. (1994), '^{29}Si and ^{27}Al MAS-NMR of NaOH activated blast furnace slag' *J. Am. Ceram. Soc.* **77** (9), pp. 2363–2368

Shi C., Krivenko P.V., Roy D. (2006), *Alkali-activated cements and concretes* Taylor & Francis, Abingdon, UK

Shi C., Fernández Jiménez A., Palomo A. (2011), 'New cements for the 21st century, The pursuit of an alternative to portland cement' *Cem. Con. Res.* **41** pp. 750–763

Škvára F., Jilek T., Kopecky L. (2005), 'Geopolymer materials based on fly ash' *Ceramics-Silikáty* **3** pp. 195–204

Škvára F. (2007), 'Alkali activated material – Geopolymer' Proceedings of 2007 International Conference on Alkali Activated Materials. Research, Production and Utilization, Prague, Czech Republic pp. 661–676

Sumajouw M.D.J., Rangan B.V. (2006), 'Low-calcium fly ash-based geopolymer concrete: reinforced beams and columns' Research Report GC3, Faculty of Engineering, Curtin University of Technology Perth, Australia

Swamy R.N. (1992), in R.N. Swamy (Ed.), Blackie and Son Ltd, *The alkali–silica reaction in concrete*, Blackie, New York

Taylor H.F.W. (1997), *Cement chemistry*, Academic Press Ltd, London

Temuujin J., Van Riessen A., Williams R. (2009), 'Influence of calcium compounds on the mechanical properties of fly ash geopolymer pastes' *J. Hazard. Mater.* **167** pp. 82–88

Temuujin J., Van Riessen A., Mackenzie K.J.D. (2010), 'Preparation and characterisation of fly ash based geopolymer mortars' *Construction Building Mater.* **24** pp. 1906–1910

Van Deventer J.S.J., Feng D., Duxson P. (2010), 'Dry mix cement composition, methods and systems involving same' US Patent 7691,198 B2

van Deventer J.S.J., Provis J.L., Duxon P. (2011), 'Technical and commercial progress in the adoption of geopolymer cement', *Minerals Engin.* **29** pp. 89–104

Van Jaarsveld J.G.S., Van Deventer J.S.J., Lukey G.C. (2002), 'The effect of composition and temperature on the properties of fly ash and kaolinite-based geopolymers' *Chem. Eng. J.*, **89** pp. 63–73

Wang S.D., Scrivener K.L. (1995), 'Hydration products of alkali activated slag cement' *Cem. Con. Res.* **25** (3) pp. 561–571

Wang S.D., Pu X.C., Scrivener K.L., Pratt P.L. (1995), 'Alkali-activated slag: a Review of Properties and Problems' *Cem. Con. Res.* **17** (27) pp. 93–102

White C.E., Provis J.L., Proffen T., van Deventer J.S.J. (2011), 'Molecular mechanisms responsible for structural changes occurring during geopolymerization: Multiscale simulation', *AIChE J.* **58** pp. 2241–2253

Winnefeld F., Leemann A., Lucuk M., Svoboda P., Neuroth M. (2010) 'Assesment of phases formation in alkali activated low and high calcium fly ashes in building materials' *Construction Building Mater.* **24** pp. 1086–1093

Wu X., Roy D.M., Langton C.A. (1983), 'Early stage hydration of slag-cement', *Cem. Con. Res.* **13** pp. 277–286

Xu H., Provis J.L., van Deventer J.S.J., Krivenko P.V. (2008), 'Characterization of aged slag concretes', *ACI Mater. J.* Vol. **105** (2) pp. 131–139

Yip C.K.B., Van Deventer J.S.J. (2002), CD Proceedings of geopolymers 2002, Sofitel, Melbourne, Australia, pp. 28–29.

Yip C.K., Lukey G.C., Deventer J.S.J. (2005), 'The coexistence of geopolymeric and calcium silicate hydrate at the early stage of alkaline activation' *Cem. Con. Res.* **35** pp. 1688–1697

Yodmunee S., Yodsudjai W. (2006), 'Study on corrosion of steel bar in fly ash-based geopolymer concrete' International Conference on Pozzolan, Concrete and Geopolymer, Khon Kaen, Thailand, pp. 189–194

Zhang Y., Wei S. (2006), 'Fly ash based geopolymer concrete' *Indian Con. J.* **80** pp. 20–24.

18
Sulfoaluminate cement

M. A. G. ARANDA and A. G. DE LA TORRE,
University of Málaga, Spain

DOI: 10.1533/9780857098993.4.488

Abstract: Yeelimite-containing, $Ca_4Al_3O_{12}(SO_4)$, cements are attracting considerable interest for their inherently low CO_2 emissions during fabrication, and also because they offer the possibility of tailored properties and performances in the resulting mortars and concretes. Here, we initially disentangle the nomenclature and compositions of these cements and, chiefly, distinguish between yeelimite-rich calcium sulfoaluminate cements (CSA, Al_2O_3 content ranging between 30 and 40 wt%) and belite-rich yeelimite-intermediate (BCSAF, Al_2O_3 content ~15 wt%) cements, also known as sulfobelite. Belite-active BCSAF cement may become an alternative to Portland cement for large-scale structural applications, although more research is still needed. Clinkering, hydration and durability issues of pastes, mortars and concretes derived from yeelimite-containing cements, and their blended cements, are discussed.

Key words: calcium sulfoaluminate (CSA), BCSAF, yeelimite, Klein salt, calcium sulfoaluminate, sulfobelite, eco-cements, CO_2 emission reductions.

18.1 Introduction

Calcium sulfoaluminate (CSA) cements are considered environmental friendly materials since their manufacturing process releases less CO_2 into atmosphere than that of ordinary Portland cement (OPC) (Sharp et al., 1999; Gartner, 2004). These binders may have quite variable compositions, but all of them contain high amounts of yeelimite, also called Klein's salt, CSA or tetracalcium trialuminate sulfate ($C_4A_3\underline{S}$) (Odler, 2000). Cement nomenclature will be used hereafter: C = CaO, M = MgO, S = SiO_2, A = Al_2O_3, F = Fe_2O_3, $\underline{S}$ = SO_3, T = TiO_2 and H = H_2O. The label 'eco' is assigned to CSA cements due to the CO_2 emission savings. CO_2 emissions from cement manufacturing industry can be classified into two main categories: those coming from raw materials and those from the operation processes. The former are easy to be estimated as compositions of raw material and products are well known (Gielen and Tanaka, 2006; Alaoui et al., 2007). Table 18.1 gives CO_2 emissions related to the production of cement components calculated according to the stoichiometry of chemical reactions, for example, $3CaCO_3 + SiO_2 \rightarrow Ca_3SiO_5 + 3CO_2$. Therefore, the production of 1 t of OPC clinker composed of 60 wt% of C_3S, 20 wt% of

C_2S, 10 wt% of C_3A and 10 wt% of C_4AF released 0.54 t of CO_2. In the next section, yeelimite-containing cement nomenclature and phase assemblages will be properly clarified; however, in order to illustrate the reduction of CO_2 emissions, a CSA clinker composed by 50 wt% of $C_4A_3\underline{S}$, 25 wt% of C_2S, 10 wt% C_4AF and 15 wt% of $C\underline{S}$ will be considered (Quillin, 2007). Table 18.1 gives CO_2 emissions from the formation of yeelimite phase according to the reaction: $3CaCO_3 + 3Al_2O_3 + C\underline{S} \rightarrow C_4A_3\underline{S} + 3CO_2$. Given the CSA phase assemblage shown above, the formation of 1 t of this clinker releases 0.27 t of CO_2. Consequently, a reduction in carbon dioxide emissions of 49% is calculated due to raw material decomposition.

Cement production, however, is an energy-demanding process and it is important to quantify CO_2 emissions coming from fuel consumption to reach clinkering temperatures in the kilns. This is not an easy task as it is directly related to the type of processing equipment and the specific fuel chosen. Gartner (2004) published an estimation of 0.28 t of CO_2 per ton of clinker produced assuming that good quality of bituminous coal is used and taking into account energy efficiency of modern kilns. McCaffrey (2002) stated that 0.34 t of CO_2 per ton of clinker produced are released into atmosphere due to the burning of coal in the kiln. The reduction in CO_2 emissions from the burning of the fuel can be achieved by different strategies (Gartner, 2004; Juenger et al., 2011) including the reduction of clinkering temperature. This is the case of CSA clinker production where the operating temperature can be reduced to ~1250 °C, ~200 °C lower than that of OPC production (Phair, 2006), with a concomitant reduction of up to 0.04 t of CO_2 per ton of CSA clinker produced.

CO_2 emissions derived from electricity consumption for grinding in OPC clinker production are not negligible. In fact, 0.09 t of CO_2 per ton of clinker produced are calculated, assuming that fossil fuels are used to produce the electricity (McCaffrey, 2002). Energy savings also occur in the grinding of CSA clinker compared to OPC, since the lower firing temperatures lead to a clinker which is generally easier to grind. Therefore, reduction on the electricity consumption yields a depletion of up to 0.02 t of CO_2 emissions.

All together, the production of 1 t of OPC clinker releases a maximum of 0.98 t of CO_2. The production of 1 t of CSA clinker leads to a reduction,

Table 18.1 CO_2 emissions per ton of component produced, considering $CaCO_3$ as calcium source

Component	t CO_2/t of component
C_3S	0.58
C_2S	0.51
C_3A	0.49
C_4AF	0.36
$C_4A_3\underline{S}$	0.22

which will depend on the composition (see next), but it can range between 0.25 and 0.35 t of CO_2.

CSA cements were developed in 1970s mainly in China (Zhang et al., 1999; Sui and Yao, 2003) and they are/were known as 'the third series cement of China'. These commercial materials contain 50–80 wt% of $C_4A_3\underline{S}$ and 30–10 wt% of C_2S and are used mainly where rapid setting, early strength or shrinkage compensation is required. However, structural applications were also tested (Zhang et al., 1999 and references therein). The terminology for yeelimite-containing cements is not uniform. Several acronyms have been used for these cements and they are gathered in Table 18.2. In the last 15 years, the interest in CSA (and related) cements has increased due to the potential environmental benefits of their productions. The uses of these cements will be discussed in the next sections.

18.2 Types of calcium sulfoaluminate cements

Yeelimite-containing cements may have a very wide range of phase assemblages that need to be clarified. These materials may be classified according to the content of the main crystalline phase. Here, we propose to unify the terminology used for these cements (see Table 18.2). In a later stage, the

Table 18.2 Proposed acronyms for yeelimite-containing cements. Other acronyms in use are also given

Acronym	Definition	Main phase	Second and other phase(s)
CSA	Calcium sulfoaluminate cement	$C_4A_3\underline{S}$	C_2S (C_4AF, $C\underline{S}$, CT, ...)
BCSAF	Iron-rich belite calcium sulfoaluminate cement	C_2S	$C_4A_3\underline{S}$ (C_4AF, CT, ...)
BCSAA	Aluminum-rich calcium sulfoaluminate cement	C_2S	$C_4A_3\underline{S}$ ($C_{12}A_7$, CA, ...)
ACSA	Alite calcium sulfoaluminate cement	$C_4A_3\underline{S}$	C_3S (C_2S, ...)
SAC	Sulfoaluminate cement	analogous to CSA	Zhang et al. (1999)
BSA	Belite sulfoaluminate	analogous to CSA	Odler (2000)
CSAC	Calcium sulfoaluminate cement	analogous to CSA	Glasser and Zhang (2001)
SCC	Sulfoaluminate cement clinker	analogous to CSA	Fu et al. (2003)
FAC	Ferroaluminate cement	Iron-reach CSA	Zhang et al. (1999)
SFAB	Sulfoferroaluminate belite	Iron-reach CSA	Odler (2000)
SAB	Sulphoaluminate belite	analogous to BCSAF	Janotka et al. (2007)
BSAF	Belite sulphoaluminate ferrite	analogous to BCSAF	Odler (2000)

ranges for mineralogical compositions should also be defined and unified. This is needed in order to standardize these cements.

The term calcium sulfoaluminate (CSA) should be reserved for those clinkers/cements with high $C_4A_3\underline{S}$ contents. They may be prepared from CSA clinkers containing $C_4A_3\underline{S}$ as the main phase ranging between 50 and 80 wt% (Sahu and Majling, 1993; Zhang et al., 1999; Odler, 2000; Glasser and Zhang, 2001). These clinkers may also have minor phases such as C_2S, CT, C_4AF, $C\underline{S}$ and others. These are the 'Chinese cements' and they cannot replace OPC as the percentage of expensive aluminum source needed in their productions is quite high (see Table 18.3). The Al_2O_3 and SO_3 contents in CSA usually range between 30–40 and 8–14 wt%, respectively (see Table 18.3). Al_2O_3 contents will be further discussed below.

A new type of yeelimite-containing cement is emerging due to their environmental benefits. The term belite calcium sulfo-aluminate (BCSA) is reserved for cements arising from clinkers with C_2S (belite) as the main phase and intermediate $C_4A_3\underline{S}$ contents. These cements, also known as sulfobelite, are prepared from clinkers containing more than 40–50 wt% of C_2S and 20–30% $C_4A_3\underline{S}$. The most common formulation of BCSA clinkers consists of β-C_2S, $C_4A_3\underline{S}$ and C_4AF (Janotka and Krajci, 1999; Adolfsson et al., 2007; Janotka et al., 2007). These are iron-rich BCSA cements, also termed BCSAF, and they are produced at temperatures close to 1250°C. These cements are being studied with the final aim of replacing OPC as the aluminum demand is much smaller. The Al_2O_3 and SO_3 contents in BCSAF usually range between 14–17 and 3.5–4.5 wt%, respectively (see Table 18.3). However, it must be highlighted, to avoid confusion, that BCSAF clinkers do not belong to 'the third series cement of China', which have much higher amounts of yeelimite.

In order to further enhance mechanical strengths at very early ages, C_4AF phase may be substituted by $C_{12}A_7$; however, the clinkering temperature should be increased to ~100 °C. Moreover, higher amounts of expensive bauxite (or another aluminum-rich source) are needed. This formulation corresponds to aluminum-rich BCSA clinkers (or BCSAA) with $C_{12}A_7$ and CA as additional phases (Martin-Sedeño et al., 2010; and references therein).

Finally, there is another possible general phase assemblage with simultaneous presence of C_3S and $C_4A_3\underline{S}$ phases. These cements may be termed alite calcium sulfoaluminate (ACSA). Other phases which may appear in these clinkers are C_2S, C_4AF and C_3A (Abdul-Maula and Odler, 1992; Odler and Zhang, 1996; Zhang and Odler, 1996).

18.2.1 Commercial yeelimite-containing cements

Commercial CSA cements with large amounts of yeelimite have been manufactured and used on a large scale in China (Zhang et al., 1999) since

Table 18.3 Mineralogical and elemental (expressed as oxide in weight percentage) compositions for commercial CSA clinkers and large-scale laboratory-prepared BCSAF clinkers

Cement Phase/composition	I[1]	II[2]	III[3]	IV[4]	V[5]	VI[6]	VII[7]	VIII[8]	IX[9]	X[10]	XI[11]	XII[12]	XIII[13]
Type	CSA	CSA	CSA	CSA	CSA	CSA	CSA	CSA	CSA	BCSAF	BCSAF	aBCSAF	aBCSAF
$C_4A_3\underline{S}$	69.5	65.6	56.2	62.8	58	68.5	53.5	57.4	73.5	27.5	20.2	29	31.1
α'_H-C_2S	9.4	–	21.4									?	56.7
β-C_2S	7.7	16.0	9.7			15.9	21.2	25.6	16.1	56.3	50.3	51	–
C_4AF	–	2.4			4	–	16.3	6.6		7.9	19.5	19	10.1
CT	3.5	9.3	3.5	5.7	4	2.9			6.9				2.1
$C\underline{S}$	9.0	–	6.3			0.5	9.0	3.3	1.0	4.4	9.7		–
CaO	41.6	41.9	44.1	36.1	38.0	42.8	44.2	45.3	39.2	52.58	48.12	51.5	51.0
Al_2O_3	33.6	33.9	27.3	45.0	47.4	39.2	30.4	28.9	37.4	19.69	15.25	16.6	17.0
SiO_2	6.5	8.2	9.0	4.5	3.6	5.5	7.5	11.0	4.7	15.45	17.17	17.2	16.5
SO_3	14.0	8.8	12.2	8.6	7.5	9.0	6.5	8.9	8.8	6.12	7.88	4.4	3.7
Fe_2O_3	0.89	2.37	2.60	1.5	1.4	1.2	8.6	3.7		2.60	6.16	7.6	6.2
B_2O_3												1.4	2.4
Na_2O	0.09	0.08	1.40	0.07		0.03	0.04					0.7	1.0
K_2O	0.39	0.25	0.30	0.35	0.16	0.14	0.17					0.1	0.3
MgO	0.68	2.73	1.50	0.91	0.3	1.5	0.59	1.45	1.7	1.50	1.66	0.2	0.97
TiO_2	1.48	1.50	1.30	2.23	2.2	1.67	1.08		1.6			0.4	0.62
Reference	[a]	[a]	[a]	[b]	[c]	[d]	[e]	[f]	[g]	[h]	[i]	[j]	[a]

[1] I, ALIPRE®, also contains 0.5 wt% of MgO and 0.5 wt% of $Na_2Si_2O_5$.
[2] II, CS10 commercialized in Europe by BELITH, also contains 2.2 wt% of MgO and 4.6 wt% of akermanite.
[3] III, S.A.cement, also contains 1.1 wt% of MgO and 1.9 wt% of CA.
[4] IV, unreported name, also contains 18.3, 8.1, 3,1, 1.8 and 0.2 wt% of C_2AS, CA, CA_2, MA and M.

[5] V, Rockfast 450, also contains 17, 16 and 1 wt% of, CA, C_2AS and $C_{12}A_7$.
[6] VI, KTS 100 commercialized in Europe by Belitex, also contains 9.5, 0.5 and 1.2 wt% of $C_{12}A_7$, S and F.
[7] VII, TS commercialized in Europe by Belitex, does not contain additional phases.
[8] VIII, R·SAC 42.5, also contains 9.5, 0.5 and 1.2 wt% of $C_{12}A_7$, S and F.
[9] IX, unreported name, also contains 0.6, 1.7 and 0.2 wt% of $C_{12}A_7$, M and S.
[10] X is a laboratory-prepared BCSAF clinker in very large quantities.
[11] XI seems to be a laboratory-prepared BCSAF clinker in very large quantities.
[12] XII, AETHER™, is an industrially fabricated active BCSAF clinker (still not commercial). So, these data are gathered from the analytical results reported from Lafarge. The different polymorphs of C_2S were not reported.
[13] XIII is a laboratory-prepared active BCSAF clinker.
References: [a] Álvarez-Pinazo et al. (2012); [b] Pelletier-Chaignat et al. (2012); [c] Paglia et al. (2001); [d] Berger et al. (2011a); [e] Coumes et al. (2009); [f] N. Sun et al. (2011b); [g] Kuryatnyk et al. (2010); [h] Janotka and Krajci (1999); [i] Janotka et al. (2007); [j] G. S. Li et al. (2007).

the 1970s with special applications such as rapid strength development used in precast concrete and at moderate curing temperatures (Glasser and Zhang, 2001; Quillin, 2001), self-stressing materials (Péra and Ambroise, 2004; Georgin et al., 2008) or with expansive properties for shrinkage compensating concrete (Chen et al., 2012). These materials have also been studied for radioactive element encapsulation in high-density cement pastes (Zhou et al., 2006; Cau Dit Coumes et al., 2009; Q. Sun et al.; 2011a).

Classification of yeelimite-containing cements is not an easy task as EN regulation standards do not yet provide any contributions helpful to do so and moreover, the properties and uses of this kind of binder are influenced by many factors: (i) the chemical and mineralogical composition of the clinker, (ii) the amount and type of sulfate carrier, (iii) the water to cement ratio, or (iv) the blending with other binders, for instance, OPC. CSA cements present a wide variety of mineralogical compositions (Zhang et al. 1999; Quillin, 2007); see Table 18.3. Those binders with high Al_2O_3 content gained 90% of their 8 day strength after 1 day of hydration, mainly due to high contents of $C_4A_3\underline{S}$, $C_{12}A_7$ and CA. On the other hand, materials with higher amounts of SiO_2 developed greater mechanical strengths after 28 days of hydration, being C_2S the responsible for this behavior. Furthermore, the elemental and phase composition is quite broad as shown in Table 18.3.

Nowadays, there are some commercial CSA clinkers/cements being produced and used at large scales for special applications, e.g. S.A. cement from Buzzi-Unicem, ALIPRE® 2009 from Italcementi Group, and CSA cement (model number 62.5, 72.5, 82.5, 92.5) from Tangshan Polar Bear Building Materials, China. Table 18.3 gives elemental and mineralogical compositions of some commercial CSA binders. There are numerous commercial CSA clinkers, but this is not the case for BCSAF clinkers. For this type of clinker, the analytical results shown in Table 18.3 are those compiled from large-scale laboratory preparations. To the best or our knowledge, today there are no commercial BCSAF clinkers/cements. CSA clinkers are mixed with different types of calcium sulfate carriers and they are also commonly blended with OPCs. Final performances depend heavily on the amount of each component (Marchi and Costa, 2011), yielding a variety of materials with applications such as Portland accelerator, fast-setting mortars, low-shrinkage and expansive mortars and concretes. Some results have been published about the use of this binder in concrete technology (Buzzi et al., 2011). This concrete exhibited rapid strength development, good resistance to freeze–thaw and excellent resistance to sulfate attack; however, durability under carbonation is still under investigation.

There was an attempt to produce a BCSAA cement, called Porsal (Viswanathan et al., 1978; Wolter, 2005), with C_2S and $C_4A_3\underline{S}$ as the main phases, and CA and $C_{12}A_7$ as secondary phases. Cements from this chemical system would combine calcium aluminate cement and sulfoaluminate cement performances. However, there is lack of information concerning the large-scale production of these cements. On the other hand, some laboratory studies state the feasibility of the production of this type of material (Pliego-Cuervo and Glasser, 1979; Valenti et al., 2007; Martín-Sedeño et al., 2010) at low temperatures with the suitable phase assemblage and even using aluminum-rich waste materials for cost reduction.

Popescu et al. (2003) reported an industrial trial of low-energy belite-based cements in which a BCSA cement was included. This study stated the environmental profits of belite-rich materials and even concluded that BCSA developed higher mechanical strengths than OPC at very large hydration ages. However, this BCSA contained only ~12 wt% of $C_4A_3\underline{S}$; thus mechanical strengths at early ages were much lower than those of a typical OPC.

Recently a new environmentally friendly type of cement containing calcium sulfoaluminate phase has emerged with the aim of substituting OPC. These are BCSAF cements; see Table 18.3. The main technological disadvantage is the low mechanical strengths at intermediate ages as the hydration of belite is rather slow. However, this drawback is being overcome by the activation of belite (Gartner and Li, 2006; Cuberos et al., 2010; Morin et al., 2011). The production of active BCSAF cements involves the stabilization of highly reactive C_2S polymorphs, i.e. β-modified form and α-forms, as they react faster with water. It is known that stoichiometric C_2S has five polymorphs (Taylor, 1997), γ, β, α'_L, α'_H and α, on heating, with β-C_2S being the form that commonly prevails in OPC and in BCSAF and BCSAA without activation. Physical and chemical properties of phases can be altered by introducing defects or strains in crystalline structures. Moreover, these defects can even stabilize high-temperature forms of C_2S at room temperature (Ghosh et al., 1979; Nettleship et al., 1992). The different type of defects can be produced by the addition of foreign elements to form solid solutions (Jelenic et al., 1978) or by specific thermal treatments (Fukuda and Ito, 1999).

There are many studies concerning the chemical stabilization of β-C_2S by foreign ions such as SO_3, B_2O_3, Cr_2O_3 Na_2O, K_2O, BaO and MnO_2 (Pritts and Daugherty, 1976; Kantro and Weise, 1979; Matkovic et al., 1981; Fierens and Tirlocq, 1983; Ziemer et al., 1984; Benarchid et al., 2004). These investigations concluded that the hydration reactivity of stabilized β-C_2S depends on preparation parameters that influence particle and crystallite size. These parameters include temperature, type and amount of stabilizer, and fineness of the final ground product. The stabilization of α′ forms by introduction of foreign oxides, such as MgO, P_2O_5, K_2O, BaO and SO_3, has also been studied (Bensted, 1979; Fukuda et al., 2001; Park, 2001). These works stated that hydraulic properties were increased when compared with the materials without foreign ions. The stabilization of α′-belite forms in a cement matrix can be attained by introducing minor elements, such as alkaline oxide, boron or phosphor, in raw materials (G. S. Li et al., 2007;

Morsli et al., 2007a, 2007b; Wesselsky and Jensen, 2009; Cuberos et al., 2010; Morin et al., 2011). Belite-activated BCSAF, aBCSAF, cements have been patented by Lafarge (Gartner and Li, 2006) with ~20–30 wt% of Klein's salt and α forms of belite, due to addition of minor elements, such as B_2O_3 and Na_2O added as borax or P_2O_5, among others. These minor elements promote the stabilization of α′ forms of belite and the distortion of β form. aBCSAF prepared with 2.0 wt% of borax developed comparable mechanical strengths to those of OPC. Two examples of the elemental and mineralogical compositions of aBCSAF are also given in Table 18.3. Moreover, an industrial trial to produce 2500 tons of this aBCSAF cements has been carried out by Lafarge in 2011 under the AETHER™ project (http://www.aether-cement.eu/). This material develops a mechanical strength of 25 MPa at one day and the resistances seem to be even higher than those developed by type I 52.5 OPC after 7 days (Walenta and Comparet, 2011).

We would like to highlight that the compositional variability is much larger for CSA than for BCSAF clinkers (see Table 18.3). Research efforts scale exponentially with the variability of the systems. This is rather important as there are several additional parameters to be investigated for tailoring the mortar/concrete properties, as it will be discussed below. We can note two of the utmost importance: (i) the amount and source (solubility) of the sulfate carrier; and (ii) the water-to-binder ratio. Hence, the research needs for optimizing/tailoring CSA cements are much larger than those for aBCSAF analogous. The initial compositional variability of aBCSAF clinkers is much smaller with a typical phase assemblage of 55 wt% of α'-C_2S, 25 wt% of $C_4A_3\underline{S}$, 15 wt% of C_4AF and about 5 wt% of minor phases like CT. In order to highlight the compositional differences between CSA and aBCSAF clinkers, Fig. 18.1 displays the laboratory X-ray powder diffraction Rietveld plots for two archetypal samples. The selected commercial CSA clinker contains ~70 wt% of $C_4A_3\underline{S}$ and ~15 wt% of C_2S, meanwhile the laboratory-prepared aBCSAF clinker contains ~55 wt% of C_2S and ~30 wt% of $C_4A_3\underline{S}$.

To finish this section, we would like to summarize and clarify the expected uses of yeelimite-containing cements. The Al_2O_3 contents of CSA, BCSAA and BCSAF clinkers are variable but usually range between 30–40, 20–24 and 14–17 wt%, respectively. These values are much higher than that typical of an OPC clinker, ~5 wt%. The high demand of aluminum needed for CSA and BCSAA clinker fabrications makes these cements inappropriate for replacing OPC for large-scale inexpensive applications. Therefore, CSA cements (with prices ranging 250–350 €/ton of clinker) are being used for special applications but they are not intended to replace OPC (with prices close to 100 €/ton of clinker). However, the relatively lower aluminum content of belite-active BCSAF, ~15 wt%, coupled to the much lower CO_2 emissions (approximately 30–35% reduction) and higher gypsum-demand

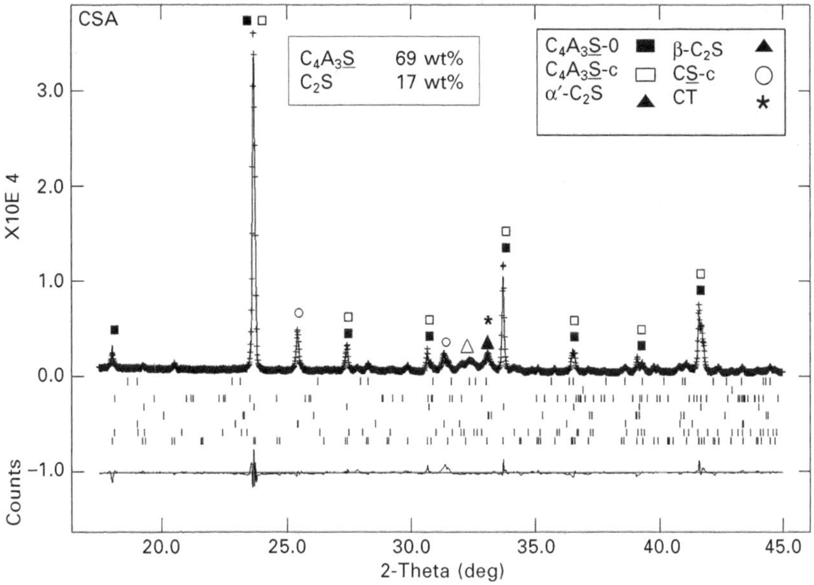

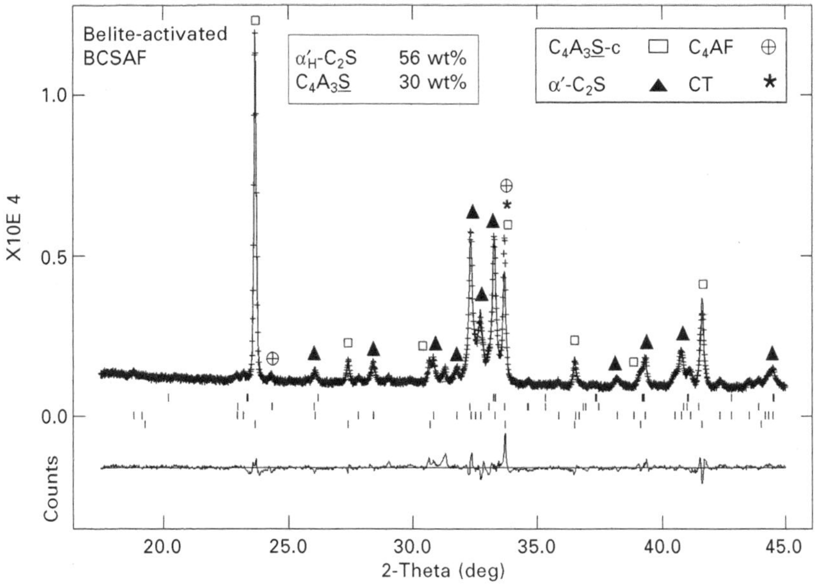

18.1 Selected range of the laboratory X-ray powder diffraction Rietveld plots for a typical CSA clinker (top) and a typical belite-activated BCSAF clinker (bottom). Crosses are the experimental scan, solid line is the calculated pattern and the line on the bottom is the difference curve. The major peaks for each phase are labeled as shown in the insets. Adapted from Álvarez-Pinazo et al. (2012).

(close to 10–15 wt%) make BCSAF a potential alternative to OPC with expected prices only slightly larger than those of OPC cements.

Very much related with the above issue, bauxite cost and scarcity (Scrivener and Kirkpatrick, 2008; Juenger et al., 2011; Schneider et al. 2011) are serious problems. Bauxite is necessary in large amounts for CSA and BCSAA fabrications and this will certainly prevent their applications for large-scale uses. However, for BCSAF the primary aluminum source may be aluminum silicates and bauxites or aluminum-rich wastes are used as correctors to reach the right Al_2O_3 content. Therefore, bauxite cost and availability are not that critical for BCSAF. Furthermore, the current value of $C_4A_3\underline{S}$ content, ~30 wt%, implies ~15–16 wt% of Al_2O_3 in the clinker. Research is ongoing in order to develop belite-active BCSAF with ~20 wt% of $C_4A_3\underline{S}$ which would lead to ~10 wt% of Al_2O_3 but cement performances must be maintained.

18.2.2 Laboratory-prepared yeelimite-containing cements

Mehta (1977) presented some compositions ranging ~20 wt% of $C_4A_3\underline{S}$, 25–45 wt% of C_2S, 15–40 wt% of C_4AF and 15–20 wt% of anhydrite. Those materials containing higher values of C_2S gained mechanical strengths up to 50 MPa at 28 days. Most laboratory studies were mainly devoted to revealing the relationship between mineralogical composition of clinkers and some final performances. This fact becomes even more relevant if waste materials were included in the raw mixtures for producing CSA and BCSAF clinkers (Beretka et al., 1993; Arjunan et al., 1999; Katsioti et al., 2006; Seluck et al., 2010). These studies concluded that the use of waste materials such as bag house dust, low-calcium fly ash, phosphogypsum or scrubber sludge successfully yielded to the preparation of CSA or BCSA clinkers at low temperatures, i.e. 1200–1300 °C. Moreover, final performances, i.e. setting times, consistency, expansion and mechanical strengths were similar to those obtained with the cements prepared using reagent grade raw materials.

Senff et al. (2011) presented the use of the design of experiments for the formulation of clinkers with mineralogical composition ranging 30–70 wt% of C_2S, 20–60 wt% $C_4A_3\underline{S}$ and 10–25 wt% C_4AF and mathematical models to predict properties. They concluded that mechanical strength at early ages strongly increases with higher amounts of $C_4A_3\underline{S}$, as expected. However, these authors do not deal with activation of BCSAF. aBCSAF materials were successfully prepared in the laboratory (Cuberos et al., 2010; Aranda et al., 2011; Álvarez-Pinazo et al., 2012). In these works, the main goal was the quantitative demonstration of α'_H-C_2S polymorph stabilization by borax addition to BCSAF clinkers, and that this polymorph hydrates much faster than β-C_2S polymorph.

Figure 18.2 gives Rietveld plots of two BCSAF, (a) without any activator

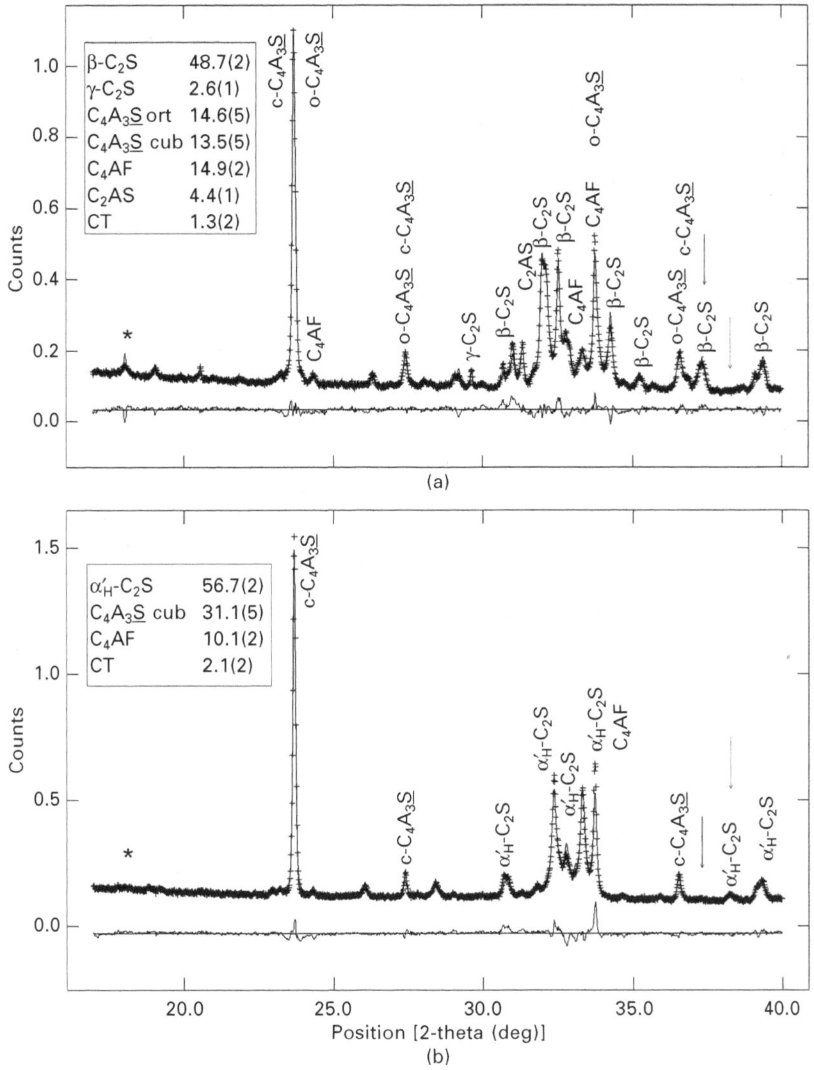

18.2 Selected range of the laboratory X-ray powder diffraction Rietveld plots for a typical BCSAF clinker without activation (a) and borax-activated BCSAF clinker (b). Crosses are the experimental scan, solid line is the calculated pattern and the line on the bottom is the difference curve. The Rietveld quantitative phase analysis results are shown in the insets. The major peaks for each phase are also labeled. Solid and dotted arrows highlight the presence of β-C_2S and α'_H-C_2S, respectively. Asterisk indicates the angular range indicative of cubic or orthorhombic $C_4A_3\underline{S}$. Adapted from Álvarez-Pinazo et al. (2012).

added and (b) with 2.0 wt% of B_2O_3 added as borax. This figure also includes Rietveld quantitative phase analyses, showing that the activation was achieved as BCSAF with 2.0 wt% of B_2O_3 only contained α'_H-C_2S polymorph. It has to be highlighted that the effectiveness of addition of B_2O_3 with no other co-dopants for stabilizing α' forms is poor (Wesselsky and Jensen, 2009; Morin et al., 2011; Cuesta et al., 2012), thus it is mandatory to add additional foreign elements such as sodium. On the other hand, the addition of minor elements in raw mixtures may also affect to $C_4A_3\underline{S}$ polymorphism. This phase presents two polymorphs, cubic (Saalfeld and Depmeier, 1972) and orthorhombic (Calos et al., 1995). BCSAF clinker without activation presents a mixture of both polymorphs, meanwhile the activation promoted the stabilization of the cubic form, asterisk in Fig. 18.2 (Álvarez-Pinazo et al., 2012). The influence on hydration of the polymorphism of $C_4A_3\underline{S}$ is currently under study.

ACSA cements have also been investigated with the same aim of enhancing final performance of clinkers with 20–30 wt% of $C_4A_3\underline{S}$. However, this clinker preparation seems difficult because the optimum temperatures for the synthesis of alite and yeelimite differ considerably. While C_3S needs at least 1350 °C to be formed and the reaction between free lime and C_2S is promoted by the apparition of melting phases (De la Torre et al., 2007), yeelimite phase decomposition takes place above 1350 °C. Nevertheless, the addition of a small amount of CaF_2 to raw mixture allows the coexistence of both phases at temperatures between 1250 and 1300 °C (J. Li et al., 2007; Yanjun et al., 2007). Other authors have prepared ACSA clinkers by jointly adding fluorite and other minor elements such as Cu, Ti, Mn, Ba or Mg (Liu and Li, 2005; Lingchao et al., 2005; Ma et al., 2006; Liu et al., 2009; Lili et al., 2009). Such clinkers exhibit a significantly improved grindability compared with OPC clinker. Mechanical strengths developed by these cements strongly depend on the amount of added gypsum.

18.3 Calcium sulfoaluminate clinkering

Yeelimite-containing clinkers are usually produced by mixing limestone, clay and/or bauxite and gypsum, as sources of calcium, silicon/aluminum and sulfur, respectively. Silica-rich bauxites and iron ores are used for BCSAF preparations. The availability of waste materials containing high amounts of alumina (such as fly ashes and aluminum anodization muds) or sulfates (such as phosphogypsum and flue gas desulfurized gypsum) is a key issue for the economic production of CSA cements (Singh and Pradip, 2008; Pace et al., 2011). Even aluminum-poor wastes, such as municipal solid waste incineration fly ash, can be used as raw materials for CSA preparations (Wu et al., 2011). CSA clinkers are prepared in rotary kilns at temperatures of 1250–1350 °C. Clinkering reactions slightly depends on the composition of raw mixture. Research in BCSAF clinker preparation has increased in the last decade (Sahu and Majling, 1994; Arjunan et al, 1999; De la Torre et al., 2011a, 2011b; Chen et al., 2011). These studies concluded that up to 900 °C H_2O and CO_2 are released into atmosphere due to dehydration, dehydroxylation and decarbonation of raw material. Below this temperature, phases arising from the decomposition of the raw materials, such as free lime, β-SiO_2 or anhydrite coexist with a bad-crystallized/amorphous aluminum-rich phase.

Figure 18.3 shows a selected range of synchrotron X-ray powder diffraction (SXRPD) raw data (λ = 0.30 Å) at various temperatures from an *in situ* clinkering study of an active BCSAF clinker. The diffraction peaks, arising from different phases, are labeled in order to highlight the transformations that are taking place on heating. Solid state formation of yeelimite has been proved to start over ~1000 °C; open circle in Fig. 18.3. Recent reports (De la Torre et al., 2011a, 2011b) have presented clinkering reactions of BCSAF clinkers with and without activators. In these studies, high-resolution SXRPD was used to follow chemical reactions with temperature. BCSAF clinkers with 50 wt% C_2S, 30 wt% of $C_4A_3\underline{S}$ and 20 wt% of C_4AF expected mineralogical phase assemblage were prepared *in situ* by mixing ~61 wt% of calcite, ~25 wt% of kaolin, ~4 wt% of alumina, ~6 wt% of gypsum and

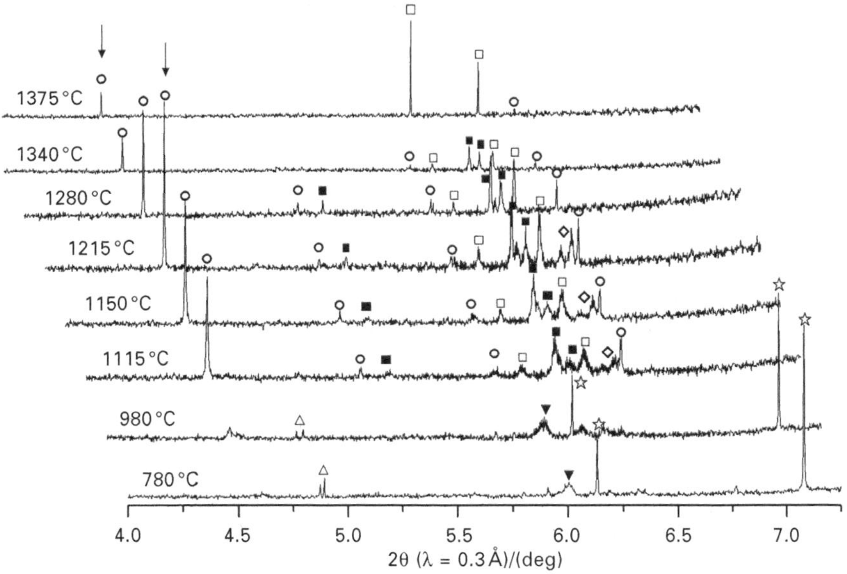

18.3 3D-view of a selected range of the synchrotron X-ray powder diffraction raw patterns for a BCSAF clinker formation collected on heating, where the symbols highlight CaO (star), β-SiO_2 (solid triangle), $CaSO_4$ (open triangle), $C_4A_3\underline{S}$ (open circle), α'_H-C_2S (solid square), C_4AF (rhombus) and α-C_2S (open square). Adapted from De la Torre et al. (2011a).

~4 wt% iron oxide. These authors concluded that as temperature increases, free lime reacts with anhydrite and amorphous aluminum/iron-rich phase to form $C_4A_3\underline{S}$ and C_4AF. Belite is also formed below 900 °C, being the α'_H-C_2S the polymorph identified at that temperature. In those experimental conditions, free lime had completely reacted at ~1100 °C; star in Fig. 18.3. The system CaO–SiO_2–Al_2O_3–Fe_2O_3–SO_3 contains 18 phases (Sahu and Majlin, 1993) related to BCSAF cements. However, it is possible to control final phase assemblage and to avoid the formation of some non-hydraulic phases such as C_2AS or $C_5S_2\underline{S}$ (Li et al., 2001). C_2AS phase is decomposed to give belite at high temperature, i.e. over 1100 °C. However if the amount of CaO concentration is not enough to react with SiO_2 to give C_2S, C_2AS may still may be present in final clinker.

Clinkering temperature of BCSAF is one important factor affecting final phase assemblage. This is due to the phase stability field of yeelimite phase at high temperature. It has been demonstrated that a certain temperature phase assemblage is closed to the targeted one, for instance, at 1215 °C mineralogical composition of the mix was 49.1(5) wt% of C_2S, 28.1(5) wt% of $C_4A_3\underline{S}$ and 22.8(4) wt% of C_4AF determined by analyzing SXRPD patterns by the Rietveld method, the target composition being 50 wt% of C_2S, 30 wt% of $C_4A_3\underline{S}$ and 20 wt% of C_4AF. It has to be borne in mind that the presence of minor elements such as alkaline oxide or boron in raw mixes will decrease clinkering temperatures, as well as melting points of all the phases. For instance, the addition of 2 wt% of borax to the raw materials decreases the minimum clinkering temperature, i.e. the temperature at which phase assemblage match the expected one, from 1300 °C down to 1150 °C (De la Torre et al., 2011a, 2011b). There have been some studies concerning the thermal formation and stability of this phase (Ali et al., 1994; Puertas et al., 1995; De la Torre et al., 2011b). These studies used as raw material reagent grade $CaCO_3$, $CaSO_4$ and Al_2O_3, thus mechanism of formation may be slightly different from that occurring when using calcite, kaolin and gypsum. After decarbonation, above 900 °C, some calcium aluminates, mainly CA or $C_{12}A_7$ are formed as intermediate phases and finally they react with $C\underline{S}$ to form yeelimite. At high temperatures, $C_4A_3\underline{S}$ melts/decomposes. The arrows in Fig. 18.3 highlight the diminution on the absolute integrated intensities of yeelimite diffraction peaks above 1300 °C due to melting and/or decomposition of this phase.

The addition of foreign elements also promotes the decrease of polymorphic transformation temperatures. For BCSAF clinkers the α'_H-$C_2S \rightarrow \alpha$-C_2S polymorphic transformation is relevant. The addition of 2 wt% of B_2O_3 lowers the temperature of this transformation from ~1370 °C to ~1000 °C, and this effect is important for the stabilization of high temperature forms of C_2S.

An industrial trial to produce about 5500 tons of BCSAF (AETHER™

from Lafarge) was carried out in early 2011 (Walenta and Comparet, 2011). Clinkering in a Portland industrial kiln was proven with lower operating temperatures (1225–1300 °C). Furthermore, the reduction of CO_2 emissions of 25–30% for BCSAF cement was also proven when compared with the emissions of CEM (I) type OPC cement. Temperature in the clinkering zone was a key parameter as too low temperatures gave under burnt binder with high free lime and $C_{12}A_7$ contents, and too high temperatures may give kiln blockage, loss of grindability and $C_4A_3\underline{S}$ decomposition with high SO_2 emissions. However, optimum clinkering temperature resulted in NO_x emissions being much lower than those of OPC fabrication, and SO_2 measured industrial emissions at the same level than for OPC production.

18.4 Hydration of calcium sulfoaluminate cements

In previous sections, yeelimite-containing cements have been mainly classified as CSA or BCSAF. However, from the point of view of hydration the first process is common in both types of binder. Once water is added to cement, the first effect that occurs is the wetting of the cement and the start of yeelimite dissolution. However, only part of this exothermic process is usually recorded in a calorimeter owing to the time required for the mixing procedure and stabilization time, point (1) in Fig. 18.4. Immediately after wetting, the following reaction takes place:

$$C_4A_3\underline{S} + 2\ C\underline{S}H_2 + 34\ H \rightarrow C_6A\underline{S}_3H_{32} + 2\ AH_3 \qquad [18.1]$$

Chemical reaction [18.1] corresponds to the formation of ettringite ($C_6A\underline{S}_3H_{32}$), also known as AFt, by the reaction of $C_4A_3\underline{S}$ with soluble calcium sulfate and water. This reaction also takes place with anhydrite or bassanite although at different rates. Formation of ettringite will continue while calcium sulfates are present; signal (2) in Fig. 18.4. If there is not enough sulfates to react with yeelimite phase, reaction [18.2] will start:

$$C_4A_3\underline{S} + (16 + x)\ H \rightarrow C_4A\underline{S}H_{(10+x)} + 2AH_3 \qquad x = 0,2,4 \qquad [18.2]$$

Reaction [18.2] corresponds to the hydration of yeelimite to form an AFm-type phase with different water contents, depending on the time of hydration, water-to-cement W/C ratio and initial cement phase assemblage. This is also an exothermic reaction which takes place once calcium sulfates are consumed, signal (3) in Fig. 18.4 (Odler, 2000). These two reactions may take place in the early hydration of CSA and BCSAF cements. This is shown in Fig. 18.5 where cryo-scanning electron microscopy photographs of BCSAF pastes are shown. Pastes from BCSAF cements with 20 wt% of gypsum showed a lot of acicular-hexagonal AFt crystals even at very early hydration ages, reaction [18.1]. Conversely, pastes from BCSAF cements with (only) 5 wt%

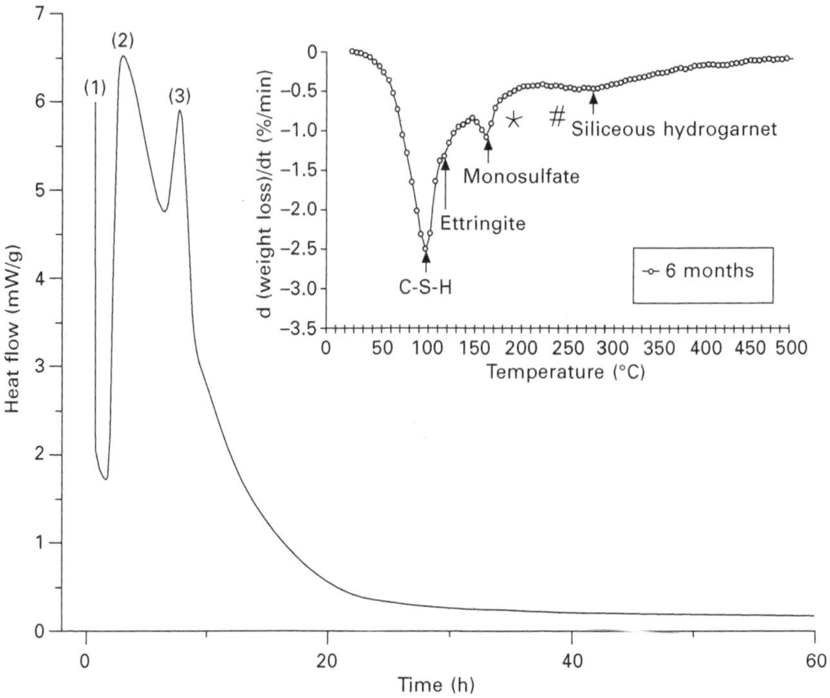

18.4 Heat flow curve for a BCSAF cement with ~45 wt% of C_2S, ~27 wt% of $C_4A_3\underline{S}$, ~18 wt% of C_4AF and ~10 wt% of gypsum (w/c=0.5). Inset shows ATD curves of a similar BCSAF. Adapted from Morin et al. (2011).

of gypsum showed, in addition to AFt crystals, thin-layered AFm crystals, due to reaction [18.2].

It must be kept in mind that there are several types of CSA and BCSA cements. Thus it is not possible to describe every reported hydration reaction. However, the most important reactions of most common minor phases are detailed below. Minor phases also react with water or calcium sulfate and somewhat contribute to rapid hardening, according to the following reactions (Meller et al., 2004; Bessy, 1938):

$$C_4AF + 3C\underline{S}H_2 + 30H \rightarrow C_6A\underline{S}_3H_{32} + FH_3 + CH \qquad [18.3]$$

$$C_{12}A_7 + 12C\underline{S}H_2 + 113H \rightarrow 4\ C_6A\underline{S}_3H_{32} + 3AH_3 \qquad [18.4]$$

$$3CA + 3C\underline{S}H_2 + 32H \rightarrow C_6A\underline{S}_3H_{32} + 2AH_3 \qquad [18.5]$$

All these reactions yield AFt and take place during the first hours of hydration, thus signal (2) in Fig. 18.4 is also due to the heat release by these reactions, if these minor phases are present. Depending on the availability of calcium sulfates again, there are alternative reactions for the direct hydration of these phases:

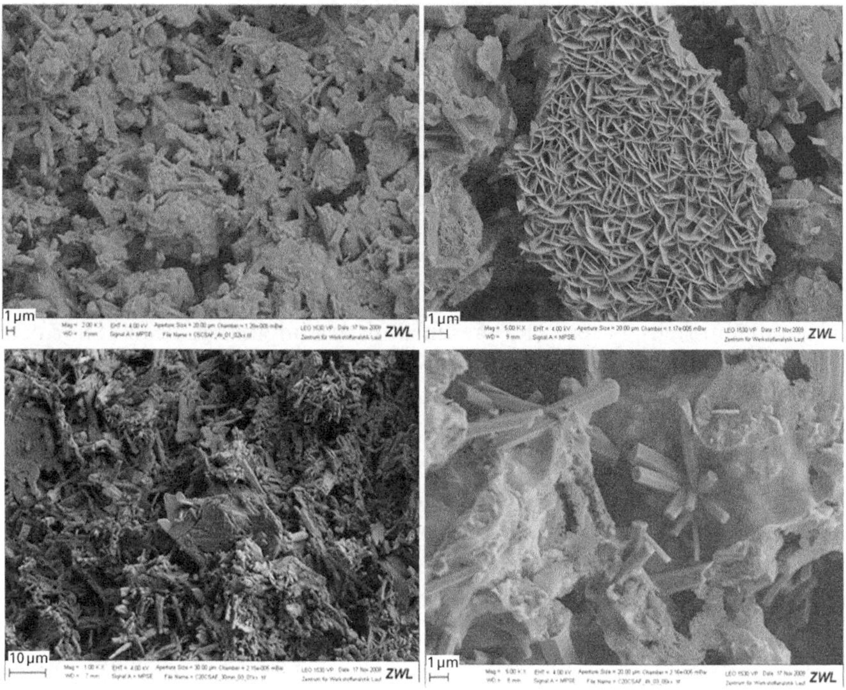

18.5 Cryo-scanning electron microscopy study for BCSAF pastes. Top images are taken for a BCSAF-5wt% $C\underline{S}H_2$ paste after 260 min of hydration: (left) photograph showing acicular-hexagonal AFt crystals; (right) photograph showing thin-layered AFm crystals. Bottom images are taken for a BCSAF-20wt% $C\underline{S}H_2$ paste: (left) photograph after 30 min of hydration showing already a lot of ettringite; (right) photograph after 260 min of hydration showing well-developed hexagonal AFt crystals.

$$C_4AF + 16H \rightarrow 2C_2(A_{0.5}F_{0.5})H_8 \qquad [18.6]$$

$$C_{12}A_7 + 51H \rightarrow 6C_2AH_8 + AH_3 \qquad [18.7]$$

$$2CA + 11H \rightarrow C_2AH_8 + AH_3 \qquad [18.8]$$

Most of these reactions also yield aluminum or iron hydroxides (AH_3 or FH_3) which are initially amorphous as they cannot be directly detected by X-ray powder diffraction (XRPD), but its occurrence may be confirmed by thermal analysis techniques (Cuberos et al., 2010; Martín-Sedeño et al., 2010; Morin et al., 2011). Later, these aluminum-rich amorphous phases may crystallize as gibbsite.

In BCSAF, hydration of C_2S is of special interest. Independently of the rate of reaction of α- and β-polymorphs, the hydration reaction would be the same as that occurring in OPC:

$$C_2S + (x + 2 - y)\ H \rightarrow C_ySH_x + (2 - y)\ CH \quad [18.9]$$

Reaction [18.9] corresponds to the direct hydration of belite to form the amorphous gel C_ySH_x and portlandite. However in BCSA pastes, belite coexists with aluminum-rich amorphous hydrates, promoting the formation of stratlingite (Palou et al., 2005; Cuberos et al., 2010; Gartner and Macphee, 2011):

$$C_2S + AH_3 + 5H \rightarrow C_2ASH_8 \quad [18.10]$$

This reaction consumes the amorphous AH_3 formed by the hydration reactions of aluminum-rich phases. The presence of stratlingite has been confirmed in BCSAF pastes by XRPD and ATD techniques (Cuberos et al., 2010; Aranda et al., 2011; Morin et al., 2011). This reaction may play a key role in the increase of early-ages mechanical strengths, and it is produced at a larger pace if high temperature polymorphs of belite are stabilized. Reaction [18.9] may take place at later ages, if aluminum hydroxide is depleted. The inset in Fig. 18.4 shows ATD signals of a BCSAF paste hydrated for 6 months (Morin et al., 2011). These data confirm the presence of CSH gel and the # symbol highlights the absence of the endothermic effect of AH_3 decomposition.

There are other reactions which have to be taken into account. Some studies have confirmed the presence of katoite phases, also known as siliceous hydrogarnet, with C_3ASH_4 being the stoichiometry of one member of the series, $C_3(A,F)S_xH_{6-2x}$ (iron-free, $x = 1$). A/F ratio is still unknown in these katoites and x is probably not greater than 1 (Gartner and Macphee, 2011). The proposed formation mechanism is detailed below:

$$C_2S + 1/2C_4AF + 5H \rightarrow C_3(A,F)SH_4 + CH \quad [18.11]$$

Reaction [18.11] justifies the consumption of ferrite and releases to solution portlandite, which is not detected by XRD or ATD. Consequently, this portlandite may be consumed by stratlingite as follows, producing larger quantities of katoite:

$$C_2ASH_8 + CH \rightarrow C_3ASH_4 \quad [18.12]$$

The inset in Fig. 18.4 shows the endothermic effect of siliceous hydrogarnet or katoite decomposition and the asterisk symbol is highlighting the absence of stratlingite after 6 months of hydration (Morin et al., 2011). Katoite formation by reaction [18.12] may be of interest for the durability of mortars and concretes based on BCSAF as portlandite is consumed. Finally, the effect of temperature on the hydration of CSA cements at very early ages has also been studied (Zhang and Glasser, 2002).

The understanding of the hydration reactions is key for tailoring the properties of CSA and BCSAF mortars and concretes, but the microstructure development is also of the utmost importance. Very recently, the early hydration (less than a day) of three cementing materials (OPC, CSA and an OPC/CSA blend) has been investigated *in situ* by synchrotron X-ray tomographic microscopy (Gastaldi et al., 2012). The reaction kinetics and the morphological evolution of mineral phases of the three investigated cements were followed, but chiefly, it was possible to determine the three-dimensional evolution of the pore networks with a spatial resolution close to a 1 μm.

The setting times of CSA and BCSAF cements are shorter than those of OPCs and depend mainly on their yeelimite content and the amount and reactivity of the added calcium sulfate. The reported typical values range between 30 min and 4 h (Zhang et al., 1999; Glasser and Zhang, 2001; G. S. Li et al., 2007; Marchi and Costa, 2011). Furthermore, the setting time can be adapted with the use of retarders (Zhang and Glasser, 2005) and also by the interaction with the filler (Pelletier-Chaignat et al., 2012).

18.4.1 Role of sulfate content

The amount and the reactivity of calcium sulfate added to CSA or BCSAF clinkers plays a key role in controlling the hydration of yeelimite and the final performances of pastes/mortars/concretes (Glasser and Zhang, 2001; Pera and Ambroise, 2004). The solubility of the calcium sulfate added to the clinker influences the ratio of formation of the main hydrate phases, i.e. AFt and AFm (Sahu et al., 1991; Winnefeld and Barlag, 2009; Winnefeld and Barlag, 2010). Increasing amounts of calcium sulfate provides the calcium and sulfate ions necessary for the formation of higher amounts of ettringite (see Fig. 18.5). This may lead to an expansive behavior of the mixture due to the formation of AFt within the hardened matrix. This risk is of especial relevance when hydrating CSA, as these cements may contain very large amounts of $C_4A_3\underline{S}$ (Chen et al., 2012). On the other hand, lower amounts of calcium sulfate promote the formation of AFm phases, once sulfate ions have depleted (see Fig. 18.5), resulting in reduced fluidity and increased water demand and may lead to premature setting.

Other important issue is the type of calcium sulfate carrier added to CSA or BCSAF clinker: gypsum, bassanite or anhydrite. Setting behavior can be controlled by the use of a reactive calcium sulfate like bassanite or gypsum which present much higher dissolution rate than anhydrite. Early strength properties are also enhanced by using gypsum (over anhydrite) as sulfate carrier (Péra et al., 2003; Berger et al., 2011a; Marchi and Costa, 2011; Pelletier-Chaignat et al., 2011). On the other hand, low-reactive calcium sulfate, like some anhydrites, causes a lack of calcium and sulfate ions in the pore solution, leading to a delay in ettringite formation and strength development. The slow dissolution kinetic of anhydrite favors the formation of well-crystallized ettringite particles that leads to a rapid paste, plasticity loss, i.e. lower setting time, and to the formation of a less dense paste lowering the early-age compressive strengths (Marchi and Costa, 2011) but higher 28 day compressive strengths (Winnefeld and Barlag, 2009). On the other hand, CSA has also been used to improve the water resistance of plaster of Paris as it counteracts the high solubility of bassanite (Kuryatnyk et al., 2010).

18.4.2 Role of W/C ratio

The W/C ratio affects the microstructure of pastes/mortars/concretes, providing water to hydrate phases and the porosity (space) where hydration products can precipitate. The ratio between porosity and hydration degree strongly affects final concrete performance. At low W/C ratio, CSA or BCSAF will develop a denser pore structure, as space available for hydration products formed is smaller. Moreover, pastes with low W/C ratio can undergo self-desiccation, as ettringite formation requires huge amounts of water. This effect can be critical for expansion properties as large amounts of cement particles remain unhydrated after setting. This can cause expansion if cement is later exposed to external water from the environment, by the formation of secondary ettringite from the reactivity of anhydrous phases (Beretka et al., 1996). However, very dense frameworks make water diffusion to the interior of the mortars and concretes difficult. On the other hand, the use of a high W/C ratio makes cements dimensionally stable, even with high amounts of yeelimite because the water in the system is enough for $C_4A_3\underline{S}$ to fully react at early ages, but the microstructure of the paste/mortars is more porous.

Porosity of pastes is directly affected by W/C ratio. Low W/C ratios lead to low porosities, enhancing mechanical strengths, although in the first hours of hydration there is a loss of plasticity. High W/C ratios lead to high porous microstructures which may result in lower mechanical strengths (Berteka et al., 1996; Bernardo et al., 2006; Marchi and Costa, 2011). It should be underlined that very different W/C ratios have been used in the hydration reactions of CSA. Reaction [18.1] theoretically requires a water-to-binder mass ratio of 0.64. However, this water-to-binder mass ratio also depends on the minor phases as well as the final performances of the paste/mortar/concrete to be produced.

It must be noted that due to the large compositional variability of CSA cements, one general theoretical water demand for full hydration cannot be established. However, for a given CSA clinker phase assemblage and sulfate content, a water-to-binder mass ratio for full hydration can be calculated. This value may not be very far from 0.60. From the experimental point of view quite different water-to-binder mass ratios have been employed to study the hydration of CSA cements. These values range from quite low ratios 0.35–0.45 (Zivica, 2000, 2001; Fu et al., 2003; Canonico et al., 2007) to quite large ones 0.70–0.80 (Winnefeld and Barlag, 2009; Lura et al., 2010; Winnefeld and Lothenbach, 2010; Winnefeld et al., 2011). However, the most frequently used ratios range from 0.5 to 0.6 (Bernardo et al., 2006; Valenti et al., 2007; Gastaldi et al., 2007; Alesiani et al., 2008; Telesca et al., 2011; Buzzi et al., 2011). Finally, there are studies varying this critical ratio. For instance, Chen et al. (2012) studied water-to-binder ratios of 0.30, 0.45 and 0.70.

On the other hand, the variability of the water-to-cement mass ratio in BCSAF cements is much smaller. Most studies used W/C = 0.5 with more variability in the amount of sulfate carrier. For instance, W/C = 0.5 and 10 wt% of $C\underline{S}$ was initially reported (Adolfsson et al., 2007; Janotka et al., 2007). G. S. Li et al. (2007) used W/C = 0.5, and 8 and 12 wt% of $C\underline{S}H_2$, for mortars and pastes, respectively. Recently, Morin et al. (2011) studied the hydration reactions of AETHER^{TM} using a W/C mass ratio of 0.5 and 5.3 wt% of $C\underline{S}$. Aranda et al. (2011) also used a W/C ratio of 0.5 but adding 10 wt% of $C\underline{S}H_2$. Finally, Juenger and Chen (2011) used W/C = 0.45 and variable amounts of sulfate carriers.

18.4.3 Admixtures

To the best of our knowledge, there is little work on the role of additions like blast furnace slag or pulverized fly ash to partly replace CSA and BCSAF cements. We can highlight two works blending yeelimite-containing cements with mineral admixtures like fly ash, silica fume, and blast furnace slag to study pozzolanic reactions (Zivica, 2000, 2001). Unfortunately, the water-to-cement mass ratio used was quite small (ranging between 0.32 and 0.41), hence, full hydration of the pastes did not take place. On the other hand, the beneficial effects of limestone filler with CSA cement have been very recently reported (Pelletier-Chaignat et al., 2012). The hydration reactions have been studied for a fixed W/C mass ratio of 0.8, but two $C\underline{S}H_2$/CSA mass ratios (0.19 and 0.40) and two temperatures (5 and 20 °C) were tested. The compressive strengths were higher for limestone than for the same level of substitution with an inert filler (quartz in this study).

There is also a situation where the waste to be safely disposed within concrete may also help as an addition which influences the hydration reaction. Municipal solid wastes were blended with CSA and the compressive strength, pore structure, hydration phases, and leaching behavior of Zn and Pb were characterized (Qian et al., 2008).

18.4.4 Uses of CSA cements to stabilize wastes

Intermediate-level and low-level radioactive wastes are usually encapsulated in blended OPC cements containing high amounts of blast furnace slag or pulverized fly ash. Unfortunately, high alkalinity of these systems may lead to high corrosion rates with reactive metals within such wastes releasing hydrogen and forming expansive corrosion products. CSA cements are being tested as alternative encapsulation binders mainly due to their highly dense matrixes and lower alkalinities (Zhou et al., 2006). Hayes and Godfrey (2007) used commercial CSA (Rockfast, see Table 18.3) which theoretically would need 37 wt% of gypsum and 0.54 water-to-cement mass ratio for full hydration. These authors varied the water-to-binder mass ratio from 0.50 to 0.65, and different gypsum and pulverized fly ash contents with very encouraging results. Finally, OPC/CSA blend cements have also been tested, since the setting time is shortened which is useful for nuclear waste encapsulation (Cau Dit Coumes et al., 2009).

CSA cements have been also investigated to solidify and stabilize different type of wastes. We can underline three examples: (i) wastes containing large amounts of soluble zinc chloride, which is a strong inhibitor of OPC cement hydration (Berger et al., 2011b, 2011c); (ii) galvanic sludge with high amounts of chromium which was stabilized in blended CSA–bottom ash cement (Luz et al., 2006, 2009); and (iii) heavy metals in general, Cr, Pb, Zn, Cd (Peysson et al., 2005). The valorization of phosphogypsum (a side-product in the fabrication process of phosphoric acid) as a binder by adding CSA has also been studied (Kuryatnyk et al., 2008).

18.4.5 Plastic/rheological properties of mortars and concretes

High mechanical strength, dimensional stability and durability are the most important final performances of cement-based materials. In order to achieve these performances, the control of the rheological properties of the fresh products is needed. Rheology of fresh cement is often characterized by its workability. The workability of cement pastes is a performance index representing how easily it is mixed or transported, and also showing the ability of filling voids. Compactibility, finishability and resistance against segregation are also important parameters (Hanehara and Yamada, 2008). In yeelimite-containing systems, the first hydration hour(s) are very important as large amounts of ettringite are formed and plasticity/workability strongly depends on this reaction. There are few studies concerning how to improve workability of CSA cements. Additives to enhance workability will act by slowing down the hydration rate and dispersing CSA particles over a longer time to enlarge setting time (Chang et al., 2009). Moreover, CSA pastes with additives may present excellent dispersibility at low W/C ratios and, consequently, with improved mechanical performances (García-Maté et al., 2012). Accelerating admixtures may also affect, in addition to the setting time, to the compressive strength development and to the rheological properties.

Some initial studies on the role of some accelerators for CSA have been undertaken (Paglia et al., 2001).

Cements with very large amounts of yeelimite may present shrinkage problems, and this effect is strongly correlated with porosity. In order to prevent water evaporation causing drying shrinkage, it is advisable to decrease the surface tension of water in paste pores, thereby lowering the capillary tension within the pore structure. As discussed above, CSA cements require larger amounts of water due to $C_4A_3\underline{S}$ hydration, thus evaporation of water is also facilitated. The use of some shrinkage-retardant admixtures will modify pore structure in order to prevent the release of water (Ambroise et al., 2009).

18.4.6 Performances of blended CSA/OPC cements

Blending cements may enhance performances and achieve new properties, and moreover valorize waste materials. OPC with CSA cement as additive are classified as type K cements which present expansive properties (Klein, 1966). The fundamental behavior of these cements is based on the following reaction:

$$C_4A_3\underline{S} + 8C\underline{S}H_2 + 6CH + 74H \rightarrow 3C_6A\underline{S}_3H_{32} \qquad [18.13]$$

In reaction [18.13], yeelimite is combined with gypsum and portlandite to yield ettringite. Portlandite is released to the system mainly by C_3S hydration (from the OPC). The presence of portlandite in the system makes the reaction pathway [18.13] more favorable than [18.1]. Consequently, each unit of $C_4A_3\underline{S}$ would produce three moles of ettringite instead of one, resulting in a larger potential for expansion (Kurdowski and Thiel, 1981). Moreover, the crystal size of ettringite is also affected by the presence of portlandite in the hydrating environment. In fact, ettringite crystals coming from reaction [18.13] are significantly smaller than those produced by the hydration of yeelimite with gypsum, reaction [18.1], causing a different microstructure of the hydrating paste (Mehta, 1973; Gastaldi et al., 2011). CSA/OPC (80/20 mass ratio) blends with limestone filler and superplasticizer were studied for self-leveling applications and glass-fiber-reinforced composites (Pera and Ambroise, 2004). Finally, CSA has also been used to improve the strength of high-grade slag cement composed of granulated blast-furnace slag, OPC and anhydrite. The only studied parameter was the gypsum ratio which led to different hydrates (Michel et al., 2011).

BCSAF cement performances may also be enhanced by its blending with OPC (Janotka and Krajci, 1999; Janotka et al., 2003; Pelletier et al., 2010, Pelletier-Chaignat et al., 2011). BCSAF cements may present very short initial setting times, however, by blending with 15 wt% of OPC setting time is sufficiently enlarged for mortar-making technological procedures.

Moreover, mortars with this blending develop higher mechanical strengths than (non-activated) BCSAF mortars.

18.4.7 Alkalinity of calcium sulfoaluminate pastes, mortars and concretes

It must be remembered that the critical pH value for steel reinforcement passivation layer is slightly above 11.5. Below this pH level, the γ-Fe_2O_3 protective oxide film on the steel surface starts to decompose. Therefore, the alkalinity nature of the pastes from yeelimite-containing cements is a key issue. Furthermore, we have to distinguish between alkalinity from CSA and aBCSAF cements as their mineralogical phase assemblages are quite different.

Data on the pore solution composition and alkalinity of CSA cements is very scarce in literature. Moreover, the large compositional variability in CSA cements (see Table 18.3), may lead to large differences in pore solution properties. A full data set at realistic W/C ratios was reported for a CSA clinker (Andac and Glasser, 1999). Pore solution chemistry was determined at hydration times from 1 to 60 days at W/C ratios between 0.5 and 0.8. These authors found a rapid release of soluble alkalis from the clinker accompanied by a high pH, ~13. Recently, Winnefeld and Lothenbach (2010) reported a very comprehensive pore solution chemistry and alkalinity study of the hydration of two CSA cements, at relevant conditions: W/C ratios of 0.72 and 0.80 and calcium sulfate contents of ~20 wt% gypsum and anhydrite, respectively. The pore solution of both cements is dominated during the first hours of hydration by potassium, sodium, calcium, aluminum and sulfate; the pH values being relatively low, between 10.5 and 11.0. After approximately one day, when the calcium sulfate is depleted, the sulfate concentration drops by a factor of 10, and pH is increased to maintain electroneutrality to close to 11.8. At 28 days of hydration, both cements gave much higher pH values, 12.7 and 12.8.

As shown above, alkalinity of CSA cements is about 1 pH unit lower than that of Portland cements at early ages. However, the steel reinforcement seems to be protected from corrosion as derived from the measured pH values after one day of hydration, see above, and as measured experimentally after 14 years of service (Glasser and Zhang, 2001). Slightly lower alkalinity also seems to be favorable for the alkali aggregate reaction (Zhang et al., 1999). On the other hand, non-active BCSAF mortars may present limited ability for protection against steel corrosion, due to their lower pH values (Kalogridis et al., 2000). These authors used a W/C ratio of 0.45 for laboratory-prepared BCSAF cements where anhydrite was not added but formed during clinkering. The measured pH values were close to 6 but these values have not been reproduced by other authors. Janotka et al. (2003) reported pH values of

10.3–10.5, which were insufficient for full steel passivation. However, the same authors prepared blended BCSAF (with 15 wt% of OPC) which showed comparable steel protection to pure OPC. In any case, the studied BCSAF clinkers were far from optimized with Al_2O_3 contents close to 7.3 wt% and so, $C_4A_3\underline{S}$ contents as low as 5–7 wt%.

Pore solution data for aBCSAF cements has been reported (G. S. Li et al., 2007). However, these data were determined at a very high W/C ratio, 20. Elemental concentration evolutions within the first 24 hours were reported (Al, Ca, S and Na + K) but pH values were not given. Passivation studies of aBCSAF mortars are lacking, but the high reactivity of belite may slightly increase the pH of the free water which is key for its technological success. Much more research is needed concerning the alkalinity of aBCSAF mortars and their potential for steel passivation.

18.5 Durability of calcium sulfoaluminate concretes

Properly produced mortars and concretes from yeelimite-containing cements are durable binders. However, depending in the environment in which these mortars or concretes are serving, degradation processes should be taken into account. Degradation may occur through a variety of chemical processes, i.e. carbonation, sulfate attack or chlorine diffusion. All these effects need ion diffusion through porous microstructure. The CSA or BCSAF cement hydration process consumes the available water in a short time, which stops the evolution of total porosity. Moreover, the large amount of hydration products generated at very early stages decreases the interior pore space (Bernardo et al., 2006; García-Maté et al., 2012), consequently, resistance to diffusion is enhanced. A durability study with two commercial CSA cements (W/C mass ratio of 0.56) was reported and the characterization techniques included carbonation (and accelerated carbonation), sulfate resistance, chloride ingress, in addition of compressive strength and expansion (Quillin, 2001). These concretes showed excellent sulfate resistance and slightly higher carbonation rates than OPC concretes.

Atmospheric carbon dioxide is able to react with hydration products by dissolution in the pore solution of the pastes, and so increasing CO_3^{2-} ion concentration. This phenomenon will cause three effects: (i) lowering of pH value, (ii) precipitation of $CaCO_3$ by the reaction of carbonate ions with Ca^{2+} ions and (iii) possible deterioration of ettringite. As mentioned before, porosity (or W/C ratio) determines the degree of resistance to carbonation. Zhang and Glasser (2005) reported a study about two Chinese concretes: (i) a normal strength concrete sample from a crane column made during the cold winter of 1982 with hot mix water with W/C = 0.55–0.60 and by adding 1 wt% of $NaNO_2$ as anti-freeze; and (ii) a high-strength concrete sample from a pile made in summer 1993 with W/C ~ 0.3 and by adding unstated superplasticizer and retarder.

The normal strength concrete was carbonated in a comparable rate to OPC concretes and ettringite persisted after more than 20 years of service. On the other hand, decreasing W/C ratio (high-strength concrete) increases resistance to carbonation as there is no free water available to dissolve CO_2. Other authors (Sharp et al., 1999) stated that carbonation appears to be more rapid than in Portland cement concretes, leading to partial decomposition of ettringite, which may cause a moderate strength loss. Summarizing, carbonation is not an important problem for CSA concretes if low W/C ratios are employed (Zhang and Glasser, 2005). However, carbonation studies for aBCSAF pastes/concretes are lacking.

Sulfate diffusion into mortar matrixes may cause expansion due to delay ettringite formation. Sulfate ions may diffuse and react with calcium ions, yielding gypsum precipitation. However, this gypsum requires unreacted yeelimite and water to produce ettringite, and the latter is the limiting reactant. After some days of hydration water may be exhausted and unless there is another deterioration mechanism, such as leaching (Berger et al., 2011d), expansion due to delayed ettringite formation is not significant (Glasser and Zhang, 2001). Moreover, BCSAF cements were shown to be excellent sulfate-resistant binders due to the absence of C_3A in the phase assemblage.

18.6 Future trends

CSA cements have been proved to be suitable for tailored properties by adjusting the sulfate carrier, the water-to-cement ratio and by blending. New applications will emerge in specific niches but it does not seem probable that CSA may be an alternative to the large-scale uses of OPC (due to their unavoidable higher prices). On the other hand (active) BCSAF cements are a very promising alternative to OPC with inherent environmental benefits. The development of these new cements for large-scale production and use should be accompanied by meeting some key objectives. Carbon dioxide emission depletion is ensured by the partial replacement of calcite by calcium sulfate. However energy consumption during the manufacturing process needs to be as low as possible, as well as other emissions which have to be carefully controlled. Taken together, CO_2 emission reductions close to 30% are expected when compared active BCSAF clinkers and OPC clinkers productions (for equal mechanical strengths). The viability of the use of by-products or wastes as raw materials needs to be tested, firstly for the proper manufacturing process and secondly for the time on service to be as effective as possible. The (beneficial) effects of admixtures such as blast furnace slag and pulverized fly ash must be studied. This is a key bottleneck, as there is no reason to invest in BCSAF clinkers for reducing CO_2 emissions in the fabrication process, if CO_2 emission reduction in blending cannot be achieved at the same level as it is currently possible for OPCs. We also underline the lack of studies about the alkalinity properties and carbonation studies for belite-active BCSAF cement pastes. Pore solution chemistry, pH evolution and carbonation tests should be measured as a function of W/C ratio, sulfate contents and hydration temperature. Finally, BCSAF concretes should be consistent with health and safety legislation, where more research and work is needed.

18.7 Acknowledgements

This work has been supported by Spanish MINECO through the MAT2010-16213 research grant which is co-funded by FEDER. We also thank Junta de Andalucía (Spain) for funding through the FQM-113 project.

18.8 References

Abdul-Maula S and Odler I (1992), 'SO_3-rich Portland cements: Synthesis and strength development', *Mat Res Soc Symp Proceedings*, **245**, 315–320.

Adolfsson D, Menad N, Viggh E and Bjorkman B (2007), 'Hydraulic properties of sulphoaluminate belite cement based on steelmaking slags', *Adv Cem Res*, **19**, 133–138.

Alaoui A, Feraille A, Steckmeyer A and Le Roy R (2007), 'New cements for sustainable development', *Proceedings of 12th International Congress on the Chemistry of Cement*, Montreal, Canada.

Alesiani M, Pirazzoli I, Maraviglia B and Canonico F (2008), 'NMR and XRD study on calcium sulfoaluminate cement', *Appl Magn Reson*, **35**, 33–41.

Ali M M, Gopal S and Handoo S K (1994), 'Studies on the formation kinetics of calcium sulphoaluminate', *Cem Concr Res*, **24**(4), 715–720.

Álvarez-Pinazo G, Cuesta A, García-Maté M, Santacruz I, De la Torre A G, León-Reina L and Aranda M A G (2012), 'Rietveld quantitative phase analysis of Yeelimite-containing cements', *Cem Concr Res*, **42**, 960–971.

Ambroise J, Georgin J F, Peysson S and Péra J (2009), 'Influence of polyether polyol on the hydration and engineering properties of calcium sulphoaluminate cement', *Cem Concr Comp*, **31**, 474–482.

Andac M and Glasser F P (1999), 'Pore solution composition of calcium sulfoaluminate cement', *Adv Cem Res*, **11**, 23–26.

Aranda M A G, Cuberos A J M, Cuesta A, Álvarez-Pinazo G, De la Torre A G, Schollbach K and Pöllmann H (2011), 'Hydrating behaviour of activated belite sulfoaluminate cements', *Proceedings of the 13th International Congress on the Chemistry of Cement*, Madrid, Spain.

Arjunan P, Silsbee M R and Roy D M (1999), 'Sulfoaluminate-belite cement from low-calcium fly ash and sulfur-rich and other industrial by-products', *Cem Concr Res*, **29**, 1305–1311.

Benarchid M Y, Diouri A, Boukhari A, Aride J, Rogez J and Castanet R (2004),

'Elaboration and thermal study of iron-phosphorus-substituted dicalcium silicate phase', *Cem Concr Res*, **34**, 1873–1879.

Bensted J (1979), 'Some hydration studies of α-dicalcium silicate', *Cem Concr Res*, **9**, 97–101.

Beretka J, De Vito B, Santoro L, Sherman N and Valenti G L (1993), 'Hydraulic behavior of calcium sulfoaluminate-based cements derived from industrial process wastes', *Cem Concr Res*, **23**, 1205–1214.

Beretka J, Marroccoli M, Sherman N and Valenti G L (1996), 'The influence of C_4A_3S content and W/S ratio on the performance of calcium sulfoaluminate-based cements', *Cem Concr Res*, **26**, 1673–1681.

Berger S, Cau Dit Coumes C, Le Bescop P and Damidot D (2011a), 'Influence of a thermal cycle at early age on the hydration of calcium sulphoaluminate cements with variable gypsum contents', *Cem Concr Res*, **41**, 149–160.

Berger S, Cau Dit Coumes C, Le Bescop P and Damidot D (2011b), 'Stabilization of $ZnCl_2$-containing wastes using calcium sulfoaluminate cement: Cement hydration, strength development and volume stability', *J Haz Mat*, **194**, 256–267.

Berger S, Cau Dit Coumes C, Le Bescop P and Damidot D (2011c), 'Stabilization of $ZnCl_2$-containing wastes using calcium sulfoaluminate cement: Leaching behaviour of the solidified waste form, mechanisms of zinc retention', *J Haz Mat*, **194**, 268–276.

Berger S, Cau Dit Coumes C, Le Bescop P, Aouad G and Damidot D (2011d), 'Leaching of calcium sulfoaluminate cement-based materials: experimental investigation and modelling', *Proceedings of the 13th international Congress on the Chemistry of Cement*, Madrid, Spain.

Bernardo G, Telesca A and Valenti G L (2006), 'A porosimetric study of calcium sulfoaluminate cement pastes cured at early ages', *Cem Concr Comp*, **36**, 1042–1047.

Bessy G I (1938), 'The calcium aluminate and silicate hydrates', *Symp on the Chem. of Cements*, Stockholm.

Buzzi L, Canonico F and Schäffel P (2011), 'Investigation on high-performance concrete based on calcium sulfoaluminate cement', *Proceedings of the 13th International Congress on the Chemistry of Cement*, Madrid, Spain.

Calos N J, Kennard C H L, Whittaker A K and Davis R L (1995), 'Structure of calcium aluminate sulfate $Ca_4Al_6O_{16}S$', *J Solid State Chem*, **119**, 1–7.

Canonico F, Bernardo G, Buzzi L, Paris M, Telesca A and Valenti G L (2007), 'Microstructural investigations on hydrated high-performance cements based on calcium sulfoaluminate', *Proceedings of the 12th International Congress of Cement Chemistry*, Montreal.

Cau Dit Coumes C, Courtois S, Peysson S, Ambroise J and Péra J (2009) 'Calcium sulfoaluminate cement blended with OPC: a potential binder to encapsulate low-level radioactive slurries of complex chemistry', *Cem Concr Res*, **39**, 740–747.

Chang W, Li H, Wei M, Zhu Z, Zhang J and Pei M (2009), 'Effects of polycarboxylic acid based superplasticiser on properties of sulphoaluminate cement', *Mater Res Innovations*, **13**, 7–10.

Chen I A, Hargis C W and Juenger M C G (2012), 'Understanding expansion in calcium sulfoaluminate-belite cements', *Cem Concr Res*, **42**, 51–60.

Chen Y L, Lin C J, Ko M S, Lai Y C and Chang J E (2011), 'Characterization of mortars from belite-rich clinkers produced from inorganic wastes', *Cem Concr Com*, **33**, 261–266.

Cuberos A J M, De la Torre A G, Álvarez-Pinazo G, Martín-Sedeño M C, Schollbach

K, Pöllmann H and Aranda M A G (2010), 'Active iron-rich belite sulfoaluminate cements: Clinkering and hydration', *Env Sci Tech*, **44**, 6855–6862.

Cuesta A, Losilla E R, Aranda M A G, Sanz J and De la Torre A G (2012), 'Reactive belite stabilization mechanisms by boron-bearing dopants', *Cem Concr Res*, **42**, 598–606.

De la Torre A G, Morsli K, Zahir M and Aranda M A G (2007), 'In-situ synchrotron powder diffraction study of active belite clinkers', *J Appl Cryst*, **27**, 892–900.

De la Torre A G, Cuberos A J M, Alvarez-Pinazo G, Cuesta A and Aranda M A G (2011a), 'In situ powder diffraction study of belite sulfoaluminate clinkering', *J Synch Rad*, **18**, 506–514.

De la Torre A G, Cuberos A J M, Alvarez-Pinazo G, Cuesta A and Aranda M A G (2011b), 'In-situ clinkering study of belite sulfoaluminate clinkers by synchrotron x-ray powder diffraction', *Proceedings of the 13th International Congress on the Chemistry of Cement*, Madrid, Spain.

Fierens P and Tirlocq J (1983), 'Nature and concentration effect of stabilizing elements of beta-dicalcium silicate on its hydration rate', *Cem Concr Res*, **13**, 267–276.

Fu X, Yang C, Liu Z, Tao W, Hou W and Wu X (2003), 'Studies on effects of activators on properties and mechanism of hydration of sulphoaluminate cement', *Cem Concr Res*, **33**, 317–324.

Fukuda K and Ito S (1999), 'Improvement in reactivity and grindability of belite-rich cement by remelting reaction', *J Am Cer Soc*, **82**, 2177–2180.

Fukuda K, Takeda A and Yoshida H (2001), 'Remelting reaction of α-Ca_2SiO_4 solid solution confirmed in Ca_2SiO_4–$Ca_{12}Al_{14}O_{33}$ pseudobinary system', *Cem Concr Res*, **31**, 1185–1189.

García-Maté M, Santacruz I, De la Torre A G, León-Reina L and Aranda M A G (2012), 'Rheological and hydration characterization of calcium sulfoaluminate cement pastes', *Cem Concr Comp*, **34**, 684–691.

Gartner E (2004), 'Industrially interesting approaches to 'low-CO_2' cements', *Cem Concr Res*, **34**, 1489–1498.

Gartner E and Li G (2006), 'High-belite sulfoaluminate clinker: fabrication process and binder preparation', World Patent Application WO2006/018569 A2

Gartner E and Macphee D E (2011), 'A physico-chemical basis for novel cementitious binders', *Cem Concr Res*, **41**, 736–749.

Gastaldi D, Boccaleri E, Canonico F and Bianchi M (2007), 'The use of Raman spectroscopy as a versatile characterization tool for calcium sulphoaluminate cements: A compositional and hydration study', *J Mater Sci*, **42**, 8426–8432.

Gastaldi D, Canonico F, Capelli L, Bianchi M, Pace M L, Telesca A and Valenti G L (2011), 'Hydraulic behaviour of calcium sulfoaluminate cement alone and in mixture with Portland cement', *Proceedings of the 13th International Congress on the Chemistry of Cement*, Madrid, Spain.

Gastaldi D, Canonico F, Capelli L, Boccaleri E, Milaneso M, Palin L, Groce G, Marone F, Mader K and Stampanoni M (2012), 'In situ tomographic investigation on the early hydration behaviors of cementing systems', *Cons Build Mat*, **29**, 284–290.

Georgin J F, Ambroise J, Péra J and Reynouard J M (2008), 'Development of self-leveling screed based on calcium sulfoaluminate cement: Modelling of curling due to drying', *Cem Concr Com*, **30**, 769–778.

Ghosh S N, Rao P B, Paul A K and Raina K J (1979), 'The chemistry of the dicalcium silicate mineral', *J Mater Sci*, **14**, 1554–1566.

Gielen D and Tanaka K (2006), 'Energy efficiency and CO_2 emission reduction potentials and policies in the cement industry: Towards a plan of action', *Proceedings of the*

IEA/WBCSD Workshop on Energy Efficiency and CO_2 Emission Reduction Potentials and Policies in the Cement Industry, Paris.

Glasser F P and Zhang L (2001), 'High-performance cement matrices based on calcium sulphoaluminate-belite compositions', *Cem Concr Res*, **31**, 1881–1886.

Hanehara S and Yamada K (2008), 'Rheology and early age properties of cement systems', *Cem Concr Res*, **21**, 175–195.

Hayes M and Godfrey I H (2007), 'Development of the use of alternative cements for the treatment of intermediate level waste', *Proceedings of the Waste Management Symposium*, Tucson, Arizona.

Janotka I and Krajci L (1999), 'An experimental study on the upgrade of sulfoaluminate-belite cement systems by blending with Portland cement', *Adv Cem Res*, **11**, 35–41.

Janotka I, Krajci L, Ray A and Mojumdar S C (2003), 'The hydration phase and pore structure formation in the blends of sulfoaluminate-belite cement with Portland cement', *Cem Concr Res*, **33**, 489–497.

Janotka I, Krajci U and Mojumdar S C (2007), 'Performance of sulphoaluminate-belite cement with high C4A3$ content', *Cer Silik*, **51**, 74–81.

Jelenic I, Bezjak A and Bujan M (1978), 'Hydration of B_2O_3-stabilized α'- and β-modifications of dicalcium silicate', *Cem Concr Res*, **8**, 173–180.

Juenger M and Chen I (2011), 'Composition–property relationships in calcium sulfoaluminate cements', *Proceedings of the 13th International Congress on the Chemistry of Cement*, Madrid, Spain.

Juenger M C G, Winnefeld F, Provis J L and Ideker J H (2011), 'Advances in alternative cementitious binders', *Cem Concr Res*, **41**, 1232–1243.

Kalogridis D, Kostogloudis G Ch, Ftikos Ch and Malami C (2000), 'A quantitative study of the influence of non-expansive sulfoaluminate cement on the corrosion of steel reinforcement', *Cem Concr Res*, **30**, 1731–1740.

Kantro D L and Weise C H (1979), 'Hydration of various beta-dicalcium silicate preparations', *J Am Cer Soc*, **62**, 621–626.

Katsioti M, Tsakiridis P E, Leonardou-Agatzini S and Oustadakis P (2006), 'Examination of the jarosite–alunite precipitate addition in the raw meal for the production of sulfoaluminate cement clinker', *J Haz Mat*, **B131**, 187–194.

Klein A (1966), 'Expansive and shrinkage-compensated cements', US Patent 3251701.

Kurdowski W and Thiel A (1981), 'On the role of free calcium oxide in expansive cements', *Cem Concr Res*, **11**, 29–40.

Kuryatnyk T, Angulski-Da-Luz C, Ambroise J and Pera J (2008), 'Valorization of phosphogypsum as hydraulic binder', *J Haz Mat*, **160**, 681–687.

Kuryatnyk T, Chabannet M, Ambroise J and Pera J (2010), 'Leaching behaviour of mixtures containing plaster of Paris and calcium sulphoaluminate clinker,' *Cem Concr Res*, **40**, 1149–1156.

Li G S, Walenta G and Gartner E M (2007), 'Formation and hydration of low-CO_2 cements based on belite, calcium sulfoaluminate and calcium aluminoferrite', *Proceedings of the 12th International Congress of the Cements Chemistry*, Montreal, Canada.

Li H, Agrawal D K, Cheng J and Silsbee M R (2001), 'Microwave sintering of sulphoaluminate cement with utility wastes', *Cem Concr Res*, **31**, 1257–1261.

Li J, Ma H and Zhao H (2007), 'Preparation of sulphoaluminate-alite composite mineralogical phase cement clinker from high alumina fly ash', *Key Eng Mat*, **334–335**, 421–424.

Lili R, Xiaocun L, Tao Q, Lian L, Deli Z and Yanjun L (2009), 'Influence of MnO_2 on the burnability and mineral fornation of alite-sulphoaluminate cement clinker', *Sil Ind*, **74**, 183–187.

Lingchao L, Jun C, Xin C, Hanxing L and Runzhang Y (2005), 'Study on a cementing system taking alite-calcium barium sulphoaluminate as main minerals', *J Mat Sci*, **40**, 4035–4038.

Liu X and Li Y (2005), 'Effect of MgO on the composition and properties of alite-sulphoaluminate cement', *Cem Concr Res*, **35**, 1685–1687.

Liu X C, Li B L, Qi T, Liu X L and Li Y J (2009), 'Effect of TiO_2 on mineral formation and properties of alite-sulphoaluminate cement', *Mat Res Innov*, **13**(2), 92–97.

Lura P, Winnefeld F and Klemm S (2010), 'Simultaneous measurements of heat of hydration and chemical shrinkage on hardening cement pastes', *J Therm Anal Calorim*, **101**, 925–932.

Luz C A, Rocha J C, Cheriaf M and Pera J (2006), 'Use of sulfoaluminate cement and bottom ash in the solidification/stabilization of galvanic sludge', *J Haz Mater*, **B136**, 837–845.

Luz C A, Rocha J C, Cheriaf M and Pera J (2009), 'Valorization of galvanic sludge in sulfoaluminate cement', *Const Build Mat*, **23**, 595–601.

Ma S, Shen X, Gong X and Zhong B (2006), 'Influence of CuO on the formation and coexistence of $3CaO{\cdot}SiO_2$ and $3CaO{\cdot}3Al_2O_3{\cdot}CaSO_4$ minerals', *Cem Concr Res*, **36**, 1784–1787.

Marchi M and Costa U (2011), 'Influence of the calcium sulfate and W/C ratio on the hydration of calcium sulfoaluminate cement', *Proceedings of the 13th International Congress on the Chemistry of Cement*, Madrid, Spain.

Martín-Sedeño M C, Cuberos A J M, De la Torre A G, Álvarez-Pinazo G, Ordónez L M, Gateshki M and Aranda M A G (2010), 'Aluminum-rich belite sulfoaluminate cements: clinkering and early age hydration', *Cem Concr Res*, **40**, 359–369.

Matkovic B, Carin V, Gacesa T, Halle R, Jelenic I and Young J F (1981), 'Influence of $BaSO_4$ on the formation and hydration properties of calcium silicates: i, doped dicalcuim silicates', *Am Ceram Soc Bull*, **60**, 825–829.

McCaffrey R (2002), 'Climate change and the cement industry', *Global Cement and Lime Magazine Environmental Special Issue*, 15–19.

Mehta P K (1973), 'Mechanism of expansion associated with ettringite formation', *Cem Concr Res*, **3**, 1–6.

Mehta P K (1977), 'High iron oxide hydraulic cement', US patent 4036657.

Meller N, Hall C, Jupe A C, Colston S L, Jacques S D M, Barnes P and Phipps J (2004), 'The paste hydration of brownmillerite with and without gypsum: A time resolved synchrotron diffraction study at 30, 70, 100 and 150 °C', *J Mater Chem*, **14**, 428–435.

Michel M, Georgin J F, Ambroise J and Péra J (2011), 'The influence of gypsum ratio on the mechanical performance of slag cement accelerated by calcium sulfoaluminate cement', *Cons Build Mat*, **25**, 1298–1304.

Morin V, Walenta G, Gartner E, Termkhajornkit P, Baco I and Casabonne J M (2011), 'Hydration of a belite-calcium sulfoaluminate-ferrite cement: AetherTM', *Proceedings of the 13th International Congress on the Chemistry of Cement*, Madrid, Spain.

Morsli K, De la Torre A G, Stöber S, Cuberos A J M, Zahir M and Aranda M A G (2007a), 'Quantitative phase analysis of laboratory active belite clinkers by synchrotron powder diffraction', *J Amer Cer Soc*, **90**, 3205–3212.

Morsli K, De la Torre A G, Zahir M and Aranda M A G (2007b), 'Mineralogical phase analysis of alkali and sulfate bearing belite rich laboratory clinkers', *Cem Concr Res*, **37**, 639–646.

Nettleship I, Slavick K G, Kim Y J and Kriven W M (1992), 'Phase transformation in dicalcium silicate: I, Fabrication and phase stability of fine-grained β-phase', *J Am Cer Soc*, **75**, 2400–2406.

Odler I (2000), *Special Inorganic Cements*, Taylor and Francis, London.

Odler I and Zhang H (1996), 'Investigations on high SO_3 Portland clinkers and cements I. Clinker synthesis and cement preparation', *Cem Concr Res*, **26**, 1307–1313.

Pace M L, Telesca A, Marroccoli M and Valenti G L (2011), 'Use of industrial byproducts as alumina sources for the synthesis of calcium sulfoaluminate cements', *Env Sci Tech*, **45**, 6124–6128.

Paglia C, Wombacher F and Bohni H (2001), 'The influence of alkali-free and alkaline shotcrete accelerators within cement systems I. Characterization of the setting behavior', *Cem Concr Res*, **31**, 913–918.

Palou M, Majling J, Dovál M, Kozanková J and Mojumdar S C (2005), 'Formation and stability of crystallohydrates in the non-equilibrium system during hydration of SAB cements', *Cer Silik*, **49**, 230–236.

Park C K (2001), 'Phase transformation and hydration of dicalcium silicate containing stabilizers', *J Cer Soc Jap*, **109**, 380–386.

Pelletier L, Winnefeld F and Lothenbach B (2010), 'The ternary system Portland cement–calcium sulphoaluminate clinker–anhydrite: Hydration mechanism and mortar properties', *Cem Concr Comp*, **32**, 497–507.

Pelletier-Chaignat L, Winnefeld F, Lothenbach B, Le Saout G, Müller C J and Famy C (2011), 'Influence of the calcium sulfate source on the hydration mechanism of Portland cement–calcium sulphoaluminate clinker–calcium sulfate binders', *Cem Concr Comp*, **33**, 551–561.

Pelletier-Chaignat L, Winnefeld F, Lothenbach B and Müller C J (2012), 'Beneficial use of limestone filler with calcium sulphoaluminate cement', *Cons Build Mat*, **26**, 619–627.

Péra J and Ambroise J (2004), 'New applications of calcium sulfoaluminate cement', *Cem Conc Res*, **34**, 671–676.

Péra J, Ambroise J, Holard E and Beauvent G (2003), 'Influence of the type of calcium sulfate on the properties of calcium sulfoaluminate cement', *Proceedings of the 11th International Congress on the Chemistry of Cement*, Durban, South Africa, 1129–1135.

Peysson S, Pera J and Chabannet M (2005), 'Immobilization of heavy metals by calcium sulfoaluminate cement', *Cem Concr Res*, **35**, 2261–2270.

Phair J W (2006), 'Green chemistry for sustainable cement production and use', *Green Chem*, **8**, 763–780.

Pliego-Cuervo Y B and Glasser F P (1979), 'The role of sulfates in cement clinkering: subsolidus phase relations in the system $CaO–Al_2O_3–SiO_2–SO_3$', *Cem Concr Res*, **9**, 51–56.

Popescu C D, Muntean M and Sharp J H (2003), 'Industrial trial production of low energy belite cement', *Cem Concr Comp*, **25**, 689–693.

Pritts M and Daugherty K E (1976), 'The effect on stabilizing agents on the hydration rate of β-C_2S', *Cem Concr Res*, **6**, 783–796.

Puertas F, Blanco-Varela M T and Giménez-Molina S (1995), 'Kinetics of the thermal decomposition of $C_4A_3\underline{S}$ in air', *Cem Concr Res*, **25**, 572–580.

Qian G R, Shi J, Cao Y L, Xu Y F and Chui P C (2008), 'Properties of MSW fly ash–calcium sulfoaluminate cement matrix and stabilization/solidification on heavy metals', *J Haz Mat*, **152**, 196–203.

Quillin K (2001), 'Performance of belite–sulfoaluminate cements', *Cem Concr Res*, **31**, 1341–1349.

Quillin K (2007), *Calcium sulfoaluminate cements: CO_2 reduction, concrete properties and applications*, BRE Report 496, IHS-BRE Press, Garston, Watford, UK.

Saalfeld H and Depmeier W (1972), 'Silicon-free compounds with sodalite structure', *Kristall Technik*, **7**, 229–233.

Sahu S and Majling J (1993), 'Phase compatibility in the system $CaO–SiO_2–Al_2O_3–Fe_2O_3–SO_3$ referred to sulphoaluminate belite cement clinker', *Cem Concr Res*, **23**, 1331–1339.

Sahu S and Majling J (1994), 'Preparation of sulphoaluminate belite cement from fly ash', *Cem Concr Res*, **24**, 1065–1072.

Sahu S, Havlica J, Tomkovh V and Majling J (1991), 'Hydration behavior of sulphoaluminate belite cement in the presence of various calcium sulfates', *Thermochim Acta*, **175**, 45–52.

Schneider M, Romer M, Tschudin M and Bolio H (2011), 'Sustainable cement production-present and future', *Cem Concr Res*, **41**, 642–650.

Scrivener K L and Kirkpatrick R J (2008), 'Innovation in use and research on cementitious material', *Cem Concr Res*, **38**, 128–136.

Seluck N, Soner I and Seluck E (2010), 'Synthesis of special cement with fluidised bed combustion ashes', *Adv Cem Res*, **22**, 107–113.

Senff L, Castela A, Hajjaji W, Hotza D and Labrincha J A (2011), 'Formulations of sulfobelite cement through design of experiments', *Cons Build Mat*, **25**, 3410–3416.

Sharp J H, Lawrence C D and Yang R (1999), 'Calcium sulphoaluminate cements – Low-energy cements, special cements or what?', *Adv Cem Res*, **11**, 3–13.

Singh M and Pradip P C K (2008), 'Preparation of calcium sulphoaluminate cement using fertiliser plant wastes', *J Haz Mat*, **157**, 106–113.

Sui T and Yao Y (2003), 'Recent progress in special cements in China', *Proceedings of 11th International Congress on the Chemistry of Cement*, Durban, South Africa.

Sun Q, Li J and Wang J (2011a), 'Effect of borate concentration on solidification of radioactive wastes by different cements', *Nucl Engin Design*, **241**, 4341–4345.

Sun N, Chang W, Wang L, Zhang J and Pei M (2011b), 'Effects of the chemical structure of polycarboxy-ether superplasticizer on its performance in sulphoaluminate cement', *J Disp Sci Tech*, **32**, 795–798.

Taylor H F W (1997), 'Portland cement and its major constituent phases', pp. 1–28 in *Cement Chemistry*, Thomas Telford Ltd, London.

Telesca A, Marroccoli M, Pace M L and Valenti G L (2011), 'Calcium sulfoaluminate cements obtained from bauxite-free raw mixes', *Proceedings of the 13th International Congress on the Chemistry of Cement*, Madrid, Spain.

Valenti G L, Marroccoli M, Montagnaro F, Nobili M and Telesca A (2007), 'Synthesis, hydration properties and environmental friendly features of calcium sulfoaluminate cements', *Proceedings of the 12th International Congress of Cement Chemistry*, Montreal, Canada.

Viswanathan V N, Raina S J and Chatterjee A K (1978), 'An exploratory investigation on Porsal cement', *World Cem Tech*, **May/June**, 109–118.

Walenta G and Comparet C (2011), 'New cements and innovative binder technologies BCSAF cements – recent developments', ECRA-Barcelona-2011. Presentation available at http://www.aether-cement.eu/press-room/publications/aether-cement-ecra-barcelona-presentation -2011-05-05.html

Wesselsky A and Jensen O M (2009), 'Synthesis of pure Portland cement phases', *Cem Concr Res*, **39**, 973–980.

Winnefeld F and Barlag S (2009), 'Influence of calcium sulfate and calcium hydroxide on the hydration of calcium sulfoaluminate clinker', *ZKG Int*, **12**, 42–53.

Winnefeld F and Barlag S (2010), 'Calorimetric and thermogravimetric study on the influence of calcium sulfate on the hydration of yeelimite', *J Therm Anal Calorim*, **101**, 949–957.

Winnefeld F and Lothenbach B (2010), 'Hydration of calcium sulfoaluminate – Experimental findings and thermodynamic modelling', *Cem Concr Res*, **40**, 1239–1247.

Winnefeld F, Ben-Haha M and Lothenbach B (2011), 'Hydration mechanisms of calcium sulfoaluminate cements assessed by scanning electron microscopy and thermodynamic modelling', *Proceedings of the 13th International Congress on the Chemistry of Cement*, Madrid, Spain.

Wolter A (2005), 'Belite cements and low energy clinker', *Cem Inter*, **3**, 106–117.

Wu K, Shi H and Guo X (2011), 'Utilization of municipal solid waste incineration fly ash for sulfoaluminate cement clinker production', *Waste Manag*, **31**, 2001–2008.

Yanjun L, Xiaocun L and Lili R (2007), 'Study of the properties of compounded alite-sulphoaluminate cement and portland cement', *Silicate Industriels*, **72**, 15–18.

Zhang H and Odler I (1996), 'Investigations on high SO_3 Portland clinkers and cements II. Properties of cements', *Cem Concr Res*, **26**, 1315–1324.

Zhang L and Glasser F P (2002), 'Hydration of calcium sulfoaluminate cement at less than 24 h', *Adv Cem Res* **14**, 141–155.

Zhang L and Glasser F P (2005), 'Investigation of the microstructure and carbonation of CSA-based concretes removed from service' *Cem Concr Res* **35**, 2252–2260.

Zhang L, Su M Z and Wang Y M (1999), 'Development of the use of sulfo-and ferro-aluminate cements in China', *Adv Cem Res*, **11**, 15–21.

Zhou Q, Milestone N B and Hayes M (2006), 'An alternative to Portland cement for waste encapsulation – the calcium sulfoaluminate cement system', *J Haz Mater*, **136**, 120–129.

Ziemer B, Altrichter B and Jesenak V (1984), 'Effect of SO_3 on formation and hydraulic reactivity of belite', *Cem Concr Res*, **14**, 686–692.

Zivica V (2000), 'Properties of blended sulfoaluminate belite cement', *Const Build Mat*, **14**, 433–437.

Zivica V (2001), 'Possibility of the modification of the properties of sulfoaluminate belite cement by its blending', *Const Build Mat*, **45**, 24–30.

19
Reactive magnesia cement

A. AL-TABBAA, University of Cambridge, UK

DOI: 10.1533/9780857098993.4.523

Abstract: This chapter presents an overview of reactive magnesia cements which have recently emerged as a viable alternative to Portland cement, with both technical and sustainability advantages. Details of research work to date on the characterisation and properties of a range of different origin materials as well as a range of applications are presented.

Key words: reactive magnesia cement, construction products, environmental applications, properties.

19.1 Introduction

Magnesium is the eighth most abundant element in the earth's crust, at ~2.3% by weight, present in a range of rock formations such as dolomite, magnesite and silicate. Magnesium is also the third most abundant element in solution in seawater, with concentrations of ~1300ppm. Magnesia (magnesium oxide, MgO) is mainly produced from the calcination of magnesite in a process similar to the production of lime from limestone. A smaller proportion of the world's MgO production comes from seawater and brine sources. The principal phases of magnesium in seawater are chlorides and sulphates and the production process is initiated by the addition of a strong base to facilitate the precipitation of magnesium hydroxide ($Mg(OH_2)$), followed by thermal decomposition (e.g. Gilbert, 1951). The current global production of MgO is ~14 million tonnes annually (USGS, 2012), compared with that of Portland cement (PC) of over 2.6 billion tonnes, with current costs of around ~£200/tonne for reactive MgO (calcined), compared to ~£70/tonne for PC.

MgO is manufactured in three main relevant grades (Shand, 2006):

- dead burned MgO, or periclase: manufactured at temperatures of ~1400–2000 °C, constitutes ~60% of MgO production, has the least reactivity, highest crystallinity and lowest surface area, is used extensively in refractory applications and is the type that is problematic in Portland cement;
- hard burned MgO: manufactured at temperatures of ~1000–1400 °C, has intermediate properties and is used in animal feeds and fertilizers and has been extensively used as expansive additive in concrete dams in China;

- light burned MgO, also called caustic magnesia, caustic-calcined magnesite (CCM) or reactive magnesia: manufactured at temperatures of ~700–1000 °C, has the highest reactivity, least crystallinity and highest surface area and its applications include plastics, rubber, paper and pulp processing, adhesives and acid neutralisation to name a few and is the main ingredient of the cement presented in this chapter.

MgO has been used in the formation of cements or as an additive in concrete since the mid-nineteenth century when Sorel in 1867, shortly after the invention of PC, developed magnesium oxychloride cements (MOC), also called Sorel cements, by combining light burned MgO with magnesium chloride solution (Sorel, 1867). Despite many technical advantages, Sorel cements suffer from poor water resistance, which has prevented their widespread use. Since then other similar chemically bonded cements have emerged including magnesium phosphate cements, which are the results of the reaction between MgO, either hard burned or dead burned, and acid phosphate salts (Wagh, 2004), and which have applications as rapid hardening repair cements. Since the mid-1960s hard burned MgO has been used for shrinkage compensation in concrete dams in China (Du, 2006). Much more recently, reactive MgO cements (Harrison, 2008), which are blends of light burned MgO and PC, emerged as a more sustainable alternative to PC and with anticipated superior technical performance. Extensive coverage of the chemistry and technology of magnesia is given in Shand (2006).

19.2 Overview, history and development of reactive magnesia cements

Reactive MgO cement formulations were developed and patented just over a decade ago by the Australian scientist John Harrison (2003, 2008) and are blends of PC and reactive MgO (sometime also incorporating fly ash) in different proportions, depending on the intended application, ranging from structural concrete to porous masonry units. They have been developed with strong emphasis on a range of sustainability advantages over PC and have received significant publicity including coverage in the *New Scientist* (Pearce, 2002) and the *Guardian* (Dyer, 2003). Advantages include: (i) manufacture at a much lower temperatures (650–800 °C), potential complete recyclability of the MgO, (iii) uptake of significant quantities of CO_2 and (iv) insensitivity to impurities. Three different composition ranges have been proposed (Harrison, 2003):

- tec-cements with MgO << PC mainly for concrete applications to enhance durability and strength through increased density, reduced permeability, internal dryness and long-term pH and volume stability;
- enviro-cements with MgO ~ PC for enhanced waste immobilisation

capacity through lower solubility and mobility of brucite and its layered structure as well as its lower stable long-term pH;
- eco-cements with MgO >> PC for porous block applications through uptake of CO_2 and subsequent carbonation to a range of strength providing hydrated magnesium carbonates.

Reactive MgO hydrates at a similar rate to PC and should not be confused with dead burned MgO, the latter being the result of a much higher calcination temperature resulting in a much lower reactivity leading to delayed hydration and cracking, as would be the case for MgO impurities within PC during the clinkering stage (~1450 °C).

The inventor John Harrison has contributed significantly to the understanding of the chemistry, interactions and performance of these three PC–MgO systems (Harrison, 2012). Significant publications in the technical literature have resulted from related work performed at the University of Cambridge over the past nine years which has investigated a whole host of reactive MgO systems. This has so far included MgO alone and in blends with PC, slags, fly ash, microsilica, limestone cement, zeolite and other magnesium-based compounds (namely brucite, magnesite, hydrated magnesium carbonates and silicates) as well as with a range of chemical admixtures. Reactive MgO has also been investigated with a range of aggregates, including natural, fly ash-based and limestone. Characterisation and fundamental properties as well as performance in a range of applications for those systems, including concrete, porous blocks, pervious concrete, ground improvement, waste and contaminant immobilisation, soil and groundwater remediation, performance in aggressive and extreme environments and carbon capture and storage, were investigated.

19.3 Characterisation and properties

Extensive characterisation studies have been conducted on over 20 commercially available MgOs from around the world, from different sources and manufacturing processes, and from commercial trials in China as well as from small-scale laboratory production studies covering a range of compositions (Jin, forthcoming). It should be pointed out that most of those MgO commercially produced from the calcination of magnesite are likely to have been calcined at temperatures at the higher end of the reactive MgO range of ~1000 °C hence their reactivity is likely to be close to that of hard burned MgOs. The MgO content was found to vary between 60% and 99.6% where the synthetically produced (seawater and chemical precipitation) were at the higher purity end. The main impurities of the calcined MgOs were CaO and SiO_2, which are common minerals of rock, while those of the synthetic MgOs were CaO, Cl and SO_3. The CaO content varied between 0.15% and

6.9% and the loss of ignition (LOI) between 0.8% and 7%. The surface areas measured were between 2 and 148 m^2/g and the average particle size between 1.8 and 35 μm. The reactivity of these MgOs, using the citric acid test, varied significantly from as little as 9 seconds (extremely reactive) to >2.5 hours (highly unreactive). The equilibrium pH of the MgOs also varied significantly between 10.0 and 12.5, mainly related to the CaO content. A degree of hydration of 40–80% was observed with a limit of 80–95%; the degree and rate of hydration were enhanced by the presence of certain hydration agents similar to those used with PC. Commercial trial production of magnesia from magnesite in China under controlled calcination conditions demonstrated that highly reactive MgO can be produced which is far more reactive than those currently available commercially in China (Li, 2012).

Fundamental studies of reactive MgO, alone and in blends, using pastes and mortars, in terms of hydration behaviour, microstructure and carbonation behaviour (Vandeperre et al., 2006, 2007, 2008a, 2008b; Vandeperre and Al-Tabbaa, 2007; Liska et al., 2006, 2008; Liska, 2009; Unluer, 2012; Li, 2012; Jin, forthcoming) showed that in the presence of water and under ambient CO_2 levels, MgO hydrates to form $Mg(OH)_2$, or brucite, according to:

$$MgO + H_2O \rightarrow Mg(OH)_2 \qquad [19.1]$$

The solubility of brucite in water is quite low at 0.009 g/L at 18 °C and the equilibrium pH of a saturated solution of pure magnesium hydroxide is 10.5. Brucite has a layered structure and its morphology varies depending on the magnesium source and formation conditions (e.g. Henrist et al., 2003; Gao et al., 2008). The binding ability of brucite as a cement was found to be limited. Brucite is far less soluble and far less mobile than Portlandite and its pH is stable in the long-term and is hence far less susceptible to attack by aggressive chemicals.

In the presence of sufficient CO_2 in the curing atmosphere (5–20%) and water, brucite was found to carbonate to form one or more hydrated magnesium carbonates:

$$Mg(OH)_2 + CO_2 + 2H_2O \rightarrow MgCO_3{\cdot}3H_2O \text{ (nesquehonite)} \qquad [19.2]$$

and/or

$$5Mg(OH)_2 + 4CO_2 + H_2O \rightarrow Mg_5(CO_3)_4(OH)_2{\cdot}5H_2O \text{ (dypingite)} \qquad [19.3]$$

and/or

$$5Mg(OH)_2 + 4CO_2 \rightarrow Mg_5(CO_3)_4(OH)_2{\cdot}4H_2O \text{ (hydromagnesite)} \qquad [19.4]$$

These hydrated magnesium carbonates form well-ramified networks of massive dense crystals, with different morphologies depending on the conditions of their formation, with a very effective binding ability. They are metastable compounds and can undergo transformation to less hydrated forms depending on the conditions they are exposed to including elevated temperature, CO_2 partial pressure, pH and water activity (Dell and Weller, 1959; Davies and Bubela, 1973; Hänchen et al., 2008). Both the hydration and carbonation reactions are expansive with solid volume increases from the MgO to brucite of 2.2-fold and from brucite to the hydrated magnesium carbonates above of 1.8–3.1-fold depending on the carbonate formed.

19.4 Performance in paste blends

Compared to PC, the water demand of reactive MgO is quite high in the region of 0.45–0.70. In PC–MgO paste blends, both PC and MgO were found to hydrate mainly independently forming a blend of their own hydration products, typical of their own raw materials, but did interact such that the setting time was prolonged compared to that of the individual materials, due to the common ion effect (Jeffery et al., 1989). In some cases a limited formation of hydrotalcite was seen as a result of interaction of MgO with PC and fly ash (Vandeperre et al., 2008b). As the MgO content in the blends increased, the ability to obtain densely packed mix decreased. With MgO content of up to 15% in PC–MgO pastes, the MgO hydration rate was found to vary in the range of 50–95%, between MgOs from different sources and different calcination conditions, and seems to correlate reasonably well with the reactivity and surface area. Some strength enhancement, up to 5%, was observed for some of the MgOs used and this was usually in the MgO content range of ~4–7% (Li, 2012).

As metal alkalis, such as those of Na and Ca, are commonly used as activators for slags (Shi et al., 2006), reactive MgO was also investigated for this application. Reactive MgO, in 2.5–20% content, showed effective activation of a range of slags, with the performance being primarily and strongly affected by the slag used and the reactivity of the MgO (Li, 2012; Jin, forthcoming; Yi et al., 2012). The early age degree of hydration of MgO in slag–MgO pastes was found to be ~50% higher than in PC–MgO pastes (Li, 2012). The main hydration products found were calcium silicate hydrates (CSH), ettringite and hydrotalcite, as well as some portlandite and brucite depending on the mix composition and hydration reactions. The presence of Mg-rich CSH and/or separate magnesium silicate hydrate (MSH) phases intermixed with the CSH is also being investigated in different mix compositions. Dense microstructure was observed generally supporting high compressive strengths observed for the pastes of up to 35 MPa with 5% MgO content. These findings correlate well with the hydration behaviour of slags reported in earlier findings (e.g. Wang and Scrivener, 1995; Shi et al., 2006) and more recent findings specific to slag–MgO blends (Shen et al., 2011) and concentrate on the impact of different source materials. Given the higher shrinkage of alkali-activated slag cements compared with PC cements, the expansive MgO reactions have been found, in the appropriate dosages, to compensate for such shrinkage (Yi et al., 2012; Jin, forthcoming).

The formation of MSH as the reaction product between reactive MgO and microsilica at room temperature confirmed the findings of others (Wei et al., 2006, 2011). Ongoing studies, using different source raw materials, giving a pH range of the blends of between 9.0 and 11.7, further showed that the reaction rate is significantly affected by the characteristics of the MgO, in particular its reactivity and pH (Jin, forthcoming) where the reactivity indicated the amount of brucite available for reaction and the pH determined the solubility of the microsilica in water. The characteristics of the microsilica were found to have less of an effect, and the ratio of Mg/Si governed the resulting pH and the reaction mechanisms, leading to different formation characteristics of the resulting MSH. Compressive strength values of over 20 MPa were obtained with a 30% microsilica content paste consistent with the findings of Wei et al. (2006) who reported values of ~60 MPa for mortar specimens.

The above work underpins ongoing extensive research investigations into binary and tertiary blends containing MgO for a wide range of applications, some of which are presented in the following section.

19.5 Performance for different applications

The performance of reactive MgO has been investigated in a number of different applications including porous blocks, both in the laboratory as well as commercial full-scale trials (Liska and Al-Tabbaa, 2008, 2009; Liska et al., 2008, 2012a, 2012b; Liska, 2009; Unluer and Al-Tabbaa 2011, 2012; Karthika, 2011; Unluer, 2012), concrete (Li et al., 2010; March, 2011; Tsang, 2010; Li, 2012), ground improvement (Jegandan et al., 2010; Jegandan, 2010; Al-Tabbaa et al., 2011; Yi et al., 2012) and in a range of aggressive and extreme environments (Iyengar, 2008; Iyengar and Al-Tabbaa, 2008; Kogbara et al., 2010; Al-Tabbaa et al., 2011; Kogbara; 2011; Mackay, 2012; Lim, 2012). An overview of the research work in each of those applications is presented in this section. In order to understand the performance mechanisms of reactive MgO in the majority of these applications, no additives were used in the initial stages of the work. It is, however, clear, as in the use of PC, that additives and admixtures play a major role in significantly enhancing the performance of the cement. Hence the applicability of the typical range of such additives and admixtures that are commonly used with PC is currently being investigated with reactive MgO (Jin, forthcoming).

19.5.1 Porous blocks

The significant contribution of reactive MgO in porous blocks has been in the use of MgO alone as the binder and through its subsequent carbonation. When reactive MgO porous blocks are cured at ambient conditions, the compressive strength developed was found to be quite low, around 20–30% of that of PC blocks, even after a year of curing (Liska, 2009; Unluer, 2012). This is mainly due to the presence of brucite, as a result of the hydration of the MgO, and while some limited carbonation products were observed, their presence was not sufficient to impact the strength development. However, such semi-dry pressed laboratory-scale blocks, with 10% MgO as the cement, and cured in elevated CO_2 environment (up to 20% CO_2), achieved compressive strength values two to three times those of corresponding PC blocks (Liska, 2009; Unluer, 2012) in which almost 100% carbonation was achieved following optimisation studies (Unluer and Al-Tabbaa, 2011, 2012; Unluer, 2012). Abundant presence of the hydrated magnesium carbonates in equations 19.2–19.4 was clearly seen in the microstructure and it was clear that in the high-strength blocks that the presence of nesquehonite dominated over the other carbonate forms. Typical micro-images from scanning electron microscopy are shown in Fig. 19.1, both of the highly carbonated blocks (both laboratory and from commercial trials, described below) as well as those from ambient curing.

Commercial trials were conducted at the Forticrete plant in Dewsbury, Yorkshire, UK, to produce full size porous blocks (440 × 210 × 100/200 mm) using four different aggregate types (natural, Lytag (lightweight ash aggregates), ash mix and limestone, the latter two also containing some crushed glass) and using MgO alone as the cement at 10% content (Liska et al., 2012a, 2012b) which were cured in existing high CO_2 chambers on site. Typical photographs from this process and of the resulting blocks are shown in Figs 19.2 and 19.3. The block production process used was exactly the same as that used for the PC blocks, which usually uses 10% cement content and results in blocks with design compressive strength of 7 MPa. The compressive strength of the resulting MgO blocks ranged from 8.5 to 21 MPa, the highest was for the natural aggregates and the lowest for the ash mix-crushed glass aggregates. Carbonation was found to be uniform throughout the blocks, and quantification of the degree of carbonation showed a degree of carbonation range of 50–100%, with the limestone aggregates blocks, in which the MgO content was only 5%, achieving full carbonation. The trials verified the potential scale-up of such a production process.

Optimisation studies to enhance the degree of carbonation and hence the compressive strength of porous blocks were conducted addressing the various contributing components including (i) block composition (MgO content, water/cement ratio, aggregate types and particle size distributions), (ii) curing

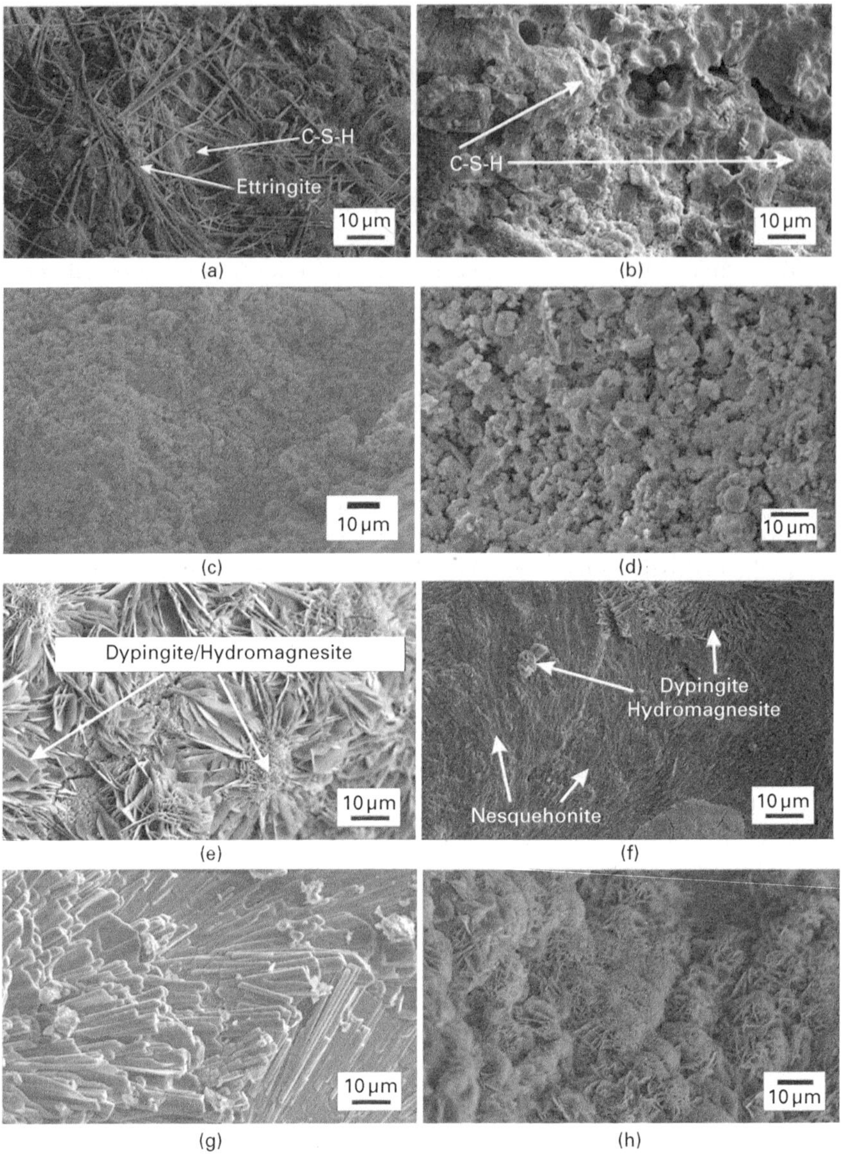

19.1 Scanning electron micro-images of various porous blocks tested (Liska et al., 2008, 2012b): (a, b) PC–natural aggregates and PC–Lytag respectively showing the typical PC hydration products, (c, d) MgO–natural aggregates and MgO–Lytag respectively under natural curing showing mainly brucite (e, f) MgO–natural aggregates, carbonated, (g, h) MgO–Lytag, carbonated, where images (e–h) clearly show the range of hydrated magnesium carbonates present.

(a) (b) (c)

19.2 Production of the MgO porous blocks at the Forticrete plant (Liska et al., 2012a): (a) mixing with manual application of the MgO, (b) fresh blocks on the conveyer belt from the plant and (c) the curing chamber with elevated CO_2 environment.

conditions (elevated CO_2 conditions, relative humidity, wet–dry cycling and ambient curing) and (iii) use of additives and admixtures (mineral admixtures and chemical admixtures) (Unluer and Al-Tabbaa, 2011, 2012; Unluer, 2012). Porous blocks with only 4% carbonated MgO were found to achieve the strength of the 10% PC blocks. The use of gap grading enhanced the strength by over 60%. Ambient curing with those enhanced conditions significantly increased the compressive strengths at 6 months but were still much lower than those of the PC mixes. The use of mineral admixtures showed that the incorporating of other magnesium-based compounds (brucite, silicates and carbonates) did in some cases enhance the performance. In particular, the inclusion of 20% of one particular heavy hydrated magnesium carbonate led to strength increase of over 20%. Frequent wet and dry cyclic curing also led to significant strength increase of over 30% compared to constant high relative humidity curing (Karthika, 2011). Significant enhancement of

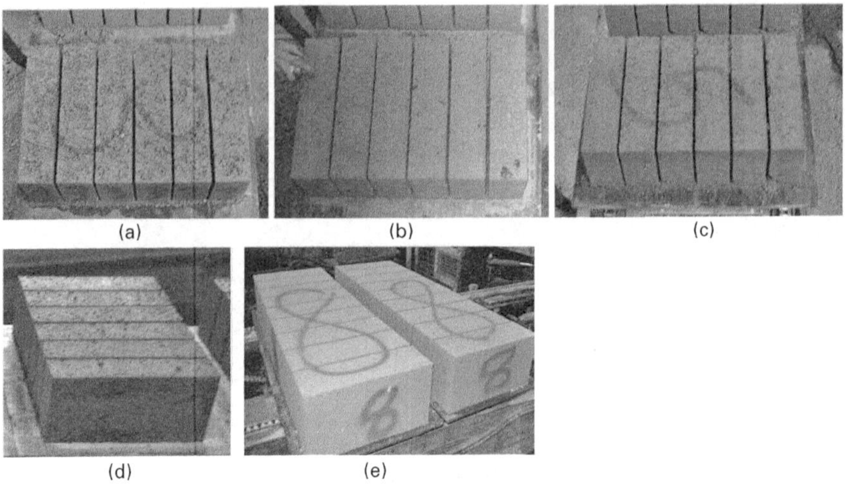

19.3 Freshly produced MgO porous blocks with different aggregate type and PC blocks as controls (Liska et al., 2012a): (a) PC with Lytag, (b) MgO with natural aggregates, (c) MgO with Lytag, (d) MgO with ash mix and crushed glass aggregates and (e) MgO with limestone and crushed glass aggregates (larger blocks).

the degree of carbonation, with almost complete carbonation was achieved with those optimisation studies (Unluer, 2012). Numerical modelling of the carbonation process and particle packing calculations are being used for further optimisation of these systems.

19.5.2 Concrete

In PC concrete formulations, using either PC alone or with 20% ash and/or slag content, as is common practice in China, the impact of up to 20% of more than ten different commercially available reactive MgOs, showed in general very little sign of strength enhancement (Li, 2012). However, one particular MgO, which was produced as part of a controlled commercial trials process in Xiuyan Deman Magnesium company in China, under controlled calcination conditions, resulted in up to 27% increase in compressive strength with the optimum of 4% reactive MgO content. All MgO contents led to an increase in strength with the 15% MgO content leading to up to 10% increase (Li, 2012). Despite the insignificant change to strength caused by the majority of the MgOs tested, there were indications that there was a reduction in permeability with the addition of a small percentage of MgO (Li, 2012) but more detailed investigation is being performed to verify this.

Investigations of the effect of MgO on PC-slag and PC-ash blends, in comparison to PC concrete, presented evidence of the presence of different hydration products in different compositions although again with minimal contribution to strength development (Tsang, 2010; March, 2011; Li, 2012). Incorporation of microsilica, at 10% content, in the cement blends above led to strength increases, which varied with curing age, when MgO was included at ~5% content compared to the effect of microsilica on the control without the MgO (March, 2011). As an activator for slag on its own, a 5% MgO content produced concrete with strengths of up to 28MPa (Li, 2012). Carbonated MgO and slag–MgO blends showed considerable promise in previous concrete applications (Hargreaves, 2010). The pH of those blends investigated is usually high enough to provide the alkalinity necessary to maintain the steel reinforcement passivation layer. Further detailed investigations of these blended compositions and applications are being conducted (Abdollazadeh, forthcoming).

The expansion behaviour of different commercial reactive MgOs was investigated in comparison to that of a commercial MgO expansive additive from China (Li, 2012). The results showed that a similar degree of expansion and trend with increasing MgO content confirming that the majority of commercially available reactive MgOs are close to the hard burned MgO end of the calcination process and properties. Reactive MgO is generally more expansive and expansion takes place much earlier on in the hydration process, leading to densification of the concrete matrix. Optimisation of shrinkage compensation of MgO in concrete is currently being investigated in detail (Lau, forthcoming).

19.5.3 Ground improvement

As PC, lime and slag–lime blends are frequently used for ground improvement applications, the potential for MgO in comparison was investigated in which carbonated MgO was compared with PC, and slag–MgO with slag–lime blends as well as with PC (Jegandan, 2010; Jegandan et al., 2010; Al-Tabbaa et al., 2011; Yi et al., 2012). The range of compositions and performance of those systems especially when the soil is sand, are similar to those of porous blocks and mortars. Studies using laboratory prepared samples as well as laboratory-scale *in situ* field application simulations showed that carbonated MgO, applied at 5 and 10% content, outperformed PC. It was possible to carbonate dry MgO mixed with sand by pumping gaseous CO_2 through the treated sand and carbonation occurred in under 3 hours. The compressive strength values achieved were in the same range as those achieved by corresponding PC–sand at 28 days. Increasing the CO_2 pressures speeded up the carbonation process although the final strength value achieved was always roughly the same and curing in a 20% CO_2 incubator also reached the same strength but in a longer curing period of 2 days.

Slag–MgO blends with MgO content of 5–50% and at up to 15% binder addition to sandy soils were tested both in the laboratory as well as in full-scale field application (Jegandan et al., 2010; Al-Tabbaa et al., 2009, 2011; Yi et al., 2012). The range of 10–20% MgO content was found to give the highest strength values and with the wider range of 5–30% reaching strength values similar to those of the corresponding PC binder. The elastic stiffness and strain at failure values obtained were in the same range as those of the PC binders. The slag–MgO binder outperformed slag–lime with strength values of up to four times higher being achieved. The permeability of the slag–MgO stabilised soils with MgO content, in the range of 5–30%, were over two orders of magnitude lower those of slag alone or with the higher MgO content blends, with the lowest permeability being around the 10% MgO content. In addition, those values were one order of magnitude lower than the permeability of the corresponding PC–sand mixes. This confirms a high degree of interaction between the MgO and slag and the microstructure confirmed the formation of hydration products that filled the sand pore spaces. Comparison between the carbonated MgO and the slag–MgO blends showed that both systems provided comparable results at 28 days for the same binder content addition, although the latter produced enhanced strength values in the longer term while the former would provide a lower carbon footprint binder through the CO_2 sequestration process. MgO alone and in blends with slag, in 5–20% content, and in blends with PC–ash, in 40% content, were tested as part of large field trials in the summer of 2011 (Al-Tabbaa et al., 2011, 2012a, 2012b) and have so far performed very well and in many case outperformed PC. Further detailed investigation of these alkali-activated binder and grout systems as well as investigation of their potential expansive self-healing properties is currently underway (Litina, forthcoming).

19.5.4 Performance in aggressive and extreme environments

Cements are employed in a wide range of aggressive and extreme environments including aggressive physical and chemical environments (acids, sulphates, freeze–thaw cycling), wastes and soil contamination, as well as extreme conditions, such as CO_2-rich, high salinity and high temperatures and pressures, as encountered in oilwells, geological sequestration and geothermal applications to name a few. The significant advantages of MgO in this respect is the unique properties of brucite as well as the hydrated magnesium carbonates as stated earlier in this chapter.

In terms of resistance to aggressive chemical environments, it is well established that PC does suffer to some degree in this respect. Investigations conducted using MgO-only porous blocks (Liska, 2009) showed superior resistance to acid (HCl) and sulphate ($MgSO_4$) attack, when cured both in ambient conditions, hence mainly through the brucite formed, and those cured under elevated CO_2 conditions, hence through the carbonation products (Liska and Al-Tabbaa, 2012). The ambient cured MgO blocks showed the highest resistance, and although they had lower initial strength, they showed no change in the strength even after 1 year of exposure, as the formed brucite did not enter into any detrimental reactions with the acid or sulphate. The carbonated MgO blocks showed good resistance, which although not as good as that of the ambient cured blocks, was much better than the performance of the corresponding PC blocks. There were no reaction products visible and it is possible that some of the nesquehonite formed converted to hydromagnesite/dypingite, which have lower strengths, leading to the partial loss in strength observed. The corresponding PC blocks showed 30% and 100% loss in strength due to acid and sulphate attack from their extensive reaction with the cement hydration products. Blending the PC with MgO presented some advantages particularly in terms of reducing the sulphate attack. Resistance to sulphuric and hydrochloric acids by blended cement concrete with up to 10% reactive MgO content showed that PC–slag concrete performed far better than PC and PC–ash concretes. The presence of 4–7% MgO enhanced the resistance of the PC–slag concrete to both acids (Tsang, 2010) for which the microstructure was denser, with brucite forming both on the outside and inside of the samples surrounding the portlandite and with very little ettringite being present.

In terms of waste and contaminant immobilisation, cement and lime are commonly used for heavy metals in particular, through the resulting high pH of ~12–13 in the pore solution, leading to the precipitation of the metal hydroxides, as well as some incorporation within hydration products. However, as the solubility of many heavy metals is lowest at pH of ~10, which is closer to that of brucite, MgO has been found to reduce metal concentrations much further (Iyengar and Al-Tabbaa, 2008; Al-Tabbaa et al., 2009, 2012b, 2012c). In addition, the lower pH will provide a far more appropriate environment for microbial activities which will facilitate biodegradation of organic contaminants (Al-Tabbaa et al., 2007; Kogbara et al., 2010; Kogbara, 2011). Concentrations of heavy metals lower than their solubility limits were found to leach out in some tests, suggesting factors, other than pH, play a major role in the immobilisation process (Iyengar, 2008). In order to better understand the performance mechanisms, the behaviour of MgO in comparison with PC, in contaminated aqueous solutions as well as pastes, was investigated. In general, the MgO performed better, and in some cases with up to three orders of magnitude reduced leachability. The results showed that MgO and PC do not significantly interact in metal removal, and that their effect is metal-specific. The hydration of the MgO was found to be accelerated by some heavy metals and retarded by others. It was also found that the metals were incorporated into the formed brucite. A detailed investigation of the immobilisation mechanisms of MgO and MgO blends is currently underway (Wang, forthcoming).

MgO and brucite have also been employed in nuclear waste containment (Xiong and Lord, 2008). Periclase type MgO, is the only engineered barrier certified by the USEPA (Environmental Protection Agency) for emplacement in the USA Waste Isolation Pilot Plant (WIPP) used for the permanent disposal of defence-related transuranic waste (US DOE, 2009), and brucite is employed as an engineered barrier in the Asse repository in Germany (Xiong and Lord, 2008). As a result extensive research has been conducted on the performance of periclase and brucite in this context in relevant geological, usually bedded salt, formations in terms of their hydration of the former and carbonation of the latter (Xiong and Lord, 2008). The low pH cements mentioned above were also investigated with nuclear waste immobilisation in mind (Iyengar, 2008) and are currently being investigated further in a large industry–academia collaboration.

A number of patented special cement formulations have been developed for applications in specific extreme environments, such as CO_2-rich environments, high salinity, extreme acidic conditions, high temperatures and high pressures to name a few, as encountered in many sectors including civil, petroleum and chemical processing (Schlumberger, 2008; Halliburton, 2009). In geothermal applications, such as oil production, geothermal energy generation and geological CO_2 sequestration, the long-term integrity of the cement linings in the respective wellbore shafts has always been a major concern (Nygaard, 2010), with the recent Deepwater Horizon oil spill being such an example (Anon, 2010). The expansive nature of MgO hydration and carbonation and the durability of the resulting products implies potential significant enhancement of thermal shrinkage compensation and resistance to such aggressive chemical environments in such applications. Research work is ongoing to assess potential advantages in the incorporation of reactive MgO in such cements and comparing their performance with commercially available CO_2-resistant cements (Mackay, 2012; Lim, 2012).

19.6 Sustainable production of reactive magnesia cement

The two current production routes of MgO, which are through the calcination of magnesium carbonates and from seawater, are currently not particularly sustainable. For the magnesite calcination process, the total theoretical energy requirement is 2415 MJ/t (Shand, 2006) and, in theory, the production of 1 t MgO would require 2.08 t of $MgCO_3$ and would result in the emission of 1.4 t of CO_2. The corresponding values for the production of PC are 1760 MJ/t (Taylor, 1990), with 1.5 t raw material required for the production of 1 t PC resulting in emissions of 0.85 t CO_2. The production route for MgO from seawater is currently a very energy-intensive process with a total theoretical energy requirement of 6700 MJ/t (Shand, 2006). The potential for using renewable energy sources for the low-temperature calcination process (~700 °C) and the significant CO_2 sequestration within porous blocks would significantly reduce the emissions above.

A number of research and development projects and commercial initiatives, which are directly or indirectly addressing more environmentally sustainable production of reactive MgO, have been reported and are ongoing. Examples include:

- development of a closed system low-temperature kiln (without releases to the atmosphere) in which grinding and calcination of $MgCO_3$ occur simultaneously; this system makes use of the heat generated in grinding to assist with calcination, thereby giving up to 30% better efficiency (Harrison and Cuff, 2006; Harrison, 2009);
- production of nesquehonite, through a low-energy low-maintenance precipitation process, from seawater or brine, including waste brines from desalination plants or produced water, through CO_2 sequestration (Ferrini et al., 2009; Constantz et al., 2009; Mignardi et al., 2011; Hassan forthcoming);
- production of reprocessed MgO from the hydroxide/carbonate slurry carbon capture cycle technology used in the leaching of MgO from low-grade $MgCO_3$ and dolomites (Fernandez et al., 1999; Butt et al., 1996; Bearat et al., 2002);
- precipitation of brucite from magnesium silicates through hydrochloric acid or sodium hydroxide digestion processes as intermediate stages within the mineral carbonation process of magnesium silicates (IPCC, 2004; Nduagu, 2008; Zevenhoven and Fagerlund, 2011);
- precipitation of hydrated magnesium carbonates from magnesium silicates through mineral carbonation at high pressures and temperatures (O'Connor et al., 2005; Huijgen and Comans, 2005; Sipilä et al., 2008)
- precipitation of magnesium carbonates through mineral carbonation of MgO-rich industrial waste streams such as slags (Stolaroff et al., 2005; Teir et al., 2007; Bobicki et al., 2012).

Current work at the University of Cambridge is investigating the second initiative above, producing nesquehonite from a range of different source brines and waste brines and assessing the performance of the resulting MgOs produced from different reactions followed by different calcinations conditions (Hassan, forthcoming; Jin, forthcoming). The produced nesquehonites with those from different reactions, e.g. Wang et al. (2008) and Cheng and Li (2009) are also being compared. The work is also performing carbon footprint calculations for these processes and comparing them with the processes from a number of the other initiatives above (Hassan, forthcoming).

It is expected that some of the above developments will lead to commercial production of MgO in much reduced or even carbon negative processes.

However, as most of these technologies are still at the research stage, there is currently very little information available to allow comparisons in terms of energy consumption and CO_2 emissions as well as commercial viability and cost although some work on this is on-going (Zevenhoven et al., 2011). Detailed life-cycle analyses investigating all emissions and environmental impacts, similar to those performed by Lei et al. (2011) and Van der Heede and De Belie (2012) on current cements, will be required. A number of the initiatives above are strongly linked to global CO_2 mineral sequestration initiatives for carbon management and there has been a recent trend towards scale-up and demonstration projects suggesting that large-scale applications will follow in the foreseeable future (Zevenhoven et al., 2011).

19.7 Future trends

As a recently developed material, reactive MgO in its wide-ranging promising blends and applications has seen significant research advances being made in the past few years. However, there is a significant amount of research work still to be undertaken which will require extensive further investigations and field validations before any commercial large scale use is feasible. Hence future research trends will see continuation of the extensive on-going initiatives in the investigation, testing and validations of a whole host of properties of reactive MgOs including in particular long-term performance and durability in order to understand with confidence the in-service performance of such cements. And just as important will be the investigations of the global availability of the raw materials with suitable low costs and low environmental impacts as well as assessment of the life cycle of reactive MgO cements in comparison with Portland cements. In addition, the current, very small annual global production rate of MgO will need to significantly increase. And finally, if the production of reactive MgO is largely associated with global carbon management schemes then it is likely that life-cycle analyses will result in favourable figures for their carbon footprint and their production will significantly increase as a result.

19.8 References

Abdollazadeh, A (forthcoming), Characteristics and performance of pervious concrete for green infrastructure. Forthcoming PhD Thesis, University of Cambridge, Cambridge, UK.

Al-Tabbaa, A, Harbottle, MJ and Evans, CW (2007), Robust sustainable technical solutions. Chapter 10 in *Sustainable Brownfield Regeneration: Liveable Places from Problem Spaces* (Dixon, Lerner and Raco Eds), Blackwell Publishing.

Al-Tabbaa, A, Liska, M, Jegandan, S and Barker, P (2009), Overview of project SMiRT for integrated remediation and ground improvement. International Symposium on Soil Mixing and Admixture Stabilisation, Okinawa, May 2009.

Al-Tabbaa, A, Barker, P and Evans, CW (2011), Recent innovations in soil mix technology for land remediation. *Institution of Civil Engineers Journal of Ground Improvement*, **164**(3), 127–137.

Al-Tabbaa, A, Liska, M, Ouellet-Plamondon, C, Jegandan, S, Shrestha, R, Barker, P, McGall, R and Critchlow, C (2012a), Soil mix technology for integrated remediation and ground improvement: From laboratory work to field trials. Proc. 4th International Conference in Grouting and Deep Mixing, New Orleans, February.

Al-Tabbaa, A, Liska, M, McGall, R and Critchlow, C (2012b), Soil mix technology for integrated remediation and ground improvement: Field trials. Invited Lecture, International Symposoium in Recent Research, Advances and Execution Aspects of Ground Improvement Works, IS-GI 2012, Brussels, May.

Al-Tabbaa, A, Liska, M, McGall, R, Critchlow, C and Sweeney, R (2012c), Soil mix technology for integrated remediation and ground improvement: Design & field performance, International Conference on Ground Improvement and Ground Control: Transport Infrastructure Development and Natural Hazard Mitigation (ICGI 2012), Wollongong, Australia, October.

Anon (2010), Cement chemistry partly to blame for BP oil spill. *Royal Society of Chemistry – Chemistry World News*, September.

Bearat H, Mckelvy MJ, Chizmeshya AVG, Sharma R and Carpenter W (2002), Magnesium hydroxide dehydroxylation/carbonation reaction processes: implication for carbon dioxide mineral sequestration. *Journal of the American Ceramic Society*, **85**(4), 728–742.

Bobicki, ER, Liu, Q, Xu, Z and Zeng, H (2012), Carbon capture and storage using alkaline industrial wastes. *Progress in Energy and Combustion Science*, **38**, 302–320.

Butt DP, Lackner KS, Wendt CH, Conzone S, Kung H, Lu Y-C and Bremster JK (1996), Kinetics of thermal dehydroxylation and carbonation of magnesium hydroxide. *Journal of the American Ceramic Society*, **79**(7), 1892–1898.

Cheng, W and Li, Z (2009), Precipitation of nesquehonite from homogeneous supersaturated solutions. *Crystal Research Technology*, **44**(9), 937–947.

Constantz, BR, Youngs, A and Holland, TC (2009), Methods of sequestering CO_2. US Patent No. 20090169452A1.

Davies, PJ and Bubela, B (1973), The transformation of nesquehonite into hydromagnesite, *Chemical Geology*, **12**, 289–300.

Dell, RM and Weller, SW (1959), The thermal decomposition of nesquehonite $MgCO_3{\cdot}H_2O$ and magnesium ammonium carbonate $MgCO_3{\cdot}(NH_4)_2CO_3{\cdot}4H_2O$. *Transactions of the Faraday Society*, **55**, 2203–2220.

Du, C (2006), A review of magnesium oxide in concrete, *Cement and Concrete World*, **58**, 53–63.

Dyer O (2003), A rock and a hard place, Eco-cement yet to cover ground in the building industry. The *Guardian*, 28 May.

Fernandez AI, Chimenos JM, Segarra M, Fernandez MA and Espiell F (1999), Kinetic study of carbonation of MgO slurries. *Hydrometallurgy*, **53**(2), 155–167.

Ferrini, V, De Vito, C and Mignardi, S (2009), Synthesis of nesquehonite by reaction of gaseous CO_2 with Mg chloride solution: Its potential role in the sequestration of carbon dioxide. *Journal of Hazardous Materials*, **168**, 832–837.

Gao, Y, Wang, H, Su, Y, Shen Q and Wang, D (2008), Influence of magnesium source on the crystallization behaviours of magnesium hydroxide. *Journal of Crystal Growth*, **310**, 3771–3778.

Gilbert, F. (1951), *Production of magnesia from seawater and dolomite*. British Periclase Company, Hartlepool, UK.

Halliburton (2009), WellLife III Cementing service for CO_2 environment, http://www.halliburton.com/public/cem/contents/Brochures/web/WellLife%20III%20CO2%20Brochure.pdf.

Hänchen, M, Prigiobbe, V, Baciocchi, R and Mazzotti, M (2008), Precipitation in the Mg–carbonate system – effects of temperature and CO_2 pressure. *Chemical Engineering Science*, **63**, 1012–1028.

Hargreaves, A (2010), Reactive magnesia cement pervious concrete. Final Year Undergraduate Project Report, University of Cambridge, Cambridge, UK.

Harrison, AJW (2003), New cements based on the addition of reactive magnesia to Portland Cement with or without added pozzolan. Proceedings of the CIA Conference: Concrete in the Third Millennium, CIA: Brisbane, Australia, see http://www.tececo.com/, as well as other publications on the same website.

Harrison, AJW (2008), Reactive magnesium oxide cements. United States Patent 7347896.

Harrison, AJW (2009), Tec-Eco Tec-Kiln, http://www.tececo.com/ products.tec-kiln.php/.

Harrison, AJW (2012), Various papers, presentations and reports at www.tececo.com.

Harrison, AJW and Cuff, C (2006), The CarbonSafe Alliance. Submission to the Australian Enquiry into Geosequestration Technology, http://www.tececo.com/files/political%20documents Aus House Reps Standing Committee Science Innovation Geosequestration Inquiry 10906.pdf.

Hassan, D (forthcoming), Sustainable production of reactive MgO. Forthcoming PhD Thesis, University of Cambridge, Cambridge, UK.

Henrist, C, Mathieu, J-P, Vogels, C, Rulmont, A and Cloots, R (2003), Morphological study of magnesium hydroxide nanoparticles precipitated in dilute aqueous solutions. *Journal of Crystal Growth*, **249**, 321–330.

Huijgen, WJJ and Comans, RNJ (2005), *Carbon dioxide sequestration by mineral carbonation – Literature review update 2003/2004*. Report No. ECN-C—05-022. Energy research Centre of the Netherlands.

IPCC (2004), Mineral carbonation and industrial uses of carbon dioxide. In the *Intergovernmental Panel on Climate Change Special Report on Carbon Dioxide Capture and Storage*, IPCC, Geneva, Switzerland, pp. 321–337.

Iyengar SR (2008), Application of two model magnesia-based binders in stabilisation/solidification Treatment Systems. PhD Thesis, University of Cambridge, UK.

Iyengar SR and Al-Tabbaa A (2008), Application of two novel magnesia-based cements in the stabilization/solidification of contaminated soils. GeoCongress 2008: The Challenge of Sustainability in the Geoenvironment, New Orleans, ASCE Publication, pp. 716–723.

Jeffery, GH, Bassett, J, Mendham, J and Denney, RC (1989), *Vogel's textbook of quantitative chemical analysis*, John Wiley & Sons, New York.

Jegandan, S (2010), Ground improvement with conventional and novel binders. PhD Thesis, University of Cambridge, Cambridge, UK.

Jegandan, S, Liska, M. Osman, A. A-M and Al-Tabbaa, A (2010), Sustainable binders for soil stabilisation. *ICE Journal of Ground Improvement*, **163**(1), 53–61.

Jin, F (forthcoming), Characterisation and performance of different source reactive MgOs and MgO blends. Forthcoming PhD Thesis, University of Cambridge, Cambridge, UK.

Karthika, S (2011), Low carbon MgO-based construction products, Final year undergraduate project report, University of Cambridge, UK.

Kogbara, RB, Al-Tabbaa, A and Iyengar, SR (2010), Utilisation of magnesium phosphate cements to facilitate biodegradation within a stabilised/solidified contaminated soil. *Water Air and Soil Pollution*, **216**, 411–427.

Kogbara, RB (2011), Process envelope for and biodegradation within stabilised/solidified contaminated soils. PhD Thesis, University of Cambridge, UK.

Lau, WY (forthcoming), Shrinkage reduction of MgO concrete. Forthcoming PhD Thesis, University of Cambridge, UK.

Lei, Y, Zhang, Q, Nielsen, C and He, K (2011), An inventory of primary air pollutants and CO_2 emissions from cement production in China, 1990–2020. *Atmospheric Environment*, **45**, 147–154.

Li, X (2012), Mechanical performance and durability of reactive magnesia cement concrete. PhD Thesis, University of Cambridge, Cambridge, UK.

Li, X, Bie, A, Tsang, SS, Al-Tabbaa, A and Qian, J (2010), Initial study on mechanical properties and durability of reactive MgO cement concrete. International Conference on Structural Faults and Repair, Edinburgh, June, Engineering Technics Press.

Lim, WJ (2012), MgO cements under CO_2 and geothermal conditions. Final Year Undergraduate Research Project Report, University of Cambridge, UK.

Liska, M (2009), Characterisation and properties of reactive magnesia cement in porous blocks. PhD Thesis, University of Cambridge, Cambridge, UK.

Liska, M and Al-Tabbaa, A (2008), Performance of magnesia cements in pressed masonry units with natural aggregates: Production parameters optimisation. *Construction and Building Materials*, **22**, 1789–1797.

Liska, M and Al-Tabbaa, A (2009), Ultra green construction: Reactive magnesia masonry blocks. *ICE Journal of Waste and Resources Management*, **162**(4), 185–196.

Liska, M and Al-Tabbaa, A (2012), Performance of reactive magnesia cement in porous blocks in HCl and $MgSO_4$ solutions. *Advances in Cement Research*, **24**, 221–232.

Liska, M, Vandeperre, LJ and Al-Tabbaa, A (2006), Mixtures of pulverised fuel ash, Portland cement and magnesium oxide: Characterization of pastes and setting behaviour. Proceedings of WASCON 2006, Apollo Graphic Production, Belgrade, pp. 563–573.

Liska, M, Vandeperre, L and Al-Tabbaa, A (2008), Influence of carbonation on the properties of reactive magnesia cement-based pressed masonry units. *Advances in Cement Research*, **20**(2), 53–64.

Liska, M, Al-Tabbaa, A, Carter, K and Fifield, J (2012a), Scaled-up commercial production of reactive magnesia cement pressed masonry units. Part I: Production. *Proceedings of the ICE Journal of Construction Materials*, **165**, 211–223.

Liska, M, Al-Tabbaa, A, Carter, K and Fifield, J (2012b), Scaled-up commercial production of reactive magnesia cement pressed masonry units. Part II: Properties of the blocks. *Proceedings of the ICE Journal of Construction Materials*, **165**, 225–243.

Litina, C (forthcoming), Green grouts. Forthcoming PhD Thesis, University of Cambridge, UK.

Mackay, W (2012). Advanced CO_2 resistant cements for geological storage of CO_2. Final Year Research Project Report, University of Cambridge, UK.

March, A (2011). Effects of additives on the performance of MgO concrete. Final year undergraduate project report, University of Cambridge, UK.

Mignardi, S, De Vito, C, Ferrini, V and Martin, RF (2011), The efficiency of CO_2 sequestration via carbonate mineralization with simulated wastewaters of high salinity. *Journal of Hazardous Materials*, **191**, 49–55.

Nduagu, E (2008), Mineral carbonation: Preparation of magnesium hydroxide [$Mg(OH)_2$] from serpentine rock, MSc(Eng.) Thesis, Åbo Akademi University, Finland.

Nygaard, R (2010). Well design and well integrity, Wabamun area CO_2 sequestration project (WASP), University of Calgary, Energy and Environment Systems Group, http://www.ucalgary.ca/wasp/Well%20Integrity%20Analysis.pdf.

O'Connor, WK, Dahlin, DC, Rush, GE, Gerdemann, SJ, Penner, LR and Nilsen, RP (2005). Aqueous mineral carbonation: Mineral availability, pretreatment, reaction parametrics, and process studies, DOE/ARC-TR-04-002. Albany Research Center, Albany, OR.

Pearce, F (2002), Green foundations. *New Scientist*, **175**(2351), 39–41.

Schlumberger (2008). EverCRETE: CO_2-resistant cement for long term zonal isolation, http://www.slb.com/~/media/Files/cementing/product_sheets/evercrete_ps.ashx

Shand, MA (2006), *The chemistry and technology of magnesia*. Wiley, New York.

Shi, C, Krivenko, PV and Roy, D (2006). *Alkali-activated cements and concretes*. Taylor & Francis, London.

Shen, W, Wang, Y, Zhang, T, Zhou, M, Li, J and Cui, X (2011), Magnesia modification of alkali-activated slag fly ash cement. *Journal of Wuhan University of Technology – Mater. Sci. Ed.*, **26**(1), 121–125.

Sipilä, J, Teir, S and Zevenhoven, R (2008), *Carbon dioxide sequestration by mineral carbonation – Literature review update 2005–2007*. Report VT 2008-1. Åbo Akademi University, Finland.

Sorel, S (1867), On a new magnesium cement, *Cognitive Science*, **65**, 102–104.

Stolaroff JK, Lowry GV and Keith DW (2005), Using CaO- and MgO-rich industrial waste streams for carbon sequestration. *Journal of Energy, Conservation & Management*, **46**(5), 687–699.

Taylor, FHW (1990), *Cement chemistry*. Academic Press, New York.

Teir, S, Eloneva, S, Fogelholm, C-J and Zevenhoven, R (2007), Carbonation of minerals and industrial by-products for CO_2 sequestration. Proceedings of 3rd International Green Energy Conference, Vasteras, Sweden, CD-ROM paper ID-129.

Tsang, SS (2010), The performance of MgO concrete in acid solutions. Final year undergraduate project report, University of Cambridge, UK.

Unluer, C (2012), Enhancing the carbonation of reactive magnesia cement-based porous blocks PhD thesis, University of Cambridge, UK.

Unluer, C and Al-Tabbaa, A (2011), Green construction with carbonating reactive magnesia porous blocks: effect of cement and water contents. Proc 2nd Int. Conf. on Future Concrete Dubai, December.

Unluer, C and Al-Tabbaa, A (2012), Effect of aggregates size distribution on the carbonation of reactive magnesia based porous blocks. 18th Annual International Sustainable Development Research Conference, Hull, June.

US DOE (2009). Title 40 CFR Part 191 Subparts B and C Compliance Recertification Application for the Waste Isolation Pilot Plant, Appendix MgO-2009, Magnesium oxide as an engineered barrier, US DOE Carlsbad Field Office, New Mexico.

USGS (2012), Minerals Information: Magnesium Statistics and Information, US Geological Survey. See http://minerals.usgs.gov/minerals/pubs/commodity/magnesium/mcs-2012-mgcom.pdf.

Van den Heede, P and De Belie, N (2012), Environmental impact and life cycle assessment (LCA) of traditional and 'green' concretes: Literature review and theoretical calculations. *Cement and Concrete Composites*, **34**, 431–442.

Vandeperre, LJ and Al-Tabbaa, A (2007), Accelerated carbonation of reactive MgO cements. *Advances in Cement Research*, **19**(2), 67–79.

Vandeperre, LJ, Liska, M and Al-Tabbaa, A (2006), Mixtures of pulverised fuel ash, Portland cement and magnesium oxide: Strength evolution and hydration products. Proceedings of WASCON 2006, Apollo Graphic Production, Belgrade, pp. 539–550.

Vandeperre, LJ, Liska, M and Al-Tabbaa, A (2007), Reactive magnesium oxide cements: properties and applications. *Proceedings of an International Conference on Sustainable Construction Materials and Technologies*, Taylor & Francis, London, pp. 397–410.

Vandeperre, LJ, Liska, M and Al-Tabbaa, A (2008a), Hydration and mechanical properties of magnesia, pulverised fuel ash and Portland cement blends. *ASCE Journal of Materials in Civil Engineering*, **20**(5), 375–383.

Vandeperre, LJ, Liska, M and Al-Tabbaa, A (2008b), Microstructures of reactive magnesia cement blends. *Journal of Cement & Concrete Composites*, **40**(8), 706–714.

Wagh, AS (2004), *Chemically bonded phosphate cements*, Elsevier, London.

Wang F (forthcoming), Novel binders for contaminated land remediation and their immobilisation mechanisms, Forthcoming PhD Thesis, University of Cambridge, UK.

Wang, SD and Scrivener, KL (1995), Hydration products of alkali activated slag cement. *Cement and Concrete Research*, **25**(3), 561–571.

Wang, Y, Li, Z and Demopoulos, GP (2008), Controlled precipitation of nesquehonite ($MgCO_3.3H_2O$) by the reaction of $MgCl_2$ with $(NH_4)_2CO_3$. *Journal of Crystal Growth*, **310**, 1220–1227.

Wei, J, Chen, Y and Yongxin, L (2006), The reaction mechanism between MgO and microsilica at room temperature. *Journal of Wuhan University of Technology – Mater. Sci. Ed.*, **21**(2), 88–91.

Wei, J, Yu, Q, Zhang, W and Zhang, H (2011), Reaction products of MgO and microsilica cementitious materials at different temperatures. *Journal of Wuhan University of Technology – Mater. Sci. Ed.*, **26**(4), 745–748.

Xiong, Y and Lord, AS (2008), Experimental investigations of the reaction path in the MgO-CO_2–H_2O system in solutions with various ionic strengths, and their applications to nuclear waste isolation. *Applied Geochemistry*, **23**, 1634–1659.

Yi, Y, Liska, M and Al-Tabbaa, A (2012), Initial investigation into the use of GGBS-MgO in soil stabilisation. Proc 4th Int. Conf. Grouting and Deep Mixing, New Orleans, February.

Zevenhoven, R and Fagerlund, J (2011), Mineral sequestration for CCS in Finland and abroad, World Renewable Energy Congress, Linköping, May.

Zevenhoven, R, Fagerlund, J and Songok, JK (2011), CO_2 mineral sequestration: developments towards large-scale application. *Review, Greenhouse Gases Science and Technology*, **1**, 48–57.

20
Nanotechnology for eco-efficient concrete

M. S. KONSTA-GDOUTOS, Democritus University of Thrace, Greece

DOI: 10.1533/9780857098993.4.544

Abstract: This chapter discusses the development of a materials science approach with an application to nanotechnology to optimize the processing and micro/nanoscale structure of cement-based materials reinforced with nanosized fibers and carbon nanotubes. The dispersion of multi-walled carbon nanotubes (MWCNTs) and carbon nanofibers (CNFs) for use in cementitious composites and specifically the effect of ultrasonication energy are discussed in detail. The nanomechanical properties of the CNT/CNF cementitious nanocomposites are examined through nanoindentation experiments. Additionally, the excellent reinforcing ability of both the MWCNTs and CNFs is demonstrated. Finally, the effect of the addition of silica nanoparticles on the degradation by calcium leaching is discussed and explained.

Key words: carbon nanotubcs, carbon nanofibers, nanoindentation, nanomechanical properties, nanosilica, calcium leaching.

20.1 Introduction

Nanoparticles, carbon nano-fibers (CNFs) and nanotubes (CNTs) offer a potential for developing new nano-enhanced materials/coatings for dramatically improving the environmental and energy performance of structural materials such as concrete. Nanomaterials can improve the strength, durability and adaptability of concrete; reduce material toxicity; improve insulation; and protect surfaces by applying new processes such as photocatalysis.

The fundamental properties of cementitious materials, such as concrete, are affected by the material properties at the nanoscale. The construction industry and research community currently investigates the changes in nanoscale properties with the addition of different mineral admixtures and CNFs/CNTs, as well as optimized processing techniques targeting the development of new materials with specific properties. One major area of focus is to develop nano-engineered materials with improved properties and to investigate the changes in the nanostructure, fracture properties, transport properties and durability of cement-based nanocomposites reinforced with highly dispersed CNTs and CNFs and/or with the addition of nanosilica. Silica nanoparticles can be of a great benefit to the transport properties of concrete since not only are the porosity and the portlandite content of the cementitious matrix reduced, they also trigger the modification of the CSH gel towards longer silicate chains. As a result, silica nanoparticles appear to greatly delay the degradation rate of the calcium leaching process in high-performance concretes, while the mechanical properties are enhanced (Gaitero et al. 2008).

All the above-mentioned topics present excellent examples of nanoscale additions that optimize the processing, modify the nanostructure of the matrix and develop nano-engineered materials with improved macroscopic properties and desired performance.

20.2 Nano-modification of cement-based materials

CNTs and CNFs are leading the way into nanotechnology. The concept of creating multi-phase, high-performance nanocomposites, is currently under development in a wide variety of matrices, with the emphasis to date being on polymers.

Cement-based materials, such as concrete, are typically characterized as quasi-brittle materials that exhibit low tensile strength and low strain capacity. Typical reinforcement of cementitious materials is usually done at the millimeter scale and/or at the microscale using macrofibers and microfibers, respectively. However, while microfibers delay the development of formed microcracks, they do not stop their initiation. The incorporation of fibers at the nanoscale will allow the control of the matrix flows and cracks at the nanoscale level and essentially create a new generation of a 'crack-free material' (Shah et al. 2009; Konsta-Gdoutos et al. 2010a).

CNTs are considered one of the most beneficial nanomaterials for nano-reinforcement. CNTs are long, hollow cylindrical structures with the length extending from few tens of nanometers to several micrometers and the outer diameters ranging from 2.5 to 30 nm. CNTs can be roughly divided into two categories: the single-walled CNTs (SWCNTs), and the multi-walled CNTs (MWCNTs). The SWCNT is a single graphite sheet that has been rolled into a hollow cylinder. MWCNTs consist of many concentric graphite layers, held in place by van der Waals forces. Their unique mechanical, electrical and chemical properties make them a revolutionary candidate for reinforcement of composite materials. The Young's modulus of an individual nanotube should be around 1 TPa and its density is about 1.33 g/cm^3 (Salvetat et al. 1999). Molecular mechanic simulations suggested that CNT fracture strains were between 10% and 15%, with corresponding tensile stresses on the order of 65–93 GPa (Belytschko et al. 2002). Their aspect ratios are generally beyond 1000.

20.3 Dispersion of multi-walled carbon nanotubes (MWCNTS) and carbon nanofibers (CNFs) for use in cementitious composites

The potential of using CNTs and CNFs as reinforcement for cementitious materials has not been fully realized, mainly because of the difficulties in processing. The two major challenges associated with the incorporation of CNTs and CNFs in cement-based materials are poor dispersion and cost. CNTs and CNFs tend to adhere together due to van der Waals intermolecular attraction, and it is particularly difficult to separate them individually (Groert 2007). To achieve good reinforcement in a composite, it is critical to have uniform dispersion of the nanoscale fibers within the matrix (Xie et al. 2005).

Attempts have been made to add CNTs and CNFs in cementitious matrices at an amount mainly ranging from 0.5 to 2.0 wt% (by weight of cement). Prior work on CNTs in liquid dispersions has focused on pre-treatment of the nanotube's surface via chemical modification. Makar and coworkers (Makar and Beaudin 2004; Makar et al. 2005) reported an ethanol/sonication technique for dispersing 2.0 wt% single walled CNTs (SWCNTs) in cement. The results obtained from scanning electron microscopy (SEM) and Vickers hardness measurements indicate that SWCNTs might affect the early hydration progress, producing higher hydration rates. Li et al. (2005) employed a carboxylation procedure to improve the bonding between 0.5 wt% multiwalled CNTs (MWCNTs) and cement matrix and obtained a 25% increase in flexural strength and a 19% increase in compressive strength. Saez de Ibarra et al. (2006) used gum arabic as a dispersing agent and measured the stiffness of cement samples reinforced with MWCNTs and SWCNTs using an atomic force microscopy (AFM) nanoindentation technique. They reported modest gains in the Young's modulus. Cwirzen et al. (2008, 2009) investigated the mechanical properties of cement matrices reinforced with different concentrations of MWCNTs. The results showed no increase in the flexural strength and a slight increase in compressive strength of the cement paste with the addition of CNTs. More recently, Collins et al. (2012) examined the use of air entrainer, styrene butadiene rubber, polycarboxylates, calcium naphthalene sulfonate, and lignosulfonate formulations for the dispersion of MWCNTs. It was found that the compressive strength of ordinary Portland cement (OPC) paste was improved 25% only when a polycarboxylate admixture was used at low water to cement ratio (W/C = 0.35).

Similarly to CNTs, the potential of using CNFs as reinforcement in cementitious matrices is still being explored. Sanchez and coworkers (Sanchez and Ince 2009; Sanchez et al. 2009) have investigated the effect of CNFs at different concentrations (0.005%, 0.02%, 0.05%, 0.5% and 2%) on the microstructure and mechanical properties of cement matrix with 10% silica fume. They also studied the effect of 0.5% CNFs surface treated with nitric acid. It was found that silica fume and surface treatment may have facilitated CNF dispersion and probably improved the interfacial interaction between CNFs and cement paste. However, dispersion was not complete since the CNFs were found to form agglomerates in various areas in the matrix, which effectively hampered any increase in the compressive and splitting tensile strength sought through the additive.

20.3.1 Effect of ultrasonication energy

Earlier studies conducted by Konsta-Gdoutos et al. (2010b) have focused on solving the two major challenges associated with the incorporation of nanoscale fibers in cementitious material: poor dispersion and cost. Experimental results have shown that by using a surfactant and applying ultrasonic energy, CNTs and CNFs can be effectively dispersed in an aqueous solution (Fig. 20.1) (Konsta-Gdoutos et al. 2010a). Ultrasonication is a common physical technique used to disperse CNTs into base fluids. Ultrasonic processors convert line voltage to mechanical vibrations. These mechanical vibrations are transferred into the liquid by the probe creating pressure waves. This action causes the formation and violent collapse of microscopic bubbles. This phenomenon, referred to as cavitation, creates millions of shock waves, as well as elevated pressure and temperature into the liquid. The cavitational collapse last only a few microseconds. Although the amount of energy released by each individual bubble is small, the cumulative effect causes extremely high levels of energy to be released. Objects and surfaces within the cavitation field are dispersed.

In a typical procedure, CNT suspensions were prepared by mixing the MWCNTs in an aqueous solution containing different amounts of surfactant and the resulting dispersions were sonicated at room temperature. Constant energy was applied to the samples using a 500 W cup-horn high-intensity

(a) (b)

20.1 SEM image of cement paste fracture surfaces reinforced with undispersed and dispersed MWCNTs (a) and (b), respectively.

ultrasonic processor with a cylindrical tip and temperature controller. The sonicator was operated at an amplitude of 50% so as to deliver an energy of 1900–2100 J/min at cycles of 20 s in order to prevent overheating of the suspensions.

20.3.2 Effect of ultrasonication energy on the rheological characteristics

The influence of MWCNT dispersion on the rheological properties of cement paste samples was investigated. The rheological characteristics of the samples with the sonicated dispersions were measured using a Haake Rheostress 150 rheometer with a 20 mm concentric cylinder measurement system. A steady stress protocol similar to the one proposed by Yang et al. (2006) was applied to determine the viscosity dependence on stress. Cement paste was placed in the rheometer immediately after mixing. Each sample was presheared at 100 s^{-1} for 200 s, and then allowed to rest for 200 s. A low stress of 4.5 Pa was applied to the sample, and the stress was increased stepwise using the protocol shown in Fig. 20.2. Initial stresses were chosen to be higher than the yield stress of the material. The samples were held at each stress condition for 40 s. The holding time was necessary to ensure that an equilibrium flow had been reached. Apparent viscosity (η) as a function of time was monitored and recorded during the test. The viscosity at each shear stress was obtained by averaging the values at the last 10 seconds that corresponded to the equilibrium region.

The effect of ultrasonic energy on the dispersion of the CNTs was investigated by measuring the rheological properties of cement paste samples reinforced with MWCNTs under steady shear stress. Rheology is a method commonly used to study the microstructure in nanotube dispersions. Under low shear stress, CNT agglomerates control the viscosity of cement nanotube

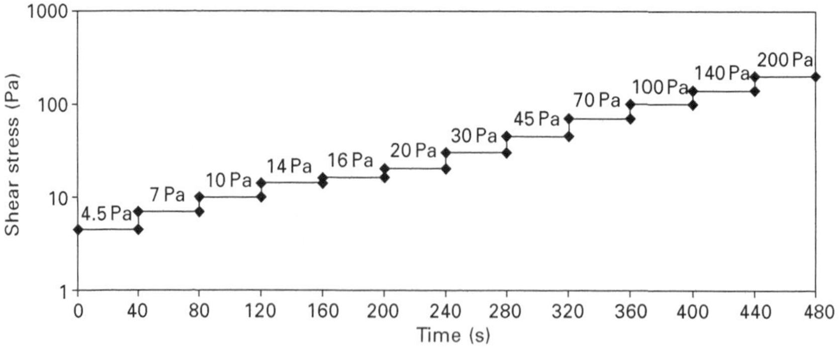

20.2 Rheology measuring protocol.

suspensions. Therefore, suspensions with larger-scale agglomerates exhibit higher viscosity (Yang et al. 2006).

Four different CNT aqueous-surfactant dispersions were studied with a surfactant to CNTs weight ratio of 1.5, 4.0, 5.0 and 6.25, respectively. The CNT content was kept constant in the solution at an amount of 0.16 wt% of water. Two mixes were prepared for each surfactant to CNT ratio, one with the use of ultrasonic energy and one without. Then the dispersions were mixed with cement and the rheological properties of the cementitious composites samples were investigated using the mixing and measuring protocol described previously.

Figure 20.3 shows the rheological behavior of the cementitious composites (W/C = 0.5) with a surfactant to CNTs ratio of 6.25 that were treated with and without the use of ultrasonic energy. The results are compared to the plain cement paste containing the same amount of surfactant (CP + SF). All samples exhibit the typical shear thinning response of cement paste. At low shear stress the viscosity is high, while at high shear stress (>70 Pa) the viscosity is decreasing and reaches a 'plateau' region in which the fluid seems to have a constant viscosity. The non-sonicated dispersion exhibits viscosity at low stress (14 Pa) of up to 0.13 Pa s while the sonicated dispersion exhibits a viscosity of 0.09 Pa s which is very close to the viscosity of plain cement paste (0.07 Pa s). As expected, at low stress conditions the application of ultrasonic energy controls the dispersion of the CNTs. Under high shear stress (>70 Pa) the agglomerates can be broken down by the fluid motions so the viscosities of the suspensions with and without sonication are similar. Analogous results, shown in Table 20.1, were obtained with a surfactant to CNTs ratio of 1.5, 4.0 and 5.0.

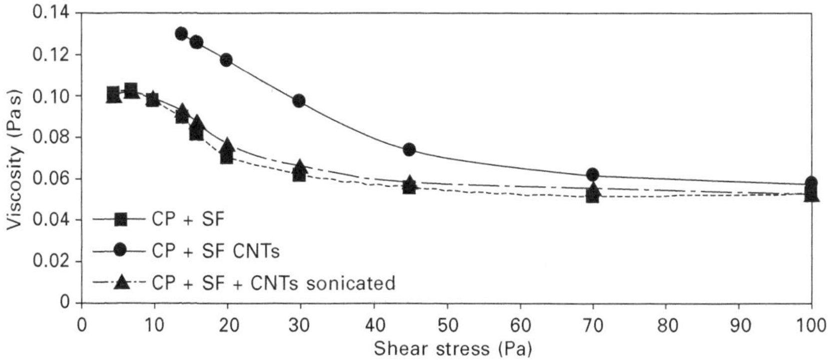

20.3 Steady shear viscosity of cement paste samples W/C = 0.5 reinforced with CNTs suspensions at a surfactant to CNTs weight ratio of 6.25.

Table 20.1 Viscosities (Pa s) at low shear stress condition

Surfactant CNTs weight ratio	1.5	4.0	5.0
CP + SF	0.86	0.27	0.20
CP + SF + CNTs	0.94	0.39	0.26
CP + SF + CNTs sonicated	0.89	0.29	0.20

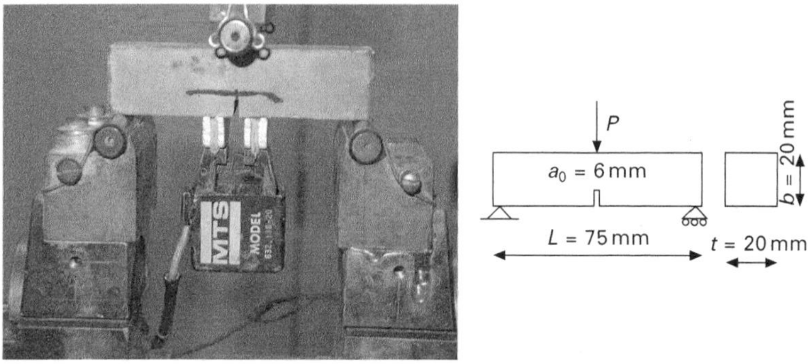

20.4 Three-point bending test set-up and configuration.

20.4 Mechanical properties

The mechanical performance of the MWCNT cementitious nanocomposites was evaluated by fracture mechanics tests. Beam specimens of 20 × 20 × 80 mm, prepared using the mixing protocol described previously, were tested by three-point bending at the age of 3, 7 and 28 days, as illustrated in Fig. 20.4. Three replications were made for each nanocomposite tested. Before testing, a 6 mm notch was introduced to the specimens using a water-cooled diamond saw. The test was performed using an 89 kN, MTS servo-hydraulic, closed-loop testing machine. A clip gauge was used to measure the crack mouth opening displacement (CMOD). The CMOD was used as a feedback to produce a stable fracture at a rate of 0.012 mm/min. The load and the CMOD were recorded during the test. The ASTM C348 was followed to determine the average values of the flexural strength and Young's modulus. The specimens that result in strengths differing by more than 10% from the average value of all test specimens made from the same sample and tested at the same period were not considered in determining the flexural strength. After discarding the strength values, if fewer than two strength values were left for determining the flexural strength at any given period a retest was made. Following the above-mentioned procedure ensures that the variations of the test results are not significant and do not affect the conclusions. The Young's modulus was calculated from the load versus CMOD results using the two-parameter fracture model by Jenq and Shah (Shah et al. 1995).

20.4.1 Effect of surfactant

The average values of maximum fracture load for cement paste samples containing MWCNTs with different surfactant-to-MWCNT ratios for 28 days are plotted in Fig. 20.5. It is observed that samples treated with different amounts of surfactant exhibit higher fracture load than the sample with no surfactant. The samples with surfactant to MWCNT ratio of 4.0 give a higher average load increase at all ages. Surfactant-to-MWCNT ratios either lower or higher than 4.0 produce specimens with less load increase. A possible explanation could be that at lower surfactant-to-MWCNT ratios, fewer surfactant molecules are absorbed to the carbon surface and the protection from agglomeration is reduced. At higher surfactant-to-MWCNT ratios, bridging flocculation can occur between the surfactant molecules. Too much surfactant in the aqueous solution causes the reduction of the electrostatic repulsion forces between the MWCNTs (Yu et al. 2007). The results indicate that for effective dispersion, there exists an optimum weight ratio of surfactant to MWCNTs close to 4.0.

Table 20.2 compares the flexural strength increase of cement paste nanocomposites achieved in this study with results of mortars and cement paste nanocomposites reported in the literature, relative to the concentration

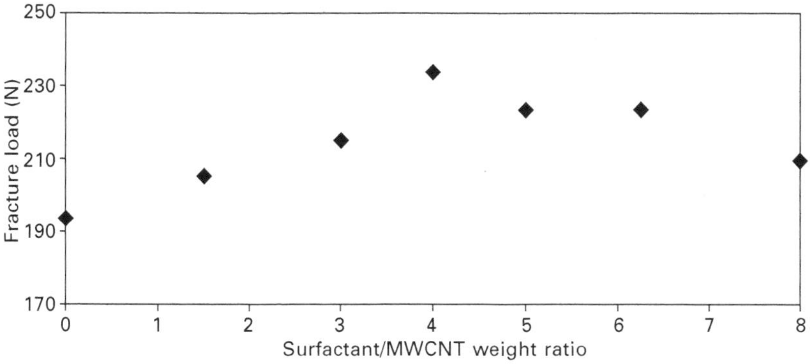

20.5 Fracture load of 28-day W/C = 0.5 cement paste reinforced with MWCNTs 0.08 wt%.

Table 20.2 Comparison of flexural strength increase in cementitious nanocomposites reinforced with CNTs

Researchers	Nanocomposites	Amount of CNTs (wt% of cement)	Flexural strength increase (%)
Li et al. (2005)	Mortars	0.50	25
Cwirzen et al. (2008)	Cement paste	0.042	10
Konsta-Gdoutos [present work]	Cement paste	0.08 and 0.048	25

of CNTs used. It is observed that a similar gain (25%) in the flexural strength of mortars nanocomposites was obtained by Li et al. (2005), using ten times higher concentration of CNTs (0.5 wt% of cement). Comparing the results of this study with the results obtained by Cwirzen et al. (2008) it is observed that, despite the similar CNTs concentration used for the preparation of cement paste nanocomposites, the flexular strength increase for Cwirzen et al. was 10%. The extent of the flexural strength improvement of the nanocomposites in this work was attributed to the successful nanofiber dispersion. Effective dispersion of the CNTs in the cement matrix resulted in the reduction of the fiber-free area in the material and in an increase in the mechanical performance of the nanocomposite. By achieving effective dispersion of small amounts of CNTs, the reinforcing effect of CNTs in the matrix increases and the cost of the nanocomposite is reduced.

20.4.2 Reinforcing efficiency of CNT/CNF in cementitious composites

The flexural strength rate of the nanocomposites reinforced with either MWCNTs or CNFs up to the age of 28 days of hydration is shown in Fig. 20.6. The samples reinforced with nanofibers at all ages exhibit higher flexural strength than plain cement paste. In particular, an increase up to 25% is achieved when MWCNTs are utilized. The use of CNFs results in an increase of the flexural strength up to 45%. Comparing the two nanocomposites, it is observed that despite the fact that MWCNTs exhibit a higher aspect ratio (due to their smaller diameter and larger amounts of nanotubes) it reinforces the cement matrix since the concentration of the fibers is constant, CNFs provide the matrix with the ability to carry higher flexural loads at lower

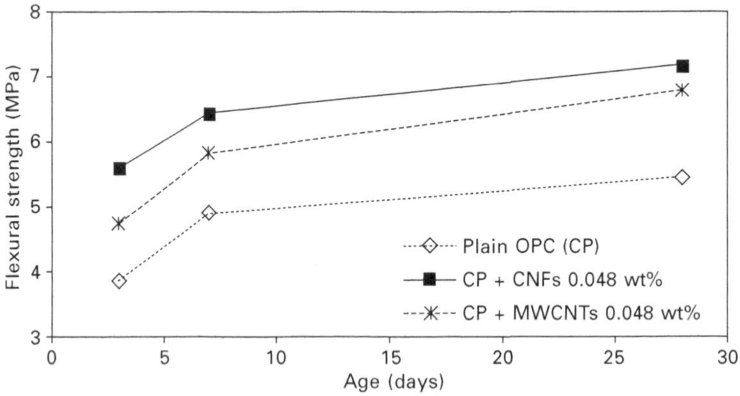

20.6 Flexural strength of plain cement paste (W/C = 0.5) and cement paste reinforced with 0.048% by weight of cement MWCNTs or CNFs.

strains. A possible explanation could be that the bonding between the CNFs and the matrix is enhanced due to the unusual outer surface texture of the carbon nanofibers. The CNFs used in this study exhibit graphite planes which extend beyond the diameter of the nanofiber and are present along the circumference of the fiber. These edges probably help anchor the fiber in the matrix, preventing interfacial slip and enabling more sufficient load transfer across nanocracks and pores.

The Young's modulus of the samples reinforced with either MWCNTs or CNFs at the age of 3, 7 and 28 days is illustrated in Fig. 20.7. Similar to the flexural strength results, samples reinforced with nanofibers clearly exhibit improved Young's modulus over OPC specimens. Specifically, an increase of the Young's modulus of 44% to 50% over plain cement specimens is achieved with the use of MWCNTs. In the case of the samples with CNFs, an increase of at least 50% is observed.

20.5 Mechanical properties at the nanoscale

The nanomechanical properties of the MWCNT/ CNF cementitious composites were investigated using a triboindenter. The triboindenter is a special type of nanoindenter that combines nanoindentation to determine the local properties of the material at the nanoscale, with high-resolution *in situ* scanning probe microscopy (SPM) imaging that allows pre- and post-test observation of the sample (Fig. 20.8). A Berkovich tip with a total included angle of 142.3 degrees was used for indentation and SPM imaging. Multiple cycles of partial loading and unloading were used to make each indention, eliminating creep and size effects (Nemecek 2009). The Oliver and Pharr method was used to determine the mechanical properties, where the indentation modulus is calculated from the final unloading curve (Oliver and Pharr 1992). Prismatic

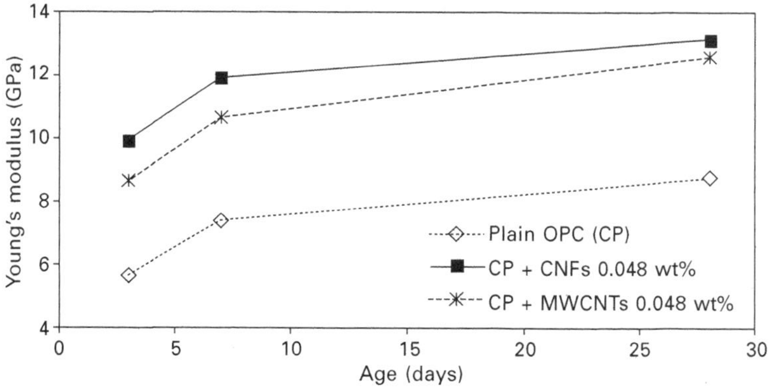

20.7 Young's modulus of plain cement paste (W/C = 0.5) and cement paste reinforced with 0.048% by weight of cement MWCNTs or CNFs.

20.8 Triboindenter.

specimens of $25.4 \times 6.35 \times 6.35$ mm were prepared and cured in water saturated with lime for 28 days. After curing, specimens were kept in acetone. Before testing, thin sections of approximately 5 mm were cut out of the specimens and mounted using an adhesive (softening temperature 71 °C) on a metal sample holder for polishing. Polishing is very important for nanoindentation. Eliminating the sample roughness without any damage is necessary for the determination of reliable local mechanical properties (Metaxa et al. 2009). Samples were polished using silicon carbide paper discs of gradation 22, 14, 8 and 5 μm and diamond lapping films of gradation 6 and 3 μm. Water was used in the first two and last two gradations. At every step an optical microscope was used to check the effectiveness of the polishing. As a final step the polished samples were ultrasonically cleaned in water for 1 minute using a bath sonicator to remove polishing debris.

An environmental scanning electron microscope (ESEM) at low vacuum mode was used to further investigate the effectiveness of the polishing procedure and find representative areas of the samples. Before nanoindentation, representative areas of the nanocomposites were also imaged using the Berkovich tip of the triboindenter to provide surface information at millimeter to nanometer scale. Nanoindentation was performed in a 12×12 grid (10 μm between adjacent grid points). This procedure was repeated in at least two different areas on each sample. Some of the nanoindentation data were discarded due to the irregular nature of the load–displacement plot, which could be due to the presence of large voids or cracking in the material (Metaxa et al. 2009).

Nanoindentation tests were performed on 28 days cement paste samples reinforced with 0.08 wt% short CNTs. Figure 20.9 shows the probability plot

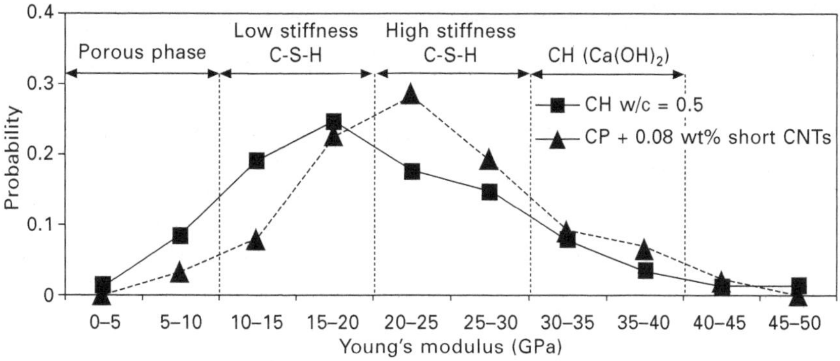

20.9 Probability plots of the Young's modulus of 28-day cement paste and cement paste reinforced with 0.08 wt% short CNTs with w/c = 0.5.

of the Young's modulus of plain cement paste and cement paste reinforced with MWCNTs (w/c = 0.5). Similar experiments, using the same type of cement and testing procedure, were carried out by Mondal (2008) on 28-day cement paste specimens (w/c = 0.5), who used a peak analyzing protocol to fit four normal distributions to the probability plot of the Young's modulus corresponding to the porous phase, low stiffness C-S-H, high stiffness C-S-H and calcium hydroxide. The same analyzing protocol was used by Constantinides and Ulm (2007) on the nanoindentation results of 5 months white cement paste specimens (w/c = 0.5).

The probability plot of Young's modulus of cement paste reinforced with MWCNTs is in agreement with the probability plot of Mondal and the frequency plot of Constantinides and Ulm. However, while the peak of the distribution of the nanoindentation modulus of both Mondal and Constantinides and Ulm and the plain cement paste falls in the area of 15–20 GPa and represents the low stiffness C-S-H, the peak of the probability plot of the Young's modulus of the MWCNTs nanocomposite was found to be in the area of 20–25 GPa which represents the high stiffness C-S-H. This result suggests that the incorporation of CNTs increased the amount of high stiffness C-S-H.

Similar results were obtained with nanosilica cement composites by Gaitero (Gaitero 2008). Furthermore, the area between 0 and 15 GPa is attributed to a material region for which the indentation response is dominated by high porosity. Comparing the probability plot of the Young's modulus of the MWCNTs nanocomposite with the plain cement paste it is observed that the probability of Young's modulus below 10 GPa is reduced from 0.10 to 0.03 in the case of the samples with MWCNTs. These results suggest that MWCNTs reduce the nanoporosity of cement paste by filling the gaps between the C-S-H gel. This result also agrees well with the experimental results of Li et al. (2005).

20.6 Calcium-leaching with nanosilica particles addition

Much of the research to date on nanomodification of concrete with nanoparticles has been with nano-silica (nano-SiO_2) (Bjornstrom et al. 2004; Ji 2005, Jo et al. 2007; H. Li et al. 2004a, 2006; D.F. Lin et al. 2008; K.L. Lin et al. 2008; Nazari and Riahi 2010, 2011a,b; Qing et al. 2007, 2008; Sobolev et al. 2009) and nano-titanium oxide (nano-TiO_2) (Chen et al. 2012; H. Li et al. 2006, 2007). A few studies exist on including nano-iron oxide (nano-Fe_2O_3) (H. et al. Li 2004a), zinc oxide (Nazari and Riahi 2011c), nano-alumina (nano-Al_2O_3) (Z. Li et al. 2006; Nazari and Riahi 2011d, 2011e), and nanoclay particles (Chang et al. 2007; Kuo et al. 2006). SEM images of four different types of nano and microsilica are depicted in Fig. 20.10 (Quercia et al. 2012). Nanosilica was shown to accelerate the hydration reactions of C_3S and to improve concrete workability and strength (Givi et al. 2011; Jalal et al. 2012; H. Li et al. 2004b; Sobolev and Ferrada-Gutiérrez 2005; Sobolev et al. 2009; Zhang and Islam 2012). Also it was found to increase resistance to water penetration (Ji 2005) and to help control the leaching of calcium (Gaitero et al. 2008).

Calcium leaching is a degradation process concerning the progressive dissolution of the cement paste as a consequence of the migration of calcium ions to the aggressive solution. The kinetics of the process depend of on several parameters with the porosity (the natural path for chemical attack), the composition of the paste (each phase degrades at a different rate), and the nature of the aggressive solution (generally soft water) being the most important ones. Although the phenomenon of calcium leaching is nowadays reasonably well understood, a way of stopping or, at least, reducing it has not been found yet. Silica, generally in the form of silica fume, has long been used to improve the performance of the cement paste in terms of strength and refinement/reduction of the porosity. Furthermore, this takes place by mean of a pozzolanic reaction that results in the reduction of the amount of portlandite, the hydrous phase most severely affected by calcium leaching. Silica fume is composed of silica particles with diameters over 100 nm, i.e. microsilica, whereas silica nanoparticles have typical diameters well below 100 nm. The use of silica nanoparticles has several advantages in comparison to other types of silica (Qing et al. 2007). As a consequence of their reduced size the nanoparticles act as nucleation site for the growth of hydration products, accelerating the hydration rate. Furthermore, their large surface area and purity (up to 99 %), in addition to their amorphous nature, provide them with great reactivity.

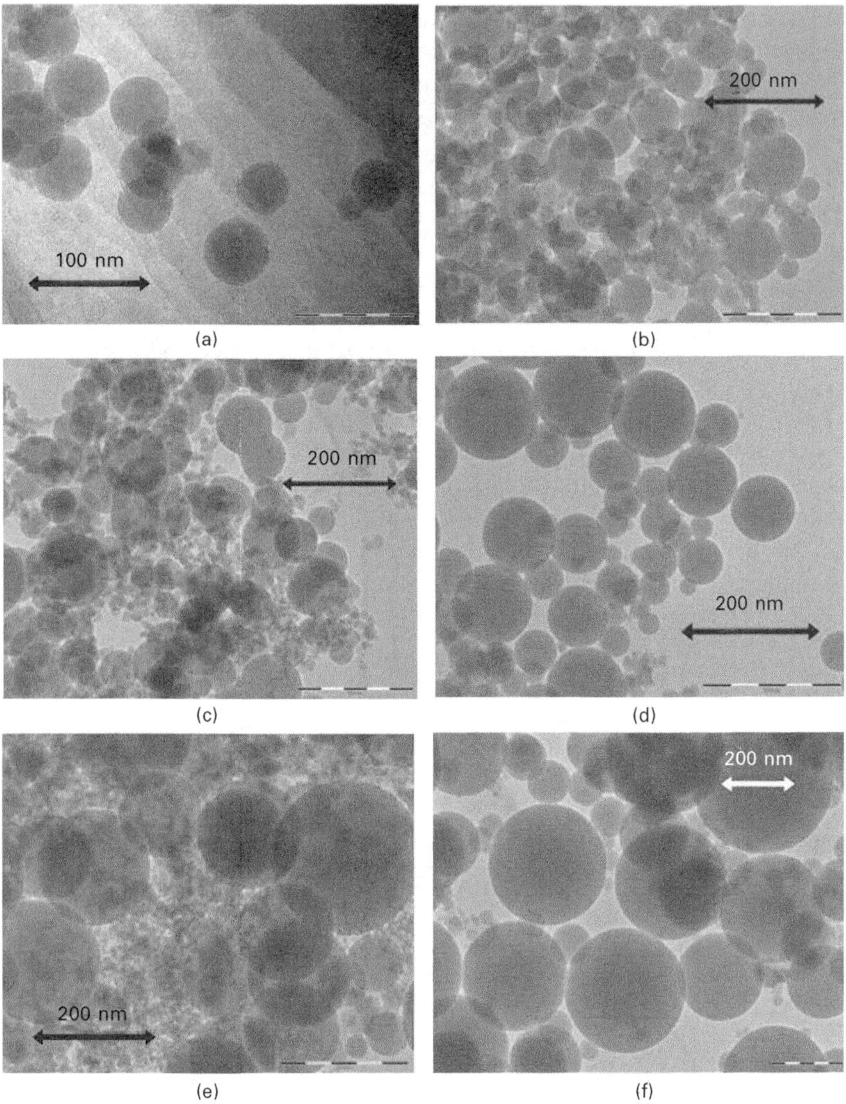

20.10 SEM images of (a, b) two commercial colloidal nanosilica suspensions with different properties, (c) commercial nanosilica fume in powder form, (d) standard microsilica fume in slurry form and (e, f) two experimental synthetic pyrogenic silica samples with different specific Brunauer–Emmett–Teller (BET) surface area (Quercia et al. 2012).

Samples of four different types of commercial silica nanoparticles, three colloidal dispersions (CS1, CS2, CS3) and one type of dry powder (ADS)), at a concentration of 6 wt.% were prepared based on the procedure discussed in Gaitero et al. (2008, 2009). The samples were cured in a saturated lime solution at room temperature for 28 days (t_0). Afterwards, they were moved into a bath containing a 6 M ammonium nitrate solution where they stayed for 9, 21, 42 or 63 days for the accelerated calcium leaching (t_1, t_2, t_3 and t_4 respectively) procedure.

The compressive strength and the porosity results are shown in Fig. 20.11. The addition of silica nanoparticles to the cement paste increased the overall strength of the paste and helped retain it along the degradation process. Comparison between the different types of additions used revealed that the colloidal dispersions were much more effective reducing the effects of the degradation than the ADS. It is also observed that a very good correlation exists between the macroscopic strength and the porosity evolution. At t_0, the reference specimen (REF) was the one with the highest porosity, followed by ADS and then the pastes with colloidal silica. As soon as the degradation began (t_1), the total pore volume increased dramatically in REF followed by ADS, while the rest of the specimens were barely affected until t_2. This behavior can be attributed to the better dispersion of colloidal silica throughout the paste which resulted in a more homogeneous porosity and

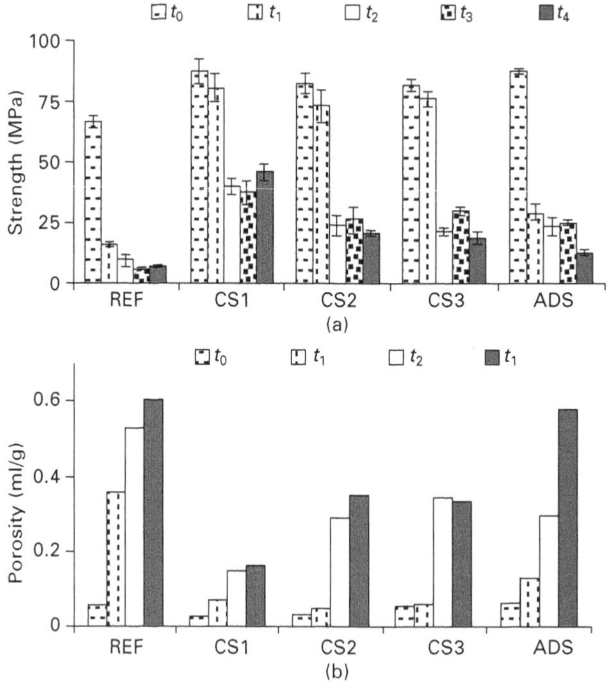

20.11 (a) Compressive strength and (b) total pore volume, measured by mercury intrusion porosimetry, as a function of the degradation time.

portlandite distribution compared to ADS. After t_2, the increase in porosity was more progressive in all the pastes.

X-ray powder diffraction spectra proved that the great changes in porosity and strength undergone by the specimens during the early stages of the degradation process (t_1 for REF and ADS, and t_2 for CS_1, CS_2, and CS_3) could be attributed almost entirely to the dissolution of portlandite (Fig. 20.12). Therefore, the reduction in the amount of portlandite, during the curing process because of the reaction with the silica nanoparticles, contributed significantly to reduce the negative consequences of calcium leaching. Portlandite dissolution and the consequent porosity increase were also accompanied by a complete hydration of all the anhydrous cement present at the paste. The slower degradation observed afterwards was a consequence of the more progressive dissolution of other phases like ettringite or the C-S-H gel itself.

^{29}Si magic angle spinning nuclear magnetic resonance (MAS-NMR) provided information about the changes taking place in the atomic structure of the C-S-H gel (Gaitero et al. 2006). According to Fig. 20.13, the addition of silica nanoparticles resulted in an increase in the average segment length. As soon as the degradation began, the average segment length was sharply reduced for all samples. Its value, however, was always greater for the samples with nanosilica than the reference samples. Furthermore, as a consequence of the calcium loss, chains began to merge together resulting in an increase of the degree of polymerization. It is observed that, during calcium loss, the samples with nanosilica exhibit lower degree of polymerization. There is a good correlation between the average segment length and the degree of polymerization. Samples with higher segment length exhibit lower degrees

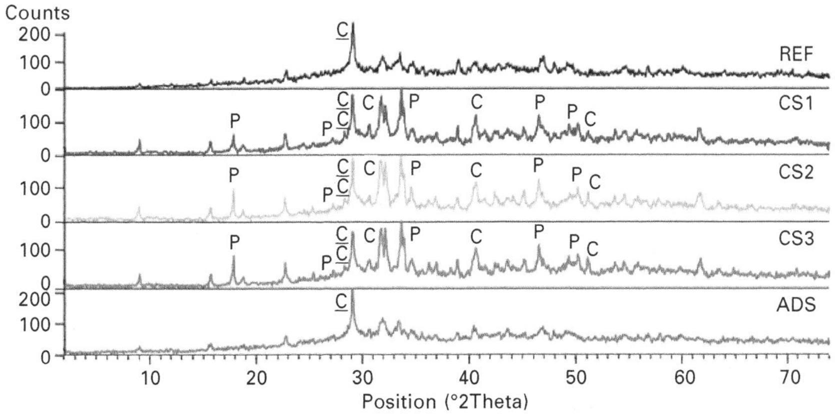

20.12 X-ray powder diffraction spectra after 9 days of degradation (t_1). C: Clinker, C̲: Calcite, P: Portlandite.

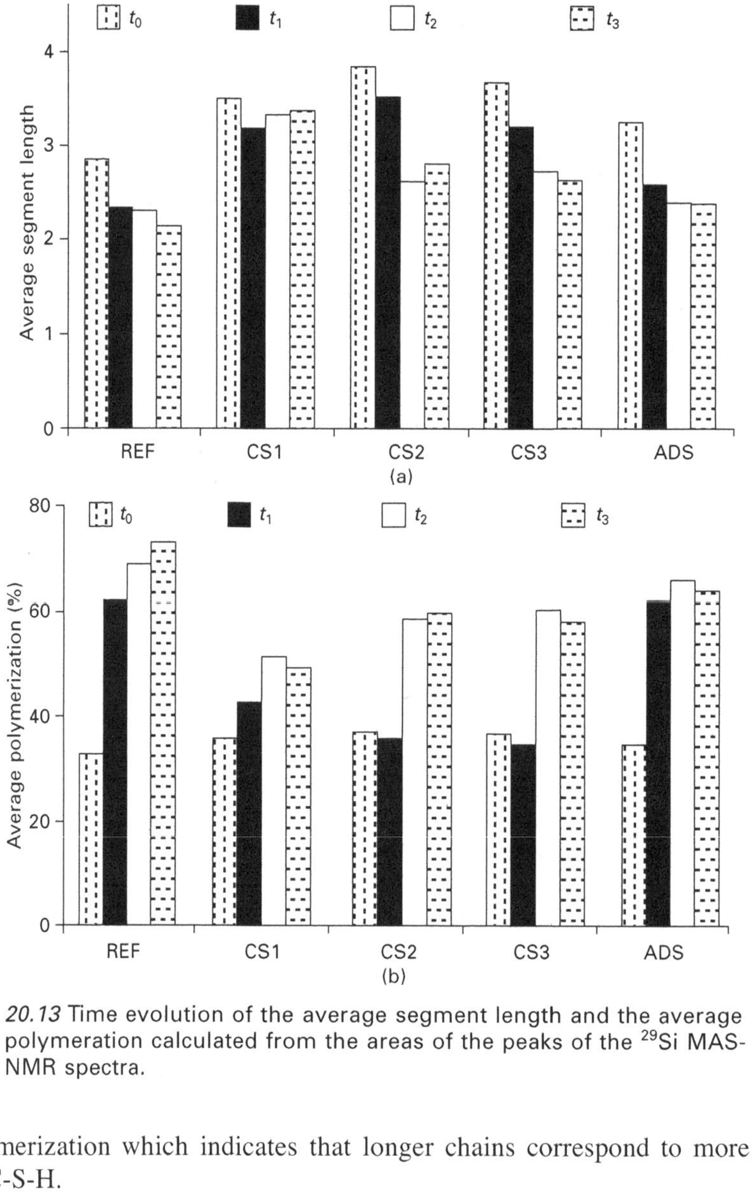

20.13 Time evolution of the average segment length and the average polymeration calculated from the areas of the peaks of the ^{29}Si MAS-NMR spectra.

of polymerization which indicates that longer chains correspond to more stable C-S-H.

20.7 Future trends

Nanotechnology is bringing a revolutionary performance improvement to concrete. Recent advances in the field of nanotechnology have led to the development of advantageous nanoscale fibers for nanoreinforcement, making possible the development of new multifunctional high performance and advanced sensing materials that could be used effectively as construction materials and at the same time, act as sensors to monitor the health of the structures. However, despite these exciting developments the study of nanotechnology for green concrete and the envisioned benefits are still young science:

- We still need to advance methods of dispersion. To achieve good reinforcement in a composite, it is critical to have uniform dispersion of the nanoscale fibers within the matrix. So far the major challenge associated with the incorporation of CNTs and CNFs in cement-based materials is dispersion.
- The effect of nanoscale fibers on the electrical conductivity of CNTs/CNFs nanocomposites with cement paste and mortar matrix so as to specify the optimum types of nanoscale fibers, and the study of the piezoresistive response of the nanocomposites and the capability of CNTs/CNFs to be used as sensors in concrete are currently considered key research topics for the application of cementitious nanocomposites in the construction industry.
- SiO_2 nanoparticles can definitely improve the environmental and mechanical performance of concrete. However, further research is needed to bridge the gap between potential benefits, reducing pore size distribution and increasing shrinkage. Recent studies suggest that the use of nano-SiO_2 increased the 7-day shrinkage by 80% compared with an equivalent mix with micro-silica (Senff et al., 2010).

20.8 Acknowledgements

The author would like to acknowledge the support of the National Strategic Reference Framework (NSRF) – Research Funding Program: 'Thales – Democritus University of Thrace – Center for Multifunctional Nanocomposite Construction Materials' (MIS 379496) funded by the European Union (European Social Fund – ESF) and Greek national funds through the Operational Program 'Education and Lifelong Learning'.

20.9 References

Belytschko, T; Xiao, S.P; Schatz, G.C.; Ruoff, R. (2002), 'Atomistic simulations of nanotube fracture', *Physical Review B* **65** (23), 235430–235437.

Bjornstrom, J.; Martinelli, A.; Matic, A.; Borjesson, L.; Panas, I. (2004), 'Accelerating effects of colloidal nano-silica for beneficial calcium–silicate–hydrate formation in cement', *Chemical Physics Letters* **392** (1–3), 242–248.

Chang, T.-P.; Shih, J.-Y.; Yang, K.-M.; Hsiao, T.-C. (2007), 'Material properties of Portland cement paste with nano-montmorillonite', *Journal of Materials Science* **42** (17), 7478–7487.

Chen, J.; Kou, S-C.; Poon, C-S. (2012), 'Hydration and properties of nano-TiO_2 blended cement composites', *Cement and Concrete Composites* **34**, 642–649.

Collins, F.; Lambert, J.; Duan, W.H. (2012), 'The influences of admixtures on the dispersion, workability, and strength of carbon nanotube–OPC paste mixtures', C*ement and Concrete Composites* **34**, 201–207.

Constantinides, G.; Ulm, F.-J. (2007), 'The nanogranular nature of C–S–H', *Journal of the Mechanics and Physics of Solids* **55** (1), 64–90.

Cwirzen, A.; Habermehl-Chirzen, K.; Penttala, V. (2008), 'Surface decoration of carbon na-no-tubes and mechanical properties of cement/carbon nanotube composites', *Advances in Cement Research* **20**, 65–73.

Cwirzen, A.; Habermehl-Chirzen, K.; Nasibulin, A.G.; Kaupinen, E.I.; Mudimela, P.R.; Penttala, V. (2009), 'SEM/AFM studies of cementitious binder modified by MWCNT and nano-sized Fe needles', *Materials Characterization* **60**, 735–740.

Gaitero, J.J. (2008), 'Multi-scale study of the fiber-matrix interface and calcium leaching in high performance concrete', PhD thesis, Labein-Technalia.

Gaitero, J.J.; Saez de Ibarra, Y.; Erkizia, E.; Campillo, I. (2006), 'Silica nanoparticle addition to control the calcioum-leaching in cement-based materials', *Physica Status Solidi (A)* **203**, 1313–1318.

Gaitero, J.J.; Campillo, I.; Guerrero, A. (2008), 'Reduction of the calcium leaching rate of cement paste by addition of silica nanoparticles', *Cement and Concrete Research* **38**, 1112–1118.

Gaitero, J.J.; Zhu, W.; Campillo, I. (2009), 'Multi-scale study of calcium leaching in cement pastes with silica nanoparticles', in: Bittnar Z., Bartos P.J.M., Nemecek J., Smilauer V., Zeman J., editors. *Nanotechnology in construction: proceedings of the NICOM3 (3rd International Symposium on Nanotechnology in Construction)*, Springer, 193–198.

Givi, A.N.; Rashid, S.A.; Aziz, F.N.A.; Salleh, M.A.M (2011), 'The effects of lime solution on the propertiesof SiO_2 nanoparticles binary blended concrete', *Composites: Part B* **42**, 562–569.

Groert, N. (2007), 'Carbon nanotubes becoming clean', *Materials Today* **10**(1–2), 28–35.

Jalal, M.; Mansouri, E.; Sharifipour, M.; Pouladkhan, A.R. (2012), 'Mechanical, rheological, durability and microstructural properties of high performance self-compacting concrete containing SiO_2 micro and nanoparticles', *Materials and Design* **34**, 389–400.

Ji, T. (2005), 'Preliminary study on the water permeability and microstructure of concrete incorporating nano-SiO_2', *Cement and Concrete Research* **35** (10), 1943–1947.

Jo, B.-W.; Kim, C.-H.; Tae, G-H.; Park J-B (2007), 'Characteristics of cement mortar with nano-SiO_2 particles', *Construction and Building Materials* **21** (6), 1351–1355.

Konsta-Gdoutos, M.S.; Metaxa, Z.S.; Shah, S.P (2010a), 'Multi-scale mechanical and fracture characteristics and early-age strain capacity of high performance carbon nanotube/cement nanocomposites', *Cement and Concrete Composites* **32** (2), 110–115.

Konsta-Gdoutos, M.S.; Metaxa, Z.S.; Shah, S.P. (2010b), 'Highly dispersed carbon nanotubes reinforced cement based materials', *Cement and Concrete Research* **40** (7), 1052–1059.

Kuo, W.-Y.; Huang, J.-S.; Lin, C.-H. (2006), 'Effects of organo-modified montmorillonite on strengths and permeability of cement mortars', *Cement and Concrete Research* **36** (5), 886–895.

Li, G.Y.; Wang, P.M.; Zhao, X., (2005), 'Mechanical behavior and microstructure of cement composites incorporating surface-treated multi-walled carbon nanotubes', *Carbon* **43**, 1239–1245.

Li, H.; Xiao, H-G.; Ou, J-P. (2004a), 'A study on mechanical and pressure-sensitive properties of cement mortar with nanophase materials', *Cement and Concrete Research* **34** (3), 435–438.

Li, H.; Xiao, H-G.; Yuan, J.; Ou, J. (2004b), 'Microstructure of cement mortar with nano-particles', *Composites Part B: Engineering* **35** (2), 185–189.

Li, H.; Zhang, M-H.; Ou, J-P. (2006), 'Abrasion resistance of concrete containing nano-particles for pavement', *Wear* **260** (11–12), 1262–1266.

Li, H.; Zhang, M-H.; Ou, J-P. (2007), 'Flexural fatigue performance of concrete containing nano-particles for pavement', *International Journal of Fatigue* **29** (7), 1292–1301.

Li, Z.; Wang, H.; He, S.; Lu, Y.; Wang M. (2006), 'Investigations on the preparation and mechanical properties of the nano-alumina reinforced cement composite', *Materials Letters* **60** (3), 356–359.

Lin, D.F.; Lin, K.L.; Chang, W.C.; Luo, H.L.; Cai, M.Q. (2008), 'Improvements of nano-SiO_2 on sludge/fly ash mortar', *Waste Management* **28** (6), 1081–1087.

Lin, K.L.; Chang, W.C.; Lin, D.F.; Luo, H.L.; Tsai, M.C. (2008), 'Effects of nano-SiO_2 and different ash particle sizes on sludge ash–cement mortar', *Journal of Environmental Management* **88** (4), 708–714.

Makar, J.M.; Beaudin, J.J. (2004), 'Carbon nanotubes and their applications in the construction industry', in *Nanotechnology in Construction, Proceedings of the 1st International Symposium on Nanotechnology in Construction*, 331–341.

Makar, J.M.; Margeson, J.; Luh, J. (2005), 'Carbon nanotube/cement composites – Early results and potential applications', in *Proceeding of the 3rd International Conference on Construction Materials: Performance, Innovations and Structural Implications*, 1–10.

Metaxa, Z.S.; Konsta-Gdoutos, M.S.; Shah, S.P. (2009), 'Carbon nanotubes reinforced concrete', ACI Special Publications 267: *Nanotechnology of Concrete: The Next Big Thing is Small*, SP-267-2, 11–20.

Mondal, P. (2008), 'Nanomechanical properties of cementitious materials', PhD thesis, Northwestern University.

Nazari, A.; Riahi, S. (2010), 'Microstructural, thermal, physical and mechanical behavior of the self compacting concrete containing SiO_2 nanoparticles', *Materials Science and Engineering A* **527**, 7663–7672.

Nazari, A.; Riahi, S. (2011a), 'The role of SiO_2 nanoparticles and ground granulated blastfurnace slag admixtures on physical, thermal and mechanical properties of selfcompacting concrete', *Materials Science and Engineering A* **528**, 2149–2157.

Nazari, A.; Riahi, S. (2011b), 'The effects of SiO_2 nanoparticles on physical and mechanical properties of high strength compacting concrete', *Composites: Part B* **42**, 570–578.

Nazari, A.; Riahi, S. (2011c), 'The effects of zinc dioxide nanoparticles on flexural strength of self-compacting concrete', *Composites: Part B* **42**, 167–175.

Nazari, A.; Riahi, S. (2011d), 'Al_2O_3 nanoparticles in concrete and different curing media', *Energy and Buildings* **43**, 1480–1488.

Nazari, A.; Riahi, S. (2011e), 'Abrasion resistance of concrete containing SiO_2 and Al_2O_3 nanoparticles in different curing media', *Energy and Buildings* **43**, 2939–2946.

Nemecek J. (2009), 'Creep effects in nanoindentation of hydrated phases of cement pastes', *Materials Characterization* **60**, 1028–1034.

Oliver, W.C.; Pharr, G.M. (1992), 'An improved technique for determining hardness and elastic modulus using load and displacement sensing indentation experiments', *Journal of Materials Research* **7**, 1564–1583.

Qing. Y.; Zenan, Z.; Deyu, K.; Rongshen, C. (2007), Influence of nano-SiO_2 addition on properties of hardened cement paste as compared with silica fume', *Construction and Building Materials* **21** (3), 539–545.

Qing, Y.; Zenan, Z.; Li, S.; Rongshen, C. (2008), 'A comparative study on the pozzolanic activity between nano-SiO_2 and silica fume', *Journal of Wuhan University of Technology – Materials Science Edition* **21** (3), 153–157.

Quercia, G.; Hüsken, G.; Brouwers, H.J.H. (2012), 'Water demand of amorphous nano silica and its impact on the workability of cement paste', *Cement and Concrete Research* **42**, 344–357.

Saez de Ibarra, Y.; Gaitero, J.J.; Erkizia, E.; Campillo, I. (2006), 'Atomic force microscopy and nanoindentation of cement pastes with nanotube dispersions', *Physica Status Solidi (a)*, **6**, 1076–1081.

Salvetat, J.P; Bonard, J.M.; Thomson, N.H.; Kulik, A.J.; Forro, L.; Benoit, W. et al. (1999), 'Mechanical properties of carbon nanotubes', *Applied Physics A* **69** (3), 255–260.

Sanchez, F.; Ince, C. (2009), 'Microstructure and macroscopic properties of hybrid carbon nanofiber/silica fume cement composites', *Composites Science and Technology* **69**, 1310-1318.

Sanchez, F.; Zhang, L.; Ince, C. (2009), 'Multi-scale performance and durability of carbon nano-fiber/cement composites', in: Z. Bittnar, P.J.M. Bartos, J. Nemecek, V. Smilauer, J. Zeman (Eds.), *Nanotechnology in construction 3, Proceedings of the Third International Symposium on Nanotechnology in Construction*, Springer, 345–350.

Senff, L.; Hotza, D.; Repette, W.L.; Ferreira, V.M.; Labrincha, J.A. (2010), 'Mortars with nano-SiO_2 and micro-SiO_2 investigated by experimental design', *Construction and Building Materials* **24**(8), 1432–1437.

Shah, S.P; Swartz, S.E; Ouyang, C. (1995), *Fracture Mechanics of Concrete: Application of fracture mechanics to concrete, rock and other quasi-brittle materials*', John Wiley and Sons, New York.

Shah, S.P.; Konsta-Gdoutos, M.S.; Metaxa, Z.S.; Mondal, P. (2009), 'Nanoscale modification of cementitious materials', in: Z. Bittnar, P.J.M. Bartos, J. Nemecek, V. Smilauer, J. Zeman (Eds.), *Nanotechnology in construction 3, Proceedings of the Third International Symposium on Nanotechnology in Construction*, Springer, 125–130.

Sobolev, K.; Ferrada-Gutiérrez, M. (2005), 'How nanotechnology can change the concrete world: Part 1', *American Ceramic Society Bulletin* **84** (10), 14–17.

Sobolev, K.; Flores, I.; Torres-Martinez, L.M.; Valdez, P.L.; Zarazua, E.; Cuellar, E.L. (2009), 'Engineering of SiO_2 nanoparticles for optimal performance in nano cement-based materials', in: Bittnar Z., Bartos P.J.M., Nemecek J., Smilauer V., Zeman J., editors., *Nanotechnology in Construction: Proceedings of the NICOM3 (3rd International Symposium on Nanotechnology in Construction)*, Springer, 139–148.

Xie, X.L.; Mai, Y.W.; Zhou, X.P. (2005), 'Dispersion and alignment of carbon nanotubes in polymer matrix: a review', *Materials Science and Engineering Reports* **49** (4), 89–112.

Yang, Y.; Grulke, A.; Zhang, G.Z.; Wu, G. (2006), 'Thermal and rheological properties of carbon nanotube-in-oil dispersions', *Journal of Applied Physics* **99** (11), 114307.

Yu, J.; Grossiord, N.; Koning, C.E.; Loos, J. (2007), 'Controlling the dispersion of multi-wall carbon nanotubes in aqueous surfactant solution', *Carbon* **45** (3), 618–623.

Zhang, M-H.; Islam, J. (2012), 'Use of nano-silica to reduce setting time and increase early strength of concretes with high volumes of fly ash or slag', *Construction and Building Materials* **29**, 573–580.

21

Biotechconcrete: An innovative approach for concrete with enhanced durability

F. PACHECO-TORGAL, University of Minho, Portugal and J. A. LABRINCHA, University of Aveiro Portugal

DOI: 10.1533/9780857098993.4.565

Abstract: The use of sealers is a common way of contributing to concrete durability. However, the most common ones are based on organic polymers which have some degree of toxicity. The Regulation (EU) 305/2011 related to the Construction Products Regulation (CPR) emphasizes the need to reduce hazardous substances. Therefore, new low-toxicity forms to increase concrete durability are needed. Recent investigations in the field of biotechnology show the potential of bioinspired materials in the development of low toxic solutions. This chapter reviews current knowledge on the use of bacteria for concrete with enhanced durability. It covers the use of bacteria in concrete admixtures and also biomineralization in concrete surface treatments. Results from practical applications exposed to environmental conditions are still needed in order to confirm the importance of this new approach.

Key words: concrete durability, sealers, Construction Products Regulation (CPR), hazardous substances, biomineralization.

21.1 Introduction

With an annual production of about 10 km^3/year (Gartner and Macphee, 2011), Portland cement concrete is the most used construction material on Earth, the majority of which is used in the execution of reinforced concrete structures. However, many concrete structures face premature degradation problems, in fact, in the USA alone about 27% of all highway bridges are in need of repair or replacement. Also, the corrosion deterioration costs due to deicing and sea salt effects are estimated at over 150 billion dollars (Davalos, 2012). The reasons for such a low durability performance have to do with the fact that some of them were built decades ago, when little attention was given to durability issues (Hollaway, 2011), but also because that concrete has a high permeability.

This allows water and other aggressive elements to enter, leading to carbonation and chloride ion attack resulting in corrosion problems. The importance of concrete durability in the context of the eco-efficiency of construction materials has been rightly put by Mora (2007), when he stated that increasing its durability from 50 to 500 years would mean a reduction of its environmental impact by a factor of 10. The importance of concrete durability for their eco-efficiency has also recently been recognized by other authors (Flatt et al., 2012). Concrete structures with low durability require frequent maintenance and conservation operations or even integral replacement, being associated with the consumption of raw materials and energy. Concrete durability means, above all, minimizing the possibility of aggressive elements entering the concrete, under certain environmental conditions, for any of the following transport mechanisms: permeability, diffusion or capillarity.

The use of concrete surface treatments with waterproofing materials (also known as sealers) to prevent the access of aggressive substances is a common way of contributing to concrete durability. However, the most common surface treatments use organic polymers (epoxy, siloxane, acrylics and polyurethanes) all of which have some degree of toxicity. Polyurethane is obtained from the isocyanates, known worldwide for their tragic association with the Bhopal disaster. The production of polyurethane also involves the production of toxic substances such as phenol and chlorofluorocarbons. Besides, chlorine is associated with the production of dioxins and furans that are extremely toxic and also bio-cumulative. Several scientist groups already suggest that chlorine industrial-based products should be prohibited (Pacheco-Torgal and Jalali, 2011a).

Recently, the European Union approved the Regulation (EU) 305/2011 related to the Construction Products Regulation (CPR) that will replace the current Directive 89/106/CEE, already amended by Directive 1993/68/EEC, known as the Construction Products Directive (CPD). A crucial aspect of the new regulation relates to the information regarding hazardous substances (Pacheco-Torgal et al., 2012). New low-toxicity materials and techniques that increase concrete durability are therefore needed. An innovative approach to solve this and other current technological problems faced by human society which can encompass a new way of perceiving the potential of natural systems (Martin et al., 2010). The continuous improvement of these systems carried out over millions of years has been leading to materials and 'technologies' with exceptional performance and that are fully biodegradable.

Recent nanotechnology achievements regarding the replication of natural systems may provide a solution to solve the aforementioned problem (Wegner et al., 2005; Pacheco-Torgal and Jalali, 2011b). Analysis of bioinspired materials requires knowledge of both biological and engineering principles, thus constituting a new research area that can be termed biotechnology. Although this area has rapidly emerged at the forefront of materials research, the fact is that the study of biological systems as structures dates back to the early part of the twentieth century with the work of D'Arcy W. Thompson, first published in 1917 (Chen et al., 2012). In this context biotechnology seems to be able to provide a solution to concrete durability enhancement by means of biomineralization, a phenomenon by which organisms form minerals and first used for crack repair by Gollapuddi et al. (1995).

21.2 Bacteria mineralization mechanisms

Bacteria are relatively simple, unicellular organisms. There are typically 40 million bacterial cells in a gram of soil and a million bacterial cells in a milliliter of fresh water; in all, there are approximately five nonillion (5×10^{30}) bacteria on Earth, forming much of the world's biomass. Under optimal conditions, bacteria can grow and divide extremely rapidly and bacterial populations can double as quickly as every 9.8 min (Siddique and Chhal 2011). Biomineralization is defined as a biologically induced precipitation in which an organism creates a local micro-environment, with conditions that allow optimal extracellular chemical precipitation of mineral phases, like calcium carbonate ($CaCO_3$) (Hamilton, 2003). Decomposition of urea by ureolytic bacteria is one of the most common pathways to precipitate $CaCO_3$. The microbial urease enzyme hydrolyzes urea to produce dissolved ammonium, dissolved inorganic carbon and CO_2. Furthermore, the ammonia released in the surroundings subsequently increases pH, leading to accumulation of insoluble $CaCO_3$ in a calcium-rich environment. Figure 21.1. shows a

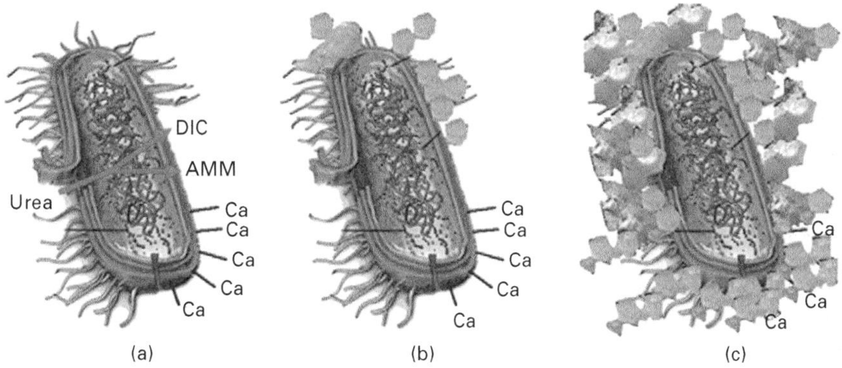

21.1 Simplified representation of the events occurring during the ureolytic-induced carbonate precipitation. Calcium ions in the solution are attracted to the bacterial cell wall due to the negative charge of the latter. Upon addition of urea to the bacteria, dissolved inorganic carbon (DIC) and ammonium (AMM) are released in the microenvironment of the bacteria (a). In the presence of calcium ions, this can result in a local supersaturation and hence heterogeneous precipitation of calcium carbonate on the bacterial cell wall (b). After a while, the whole cell becomes encapsulated (c) (De Muynck et al., 2010).

simplified representation of the events occurring during the ureolytic induced carbonate precipitation.

Ramachandran et al. (2001) reported that a high pH hinders the growth of bacteria. They also state that the optimum pH for the growth of *Bacillus pasteurii* is around 9. According to Arunachalam et al. (2010) *Bacillus sphaericus* bacteria-induced calcium carbonate precipitation at pH = 8 yields the highest results. Urease-catalyzed ureolysis is also temperature dependent and the optimum temperature ranges from 20 to 37 °C. In fact, increasing the temperature from 15 to 20 °C increased the rate of ureolysis, k_{urea}, 5 times and 10 times greater than k_{urea} at 10 °C (Okwada and Li, 2010). The same authors reported that k_{urea} is more dependent on the bacterial cell concentration than initial urea concentration so long as there is enough urea to sustain the bacteria. At 25 mM Ca^{2+} concentration, increasing bacterial cell concentration from 10^6 to 10^8 cells mL^{-1} increased the $CaCO_3$ precipitated by over 30%. However, when Ca^{2+} concentration was increased 10-fold, the $CaCO_3$ precipitated increased by over 100% irrespective of initial urea concentration.

21.3 Bacterium types

Different bacteria lead to very different calcium carbonate precipitation results. Table 21.1 summarizes the species used by different authors.

Table 21.1 Bacterium types

References	Bacterium type
De Muynck et al. (2008a, 2008b) Van Tittelboom et al. (2010) Arunachalam et al. (2010) Wang et al. (2012a) Wang et al. (2012b)	*Bacillus sphaericus* LMG 225 57
Ramachandran et al. (2001) Okwadha and Li (2010) Okwadha and Li (2011)	*Sporosarcina pasteurii* ATCC 11859
Chahal et al. (2012) Grabiec et al. (2012)	*S. pasteurii (formerly Bacillus pasteurii)*
Reddy et al. (2010) Afifudin et al. (2011)	*Bacillus subtilis*
Achal et al. (2009) Achal et al. (2011c)	Phenotypic mutant of *S. pasteurii* (Bp M-3)
Ramachandran et al. (2001)	*Pseudomonas aeruginosa* ATCC 27853
Jonkers et al. (2010)	*Bacillus pseudofirmus* DSM 8715
Ghosh et al. (2005, 2009)	*Shewanella* sp.
Achal et al. (2011a)	*Bacillus megaterium* ATCC 14581
Achal et al. (2011b)	*Bacillus* sp. CT-5
Wiktor and Jonkers (2011)	*Bacillus alkalinitrilicus*

Ramachandran et al. (2001) found that the contribution of *Pseudomonas aeruginosa* related to calcium carbonate precipitation was insignificant. Ghosh et al. (2005) mentioned that the use of *Escherichia coli*. showed no evidence of biomineralization. Achal et al. (2009) developed a phenotypic mutant of *Sporosarcina pasteurii*. The mutant, Bp M-3, was found to be more efficient in improving urease activity and was able to survive at very high pH values. In the majority of the studies, ureolytic bacteria of the genus *Bacillus* were used as the agent for the biological production of calcium carbonate-based minerals. The mechanism of calcium carbonate formation by these bacteria is based on the enzymatic hydrolysis of urea to ammonia and carbon dioxide.

A potential drawback of this reaction mechanism is that for each carbonate ion two ammonium ions are simultaneously produced. This may increase the risk of reinforcement corrosion (Jonkers et al., 2010). Besides atmospheric ammonia is recognized as a pollutant that contributes to several environmental problems (De Muynck et al., 2010).

21.4 Using bacteria as admixture in concrete

Ramachandran et al. (2001) found that using *Bacillus pasteurii* has a positive influence on the performance of cementitious composites. Ghosh et al. (2009) reported that anaerobic hot spring bacterium leaches silica and helps in the formation of new silicate phases that fill the micro-pores. They also mention that a concentration of 10^5 cells/ml optimizes the microstructure of cementitiuous composites. In order to overcome the problem of excessive ammonia production associated with the use of genus *Bacillus,* Jonkers et al. (2010) used bacterial spores (*B. cohnii*). They reported a loss of bacteria linked to the continuing decrease in the matrix pore diameter sizes with the progress of concrete curing age. In order to avoid bacterial loss, these authors suggest their encapsulation prior to addition to the concrete mixture, or else, the addition of air-entraining agents. Reddy et al. (2010) reported that the use of *Bacillus subtilis* bacteria for a cell concentration of 10^5 cells per ml of mixing water increases the concrete resistance to sulfuric acid attack. For the same bacteria an optimum concentration of 10^6 cells per ml was reported by other authors (Afifudin et al., 2011). Van Tittelboom et al. (2010) confirm that the use of bacteria can help to reduce the water permeability of concrete; however, they mention that the highly alkaline pH of concrete hinders the growth of the bacteria. To overcome this problem they immobilized the bacteria in silica gel. Other authors had already suggested the use of polymer encapsulation (Bang et al., 2001). According to Achal et al. (2011a) fly ash concrete containing *Bacillus megaterium* cells absorbed nearly 3.5 times less water than the control concrete. They also found that the permeability of the concrete with bacterial cells was lower than that of the control concrete.

Wiktor and Jonkers (2011) reported that the combined effect of viable bacterial spores plus calcium lactate embedded in porous clay capsules significantly enhanced mineral precipitation on crack surfaces further resulting in the healing of cracks (Fig. 21.2) with a maximal width of 0.46 mm. They also mentioned that since bacteria consume oxygen, it may provide an additional benefit associated with the potential to inhibit reinforcement corrosion. Chahal et al. (2012) studied the influence of *Sporoscina pasteuri* bacteria on fly ash concrete. The optimum performance was achieved for a 10^5 cells/ml of bacteria concentration. These authors reported a four times reduction in water absorption and a eight times reduction in chloride permeability due to calcite deposition (Fig. 21.3). Wang et al. (2012a) suggest the use of diatomaceous earth to protect *Bacillus sphaericus* from the high pH of the concrete matrix.

These authors report that the bacteria immobilized in diatomaceous earth had much higher ureolytic activity (12–17 g/L urea was decomposed within 3 days) than that of non-immobilized bacteria (less than 1 g/L urea was decomposed within the same time span) in cement slurry. Being that

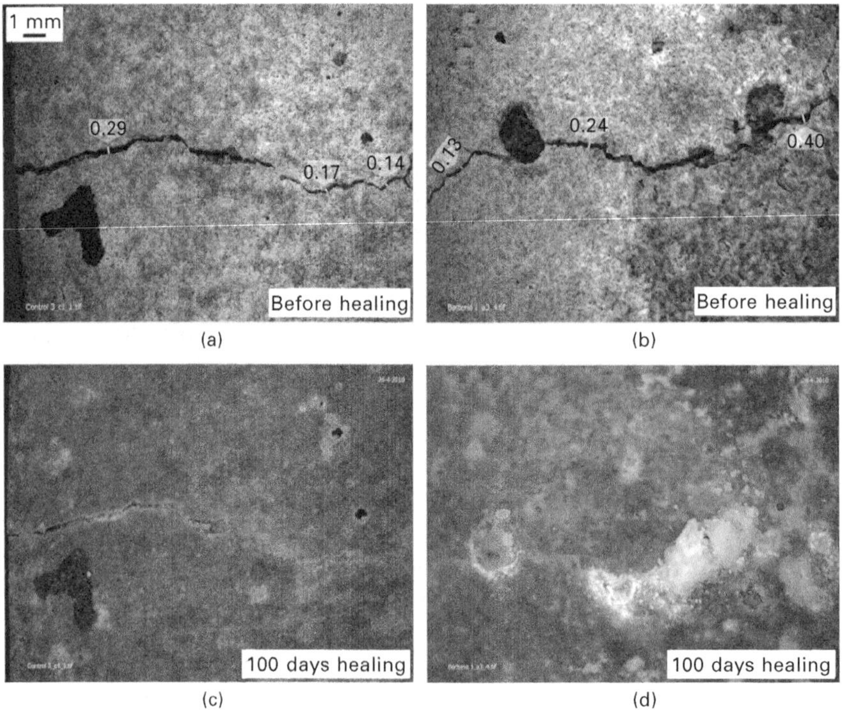

21.2 Stereomicroscopic images of crack-healing process in control mortar specimen before (a) and after (c) 100 days healing, in bio-chemical agent-based specimen before (b) and after (d) 100 days healing (Wictor and Jonkers, 2011).

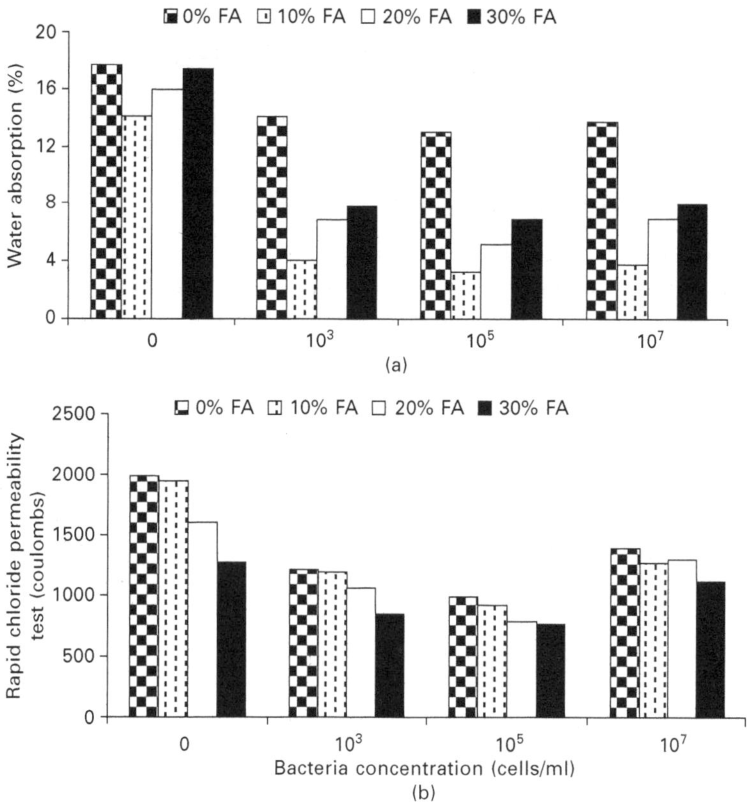

21.3 Effect of *S. pasteurii* on (a) water absorption of concrete at 7 days and (b) chloride permeability of fly ash (FA) concrete at 28 days (Chahal et al., 2012).

the optimal concentration of diatomaceous earth for immobilization was 60% (w/v, weight of diatomaceous earth/volume of bacterial suspension). Wang et al. (2012b) compared the performance of two different techniques (silica gel and polyurethane) to protect bacteria when immobilized inside concrete. The silica gel technique uses Levasil®200/30% sol with a specific surface area of 200 m^2/g and a solid content of 30% was used to embed bacterial cells. The immobilization of bacteria into polyurethane uses a two-component polyurethane MEYCO MP 355 1 K (BASF), to encapsulate bacterial cells.

The incorporation of the bacteria into mortar specimens is made by glass tubes with a length of 40 mm and an inner diameter of 3 mm. Experimental results show that the silica gel immobilized bacteria exhibited a higher activity than polyurethane immobilized bacteria, and hence, more $CaCO_3$ precipitated in silica gel (25% by mass) than in polyurethane (11% by mass), based on thermogravimetric analysis. However, cracked mortar specimens healed by polyurethane immobilized bacteria had lower water permeability coefficient (10^{-10}–10^{-11} m/s) compared with specimens healed by silica gel immobilized bacteria, which showed a water permeability coefficient of 10^{-7}–10^{-9} m/s (Fig. 21.4).

21.5 Concrete surface treatment

De Muynck et al. (2008a) compared the durability (concerning capillary water uptake and gas permeability) of concrete when its surface was treated with pure and mixed cultures of ureolytic bacteria. They concluded that the type of bacterial culture and the medium composition had a profound impact on $CaCO_3$ crystal morphology, being that the use of pure cultures resulted in a more pronounced decrease in the uptake of water. They also concluded that the durability performance obtained with cultures of the species *B. sphaericus* was comparable to the ones obtained with conventional water repellents (silanes, siloxanes). (Fig. 21.5).

De Muynck et al. (2008b) studied different durability parameters (carbonation, chloride penetration, and freezing and thawing), confirming that the biodeposition treatment showed a similar protection towards degradation processes when compared with some of the conventional surface treatments under investigation. They also mention the need for investigations regarding the durability of the treatment under acidic media. They further mentioned that biological generated calcite is less soluble than the one inorganically

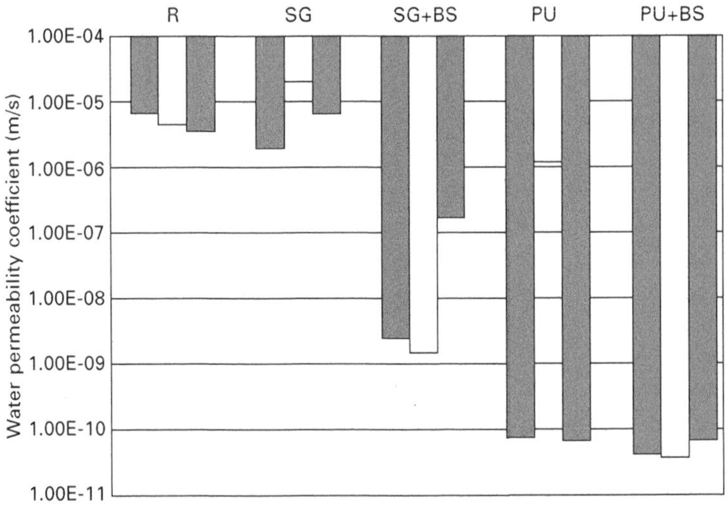

21.4 Water permeability of cracked cylinders. (R = reference, SG = silica gel, BS = bacterial suspension, PU = polyurethane immobilized bacteria) (Wang et al., 2012b).

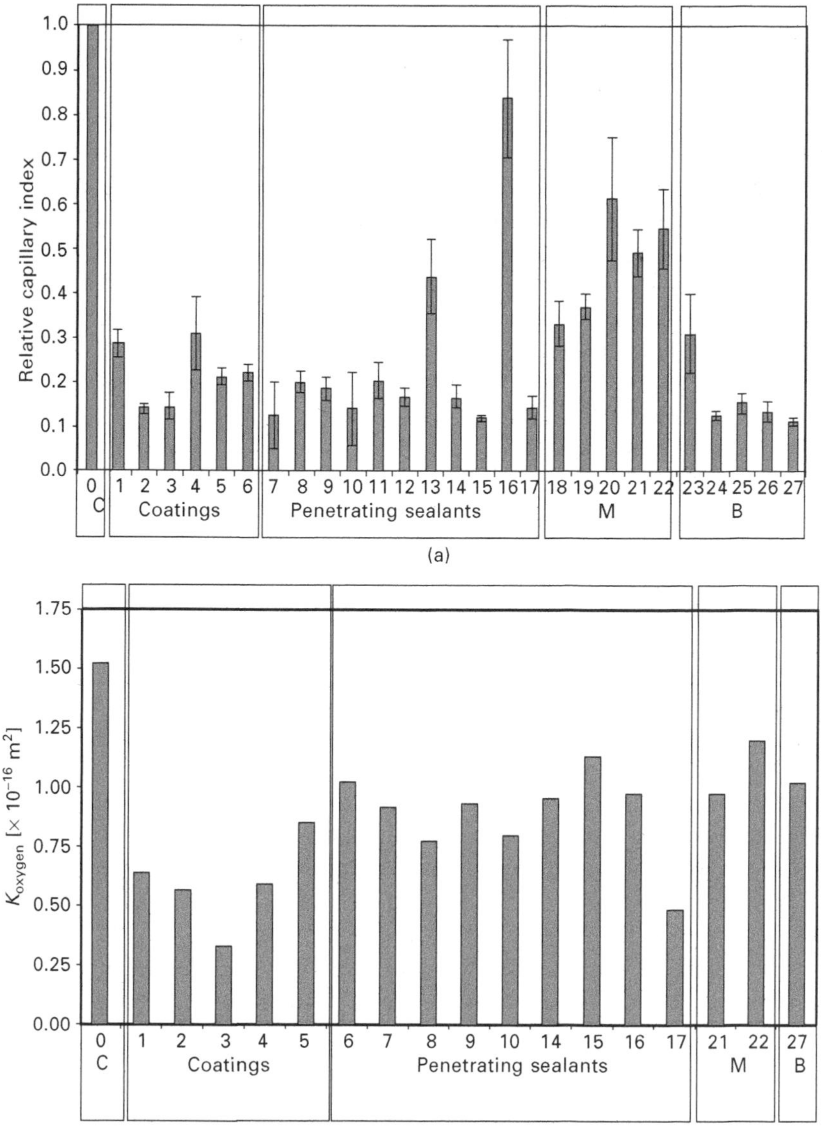

21.5 Concrete durability: (a) capillary water suction results expressed as the relative capillary index; (b) Permeability towards oxygen of treated and untreated concrete specimens expressed as specific permeability coefficient K to oxygen.

precipitated, thus suggesting a higher performance. Okwada and Li (2011) used bacterium *S. pasteurii* strain ATCC 11859 to create a biosealant on a polychlorinated biphenyl (PCB)-contaminated concrete surface, reporting a reduction on water permeability by one to five orders of magnitude. They also state that the treated concrete had a high resistance to carbonation. Achal et al. (2011b) mention a six times reduction in water absorption due to the microbial calcite deposition. In a different study the same authors (Achal et al., 2011c) used a phenotypic mutant of *Sporosarcina pasteurii* (Bp M-3) with improved urease activity also reporting a significant reduction in water absorption, permeability and chloride permeability. Li and Qu (2012) confirm that bacterially mediated carbonate precipitation on concrete surface reduces capillary water uptake, leading to the carbonation rate constant to be decreased by 25~40%. Nevertheless, the cost of biodeposition treatment still remains a major drawback to be overcome being dependent on the price of the microorganisms and the price of the nutrients (5–10 € per m^2), which is far from being cost-efficient (De Muynck et al. 2010). More recently Achal et al. (2011c) used corn steep liquor, an hazardous industrial effluent, as a nutrient source reporting a biodeposition cost of just (0.3–0.7 €) per m^2.

21.6 Future trends

Further investigation in the field of bacteria based concrete is still needed:

- Which calcite-producing bacteria are more efficient in highly alkaline environments?
- Can air-entraining agents be effective in preventing bacteria loss associated with reduction in pore size?
- Which is the more eco-efficient encapsulation method?
- Will biologically deposited calcite endure the test of time?
- Can biomineralization be made cost-efficient?
- What are the environmental implications related to the use of corn steep liquor as a nutrient source?
- Are there any health implications involved in the use of bacteria?
- Which is the life-cycle assessment of biotechconcrete?

21.7 References

Achal, V., Mukherjee, A.; Basu, P.; Reddy, M. S. (2009) Strain improvement of Sporosarcina pasteurii for enhanced urease and calcite production. *J. Ind. Microb. Biotechnol.* **36**, 981–988.

Achal, V.; Pan, X.; Ozyurt, N. (2011a) Improved strength and durability of fly ash-amended concrete by microbial calcite precipitation. *Ecol. Engi.* **37**, 554–559.

Achal, V.; Mukherjee, A.; Reddy, M. (2011b) Microbial concrete: Way to enhance the durability of building structures. *J. Mater. Civil Eng.* **23**, 730–734.

Achal, V.; Mukherjee, A.; Reddy, M. (2011c) Effect of calcifying bacteria on permeation properties of concrete structures. *J. Ind. Microbiol. Biotechnol.* **38**: 1229–1234.

Afifudin, H.; Hamidah, M.; Hana, H.; Kartini, K. (2011) Microorganism precipitation in enhancing concrete properties. *Appl. Mech. Mater.* **99–100**, 1157.

Arunachalam, K.; Sathyanarayanan, K.; Darshan, B.; Raja, R. (2010) Studies on the characterisation of biosealant properties of *Bacillus sphaericus*. *Int. J. Eng. Sc. Technol.* **2**, 270–277.

Bang, S.; Galimat, J.; Ramakrishan, V. (2001) Calcite precipitation induced by polyurethane-immobilized *Bacillus pasteurii*. *Enzyme Microbiol. Technol.* **28**, 404–409.

Chahal, N.; Siddique, R.; Rajor, A. (2012) Influence of bacteria on the compressive strength, water absorption and rapid chloride permeability of fly ash concrete. *Const. Build. Mater.* **28**, 351–356.

Chen, P.-Y.; McKittrick, J.; Meyers, M. (2012) Biological materials: Functional adaptations and bioinspired designs. *Prog. Mater. Sci.* (in press)

Davalos, J.F. (2012) Advanced materials for civil infrastructure rehabilitation and protection. Seminar at The City College of New York, New York.

De Muynck, W.; Cox, K.; De Belie, N.; Verstraete, W. (2008a) Bacterial carbonate precipitation as an alternative surface treatment for concrete. *Const. Build. Mater.* **22**, 875–885.

De Muynck, W.; Debrouwer, De Belie, N.; Verstraete, W. (2008b) Bacterial carbonate precipitation improves the durability of cementitious materials. *Cem. Concr. Res.* **38**, 1005–1014

De Muynck, W.; De Belie, N.; Verstraete, W. (2010) Microbial carbonate precipitation in construction materials: A review. *Ecol. Eng.* **36**, 118–136.

Flatt, R.; Roussel, R.; Cheeseman, C.R. (2012) Concrete: An eco-material that needs to be improved. *J. Europ. Ceram. Soc.* (in press).

Gartner, E.; Macphee, D. (2011) A physico-chemical basis for novel cementitious binders. *Cem. Concr. Res.* **41**, 736–749.

Ghosh, P.; S. Mandal, S.; Chattopadhyay, B.; Pal, S. (2005) Use of microorganisms to improve the strength of cement mortar. *Cem. Concr. Res.* **35**, 1980–1983.

Ghosh, P.; Biswas, M.; Chattopadhyay, B.; Mandal, S. (2009) Microbial activity on the microstructure of bacteria modified mortar. *Cem. Concr. Comp.* **31**, 93–98.

Gollpudi, U.; Knutson, C.; Bang, M.; Islam, M. (1995) A new method for controlling leaching through permeable channels. *Chemosphere* **30**, 695–705.

Grabiec, A.; Klama, J.; Zawal, D.; Krupa, D. (2012) Modification of recycled concrete aggregate by calcium carbonate biodeposition. *Constr. Build. Mat.* **34**, 145–150.

Hamilton, W.A., 2003. Microbially influenced corrosion as a model system for the study of metal–microbe interactions: a unifying electron transfer hypothesis. *Biofouling* **19**, 65–76.

Hollaway LC. (2011) Key issues in the use of fibre reinforced polymer (FRP) composites in the rehabilitation and retrofitting of concrete structure. In Karbhari VM & Lee LS (Eds.), *Service life estimation and extension of civil engineering structures*. Woodhead Publishing Limited, Cambridge, UK.

Jonkers, H.; Thijssen, A.; Muyzer, G.; Copuroglu, O.; Schlangen, E. (2010) Application of bacteria as self-healing agent for the development of sustainable concrete. *Ecol. Eng.* **36**, 230–235.

Li, P.; Qu, W. (2012) Microbial carbonate mineralization as an improvement method for durability of concrete structures. *Adv. Mater. Res.* **365**, 280–286.

Martin, J.; Roy, E.; Stewart, A.; Fergunson, B. (2010) Traditional ecological knowledge (TEK): Ideas, inspiration, and designs for ecological engineering. *Ecol. Eng.* **36**, 839–849.

Mora, E. (2007) Life cycle, sustainability and the transcendent quality of building materials. *Build. Environ.* **42**, 1329–1334.

Okwadha, G.; Li, J. (2010) Optimum conditions for microbial carbonate precipitation. *Chemosphere* **81**, 1143–1148.

Okwadha, G.; Li, J. (2011) Biocontainment of polychlorinated biphenyls (PCBs) on flat concrete surfaces by microbial carbonate precipitation. *J. Environ. Manage.* **92**, 2860–2864.

Pacheco-Torgal, F.; Jalali, S. (2011a) Toxicity of building materials. A key issue in sustainable construction. *Int. J. Sustain. Eng.* **4**, 281–287.

Pacheco-Torgal, F.; Jalali, S. (2011b) Nanotechnology: Advantages and drawbacks in the field of building materials. *Const. Build. Materi.* **25**, 582–590.

Pacheco-Torgal, F.; Jalali, S.; Fucic, A. (2012) *Toxicity of building materials*. Woodhead Publishing Limited, Cambridge, UK.

Ramachandran, S.; Ramakrishnan, V.; Bang, S. (2001) Remediation of concrete using micro-organisms. *ACI Mater. J.*, **98**, 3–9.

Reddy, S.; Rao, M.; Aparna, P.; Sasikala, C.(2010) Performance of standard grade bacterial (*Bacillus subtilis*) concrete. *Asian J. Civil Eng. (Building and Housing)* **11**, 43–55.

Siddique, R.; Chahal, N. (2011) Effect of ureolytic bacteria on concrete properties. *Const. Build. Mater.* **25**, 3791–3801.

Van Tittelboom, K.; De Belie, N.; De Muynck, W.; Verstraete, W. (2010) Use of bacteria to repair cracks in concrete. *Cem. Con. Res.* **40**, 157–166.

Wang, J.; De Belie, N.; Verstraete, W. (2012a) Diatomaceous earth as a protective vehicle for bacteria applied for self-healing concrete. *J. Ind. Microbiol. Biotechnol.* **39**, 567–577.

Wang, J.; Van Tittelboom, K.; De Belie, N.; Verstraete, W. (2012b) Use of silica gel or polyurethane immobilized bacteria for self-healing concrete. *Const. Build. Mater.* **26**, 532–540.

Wegner, T.; Winandy, J.; Ritter, M. (2005) Nanotechnology opportunities in residential and non-residential construction. In: 2nd International Symposium on Nanotechnology in Construction, 13–16 November 2005, Bilbao, Spain Bagneux, France: RILEM

Wiktor, V.; Jonkers, H. (2011) Quantification of crack-healing in novel bacteria-based self-healing concrete. *Cem. Con. Comp.* **33**, 763–770.

Index